Handbook of Fillers and Reinforcements for Plastics

Handbook of Fillers and Reinforcements for Plastics

Edited by

Harry S. Katz
Utility Research Co.
Montclair, New Jersey

John V. Milewski
Los Alamos Scientific Lab.
Los Alamos, New Mexico

VNR VAN NOSTRAND REINHOLD COMPANY
NEW YORK CINCINNATI ATLANTA DALLAS SAN FRANCISCO
LONDON TORONTO MELBOURNE

Van Nostrand Reinhold Company Regional Offices:
New York Cincinnati Atlanta Dallas San Francisco

Van Nostrand Reinhold Company International Offices:
London Toronto Melbourne

Library of Congress Catalog Card Number: 77-22335
ISBN: 0-442-25372-9

Manufactured in the United States of America

Published by Van Nostrand Reinhold Company
135 West 50th Street, New York, N.Y. 10019

Published simultaneously in Canada by Van Nostrand Reinhold Ltd.

15 14 13 12 11 10 9 8 7 6 5 4 3 2 1

Library of Congress Cataloging in Publication Data

Main entry under title:

Handbook of fillers and reinforcements for plastics.

Includes bibliographical references and index.
1. Plastics–Additives–Handbooks, manuals, etc. 2. Fillers (in paper, paint, etc.)–Handbooks, manuals, etc. I. Katz, Harry S. II. Milewski, John V.
TP1142.H36 668.4 77-22335
ISBN 0-442-25372-9

Foreword

Plastics are playing an increasingly important role in our daily lives—in our homes, businesses, and environment. Until recently, their plentiful supply and low cost had been taken for granted. However, the oil shortage and subsequent materials shortages, plus escalating prices for monomers and polymers, have resulted in a rude awakening to the fact that new initiatives must be exerted in our field.

A more important role must now be accorded to the increased and more efficient use of fillers and reinforcements as a means for stretching the resin supply and lowering the cost of molding compounds.

In my past years of plastics engineering, I have strived to bring to the industry a greater awareness of the proper materials, design, fabrication methods, and economics. I consider this *Handbook* a giant step forward in that direction, and believe that it will accelerate the proper use of fillers and reinforcements and therefore be beneficial to the plastics industry.

J. Harry DuBois
Morris Plains, New Jersey

Acknowledgments

The preparation of this *Handbook* has involved the advice, encouragement, and aid of many persons, and the editors are grateful to each of these individuals. In particular, we are indebted to the editorial staff of Van Nostrand Reinhold, Mr. Ashak M. Rawji and Ms. Alberta Gordon, for their cooperation, and to Mr. Tom Ferrigno for his constructive comments and input. We are certain that each chapter contributor will join us in special gratitude to our wives and families for their indulgence during the many hours spent on this project.

John V. Milewski
Harry S. Katz

Preface

Until recently, the plastics industry lived in a fool's paradise, where resins were low in cost and plentiful, a condition that was expected to persist. Then came the rude awakening during the oil embargo of 1973, which caused a shortage of raw materials and resins, and initiated a series of escalations in the prices of polymers. This situation led to an increased interest in the use of fillers and reinforcements as a means of reducing the price of molding compounds and extending the supply of resins. The editors were among those who considered it desirable, for many reasons, to make more extensive use of fillers and reinforcements, and realized that this goal would be aided by a unified compilation of information and data that would enable the rapid choice of a satisfactory filler or reinforcement. At the present time, the choice of a filler or reinforcement usually involves many contacts with materials suppliers, compounders, design engineers, and molders in order to select candidate materials and formulations.

During recent years, while this manuscript was being prepared, there has been a great increase in technical magazine articles related to fillers and reinforcements, and other factors that indicated a growing need for this type of *Handbook*. We hope that the consensus of opinion will be that this is a timely and useful compilation, which will serve as a standard reference of information for everyone involved in the plastics industry.

The Introduction, Chapter 1, discusses basic reasons for the use of fillers and reinforcements, and also describes the organization of each chapter. There are subjects in this *Handbook* that have not been treated in depth elsewhere, such as detailed information on new materials including Kevlar Filaments, ribbon reinforcements, microfibers, Basalt Fibers, and flakes, and the theoretical considerations for the selection of mineral fillers and fiber combinations. The editors are thankful to the expert contributors of the chapters in this *Handbook* and to the many companies and individuals who contributed information and data.

This *Handbook* is directed toward all individuals involved in the production, design, or specification of a molded end product. This includes design engineers, materials scientists, polymer chemists, compounders, and molders. The editors will welcome comments from every profession so that future editions of the *Handbook* will provide the information required for more efficient utilization of fillers and reinforcements.

HARRY S. KATZ
JOHN V. MILEWSKI

Contents

Handbook of Fillers and Reinforcements for Plastics

Section I

1 Introduction

John V. Milewski and Harry S. Katz
Los Alamos Scientific Lab.
Los Alamos, New Mexico
Utility Research Co.
Montclair, New Jersey

Growing industrial activities create a continual demand for improved materials that satisfy more stringent requirements, such as higher tensile strength, modulus, thermal conductivity, heat distortion temperature, and lower thermal expansion and cost. These requirements, which often involve a combination of many difficult-to-attain properties, can often be satisfied by utilizing a composite material, whose constituents act synergistically to solve the needs of the application. As we cross the threshold into the "Composite Materials Age," it becomes increasingly important to understand the properties, performance, costs, and potential of the available composite materials.

Fillers and reinforcements have always played an important role in the plastics industry. The early growth of the phenolic plastics industry would not have been possible without the enhancement of properties by the use of fillers and reinforcements. The later commodity resins, such as polyvinyl chloride, polystyrene, polyethylene, and polypropylene, have properties that meet the requirements of high volume end uses; thus they have been sold and used as essentially pure resins. However, recent price escalations, combined with the sporadic and projected shortages of resins and petroleum feed stocks, have established the urgent need for efficient and widespread utilization of fillers and reinforcements. The composite systems afford a means of extending the available volume of resins while improving many properties. These improvements in properties are often associated with economic advantages such as lower raw material cost, faster molding cycles as a result of increased thermal conductivity, and fewer rejects due to warpage.

At present, most molded products do not contain any filler or reinforcement, in spite of the fact that the judicious choice of a filler or reinforcement can result in a lower cost product with equivalent or improved properties. Recent technological advances, such as improved dispersion and the often-neglected factor of filler and reinforcement packing, have not been widely applied, and as a consequence, much current information on filled or reinforced plastics indicate poorer properties than the true potential of these composites. Also, there have been recent advances in the equipment and procedures for compounding and molding highly filled polymers. Currently, there are few valid technical justifications for the use of an unfilled resin in most molded products, from a consideration of either properties or moldability. Moreover, as this technology advances, there will be increased benefits from the use of fillers and reinforcements.

Resin production statistics for 1975-76 are shown in Table 1-1. These figures indicate an annual production of about 13,300,000 metric tons or 30 billion lb. Table 1-2 lists the annual usage of most major fillers and reinforcements. This estimate does not include glass fibers, carbon-graphite filaments, and a number of other reinforcements. Even with these items added, the figures would indicate that the usage of fillers and reinforcements is less than 10% of the resin volume. This is much less than their potential usage. For example, current fiberglass reinforced injection molding compounds contain up to 40% by weight glass, while calcium carbonate filler has been used at 40 wt % in PVC pipe extrustion compounds. Moreover, the improvements in coupling agents, compounding procedures, and molding are leading to the effective use of higher filler concentrations. Good molding compounds that contain 75 parts of filler per hundred parts resin may be common in the near future.

Table 1-1. Production statistics for 1975-1976

	1,000 METRIC TONS	
Material	1975	1976
Polyethylenes	3,196	4,042
LDPE and copolymers	2,148	2,625
HDPE	1,048	1,417
Vinyls	1,700	2,103
Other	320	390
Polystyrene and copolymers	1,875	2,285
Polypropylene and copolymers	861	1,178
Phenolics	465	600
Polyesters	350	430
Aminoplastics (UF and MF)	369	450
Acrylics	193	222
Alkyd resins[a]	280	330
Coumarone and petroleum resins	113	136
Cellulosics	56	70
Epoxies	85	110
Polyamides	65	95
Polyurethanes (excluding foams and elastomers)	614	738
Engineering plastics[b]	130	170
Total	10,672	13,353

[a]Includes captive consumption; about 50%.
[b]Includes phenylene oxide-based resin, polyacetal, polybutylene, fluoroplastics, polycarbonate, silicones, thermoplastic polyesters, thermoplastic urethanes, and others. (*Courtesy Modern Plastics, January, 1977.*)

Figure 1-1, which appeared in *Modern Plastics*, October 1975, shows an estimate of the growth rate of fillers and reinforcements as compared with plastics. A more rapid growth for fillers and reinforcements is predicted, but the growth rate may be much higher than indicated, due to the following facts: first, the rate of increase in costs of fillers will be lower than that of resins, since most fillers are not as energy dependent or derived from dwindling petroleum stocks as is the case with commodity resins; and second, more processors are learning to work with filled or reinforced compounds. More efficient processing machinery is being developed and becoming available. Thus, there is an increasing basis and impetus for processors to realize the economic advantages and greater profits resulting from the lower raw materials cost.

During the period from 1973 to 1976, the prices of commodity resins, such as polyethylene and polyester resins, have approximately doubled. During this same time, the price of commercial fillers and reinforcements have increased at a significantly lesser rate, on the order of 25 to 50%. Typical examples are shown in Table 1-3. The important point is that the price spread between resin cost and the cost of a filler or reinforcement is increasing, so

Modern Plastics, October, 1975.)

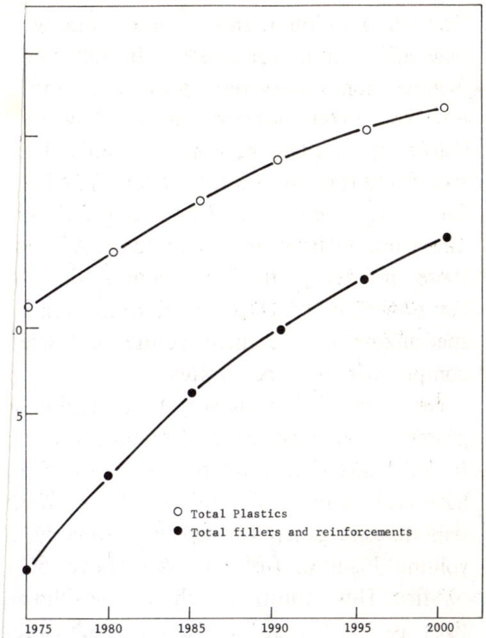

g. 1-1. How fillers and reinforcements will grow mpared with plastics as a whole. (*Courtesy Modern astics, October 1975.*)

there will be an even greater economic advantage in the future use of low-cost fillers and reinforcements.

Until recently, large suppliers of resin would not become involved in the development or sale of a filled compound. Apparently, they considered the potential sales volume and profit too low to justify the compound development cost. However, major suppliers have begun to market a number of mineral-reinforced polyamide polymers. These materials contain about 40% by weight of the filler. Reports and advertisements in the literature on this type of product indicate the following: the composite has greater strength than nylon and offers a unique combination of properties and low cost; mold cycles may be 20 to 40% faster than most unreinforced thermoplastics and the parts have high stiffness and low warpage; and the parts can be painted with no pretreatment. The material has been suggested as a replacement for die-cast

Table 1-3. Price changes of resins, fillers, and reinforcements.

	PRICE	
Material	1973	1976
Commodity resins		
Polypropylene	17¢/lb	28¢/lb
Polyethylene	14¢/lb	30¢/lb
Polyester	20¢/lb	37¢/lb
Engineering resins		
Nylon	60¢/lb	100¢/lb
Polycarbonate	75¢/lb	108¢/lb
Fillers and reinforcements		
Calcium carbonate		
fine	$50/ton	$65/ton
intermediate	$32/ton	$36/ton
coarse	$18/ton	$22/ton
Talc		
fine	$68/ton	$89/ton
intermediate	$46/ton	$60/ton
coarse	$40/ton	$55/ton
Wollastonite		
F-1	2.4¢/lb	3.2¢/lb
P-1	1.9¢/lb	2.4¢/lb
Fiberglass roving	35–40¢/lb	43–52¢/lb

SOURCE: *Chemical Market Reporter*, and Industrial Sources.

zinc and aluminum, and will undoubtedly open new applications for plastics. In an analogous manner, commodity resin producers, who have not yet taken vigorous action toward the marketing of filled versions of their polymers, will surely turn toward this direction as a means for entering new markets and for establishing a direct line to large volume molders. At present, these molders either compound their own composites or purchase them from companies specializing in the manufacture and sales of compounds and concentrates.

The term "fillers" is very broad and encompasses a very wide range of materials. In order to limit this Handbook to a practical size, we have arbitrarily defined fillers as the solid materials that are generally used in reasonably large volume loadings (usually well above 5%) in plastics. This definition excluded the solid additives used in low concentrations, such as lubricants, pigments, catalysts, anti-static agents, and all liquid extenders or plasticizers. These are important formulation ingredients, but are beyond the scope of this Handbook.

Since widespread utilization of fillers and reinforcements is becoming more urgent, it appeared desirable to compile a Handbook that would be a basic source of information on these materials.

This Handbook is intended to guide the reader in the selection and optimum use of fillers and reinforcements, and to furnish information on suppliers and specific product grades. Material suppliers usually furnish only basic data, without any theory or practical know-how regarding the effective use of the material. Also, a supplier's data sheet will not include information on competitve companies or materials. This Handbook lists major suppliers and their products and compares competitive materials where significant differences are apparent. This should enable the user to make a more judicious selection of a suitable material and supplier.

During recent years, there have been frequent price increases caused by inflation and materials shortages. In the case of the high performance filaments, carbon-graphite and boron, there have been price reductions due to increased sales volumes and improved manufacturing methods. In spite of price changes that will undoubtedly occur, it was considered important to present current prices so that the reader could have an indication of relative costs of materials, as well as grades of the same material. Prior to any precise cost analysis, the reader should obtain a price quote from the supplier.

Some subjects which have not been treated in depth elsewhere are information on new materials, such as Kevlar, Basalt Fibers, flakes, ribbon reinforcements, and microfibers, and the theoretical considerations for selection of mineral fillers and fiber combinations.

All theoretical treatments have been kept to a minimum, especially in areas already presented in great detail in other books. However, the new science of fiber-sphere packing is covered in some depth because this information has never before appeared in a book, and this subject is considered extremely important to the future of filled and reinforced plastics.

The book is divided into 12 sections: Section I is the Introduction; Section II, a guide to the selection and use of mineral fillers and reinforcements in three chapters. The first chapter of Section II deals with fillers and how to select and use them efficiently. This is one of the most comprehensive works in this field, covering in detail all important aspects of the use of fillers. The next chapter presents fundamentals of fiber-reinforced composites. This subject has been covered in great depth elsewhere, as referenced. Therefore, this presentation is a brief, concise review of the subject in easily understood terminology. Chapter 4 in Section II presents the new technology of packing combinations of fillers and reinforcements and explains why, in many cases, synergistic effects can result if the proper combinations of fillers and fibers are made. Detailed and theoretical explanations regarding the phenomena of packing are given, as well as concrete examples of how this technology can and should be applied to formulating combinations of fibers and spheres to produce theoretically and economically efficient composites.

Section III covers the mineral fillers. Great

attention is given to calcium carbonate, the most widely used filler. A large amount of very practical, detailed information on formulations with every major resin system is given. The other chapters in this section describe currently utilized mineral fillers, including clay, feldspar, silica, and talc.

Section IV contains 2 chapters, one on metallic and conductive fillers, the other on magnetic fillers.

Section V presents significant information on flame retardants, with separate chapters on the two major flame retardants, antimony oxide and alumina trihydrate. This fast growing area is covered in the depth and clarity needed to permit someone unfamiliar with this subject to understand the use and selection of these flame retardants.

Section VI covers carbon fillers and organic fillers, many of which are used in very large volume. Carbon, as a filler for plastics, is presented in Chapter 16. Chapter 17 presents a comprehensive description of the wide variety of organic fillers.

Section VII contains 2 chapters on spherical filler reinforcements, both solid and hollow. All aspects of sphere-filled composites are given, including discussions of glass, ceramic, plastic, and metallic coated spheres.

Section VIII, Chapter 20, is a detailed presentation of the expanding field of flake reinforcements, principally mica. Section IX, Chapter 21, describes the technology of ribbon reinforcements and what can be expected from this type of lesser known reinforcement, which has potential advantages over fiber reinforcement in some applications.

Section X is made up of 4 chapters on short fiber reinforcements. These include, in detail, practical information on the use of such established materials as Wollastonite and asbestos, as well as new and futuristic ideas on the short fibers known as microfibers and whiskers. The latter material, although not readily available at the present time, has tremendous potential as a superior reinforcement; this was considered sufficient justification for its inclusion in this Handbook.

Section XI contains 6 chapters on continuous filament reinforcements, such as the high volume fiber glass. Other chapters describe the aramid fiber, Kevlar; the two major aero-space filaments, carbon-graphite and boron; experimental ceramic fibers; and, finally, metal filaments. Comprehensive coverage is given to all aspects of these continuous filaments, which are the reinforcements used in the highest strength composites currently produced commercially.

As noted above, Sections III through XI present all of the significant materials used as fillers or reinforcements. The editors had presented to the authors of these chapters a suggested chapter outline that might be used as a standard format for each of these chapters. However, it was apparent that each author required some latitude for variation from the standard format. In spite of some of these differences, the reader will note that most chapters have the same basic format for presenting the data, and this will facilitate locating specific information on a particular material. Also, each author had the prerogative of defining the terms used in this chapter and presenting them as a glossary at the end of the chapter. This was considered important, since many terms can be ambiguous in the various divisions of the plastics industry.

It seemed desirable to have a Contents included for each chapter to assist the reader in locating specific information.

Section XII contains a practical chapter on procedures, equipment, and utilization technology. The filler or reinforcement cannot be used effectively if the user does not know how to blend and mix the filler/reinforcements with the polymer and proceed with proper molding methods and equipment. The chapter describes available processing equipment and lists suppliers.

The field of fillers and reinforcements is growing rapidly and encompasses a wide variety of materials. This Handbook has been designed to present a complete and orderly discussion of pertinent information in this field. However, there have been some arbitrary choices and, undoubtedly, some omissions. Nevertheless, we are confident that the reader will find this a useful and rewarding document. In addition, it is hoped that this presentation will act as a catalyst to inspire the increased use of fillers and reinforcements.

The editors anticipate that there will be a revision of this Handbook about 5 years after this first edition. Therefore, comments and suggestions on more efficient formats, additional data on current materials, and information on new materials would be appreciated.

Section II
Guide to the Selection and Usage of Fillers and Reinforcements

2
Principles of Filler Selection and Use

Thomas H. Ferrigno
Improde
Trenton, New Jersey

CONTENTS

1. INTRODUCTION

This chapter deals with the essentials of filler properties and their relationships to plastics properties. The presentation is intended to familiarize the reader with those properties of fillers which have the most influence on the plastics-filler system. As such, only those properties of fillers which are related to their use in plastics are described. The reader is directed to references providing in-depth studies of particular aspects of fillers properties where such information may be useful for the evaluation and thorough understanding of new fillers and the delineation of certain subtleties.

Reliance has been placed on the reported properties of commercial fillers to the extent that they help to provide information on their use. In the many cases where the technical data on commercial fillers is inadequate, tables and check lists are provided to place each filler in its proper perspective. As a further aid, simplified tests and related values are described so that a property of a filler may be assessed to determine its suitability in a particular application.

For the most part, fillers have been considered, especially by their manufacturers, to be cheap extenders or diluents, and have not been used correctly to obtain the full measure of their benefits. It is hoped that the method presented here will provide the tools necessary for the understanding and use of fillers. If their primary use is to reduce the cost of a compound, it is our purpose to show the most efficient way to this end. Many cases are discussed and illustrated where the highest concentration of filler, and therefore maximum cost reduction, is highly advantageous. Conversely, the limitations of filler use are also explained, in the section on systemic effects of fillers. Thus, each aspect, both pro and con, has been treated so that filler selection and use may be understood and practiced by those involved in plastics formulating and compounding.

2. PRIMARY PROPERTIES OF FILLERS

2.1 The Particle

Filler selection is primarily determined by the particle size distribution, the particle shape, and, as a consequence of both, the manner in which the particles pack together. First, an examination of the particle as the primary unit of the filler is required.

A general classification of filler particles is presented in Table 2-1. The classes are based on a somewhat arbitrary classification according to surface area. This classification is based on two primary properties of fillers; particle size and

Table 2-1. Particle Characteristics.

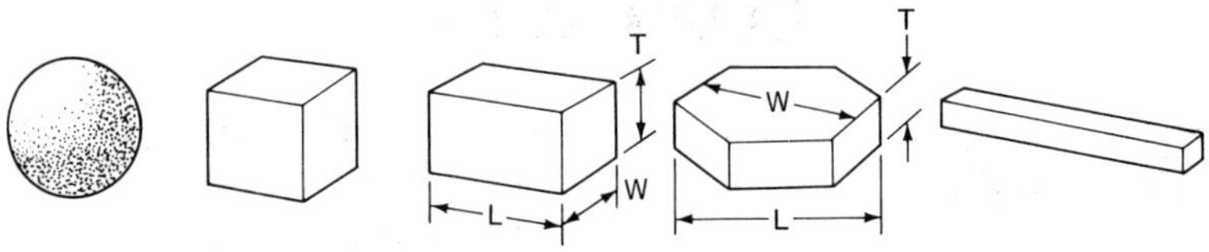

Particle Class	Sphere	Cube	Block	Flake	Fiber
Idealized shape class			(figures for Table 2-1)		
Descriptor[a]	spheroidal[b]	cubic[c] prismatic rhombohedral	tabular prismatic pinacoid irregular	platy flaky[d]	acicular elongated fibrous
Shape ratios;					
length (L)	1	~1	1.4–4	1	1
width (W)	1	~1	1	<1	<1/10
thickness (T)	1	~1	1–<1	$\frac{1}{4}$–1/100	<1/10
Sedimentation diameter[e]	1	esd	esd	esd[f]	esd[f]
Surface area equivalence[g]	1	1.24[h]	1.26–1.5[i]	1.5–9.9[j]	1.87 for 1/10 2.3 for 1/20[k]
Examples	glass spheres microspheres	calcite[l] feldspar	calcite feldspar silica barite nephelite	kaolin mica talc graphite hydrous alumina	wollastonite tremolite wood flour

[a] Preferred to particle class since this is based on relative surface area. First descriptor is preferred.
[b] In the sense that a spheroid approaches a true sphere.
[c] Generally distorted cubes; more nearly prismatic.
[d] Generally having the nature of hexagonal platelets, as illustrated.
[e] According to Stokes' law, esd = equivalent spherical diameter or the diameter of a sphere having the same volume as that of the particle.
[f] Must be modified for dissymetry of greater than 4–1, maximum to minimum particle dimensions, see Reference (1), pp. 20, 61.
[g] Equivalent to a spherical diameter of 1; an approximation of the area when the particle has a volume equivalent to an esd of 1.
[h] About the same for cubic and prismatic shapes.
[i] For lengths of 1.4–4, respectively.
[j] Based on hexagonal platelets as follows:

length/thickness	area factor
4/1	1.47
6/1	1.78
8/1	2.09
10/1	2.34
100/1	9.88

[k] For a square cross-section.
[l] There are more than 300 crystal shapes for calcite alone, but generally it is an irregular, low surface-area particle.

surface area, both of which are directly measurable and serve as a basis for systematizing filler functions. Most fillers, and those most commonly used, are minerals which are ground rock or ores otherwise processed to obtain the material in particulate form. As such, they do not cleave, part, or fracture uniformly and their particles are generally irregular shapes. Certain fabricated fillers, such as glass spheroids, precipitated silicates, and calcium carbonates, will have more nearly precise forms. Kaolinite has the nature of hexagonal platelets, for example, but very rarely will a particle be found that is a *true* hexagonal platelet.

There are many types of fillers that contain variously shaped particles having greater surface areas to esd ratios than those indicated in Table 2-1. An example of heterogeneity is shown in Fig. 2-1, wherein the smallest (thickness) to largest (diameter) dimensions of kaolinite were obtained by microscopic methods. It may be seen that at about 13 μ, the particles are stacks of bound platelets equal in two dimensions.

A classical illustration of indescribable particle shape may be seen in fotomicrographs of diatomite. The skeletal remains of diatoms are frequently pierced by sub-micron sized holes, have dished sections, are of toroid, oblate, spoked-wheel, and other forms which, to say the least, are indescribable by any simple classification. When diatoms are ground or subsequently processed, the fragments have an even greater variety of shapes.

In effect, then, the shapes of most real particles are too variant to be dealt with directly. Taken together, particle size and surface area provide the only means for classification of particles and their effects in filler-binder systems.

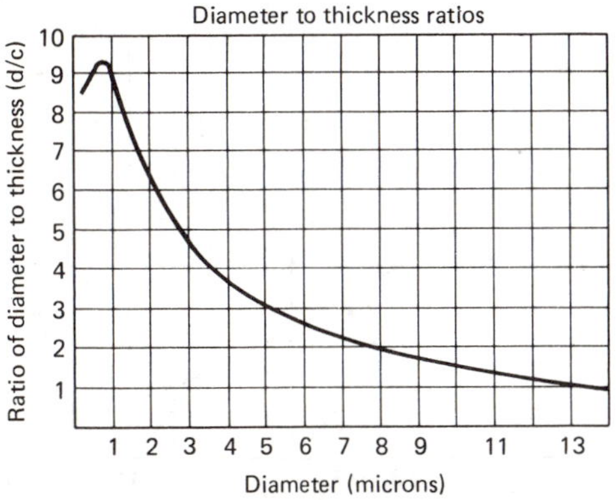

Fig. 2-1. Comparison of diameter (largest dimension) to thickness (smallest dimension) of kaolinite particles.[2]

In view of the above discussion, it is appropriate that the standard definition for particle size is equivalent spherical diameter (esd), or the diameter of a sphere having the same volume as that of the particle. Stokes' law is utilized in several sedimentation methods for determining particle size distribution. The various methods and adaptations for particle size analysis are better left to an authoritative source. Reference 1 is such a source and is recommended.

In regard to particle size measurement, microscopic techniques are usually the final criteria, but such methods are arduous and time consuming. Microscopy is a valuable tool for assessing particle shapes and their contributions to filler functions.

Particle size is also measured by sieving; dry methods being used for large particles, and wet methods for analysis below about 100 mesh (149 μ). Woven wire sieves are used for +37 μ particles, and flat, electroformed micromesh sieves are used with mesh openings of from 120–10 μ. It is apparent that woven wire will not provide accurate values for the smallest dimension of a particle passing through the mesh due to variation in wire diameter, distortion, angularity of the windows, wear, and other factors. The micromesh sieves are much more precise, but both require standardization with particles of known size.

Sieves are indispensable for measuring particles larger than about 40 μ since sedimentation methods are imprecise above this value when water is used as the sedimenting medium. Values obtained by sieving must be modified to conform with those obtained by methods yielding values as esd, unless, of course, the particles are spheroids. Sieves are subject to considerable inaccuracies because of mechanical factors and even new ones should be standardized. As mentioned previously, most filler particles are quite irregular, so the smallest dimension which will pass through the sieve opening must be considered. This will be discussed in detail in the next section.

2.2 Particle Size Distribution

Figure 2-2 illustrates the most common method of particle size distribution graphing which describes a single species; all particles have essentially the same density, and therefore, the percent finer by weight and volume are equal. Although a filler of very broad distribution was chosen and values approach zero, it is a common failing of suppliers' literature that the entire distribution is not reported. It shall be emphasized later that these missing data do not allow adequate assessment of the particle size distribution for surface area, packing values, and the true nature of the distribution. Such has been the case with filler literature, and although the effort to provide complete data is great in some cases, there is greater justification for providing complete information.

Figure 2-3 illustrates the data of Fig. 2-2 presented as a log probability graph. This graphical method is preferred by authorities[1] since it exaggerates the coarse and fine ends of the distribution. Absolute values are not obtainable, but the log probability graph is constructed so that the data can be compared with a normal distribution, which, on such a graph, is a straight line. The graph of particle size by number is also a straight line, parallel to and below that by weight. Deviation from the straight line also characterizes the history and properties of the filler. The graph of kaolin illustrated is almost a straight line, indicating that little has been done in processing to alter the natural distribution. Almost all fillers are fractionated to remove the coarser particles; the result is the curve turning toward the abscissa in the range of 90% finer. The slight deviation from the straight line for this kaolin indicates that a small amount of particles between 44 and 10 μ was removed in processing. At the fine end of the curve, the slight deviation from the straight line below about 0.4 microns may be, and frequently is, due to inaccuracies in the method of particle size determination.

Log probability paper of this type (Keuffel & Esser SP 46-8040) is complete to the 99.99 and 0.01% values. Figure 2-3 is extended to the 99.9% value only. The coarseness parameter, as microns at 99.9% finer, should be the standard, since it adequately describes the thickness of a film into which the largest particles of a distribution will fit. Certain tests, such as the fineness of dispersion test of ASTM D 1210 (also called Hegman) delineate the size of particles or agglomerates appearing in a film. When a coarseness parameter of 99.9% finer is specified, this will precisely coincide with the fineness of dispersion value in microns for cubic and tabular particles. Platy particles are more likely to provide fineness of dispersion values somewhat lower than the 99.9% coarseness parameter, since the measure of particle size is in esd and the plate or flake thicknesses are much less than the esd value. Nevertheless, the 99.9% coarseness parameter provides a more

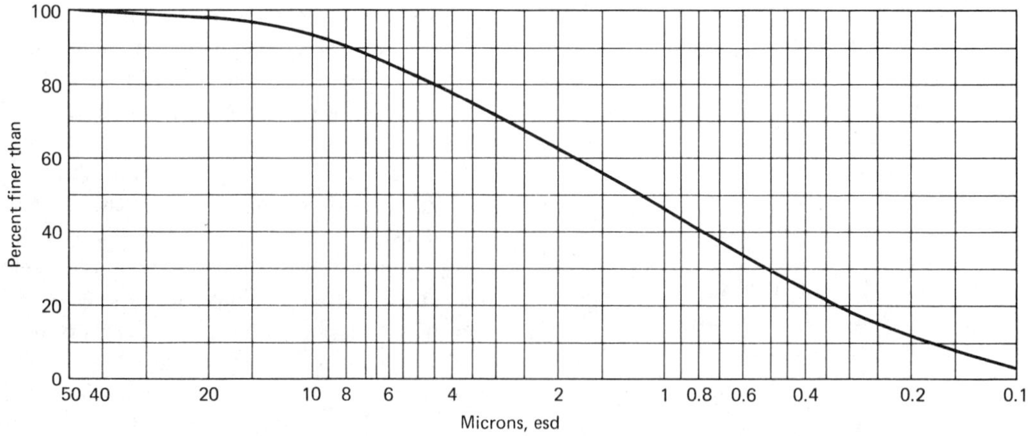

Fig. 2-2. Particle size distribution of Huber 65A kaolin, semi-log graph.

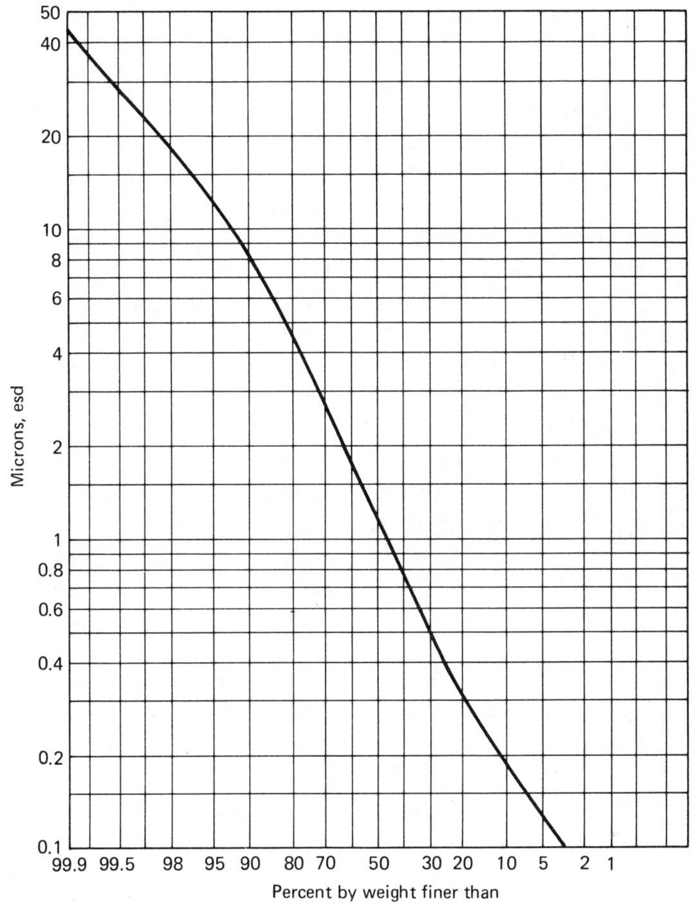

Fig. 2-3. Particle size distribution of Huber 65A kaolin, log-probability graph.

precise specification than the subjective values generally reported. When this test is used as a measure, or for control, of coarseness parameter, it requires that the filler be completely dispersed.

Sieve test data are frequently and incorrectly used to extend the particle size curve. For reasons explained above, it is obvious that mixed data are presented by this means. Rather than esd, sieve data provide minimum geometric dimension values for a particle passing through a sieve opening. A comparison of esd and sieve data may be obtained by extending the esd curve by drawing a tangent to the curve in the range of the most accurate values; below about 20 microns for water suspension data. Ratios of sieve versus esd values at the same percent finer usually fall within a narrow range. For example, esd values for tabular particles are usually 1.2-1.4 times larger than those for sieve values. Calculating the equivalent spherical volume based on the minimum sieve dimension and uniformly shaped particles provided the approximate shape ratios and surface areas of Table 2-1.

Average particle size by weight and number are useful for evaluating specific properties of fillers, such as the standard deviation and skewed, bimodal, or heterogeneous size distributions. These are properties which also should be reported by filler suppliers in order to aid in the selection and use of their products. Calculation of these parameters by the filler user is laborious and demanding and best left to the specialist in particle size analysis. Where certain specific characteristics in the particle size distribution of a filler are indicated, these will be noted in the tables of sections 2-5.1 with reference to section 2.4.

Almost all mineral fillers are fractionated products, that is, some portion of the parent distribution, usually the coarse end, has been removed. Quality control of these fillers is usually exercised by controlling the coarseness and average particle size. Control of the coarseness is essential, as explained above, but average particle size can be misleading, as considerable deviation can occur below this point. In all, emphasis will be placed on the entire particle size distribution, since differences are reflected in rheological effects, abrasiveness and abrasion resistance, particle packing relationships, dispersability, optical effects, chemical resistance, and strength properties.

2.3 Surface Area

Surface area is one of the most important of filler properties. Many effects of fillers are surface area dependent, particularly where surfactants, dispersants, surfiers (surface modifiers), polar polymers, etc., are sorbed by or reacted with filler surfaces. The chemistry and physics of solid surfaces is highly complex, so a basic text is suggested.[3]

Similar esd particle size distributions give no indication of filler surface area without considering shape factors. Recourse to direct surface area measurement is necessary if precise values are required. The standard method is the isothermal adsorption of nitrogen gas based on the BET equation.[4] Another lesser known, but considerably simpler method, relies on methylene blue adsorption on minerals having very low solubilities in acidic water.[5]

Frequently, surface area data are not given in suppliers' literature, and it becomes necessary to obtain an estimate. Comparison with the same type of filler having a similar particle size distribution can be done by referring to the table of section 5.1. Where this fails, it may be necessary to compute a value by one of two methods; the Gates diagram or a computation based on esd.

The Gates diagram[6] is a graph of the reciprocal of particle size versus percent finer on arithmetic graph paper. The measured area beneath the curve represents surface area. When using such a method, the data should be complete from 100 to very nearly zero percent finer

Table 2-2. Data for Gates diagram plotting and surface area calculations.

COARSER FILLER			FINER FILLER		
% Finer	Microns, esd	1/μ	% Finer	Microns, esd	1/μ
99.9	40	0.025	99.9	20	0.05
97	30	0.033	98	15	0.067
87	25	0.04	91	10	0.1
55	10	0.1	83	8	0.125
45	8	0.125	71	6	0.167
34	6	0.167	63	5	0.2
29	5	0.2	51	4	0.25
24	4	0.25	39	3	0.333
18	3	0.333	24	2	0.5
12	2	0.5	16	1.5	0.67
9	1.5	0.67	9	1	1
6	1	1	0	0.5	2
0	0.5	2			
Area beneath curve, Fig. 2-4, square units 78					
Known surface area, M^2/g 1–1.3					
Surface area factor				140/78 = 1.79	
Calculated surface area, m^2/g				1.8–2.3	

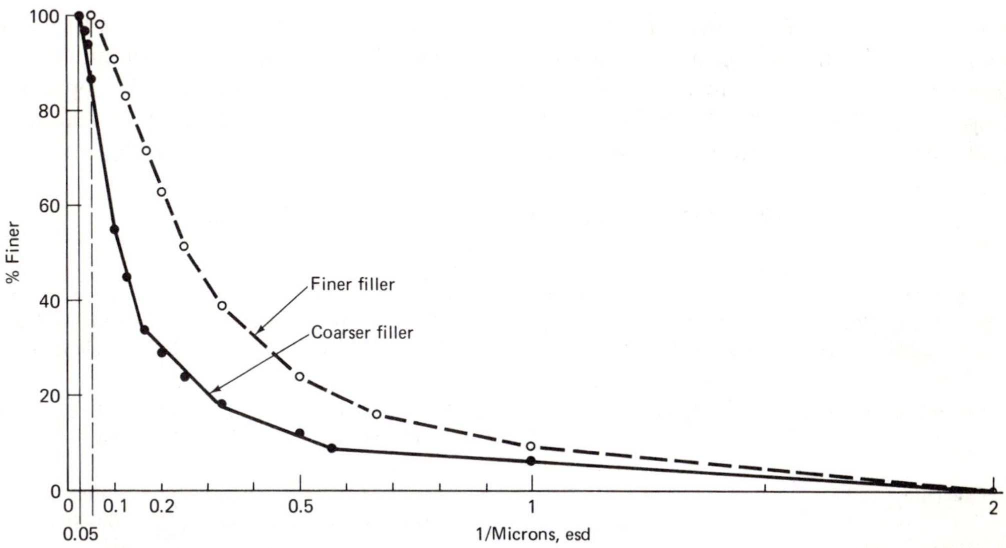

Fig. 2-4. Reciprocal of particle size versus percent finer for data of Table 2-2; Gates diagram for surface area under the curves.

Table 2-3. Calculation of surface area.

Size Distribution	dW, grams	$\overline{X}_m$, esd μ	$dW/\overline{X}_m$
100–97	0.03	35	0.0008
–94	0.03	28	0.0011
–87	0.07	23	0.0030
–55	0.32	15	0.0213
–45	0.10	9	0.0111
–34	0.11	7	0.0157
–29	0.05	5.5	0.0091
–24	0.05	4.5	0.0111
–18	0.06	3.5	0.0171
–12	0.06	2.5	0.0240
– 9	0.03	1.8	0.0167
– 6	0.03	1.3	0.0231
– 0	0.06	0.8	0.0750
Total	1.00		$\sum \frac{dW}{\overline{X}_m} = 0.2291$

$$SA = \frac{K}{\rho}\left(\sum \frac{dW}{\overline{X}_m}\right)$$

where

SA = surface area, m^2/g

K = a proportionality constant for particle shape:

- spheres = 6
- blocky and tabular = about 12
- platy = about 18
- flaky, thin = about 30
- acicular, fibrous, short = about 18–30

dW = weight increment, grams

ρ = density, g/cm^3

$\overline{X}_m$ = average particle size of weight increment as esd

For the coarser filler of Table 2-2, $\frac{12}{2.6}$ (0.2291) = 1.06 m^2/g, which compares with the determined values of 1–1.3 m^2/g

and the *same* type of filler having a known surface area must be plotted. Taking a proportionality of the measured areas under the curves versus the known surface area then provides a reliable estimate. This method is illustrated in Table 2-2 and Fig. 2-4.

Computational values are based on the areas of equivalent spheres. Unless spheres are the subject of calculation, it is obvious that for all other particles, a proportionality constant related to particle shape must be applied. Calculation is simplified by modern data processors and provides a reasonable estimate of surface area, provided the deviation from sphericity for the particles is not extreme. Table 2-3 illustrates the method. It should be noted that the ratios of surface areas in Table 2-1 versus esd are for *regularly* shaped particles. Few filler particles are regularly shaped, so factors for this deviation were empirically derived from known values of surface area for shape classes. Differences between the actual and calculated surface areas will be modest for low area fillers, but for others, there may be differences greater than two or three times.

2.4 Particle Packing

Maximum packing systems. In all the literature on this subject, the most significant is that of C.C. Furnas.[8] It is the most practical method devised because it is based on the determined void volumes of specific size ranges of uniformly shaped particles. For simplicity and conformance with sieve sizes, he chose "one" size ranges of $\sqrt{2}$ (or 1-1.414). By determining the void volume of a size range, the shape factors referred to previously are eliminated or compensated for in the calculations. In this method, specific sizes suitable to a particular application are selected. The steps of Table 2-4 illustrate the method for stepwise, or discontinuous,

Table 2-4. Stepwise grading of particles for maximum packing.

Steps	Examples
1. Select sizes; "one" size has the range $1 - \sqrt{2}$. d_1 = average esd of largest size; d_s = average esd of smallest size.	1. "One" size is through 16, on 20 mesh for d_1, or 1190 to 840 μ, 840 $\sqrt{2}$ = 1190. Smallest size is 0.5 μ.
2. Determine void volumes, V, of d_1 and d_s.	2. Tamped volume of d_1 can be used. Oil absorption by volume using ASTM D-281 for finest.
3. Calculate K ratio $= \dfrac{d_s}{d_1}$.	3. Average of 1190 to 840 μ is 1000 μ. Finest size is narrow range 0.5 μ. K ratio $= \dfrac{0.5}{1000} = 0.0005$.
4. From Fig. 2-5, determine number of components for K ratio.	4. If d_1 and d_s have 50% voids, number of component sizes = 3.05; use 3 for practical purposes.
5. Determine intermediate size; $d_2 = \sqrt{d_1\, d_s}$.	5. For symmetrical distribution $d_2 = \sqrt{0.5 \times 1000}$ = 22.4 μ or 19 to 27 μ for "one" size ($d_2/1.189$ = $19 \times \sqrt{2} = 27$.)
For component sizes which are not whole numbers, size ratio $R_s = K^{1/n}$ where K = K ratio and n = number of intervals or one less than component sizes.	$R_s = 0.0005^{1/2.05}$, this is an X^y function and = 0.0245; d_1 = 1000 μ and 1000 × 0.0245 = 24.5 μ = d_2. For this example, d_2 = 22.4 μ may be used without significant effect on final volume (see Step 7).
6. Determine void volume of d_2.	6. Use appropriate method, d_2 must be in the range of $\sqrt{2}$ for average esd.
7. Calculate absolute volumes of each component: $S = \underset{d_1}{\dfrac{1}{1+V_1}} + \underset{d_2}{\dfrac{V_1}{1+V_2}} + \underset{d_3}{\dfrac{(f)\,V_2^2}{1+V_3}} + \underset{d_4}{\dfrac{(f)\,V_3^3}{1+V_4}}$ where: S = total absolute volume of solid material; V_1 = fractional voids of d_1, etc.; (f) = fractional proportion of last component, if any.	7. Say void volumes of all components = 50%; $S = \underset{d_1}{\dfrac{1}{1+0.5}} + \underset{d_2}{\dfrac{0.5}{1+0.5}} + \underset{d_3}{\dfrac{(0.5)^2}{1+0.5}}$ $= 0.667 + 0.333 + 0.167$ $= 1.167$. If 3.05 components were used, an additional term; d_4 would be used, but the fourth term would be only $\dfrac{0.05\ V^3}{1+V_4} = \dfrac{0.05\ (0.5)^3}{1+0.5} = 0.004$ additional volume.
8. Calculate packing fraction, P_f, or the volume that the solid particles occupy of the total volume; $P_f = S - SV_1^2$.	8. $P_f = 1.167 - 1.167 \times 0.5^2$ $= 0.875$, 87.5%; $V_v = 1 - 0.875$ $= 0.125$, 12.5%
Void volume $V_v = 1 - P_f$.	
9. For combinations by weight, multiply each component volume for d_1, d_2, etc., by its specific gravity (see Step 7).	9. Using limestone as an example: $d_1 = 2.71 \times 0.667 = 1.81$ 57.3%; $d_2 = 2.71 \times 0.333 = 0.90$ 28.5%; $d_3 = 2.71 \times 0.167 = 0.45$ 14.2%; total 3.16

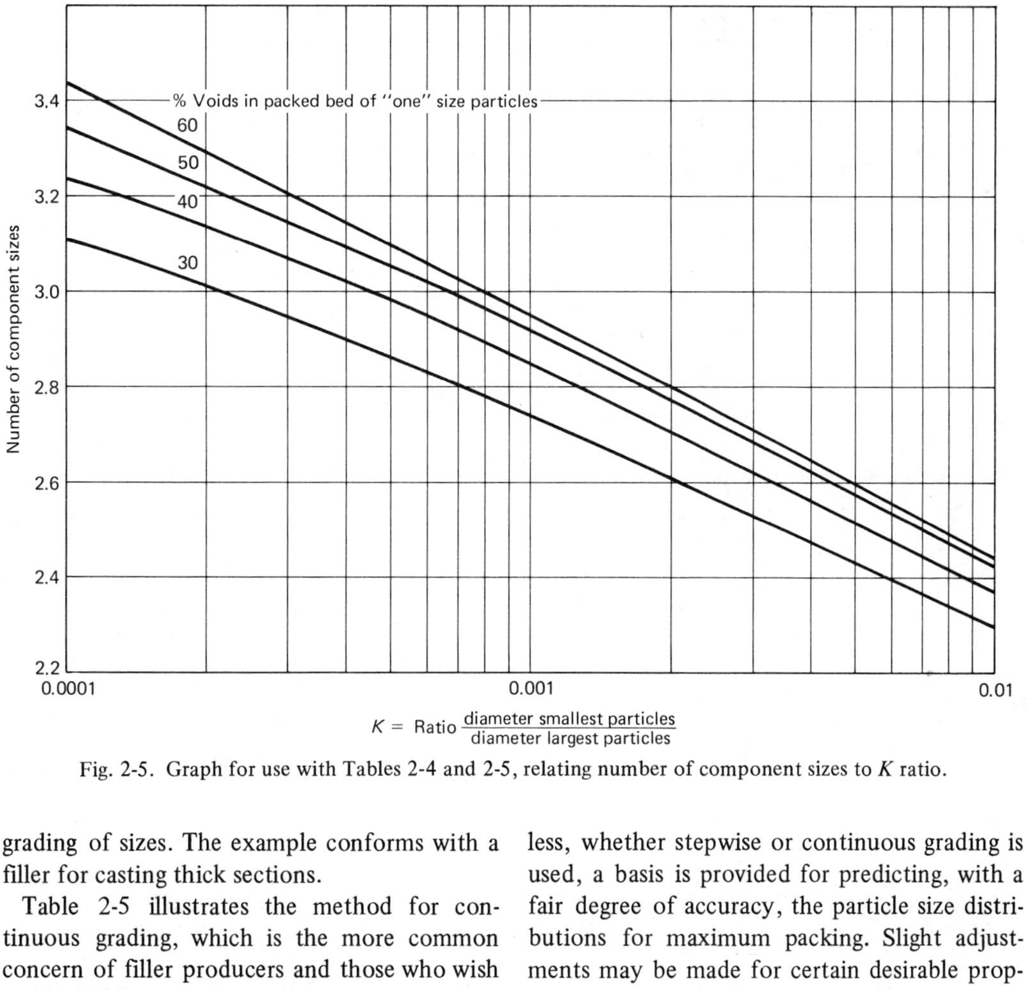

Fig. 2-5. Graph for use with Tables 2-4 and 2-5, relating number of component sizes to *K* ratio.

grading of sizes. The example conforms with a filler for casting thick sections.

Table 2-5 illustrates the method for continuous grading, which is the more common concern of filler producers and those who wish to blend fillers to obtain the minimum voids to be occupied by binders.

It may be noted that the system is actually based on the fact that the coarsest particles pack to produce the gross volume of the system. As succeeding finer material is added, it occupies the voids between the largest particles such that the total volume is not expanded. Each additional finer material also has a void volume which, in turn, may be occupied by a still finer material. The resulting geometric progression then depends on the least voids left in the system and on very wide ranges of particles sizes to achieve a high degree of packing.

In practice, one may combine different types of fillers having somewhat, but not radically, different particle shapes. In these cases, the calculations may not agree with the results obtained from the fabricated mixtures. Nonetheless, whether stepwise or continuous grading is used, a basis is provided for predicting, with a fair degree of accuracy, the particle size distributions for maximum packing. Slight adjustments may be made for certain desirable properties, but the entire process is far removed from a method that is entirely trial and error.

Minimum packing systems. Many systems do not benefit from maximum filler content. Although examples and applications of packing effects will be given in sections 4 and 5, it is apparent that one may wish to use a filler merely to modify the viscosity of a compound. Thus, the poorest packing and minimum filler content may be indicated. If maximum packing is obtained with a particular distribution of particles sizes, it follows that minimum packing is obtained by one particle size. Absolute minimum filler packing is then obtained with the finest, highest surface area particles, requiring the greatest concentration of binder to wet and disperse them. Particles having very high aspect ratios have been used for this purpose for some time. Fibers, or long, needle-shaped

Table 2-5. Continuous grading of particles for maximum packing.

Steps	Examples
1. Select size range to be used. Obtain "one" size material differing by $\sqrt{2}$ for minimum to maximum and determine void volume for several fractions.	1. Filler chosen in 358 to 0.5 μ range. For $\sqrt{2}$ "one" sizes: (see below)

−mesh	40	70	140	230
+mesh	50	100	200	325
Ave. μ	358	180	90	53
% voids	44.4	45.4	42.5	42.3

% void volume by oil absorption, average, V, 43.6% or 0.44.

Steps	Examples
2. Calculate K ratio $= \frac{d_s}{d_1}$ where d_s = average esd of smallest size and d_1 = average esd of largest size.	2. K ratio $= \frac{0.5}{358} = 0.0014$
3. Prepare table for number of sizes differing by $\sqrt{2}$ for range selected.	3. (see below)

Size $\times\sqrt{2}$	Step 5, volume of each size X 1.081	% individual volume	% cumulative volume finer than
0.5	1.000	2.16	2.16
0.71	1.081	2.34	4.50
1.00	1.168	2.52	7.02
1.41	1.263	2.73	9.75
2.00	1.366	2.95	12.70
2.83	1.476	3.19	15.89
4.00	1.596	3.45	19.34
5.66	1.725	3.73	23.07
8.00	1.865	4.03	27.10
11.3	2.016	4.36	31.46
16.0	2.179	4.71	36.17
22.6	2.355	5.09	41.26
32.0	2.546	5.50	46.76
45.0	2.752	5.95	52.71
64.0	2.975	6.43	59.14
90.5	3.216	6.95	66.09
128	3.477	7.51	73.60
256	4.063	8.78	90.50
362 ≅ 358	4.392	9.49	99.99
20 sizes	46.270	99.99	
19 intervals			

Steps	Examples
4. For the ratio of weight or volume in a continuous series for particles having the same true density; $r = \frac{1}{V^{n/m}}$ where r = factor or ratio of one size to the next larger; V = fractional void volume; m = number of size intervals; n = number of component size intervals; one less than the number of component sizes from Fig. 2-5 for K ratio.	4. From Fig. 2-5, interpolating for V = 0.44 or 44%, number of component sizes = 2.81, 2.81 − 1 = n, 1.81. m = 19 from Step 2; $r = \frac{1}{0.44^{1.81/19}} = 1.081$
5. Multiply each size by r beginning with 1 for the finest size.	5. See table of Step 3 for volume or weight of particles and % composition.

Table 2-5. (*Continued*)

Steps	Examples
6. Calculate V_v, fractional voids in the continuous distribution: $$V_v = \frac{r^{\log d} - r^{\log d_s}}{r^{\log d_1} - r^{\log d_s}}$$ where: d = the size interval, $\sqrt{2}$ or 1.414; d_s = diameter of smallest particle size; d_1 = diameter of largest particle size; r = ratio or factor of sizes from Step 3.	6. $$V_v = \frac{1.081^{\log \sqrt{2}} - 1.081^{\log 0.5}}{1.081^{\log 358} - 1.081^{\log 0.5}} = \frac{1.0118 - 0.9768}{1.2200 - 0.9768} = \frac{0.0350}{0.2432} = 0.1439 \text{ or } 14.39\%$$
7. Calculate $P_f = 100 - V_v$	7. $P_f = 100 - 14.39 = 85.61\%$

particles, are very effective since they do not readily orient side by side but form a voluminous mass occupying a large volume at rest. For a detailed discussion of fiber and fiber-sphere packing, see Section II, Chapter 4.

General considerations. Oil absorption by volume may be used as a guide for the packing properties of all fillers as single components of a system by ASTM D 281. For mixtures of fillers and pigments, the blends may be evaluated by the same method. Determination of the best size ranges to employ for either minimum or maximum packing may then be done by using the Furnas calculations above. Plotting these distributions against the ideal on semi-log graph paper as percent finer versus esd particle size will give a graphical comparison of those distributions which conform, more or less, to the ideal maximum packing.

It should be noted that esd particle size is used as the basis for all of the above discussion. In many cases, peculiar packing effects will be obtained with admixtures of particles having divergent shape factors or where a single type of filler varies in shape factor according to its esd particle size (see Table 2-1).

2.5 Chemical Composition

Chemical composition is a primary property of fillers and is an essential consideration for their use in many systems. Essentially, chemical reactivity is the chief concern of filler users. Reactivity is a surface effect due to composition. Bulk effects due to chemical composition are properly placed in the Physical Properties section of this chapter. All reactivity proceeds from the surface of the filler, so it is easy to be mislead by apparent equivalence. For example; all silicates contain silica and *might* be considered to have equivalently good acid and poor alkali resistance. This is *not* the case, since chemical composition does not indicate the relative positions of modifying elements in the crystal lattice or at the surface of the particle. Crystal structure and bond strength relationships are dealt with by Pauling[10] and are much too detailed to be discussed here. The data of Table 2-6 indicate the relative differences between silicates and other filler classes, sufficient to provide a useful classification.

In each of the sections below dealing with specific chemical classes of fillers, reference will be made to the influence of filler surface and, where appropriate, of bulk composition on reactivity.

Properties of fillers are fully categorized in section 5.1 for convenient reference. The principles underlying systemic effects will be discussed here to provide a general understanding of class differences. Mineral classification has been systematized by Hurlbut[9] and is used in Table 2-6.

Oxides. Corundum, or alumina, the oxide of aluminum, is the only mineral or synthetically derived oxide sometimes used as a filler. Alumina is Al_2O_3, but the so-called hydrated alumina of commerce is $Al(OH)_3$, or gibbsite, which loses about 35% water on heating to 300°C or higher, and therefore provides flam-

Table 2-6. Chemical Composition Classification of Fillers[a]

Chemical Class	Derivation	Types	CHEMICAL RESISTANCE[b] Acid	Alkali	Other
Oxide	Mineral	Alumina	G	G	
		Gibbsite (alumina trihydrate)	G	G	
Salt	Mineral, synthetic	Calcium carbonate	P	F	Water, sol.
		Barium sulfate, barite	E	E	
	Animal	Aragonite, calcium carbonate (oyster shell)	P	F	Water, sol.
Silicate[c]	Mineral				
neso-		Zirconium silicate, zircon	E	E	
ino-		Calcium silicate, wollastonite	P	F	Water, sol.
		CaMg silicate, tremolite	F	G	
phyllo-		Alumino silicate, kaolinite	G	G	
		K alumino silicate, mica	G	G	
		Mg silicate, talc	G	G	
		Mg silicate, serpentine (asbestos)	G	G	
		Alumino silicate, pyrophyllite	G	G	
tecto-		Silica	E	P	
		Hydrous silica, opal	E	G	
		Na, K alumino silicate feldspar, nepheline	G	G	
	Synthetic	Glass (microbeads)	G	P	
		Calcium silicate, precipitated	P	F	
Elements	Mineral, synthetic	Crystalline carbon, graphite	E	E	
	Synthetic	Metals	P	E, P for aluminum	
Organic	Vegetable	Coal (anthracite)	E	F	Volatiles
		Wood, bark, cork, nutshell flours	P	P	Reactive with acids, salts

[a] Types used for prime purposes other than as fillers are not included.
[b] E = excellent, G = good, F = fair, P = poor.
[c] Prefixes from the Greek characterize the silica tetrahedral arrangements in the crystal lattice; soro for group, neso for island (isolated), cyclo for ring, ino for chain or thread, phyllo for sheet, and tecto for framework. Soro- and cyclosilicates have not been included since no commercial fillers exist for either class, but they may be present as impurities in principal types.

mability retardation. Due to the amphoteric nature of aluminum, alumina and gibbsite are reactive with acids and alkalis and have highly active surfaces. The hydroxyls on the surface of gibbsite may react as acids (proton donors) or as alkali hydroxyls. This versatility gives rise to a great variety of surface treatments suitable for particular effects, according to the matrix in which it is used as a filler.

Salts. Products of acids and bases are salts. The most widely used filler, calcite, or $CaCO_3$, is a salt differing considerably from barite, $BaSO_4$. Over 300 different crystal forms of calcite have been identified, but the item of commerce is generally derived from limestone, a sedimentary rock mass. When metamorphosed, limestones are called marble. Typical of a weak acid and a weak base, calcium carbonate has very little resistance to even weak organic acids and has a fairly high water solubility for a filler, making it reactive with a great variety of chemical reagents.

Aragonite is a higher density, harder, and higher refractive index crystal form of calcium carbonate, the main constituent of mollusk shells and available as oyster shell flour. It is also somewhat less reactive than calcite. Precipitated calcium carbonates are mixtures of both calcite and aragonite.

Dolomite, $CaMg(CO_3)_2$, is sometimes used as a filler but is more reactive than calcite due to the presence of magnesium. All calcium carbonates contain some magnesium, usually as dolomite.

Anhydrite is anhydrous calcium sulfate and has the failing of high water solubility, acidic reaction in water, and reactivity with many binders. Terra alba is the mineral name for fully hydrated calcium sulfate or gypsum, and is similar to anhydrite.

Barite, the mineral, and blanc fixe, the precipitated form of barium sulfate, lie at the extreme end of the scale for salts. Both forms have excellent chemical resistance, low solubility, and bond very well to metals through surface sulfate groups.

Silicates. The most complicated of fillers are those derived from several mineralogical processes resulting in an extremely wide variety of types. Ore deposits large enough to be commercially exploitable exist for only a few. Most ores contain impurities or different minerals which may detract from the properties of the primary mineral component. Thus, processes designed to obtain kaolin from an ore involve first removing contaminants such as silica, graphite, vegetable matter, and others with simple methods. Highly refined methods are used to remove ilmenite and other iron compounds, the discoloring agents. In a similar manner, but with the products reversed, kaolin and other clays are removed from silica. This is to emphasize that uniform chemical composition for the silicates and other minerals frequently leaves much to be desired. Only for certain ores and processes can it be said that the filler is *essentially* one mineral or functionally similar species. The source and processing of the filler are very important and frequently proprietary. Where a filler is described as a chemical class in Table 2-6, its properties are those of the practically pure materials. For these reasons, it is emphasized that "local" materials be carefully evaluated or that the supplier's literature be searched for clues to impurities which can be reactive with, or otherwise deleterious to, the polymer or combinations of ingredients in a compound.

The silicates vary widely in their characteristics, but, as a rule, all have good resistance to acids (exception; wollastonite) and poor to excellent resistance to alkalis, depending on the disposition of the silica tetrahedra in the crystal lattice.

Synthetic silicates which are precipitated products are made with sodium silicate solutions and insolubilizing cations such as calcium. Their compositions differ considerably from natural products, and, due to small particle sizes and high surface areas, they have poorer acid and alkali resistances.

Glass spheroids are also synthetic silicates, principally soda-lime glass. Since glass is a supercooled liquid, it does not benefit from the bonding or regular compositional order of crystalline, natural fillers. Certain special types are formulated for better resistance properties, but generally, glasses are inferior to natural silicate fillers for acid and alkali resistance.

Elements. Various metals and alloys are processed for flake, shot (spheroids), or irregular particles. All have poor acid resistances typical of metals, with aluminum having the additional failing of poor alkali resistance. Their uniformities are excellent as all are made from melts and chemical analyses are precisely controlled.

Graphite is crystalline carbon formed by mineralogical processes (inorganic), is very uniform, and provides very good acid and alkali resistance.

Organics. Coal, as anthracite, is a hard form of carbon containing aromatic compounds of volatile, low molecular weights and high molecular weight "tars" with a host of oxygenated compounds between. In addition, water, sulfur, sulfur containing compounds, and "ash" (which is essentially feldspar) add to the complexity of the material. It is frequently very difficult to devolatilize and thus may cause "gasing" in melt processing of plastics, as well as reactivity with a number of polymers. Acid resistance is generally good and alkali resistance poor.

Wood, nutshell, bark, and cork flours are all similar for their lignin and cellulose contents, but are neither consistent nor uniform in their analyses. Their structures are porous to varying degrees, enabling reaction with contained acids, tannins, waxes, salts, etc. Chemical and solvent resistances are poor.

General considerations. Chemical composition also derives a variety of surface activities and effects. Alumina is catalytically surface-active under many circumstances. Similarly, when kaolin is calcined to produce calcined kaolin, it too has alumina surfaces. The surface is not always an indicator of the material within and vice versa. Because of this, tests are employed to determine the nature of the particles according to the media in which they are dispersed. Electrophoretic mobility, zeta potential, and pH reflect surface effects in aqueous media. In non-aqueous media, conditions of surface activity frequently depend on sorbed surface water. Flocculation and dispersion are thus very complex functions of chemical composition which are too involved for lengthy discussion here. In section 4, on the systemic effects of fillers, this subject will be discussed relative to rheology and dispersion.

3. SECONDARY PROPERTIES OF FILLERS

3.1 Optical

Color is an obvious property of fillers, but is frequently misunderstood. The color of a dry powder is usually evaluated by pressing a disc-shaped compact and obtaining a reflectance value at a particular wavelength of light on the flat surface. When the full spectrum of visible light is used for the evaluation, subtle differences in the true color of the particles can be determined by calculation or by inspection of the reflectance graph, although such data is influenced by particle size. Particles reflect light, as well as refract or bend light, according to their sizes. When a pressed disc is evaluated for reflectance or "brightness" by ASTM E 97, the effect of particle size is obscured. Color is also obscured due to the high reflectance of the powder in air according to the Fresnell equation:

$$R = \frac{(N_1 - N_2)^2}{(N_1 + N_2)^2}$$

where R is the reflectance coefficient or the relative amount of light reflected, N_1 is the refractive index of the particle, and N_2 is the refractive index of the surrounding medium.

When air is the surrounding medium (refractive index = 1) of a typical mineral filler having a refractive index of 1.53, $R = 0.0438$. But when the medium is a plastics matrix having a refractive index of 1.5, $R = 0.0001$. Although this is over-simplified, almost no light is reflected by filler particles dispersed in a matrix under the above conditions. This will be discussed more fully in systemic effects of fillers, section 4.6, but it is necessary here to emphasize that optical reflectance of a filler in air will not always provide a measure of its reflectance or color in a matrix.

In regard to refractive index, we are limited to discussing clear or translucent particles which will transmit light. Few fillers are either amorphous (non-crystalline) or isometric crystals, and therefore would have but one refractive index. Only glass beads and flakes are amorphous, and even opal, which is essentially amorphous hydrated silica (diatomite), contains crystalline aggregates of cristobalite, a high temperature quartz. No example of isometric crystallizing fillers are known to be in general use, although fluorite (CaF_2) is one which may be used for certain special applications.

Since reference will be made to crystal structure of fillers in regard to optical as well as other structure-dependent properties, a general basis for crystal structure versus properties is given.

The internal structure of any crystalline material is constant. The particular elements, arrangements of their atoms, and the types of bonds between them determine the crystal morphology or shape. Atoms of the elements have different sizes and valences or abilities to combine with other elements. When atoms combine, they arrange themselves in motifs to produce electrical neutrality or the most stable configuration. This configuration of motifs is called the space lattice and is the framework on which the crystal is built. There are only 14 possible ways in which points (groups of atoms) can be arrayed in space to produce these three-dimensional units. The atoms which constitute these point groups may be arranged in only 32 ways, which results in 230 possible types of arrangements, called space groups[9]. Since the groups of atoms at these lattice points may be single atoms, molecules, or ionic clusters, they may build on one another in a much greater number of ways to produce crystals.

Only isometric lattices and isotropic substances have all points equidistant from each other. When a beam of light passes through a crystal, the velocity of the light is reduced according to the number of atoms in its path. When all atoms are equally spaced, the crystal will have only one refractive index, which is the reciprocal of the velocity of the light, regardless of the path of the light beam through the crystal. Ordinary salt is an example of an isometric crystal. More commonly, lattices are not so simply arranged, and their crystals vary considerably in the orientation of atoms or groups of atoms.

Many crystals differ in the lengths of the three axes. When the shorter *a* axes are equal and the longer *c* axis is at right angles, the *c* axis is the optic axis. Light passes along the *c* axis without being polarized, but when light passes through the crystal in any other direction, it is broken into two rays with different velocities. This produces two refractive indexes, and the difference between them is called the birefringence. Very common examples of this are quartz, with 1.553 and 1.544, and calcite, with 1.658 and 1.486 refractive indexes.

Many other crystals have two optic axes, that is, light passes through the crystal in two directions without being altered. Such crystals are said to be biaxial and have three refractive indexes.

These differences in refractive index give rise to another property of crystals which depends on the wave length of light. Since white light contains all wave lengths of visible radiation from 0.4 to 0.7 microns, a filler particle having a particular thickness in the path of the light beam will split the beam, and one portion (wavelength) will emerge out of phase with the other, with resulting interference colors produced. Interference colors depend on crystal or particle orientation, thickness, and birefringence. For example, quartz particles in plastics may have any orientation, but a sufficient number will be oriented so that some particles of 30 μ thickness will pass white light. Below this thickness, the light approaches extinction, so that at 10 μ a dark gray color is transmitted. Above 30 μ, various interference colors are produced.

Other optical properties of fillers lie on either side of the visible spectrum; X, ultraviolet, infrared, and microwave radiations. X rays are used to identify materials according to the diffraction obtained from layers of atoms. X rays have wavelengths of from 1 angstrom (0.0001 μ) to 100 angstroms, which is in the range of atomic spacings, so that they can penetrate matter via the spacings between atoms.

Ultraviolet (UV) lies in the range between X and visible radiation; 0.01-0.4 μ wavelength. Fillers such as coal and graphite absorb UV light. Organic fillers and most polymers are degraded by UV, that is, they will decompose over a protracted period of exposure. Inorganic fillers absorb, transmit, or reflect according to the crystal orientation, and many produce luminescent effects. Fluorescent minerals produce visible light when exposed to UV radiation. If the luminescence persists after exposure to UV, the material is said to be phosphorescent, although there is no precise distinction between the two. The conversion of UV to longer wavelength radiation indicates that the filler converts this destructive radiation to a less destructive form. Luminescent properties, such as fluorescence and thermoluminescence, depend on the presence of impurity ions as activators, so these properties are not dependable indicators of the ability of a filler to aid in stabilizing polymers.

3.2 Thermal

Since many plastics processes involve heating, melt processing and the like, and since filled plastics are exposed to thermal influences, there are four properties of fillers which are important. These are: thermal conductivity, specific heat, thermophysical effects, and thermochemical effects.

Thermal conductivity, K, is reported as g · cal/(sec · cm^2)(°C/cm). Inorganic fillers have values on the order of 1-8 × 10^{-3} and organic fillers have values less than 1 × 10^{-3}; however, graphite has a value of 0.1. At either end of the scale is air ($K = 6 \times 10^{-6}$) and metals (at up to $K = 0.92$ for copper). Most polymers have K values one order less than those for fillers; 1 × 10^{-4}, since their constituent atoms are free to rotate and to dissipate the effects of thermal energy, whereas mineral fillers are rigidly fixed groups of atoms.

Specific heat, or heat capacity, is the amount of heat in cal required to raise 1 g material 1 °C (cal/g) and most fillers have values of about 0.2. Water is the standard at 1 cal/g. Plastics have values in the order of 0.3-0.5. When these values are multiplied by density, to obtain cal/cm^3, the range for plastics and fillers is 0.35-0.56. Again, graphite and coal are exceptions, with volumetric specific heats of 0.2-0.25. Ambient air has a value of 0.17 cal/g, but its density is so low that it requires very little heat to raise its temperature. Metals have very high specific heats, with aluminum having a value of 0.64 cal/cm^3.

Coefficient of thermal expansion is reported as units per unit of linear dimension per °C for a specified range of temperature. The coefficient of cubical expansion is reported as the volume increase per unit volume of material per °C and is about three times that of linear expansion.

Most fillers expand differently in different directions due to nonuniform composition or crystalline structure. The value for pine wood (wood flour) is 5.4×10^{-6} parallel to the fiber and 34.1×10^{-6} in./in./°C perpendicular to the fiber. At high temperatures, and especially if the fibers are aligned, moldings containing wood flour can be expected to exert considerable expansion forces. Coefficient of expansion values for inorganic fillers lie in the range of about $1\text{-}8 \times 10^{-6}$ linear units/linear unit/°C. Metals have greater thermal expansions, about $1\text{-}3 \times 10^{-5}$ in./in./°C, and are uniform in all directions, due to their cubic (isometric) crystal lattices.

Thermochemical effects are of particular importance to fillers. Flammability of fillers is limited to organic types, with graphite, again, an exception, having oxidative stability to 1,000°C. Hydrated alumina, which is actually aluminum hydroxide, begins to lose significant amounts of hydroxyl water at 230°C and continues endothermically until water loss totals 35% at 900°C. Hydrous silicate fillers also split off hydroxyls as water, but at much higher temperatures. Limestone fillers liberate carbon dioxide beginning at about 900°C, and continues at that temperature to leave a residue of calcium oxide or lime. Aluminum pigments will burn, and other metallic fillers are effective thermal conductors to promote heating and decomposition of plastics.

3.3 Physical

Density is the most important physical property of fillers since it affects the economics of composites (section 5.2.1). General rules which apply are that densities for solid fillers vary from organics to metals in ascending order. Porous or cellular fillers fall into the lowest density range, depending on the voids contained in the fabricated part.

Hardness is a most significant property of fillers. There are a number of systems applied to categorizing material hardnesses, but the Mohs scale, developed for minerals, is useful as an abrasiveness index, as well. In essence, the Mohs scale rates one material against another for its ability to scratch or to be scratched by an adjacent member in the series, or by a knife or file. The scale is arbitrary and is not linear, as reported in Table 2-7.

Crystals have varying degrees of hardness, depending on the direction in which they are scratched and the particular crystal surface, both of which depend on the atomic orientation and spacing. Only isometric crystals, such as the cubic crystals of metals, have but one hardness value, but frequently the differences are too subtle to be evaluated precisely according to crystallographic direction.

Particle size and shape are two determinants, in addition to Mohs hardness, of filler abrasive-

Table 2-7. Mohs Scale of Hardness.

	Value	Material
Softest	1	talc, vermiculite 1.5
	2	kaolin, gypsum, tin, mica 2–2.5 serpentine asbestos 2.0–5.0
	2	fingernail
	3	calcite, barite, copper, gold, anhydrite 3–3.5
	4	aragonite, flourite, iron 4.5
	5	tremolite 5–6, wollastonite 5–5.5 steel knife, windowglass 5.5
	6	feldspar 6–6.5, nepheline 6–6.5 titanium dioxide pigment 6–6.5 file steel 6.5
	7	silica, zircon 7.5
	8	topaz
	9	corundum (alumina)
Hardest	10	diamond

NOTE: Where sufficient distinction can be determined, minimum and maximum values are given.

ness. Pure crystals of titanium dioxide have been determined to have a Mohs hardness of 7-7.5, making the most useful white pigment extremely abrasive. However, the average particle size of the pigment is about 0.25 microns, and the crystals do not have distinct sharp edges; thus, the abrasiveness is considerably less than might be expected from hardness alone.

Abrasiveness is frequently a property of contaminants or impurities in an otherwise soft filler. Unrefined kaolin may be quite abrasive due to low concentrations of silica. The silica is usually of a much larger average particle size than that of the host mineral, and therefore, increases the abrasiveness considerably.

Coefficient of friction varies with shape, surface, and bonding forces between crystals or particles. Talc, for example, has silica surfaces on either side of the platelets (crystals) with so little Van der Waals forces acting between them that they may be easily separated with the fingernail. The silica surface is also hydrophobic, so that the normal neutralization of charge by sorbed water layers present on almost all surfaces has little effect on its greasy feel and lubricating properties. Graphite has a similar nature and is an excellent lubricant, as are most platy particles. Rough, angular, and hard particle surfaces generally produce increased friction between themselves or other components of a mixture.

A final physical property of filler particles is porosity or sorptiveness. Diatomite is the highly porous opal skeleton of diatoms. Wood and nutshell flours may have natural or induced porosity due to processing. More subtle is porosity on a microscopic scale, which is frequently obvious as staining on particles. Larger particles of limestone are generally quite porous, since they are concretions of very small crystals This property may be somewhat confused, since fillers are electrostatically charged when bonds are broken in the comminuting process, and agglomerates which form as a result may appear to be absorptive, larger particles. This will be dealt with later under the section on dispersion of particles, and is also an effect of adsorption or, more properly, chemisorption, which is related to surface charge and bonding with oppositely charged particles, molecules, or ions.

3.4 Electrical

It is necessary to consider the fundamental nature of filler particles in the *absence* of water when discussing electrical properties. Virtually every surface at ambient conditions has a condensed layer of water molecules on its surface which is more or less tightly bonded, depending on the nature of the surface. For this reason, the electrical properties of fillers may differ considerably on a fundamental versus a practical basis. As a generally valid statement, all metallic bonds produce excellent electrical conductivity, whereas ionic and covalent bonds produce nonconductors. Some materials have mixed bonds; ionic and metallic, and are semi-conductors. Electrical conductivity varies with crystallographic direction. Graphite, for example, is a much better conductor parallel to the platelet than through the shorter dimension.

Fillers having hydroxyl surfaces, adsorbed free ions, mixed contaminants of polyvalent metal compounds, free salts, water, and water soluble matter all provide surface conductivity paths for electricity.

High current densities or discharges will carbonize organic fillers; thus, carbon bridges are the conductors.

In light of the above facts, it is apparent that the use of fillers can lead either to improved or adversely affected electrical properties; thus, due care must be taken in the choice of the fillers used in electrical applications.

4. SYSTEMIC EFFECTS OF FILLERS

4.1 Economics

When the use of fillers is first considered, it is usually to reduce the cost of a compound. This purpose exemplifies the regard generally held for fillers and filled systems. The term filler itself implies adulteration, but unfortunately, a more suitable term has not been devised to characterize the variety of materials used for this purpose. Nonetheless, economics is an important consideration, although there are many circumstances under which fillers exert property improvements compared to plastics without fillers.

Filler concentration is properly reported as

percent by weight of the total composition for costing purposes. The percent composition and the density relate to the cost/part; volume of part X cost/unit wt X density. Economically and technically, a filler replaces a portion of the matrix by volume, so it is very important, for cost analysis, that the true density of the fabricated part be known.

An outline for costing a filled versus an unfilled compound is given in Table 2-8. Accounting procedures vary considerably with certain segments of the plastics industry, so the purpose here is to provide a basis which is generally applicable. For example, raw material costs alone may be very deceiving. Batch sizes may increase considerably, due to the higher bulk densities of mineral fillers to offer preferred economics. Table 2-8 was arranged to show such effects in a dry-blending operation. With liquid batches, the total volume will be very nearly that of the regular mixer capacity and may be calculated from the true densities of the ingredients.

The costs indicated are approximations to be used as a guide only, but the intention is to show that costs for other than raw materials may represent more than 50% of the product cost.

Extrusion operations may be reduced in volume output due to the slower thruput rate of higher melt viscosity filled compounds. Injection molding, on the other hand, may be somewhat more efficient due to shorter chill times, because a filled compound will gain stiffness more rapidly on cooling than an unfilled compound.

Cost comparison of fillers must be done on a volume basis also; cost/unit wt X density or specific gravity. The true density of the filler in the part must be known for two reasons: the filler may be incompletely dispersed, purposely or otherwise, so that voids appear in the compound and part; and in the case of low-density, void containing fillers, such as perlite or microballoons, degradation occurs on processing to solid particles. In either case, the density of the compound must be determined to obtain the functional density of the filler.

Table 2-8. Costing Filled Versus Unfilled Compounds.

RAW MATERIALS/BATCH			UNFILLED			FILLED		
Material	$/unit wt	Density	Wt	Volume	$	Wt	Volume	$
Polymer	0.50	1.25	100	80.000	50.00	85.00	68.000	42.50
Additives	1.00	0.95	1	1.052	1.00	1.25	1.316	1.25
Filler	0.05	2.71				85.00	31.365	4.25
Sub totals			101	81.052	51.00	171.25	100.681	47.50
Density, calculated				1.246			1.701	
Density, determined				1.246			1.700	
Manufacturing								
Process yield, %				98			97	
Batch yield, wt			98.98			166.11		
Hours/batch				1.0			1.25[a]	
Direct costs, $/batch · hr[b]					20.00			25.00
Fixed costs, $/batch · hr[c]					10.00			12.50
Overhead, $/hr[d]					10.00			12.50
General expense, $/hr[e]					5.00			6.25
Total cost/batch, $					96.00			103.75
Cost/unit wt yield, $					0.9698			0.6245
Cost/unit volume, $ ($/wt X density)					1.2083			1.0622
Cost/unit volume, %					100			82.79

[a]Assuming material handling and addition of filler requires an additional 15 min.
[b]Includes labor, supervision, maintenance, power, etc.
[c]Includes depreciation, taxes, insurance, rent, etc.
[d]Includes general maintenance, employee benefits, control lab costs, warehousing, etc.
[e]Includes administration, sales, financing, packaging, research, etc.

$$\text{Filler density} = \frac{d_c F_f}{1 - (d_c F_p / d_p)}$$

where

d_c is the determined compound density
F_f is the fractional proportion by weight of the filler
F_p is the fractional proportion by weight of all ingredients except the filler
d_p is the density of all ingredients combined except the filler.

Using the above equation, the true cost of the filler/unit volume may be calculated directly; cost/unit wt X determined density in the compound. For a cost comparison of the filler in other systems, and for technical reasons, it is advantageous to know the true volume and true volume cost of the filler.

4.2 Physical

4.2.1 Modulus. The most prominent physical effect of fillers is the stiffening, or modulus increase, which they cause in composites. Figure 2-6 is an adaptation relating modulus effects of fillers to the modified Kerner equation[11]. Basic data and equations were taken from texts by Nielsen[12,25].

The equations for M, any modulus, which may be shear, Young's, or bulk are:

$$\frac{M}{M_1} = \frac{1 + ABV_f}{1 - B\psi V_f}$$

$$A = k_E - 1 \qquad B = \frac{M_2/M_1 - 1}{M_2/M_1 + A}$$

$$\psi = 1 + \left(\frac{1 - P_f}{P_f^2}\right) V_f$$

where

M, M_1, and M_2 are moduli of the composite matrix and filler, respectively
k_E is the Einstein coefficient (see Table 2-9)
V_f is the fractional filler volume
P_f is the packing fraction of the filler at maximum filler concentration.

The constant k_E accounts for the shape of the filler particles and Poisson's ratio of the matrix. Section 4.3.3 describes the rheological method for determining k_E which is more precise for fillers having varied shapes and sizes, typical of those most commonly used. Correction for Poisson's ratio of the matrix at the temperature of testing is necessary, as indicated in Table 2-9.

Solid fillers always increase the modulus of a composite, and Fig. 2-6 is a generalized version of this effect, with reference to the influential

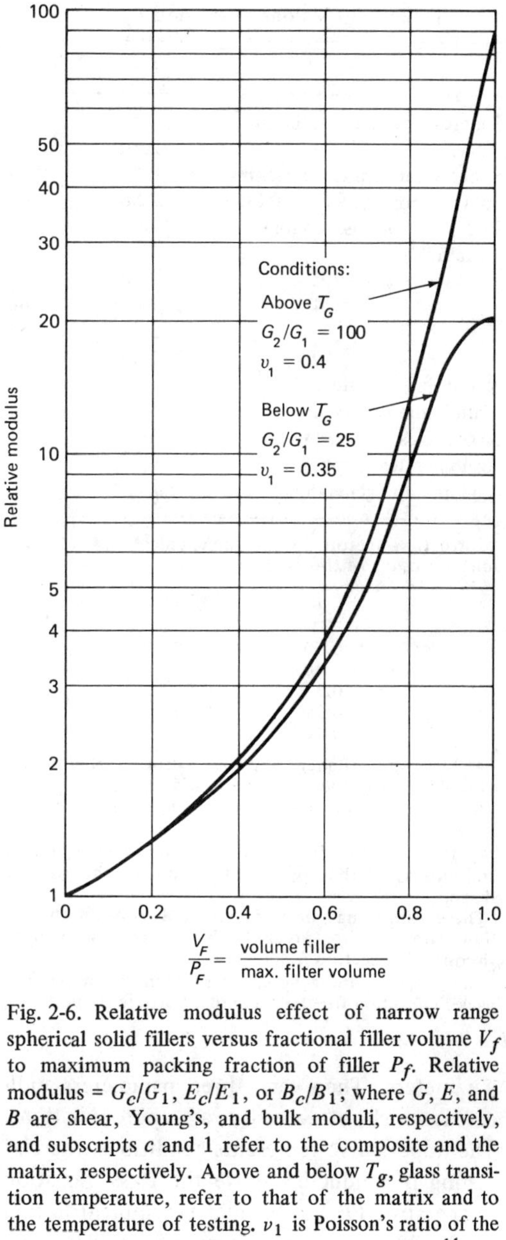

Fig. 2-6. Relative modulus effect of narrow range spherical solid fillers versus fractional filler volume V_f to maximum packing fraction of filler P_f. Relative modulus = G_c/G_1, E_c/E_1, or B_c/B_1; where G, E, and B are shear, Young's, and bulk moduli, respectively, and subscripts c and 1 refer to the composite and the matrix, respectively. Above and below T_g, glass transition temperature, refer to that of the matrix and to the temperature of testing. ν_1 is Poisson's ratio of the matrix and G_2 is the shear modulus of the filler.[11]

Table 2-9. Values of k_E and P_f[25,28].

Filler	k_E[a]	P_f
Spheres, one size, maximum packing	2.50	0.74
Spheres, random close packing	2.50	0.64
Spheres, random loose packing	2.50	0.60
Rods or ellipsoids, random packing[b]		
Aspect ratio = 2	2.58	0.68
4	3.08	0.60
6	3.80	0.53
10	5.93	0.42
Mixed sizes, irregular shapes, minimum surface area	4.00[e]	
Mixed sizes, plates, and flakes	5+[e]	
Agglomerates of spheres	$2.5/\phi_a$[c,e]	
Agglomerates generally	k_e/ϕ_p[d,f]	

[a]Correction factors for the mechanical case of Poisson's ratio, ν_1, of the matrix:

ν_1	Factor
0.50	1.00
0.40	0.90
0.35	0.87
0.30	0.84
0.20	0.80

[b]At high rates of shear and low V_f for the rheological case, these shapes tend to orient and to reduce k_E.
[c]ϕ_a is the volume fraction of the agglomerate which is spherical.
[d]ϕ_p is the volume fraction of particles in an agglomerate, and k_E is the appropriate value for the particle shape.
[e]These values may be approximated from oil absorption data or more precisely determined by intrinsic viscosity; see section 4.3.3.
[f]These values are generally less than the maximum packing fraction for the particular particle shape.

parameters. There are three parameters indicated, which, of necessity, are approximations. The upper curve represents conditions whereby testing is conducted above the glass transition temperature (T_g) of the matrix; ambient testing of flexible matrices such as polyethylene and plasticized polyvinyl chloride. The lower curve represents rigid matrices wherein the composite is tested below its T_g, as in thermosets and rigid PVC. This is a principal distinction which must be considered because of the pronounced influence of the matrix or polymer. Moduli of the filler and matrix are also very influential, since solid fillers are generally many times more rigid than polymers. Poisson's ratio, ν, the change in width divided by the change in length in tensile testing, is 0.5 for an ideal elastomer which does not change in volume when stretched. Flexible polymers, rigid polymers, and solid fillers have values of about 0.4, 0.3, and 0.3, respectively.

Due to the manner in which test specimens are usually molded, at least two surfaces of the filled composite will be matrix-rich and therefore test data do not reveal the true moduli of an ideally uniform distribution of the filler. The data of Fig. 2-6, however, are based on actually determined values of typically molded specimens. Large, narrow size-range particles will provide lower moduli for the above reason, but finer, wide size-range fillers having high packing fractions will provide about equivalent values. It should be noted that the curves of Fig. 2-6 represent completely and uniformly dispersed filler particles. Agglomerates or flocculated particles provide higher moduli, since that portion of the matrix which is isolated in the agglomerate is less free to react to stress and strain than the continuous phase at values below the maximum packing fraction P_f. Oil absorption by volume provides a reasonable guide for maximum packing fraction as previously discussed. Particle shape may have a pronounced effect on these calculations. When filler aspect ratios become greater than about 2 to 1, as with flakes and fibers (see Table 2-1), moduli increase considerably above about 0.3 relative filler volume $V_r = V_f/P_f$. In all cases, the maximum filler volume, P_f, achievable in a system governs all physical effects, since to have a uniformly dispersed system, the minimum matrix present will be that required to coat the particles and to fill the interstitial voids. In this regard, it should be noted that the concentration of matrix required to just coat the filler per unit of its surface area will vary considerably from one matrix to another and from one filler to another. For fillers, this will depend on their surface polarities, presence of water, degree of work in dispersion, presence of dispersants such as vinyl stabilizers, and surface pretreatments such as coupling agents, as well as previously mentioned factors.

An additional factor in filler effect on physical properties stands alone for its significance. It was mentioned in section 3.2 that fillers have different thermal expansions in different crystallographic directions. Since heating is most generally involved in plastics processing, differences in the coefficient of thermal expansion of the filler and matrix produce internal stresses at room temperature. These stresses produce significant loss of modulus below the T_g of the matrix[13] and is more pronounced with increasing filler concentration. The stress will vary according to filler particle orientation and reach a maximum with high aspect ratio fillers such as flakes and fibers. It follows, then, that lower residual systemic stresses will be produced by spherical fillers having coefficients of expansion matching those of the matrices. Further, it also follows that with high aspect ratio fillers, their orientation in the molded item should be such that the maximum effect on modulus is produced where it is of greatest benefit. For example, maximum flake or fiber orientation is produced in thin sections of uniform cross-section and flow. Annealing of filled items made with matrices having T_g's below room or use temperature has been observed by the author to significantly reduce thermal stresses and to result in greater uniformity of modulus throughout the item.

Finally, the presence of air bubbles or voids caused by agglomerates of undispersed filler particles or air entrainment in processing will lower the modulus, since the modulus of air or voids may be assumed to be zero[14]. This would offset the modulus increase due to solid fillers and is undoubtedly the cause of disparate data for filler effects.

4.2.2 Tensile Strength. Fillers affect tensile properties according to their packing characteristics, sizes, and interfacial bonding. The maximum volumetric packing fraction of a filler, P_f, reflects the size distribution and the shapes of the particles. At P_f, the filler particles of a composite are separated by at least the thickness of matrix sufficient to uniformly coat them. The spaces between particles are assumed to be filled with matrix and no voids or air bubbles are present. Under these conditions, for a given system, the matrix volume is at a minimum and acts as individual segments or pockets to support a tensile load. When a tensile load is applied, these matrix segments stretch and pull away from the particles, resulting in the very low strengths and elongations of highly filled composites. At the other extreme, a specimen of matrix is not restricted in its ability to stretch and to support a load and therefore would be expected to provide greater strength. Now, if two fillers having significantly differing P_f values are compared well below P_f and at the *same* filler volume in the same matrix, the one having the higher P_f will have larger spaces between particles, larger segments of matrix to support the load, and therefore will provide a higher tensile strength.

Large particles cause greater stress concentrations in the matrix than small ones. All other conditions being equal, smaller average particle sizes provide higher strengths. This effect is well documented[25].

The effect of filler to matrix bonding may be more easily understood by referring to stress whitening which occurs when a tensile specimen is elongated. At a particular stress for the filled composite, the matrix de-wets or separates from the filler surface, causing vacuoles or minute voids, which reflect light due to the refractive index differences between components; this is called the Fresnell effect. For a given critical strain and equal interfacial bonding, the immobility and greater surface of a large particle will cause more voids to form in the matrix than will occur with an equal volume of smaller particles.

Under conditions where the particles have been treated with bonding or coupling agents, the matrix is severely restricted by interfacial bonding, so tensile strengths are improved but elongations are reduced.

The ultimate in tensile reinforcement will be obtained with very fine, high packing fraction fillers having a very high order of adhesion with the matrix. Under these conditions, the particles are more likely to be displaced uniformly with the matrix as the composite is distended, so that maximum benefit may derive from the improved adhesion.

Figure 2-7 presents a generalization based on the relative tensile yield strength versus the relative filler volume. Tensile yield strength is used here, since plastics applications are limited to those at or below the yield strength. Primarily,

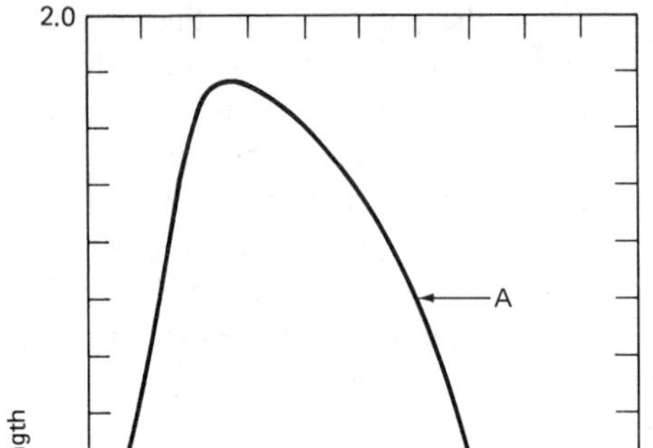

Fig. 2-7. Relative tensile yield strength = tensile yield of composite/tensile yield of matrix. V_f/P_f = relative packing volume; volume of filler in composite/maximum packing volume of the filler.

A = Above T_g of non-polar, crystallizing polymers with high aspect ratio, high surface area fillers.

B = Above T_g of non-polar, crystallizing polymers with low surface area fillers.
Below T_g of polar, rigid polymers with with reinforcing fillers; fibers, flakes, and those treated with coupling agents.

C = Below T_g of polar, rigid polymers with low surface area fillers.

D = Above T_g of polar polymers with low surface area fillers.

the T_g of the matrix and its polarity are limiting on the filler effect. Non-polar polymers which orient or crystallize with strain, such as polyethylene, exhibit remarkable tensile yield strength improvements with most fillers (curves A and B). They also exhibit poor bonding with fillers, and thus, are able to draw around the particles to orient and work-harden. Fillers which promote this effect are high aspect ratio, high surface area platy and fibrous types which, due to their greater surface per particle, exert a greater viscous drag on the polymer per unit volume of polymer to aid in molecular alignment. Curves B and C reflect differences below T_g of polar matrices due to filler type and bonding. Reinforcing fillers, which are surface modified, and high aspect ratio fillers will provide strengths above those indicated for curve C. Thus, the same parent filler of curve C may provide an effect indicated by curve B. These curves represent the temperature of testing versus the T_g of the matrix.

There are certain subtle effects frequently produced with low relative filler volumes whereby an increase in tensile strength is obtained. This appears to be due to a slight restriction of the matrix, which induces an orientation or work hardening effect in the order of about 0.05 relative filler volume.

A case which is not illustrated in Fig. 2-7 is below T_g of the matrix for non-polar, amorphous polymers. Polystyrene is such a polymer and, as may be expected, is severely degraded in tensile yield strength by rigid, non-reinforcing fillers, since its elongation is very low without fillers and any interference with interchain molecular bonding causes strength loss.

The data for Fig. 2-7 were obtained from various sources, particularly Reference (15), and calculations were based on known or determined values of P_f, using the equation:

$$\sigma_{yr} = \frac{V_{mr}\,\sigma_{ym}\,A}{\sigma_{ym}}$$

where

σ_{yr} is relative tensile yield strength; σ_y composite/σ_y matrix
σ_{ym} is tensile yield strength of the matrix
V_{mr} is relative matrix volume fraction; $V_{mr} = 1 - (V_f/P_f)$
and A is an adhesion factor.

For rigid systems below T_g of the matrix or above T_g for polar systems, A varies from 0.8 to about 1.1 for different types of fillers. This affords a convenient method for evaluating the effectiveness of one filler versus another and for evaluating the precison of the data.

For other matrixes or mixed systems, A varies according to V_{mr} and has not been investigated thoroughly for contributory factors such as draw ratio per matrix volume, adhesion to filler, Poisson's ratio, polymer crystallinity, molecular weight, and others.

Tensile yield strength requirements impose severe limitations on particulate filled systems. Only under certain circumstances, as above, will fillers be of benefit. Polymeric fibers which are drawn to orient the molecules, and thus, improve toughness can be, and are, significantly benefited by the use of fillers, which help to arrest the orientation stresses and to improve tensile strength.

4.2.3 Flexural Strength. When a force is applied at the center of the upper surface of an end-supported beam, as in ASTM D 790, the upper surface is under compression and the lower surface is in tension. Maximum stress is applied only to the surface in tension. The decreasing strain through the cross-section (from lower to upper surface) had the effect of providing somewhat higher flexural than tensile yield strengths. This is especially pronounced for matrices having elongations greater than about 5%. Inherently tough polymers may obtain higher flexural strengths with high aspect ratio fillers or solid fillers treated with coupling agents. The upper boundary of maximum effect generally conforms with curve B of Fig. 2-7. Also conforming to curve B are those matrices which are toughened by fillers (non-polar, rigid polymers with moderate elongations) or various thermoset polymers with moderate elongations.

In general, fillers reduce flexural strength in proportion to their relative packing volumes, as indicated by curve D of Fig. 2-7.

4.2.4 Elongation. All present theories fail to describe the actual, and frequent, occurrence of equal or improved strengths and elongations resulting at low V_f. This phenomenon is particularly noticeable with fine or very fine fillers and at testing conditions above the T_g of the matrix. One of the primary failings relates to inadequate data, particularly those which do not distinguish between ultimate and yield elongations. It is a rare occurrence for elongation at the yield point to be increased by a filler, but there are many examples of increased ultimate elongation at V_f values of about 5%.

The explanation for the above may be drawn from the observation that stress whitening occurs almost immediately on elongation of composites containing higher filler concentrations, or when the filler consists of large particles. In both cases, the matrix is restricted in its ability to stretch between packed particles or around large particles, and de-wetting or de-bonding of the particles causes cavitation, which appears as stress whitening. Smaller particles at low concentrations are freer to move with the matrix, and correspondingly, the matrix is freer to stretch around them. The viscous drag of small particles with good adhesion to the matrix can then produce an apparent strength improvement and greater tenacity, resulting in higher ultimate elongations. If the interfacial bond exceeds the matrix strength, the matrix is less free to draw around the particles, and lower elongations and higher strengths result. Conversely, very weak interfacial bonding causes almost immediate separation of the matrix from the particles and cavitation begins at low elongations.

Since elongation is one dimension of volume, it will vary as the cube root of volume according to the approximation:

$$\epsilon_c \cong \epsilon_m \left[1 - \left(\frac{V_f}{P_f}\right)^{1/3}\right]$$

where ϵ_c and ϵ_m are elongations of the composite and matrix, respectively, and V_f and P_f were defined above. This equation has been found to be consistent with many data. No compensation is indicated for several factors, such as adhesion of the matrix to the filler, Poisson's ratio of the matrix, k_E of the filler, and, possibly, T_g of the matrix.

4.2.5 Tear Resistance. Tear resistance involves a combination of effects, but the predominant effect is that of filler-matrix bonding in flexible composites. As in impact resistance testing, where the area under the stress-strain curve is proportional to toughness or the energy to fracture a specimen, tear resistance is a measure of crack or slit propagation. To illustrate the role

Table 2-10. Filler effect on tear resistance. (Flexible vinyl, 30 phr DOP, 50 phr filler.)

Filler	Silane[a]	TENSILE STRENGTH, PSI at 100% elongation	TENSILE STRENGTH, PSI at break[b]	Elongation at break, %	Tear, lb/in.[c]
None	None	2600	3870	210	500
9 micron[d]	None	2250	2500	160	410
9 micron[d]	Yes[e]	2400	2620	180 (?)	520
4 micron[d]	None	2340	2550	170	425
4 micron[d]	Yes[e]	3270	3270	125	510

[a] Union Carbide A-1100.
[b] ASTM D 638.
[c] ASTM D 1004, die C.
[d] Average particle size, feldspar filler.
[e] About optimum concentration for tear resistance.
(*Courtesy Union Carbide Corp. and Lawson United Feldspar and Mineral Co.*)

of filler-matrix bonding, the data of Table 2-10 indicate that without a coupling agent, tear strength is a function of particle size or surface area, but with a coupling agent, both tensile and tear strengths are considerably improved.

4.2.6 Impact Strength. A full description of impact strength may be found in Reference (25). Particularly emphasized is that impact strength is of great practical importance and that few test methods simulate actually encountered conditions of use. Fundamentally, impact strength is proportional to the area beneath the stress-strain curve at relatively high testing speeds. As stated in the previous section, both strength and elongation are contributory.

It has been established that inclusions in a matrix act as stress concentrators. Such inclusions must differ substantially from the matrix in ductility. Thus, crystallites, voids, particles, fibers, notches, and cracks all act as stress concentrators. Considering only the case of particulate fillers (not fibrous reinforcements), it is apparent that particles having greater ductility than the matrix will tend to improve impact strength. Impact modifiers are generally of this class. Most commonly used fillers are rigid, very high modulus materials with virtually no elongation, and therefore embrittle the composite. This statement must be qualified according to two factors; if a filler contributes improved cohesive strength to a composite compared to that of the matrix, and if the filler particles distribute the impact stress over a larger area perpendicular to the force of impact, then an improvement in impact strength will result. Fibers operate in this manner and, as discussed in section 4.2.5, fillers having improved adhesion to the matrix can improve tear and impact strengths of low T_g matrices. Improvement is usually limited to low filler concentrations and very fine particle size fillers. Composites tested below the T_g of the matrix are generally embrittled by fillers due to locked-in stresses.

Again, the packing fraction of the filler is an important determinant. Poorly packing fillers occupy larger volumes, and therefore, contribute greater numbers of stress concentrators or, to view the composite as a whole, they more effectively reduce the continuity of the matrix. Since it is the matrix which must absorb the shock of impact, those fillers having high packing fractions will tend to reduce impact strength much less at the same filler volume.

4.2.7 Compressive Strength. Compressive strength applies only to rigid materials and is somewhat akin to tensile strength in reverse. Deformations are considerably lower, so the filler-matrix adhesion and cohesive strength of the weakest component will govern this property, very much as indicated by Fig. 2-7. Soft, compressive fillers, such as wood and nutshell flours, will reduce compressive strengths, whereas the reverse will be true for hard, solid fillers, provided that adhesion between components is equal to or greater than the cohesive strength of the matrix.

Deformation under load (ASTM D 621) defines the strain component of compressive strength and varies with relative filler volume, as described in section 4.2.4..

4.2.8 Creep and Stress Relaxation. Creep is actually a viscoelastic property, as elucidated by Nielsen,[25] and its practical aspects are described in the *Modern Plastics Encyclopedia*.[17]

Fillers increase the relative viscosity of thermoplastics and, as expected, reduce the degree of creep or deformation over a period of time for an applied initial stress. Viscometric effects will be detailed in section 4.3. Packing fraction is a determining factor in creep and most other systemic effects. Relative packing fraction determines the proportion of free matrix, and therefore, maximum creep resistance can be expected at the maximum packing fraction, P_f.

Creep for filled systems may be estimated from relative modulus data:

$$\epsilon(t) = \epsilon_1(t)\,\frac{E_1}{E}$$

Where ϵ is the creep elongation at any time (t) for the composite, ϵ_1 is the corresponding creep of the polymer or matrix, and E and E_1 are the elastic moduli of the composite and matrix, respectively.[25] Thus, the moduli of the filled and unfilled matrix can be readily determined and compared with data for the matrix only. This would represent a great saving in time but conditions must be suitably chosen for the composite. Because elongations may be severely reduced by the filler, these calculations will be valid only up to the point at which the filler begins to dewet or debond from the matrix. At this point, catastrophic failure usually results.

It is possible to calculate moduli, as indicated in section 4.2.1, for various filler concentrations and to thereby estimate creep, as above. This would provide a very good starting point for an experimental or a candidate filler.

4.2.9 Hardness. Hardness of plastics is, by most tests, a measure of modulus of elasticity. As such, these tests do not conform with these for minerals (Mohs), metals (Brinell), coatings (Hoffman, Toucon), and others. Plastics hardness tests are convenient, quick methods for determining, to a degree, stiffness (modulus), resilience, creep, deformation under load, tear strength, compressive strength, and other properties, depending on the nature of the test and how it is applied.

Fillers which increase the moduli of composites generally will increase hardness values of thermoplastics and elastomeric thermosets. Penetration tests, such as Barcol and durometer, rather than indentation tests, such as Rockwell indentation, for rigid thermosets may not indicate the presence of filler or its type, due to the inherent toughness of the matrix, unless the filler is much less resistant to penetration (wood and nutshell flours, cork, bark, etc.).

Hardness is a function of the relative packing fraction of the filler and its true modulus. Other factors which influence hardness of filled systems are degree of dispersion, interfacial bonding, inhibition of cross-linking in thermosets, shear degradation of the polymer by the filler during processing, distribution of filler due to sample preparation, disruption or improvement in crystallite bonding, filler adsorption of stabilizers, processing aids or other additives, and probably many more.

4.2.10 Coefficient of Friction. The direct influence of fillers on coefficient of friction, $\mu = F/W$, where F is the tangential force to move one surface over another under an applied load W normal to the surface, only becomes important when the filler is at the surface. Molded surfaces are normally resin-rich, so the differential contraction between the filler just below the surface and the matrix can cause a degree of irregularity in surface smoothness. This condition is exaggerated by coarse fillers at low concentrations, and has been applied to blown film and blow-molded items to provide a degree of roughness for better printing. For items molded under pressure, surface irregularity due to filler becomes extremely difficult to measure. At filler concentrations sufficiently high to reduce shrinkage on cooling or polymerization, surfaces will be more nearly flat and thus reduce the plowing component of friction which results from surface waviness.

Where lubricating or low coefficient of friction fillers are used to reduce friction, they are effective when the surface has been worn or polished to expose the filler.

Fillers reduce the coefficient of friction if they are fine enough to effectively increase the T_g of the matrix very near the surface, since frictional heating and matrix softening are frequently responsible for high coefficients of friction.

For filled systems, the best uniformity in surface smoothness and low coefficient of friction will be obtained with those fillers providing the highest packing fractions and at concentrations approaching P_f. The limiting condition depends on the actual volume of matrix separating the filler particles, since the matrix will have a greater coefficient of expansion and contraction than the most commonly used mineral fillers, and reaches a minimum with fillers capable of being more closely packed.

Powdered Teflon and molybdenum disulfide have been used individually to reduce the coefficient of friction of composites. Such composites have been molded into gears and sleeve bearings. Teflon has the lowest coefficient of friction of any solid material. When molded into parts, such as oil-less bearings, its wear resistance is improved by a factor of 1,000 by the inclusion of fillers such as powdered bronze or graphite. The reasons for this effect are discussed below.

4.2.11 Abrasion Resistance. Abrasion resistance is usually measured by subjecting a specimen to a highly abrasive medium. Such methods as ASTM D 394, 1044, 1242, 1526, and 1630 are generally accepted for specific service conditions.

Friction is a highly influential component of abrasion, and several of the factors mentioned in the last section apply here. Three factors appear to predominate in the effects of fillers on abrasion or wear resistance; the Mohs hardness of the filler, the strength of the interfacial bond, and the relative filler packing fraction. In the first case, most abrasive media consist of silica, corundum, or the like, which are hard, tough particles, either free, bonded to a substrate, or in a soft matrix. If the filler is softer than the abrasive (lower Mohs hardness), it is readily ground or abraded. Filler-matrix bond

Table 2-11. Abrasion resistance of filled PVC Compounds.[a]

				TABER ABRASER TEST DATA	
PVC Type	Filler[b]	V_f[c]	phr Filler	Revs/100 mils[d]	% of Control
Rigid	None	0.00	0.0	2278	100
	LU-390	0.04	10.0	4444	195
	LU-390	0.12	30.0	4537	199
	York White	0.04	10.4	1450	64
	York White	0.12	31.1	1030	45
Flexible	None	0.00	0.0	1035	100
	LU-390	0.07	26.0	336	32
	LU-390	0.19	78.0	154	15
	LU-390	0.28	130.0	105	10
	York White	0.07	27.0	331	32
	York White	0.19	82.0	159	15
	York White	0.28	136.0	107	10

[a]Same basic formulation used for both with 50 phr (parts per hundred resin) of DOP added for flexible compound.
[b]LU-390 is feldspar, Mohs hardness 6–6.5. York White is limestone, Mohs hardness 3. Both fillers have almost identical particle size distributions and average particle sizes of 14 microns.
[c]V_f = fractional filler volume of composite.
[d]ASTM D 1630, CS-14 wheels.
(Courtesy Lawson-United Feldspar and Mineral Co.)

strength is significant when the abrasion resistance of the filler is greater than that of the matrix. In this case, hard particles are dislodged from the matrix, and this effect becomes apparent with larger particle size fillers, as indicated in Table 2-11. It may be noted that rigid PVC has inherently good abrasion resistance, which may be doubled by the inclusion of low filler volumes of harder and tougher mineral fillers. Limestone, however, is considerably softer than the abrasive, and therefore detracts considerably from the abrasion resistance of unfilled rigid PVC. Flexible vinyl compounds, which are only moderately resistant to the severe conditions of this test, provided the opposite effect with fillers, due to the lower tensile strength of the matrix, coupled with the poorer filler-matrix bonding. Under magnification, the voids left by the removal of the larger filler particles were apparent.

Finer particle sizes provide somewhat higher adhesion factors, and provided that the abrasion resistance of the filler is at least equal to that of the matrix, coupling agents would also be expected to yield improvements.

The abrasion resistance of a composite ultimately depends on the conditions to which it is subjected in use. Where the matrix contributes the qualities of resilience and slip, its proportion will determine the resistance of the composite to abrasion. If the filler is the main contributor to abrasion resistance, then its proportion will determine that property. Since all systems are equal at P_f, both cases are satisfied at the relative filler packing fraction; V_f/P_f, and fillers providing the highest P_f will provide the best abrasion resistance in either case.

4.3 Rheological

4.3.1 Filler Concentration, V_f and P_f. Rheology; the science of flow, is a complex subject made more so when suspensions of particles, rather than pure liquids, are involved. Green's text is still considered a basic reference.[18] More extensive, and treating all aspects of colloids, dispersions, and suspensions of particulate matter, is Mysels' text.[19]

Of the two primary effects of fillers, that of their concentrations on viscosity, will be discussed first and the role of dispersion later.

Patton[20] developed a basis for predicting the effect of filler concentration on relative viscosity, by deriving the equation:

$$\frac{1}{\log(\eta_f/\eta_0)} = \frac{1}{kV_f} - \frac{1}{kP_f} \quad \text{or}$$

$$k = \frac{d\,1/V_f}{d\,1/(\log \eta_f/\eta_0)}$$

where η_f is the viscosity in poises of the suspension of filler at the stated filler volume, η_0 is the viscosity in poises of the liquid of the suspension, d is the difference in these values, and k is a derived constant for the slope of the straight line. Table 2-12 provides the data of Fig. 2-8, adapted from Patton.

Patton has stated that η_f is the viscosity at infinite shear rate (η_∞) where the rate of shear in $\sec^{-1} = dv/dx$, and where dv, the differential velocity, and dx, the differential thickness, is sufficiently high to eliminate effects due to filler flocculation or to other sources. For very viscous media, such as molten plastics, this would be at a fairly low shear rate; below about 500 $\sec^{-1}$, since considerable deviation from linearity in shear stress-shear rate curves occurs at higher shear rates and would be further deviated by the filler. Very fluid systems would be much less problematical for developing graphs yielding the straight line of Fig. 2-8. In any case, it is probably advisable to determine viscosity at several shear rates for plotting, in which case the curve of the data of Table 2-12 would also be a straight line, but would go through the origin (Fig. 2-9) using the equation:

$$\log \frac{\eta_f}{\eta_0} = \frac{kP_fV_f}{P_f - V_f} \quad \text{where } k = \frac{d\,V_f}{d(\log \eta_f/\eta_0)}$$

Note that in this case k is different from that of Fig. 2-8. In either case, the calculation of k and the known V_f allow the calculation of P_f, which may be graphically solved in Fig. 2-8 as $1/1.65 = 0.606$. In practice, it would be advisable to use both equations to assure conformance with the other and to apply average values.

Determining P_f by this means will provide direct values for the continuous or fluid medium

Table 2-12. Computation of effects of fillers on viscosity.

V_f	$1/V_f$	η_f	η_f/η_0	$\log(\eta_f/\eta_0)$	$1/\log(\eta_f/\eta_0)$	$V_f/P_f - V_f$
0.091	11	1.942	1.295	0.1123	8.90	0.177
0.100	10	2.000	1.330	0.1260	7.95	0.198
0.125	8	2.196	1.464	0.1655	6.04	0.260

η_0 = 1.5 poise. (See text and Fig. 2-8 for derivation of P_f = 0.606.)

and will differ for each medium. Although it has been stated that oil absorption may be used as a method for determining P_f, different polarities, viscosities, molecular entanglements, etc. for the fluid medium actually used will provide necessarily different P_f values. Thus, this method is valuable for ascertaining the P_f of the filler or mixtures of fillers in a particular system.

It must be emphasized that the above equations relate to viscosities of the fluid medium, η_0, which may include volatile liquids; water or solvents. In cases where such volatiles are present, the P_f values obtained will be that for the entire system, and not for the solution solids of the system.

Viscosity determination, to be useful in these calculations, must be made with instruments yielding fundamental data. For viscous suspensions, Saybolt or Severs are perhaps the simplest of the capillary orifice instruments. Similarly, the Instron viscometer may be used for viscous melts. Such data have been successfully applied to mixed particle systems such as plastisols, using Severs, *not* Brookfield, viscosities.

4.3.2 Dispersion. Dispersion of particles in fluid media is a complicated process.[20,21] There are three principal factors concerned with the work or energy required to obtain a complete separation (dispersion) of particles in fluid media. Dispersion is the sum of the work of adhesion, immersion, and spreading of a liquid on a solid surface:

$$W_d = -6\sigma_1 (\cos\theta)$$

where the work of dispersion W_d is a function of the liquid surface tension σ_1 and the contact angle of the liquid with the solid surface θ. For simple liquids, and many solids, these values

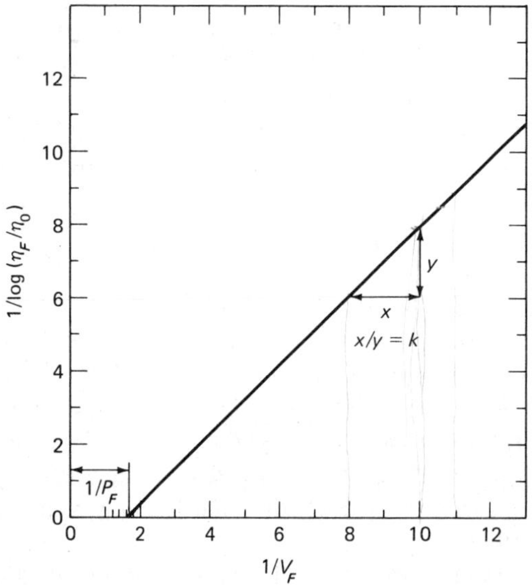

Fig. 2-8. Computation of P_f by graphical means.

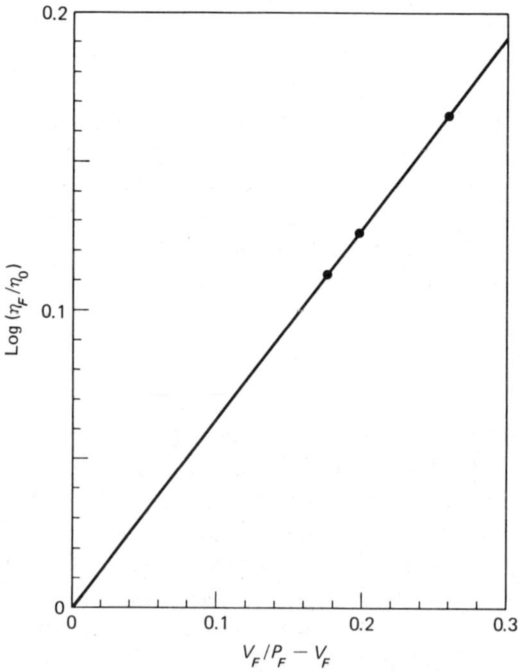

Fig. 2-9. Effect of filler volume on rheology.

may be determined quite readily, but for molten plastics, the values are very difficult to obtain. Also complicating dispersion is the viscosity of the fluid medium. As viscosity increases, the capillary forces pulling liquid into the interstices of filler agglomerates via surface tension result in slower surface wetting rates. Then, work must be performed on the agglomerate to overcome the slow dispersion process, or the filler surfaces must be pre-treated for reduction of the liquid contact angle. Spontaneous or gravitational spreading will occur if the contact angle is 0°; therefore, conditions which approach this will aid dispersion.

Estimates of maximum dispersion of fillers in liquids may be made by a titration versus the viscosity of the mixture at constant volumetric solids, as illustrated in Fig. 2-10. Although viscosity in poises is indicated as the ordinate, arbitrary units, such as torque or force for rotational viscometers, or seconds to flow through an orifice for efflux viscometers, may be used. Arbitrary units will not, however, provide a true relationship to the fundamental unit, the poise. Further, such data are only an approximation of the true state of dispersion, since one-point values do not characterize the reheology of the system as newtonian, thixotropic, plastic, pseudoplastic, or dilatant, all of which

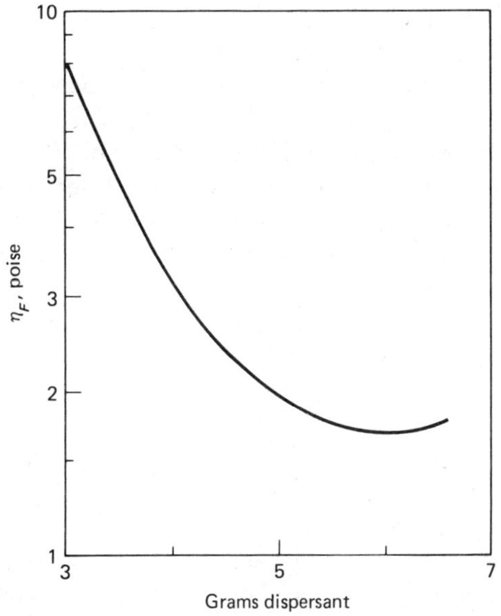

Fig. 2-10. Titration of a filler dispersion to determine minimum viscosity versus dispersant concentration.

may be a result of over- or under-dispersion of the particles, depending on the shear rate applied to the system. At high shear rates, the energy supplied to the mixture may give the apparent effect of complete dispersion, whereas at rest or at low shear rates, the mixture may "set" or gel as an indication of thixotropic or plastic rheology.

In the above discussion of packing fraction, as well as in all of section 4, complete dispersion of the filler particles is assumed and necessary for the indicated effects to be obtained. Dispersion is an often neglected factor in filler performance and one which has made much comparative data on fillers meaningless, since optimum conditions of dispersion were not employed.

Physical dispersion, or the uniform mechanical distribution of particles in viscous media, is often not difficult to obtain. However, time dependent effects of wetting and spreading of the matrix on the filler surfaces can often be neglected. In some cases, several days aging of a filled compound before fabrication results in significantly improved properties. Similarly, the use of wetting, dispersing or coupling agents should be considered for improving dispersion. Many systems contain agents which are used for one purpose but serve as dispersants, such as vinyl stabilizers. Their relative effects are sometimes very noticeable in filled plastics, where significant differences in rheology result with the substitution of one for another. A number of lubricants also fit this category. Finally, the interaction of different fillers and pigments with each other can result in nonuniform colors, rough surfaces, porosity, poor physical properties, etc., in a fabricated part.

Of the many ways to evaluate dispersion, rheological testing is the most precise. The lowest viscosity is always an indication of optimum dispersion. Note that for certain ionizing dispersants, flocculation can be induced and viscosity increased by an excess of dispersant, as indicated by Fig. 2-10. This is usually encountered in aqueous media, but can also occur with high molecular weight, polyfunctional dispersants in non-aqueous media.

Microscopic inspection of thin sections also may reveal a flocculated condition of the filler.

Very accurate specific gravity can reveal the presence of porous floccs of fillers for non-destructive testing, provided that accurate specific gravity data are obtainable for the matrix and filler, as described in section 4.1.

4.3.3 Intrinsic Viscosity and k_E. A final aspect of rheology concerns the determination of intrinsic viscosity, which is also called limiting viscosity number $[\eta]$ for fillers or k_E:

$$[\eta] = \left(\frac{\eta_{r^{-1}}}{V_f}\right)_{V_f \to 0}$$

where η_r is the relative viscosity (viscosity ratio) η/η_0, the viscosity of the suspension divided by the viscosity of the fluid medium, and V_f is the fractional filler volume.

A typical graph of the above for spheres is shown in Fig. 2-11. Absolute viscosity data are required and the extrapolation to zero filler volume provides the incremental effect of the filler on viscosity or k_E. Most fillers have ranges of sizes and shapes which tend to vary over the size range, so this is an invaluable tool to determine k_E for use in the above sections. Filler blends which may vary widely in size distribution, particle shape, and P_f also may be assigned a k_E value.

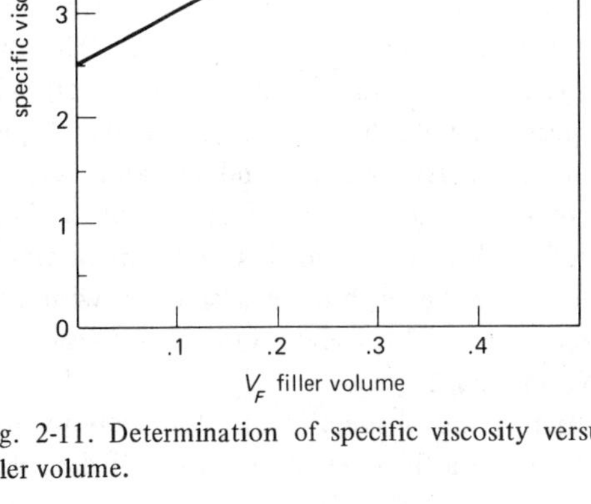

Fig. 2-11. Determination of specific viscosity versus filler volume.

4.4 Chemical

4.4.1 Corrosion Resistance. Assuming complete dispersion, it is apparent that a filler particle will be completely enveloped in a coating of matrix, with a continuous phase between particles. Chemical resistance of the system then depends essentially on the matrix. Since molded surfaces are usually filler deficient, it follows that many non-chemical resisting fillers may be used with corrosion resisting polymers. When fillers are present in blow-molded, drawn, or stretched items, the filler particles will be nearer the surface and the glosses of such items will be reduced. This indicates a very thin film of matrix covering the particles. In such cases, the transmittance properties of the matrix for the corrosive agent will be more critical and will determine the choice of filler. These conditions apply to static conditions of chemical resistance where, for example, a container is filled with a corrosive agent. Under dynamic conditions, as in pipes, pump linings, tanks, or in any application where abrasion or wear may occur, it is essential that the filler be at least as resistant to the corrosive as the matrix. This is an often overlooked factor in the long-term durability of filled compositions. Furthermore, fillers which have good chemical resistance and improve the abrasion resistance of the fabricated items are most preferred.

Those fillers which are inherently porous (wood flour, cork, bark, diatomite) may provide apparently adequate short-term water or corrosive resistance if uniformly embedded in the matrix. Air or voids due to high temperature processing provide pockets which accumulate vapors or condensate as a result of matrix permeability. Thermal cycling of such compositions will usually accentuate the degradation process, and accelerated tests of this type are essential for the evaluation of long-term durability. Due to differential expansion and contraction of the filler, matrix, and void, great stresses are induced which result in checking, cracking, and even distintegration of an item under severe thermal cycling. Chemical resistance, in general, is a primary attribute of the continuous phase, while that of the filler and its complete dispersion are secondary.[22]

4.4.2 Permeance. Solid, inorganic fillers generally reduce the permeability of plastics to liquids and gases, provided that the filler-matrix bond is unaffected by the permeance of the matrix. Fillers having significant water solubilities of about 0.04% or greater (limestone, wollastonite, anhydrite, terra alba) frequently severely reduce strengths after exposure to hydrogas or water. The soluble material at the filler-matrix interface de-bonds the matrix, and capillaries are sometimes formed which allow the composite to imbibe high concentrations of water. Since capillaries are two-way streets, such composites as limestone-filled thermosetting polyesters also exhibit a "frosted" surface, due to soluble matter exuding with the water when the surface dries.

Fillers having adequately low water solubilities or transmittance properties significantly reduce permeance according to their volume concentrations, since gases or liquids passing through the permeable matrix must take a more circuitous route. Platy fillers, such as mica, are especially effective. Plastics equilibrate to a specific absorption or permeance, so when their volumes are reduced by nonsorptive fillers, equilibrium is reached at a much slower rate.

4.4.3 Interface. Another chemical effect of fillers in composites concerns surface reactivities of the fillers or their associated impurities. First; few, if any, fillers are genuinely nonreactive (see section 2.5), and many polymers have reactive sites, chemical groups, or contain additivies which are reactive with each other and the filler surface. In many instances, the reactivity of the filler is actually chemisorption, wherein the charged organic molecules bond to the filler by Van der Waal's force.

Water which is condensed and/or chemisorbed is present on the surfaces of all pulverized matter. With fillers containing calcium and magnesium carbonate, water is the catalyst causing reaction with free acids or carboxyl groups of thermosetting polyesters, especially below 100°C. This reaction causes microscopic bubbles to form in laminates, due to the carbon dioxide liberated.

Compounding techniques which apply heat and vacuum stripping to remove water generally produce more uniform physical properties, except in instances where water is a plasticizer for the polymer, as with nylon. Epoxies are also catalyzed by trace quantities of water, and isocyanate or isocyanurate groups are reactive with water to liberate carbon dioxide, which is utilized in the foaming and gelling reaction.[23]

Silicate minerals of several classes have silanol and aluminol groups on their surfaces as a result of broken and hydrated bonds or as an inherent part of their structures. These hydroxyls are proton donors or Lewis acids. Proton acceptors, such as amines and other onium compounds, will then bond securely to the surfaces of silicate fillers. In a like manner, epoxies, which are catalyzed by proton donors, are polymerized by kaolin, fine silica, and other acidic minerals. Package or long-term stability prior to polymerization is likely to be poor with such fillers unless inhibitors or surface neutralizing agents are present.

Salts of heavy metals, such as vinyl stabilizers, will also bond to acidic fillers and, in so doing, will become deactivated so that their effectiveness is reduced. Because many polar agents are used as processing aids, stabilizers, lubricants, and the like, comparison of one filler with another in a "standard" formulation may not provide a true evaluation under conditions most suitable for a particular filler.

The complexity of interfacial phenomena is described in two recent symposia.[24]

4.4.4 Impurities. Predominant impurities causing undesirable reactions with polymers and additives are included in the tables of section 5.1. Systemic effects of impurities are complex and worthy of greater study and attention than they have been given by filler suppliers.

Impurities which can affect the stability to oxidation, temperature, ultraviolet, etc., of a composite are those which react with or cause a reaction of the polymer. They are principally inorganic compounds found as extraneous minerals in mineral fillers. A most common contaminant is manganese dioxide, which is a powerful oxidizer. Copper oxides cause depolymerization of some polymers and inhibition of polymerization of unsaturated thermosets. Vanadium oxides are catalysts for many organic and fotochemical reactions. The above and other contaminants form salts with free acids or carboxyl groups of unsaturated polyesters to produce characteristic effects and colors. Under certain conditions, a few parts per million of polymer may suffice to produce an undesirable effect.

Impurities in vegetable-derived organic fillers are even more complex, since it is frequently difficult to distinguish the impurity from the primary material. Many inorganic and organic compounds are present and analyses vary widely, so it may be difficult, at best, to isolate the cause of a particular systemic effect.

4.5 Thermal

4.5.1 Coefficient of Thermal Expansion. Coefficient of thermal expansion can only be estimated, as there are several complicating factors. First, fabrication of a composite involves shrinkage of the matrix due to thermal processing or polymerization which is greater than that of the rigid fillers on cooling. Most polymers have coefficients of expansion of about ten times greater than rigid fillers. The filler is thus under compression at room temperature, and very high residual stresses surround the filler particles.

Rigid, cross-linked polymers frequently reveal microchecks due to differential shrinkage in the polymer-rich areas of composites where coarse, narrow size range or poorly packing fillers are used. Microcracks in the composite provide slippage planes which yield coefficients of expansion considerably below those which may be expected. Exceptionally high stresses at the filler-matrix interface, induced by coupling agents, for example, also have the effect of reducing thermal expansion.

By definition, the filler packing fraction is that point at which the filler is at maximum concentration and is completely wet by and dispersed in the matrix. Thus, the filler particles are always separated by at least that film of matrix necessary to coat the particles, and residual stresses will increase as the film thickness decreases.

If the composite has been annealed to reduce or eliminate residual stresses, which may not be practical or even possible, the coefficient of expansion may then be estimated by the rule of mixtures.

The tables of section 5.1 list coefficient of expansion values for fillers to be used in the equation:

$$a = a_1 V_m + a_2 V_f$$

where the coefficients of expansion of the composite, matrix, and filler are a, a_1 and a_2, respectively, and V_m and V_f are the fractional volumes of matrix and filler, respectively.

In this and other cases, orientation of flake and fibrous fillers will have a significant effect. Depending on the manner of fabrication and the selection of the specimen used for testing, thermal expansion can be quite different in different directions. Fibrous and platy fillers severly restrict expansion in their longitudinal directions by viscous drag and mechanical restraint. Because of this, the matrix may expand to a much greater extent in the transverse direction to relieve the stress. This is typically the case when warping on heating or cooling is seen in composites. It has been mentioned in section 3.2 that crystalline fillers and others have different coefficients of expansion in different directions, which also introduces stress differentials.

4.5.2 Deflection Temperature. Tensile, flexural, and torsional tests are used to determine elongation or deflection temperature under an applied stress (ASTM D 1637, 648, 1043, and 1053). Lower stresses are applied to polymers above T_g, and higher stresses are applied below T_g. In all cases, these dynamic tests have greater value if graphs, rather than one point values, are generated. Varying the stress can tell much about the matrix-filler interaction above and below the deflection temperature. For example, fillers tend both to promote crack propagation and to build heat on flexing, causing reduced flex and fatigue lives and greater stress relaxation and creep with a rise in temperature. A series of graphs could then be used to anticipate the limits of use of filled composites. Fillers also cause internal stresses which vary according to the method of fabrication. At the beginning of these curves, the specimens may shrink slightly as the temperature is raised, indicating that orientation or residual stress may have to be relieved by annealing. The annealing process also produces higher deflection temperatures for filled plastics.

De-bonding or de-wetting of the filler is also indicated when catastrophic failure occurs in a deflection temperature test by a steep incline in the curve of deformation versus temperature. The point at which the composite fails may then be estimated, and an upper temperature limit assigned. Since moduli, as well as elongation, can be estimated below the inflection point of the curve, many practical applications are delineated.

Because of the manner in which deflection tests are conducted, deflection temperatures for amorphous polymers are nearer to their T_g's, whereas for highly crystalline polymers, deflection temperatures are nearer to their melting points. Since fillers increase moduli and reduce creep, the effect of the amorphous or crystalline matrix predominates.

Fillers increase modulus, viscosity, and—assuming that the applied stress is below the point at which significant filler de-wetting or de-bonding occurs—deflection temperature. Packing fraction is highly influential, since maximum effect will be obtained at P_f and, for a particular filler type, the relative filler volume will be controlling.

4.5.3 Thermal Conductivity. Thermal conductivity is reported as

$$k = \frac{g \cdot \text{cal}}{\sec \cdot \text{cm}^2\ (°\text{C/cm})} = \frac{H}{ta\,(dT/L)}$$

or other units based on Btu.

Assuming that a composite is void-free and water-free, and that the filler particles are uniformly dispersed, the thermal conductivity may be approximated by the rule of mixtures:

$$k = k_1 V_m + k_2 V_f$$

where k, k_1 and k_2 are the thermal conductivities of the composite, matrix, and filler, respectively, and V_m and V_f are the volumes of matrix and filler, respectively. For a sheet containing 25 volume percent limestone: $V_f = 0.25$, $k_2 = 5.6 \times 10^{-3}$; and PVC: $k_1 = 3.9 \times 10^{-4}$; and $V_m = 0.75$: $k = 16.9 \times 10^{-4}$. The filled sheet is therefore about 4 times more conductive than PVC only.

Nielsen[28] has developed a more precise method for predicting thermal conductivities of composites. Although less simple than the above, the equation provides very good agreement with experimental data:

$$\frac{K}{K_1} = \frac{1 + AB\,V_f}{1 - B\psi V_f}$$

$$A = k_E - 1$$

$$B = \frac{K_2/K_1 - 1}{K_2/K_1 + A}$$

$$\psi = 1 + \left(\frac{1 - P_f}{P_f^2}\right) V_f$$

where K, K_1 and K_2 are the thermal conductivities of the composite, matrix, and filler, respectively, k_E is the Einstein coefficient, Table 2-9, ψ is a reduced concentration factor, and B is a constant to account for the relative conductivities of the two components.

Using the prior example with $P_f = 0.7$ from oil absorption data, and $k_E = 4$ from Table 2-9;

$$A = 4 - 1 = 3$$

$$B = \frac{(5.6/0.39) - 1}{(5.6/0.39) + 3} = 0.77$$

$$\psi = 1 + \frac{1 - 0.7}{0.7^2}\,0.25 = 1.15$$

$$K = \frac{3.9 \times 10^{-4}\,(1 + 3 \times 0.77 \times 0.25)}{1 - (0.77 \times 1.15 \times 0.25)}$$

$$= 12.6 \times 10^{-4}$$

According to these calculations, the filled sheet is about three times as conductive as the PVC sheet only. For these conditions, the rule of mixtures provides a fair approximation, but will provide much higher calculated values than those actually determined, especially with increasing filler volume.

4.5.4 Specific Heat. Specific heat is reported as the calories required to raise the temperature of 1 g material 1°C. When experimental values are not available, Kapp's rule[27] may be used; the sum of the heat capacities of the constituent elements of a compound provides an approximation of the cal/g formula weight using the appropriate atomic heat capacities for liquids and solids:

Element	Liquids	Solids
C	2.8	1.8
H	4.8	2.3
O	6.0	4.0
S, P	7.4	5.4
F	7.0	5.0
Si	5.8	3.8
B	4.7	2.7
All others	8.0	6.2

For limestone of the example of section 4.5.3;

$CaCO_3$ = 40.08 + 12.01 + 3 × 16 = 100.09, and 6.2 + 1.8 + 3 × 4.0 = 20.0/100.09 = 0.20

The reported value for calcium carbonate at 20°C is 0.205 cal/g.[27]

In a thermoforming process where a sheet is heated from one side, as in the determination of K in section 4.5.3 above, thermal conductivity and specific heat may be combined to determine the heating rate of the sheet. First, correction must be made for density versus specific heat per gram when, using the example of section 4.5.3, a rigid PVC sheet contains 25 volume percent limestone versus the unfilled sheet at a thermoforming temperature of 200°C:

PVC specific heat = 0.36 × 1.3 density = 0.468 cal/cm^3
Limestone specific heat = 0.205 × 2.71 = 0.566 cal/cm^3

For the filled sheet, the rule of mixtures applies:

sp. heat by volume × V_f + sp. heat by volume × V_m = cal/°C/cm^3;

0.556 × 0.25 + 0.468 × 0.75 = 0.49 cal/°C/cm^3

For sheets having the following dimensions:

thickness = L = 0.1 cm
area = a = 40 cm × 40 cm = 1600 cm^2
volume = $L \times a$ = 160 cm^3

The total calories required to heat the sheet from ambient (20°C) to 200°C is:

for the PVC sheet, 160 cm^3 × 0.468 × 200 − 20°C = 13.5 × 10^3 cal
for the filled sheet, 160 cm^3 × 0.49 × 200 − 20°C = 14.1 × 10^3 cal

As stated, we wish an average internal temperature; L in the following equation = $\frac{1}{2}$ × 0.1 = 0.05 and the heat source is 300°C in each case:

$H = k \times a \times t \times (dt/L)$
for the PVC sheet; 13.5 × 10^3 = 3.9 × 10^{-4} × 1600 cm^2 × (300 − 200°C/0.05) t
13.5 × 10^3 = 1248 t = 10.8 sec
for the filled sheet; 14.1 × 10^3 = 12.6 × 10^{-4} × 1600 cm^2 × (300 − 200°C/0.05) t
14.1 × 10^3 = 4032 t = 3.5 sec

Therefore, although the filled sheet has a higher caloric requirement, its superior conductivity allows the sheet to reach average thermoforming temperature in only 32% of the time required for the unfilled sheet, and the cooling time will be similarly reduced.

Where exotherms are produced, as in unsaturated polyesters, similar calculations may be used to determine the temperature or time differentials between filled and unfilled compositions.

These calculations do not apply to dynamically heated systems such as occur in extruder barrels or intensive mixers, since air is entrained in the dry-blend, frictional heating is generated, sorbed water may be evaporated, and several other factors, such as lubricants, processing aid, and stabilizer effects make it very difficult to calculate the approximate effect of fillers. Practically, dry-blends always require more time or heat input to reach processing uniformity than polymers alone. For this reason, it is generally much more efficient to pre-compound filled plastics to produce pellets for feeding extruders or injection molding equipment to obtain minimal effect on the time and conditions to fabricate an item.

4.5.5 Fire Retardancy. Fire retarding applications of fillers are probably the most promising for market development and expansion.[30] Hydrous alumina is the most important filler of this type, functioning to reduce the temperature of the composite during exposure to flame, and releasing hydroxyl water beginning at 230°C, which is in the range of ignition of many organic materials. Although hydrous alumina will suffice, in itself, as a fire retardant at adequate concentration, most studies have focused on its effect with other fire retarding agents, such as phosphorous and chlorine, as part of the polymer or plasticizer.[31]

Two criteria are currently being applied to establish material resistance to fire and the human hazard involved; oxygen index and smoke generation. The former is a measure of the percent oxygen required in an atmosphere to sustain burning, and the latter is a measure of optical density or weight of smoke.

Antimony oxide is used at about 1-7% as a synergist with phosphorous and chlorine to provide fire retardance.

A recent introduction is Dawsonite, a hydrated sodium alumino carbonate which is synthetically derived, has an acicular particle shape, and may be used as a filler, reinforcement, and flame retardant.[30] Another new product is an organic filler, derived from soybeans, which has a low density and is used in conjunction with hydrous alumina.[32]

Generally, fillers containing hydroxyl water offer some advantage in conjunction with fire retarding additives, but their effects are minimal at low concentrations and at temperatures below about 1000°C. In certain thermoset plastics, limestone fillers disintegrate at about 1000°C in a fire, liberating carbon dioxide very rapidly so as to be ineffective for flame snuffing. The disintegration of the filler ruptures the composite and exposes more combustible polymer to the fire. The most important contribution of fillers in this regard is the reduction in volume of combustible matter, provided that the filler itself is noncombustible.

For detailed information on fire retardants, see Section V.

4.6 Optical

4.6.1 Color. The color of a fabricated part frequently limits the choice of filler. Ideally, pigmented (colored) systems would benefit from the choice of a filler having no effect on color. Unfortunately, glass beads and flakes, nepheline, and feldspar are the only fillers having virtually no effect on color but falling into fairly coarse particle size ranges. Pyrogenic or precipitated silicas will satisfy the very fine particle size range but are limited to very low concentrations. Hydrated alumina, highly purified kaolins, silica, and some talcs produce minor systemic color effects.

The primary reference to color effects of fillers is based on the color difference obtained with the pigmented matrix versus the pigmented matrix containing filler. Although the impurities present in mineral fillers may have a significant influence, refractive index and birefringence (section 3.1) are most influential. A simple dispersion of filler in a matrix will not yield a comparison of color effect of the filler unless the filler lends a distinct color to the system, such as wood flour, coal, etc. Fillers providing a high degree of transparency in the matrix usually appear slightly gray and have little effect on color.

All optical properties, as well as other properties, depend on the completeness of filler dispersion. White specks in a composite indicate incompletely dispersed filler particles present as agglomerates containing air. Similarly, water occluded in agglomerates may be evaporated during processing—or later—to leave voids which have the same opacity effect. The use of a hand magnifier or microscope is very helpful in determining the state of dispersion. Such effects as filtering of large filler particles on glass cloth or mat; agglomerates; non-uniform color pigment dispersion; flow lines magnified by non-uniform filler distribution; stress whitening; crazing due to resin-rich areas; surface irregularities due to excessive shrinkage and inadequate filler volume; and "chatter" marks on molded surfaces, due to dilatant flow induced by narrow particle size ranges at too-high concentrations are all frequently visually apparent and readily confirmed at low magnification.

4.62. Refractive Index. If a filler has only one refractive index and it matches that of the matrix, then the composite of the two will appear to be optically clear at or below P_f, assuming perfect wetting of the filler surfaces. Only glasses or isometric crystals satisfy this condition. Because most fillers are crystalline and have two or more refractive indexes, a match of the average to that of the matrix does not provide the ultimate in clarity. This was discussed in section 3.1, with particular reference to interference colors produced by crystalline materials according to birefringence, orientation, and thickness of the particles. These three factors interact to produce interference colors for fillers in plastics, since ranges of both particle sizes and orientations normally prevail. In one orientation only, quartz (silica) particles 30 μ thick will pass white light; thinner sections or particle thicknesses will transmit specific colors, depending on the thickness dimension. With random orientation of the particles and dimensions, which vary for each and from particle to particle, and the polarizing effect of crystals, silica filled plastics generally appear to be gray in transmitted light. Certain fillers having a uniform thickness, such as mica, may provide specific colors when carefully oriented in a part. Most fillers are not uniform crystals, so the result is generally one of graying when the average refractive index approaches that of the matrix.

When there is a difference in refractive indexes between the matrix and filler, Fresnell's law prevails (section 3.1). This applies only to essentially colorless and transparent particles. The greater the difference in refractive index, the greater will be the opacity effect. Calcite is a typical example, since its rcfractive index is 1.66 average and is sufficiently different from polymers to provide a degree of opacity.

4.7 Electrical

4.7.1 Dielectric Constant. Dielectric constant and dissipation factor are determined according to ASTM D 150. Solid fillers do not alter these values significantly. The primary sources of difference lie with water absorption, contaminants, and the like, which fillers introduce.

According to Nielsen[28], the equations of section 4.5.3 may be used with possible modification of k_E to predict permittivity (dielectric constant)[29]. Dielectric constant ϵ is the ratio of electrical flux density D to the electrical field strength E; $\epsilon = D/E$. The analagous equations are:

$$\frac{\epsilon}{\epsilon_1} = \frac{1 + ABV_f}{1 - B\psi V_f}$$

$$A = 1/A_e - 1 \cong k_E - 1 \qquad \text{(see Table 2-12)}$$

$$B = \frac{\epsilon_2/\epsilon_1 - 1}{\epsilon_2/\epsilon_1 + A}$$

$$\psi = 1 + \left(\frac{1 - P_f}{P_f^2}\right) V_f$$

where A_e is the depolarization factor and $1/A_e$ is analogous to k_E, which depends on the shape and orientation of the filler particles, and ϵ, ϵ_1, and ϵ_2 are the dielectric constants of the composite, matrix, and filler, respectively.

Using k_E values of Table 2-12, reasonable agreement may be obtained between calculated and experimentally determined values. Solutions depend on the state of agglomeration of the filler particles, which alters A and P_f significantly. Nielsen[29] suggests that A and P_f be considered as adjustable parameters based on observed states of the system (agglomeration) but for fully dispersed systems, both A and P_f should have non-ambiguous values.

4.7.2 Dielectric Strength. Dielectric strength is determined by ASTM D 149 and measures the dielectric breakdown of an insulating material by short-time or step-by-step tests in volts/0.001 in. thickness. Mineral fillers which do not disintegrate, due to the high temperatures involved, generally improve this property. Organic fillers, or those having sorbed or tightly bound free water on their surfaces, will reduce this property.

4.7.3 Arc Resistance. Arc or tracking resistance tests are determined by ASTM D 495. Electrodes are applied to a specimen and the time in seconds is taken for an applied DC voltage to arc between the electrodes. The primary cause of failure for polymers is carbonization of the surface; therefore, high mineral filler contents generally interrupt the formation of conducting carbon paths. Mineral fillers having high P_f values used at or near their maxima provide the greatest interruption of the electrical path.

4.7.4 General Electrical Properties. ASTM D 257 lists several other tests for determining electrical properties. Since most organic polymers are covalently bonded, their electrical properties will be very good. Many mineral fillers are ionic and covalently bonded; both kinds of bonds generate nonconductors. Only metallic bonded materials provide free electrons which readily transport electricity. For those electrical properties depending on volume, the calculations of section 4.7.1 will apply.

Electrical properties depend on many subtle effects of sample preparation, conditioning, uniformity, etc. The generalizations above are based on ideals which are frequently difficult to practice in this complex art. Certain processing aspects, such as the virtually complete dispersion of fillers and de-gassing to remove air and volatiles (particularly water) will provide the most predictable properties.

5. SELECTING THE FILLER TO FIT THE NEED

5.1 Classification of Commercial Fillers

Table 2-13 lists data for general classes of fillers. Many properties are not included since they do not provide a basis for selection and use. A great number of intermediate and different grades are available, so the sections of Chapter 3 should be consulted. Some data listed are nearest approximations or most probable data based on limited information. There are many grades of surfied (surface modified) fillers which provide significant improvements in all properties, largely due to improved dispersion and bonding. This table was devised to help the practitioner make a reasonable selection for calculating or anticipating the effects of a filler in a particular system based on section 4 above.

Special effects are obtainable with fillers designed for a specific purpose. Lubricating fillers are molybdenum disulfide, graphite, and teflon. Magnetic fillers are black iron oxide, powdered iron or shot, and barium and strontium ferrites. Similar in effect to microballoons or expanded perlite are glass nodules of Pittsburgh Corning, and Saran microballoons of Dow Chemical Co. Although generally used as reinforcements, cotton flock, alpha cellulose, and other fibers are used with, or as, fillers for decorative or viscosity control effects. Powdered cork and protein fillers are used for reducing resistance to compression or for sound absorption. Carbon blacks (channel and furnace) are more often regarded as pigments, but may be used in conjunction with coarser fillers to obtain better packing. Powdered bituminous coal (Austin Black) functions similarly. Fused silica is made in various particle sizes from high-purity silica for demanding electrical applications. Finally, beta eucryptite, a lithium alumino silicate, has a negative coefficient of expansion above ambient; -8×10^{-6}, and reduces the thermal expansion of composites.

5.2 Optimizing the Selection of Fillers

5.2.1 Cost Reduction. Cost reduction by the use of fillers is of primary concern to all plastics compounders. If improvements of certain properties can be obtained, then this can be an added bonus, frequently necessary to offset the increased compounding cost, inventory requirements, possibly higher product weight and density, and possibly lower output rate. The primary factor in cost reduction is filler volume concentration and its cost per unit volume. Obviously, the filler providing the highest volume concentration, P_f, in a system and having the lowest cost per unit volume versus the matrix will provide the lowest part cost. Figure 2-12 shows a graphical method for comparing matrix and filler volume costs. There is usually a point at which break-even occurs, due to cost-increasing factors mentioned above, and this must be determined for the particular process. Such a graph is easily constructed to suit the circumstances, and it provides several primary cost relationships. If maximum cost reduction is desired, several fillers can be compared in the same graph and the maximum utilizable filler volume for each will indicate the lowest. The volume of one filler necessary to obtain an equivalent cost reduction versus another also is indicated. This will then indicate concentration limits for other factors and eliminate the need to calculate the costs at intermediate concentra-

Table 2-13

CLASS		GROUND LIMESTONE				PRECIPITATED CALCIUM CARBONATE		
Producers[a] Properties	Terms	L1	L2	L3	L4	C1	C2	C3
Particle								
Size, mean	Microns, esd[b]	120	14	5	3	1.0	0.07	0.07
Range shape	Microns, esd[c]	590–20	74–0.3	35–0.2	20–0.1	2–0.1	0.5–0.1	0.5–0.1
	Class[d]	Block →						
Bulk								
Specific gravity	g/cm^3 [e]	2.71 →					2.65	2.55
Surface area	m^2/g [f]	0.3	0.9	1.2	2.2	5.3	20–26	11–17
k_E	g	4.0 →						
Packing fraction, P_f	h	0.8–0.83	0.8	0.68–0.76	0.67–0.7	0.46–0.53	0.37–0.54	0.39–0.64
Dispersibility: water	i	G →					G	N
organics	j	E →					G	E
Resistance	k	Al, W, S →						
pH		9.2–9.5 →				9–9.5	9–10.4	
Specific resistance	ohm · cm × 10^{-3} [l]	21–23 →						
Impurities		SiO_2	$MgCO_3$ →			$MgCO_3$	$CaSO_4$ →	
Thermal								
Conductivity	cal/sec, cm^2 (°C/cm) × 10^3	5.6 →						
Specific heat	cal/g°C	0.205 →				0.21 →		
Coefficient/expansion	linear unit/unit/°C × 10^6	10 →						
Optical								
Refractive index: mean	sin i/sin r	1.59 →				1.66 →		
range	–	1.49–1.66 →				1.53–1.69 →		
Color: powder	n	W →						
in DOP	o	W →						
Physical								
Young's modulus, E	dynes/cm^2 × 10^{-11}	2.6 →						
Poisson's ratio, ν	p	0.27 →						
Mohs hardness	q	2.5–3 →						
Electrical								
Dielectric constant	$\epsilon = D/E$ [r]	6.14 →						

tions. Such a solution, as indicated in the example of Fig. 2-12, may be used to calculate theoretical part density:

	Volume		*Density*	
Filler *B*	0.27	×	2.71	= 0.732
Matrix	0.73	×	1.30	= 0.949
Theoretical part density =				1.681

Where filler blends may be more efficient, the calculations of section 2.4 should be applied. The precise packing fractions are rarely obtainable, but serve as a guide to those which most nearly conform to the theoretical distributions. In most cases, it is desirable to consider fillers or combinations which provide the highest P_f, since these will provide the greatest

Table 2-13. *(Continued)*

KAOLIN					TALC		
K1	K2	K3	K4	K5	T1	T2	T3
5	1.2	0.8	0.6	1.2	6–8	5–6	1.8
45–0.1	35–0.03	20–0.03	10–0.03	35–0.4	60–0.3	50–0.1	15–0.05
Block and flake →	→	→	→	→	Block and flake →	→	→
2.58 →	→	→	→	2.63	2.8 →	→	→
6–10	10–12	14–16	18–22	8	6–8	6–10	17
4 →	→	→	→	→	4 →	→	→
0.53–0.56	0.53–0.56	0.49–0.51	0.46–0.48	0.35–0.44	0.51–0.56	0.52–0.54	0.4–0.42
P →	→	→	→	→	P →	→	→
G →	→	→	→	→	E →	→	→
Ac, Al, W, S	– →	→	→	→	Ac, Al, W, S →	→	→
3.8–52	5.3–6.0	3.8–5.4	3.8–7.	4.2–6.	9.–9.5 →	→	→
10+ →	→	→	→	20+	–	–	–
–	SiO_2	–	–	–	t	t	t
4.7 →	→	→	→	→	5.0 →	→	→
0.22 →	→	→	→	→	0.208 →	→	→
8 →	→	→	→	3	8 →	→	→
1.56 →	→	→	→	1.62	1.57 →	→	→
1.55–1.57 →	→	→	→	–	1.54–1.59 →	→	→
W–T	W–T	W	W	W	W →	→	→
T	T	T	T	W	G →	→	→
2 →	→	→	→	→	2 →	→	→
					0.3 →	→	→
0.3 →	→	→	→	→			
2 →	→	→	→	7	2	1	1
2.6 →	→	→	→	1.3	5.5–7.5 →	→	→

latitude in cost reduction as well as choice of filler volume for a specific property sought.

5.3.2 Physical Properties Improvement. In most cases, P_f is controlling, and although relative filler volume, V_f/P_f, is most important to some of the physical properties, it is preferable to first consider those fillers which have the highest P_f, because this allows more latitude in the range below P_f to obtain a specific property. Table 2-14 is designed to indicate this effect. The greatest improvement will result with the filler type indicated, but there are many circumstances where, for various technical reasons, the most preferred may be the least useful type or the most difficult to obtain. In this

Table 2-13. (*Continued*)

CLASS		MICA		FELDSPAR, NEPHELINE			WOLLAS-TONITE	PERLITE, EXPANDED
Producers[a] Properties	Terms	M1	M2	AS1	AS2	AS3	W	P
Particle								
Size, mean	Microns, esd[b]	s	5	15	8	5	6	Large
Range shape	Microns, esd[c]	–	14–?	90–0.5	40–0.5	30–0.3	44–0.1	Various
	Class[d]	Flake →		Block →			Block	Spheroid
Bulk								
Specific gravity	g/cm^3 [e]	2.82 →		2.6 →			2.9	0.15–0.2
Surface area	m^2/g [f]	–	30	0.8–0.9	1–1.3	1.8–2.3	1.1–1.4	1–2
k_E	g	5 →		4 →			4	3
Packing fraction, P_f	h	0.45	0.38	0.68–0.72	0.62–0.67	0.56	0.62	0.64
Dispersibility:								
water	i	G →		E →			E	E
organics	j	E →		E →			G	E
Resistance	k	Ac, Al, W, S →		Ac, Al, W, S →			Al, W, S	Ac, Al, W, S
pH		8–8.5 →		7–10	7–10	10	10	7.5–8
Specific resistance	ohm · cm × 10^{-3} [l]	–	–	4–15	4–15	3	Low	–
Impurities		–	–	u			$CaCO_3$	–
Thermal								
Conductivity	cal/sec, cm^2(°C/cm) × 10^3	6 →		5.6 →			6	0.02
Specific heat	cal/g°C	0.206 →		0.21 →			0.24	0.24
Coefficient/expansion	linear unit/unit/°C × 10^6	8 →		6.5 →			6.5	8.8
Optical								
Refraction index:								
mean	sin *i*/sin *r*	1.58 →		1.53 →			1.63	–
range	–	1.56–1.62 →		1.528–1.538 →			1.63–1.67	–
Color:								
powder	n	G	W	W →			W	W
in DOP	o	C	G	C →			W	W
Physical								
Young's modulus, *E*	dynes/cm^2 × 10^{-11}	3 →		3 →			3	0.01
Poisson's ratio, ν	p	0.3 →		0.3 →			0.3	0.4
Mohs hardness	q	2.5–3 →		5.5–6.5 →			5–5.5	5.6
Electrical								
Dielectric constant	$\epsilon = D/E$ [r]	2–2.6 →		6 →			6	1.5

case, compromises will be sought relative to cost and rheology factors.

Table 2-15 provides comparative effects of fillers on rheology. All processes have a maximum limitation on viscosity, which is especially critical in melt-processing such as extrusion and injection molding. In most instances, viscosity considerations form the basis for filler selection. Mixtures which are liquid or mobile at ambient conditions (plastisols, castings, and coatings) frequently require mixing to re-suspend sedimented particles. In these cases, a very fine, low P_f filler is advantageous to increase the mixture viscosity to prevent or retard settling of the larger particles. If a trixotropic effect is imparted, settling is prevented by the

Table 2-13. *(Continued)*

GLASS		SYNTHETIC SILICATES		NUTSHELL FLOUR	WOOD FLOUR	HYDROUS ALUMINA		
GB	GA	SP1	SP2	O1	O2	H1	H2	H3
25	50	0.02	0.08	30	?	40	6.5–8.5	0.7
44–4	150–10	–	0.17–0.03	177–1	–	150–5	60–0.3	3–0.05
Spheroid →		Floc →		Block	Fiber	Floc.	Floc	Plate
2.48	0.28	2.0	1.93	1.3–1.35	0.5–0.7	2.42	2.42	2.40
0.3	0.6	140–160	40	0.4–0.5	–	6	6	6–8
2.5	2.5	5(?)	5(?)	5(?)	10(?)	4	4	5
0.64	0.65	0.32	0.4	0.56	0.4	0.64	0.56	0.5
E	E	P	P	G	G	G →		
E	E	G	G	G	G	G →		
Ac, W, S	Ac, W, S	Ac, W, S	Al, W, S	None	None	Ac, Al, W, S →		
9.5	9.5	6.5–7.3	9.5	4–5	4–5	–	–	10
–	–	–	–	Low	Low	–	–	–
–	–	NaCl	NaCl	H_2O	H_2O	–	–	–
				Lignin	Lignin			
1.7	0.02	2	3	1.4	0.6	2 →		
0.27	0.24	0.2	0.2	0.43	0.42	0.19 →		
8.6	8.8	10	10	5–50	5–50	4–5 →		
1.52	–	1.46	1.45	–	–	1.57 →		
–	–	–	–	–	–	–		–
C	W	W	W	T	T	W	W	W
C	W	W	W	T	T	W	W	C
6	0.02	2	2	0.8	1	3 →		
0.23	0.4	0.3	0.3	0.35	0.3	0.3 →		
6	6	5–6	5	1	1	1–3	1	1
5	1.5	9	9	5	5	7 →		

very high viscosity at low shear rates produced by gravity acting on the larger particles.

Low viscosity matrices are generally chosen for pourable mixtures where entrained air must be removed by vibration or vacuum, either before or after the mixture is placed in the mold. Other processes are frequently benefited by higher viscosities, provided that the mixtures are void-free or air-free prior to placement. Matched die fiberglass mat molding, wherein the filler-polyester mixture is poured onto a preform or mat, is more efficient for displacing air in the mat if the fluid mixture is of moderately high viscosity. Low viscosity mixtures cause turbulance and foaming when passing through the mat, whereas high viscosity mix-

Table 2-13. (*Continued*)

CLASS		ASBESTOS	COAL	BARITE	SILICA		
Producers[a] Properties	Terms	AB	A	B	S1	S2	S3
Particle							
Size, mean	Microns, esd[b]	0.17	14	3	10	5	2
Range shape	Microns, esd[c]	–	100–2	40–0.1	74–0.5	40–0.5	15–0.1
	Class[d]	Fiber	Block	Block	Block →		
Bulk							
Specific gravity	g/cm^3 [e]	2.4–2.6	1.47	4.4–4.5	2.65 →		
Surface area	m^2/g [f]	40	1.5	0.6	0.8	1.2	1.5
k_E	g	10	3	3	4 →		
Packing fraction, P_f	h	0.6	0.6	0.65–0.75	0.59–0.67	0.52–0.55	0.49–0.52
Dispersibility:							
water	i	G	P	G	G	G	G
organics	j	G	E	G	G	G	G
Resistance	k	Al, W, S	Al, Ac, W	Al, Ac, W, S	Ac, W, S →		
pH		9–10	6–7	6–7	6.5–7	7	7
Specific resistance	ohm · cm × 10^{-3} [l]	Low	Very high	Very high	25 →		
Impurities		Fe_3O_4	Mineral	–	–	–	–
Thermal							
Conductivity	cal/sec, cm^2(°C/cm) × 10^3	5.0	0.6–0.8	6	7 →		
Specific heat	cal/g°C	0.26	0.3	0.111	0.19 →		
Coefficient/expansion	linear unit/unit/°C × 10^6	0.3	5	10	10 →		
Optical							
Refraction index:							
mean	sin *i*/sin *r*	1.53	–	1.64	1.55 →		
range	–	1.50–1.55	–	1.64–1.65	1.54–1.55 →		
Color:							
powder	n	G	B	W	W →		
in DOP	o	G	B	W	G →		
Physical							
Young's modulus, *E*	dynes/cm^2 × 10^{-11}	14.5	2	3	3 →		
Poisson's ratio, ν	p	0.2	0.3	0.3	0.3 →		
Mohs hardness	q	3–5	3	3–3.5	6.5–7 →		
Electrical							
Dielectric constant	$\epsilon = D/E$ [r]	10	3	7.3	4.3 →		

[a]See index of producers and trade names at end of footnotes.
[b]Equivalent spherical diameter; see section 2.1.
[c]Approximate from producer's literature. Range conforms with coarseness and fineness parameters of 99.9% coarser and finer by weight, esd.
[d]Idealized class from Table 2-1.
[e]ASTM D 153.
[f]Calculated when data were not available, see section 2.3.
[g]Einstein coefficient, approximate, for calculations of section 4.
[h]Approximate from oil absorption data, ASTM D 281.
[i]E = excellent, G = good, P = poor, N = non-dispersible. Dispersibility indicated without dispersant. All fillers provide excellent dispersion with correct dispersant for a particular system.
[j]Good to excellent resistance to acids, alkalis, water, and solvents, indicated by Ac, Al, W, and S, respectively.
[k]10 to 20% in water.
[l]ASTM D 2448, a measure of water solubility.
[m]Predominant minor contaminants, less than about 3%.

(*Footnotes continued*)

[n]Powder appearance; W = white, T = tan to brown, G = gray, B = black, C = clear.
[o]50% by weight in DOP (di-2-ethyl hexyl phthalate), same appearance classification as footnote (n).
[p]See section 4.2.1.
[q]See scale of section 3.3.
[r]See section 4.7.1.
[s]Available as 5-200 mesh products.
[t]Tremolite, calcium carbonate, chlorite, etc., depending on source.
[u]Feldspar—silica and mica; nepheline—emery.

L1 Casting grade limestone, 40-200, Georgia Marble Co.; C-50 or CLF 20--200, Thompson, Weinman Co.
L2 No. 1 White, Thompson, Weinman Co.; No. 10, Georgia Marble Co.
L3 Calwhite, Gama-Sperse 6451, Georgia Marble Co.; Vicron 25-11, Pfizer; Snowflake, Thompson, Weinman Co.; Camel-Tex, H. T. Campbell Co.
L4 Atomite, Thompson, Weinman Co.; Gama-Sperse 140, Georgia Marble Co.; Camel-Wite, H. T. Campbell Co.; Vicron 15-15, Pfizer.
C1 Albacar 5950, Pfizer.
C2 Multifex MM, Diamond Shamrock Co.; Calcene NC, PPG Industries.
C3 Coated with 3.5% fatty acids, Calcene TM and CO, PPG Industries; Super Multifex, Diamond Shamrock Co.
K1 Coarse, leached and fractionated, Hydrite Flat D, Georgia Kaolin Co.; ASP 400, Minerals and Chemicals Div., Engelhard Ind.
K2 Air-floated, Huber 65A, J. M. Huber Co.
K3 Fine, leached, fractionated, ASP 600, Engelhard Ind.; Huber 80, J. M. Huber Co.; Hydrite R, Georgia Kaolin Co.; Al-Sil-Ate H02, Freeport Kaolin Co.
K4 Very fine, leached, fractionated, ASP 200, Engelhard; HG 90, J. M. Huber Co.; Hydrite 10, Georgia Kaolin Co.; Al-Sil-Ate HO, Freeport Kaolin Co.
K5 Calcined kaolin, Glomax HE, Georgia Kaolin Co.; Satintone Special, Engelhard Ind.; Huber 80C, J. M. Huber Co.; Al-Sil-Ate O, Freeport Kaolin Co.; Whitetex, Freeport Kaolin Co.
T1 Beaverwhite, Cyprus Mines Corp.; 3X, International Talc Co.; Talcron CP 44-31, C. Pfizer & Co.
T2 C-400, Cyprus Mines Corp.; 325, International Talc Co.; Talcron 38-33, C. Pfizer & Co.
T3 Mistron Vapor, Cyprus Mines Corp.; 625, International Talc Co.; MP 12-50, C. Pfizer & Co.
M1 Dry ground and delaminated flake, English Mica Co., Franklin Mica Co.
M2 Fine, wet-ground and delaminated, Micro Mica C-1000, English Mica Co.; Alsibronz No. 12, Franklin Mica Co.; No. 21 Mica, Thompson, Weinman Co.
A Anthracite, 325, Keystone Filler Co.; 325, Shamokin Filler Co.
B Mineral barium sulfate, Baryta White and Foam A, De Lore Div., NL Industries; No. 22 Barytes, Thompson, Weinman Co.; No. 1 Barytes, C. Pfizer & Co.
S1 Min-U-Sil 30, PGS-Floridin Co.; 1240, Illinois Mineral Co.; 625, Malvern Minerals Co.; Gold Bond R, Tammsco Div. of Lowe's Inc.
S2 Min-U-Sil 15, PGS-Floridin Co.; Imsil A-25, Illinois Mineral Co.; S Micron, Tammsco Div. of Lowe's Inc.
S3 Min-U-Sil 5, PGS-Floridin Co.; A-10, Illinois Mineral Co.; Neosil XV, Tammsco Div. of Lowe's Inc.
AS1 Anhydrous Silicates, LU-390, Lawson-United; Minex 2, Indusmin Limited.
AS2 LU-340, Lawson-United; Minex 4, Indusmin Limited.
AS3 Minex 7, Indusmin Limited.
W Wollastonite P-4, Interpace Corp.
P Johns-Manville Corp.; Silbrice Corp.
GB Glass beads, 3000, Potters Ind., Inc.
GA Glass Microballons, B30B, 3M Co.; Q Cel, Philadelphia Quartz Co.
SP1 Synthetic precipitated silica, Hi-Sil 210, 215, 233, PPG Ind.
SP2 Synthetic precipitated calcium silicate, Silene D, PPG Ind.
O1 WF-5, Agrashell Inc.; Composition Materials Co.
O2 Chicago Wood and Plastic Products; Composition Materials Co.; International Filler Corp.; Wilner Wood Products Co.; Wood Flour, Inc.
AB 7R Asbestos, Asbestos Corp.; Bell Asbestos Mines, Limited; Carey-Canadian Mines, Limited; Johns-Manville Corp.; Uniroyal, Inc.: RG600, Union Carbide Corp.
H1 C-33, Alcoa; RH-33, Reynolds Chem.
H2 C-331, Alcoa; GHA-331, GHA-332, Great Lakes Foundry Sand Co.; SB-331, SB-332, Solem Ind.
H3 Hydral 710, Alcoa; RH-730, Reynolds Chem.

tures displace the air and force it ahead of the flow-front with less turbulance.

Plastisols, being mixtures of fluid plasticizers and dispersion resins, are suspensions of particles complicated by the addition of fillers. Packing fraction for plastisols then reflects the relative efficiency of the filler-dispersion resin packing in the fluid state. It is therefore very important to obtain particle size distributions for the dispersion and blending resins to anticipate the effect a filler will have on such a system. The rheological effects of both resin and filler apply to the conditions of Table 2-15.

Under shear-processing conditions, large particles may become non-uniformly distributed in a mixture if the matrix is too low in viscosity. This is due to the inertia and peripheral drag of large particles and the tendency for highly fluid media to flow around them. Graininess or non-uniform appearance is usually produced, and non-uniform composite properties confirm such a filler distribution.

From a practical viewpoint, improvements in physical properties begin with a selection of the polymer or matrix; filler is added to improve a property or to counter a shortcoming.

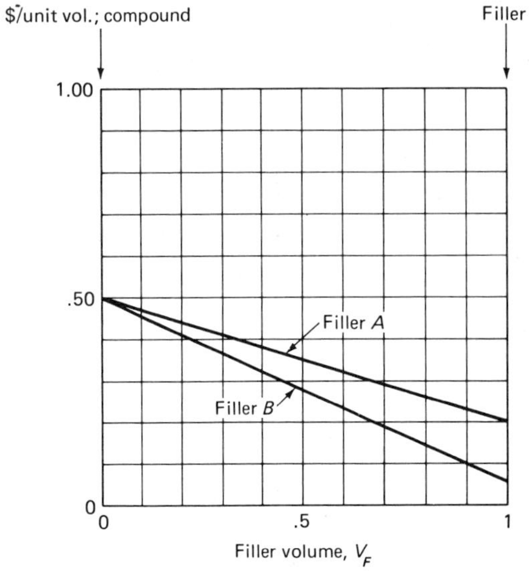

Fig. 2-12. Filler effect on compound cost.

Example:

	\$/lb ×	Density =	\$/Volume
Matrix	0.385	1.30	0.50
Filler *A*	0.74	0.27	0.20
Filler *B*	0.022	2.71	0.06

Equal cost at $0.4 V_f$ Filler *A*, 0.27 Filler *B*.

For most composites, modulus will be significantly increased, but this is not always a benefit. When such a modulus increase is detrimental, it may then become necessary to modify the matrix to counter the effect.

Table 2-16 lists some of the reasons for using fillers; properties are associated with the matrix type. It is beyond this presentation to consider all circumstances for the use of fillers, but Tables 2-14 to 2-16 provide guides which can be applied with the other sections of this chapter to make a more exacting selection of the appropriate filler for a particular purpose. Certain limitations should be kept in mind: parts requiring trimming or finishing should utilize soft (low Mohs hardness) fillers; combinations of fillers should be considered in all cases where the use of one grade or type restricts a certain improvement; similar and dissimilar types should be examined, since one limiting factor which will obviate the choice may be discovered for the system. Almost every circumstance is different, so even if improvement in only one property is desired, the system providing more secondary benefits will be preferred.

Table 2-16 may be restricted in terms of filler selection because of processing limitations. Not all processes are amenable to the types and concentrations of fillers indicated in Table 2-14, so Table 2-16 indicates less efficient types in order to obtain the most critically limiting property which, for most purposes, is viscosity.

5.2.3 Thermal Properties Modification. Section 4.5 presents means for calculating conductivity and specific heat which are suitable for estimating sheet forming requirements. These calculations also may be applied to various thermoset molding operations wherein an exotherm is produced by the polymer and becomes the heat source. Most solid fillers (except metallic) have similar specific heats and thermal conductivities; therefore, the predominant effect is due to filler volume. Low density fillers have very low conductivities and provide good insulation for the fabricated part. In processing operations, low density fillers can cause higher exotherm temperatures and long cooling periods because of the insulating effect. For this reason, low density fillers are normally used at high volume displacement of the thermoset polymer to reduce the exothermic heat, or in conjunction with solid fillers to aid in dissipating exothermic or thermoforming heat.

Deflection temperatures are generally increased by fillers due to modulus increase. Very high bonding interaction between the matrix and filler will result in greatly increased deflection temperature. The guides given above for modulus and secondary effects apply here also, but creep is more precisely evaluated and predicted by the equation of section 4.2.8. Creep data are of greater use than deflection temperature data, so the above recommendations based on modulus increase should be applied. For maximum creep resistance below the elastic limit of the composite, the highest relative filler volume, V_f/P_f, is indicated. Data are readily available in the literature[17] for calculations based on the creep values of polymers.

Coefficient of expansion is always reduced by solid and low density fillers, a fact which is

Table 2-14. Influence of fillers on physical properties of composites.

Physical Property of Composite	MAXIMUM IMPROVEMENT BY FILLER at	when V_f/P_f is	FILLER Bond/ Matrix[b]	Type[c]	Size Range[d]	RELATIVE FILLER EFFECT ON MATRIX TYPE[e] Rigid	Flexible	Reference Section
Modulus, M[a]	P_f	high	max	solid, max M	broad	least	most	4.2.1
Tensile strength	V_f/P_f	low	max	fiber	narrow	least	most	4.2.2
Flexural strength	V_f/P_f	medium	max	fiber	narrow	least	most	4.2.3
Elongation	V_f/P_f	low	min	sphere	narrow, fine	least	most	4.2.4
Tear strength	V_f/P_f	low	max	fiber	narrow			4.2.5
Impact strength	V_f/P_f	low	min	fiber	narrow, fine	most	least	4.2.6
Compressive strength	P_f	high	max	solid, max M	broad	least	most	4.2.7
Creep	P_f	high	max	fiber	broad	least	most	4.2.8
Hardness	P_f	high	max	solid, hi Mohs	broad	least	most	4.2.9
Coefficient/ friction	P_f	high	max	solid	broad	least	most	4.2.10
Abrasion resistance	P_f	high	max	solid, hi Mohs	broad	most	least	4.2.11
Density	P_f	high		sphere, low density	broad			4.1

[a]Young's, shear or bulk.
[b]Surface modification of filler may be necessary to maximize bonding.
[c]See Tables 2-1 and 2-16.
[d]Broad size range produces highest P_f.
[e]At a testing temperature below T_g of matrix for rigid, and above T_g for flexible.

Table 2-15. Rheological influences of fillers.

Filler Property	Rheological Effect
Maximum packing fraction, P_f	Essentially no mobility at P_f; below P_f, lowest viscosity for highest P_f types
Particle shape, k_E	Lowest viscosity for lowest k_E; spheres lowest, fibers highest (Table 2-9)
Particle size range	
narrow, coarse	Dilatancy under shear
wide, coarse to fine	Lowest viscosity
wide, fine	Good uniformity over different shearing conditions
narrow, fine	Highest viscosity, best uniformity of particle distribution
Particle interaction	
non-flocculating	Lowest viscosity
flocculating	Thixotropy, viscosity reduction with shear increase
Dispersion	
complete	Lowest viscosity
incomplete	Thixotropy, viscosity reduction on aging and with shear

Table 2-16. General guide to filler use.

Improvement	Matrix Type[a]	Without loss of	Matrix should be[b]	Best Filler Type[c]	Concentration[d]
Thermoplastics					
Modulus	high T_g	impact	ductile	fine, narrow range	low
Modulus	high T_g	tensile	ductile, tough	fine, fiber or flake, max bonding	low–moderate
Modulus	high T_g	flexural	tough	high P_f, max bonding	moderate–high
Abrasion resistance	high T_g	compressive strength	ductile, tough	high hardness, high P_f	high
Lower density	high T_g	compressive strength	hard, tough	glass micro-balloons	moderate–high
Modulus	low T_g	low viscosity	low melt viscosity	high P_f, wide range	low
Tensile	low T_g	impact	high elongation type	fine, narrow range, moderate bonding	low
Creep	low T_g	tensile	moderate elongation type	fine, wide range, max bonding	low–moderate
Tear resistance	low T_g	impact	moderate–high elongation type	fine, narrow range, moderate bonding	low
Thermosets					
Modulus	high T_g	impact	ductile	fine, wide range	low–moderate
Modulus	high T_g	tensile	moderate ductile	fiber or flake	moderate
Flexural	high T_g	–	moderate ductile	fiber or flake	moderate
Compressive strength	high T_g	impact	moderate ductile	fine or flake	moderate
Tear resistance	low T_g	abrasion resistance	tough	fine, hard, max bonding	low–moderate

[a]Versus temperature of testing.
[b]Matrix may require modification to obtain indicated property.
[c]Compare with Table 2-14; best choice may not be the most efficient filler type.
[d]According to relative filler volume; V_f/P_f, where high is in excess of 0.5, moderate is 0.25–0.5 and low is below 0.25.

of great practical importance. Virtually all matrices have very high coefficients of expansion, which limits their attachment or abutment to metal, ceramic, masonry, and other low-expansion materials where warping or strain with temperature change occurs.

Mechanical fastening frequently involves the use of metal fasteners. Since fillers reduce the coefficient of expansion of plastics to more nearly that of metals, composites are more compatible where temperature changes may produce joint separation due to differential expansion. Low density, crushable fillers also provide reduction in crack propagation on impact, especially in thick sections, and help to grip nails or punched inserts.

Heat-sealed and welded joints are improved by the inclusion of fillers in thermoplastics due to the improvement in conductivity, but the reduced flow of the melt may be a hindrance. Magnetic, high-frequency and inductive bonding require a magnetically susceptible filler in the thermoplastic or at the interface to generate heat. Ultrasonic bonding methods limit filler concentrations to low values, since they tend to damp or reduce the vibrational susceptibility and heat generated in the matrix. This is especially true for filler exhibiting excellent bonding with the matrix.

5.2.4 Electrical Properties Modification. Most polymers have excellent electrical properties and are considerably superior to fillers. For certain specific properties, modification with fillers is advantageous. Dielectric constant is such a property which requires low values for the filler so that less charge is stored in an alternating electrical field. Calcined kaolin and fused silica are suitable where fine particles are preferred, and glass microballoons are superior in castings.

For dielectric strength and dielectric loss, talc, alumina, and mica provide some improvement, but most fillers, excepting extremely hydrophobic types, do not greatly alter these properties of the matrix.

Volume resistivity is similar to dielectric strength for filler performance, with the most hydrophobic types preferred. In order of lowest to highest values from 10^9 ohm-cm under humid conditions, suitable fillers are talc, electrical grade mica, and hydrophobic silica. Fillers generally have resistivities below those of polymers.

Only partial conductance can be obtained with metallic fillers since the particles are separated by the insulating layers of matrix. The heating which occurs limits their use to thermoset matrices.

5.2.5 Special Requirements. Various chemical and optical properties have been mentioned in sections 4.4 and 4.6. Appearance attracts considerable attention since it is readily apparent and frequently evaluated before any other property. Gloss usually depends on mold finish or processing method, but subtle differences can be obtained with fillers. Very fine, poorly packing (low P_f) fillers allow contraction of the matrix after thermal processing to cause a patina effect. Gross gloss reduction is generally produced by medium to coarse particles having a narrow size range at concentrations near the P_f. Filled thermoplastics can be reduced in gloss by slightly drawing an extruded melt to bring the particles nearer to the surface. Similarly, drawing oriented film or sheet to slightly above the yield point causes cavitation which may be frozen by chilling to render the composite opaque. A secondary benefit is density reduction.

Whenever gloss is reduced by fillers, coefficient of friction is increased, provided that the filler particles are irregular and sufficiently harder than the matrix.

Most crystalline fillers reflect infrared radiation, and if they provide clarity in the visible range, fillers such as feldspar, nepheline, and glass can be used to reduce infrared transmittance of glazing panels.

Sound transmittance and reflectance differ according to the source of the vibration. For sound reflectance, thick sections of high density and modulus are most efficient. When sound originates at or in the composite, as by the "drumming" of waves on boat hulls, low densities, low modulus, and thick sections will reduce sound transmittance. For this reason, foam-filled boat hulls are quieter.

References

1. Irani, R. R. and Callis, C. F., *Particle Size; Measurement, Interpretation and Application,* John Wiley & Sons, New York, 1963.
2. Morris, H. H. and Schwartz, E. W., *Rubber Age* **93:** 739, Aug., 1963.
3. Mysels, K. J., *Introduction to Colloid Chemistry*, Interscience, New York, 1967.
4. *Ibid*., p. 197.
   Thomas, J. M. and Williams, B. R., *Application of Metal Vacuum Microbalance to the Study of Solid Surfaces by Physical Adsoprtion* **4,** P. M. Waters (ed.), Plenum, New York, 1965.
5. Hang, P. T. and Brindley, G. W., *Clays and Minerals* **18:** 203–212, Pergamon, New York, 1970.
6. Work and Witby, *Encyclopedia of Chemical Technology* **12:** 477–478, 1954.
7. Gray, W. A., *The Packing of Solid Particles*, Chapman and Hall, London, 1968.
8. Furnas, C. C., *Ind. and Eng. Chem.* **23,** *No. 9:* 1052, Sept., 1931.
9. Hurlbut, C. S., Jr., *Dana's Manual of Mineralogy, 18th Edition*, John Wiley & Sons, New York, 1971.
10. Pauling, L., *The Nature of the Chemical Bond, 3rd Edition*, Cornell U. Press, Ithaca, New York, 1972.
11. Lewis, T. B. and Nielsen, L. E., *J. Appl. Polym. Sci*. 14: 1449 (June 1970).
12. Nielsen, L. E., *Mechanical Properties of Polymers*, Van Nostrand Reinhold, New York, 1962.
13. Nielsen, L. E. and Lewis, T. B., *J. Polym. Sci. A-2:* 7, 1705 (1969).
14. Nielsen, L. E., *Mechanical Properties,* pp. 134–135.
15. Frissell, W. J., *Plastics Tech.:* 723 (Nov. 1956).
16. Ranney, M. W., Berger, S. E., and Marsden, J. G., *Silane Coupling Agents in Particulate Mineral Filled Composites*, 27th Reinforced Plastics Tech. and Man. Conf., Sec. of Plastics Ind. (Feb. 8–11, 1972).
17. *Modern Plastics Encyclopedia* **51,** *No. 10A* (Oct. 1974).

18. Green, H., *Industrial Rheology and Rheological Structures*, John Wiley & Sons, New York, 1949.
19. Mysels, K. J., *Introduction to Colloid Chemistry*, Interscience, New York, 1967, Chapts II-VII.
20. Patton, T. C., *Paint Flow and Pigment Dispersion*, Interscience, New York, 1964, p. 443.
21. Sonntag, H. and Strenge, K., *Coagulation and Stability of Disperse Systems*, John Wiley & Sons, New York, 1972.
22. Ferrigno, T. H., *Fillers in Corrosion Resisting poly(vinyl chloride)*, 21st Ann. Conf., Nat. Assn. Corrosion Engrs., March 15-19, 1965; *Materials Protection*, p. 13 (Oct. 1965).
23. Ferrigno, T. H., *Rigid Plastics Foams*, Van Nostrand Reinhold, New York, 1967, p. 5.
24. *Chemistry and Physics of Interfaces I (1965) and II (1971)*, American Chemical Society, Washington, D.C.
25. Nielsen, L. E., *Mechanical Properties of Polymers and Composites* 2, Marcel Dekker, New York, 1974.
26. *Ibid.*, p. 431.
27. *Lange's Handbook of Chemistry, 11th Edition*, J. A. Dean (ed.), McGraw-Hill, New York, 1973, pp. 9-119.
28. Nielsen, L. E., *Ind. Eng. Chem., Fundam.* **13**, *No. 1*, 1974.
29. Nielsen, L. E., *J. Appl. Polym. Sci.* **17**: 3819 (1973); *J. Phys., D7:* 1549 (1974).
30. Kalina, J. F., *Mod. Plastics:* 42 (May 1975).
31. Hoke, C. E., *SPE J.:* 36 (May 1973).
32. Wood, A. S., *Mod. Plastics:* 44 (June 1975).

3

Concise Fundamentals of Fiber Reinforced Composites

Harry S. Katz
Utility Research Co.
Montclair, New Jersey

Fibers have extremely high tensile strengths and moduli. There have been many definitions for a fiber, but a number of these appear to agree in defining a fiber as any material that has a high ratio of length to cross sectional area, with minimums suggested as 10:1 to 100:1, and with a maximum cross sectional area of 7.85×10^{-5} in.^{2} and a maximum transverse dimension of 0.010 in. The tensile strength of a fiber is many magnitudes greater than the strength of the same material in bulk form. For example, bulk graphite is a brittle material with a tensile strength level below 10,000 psi whereas commercial graphite fibers have tensile strengths that range between 250,000 to 400,000 psi, and laboratory research on graphite whiskers yielded fibers that had 3,000,000 psi tensile strength. As another example, nylon plastic has a tensile strength of about 12,000 psi, whereas nylon fibers have tensile strengths in the order of 120,000 psi. Also, the tensile strengths of E-glass at about 400,000 psi, boron filaments at 500,000 psi, and alumina fibers at about 150,000 psi, are far above the strength levels of the bulk materials. Thus, it is apparent that when these high strength fibers are used effectively in a matrix, composites with strength levels above those of bulk materials can be produced.

Since fibers are the highest strength materials, they are the best reinforcements for many applications, ranging from improvement of the strength of low-cost materials or composites to acceptable levels, to the production of superior structural aerospace components. Therefore, everyone involved in the production, design, specification, or use of plastics should be familiar with the fundamental characteristics of fibers as reinforcements.

Chapters 1 and 2 of this Handbook have presented many good reasons for adding a high volume fraction of low cost fillers to plastics, especially from the standpoint of stretching the resin supply. However, high particulate filler loadings will usually result in a substantial reduction of physical properties. For the many end uses where the reduced physical properties would be unacceptable, the addition of fibers to the compound formulation can be a means for increasing the properties to a satisfactorily high level. Many subtle considerations are involved in the science and art of formulating and fabricating a fiber-reinforced composite. There has been much progress in this field during the past decade, and continued advances may be anticipated. An important aspect that will form the basis for future progress in improved molding compounds is the study of the packing of fibers with spheres and particulate fillers, as presented in Chapter 4.

Commercially available fibers have a wide range of costs and properties. Graphite and boron filament costs have been in the range of \$50–200/lb, with tensile strengths up to about 500,000 psi and tensile moduli up to about 60,000,000 psi. Recently, whisker materials were available in limited quantities at over \$250/lb, with reported strengths up to 2,000,000 psi. At the lower end of the spectrum of fiber costs are glass fibers and asbestos. The cost of glass fibers is below the cost of most of the engineering plastics. Therefore, considering the fact that E-glass has a tensile strength on the order of 400,000 psi and a tensile modulus of 10,000,000 psi, it is obvious that the use of E-glass as a reinforcement can provide great improvements in physical properties without increasing the cost of engineering plastics.

The use of fibers as a reinforcement can lead to superior materials; in some cases, at a premium in cost; or they can be a means for achieving improved properties while maintaining low costs and stretching the resin supply. The choice of the fiber reinforcement will depend on the requirements of the different end products. At the present time, the highest volume fiber reinforcement is glass fiber, but the near future will undoubtedly lead to the increased use of competitive materials, such as the anticipated low cost carbon-graphite filaments, whiskers, classified asbestos, micro fibers, and boron filaments. As industrial competition becomes more sophisticated, and end uses more demanding in performance and cost, it will be essential to continue the rapidly expanding use of fiber-reinforced composites.

This chapter is intended to give the reader some insight into the principal characteristics of fiber-reinforced composites. There have been a number of excellent books that give comprehensive treatments of important topics, including precise theoretical analyses of the strength and modulus of various composite structures, crack propagation theories, and the stress analysis of advanced composite materials. These books are listed in the References at the end of this chapter.

There are three basic elements in a fiber-reinforced composite: fiber, matrix, and the fiber-matrix interface. Each of these three elements must have appropriate characteristics and function both individually and collectively in order for the composite to attain the desired superior properties.

The fiber contributes the high strength and modulus to the composite. It is the element that provides the resistance to breaking and bending under the applied stress or load.

The main roles of the matrix are to transmit and distribute stresses onto the individual fibers, and to maintain the fibers separated and in the desired orientation. The matrix also provides protection against fiber abrasion as well as fiber exposure to moisture or other environmental conditions, and causes the fibers to act as a team in resisting failure or deformation under load. The maximum service temperature of the composite is limited by the matrix.

The fiber-matrix interface is the critical factor that determines to what extent the potential properties of the composite will be achieved and maintained during usage. Localized stresses are usually highest at and near the interface, which may be the locus of premature failure of the composite. The interface must have appropriate chemical and physical features in order to provide the necessary load transfer from the matrix to the reinforcement. The use of a coupling agent can provide improved interfacial conditions. The interfacial bond must resist stresses due to factors such as the differential thermal expansion of fiber and matrix, and shrinkage of the resin during cure. Other desirable features of the matrix are resistance to liquid penetration and freedom from voids. The interface also assists the matrix in protecting the fibers.

As indicated above, the presence of a coupling agent at the interface can be the means for obtaining the optimum physical properties of the composite, and retaining these properties after environmental exposure or aging. This fact became apparent in early studies of fiberglass-reinforced polyester laminates, when the simple test of putting a test panel into boiling water for two hours showed an amazing difference in strength retention between the composite utilizing a fiberglass treated with a silane coupling agent and a control containing untreated fiberglass.

Silanes have been the most frequently used type of coupling agent. The silanes provide dual functionality in one molecule, such that one part of the molecule forms a bond to the glass filament surface while another part forms a covalent bridge with the matrix molecule. There are many different silanes, and the choice must be based upon the matrix to be used. A wide selection of silanes are offered by Dow Corning Corporation and Union Carbide Corporation. Other types of coupling agents are available. Dupont produces Volan, a methacrylato chromium complex, and Kenrich Petrochemicals Inc. produces a line of organic titanates that have shown remarkable results in some composite systems.

As composite technology has advanced, it has become virtually imperative that the properties of a structure should be predictable from a knowledge of the constituent matrix, fiber, fiber volume, and fiber orientation. This is especially required in the case of the advanced composites, where fiber and fabrication costs are high and the end product must be extremely reliable in performance, so that the trial-and-error approach cannot be tolerated. There has been good progress during the past decade in the theoretical analysis and design of composite materials, to the point where even critical aerospace structures can be designed with assurance that they will safely meet the material property requirements.

In the analysis of the theoretical properties of different types of composites, various analytical models and failure theories have been used. There are different considerations involved in the case of a randomly-oriented discontinuous fiber composite than for a continuous-filament unidirectional or angle-ply laminate.

A first approximation of the tensile modulus and strength of a fiber-reinforced composite can be obtained from the Rule of Mixtures. The modulus of a continuous filament composite with filaments oriented longitudinally can be estimated, with good precision, for the direction parallel to the fibers, by:

$$E_L = V_f E_f + V_m E_m \qquad (3\text{-}1)$$

where

E_L is the longitudinal modulus of the composite
V_f is the volume fraction of the filaments
E_f is the tensile modulus of the filaments
V_m is the volume fraction of the matrix
E_m is the tensile modulus of the matrix.

Equation 3-1 assumes that all filaments are perfectly bonded to the matrix.

Equations analogous to Eq. 3-1 may be used to express the theoretical longitudinal Poisson's ratio or the tensile strength of the composite. This prediction of tensile strength will not be as accurate as in the case of the modulus. Actual strengths are usually lower than the predicted strengths, and this ratio has been used as an indication of the fiber efficiency in the composite. In the case of some metal matrix composites, however, the indicated fiber efficiency has been greater than 100%, because of factors such as favorable residual stresses in the matrix.

In a discontinuous fiber composite, the stress along the fiber is not uniform. There are portions along the fiber's ends where the tensile stresses are less than that of a fiber which is continuous in length. This region is often called the fiber ineffective length. The tensile stress along the fiber length builds up to a maximum along the middle portion of the fiber. If the fiber is sufficiently long so that the length-to-diameter ratio, or aspect ratio, equals or exceeds the critical aspect ratio, the middle portion stress will be equal to that of a continuous length fiber. The shear stresses at the fiber-matrix interface attain maxima at the ends of the fiber, as shown in Fig. 3-1.

The critical aspect ratio $(l/d)_c$ which would result in fiber fracture at its midpoint, can be expressed as:

$$\left(\frac{l}{d}\right)_c = \frac{S_f}{2Y} \tag{3-2}$$

where

l and d are the length and diameter of the fiber
$(l/d)_c$ is the critical aspect ratio
S_f is the tensile stress of the fiber
Y is either the yield strength of the matrix in shear or the fiber-matrix interfacial shear strength. Whichever value is lower will determine the critical aspect ratio.

If the fiber is shorter than the critical length, the stressed fiber will de-bond from the matrix and the composite will fail at a low strength level. When the length is greater than the critical length, the stressed composite will lead to breaking of the fibers and high composite properties.

The rule of mixtures for discontinuous fiber composites may be expressed as:

$$S_c = V_f S_f \left(l - \frac{l_c}{2l}\right) + V_m S_m \tag{3-3}$$

where

S_c is the tensile strength of the composite
V_f and V_m are the volume fractions of the fiber and matrix
S_f and S_m are the tensile strengths of the fiber and matrix
l_c is the critical length of the fiber
l is the length of the fiber.

The rule of mixtures does not apply at low volume percentages of fibers. In order for the composite to have a higher strength than the matrix, a minimum V_f must be exceeded. This value may be 10% or greater for plastic matrix composites, but can be much lower for metal or ceramic matrix composites. Also, at high V_f, on the order of 70%, the composite properties may decrease sharply, due to practical problems such as the difficulty in avoiding fiber-fiber contact, which results in high stress points that initiate failure. If the load is compressive, the theoretical problem is more difficult, due to the possibility of fiber buckling. A number of theories have been proposed to predict the strength of discontinuous fiber composites. These theories indicate that, primarily because of high stress concentrations at discontinuities that occur at the fiber ends, the tensile strength of a discontinuous fiber composite will be from 55 to 86%, and the modulus can approach 90 to 95% of the tensile strength and modulus of the corresponding continuous filament composite.

Analytical methods for composite materials can be divided into two main types, micro and macromechanics.

Micromechanics analyses are applied to the basic filament-matrix microstructure in order to determine the strength and elastic properties from the properties and arrangement of the constituent filament and matrix materials. Micromechanics attempts to predict fracture behavior of the composite from the state of stress and strain at the micro level. Although this science has been pursued vigorously only during recent years, much progress has been

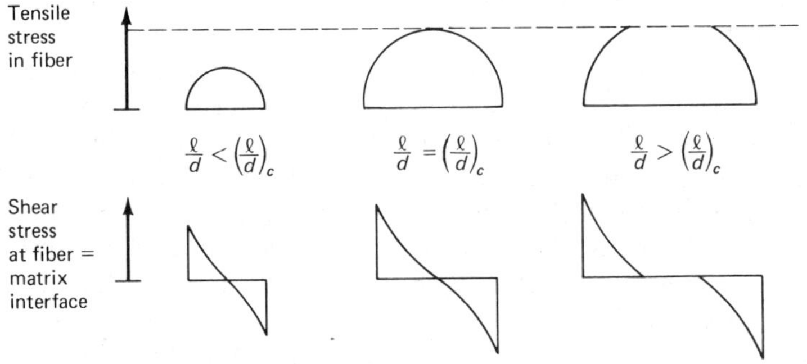

Fig. 3-1. Shear stresses at fiber-matrix interface.

made in developing precise mathematical procedures for estimating the elastic moduli and strength of composites. Computer programs have evolved that will provide this information in tabular or curve forms.

Macromechanics encompasses the analytical methods that are used to predict the strength and elastic properties of an advanced composite laminate from the known properties of the individual unidirectional monolayers. Macromechanic studies have been essential for design guidance in the rapid progress of advanced composite development in the aerospace industry.

The three failure criteria currently being utilized in estimating the strength of a composite are the maximum stress, maximum strain, and maximum energy theories.

In the maximum stress theory, failure is assumed to occur when the tensile or compressive stresses parallel or perpendicular to the filaments, or the in-plane shear stresses, reach any one of the individual ultimate strengths.

The maximum strain theory assumes that failure occurs when any one of the strains associated with the material axes reaches its maximum value.

The distortional energy or maximum energy theory assumes that failure occurs when the strain energy of distortion reaches a critical value.

Computer programs based on each of the failure criteria have been developed. Significant differences in predicted properties for a specified structure can be obtained, at the present time, when the various failure criterions are compared. The choice of which theory should be used for a particular composite structure must be substantiated by experimental fabrication and testing. The analytical program will provide information that may eliminate futile efforts and minimize the time and experimental work needed to develop a satisfactory composite structure. Future progress in micromechanics and macromechanics will further increase the capability for predicting the precise properties of proposed composite structures.

The design of a part that will be fabricated from an isotropic material, such as a structural metal, has become relatively simple and straightforward, as a result of the long history of usage and complete design data that has been generated. In contrast, composite materials and their fabrication processes are frequently unfamiliar to the designer, and the information that he can obtain may not be as complete as he might wish. In spite of these drawbacks, which are rapidly being improved, the usage of composites has been expanding in all markets, and excellent progress has been made in the design of more intricate and larger parts. Notable accomplishments have been attained in the automotive and aerospace industries. The use of sheet molding compounds has led to the design and mass production of large components with complex contours, for automobiles and other forms of transportation. The aerospace industries have led the way toward high performance composites that surpass the specific strength and stiffness of the structural metals. These advanced composites offer the designer the opportunity to produce aircraft where even the outer appearance may not resemble current aircraft design.

In order for the potential of fiber-reinforced composites to be fully realized, an interdisciplinary effort is required. There must be inputs and liaison between experts in materials, chemistry, adhesive bonding, crystal structure, metallography, stress analysis, engineering mechanics, fracture mechanics, vibration analysis, design, composite fabrication, quality control, and testing.

The design of the composite part must take into consideration the complete operational environment and the ability to meet the environmental requirements for the anticipated life of the structure. Important factors will include physical properties, fatigue and creep resistance, materials and manufacturing costs, maintenance and repairability, attachment method, and means for quality assurance.

Methodical testing procedures are an essential supporting activity for the utilization and improvement of fiber-reinforced composites. Testing has guided the research and development of each composite material and structure, and must be used for quality control in production programs. The use of small test samples as a means for predicting the performance of large structures is often necessary. In this situation, the test should be conducted by simulating, as closely as possible, the end use conditions. However, it is seldom feasible to apply to a test part the complex load conditions and extended time to which a component may be subjected. Therefore, good judgment must be exercized in the choice of tests that will provide either valid comparison data that will permit ranking of candidate materials, or meaningful data for designers and materials engineers. Standard tests, such as those proposed by the ASTM, will usually be preferable to specially designed tests that cannot be directly compared with similar testing by other investigators. In reporting the results of these tests, it is important that the exact test conditions be specified, since minor changes in test conditions can result in great differences in the data produced.

In each composites program, careful planning must include the choice of the testing procedures that will be used. A brief discussion of pertinent test methods is presented in the following paragraphs.

Flexural testing has been used frequently in composite development programs. This test is convenient and easy to perform. The data obtained can be used for grading specimens in materials evaluation, process development and improvement, and quality control. However, a true correlation between flexural properties and other mechanical properties has never been established. There are so many subtle variables affecting bending test strengths that the problem of correlating flexural data with the more basic properties can outweigh all advantages of the test when meaningful design data is required. In flexural testing, either 4-point loading or the more frequently used 3-point loading fixtures can be employed. During the application of load, one face of the sample is under compression and the other face is in tension, and the failure of fiber-reinforced composites usually occurs on the tension side. Flexural strength values for fiber-reinforced composites are usually much higher than the tensile strength values for the same material. The percentage difference in the flexural and tensile moduli is not generally as large as the difference in the strength values.

Many different tensile test procedures have been proposed for individual fibers and fiber bundles or strands. Tensile tests of FRC have been performed by use of various shaped specimens, including the dogbone configuration, NOL ring, filament-wound pressure vessels, and straight-edge panels with bonded end tabs. The choice of method is important, especially if data is to be compared with similar information from other sources in order to choose the best candidate material for a specific application. A common mistake in the reported tensile strength data is the neglect to record the exact test conditions, such as sample configuration, gauge length, and rate of jaw separation, and method of gripping the specimen.

A great deterrent in the use of advanced composites for aerospace structural applications has been the concern of designers that there was no guarantee against the use of a defective part that would cause failure of the vehicle and mission. Designers feel confident in their established knowledge of the performance of structural metals, but consider the use of NDT methods essential for the safe utilization of the new composite structural components. The procedures might preferably be called nondestructive inspection, since they provide a means for detecting minor flaws and irregularities. Great progress has been made in NDT methods during recent years, and there is now a high confidence level in the capability of detecting defective parts.

NDT procedures include, first of all, checks for dimensional accuracy and freedom from contour distortion, and visual and microscopic checks for obvious defects. Frequently used methods for further probing include x-ray, sonic transmission, pulse echo, and infra-red void detection. By the use of through-transmission ultrasonics, the ultrasonic longitudinal velocity can be correlated with the tensile and compressive moduli, as well as the ultimate tensile, compressive, and interlaminar shear strengths of boron-epoxy, carbon-carbon, and other composites, thus providing a means of physical property measurement as well as flaw detection. Other NDT methods are the use of liquid fluorescent or dye penetrants, magnetic-particle, liquid rubber penetrants; ultrasonic resonance; optical, acoustic, and laser holography; thermoluminescent coating, liquid crystals, microwave, eddy current, eddy sonic, beta back scatter, neutron radiography, and acoustic transmission or emission.

REFERENCES

General

Broutman, L. J. and Krock, R. H. (eds.), *Modern Composite Materials*, Addison-Wesley, Reading, Mass., 1967.

Holister, G. S. and Thomas, C., *Fibre Reinforced Materials*, Elsevier, New York, 1966.

Holliday, L., *Composite Materials*, Elsevier, New York, 1966.

Lubin, G. (ed.), *Handbook of Fiberglass and Advanced Plastics Composites*, Van Nostrand Reinhold, New York, 1969.

Oleesky, S. S. and Mohr, J. G., *SPI Handbook of Technology and Engineering of Reinforced Plastics/Composites*, Van Nostrand Reinhold, New York, 1973.

Nielsen, L. E., *Mechanical Properties of Polymers and Composites*, Marcel Dekker, New York, 1974.

Parratt, N. J., *Fibre-Reinforced Materials Technology*, Van Nostrand Reinhold, London, 1972.

Manson, J. A., and Sperling, L. H., *Polymer Blends and Composites*, Plenum, New York, 1976.

Schwartz, R. T. and Schwartz, H. S., *Fundamental Aspects of Fiber Reinforced Plastic Composites*, Interscience, New York, 1968.

Theory

Ashton, J. E., Halpin, J. C., and Petit, P. H., *Primer on Composite Materials: Analysis*, Technomic, Stamford, Conn., 1969.

Calcote, L. R., *The Analysis of Laminated Composite Structures*, Van Nostrand Reinhold, New York, 1969.

Tsai, S. W., Halpin, J. C., and Pagano, N. J., *Composite Materials Workshop*, Technomic, Stamford, Conn., 1968.

Levy, S. and Dubois, J. H., *Plastics Product Design Engineering Handbook*, Van Nostrand Reinhold, New York, 1977.

4

Packing Concepts in the Utilization of Filler and Reinforcement Combinations

John V. Milewski
Los Alamos Scientific Lab.
Los Alamos, New Mexico

CONTENTS

1. INTRODUCTION

Chapters 2 and 3 have explained some of the basic theory of the effect of using fillers and reinforcements in plastic composites. For years, it has been found that combinations of filler and reinforcement are complementary and, in many cases, synergistic[1,2,3,4], but until recently there has been no theory available to explain these effects or to predict and find the optimum combination of fillers and fibers. It is now proposed that the major part of the improvement is due to better packing of filler and reinforcement[5 6]. This Chapter will review the most recent information available on the subject of bimodal packing.

First, a review will be given on the bimodal packing of spheres. Then the packing of fiber at various length to diameter ratios (L/D's) will be given, followed by the bimodal packing of fibers and spheres. Next, experimental data and examples will be given to explain how to utilize this information for the design of a material containing nearly sphere-shaped particles and fibrous reinforcements.

2. BACKGROUND (PACKING)

The packing of particulate solid materials enters into many aspects of applied science, and has therefore been studied more or less continually for a number of years. This great interest in packing particles has attracted many investigators, and as Graton and Fraser[7] pointed out in 1935, it might be expected that the subject of particle packing would have been, thoroughly explored by this time. Such is not the case, however; the empirical approach adopted by many investigators has produced data for specialized applications, but has produced little precise, quantitative information, directly applicable to all packing circumstances. In 1961, McCrae and Gray[8] reported that researchers have tended to be concerned either with theoretical, unrealistic situations, or else those which are entirely empirical. The former are mainly concerned with the geometry of simplified packing situations (although restricted supporting experimentation is not uncommon) and are principally the study of systems of perfect spheres.

The author has been involved in the study of the packing of fiber and combinations of spheres and fibers. These studies have shown that there are specific conditions where the packing obtained would not have been anticipated from the prior theories that evolved from the study of packing of spheres alone.

In 1973, Milewski[5] introduced a system for the study of the packing of fibers and spheres. This original paper covered minimum packing conditions for fiber *L*/*D*'s from 4/1-15/1, and gave only general direction for obtaining maximum packing conditions. His 1974 paper[9] was much broader in scope, and identified the minimum packing conditions for fiber *L*/*D* from 4/1-37/1 plus the theoretical and experimental maximum packing conditions at various values for parameters such as percent fiber loading, fiber *L*/*D* and *R* ratios (the ratio of the diameter of the sphere to the diameter of the fiber). Also, a new parameter called critical *L*/*D* was identified, as well as the relationship of this critical *L*/*D* to fiber loading in fiber-sphere systems.

By utilizing the above information, the density of packing of various fiber-sphere compositions can be calculated, and the void volume or resin demand can be determined. This also permits the economics of any fiber-sphere system (in any chosen combination) to be calculated for wide ranges of fiber loading *L*/*D*, *R*, and resin and fiber cost. Recent high resin costs and resin shortages make these packing concepts appear worthy of serious consideration, because they point the way to more efficient utilization of systems with higher filler loadings.

3. REVIEW OF PACKING

3.1 Fiber Packing

The packing combinations that were studied in a thesis[6] at Rutgers University are illustrated in Fig. 4-1 as (a) fiber packing at various *L*/*D*'s, (b) fibers packed into fibers, (c) fibers packed into spheres, and (d) spheres packed into fibers. The experiments were run on two scales. The larger scale experiment used wooden rods; the smaller scale experiment used fiber glass. Figures 4-2 and 4-3 visually illustrate that bulk density varies with *L*/*D*. The data obtained in measuring the relative bulk volume for known *L*/*D*'s of both wooden rods and fiber glass are plotted on Fig. 4-4 and show very good volume packing agreement between the exact *L*/*D*'s of wooden rods and the numerical average *L*/*D*'s for the fiber glass.

The significance of this relationship is that it has been found that three-dimensional random packing values of fiber varies with the *L*/*D* and can be predicted from the given curve. Thus, with this relationship, the average fiber *L*/*D*'s can be determined from the bulk volume data

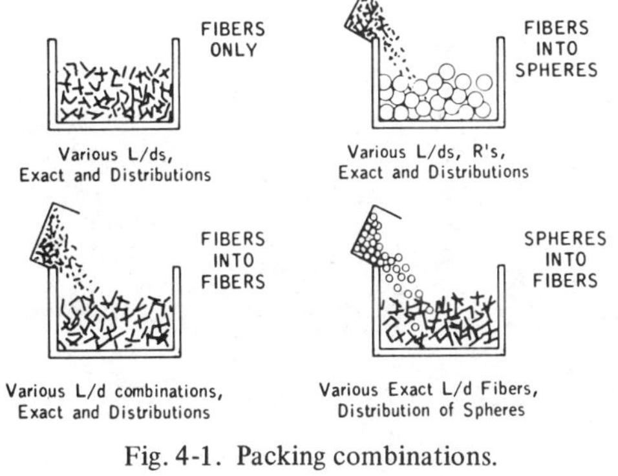

Fig. 4-1. Packing combinations.

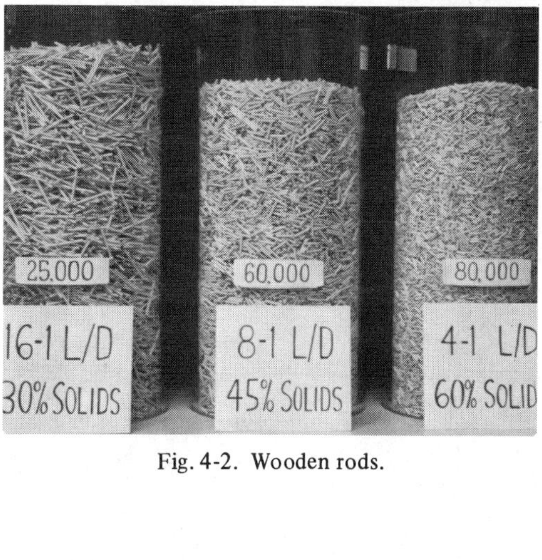

Fig. 4-2. Wooden rods.

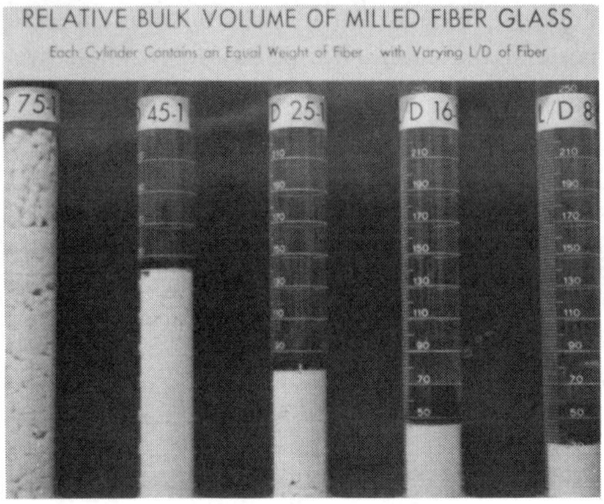

Fig. 4-3. Milled fiber glass.

of the fiber, without the time consuming expense of photographing, counting, and averaging hundreds of fibers to obtain a statistically significant sample of the fiber L/D.

3.2 Review of Bimodal Packing Theory

Before studying the packing of fibers and spheres, the bimodal packing of spheres will be reviewed so that the detailed concept of bimodal packing of fibers and spheres can be more easily understood.

3.2.1 Theoretical Maximum Density Boundary Conditions for Sphere-Sphere Packing

Figure 4-5 illustrates the densification that occurs when small spheres are added to large spheres. Maximum density is obtained when the small spheres are packed to their maximum density within the voids of the larger spheres.

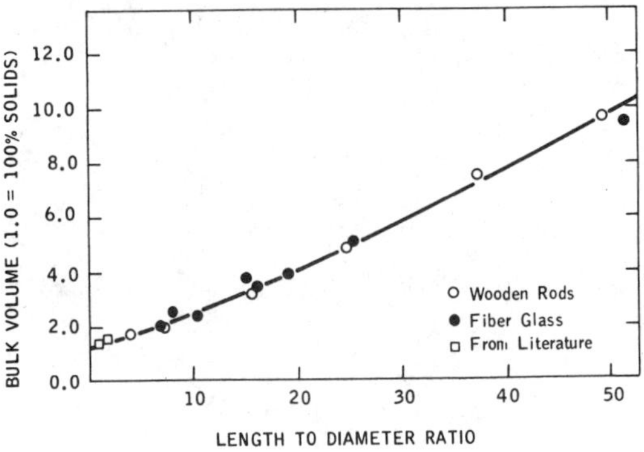

Fig. 4-4. Packing of various L/D fibers.

In Figs. 4-5A and 4-5B, each illustration represents the same volume of solid material; thus, the relative bulk volume is shown decreasing as densification occurs. In Fig. 4-5A, this results because in each step toward greater density, a large sphere is removed, and the same amount of solid material is replaced as small spheres within the voids of the remaining large spheres. Figure 4-5B illustrates how densification occurs by the opposite process, in which a number of small spheres and their associated voids are removed, and the same amount of material is replaced as one large solid sphere. The solid line in Fig. 4-6 is a theoretical packing curve for an infinite size ratio $R = \infty$. For this example, the ratio R is the diameter of the large sphere divided by the diameter of the small spheres. The maximum density point represents the condition illustrated in Figs. 4-5A and 4-5B at the extreme right. In Fig. 4-6, the composition of the mixture is shown by the horizontal scale X, the volume fraction of the large spheres, and Y, the volume fraction of the small spheres, for a total volume of unity. The left hand ordinate is the relative bulk volume and is defined such

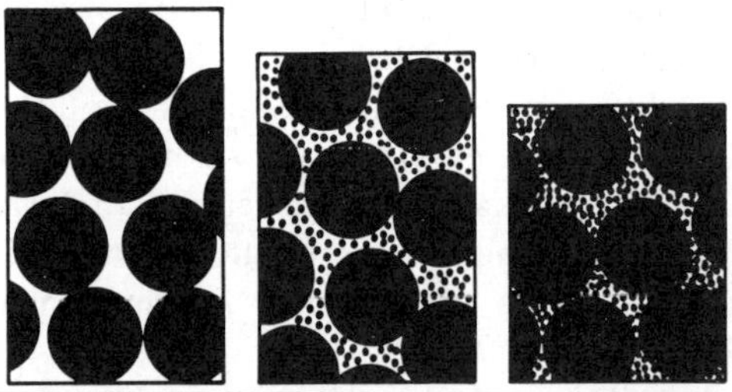

Fig. 4-5A. Addition of small spheres to large.

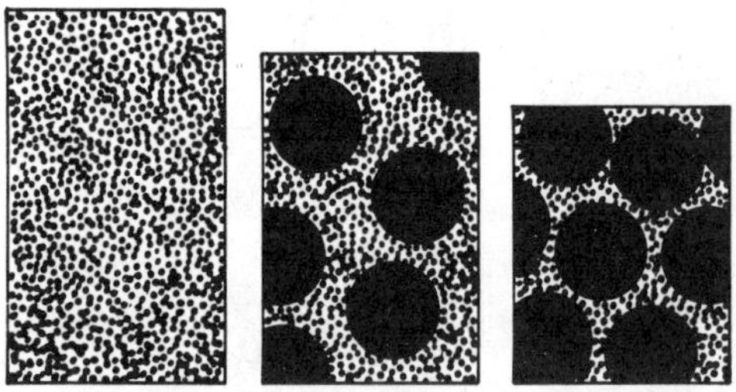

Fig. 4-5B. Addition of large spheres to small.

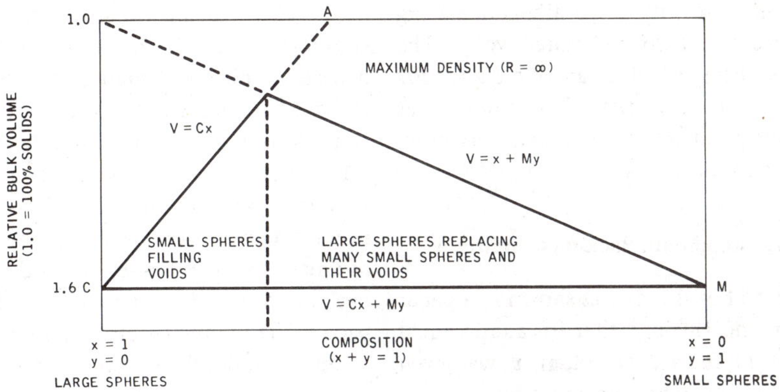

Fig. 4-6. Theoretical packing of two sphere systems.

that 1.0 is equal to 100% solid material (100 divided by the percent theoretical density). A material with a relative bulk volume of 1.6 would be 62.5% theoretically dense. Thus, C gives the experimentally determined packed volume of the large spheres, and M gives that of the small spheres. By using relative bulk volume, rather than percent theoretical density, the packing curves for the infinite size ratio become two straight lines.

3.3 Maximum Density Boundary Conditions for Sphere-Fiber Packing

The theory of bimodal packing must be modified when packing two differently shaped materials, because it is then possible to have R values less than 1. (R now being defined as the diameter of the sphere divided by the diameter of the fiber.) This leads to the development of two theoretical maximums, both having the same relative bulk volume but at different fiber loadings. This concept is illustrated in Fig. 4-7. An example is given as follows:

S = Spheres packing at 50% solids + 50% voids
Relative bulk volume = 2.0
F = Fibers packing at 25% solids + 75% voids
Relative bulk volume = 4.0
x = Volume fraction of S;
y = Volume fraction of F: $x + y = 1$

The curves are constructed in this manner: Line SE represents the filling of voids within the spheres, similar to Fig. 4-6, except that the small component is a fiber instead of other spheres. This line SE is represented by equation $V = Sx$. Line FC represents the fiber analog of Fig. 4-5B, with large diameter spheres replacing voids and small diameter fibers; this line is represented by equation $V = x + Fy$. The intersection of these two lines at point A represents the theoretical maximum for packing conditions for large R values where fibers fit within the voids of spheres.

For cases when the sphere diameter is smaller than the fiber diameter, a different construction is required. Line FD represents the void filling case with small spheres filling the voids between the fibers, and SH is again the analog

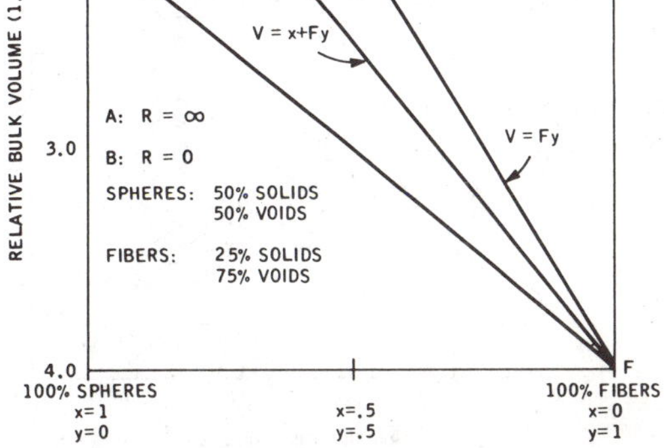

Fig. 4-7. Theoretical packing of fiber-sphere systems.

of Fig. 4-5B, but with large fibers replacing small spheres and their associated voids. The equation for these two lines are $V = Fy$ for FD and $V = y + Sx$ for SH. These lines intersect at point B. This point is the theoretical maximum packing density for $R = 0$.

3.5 Less than Maximum Packing Conditions

In reference to Fig. 4-6, it was shown in a paper by Westman and Hugill[11] that for a ratio equal to unity ($R = 1$), the packed volume v, was given by the line CM which has the equation

$$v = C_x + M_y \tag{4-1}$$

and for an infinite size ratio ($R = \infty$), the values of v are given by the line joining M with unity at the left side of the diagram, and the line CA is obtained by joining C with the zero point at the right side of the diagram, the equations of these lines being

$$v = C_x \tag{4-2}$$

and

$$v = X + M_y. \tag{4-3}$$

When the size ratio R is less than infinity, maximum density is not attained, because the small spheres begin to dilate the packing of the larger spheres before all the voids are filled. This concept is graphically illustrated in Fig. 4-8, which shows curves for several R values. At an R value of 1, the two spheres are identical and obviously no packing advantage is seen. In this case, the packing of the right component with the left follows the law of mixtures, that is, a straight line connecting the two components. This is of no consequence when the two components are spheres, but when one component is of different shape, such as a fiber, the straight line defines the worst possible packing conditions, also defining 0 packing efficiency.

In sphere-sphere packing for diameter ratios between the two limits of $R = 1$ and $R = \infty$, packing curves were developed by McGeary,[10] as illustrated in Fig. 4-8 and in the sphere-fiber packing system by Milewski[5] in the 1973 SPI paper. These curves were developed by packing of fibers and spheres over a wide range of R values (R now defined as sphere diameter divided by the fiber diameter), while varying the L/D of the fiber component. The results were viewed in two basic ways illustrated in Figs. 4-9 and 4-10.

In Fig. 4-9, the L/D of the fiber component (milled fiber glass) is held constant while R is varied. This is done with graded sizes of glass beads. The general effect is the same as that previously discussed for sphere packing, although the two components no longer have the same initial bulk volume.

Figure 4-10 shows the effect of a changing L/D at a constant R value. In the example, for $R = 10$, it can be seen that the short fibers fit easily in between the larger spheres, producing an increase in density, or decrease in bulk volume, while the longer fibers nearly follow the linear law of mixtures.

The concept of packing efficiency is defined as the maximum deviation of the bulk volume from the mixture line H_{max} divided by the

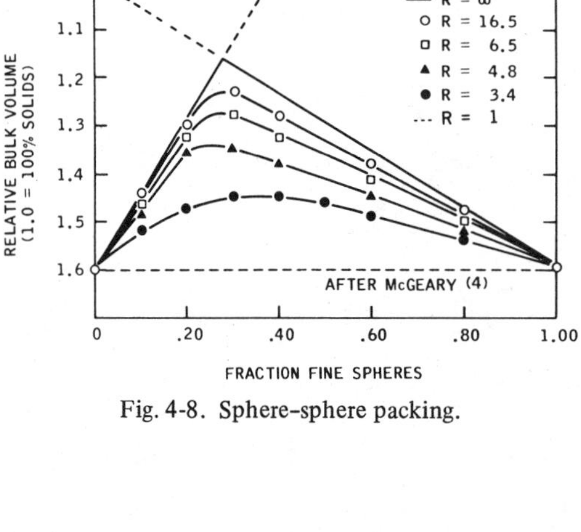

Fig. 4-8. Sphere-sphere packing.

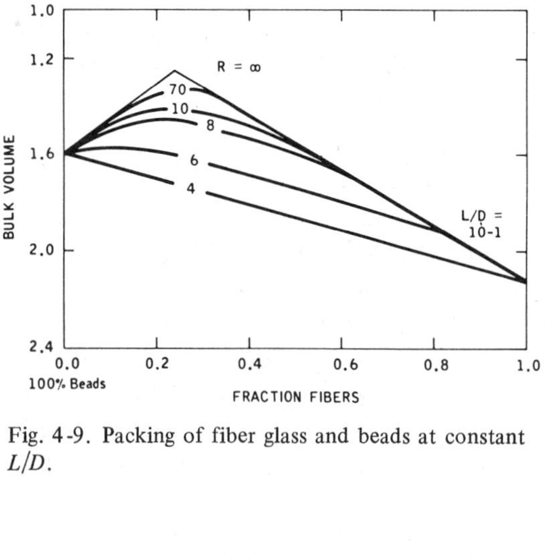

Fig. 4-9. Packing of fiber glass and beads at constant L/D.

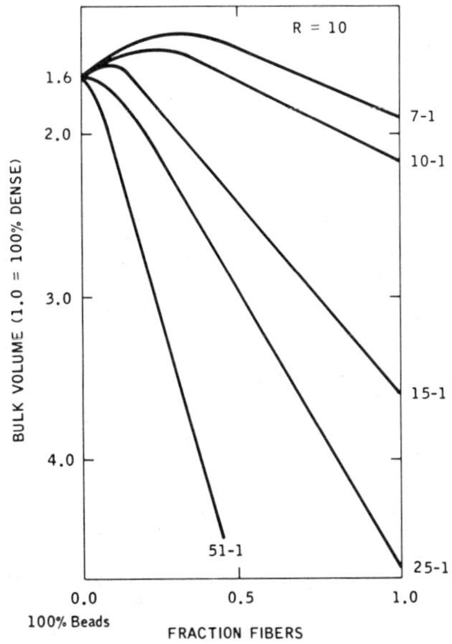

Fig. 4-10. Packing of fiber glass and beads at constant R and varying L/D's.

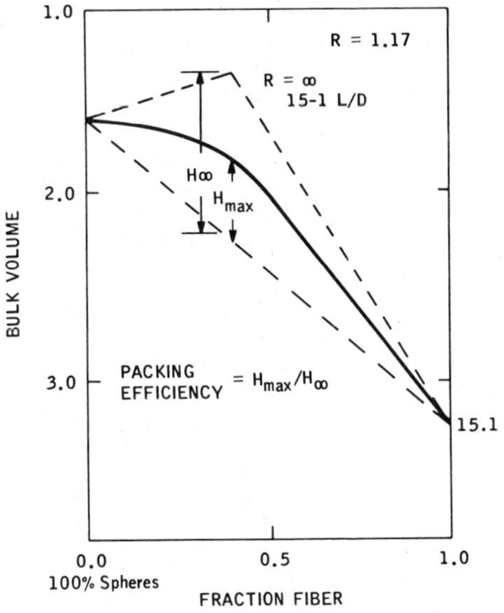

Fig. 4-11. The calculation of packing efficiency.

theoretical maximum deviation H^{∞} (see Fig. 4-11). The packing efficiency approaches zero for the long fibers and approaches 100% for the short fibers in the example of Fig. 4-10. Plots similar to this for wider ranges of fibers and spheres for various R's are shown in Fig. 4-12. Note the increased curvature, or higher packing efficiency, at the largest and smallest R's.

3.6 Experimental Minimum Packing Efficiency for Fiber Sphere Systems

Using the concept for packing efficiency, the bulk of the experimental packing work with fibers and spheres (from Fig. 4-12) can be viewed in a single graph (Fig. 4-13).

Each fiber of different L/D exhibits a minimum packing efficiency. The longer fibers show the minimum at higher R, as can be seen by the

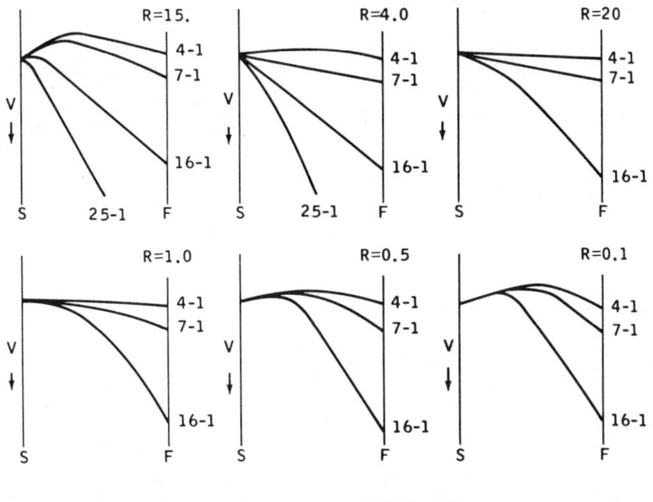

Fig. 4-12. Packing of fibers and spheres.

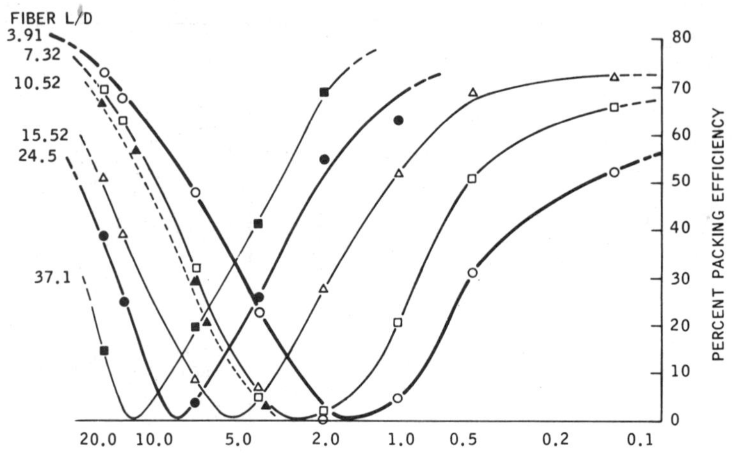

Fig. 4-13. Location of minima in packing curves.

minimum in the 37/1 L/D curve at $R = 13.5$, compared to the minimization between $R = 1$ and 2 for the 4/1 L/D fiber. The packing efficiency increases on either side of the minimum, and eventually reaches 100% as R approaches ∞, or 0, as defined by the system (Fig. 4-11). Although the maximum packing density or minimum bulk volume for either very large or very small R could have been predicted, the exact location of the minimum conditions for the different L/D fibers must be located by experiment.

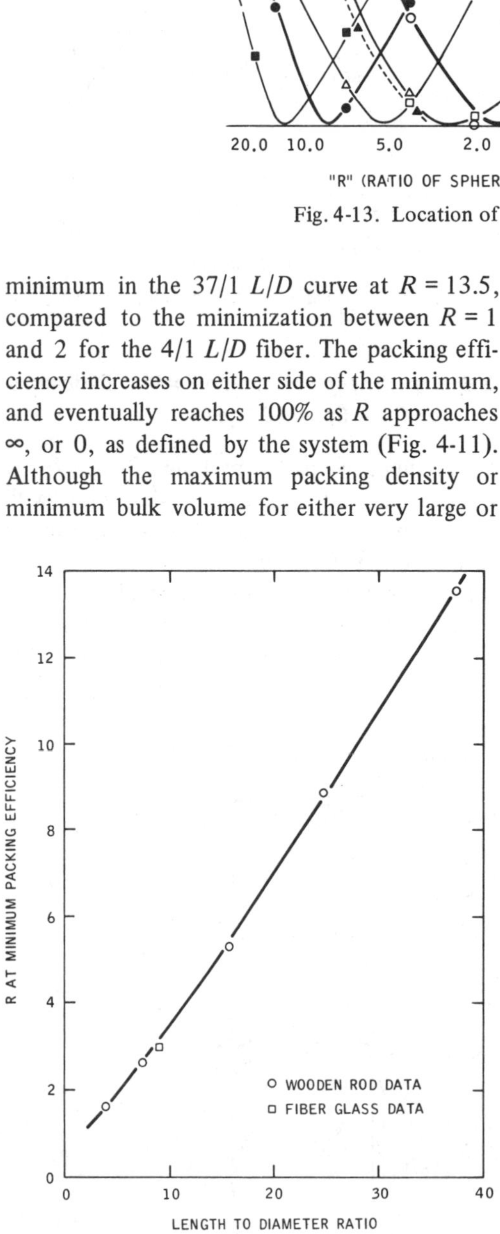

Fig. 4-14. Relationship between L/D and R at minimum packing efficiency.

From Fig. 4-13, the R values at the minimum packing conditions were determined for each fiber L/D; this relationship is illustrated in Fig. 4-14. The curve is of importance to future investigators, in that it points out what R values are to be avoided if good packing is desired, or at what R values lowest packing density can be obtained within the range of fiber L/D studied.

3.7 Theoretical Maximum Packing Densities

The theoretical maximum packing percentages were determined graphically by a method given early in bimodal packing theory for Fig. 4-7, except that the relative bulk volume data for the sphere and fiber L/D studied were used in construction of the graphs. This data is given in Table 4-1. Then experimental maximum densities were read from the data points of the packing curve for various R's and L/D's at 25, 50, and 75% fiber. These data are given in Tables 4-2 and 4-3. The 25% fiber data are plotted on Fig. 4-15. An examination of this figure tells what percent solids are the theoretical limits when packing spheres with 25% fiber of various L/D's at different R values. For example, with a fiber L/D of 13/1, it will be impossible to pack these fibers with any size spheres to a density

Table 4-1. Theoretical percent solid content for fiber-sphere packing.

% Fiber loading	R	L/D 1.00	2.00	3.91	7.31	15.51	24.49	37.10
10	$R = \infty$	68.5	68.5	68.5	68.5	68.5	68.5	61.0
	$R = 0$	64.1	64.1	64.1	64.1	64.1	64.1	64.1
20	$R = \infty$	77.0	77.0	77.0	77.0	69.0	56.2	44.3
	$R = 0$	66.7	66.7	66.7	66.7	66.7	66.7	66.7
30	$R = \infty$	87.7	87.0	83.3	77.5	60.0	46.2	34.5
	$R = 0$	69.5	69.5	69.5	69.5	69.5	68.5	45.5
40	$R = \infty$	85.5	83.3	78.7	72.0	52.9	39.1	28.6
	$R = 0$	72.5	72.5	72.5	72.5	72.5	51.3	34.2
50	$R = \infty$	82.7	80.6	74.6	67.2	47.2	34.1	24.2
	$R = 0$	76.4	76.4	76.4	76.4	62.5	41.2	27.5
60	$R = \infty$	80.0	77.5	71.0	62.9	42.7	30.2	21.0
	$R = 0$	80.0	80.0	80.0	80.0	51.8	34.2	23.0
70	$R = \infty$	77.5	74.6	67.6	59.2	39.1	27.0	23.0
	$R = 0$	84.0	84.0	84.0	72.5	44.5	29.4	19.7
80	$R = \infty$	74.6	72.0	65.0	56.2	36.0	24.5	16.7
	$R = 0$	88.5	84.0	74.6	63.3	39.0	25.8	17.5
90	$R = \infty$	72.5	69.5	62.2	53.2	33.4	22.4	15.1
	$R = 0$	78.7	74.6	66.2	56.2	34.5	23.0	15.4
100% fibers		70.5	67.1	59.5	50.3	31.1	20.7	13.8
100% spheres		61.5	61.5	61.5	61.5	61.5	61.5	61.5
Theoretical max.		88.5	87.4	84.4	80.6	73.5	69.5	66.7

Table 4-2. Fiber loading at maximum solid content of various fiber-sphere systems.

R		L/D 3.91	7.31	15.51	24.49	37.10
0.11	% Solids	74.6	73.0	67.6	61.5	61.5
	% Fiber loading	60.0	40.0	30.0	0.0	0.0
0.45	% Solids	67.1	68.0	63.7	61.5	61.5
	% Fiber loading	50.0	45.0	30.0	0.0	0.0
0.94	% Solids	61.5	61.5	61.5	61.5	61.5
	% Fiber loading	0.0	0.0	0.0	0.0	0.0
1.95	% Solids	61.5	61.5	61.5	61.5	61.5
	% Fiber loading	0.0	0.0	0.0	0.0	0.0
3.71	% Solids	64.5	61.7	61.5	61.5	61.5
	% Fiber loading	30.0	8.0	0.0	0.0	0.0
6.96	% Solids	71.0	65.0	61.5	61.5	61.5
	% Fiber loading	30.0	22.0	0.0	0.0	0.0
14.30	% Solids	75.8	73.5	63.7	61.7	61.5
	% Fiber loading	31.0	25.0	8.5	3.5	0.0
17.40	% Solids	78.1	75.1	64.9	62.5	61.5
	% Fiber loading	33.0	26.0	12.0	4.0	0.0
*	% Solids	84.7	80.6	73.5	69.5	68.0
	% Fiber loading	27.5	24.0	16.5	11.3	9.6

*Theoretical Values $R = 0$ or $R = \infty$.

Table 4-3. Experimental solid contents at 25%, 50% and 75% fiber loading for fiber-sphere packing.

Fiber L/D	Percent Fibers	*R* Value 0	0.11	0.45	0.94	1.95	3.71	6.96	14.30	17.40	∞
	25	68.5	68.5	65.4	61.7	61.0	64.5	70.0	74.6	76.4	82.0
3.91	50	76.4	74.6	67.2	61.7	60.2	64.1	67.5	72.5	74.5	75.7
	75	78.2	69.5	64.5	61.0	59.5	62.5	64.2	66.7	67.2	67.1
	25	68.5	68.5	64.5	61.0	58.5	59.9	64.5	73.5	74.6	80.6
7.31	50	76.4	71.4	67.5	58.8	55.5	56.6	58.8	65.4	67.1	67.1
	75	66.3	61.7	60.0	55.0	52.8	53.5	54.6	57.2	58.2	57.4
	25	68.5	66.7	63.7	59.9	54.6	50.3	50.5	54.1	57.5	65.0
15.52	50	61.7	55.6	51.8	50.7	45.5	42.0	42.4	44.3	44.3	48.1
	75	41.0	40.4	37.9	38.2	37.3	35.7	35.5	36.0	36.8	38.2
	25	68.5	66.5*	61.5*	55.5*	47.5	45.5	40.2	42.7	44.7	50.5
24.50	50	40.0	39.0*	38.0*	36.0*	34.0*	32.7	30.3	31.8	31.8	33.5
	75	26.4	26.3*	26.2*	25.8*	25.5*	25.2	24.3	25.0	25.6	26.2
	25	50.0	48.0*	45.0*	42.0*	39.4	37.7	33.8	33.1	39.2	41.3
37.10	50	25.7	—	—	—	—	—	22.6	22.6	22.6	25.6
	75	—	—	—	—	—	—	—	—	—	—

*Estimated values (extrapolated data)

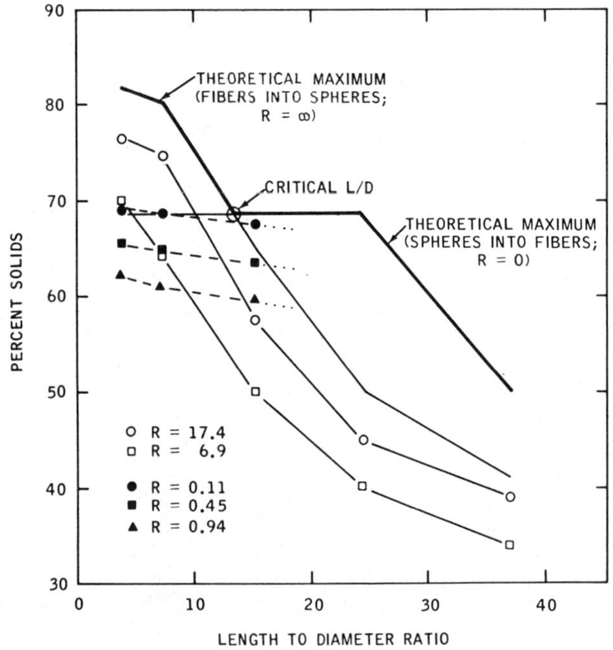

Fig. 4-15. Theoretical maximums and experimental percent solids at 25% fiber loading for various R values.

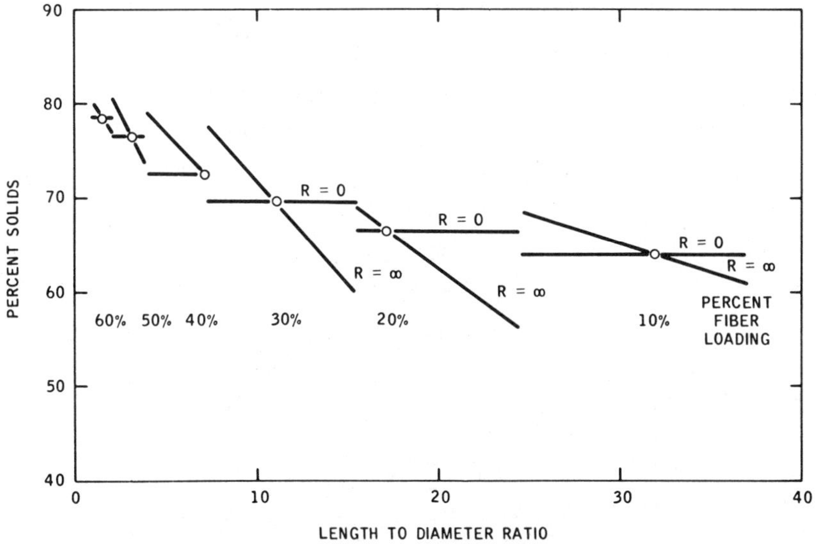

Fig. 4-16. Location of critical L/D.

greater than 68.5%, since this is more than the theoretical value. However, 74% solid packing is not only possible, but practical, and can be obtained when packing fiber with an L/D of 7/1 or less with spheres with R values of 17.4 or greater. These data also indicate that R values much less than 1 give higher actual packing densities at intermediate L/D's (greater than the critical L/D) than R values much greater than 1. For example, R of 0.11 will pack to 68% solid with a 15/1 L/D fiber, while an R of 17.4 will pack to 57.5% solid for the same 15/1 L/D fiber. This effect can also be clearly seen by the crossing of theoretical maximum curves for $R = \infty$ and $R = 0$. The points at which these two curves cross is defined as the critical L/D. Theoretically, the highest percent solids are obtained with very large R's when the L/D's are less than critical, and highest percent solids are obtained when using very small R's with fiber of L/D's greater than critical.

The location of the critical L/D was studied and determined by plotting its theoretical maxima for $R = \infty$ and $R = 0$ versus the L/D's, at various percentages fiber loading, as given in the data from Table 4-1. Figure 4-16 locates the critical L/D from the crossover points of theoretical curves for $R = 0$ and $R = \infty$ at various volume loading of fibers.

A relationship between the critical L/D's and volume percent fiber loading is shown in Fig. 4-17. When packing fibers and spheres, Fig. 4-17 will determine whether large R's or very small R's should be used if maximum theoretical solid packing is desired and the percent fiber loading and fiber L/D are given. For example, when using a 25% fiber loading with a 30/1 L/D, small R's are most advantageous, while, on the other hand, if the L/D is 10, a large R is necessary. A limit of 60% fiber loading is indicated for large R's from Fig. 4-17.

4. GENERAL APPLICATION OF PACKING CONCEPT

To approach the maximum packing condition of the fiber-sphere mixture, in general, fibers should be chosen with L/D's ranging from 1/1-100/1, and preferably about 5/1-50/1, fiber loadings should be between 5-40%, and R should be chosen to be either large or small.

To maximize packing efficiency when choosing small R values, R should be less than 2, preferably less than 0.01, but practically about 0.1. If large R values are chosen, then R should be greater than 5, preferably greater than 100, and more practically, about 20.

In a typical fiber-sphere formulation in which

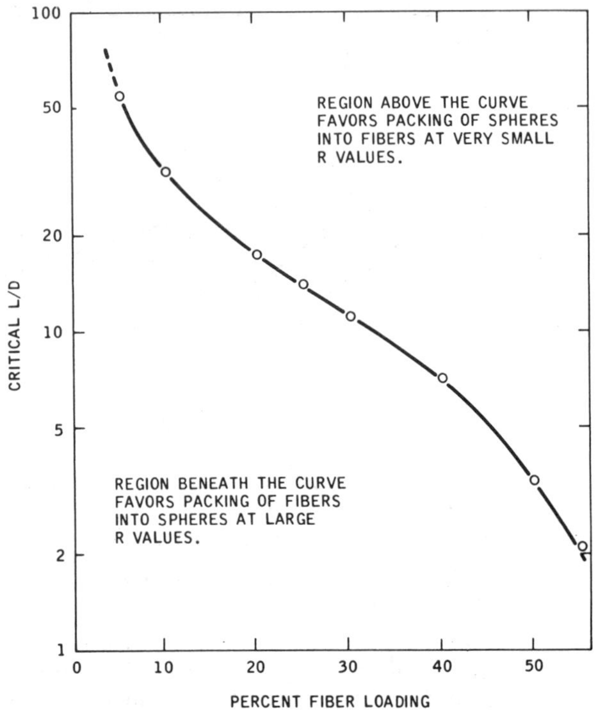

Fig. 4-17. Relationship between critical *L*/*D* and percent fiber loading.

the fiber *L*/*D* is 40 or greater with a fiber loading range of 10-80%, *R* should be chosen to be about 0.1.

In a typical fiber-sphere formulation utilizing large *R* values, for *L*/*D*'s ranging from 2/1-19/1 and fiber loadings of 1-25%, *R* at about 20 is very desirable.

In fiber loadings greater than 60%, for all *L*/*D* values, *R* should be small or about 0.1.

Where a moldable article is desired, the *L*/*D*'s preferably range from 15/1-100/1, the fiber loadings from 5-75%, and *R* should be considerably less than 5 and as small as 0.5, if practical.

5. Specific Examples of How to Apply Packing Concepts in Fiber glass-Sphere Combinations

Nearly spherical particle fillers can be used instead of perfect spheres, leaving the major packing concepts essentially valid.

Select the fiber to be used and determine its average *L*/*D* from its relative bulk volume (Fig. 4-4). For example, a commercial 1/16 in. milled fiber glass has an average *L*/*D* of about 25/1, and an average fiber diameter of about 14.0 microns. If one wishes to replace half the fiber glass with glass beads of -325 mesh (average diameter 30 μ), this would give an *R* value of about 2. From the data in Table 4-2, 50% fibers of 24.5/1 *L*/*D* and *R* value 1.94 will pack to about 33% solid volume or 67% void volume. To make a fully dense plastic composite, this 67% voids must be completely filled with resin; any lesser amount of resin would result in a partially dry mix with molded in voids.

A more practical combination would be 3 parts beads (filler) to 1 part fiber, or 25% fibers by volume; in this case, 1/32 in. milled fiber glass, which has an average *L*/*D* of about 10/1, will be considered. Table 4-2 does not contain data for 10/1 *L*/*D*, but it does contain 7.3 and 15.5/1 *L*/*D* data. Extrapolation for the 10/1 *L*/*D* data at 25% fiber loadings and *R* values of 1.95, yields 57% solids. This leaves the 43% voids to be filled with resin. If the resin is the most expensive component, this combination would be more economical than the previous mixture of fibers and beads. Since there are a wide variety of fiber and bead sizes, as well as resin costs, the optimum formulation cannot be determined without being specific to particular needs. But much can be done to reduce resin demand and percent fibers, if proper sizes of beads are chosen to match specific packing parameters.

6. SAMPLE CALCULATION OF ECONOMICS IN PACKING-FORMULATIONS

Case 1. Assume that a batch of fiber glass reinforced nylon (20/80 by volume) will be prepared, and the following components will be used:

A. Fiber glass, 1/16 milled, average L/D = 25/1, specific gravity 2.5, estimated price 50¢/lb. The theoretical percent solid content for random packing of this material, as derived from Table 4-1, is 20%. Thus, the resin demand will be 80% by volume.

B. Nylon resin, specific gravity 1.1, price 80¢/lb.

C. Composite:

20 volume percent milled fiber glass at specific gravity 2.5; 20 X 2.5 = 50 parts by weight.
80 volume percent nylon, 80 X 1.1 = 88 parts by weight.
Total batch weight = 131 lb.
Total batch cost = 50 X \$.50 = \$25 for fiber glass and 88 X \$.80 = \$70.40 for nylon; total \$95.40.
Cost/lb = \$95.40/138 = \$0.69/lb materials cost.

Case 2. Assume that a 50/50 mixture of glass fibers and glass beads will be used as a filler for nylon resin.

A. Fiber glass, 1/16 milled, average L/D = 25/1, specific gravity 2.5, fiber diameter 14.0 μ, estimated price \$0.50/lb.

B. Glass beads, -325 mesh (average diameter 28 μ), R = 2.0, specific gravity 2.5, price \$0.15/lb.

C. The theoretical solid content of a packed combination of 50-50 fiber glass (25/1 L/D) plus glass beads with R value 1.95, as shown in Table 4-3, is 34%. Thus, the resin demand will be 66%.

D. Composite:

34% mixed filler volume:
17% fiber glass X 2.5 = 42.5 parts by weight;
17% glass beads X 2.5 = 42.5 parts by weight.
66% nylon X 1.1 = 72.6 parts by weight.
Total batch weight = 157.6 lb.
Total batch cost = 42.5 X \$0.50 + 42.5 X \$0.15 + 72.6 X \$0.80 = \$85.71.
Materials cost = \$85.71/157.6 lb. = 54¢/lb.
Summary of Materials cost:
Nylon = 80¢/lb
20 wt% fiber glass = 69¢/lb
35 wt% fiber glass bead mixture = 54¢/lb

Additional costs would be associated with the blending of the fillers with the nylon resin.

It is apparent that the glass bead and fiber glass mixture can offer considerable economic advantages in materials costs. Additional advantages would be reduced shrinkage and part distortion, and a high modulus due to the higher loading. The addition of glass beads to a molding compound has frequently improved flow, and this may be a further advantage.

7. CONCLUSIONS

From an understanding of basic principles of packing, the following information can be realized:

(1) For fiber packing alone, the average L/D can be determined from its relative bulk volume (Fig. 4-4).

(2) From the relationship between the critical L/D and percent fiber loading (Fig. 4-17), a positive guide is available for selecting bead (particle filler) sizes at various combinations of fiber loading and fiber L/D.

(3) In making more exact formulations, the experimental solid contents are given in Table 4-2; this gives precisely the void volume or resin demand for a specific fiber-bead combination.

(4) For designing maximum solid filled systems, percent solid values are given over a wide range of R's and L/D's in Table 4-3.

(5) By increasing solid content (more filler and fiber), void volume or resin demand can be lessened. In cases where the resin is the most expensive component, there is a definite eco-

nomic advantage to using a better packed system of fillers and reinforcements.

8. AUTHOR'S FOOTNOTE

The author realizes that the final selection of the formulation in making a plastic composite is often directly related to the resulting physical properties and processability of the composite. This chapter suggests many maximum loading combinations of fibers and beads which should produce highly desirable properties. In general, the preferred properties are obtained with higher percents of reinforcements, as long as a void free composite can be made. Very few of these maximum packing combinations have been tried, due to the great variety of plastics, as well as of fibers and beads.

The authors welcomes questions and comments on new combinations as they are evaluated.

References

1. Ritter, J., "Sphere-Filled Plastic Composites Theory," Testing Applications 25th Annual Tech. Conf. S.P.I., Composites Div., 1970.
2. Potter's Industries, Inc., Carlstadt, N.J., Technical Bulletin #T24-1.
3. Wells, H., "Preliminary Investigation into the Use of Glass Beads in Fiberglass Reinforced Plastics," 22nd Annual Tech. Conf. S.P.I., Reinforced Plastics Div., 1967.
4. Potter's Industries Inc., Carlstadt, N.J., Tech. Bulletin #PB24-1, June 1969.
5. Milewski, J. V., "A Study of the Packing of Milled Fiberglass and Glass Beads," 28th Annual Tech. Conf. S.P.I., Reinforced Plastics Div., Feb. 1973.
6. Milewski, J. V., Ph.D. Thesis, A Study of the Packing of Fibers and Spheres," Ceramics Dept., Rutgers Univ., New Brunswick, N.J. (1973).
7. Graton, L. C. and Fraser, H. J., "Systematic Packing of Spheres–with Particular Relation to Porosity and Permeability," *J. Geology* **XLIII** (8): 785-909 (1935).
8. Macrae, J. C. and Gray, W. A., "Significance of the Properties of Materials in the Packing of Real Spherical Particles," *British J. Appl. Phys.* **12:** 164-172 (1961).
9. Milewski, J. V., Identification of Maximum Packing Conditions In the Bimodal Packing of Fibers and Spheres, 29th Annual Tech Conf. S.P.I. Reinforced Plastics Div., Feb. 1974.
10. McGeary, R. K., "Mechanical Packing of Spherical Particles," *J. Amer. Ceram. Soc.* **44** (10): 513-521 (1961).
11. Westman, A. E. R. and Hugill, H. R., "The Packing of Particles," *J. Amer. Ceram. Soc.* **13:** 767-779 (1930).

Section III
Mineral Fillers

5
Calcium Carbonate

Paul Kummer and George Crowe
Thompson-Weinman Co.
Montclair, New Jersey

CONTENTS

1. INTRODUCTION

Calcium carbonate ($CaCO_3$) fillers have high volume usage in plastics, due to their unique advantages, including the following:

- Their cost is among the lowest of the fillers.
- They are nontoxic, non-irritating, odorless.
- They are white with low refractive index, easily colored to any desired shade.
- Their softness (the standard material for Mohs 3 on the Mohs hardness scale).
- Their dryness; no water of crystallization.
- The abundant raw material supply.
- The particle sizes available; perhaps the widest range of all fillers.
- Particle size distribution is controllable for optimal packing in each polymeric system.
- They are easily coated in dry form in a high intensity mixer, to improve plastics' melt rheology.
- They are easily mixed into formulations, sometimes aiding the mixing of other ingredients.
- They are anti-plate-out agents, exerting mild scouring action during processing.
- They are acid acceptors for secondary stabilization of PVC, neutralizing chloride ions.
- They reduce shrinkage during molding and curing, as in no-shrink reinforced polyester sheet molding compounds.
- They have relatively low stiffness, even with high loading (sometimes a disadvantage).
- They are stable over a wide temperature range, converting to calcium oxide (CaO) and carbon dioxide (CO_2) between about 800-900°C.

Calcium carbonate is a polar, reactive substance, having several possible disadvantages for plastics use:

- When attacked by acids, CO_2 is evolved and soluble salts are formed. Many plastics, such as epoxy and polyester, successfully and thoroughly wet and bind $CaCO_3$ so that the compound, even though highly loaded, resists acid attack.

- When heated to between about 800-900°C, CO_2 is evolved and CaO is formed.
- In polyethylene and polystyrene, $CaCO_3$ loading tends to cause embrittlement; however, ethylene copolymers, cross-linking agents, film grades, and coupling agents are available to permit successful loading of polyethylene. Impact resistant grades of polystyrene can be loaded with $CaCO_3$.
- The trigonal crystal shape of $CaCO_3$ provides little reinforcing action, compared to strongly reinforcing materials.
- In polypropylene, talc and asbestos fillers have higher stiffness, flexural modulus, and deflection temperature than calcium carbonate; however, calcium carbonate in polypropylene has better impact resistance, possibly due to better bonding between polymer and calcium carbonate.
- Drying of calcium carbonate may be needed for moisture-sensitive applications, since the moisture content of purified calcium carbonate is usually about 0.06-0.20%.

$CaCO_3$ contrasts sharply with other high volume fillers such as clay, talc, silica, silicates, and asbestos, which may be much darker in color, stiffer in filled plastic compounds, and, in the case of silicates and clays, may contain water of crystallization or hydration. Removal of the water in clay, to form calcined clay, results in a more expensive product.

Beneficiated $CaCO_3$ description: pure, soft, white powder; low cost; abundant; retains flexibility of plasticized PVC; yet maintains stiffness and surface smoothness in glass-reinforced polyesters.

2. MANUFACTURING PROCESSES FOR $CaCO_3$

2.1 Purified Ground $CaCO_3$

This type has the highest usage for plastics, because it is purified by removal of iron and silica to minimize degradation of plastics, and ground fine enough to provide sufficiently high physical properties of filled plastics at lowest possible cost. The $CaCO_3$ is white with a fairly low refractive index, and all colors, from white plastics with TiO_2 to black plastics with carbon black, can be produced with minimum interference from the $CaCO_3$.

Production is straightforward, using equipment and procedures described in detail in Perry's Chemical Engineers' Handbook[1] and Riegel's Industrial Chemistry:[2]

"Crushing" in gyratory crusher or jaw crusher or roll; large lumps are crushed to pass through 3 in. mesh.

"Disintegrating" in pot crusher, rotary hammer mill, squirrel cage disintegrator or stamp mill, ring-roll mill, pebble mill, or edge runners to pass through 100 mesh screen.

Flotation to remove impurities.

Wet or dry grinding (example: pebble mill) autogenous or with grinding media.

Carefully controlled classification, solids separation, drying, and micronizing steps, resulting in a *median* particle diameter of about one to 10 micrometers by ASTM D422.[3] Micrometer is customarily abbreviated "μ," or "micron."

2.2 Dry Processed Coarser $CaCO_3$

For some applications, such as vinyl foam carpet backing and inexpensive dark floor tiles, fairly coarse unpurified $CaCO_3$ is used. Processing is simple, consisting of crushing, disintegrating, and pulverizing, followed by "air separation," i.e., removal by centrifugal separation of oversize particles by a counteraction "whizzer" or a multivane "deflector."

A typical example of processing might consist of a jaw crusher to a rotary hammer mill to a Raymond suction mill with dust collector and vacuum exhauster.

Finest grind: about 12 μ median diameter.
Properties: unpurified, coarse, wide size distribution, color varies with feedstock.

2.3 Precipitated $CaCO_3$ by Na_2CO_3 Process

Ernest Solvay, a Belgian, in 1864, successfully manufactured Na_2CO_3 by an ammonia process:[2,4]

$$NH_3 + H_2O + CO_2 + NaCl \rightleftharpoons NaHCO_3 + NH_4Cl$$

$$2NaHCO_3 + \text{heat} \longrightarrow Na_2CO_3 + H_2O + CO_2$$

$CaCO_3$ is used in this process as follows:

$$CaCO_3 \xrightarrow{\text{heat}} CO_2 + CaO \xrightarrow{H_2O} Ca(OH)_2$$

$$Ca(OH)_2 + 2NH_4Cl \longrightarrow 2NH_3 + 2H_2O + CaCl_2 \xrightarrow{CO_2} H_2O + 2NH_4Cl + CaCO_3$$

Naturally occurring $CaCO_3$ is a starting material and in this process precipitated $CaCO_3$ is a by-product. This is also the case in the NaOH and recarbonation processes described in the following sections. Production of the by-product is limited by amount of main product produced. Precipitated $CaCO_3$ is washed to remove soluble salts, then dried and micropulverized. The particle size is controlled by conditions of precipitation: temperature, concentration, rate and order of addition of ingredients, and amount of agitation. Particle sizes are fairly uniform and can be varied from about 0.03–0.05 μ up to about 8–10 μ. Impurities are NaCl and water. Precipitated $CaCO_3$ is used in specialized applications where purity, fineness, uniform particle size, and excellent whiteness justify their higher cost.

2.4 Precipitated $CaCO_3$ by NaOH Process

Since 1940, more NaOH has been produced electrolytically than by the "lime-soda" process described in this section. Therefore, the production of precipitated $CaCO_3$ by this process is necessarily limited. In this lime-soda process, $CaCO_3$ is heated to form CaO and reacted with water to form calcium hydroxide $Ca(OH)_2$. The $Ca(OH)_2$ is then reacted with sodium carbonate to form calcium carbonate and sodium hydroxide, as follows:

$$Na_2CO_3 + Ca(OH)_2 \rightleftharpoons 2NaOH + CaCO_3$$

The $CaCO_3$ settles out, is removed and washed to remove NaOH. Fine particle size $CaCO_3$ is produced by this method. Since Solvay process and soda-lime plants are being closed for economic reasons, less by-product $CaCO_3$ is available. Finely ground purified $CaCO_3$ is replacing the by-product material for many uses; where precipitated $CaCO_3$ is still required, it is manufactured by a more direct process described in the following section.

2.5 Precipitated $CaCO_3$ by Recarbonation

This process consists of heating $CaCO_3$ to form CO_2 + CaO that is reacted with water to form $Ca(OH)_2$. The CO_2 is saved and recombined with the $Ca(OH)_2$:

$$CaCO_3 \xrightarrow{\text{heat}} CO_2 + CaO \xrightarrow{H_2O} Ca(OH)_2 \xrightarrow{CO_2} CaCO_3.$$

The product is aragonite, with higher sp.g. than calcite. Median particle sizes are between 0.2–2.0 μ, and particle size distribution is wider than the previous two precipitated materials.

Advantages of precipitated $CaCO_3$: High purity grades for U.S.P., FDA and food contract uses; very fine particle sizes for high viscosity non-sag plastisols and polyesters.

Disadvantages: higher cost, higher energy to manufacture, ionic and water content must be taken into consideration, fineness causes excessive viscosity in some polyesters.

2.6 Surface-Coated $CaCO_3$

In general, calcite or chalk coated with stearic acid, calcium stearate, or rosin have improved rheological properties, with little if any improvement in physical properties.[5] Surface coating of 3 μ dry calcite with 0.8–1.0% stearic acid is readily accomplished in a high intensity (Henschel) mixer above the melting temperature of the stearic acid. For laboratory tests, a Ronson Cook 'n' Stir mixer set at 100°C (212°F) will easily coat 500–600 g of 3 μ calcite with 5–6 g of stearic acid. Commercially, calcium carbonate coated with stearate are used in some plastisols and rigid vinyl compounds.

A new coating material, Ken-React TTS, that improves both the melt rheology and physical properties, is discussed in section 7, **New Developments**.

2.7 $CaCO_3$ Dispersed in Liquids

For aqueous polymeric emulsions, such as vinyl and acrylic emulsions, purified calcite can be used in aqueous suspension, avoiding the use of energy to dry the filler. A disadvantage is the added cost of shipping the water content.

Precipitated $CaCO_3$ of 1 μ size is being marketed, dispersed in either dioctyl phthalate plasticizer for PVC compounds or naphthenic process oil for elastomeric compounds. Eighty parts by weight of $CaCO_3$ and 20 parts by weight of liquid are shipped.

3. Properties of $CaCO_3$

As shown in Tables 5-1–5-6, calcium carbonate is an abundant material, so stable that entire mountain ranges and coral and mollusk shell portions of seashores are composed of $CaCO_3$. In living organisms—from simple seashells, coral, and mollusks, to the most highly advanced vertebrates—structures containing $CaCO_3$ furnish defense and support, and contribute to motion and leverage. Skeletal structures, containing calcium phosphate and $CaCO_3$ far outlast all other body parts.

Table 5-1A. $CaCO_3$ chemical properties and analysis.

Commercial $CaCO_3$ production starts and ends with $CaCO_3$:

1. Precipitated $CaCO_3$ as by-product of NaOH process (Diamond Alkali);

$$CaCO_3 \xrightarrow{heat} CO_2 + CaO \xrightarrow{H_2O} Ca(OH)_2 \xrightarrow{Na_2CO_3} 2NaOH + CaCO_3.$$

2. Precipitated $CaCO_3$ as by-product of Na_2CO_3 process (Wyandotte);

Overall Na_2CO_3 process: $CaCO_3 + 2NaCl \longrightarrow Na_2CO_3 + CaCl_2$

$$CaCl_2 + CO_2 + H_2O + 2NH_3 \longrightarrow CaCO_3 + 2NH_4Cl.$$

3. Precipitation of $CaCO_3$ from "carbonated lime" (Pfizer);

$$CaCO_3 \longrightarrow CaO + CO_2 \xrightarrow{H_2O} Ca(OH)_2 + CO_2 \longrightarrow CaCO_3 + H_2O.$$

4. Wet-ground or dry-ground purified $CaCO_3$;

Grinding, impurity removal, and classification uses less energy than that required for chemical reactions.

Main chemical reactions:

1. CO_2 is evolved when heated or reacted with acids;

$$CaCO_3 \xrightarrow{900^\circ C} CaO + CO_2$$
$$CaCO_3 + 2HCl \longrightarrow CaCl_2 + H_2O + CO_2$$
$$CaCO_3 + H_2SO_4 \longrightarrow CaSO_4 + H_2O + CO_2.$$

2. In nature, dissolved calcium tends to combine with CO_2 from the air and precipitate as $CaCO_3$;

$$Ca^{++} + H_2O + CO_2 \longrightarrow CaCO_3 + 2H^+.$$

3. Life forms, from marine life to vertebrates, utilize $CaCO_3$ and calcium phosphate structures for protection, rigidity and compatibility with many chemical systems. The $CaCO_3$ structure tends to last longer than any other part of the organism.

Typical chemical composition of purified $CaCO_3$:

$CaCO_3$	98.2%	H_2O	<0.10
$MgCO_3$	1.2	MnO_2	0.0035
SiO_2	0.25	Cu	none
Al_2O_3	0.12	heavy metals	<20 ppm
Fe_2O_3	0.06	as lead	USP test

Formula: $CaCO_3$ (no water of crystallization)
Formula weight: 100.09.

Table 5-1B. Food contact packaging uses.

	Ground Limestone	Precipitated Calcium Carbonate
Assay: $CaCO_3$ after drying, min.	94.0%	98.0%
Loss on drying, max.	2.0%	2.0%
Magnesium and alkali salts, max.	3.5%	1.0%
Acid-insoluble substances, max.	2.5%	0.2%
Fluoride, max.	0.0050%	0.0040%
Heavy metals (as lead), max.	0.0040%	0.0030%
Lead, max.	0.0003%	0.0010%
Arsenic, max.	0.0003%	0.0003%
Mercury, max.	0.00005%	—
	(0.5 ppm)	—

Test methods are described for the above requirements in Food Chemicals Codex, Second Edition, 1972, National Academy of Sciences, Washington, D.C.

In nature, when the calcium is dissolved by aqueous acid, eventually the dissolved calcium recombines with carbon dioxide to form the stable calcium carbonate again.

Unique features:

1. An index of refraction close to that of many plasticizers and resins leads to minimum interference with colorants.
2. Acid acceptor serves as secondary stabilizer in PVC.
3. Purity in some grades for USP, FDA, food contract applications.
4. Temperature stability up to about 550°C.
5. Wide range of particle sizes and distributions available.
6. Readily coated for improved melt rheology when needed.
7. Retards smoke formation when PVC compounds are heated.

References 6, 7, 8, 9

Table 5-2. Physical properties of calcite, aragonite, dolomite, and $MgCO_3$: "Handbook" values.

	Calcite	Aragonite	Dolomite	Magnesite
	$CaCO_3$ most stable form	$CaCO_3$ converts to calcite	$CaCO_3$-$MgCO_3$ (45% $MgCO_3$ by wt)	$MgCO_3$
Specific gravity	2.60–2.75	2.92–2.94	2.80–2.90	3.0–3.1
Hardness, Mohs	3.0	3.5–4.0	3.5–4.0	3.5–4.5
Solubility product	0.99×10^{-8} (15°C)	—	—	2.6×10^{-5} (12°C)
Solubility product	0.87×10^{-8} (25°C)	—	—	—
Solubility-grams in 100 parts water at 18°C	0.0013	0.0019	0.032	0.0106
Solubility in aq. CO_2	100 times above	100 times above	100 times above	100 times above
Melting point	1339°C, 1025 atm. decomp. 900°C	decomp. 825°C to calcite >400°C	decomp. 730–760°C	decomp. 350°C
Formation	Limestone, marble, calc-spar. chalk, eggshells, bones (with Ca phosphate), entire mountain ranges (with dolomite)	Shells of mollusks and corals; forms above 30°C; converts to calcite above 400°C and below 30°C (metastable)		

Table 5-3. Typical Physical properties of Pulverized Limestone Association (PLA); standard grades of limestone.

	−325 Mesh High Calcium Limestone	−325 Mesh High Calcium Limestone	−325 Mesh High Calcium Limestone	−325 Mesh Dolomitic Limestone
PLA grade	P1	L1	A1	A2
Particle diameter		(ASTM D 422, E 161, and methods)		
% finer than 0.5 μ	4	5	—	—
1	18	11	4	—
2	38	26	14	5
4	65	40	—	11
5	—	—	31	—
6	80	52	—	18
8	90	63	45	26
10	96	72	52	33
15	—	89	68	47
20	—	—	79	62
25	—	—	91	76
30	—	—	96	—
35	—	—	—	90
40	—	—	—	96
% coarser than 44 μ (% on 325 mesh)	<0.01	<0.01	0.5	—
Hunter reflectance, %				
green	96.6	95.7	91.7	—
blue	95.6	95.0	89.4	—
Dry bulk, lb/ft^3				
loosely packed	45	55	70	—
tightly packed	70	87	92	—
Hardness, Mohs	3.0	3.0	3.0	3.5–4.0
pH in water, 5% slurry, 23°C.	9.0–9.5	9.0–9.5	9.5	9.0–10.0
Water soluble salts, %, 23°C.	0.06	0.06	0.06	—
Resistance in water, ohms, 23°C.	17,000–25,000	17,000–25,000	17,000–25,000	3,000–5,000
ASTM D1199 type GC, grade	I	I	II	—
ASTM D1199 max. % on 325 mesh	0.05	0.05	0.5	—
Trade names:				
Thompson-Weinman	Atomite	Snowflake	#1 White	—
Georgia Marble	Gamaco	Cal White	#10 White	—
Camel	Camel White	Camel Tex	—	—
Pfizer	Vicron 15-15	Vicron 25-11	Marblewhite 325	—
Engelhard	Chem Carb 11	Chem Carb 22	Chem Carb 55	—
Ohio Lime	—	—	—	Stonelite
Sylacauga Calcium	Microwhite 25	Microwhite 50	—	—

Table 5-4. Electrical properties of $CaCO_3$.

Dielectric constant	
Calcite, ⊥ to optic axis, 10^4 hertz, 17–22°C	8.5
Calcite, ∥ to optic axis, 10^4 hertz, 17–22°C	8.0
Dolomite ⊥ to optic axis 10^8 hertz, 17–22°C	8.0
Dolomite ∥ to optic axis, 10^8 hertz, 17–22°C	6.8
Volume resistivity at 20°C	
Marble	10^9–10^{11} ohm-cm

Table 5-4. *(Continued)*

Surface resistivity at 20°C	
Marble, 50% relative humidity	3 to 8 × 10^9 ohm-cm
Marble, 90% relative humidity	1 to 3 × 10^7 ohm-cm
Specific resistance of aq. slurry at 23°C	
5 g $CaCO_3$ (calcite)/100 ml water	17,000–25,000 ohms
5 g $CaCO_3 \cdot MgCO_3$ (dolomite)/100 ml water	3,000–5,000 ohms
pH, calcite	9–9.5
pH, dolomite	9–10
Volume resistivity for electrical insulating PVC compound: at 50°C >4 × 10^{14} ohm-cm. See Table 5-14.	

Table 5-5. Thermal properties of $CaCO_3$.

Standard *heat of formation*, $-\Delta H_f^\circ$, heat *evolved* during the formation of 100.90 grams (one mole) of
$CaCO_3$ from its elements, at 25°C equilibrium temp.
Calcite 288.45 Kg-cal/mole
Aragonite 288.49 Kg-cal/mole

Standard *free energy of formation*, $-\Delta F_f^\circ$, free energy evolved during formation of 100.90 grams of $CaCO_3$ from its elements, with each substance in its thermodynamic reference state of unit activity:
Calcite 269.78 Kg-cal/mode
Aragonite 269.53 Kg-cal/mole

Entropy of formation of $CaCO_3$ from its elements, 25°C.
Calcite S° = 22.2 cal/degree-mole
Aragonite S° = 21.2 cal/degree-mole

Equilibrium constant of formation of $CaCO_3$ from its elements, 25°C:
Calcite $K_f = 10^{197.745}$, that is, $\log_{10}$ 197.745
Aragonite $K_f = 10^{197.562}$, that is, $\log_{10}$ 197.562

(For $MgCO_3$ formation, ($-\Delta H_f^\circ$ = 266 Kg-cal/mole)
($-\Delta F_f^\circ$ = 246 cal/degree-mole)
(S° = 15.7 cal/degree-mole)
($K_f = 10^{180.3}$ or $\log_{10}$ 180.3)

Specific heat of $CaCO_3$: 0.200 to 0.214 cal required to heat 1 g 1°C (between 0–100°C)

Heat conductivity of marble 0.0071 g · cal/sec · cm^2 · 1 cm thick, 20°C

Heat conductivity of chalk 0.0022 g · cal/sec · cm^2 · 1 cm thick, 20°C

Coefficient of linear expansion, limestone C = 9 × 10^{-6}, 25 to 100°C

Coefficient of linear expansion, marble C = 11.7 × 10^{-6}, 15 to 100°C

Coefficient of linear expansion = increase in unit length for each °C increase

Formation temp: Calcite forms below 30°C, aragonate forms above 30°C.

Melting points: Calcite 1339°C 1025 atm; aragonite 825°C (decomp.).

Dissociation pressure: $CaCO_3 \rightleftharpoons CaO + CO_2$

t°C	500	600	700	800	898	900	950
P_{CO_2} mm Hg	0.11	2.33	25.3	168	760	773	1490

Table 5-6. Optical properties of $CaCO_3$.

Index of Refraction		n_D^{20}
Calcite (trigonal cryst. form)	ϵ_D	1.4864
	ω_D	1.6585
Aragonite		1.5300
orthorhombic cryst. form	α_D	1.6809
orthorhombic cryst. form	β_D	1.6854
orthorhombic cryst. form	γ_D	
axial angle and dispersion	2V = 18°11′	
optical orientation	Ax.Pl. 100	
	$X = c$	
	$Z = b$	
Dolomite (trigonal cryst. form)	ϵ_D	1.500
	ω_D	1.68174
Magnesium carbonate (trigonal cryst. form)	ϵ_D	1.509
	ω_D	1.700

Calcite, dolomite ($CaCO_3 \cdot MgCO_3$), and $MgCO_3$ have a uniaxial negative refractive index and trigonal crystalline form.

Aragonite, the λ phase of $CaCO_3$, metastable above 0°C., has a biaxial negative refractive index and an orthorhombic crystalline form.

Comparison of refractive index of SiO_2, high volume PVC plasticizers, nepheline, calcite, and rutile TiO_2:

SiO_2		~1.47–1.54
Phthalate and phosphate plasticizers		~1.45–1.56
Nepheline (Na, K) $AlSiO_4$ (HEX)	ω	1.536–1.549
Calcite (trigonal)	ω	~1.66
TiO_2, rutile (tetragonal)	ω	2.615

4. SUPPLIERS

Major suppliers and grades of $CaCO_3$ are listed in Tables 5-7–5-10. Information from "Materials and Compounding Ingredients for Rubber"[10] and from the National Paint and Coatings Association Raw Materials Index, Pigment Section[11] was used in compiling the list. In addition to the suppliers listed, there are many suppliers of air-floated unpurified $CaCO_3$ that is generally coarser and has a broader particle size distribution than the materials listed.

Screening the $CaCO_3$ through a 325 meash screen (44 μ) provides a simple method for measuring coarseness, as shown in an ASTM

Table 5-7. Suppliers of calcites, selected products, typical properties

Supplier	Purified Grades*	μ	Maximum % +325 Mesh
	CCC H-White	2.5	
Calcium Carbonate Co.	CCC G-White	5.5	0.02%
3150 Gardner Expressway	CCC Quincy-1	5	0.5
Quincy Ill. 62301	CCC Quincy-2	5	1.5
(217) 224-1100	CCC Quincy-3	8	20
	CCC Quincy-4	12	35
	CCC Quincy 1 Electro surface treated	—	0.5
	(CCC Mosher AA, A, B, C Oolitic Limestone Grades)		

Table 5-7. (*Continued*)

Supplier	Purified Grades*	μ	Maximum % +325 Mesh
Harry T. Campbell Son's Corp.	Camel-Wite	3	0.001
100 W. Pennsylvania Ave.	Camel-Tex	5	0.01
Baltimore, Md. 21204	Camel-Karb	7.5	0.5
(301) 823-7000	Camel-Kote	11.0	0.1
Engelhard Minerals and Chemicals Corp.	Chem Carb 11	3	0.01
Menlo Park	Chem Carb 22	5	0.01
Edison, N.J. 08817	Chem Carb 33	10	0.01
(201) 548-2200	Chem Carb 44	5	0.15
	Chem Carb 55	12	0.5
	Chem Carb 66	14	3
Georgia Marble Company	Gama-Sperse 80	2	0.005
Calcium Products Div.	Gama-Sperse 140	2.8	0.005
Tate, Ga. 30177	Gama-Sperse 6532	3	0.005
(404) 893-2511	Gamaco	3	0.015
	Calwhite	5	0.02
	Kalmac (Cationic Coated)	5	0.02
	Gama-Sperse 6451	6	0.008
	Wingdale White	7.5	0.2
	Gama-Sperse 255	9.5	0.02
		—	—
Pfizer, Inc.			
Pfizer Minerals, Pigments and Metals Div.	Vicron 15-15	3.2	nil
235 E. 42 Street	Vicron 25-11	5	nil
New York, N.Y. 10017	Vicron 31-6	8.9	nil
(212) 573-2834	Vicron 41-8	7.5	trace
	Vicron 45-3	17.2	trace
	Marblewhite 325	17.2	0.5
	Hi-Pflex 100	3.2	surface treated
Sylacauga Calcium Products Co.	Micro-White 25	3.2	
Box 330	Micro-White 50	6.5	
Sylacauga, Ala. 35150	Micro-White 75	9	
(205) 245-5261	Dryca-Flo 125	2.5	
	Dryca-Flo 150	6.5	
	Dryca-Flo 165	9	
	Dryca-Flo 225	12.5	
Thompson, Weinman & Co.	Atomite	3	trace
P.O. Box 130	Snowflake White	5	trace
Cartersville, Ga. 30120	Drikalite	5	0.1
(404) 382-5353	Duramite	10	trace
	Supercoat (coated)	1	nil
	Supermite (uncoated)	1	nil
	Kotamite (coated)	3	trace
Pluess-Staufer (North American) Inc.	OMYA BLS	2.3	0.005
One World Trade Center, Suite 2173	BLR/3 {surface coated with calcium stearate}		0.01
New York, N.Y. 10048			
(212) 466-0550			
	See Table 5-9 for chalk grades		

*Most suppliers also sell coarse grades

Table 5-8. Pulverized Limestone Association members.

Calcium Carbonate Co., Quincy, Illinois (address in table above)

Harry T. Campbell Sons' Co., Div. of the Flintkote Co., Towson, Maryland (address in table above)

Georgia Marble Co., 3460 Cumberland Parkway, Atlanta, Ga. 30339, (404) 432-0131

National Gympsum Co., 325 Delaware Ave., Buffalo, N.Y. 14202, (716) 852-5880

Ohio Lime Co., subsidiary of General Refractories Co., Woodville, Ohio 43469, (419) 849-2321

Pfizer, Inc., Minerals, Pigments & Metals Div., 640 N. 13th Street, Easton, Penn. 18042, (215) 253-6261

Pure Stone Co., P.O. Box 297, Marble Falls, Tex. 78654, (512) 693-4300

Sylacauga Calcium Products Co., division Moretti-Harrah Marble Co., P.O. Box 330, Sylacauga, Ala. 35150, (205) 245-5261

Thompson, Weinman & Co. (address in table above)

White Pigment Corp., 61 Main Street, Proctor, Vt. 05765, (802) 459-3311

specification for $CaCO_3$ (Table 5-11). This specification, ASTM D1199, also defines both particle size distribution width and particle size, combined in one equation. Most manufacturers prefer to control the entire distribution curve by measuring aqueous suspension density *vs* settling time at constant temperature, ASTM D422, or by a sedigraph[12] checked against fine screens (ASTM E161) down to 5 μ sieve size.

5. COST AND AVAILABILITY

5.1 Cost of $CaCO_3$ in 1975:

	Non-beneficiated Dry Ground	Beneficiated Natural	Precipitated
¢/lb	0.75–1.5	1.5–3	2.5–7
¢/in.^{3}	0.073–0.146	0.146–0.293	0.246–0.685

Future price depends mainly on changes in fuel costs for drying and energy costs for grinding and hauling the ore, as well as tax changes.

5.2 Availability

Currently, supplies are generally adequate, but are limited in some cases, due to shortages of fuel for drying and hauling, and electrical capacity for grinding.

In the future, adequate supplies should be available because more efficient grinding and drying methods are being used, and more grinding capacity is being installed. Calcite ore of good quality is available in Georgia, Alabama,

Table 5-9. Suppliers of natural chalks, grades, typical properties.

		μ	Maximum % +325 Mesh
Calcium Carbonate Co.	Mosher AA	6.5	0.5
3150 Gardner Expressway	Mosher A	7.5	2
Quincy, Ill. 62301	Mosher B	9	10
(217) 224-1100	Mosher C	12	20
Pluess-Staufer (N. Amer.) Inc.	Omyalite 90	0.7	0.01
One World Trade Center	Omyalite 90T	0.7	0.01 stearate coated
Suite 2173	Omya Hydrocarb	0.7	0.01
New York, N.Y. 10048	Omya BSH-BSH/20	2	0.01 coated
(212) 466-0550	Omya Durcalite	2.3	—
	Omya BLS	2.3	0.005 (calcite)
	Omya Micromya	2.5	0.01
	Omya BLH, BLR/3	4.5	0.01 coated (BLR/3 is calcite)
	Omya BLR/2	4.5	0.01
	Omya Violet	2.5	0.5
	Omya Yellow	4	8

Table 5-10. Suppliers of precipitated calcium carbonates.

Supplier		SpG	μ	
Basic Chemicals 845 Hanna Bldg. Cleveland, Ohio 44115 (216) 241-5000	LIQUI CARB DC LIQUI CARB RC	1.910 1.915	1.0 1.0	80% $CaCO_3$ in DOP plasticizer 80% $CaCO_3$ in naphthenic process oil
		SpG		Maximum % +325 Mesh
Mississippi Lime Co. 7 Alby Street Alton, Ill. 62002 (618) 465-7741	M-60 PPT'D $CaCO_3$	2.85	–	–
Pfizer Minerals, Pigments & Metals Div.	ALBAGLOS	2.71	.75	.05
235 E. 42 Street	ALBACAR 5960	2.71	.65	nil
New York, N.Y. 10017	ALBACAR 5950	2.71	.85	nil
(212) 573-2834	ALBACAR 5970	2.71	.95	nil
	ALBACAR 5930	2.71	1.20	nil

Vermont, California, Maryland, Massachusetts, and other states.

Beneficiation methods have been evolved and are being improved for the removal of impurities, particularly ions that would interfere with electrical volume resistivity, as well as iron and silica. The $CaCO_3$ filler industry has designed fillers to meet requirements for each major use. Some examples are listed below:

1. Electrical PVC wire coating type filler shipments are specifically controlled for low ionic content, and checked by making a standard compound and measuring electrical volume resistivity (see Table 5-14).

2. Glass fiber-reinforced sheet molding compound filler shipments are controlled for low and uniform polyester viscosity.

Future developments will be pointed directly at each use. Particle size and distribution may be tailored for best performance in each process. Coating, when needed, will meet melt rheology requirements for easy processing with optimal production rate using minimal power.

6. CURRENT APPLICATIONS

6.1 Polyvinyl Chloride (PVC) Compounds

6.1.1 Plasticized, Flexible PVC.[13-16] The highest volume uses for beneficiated $CaCO_3$ are for flexible PVC compounds, plastisols, and glass-reinforced polyester compounds. Filler

Table 5-11. ASTM D1199 specification for $CaCO_3$.

Natural ground $CaCO_3$: type GC		Precipitated $CaCO_3$: type PC
Moisture (volatiles 110°C)	0.2 % max.	Moisture (volatiles 110°C):
Mg as $MgCO_3$	3 % max.	1% max. for 0.05 μ
$CaCO_3$ + $MgCO_3$, dry basis	95 % min.	0.2% max. for 2 μ and coarser
Grade I on 325 mesh (44 μ)	0.05% max.	$CaCO_3$: 98% min., dry basis
II on 325 mesh (44 μ)	0.5 % max.	
III on 325 mesh (44 μ)	25 % max.	
IV on 200 mesh (75 μ)	30 % max.	

level for flexible PVC is usually in the range of 20–60 phr of 3 μ $CaCO_3$. (phr = parts by weight of ingredient per 100 parts resin by wt.) Many examples of filler loading *vs* physical properties are published in trade literature of PVC resin manufacturers and PVC stabilizer manufacturers,[17-21] and in books on PVC technology and plastics. See Tables 5-12 and 5-13 for typical properties of flexible PVC extrusion compounds. Finer fillers cause less of a decrease in physical properties at the same loading, while coarser fillers cause problems in thinner calendered films and coatings. Adequate physical properties at lowest cost per unit volume influences choice of filler and loading.

Flexible PVC electrical insulator compounds for extrusion onto wire are listed in Tables 5-14–5-15. The 50°C electrical compound is a Thompson-Weinman standard used for electrical volume resistivity tests. All ingredients must be low in ionic impurities to meet the required minimum volume resistivity of 4.0×10^{14} ohm-cm at 50°C by ASTM D257, using 500 volts DC 120 sec readings, and 50–70 mil thickness.

Various accelerated aging tests are available to define adequate amounts of stabilizer and lubricant for a given amount and type of filler. Forced hot air oven aging is described in ASTM D2115: milled sheets 0.040 in. thick are cut into 1 in. × 1 in. squares. At 10 or 15 min intervals, a square from each test compound is laid on a large piece of aluminum foil in a forced air oven at 177°C, to form a display showing accelerated aging *vs* time. Torque rheology measurements furnish an excellent method of determining the correct amounts of stabilizer and lubricant.[22,23] Vinyl stabilizer

Table 5-12. PVC upholstery covering, supported film or sheeting.

Formulation: (parts by weight, pbw)	
PVC: Geon 101 EP, B. F. Goodrich (BFG)***	100
DOP: Di 2 ethyl hexyl phthalate, various suppliers***	10
P72: Trimethyl glycol dicaprylate, Hatco Division W. R. Grace	27
TCP: Tricresyl phosphate, various suppliers	13
Epoxidized soybean oil	5
Barium, cadmium, zinc, phosphite, phenate stabilizer	2
Stearic acid	0.3
$CaCO_3$, Atomite	20–40

Milling and compression molding: 170°C

Properties of compression molded vinyl sheet, not supported

$CaCO_3$ Atomite, pbw	0	10	20	30	40
Tensile strength, psi	2585	2605	2500	2465	2410
Elongation, %	325	335	325	315	305
Mod., 100% elong., psi	1055	1065	1065	1140	1140
Mod., 200% elong., psi	1750	1760	1815	1805	1800
Hardness, Shore A	79	80	81	82	83
Specific gravity	1.234	1.271	1.312	1.350	1.384
Permanent set*	48	49	47	46	47
Crescent tear**	1525	1485	1500	1545	1300
Water absorption, 30°C, %	1.10	0.93	0.71	0.77	0.77
Water extraction, 30°C, %	0.40	0.43	0.46	0.49	0.47
Low temp. impact, °F ASTM D1593	-50	-40	-35	-30	-30
Plasticizer volatil. 70°C, %	0.90	0.91	1.04	1.18	0.89
Ross Flex, cut growth		No cut growth in 150,000 cycles			

*ASTM D412, %
**ASTM D624 Die A, lb/in.
***Suppliers are listed in Tables 5-41 to 5-45.

Table 5-13. PVC hose extrusion compounds.

Formulation A (pbw)	No $CaCO_3$	36.4 phr $CaCO_3$ Atomite	100 phr $CaCO_3$ Atomite
PVC/3%VAc VYNW UCC	100	100	100
DOP plasticizer	40	42.7	47.5
BaCd Zn stabilizer	2.5	2.5	2.5
Properties			
Tensile str., psi	2260	2260	1770
100% modulus, psi	1800	1690	1285
Elongation, %	300	290	270
Hardness, Shore A	96	95	95
Ross Flex, cut growth, 150,000 cyc.	20%	20%	30%
Taber abrasion loss, mg	20	63	93
Permanent set	150	125	60
Sp.G.	1.24	1.38	1.57
Formulation B (pbw)			
VYNW PVC/Ac	100	100	100
DOP plasticizer	13.3	14.2	15.9
Epoxidized soybean oil plasticizer	13.3	14.2	15.9
Chlorinated paraffin 40% Cl, Chlorowax 40	13.3	14.2	15.9
BaCd Zn stabilizer	2.5	2.5	2.5
Properties			
Tensile strength, psi	2280	2270	1845
100% modulus psi	2010	1845	1485
Elongation, %	310	300	240
Hardness, Shore A	97	97	97
Ross Flex, cut growth, %	100%	30%	40%
Ross Flex total cycles	17,000	150,000	150,000
Taber abrasion loss, mg	43	57	100
Permanent set	186	155	80
Sp.G.	1.26	1.40	1.59

and lubricant manufacturers can furnish information on the correct choice and amounts of these ingredients.

6.1.2 PVC Plastisols and Organisols. $CaCO_3$ loading is usually 20-100 phr in plastisols and organisols, and particle size is usually 3-15 μ. A wide range of particle sizes are used in this area: from coarse $CaCO_3$ in PVC carpet backing and mechanically frothed PVC foam coating under PVC tile, to ultra-fine precipitated $CaCO_3$ and coated chalks which control the thixotropic viscosity of sealants. Coarse particles lead to lower viscosity plastisols, but the coarse particles may form a sediment during storage and cause handling problems. Literature includes a book by H. A. Sarvetnick, titled "Plastisols and Organisols,[24] and technical bulletins from Firestone,[25-26] Goodyear, Goodrich, and many other material suppliers. See Tables 5-16, 5-17.

6.1.3 Rigid PVC. At present, a main large use for rigid PVC is in pipe and conduit. The PVC compound must meet ASTM D1784 specification (Table 5-18), and the pipe itself is defined by specfications of the National Sanitary Foundation,[27] the Plastic Pipe Institute,[28] and

Table 5-14. PVC electrical wire insulation compound.

Standard compound for Electrical V.R. tests at 50°C:

Formulation (pbw)	
PVC Geon 102 EP-F1 BFG	100
DIDP-E Diiso decyl phthalate, elec. grade	50
Pigment 33 Calcined Clay, Freeport Kaolin	20
$CaCO_3$, Atomite	15
Tribase E-XL Tribasic lead sulfate NL Industries	8
Stearic acid	0.25
Properties	
Tensile strength, psi	2750
Modulus, 200% elongation, psi	2400
Elongation, %	290
Ross Flex	higher than 200,000 cycles

Electrical volume resistivity, 50°C., 2-min rdg, ohm-cm 4×10^{14} minimum

Milling: 100 psig steam (170°C.) 5 min. Remove sheet at 50–60 mils, cut 4.5 in. diameter circles for VR test, condition overnight in 50°C forced air oven.

ASTM requirements. Currently 0–5 phr 2–3 μ $CaCO_3$ is used in potable water pipe; potentially up to 10–15 phr $CaCO_3$ should meet ASTM D1784 according to results shown in Table 5-18.

There are two separate reasons that 1–4 phr $CaCO_3$ is now used in rigid PVC pipe: either stearate-coated chalk is used to improve the melt rheology and smoothness of extrusion, or uncoated $CaCO_3$ serves as an anti-plate-out agent by exerting a mild scrubbing action, particularly on the extruder die.

Filled rigid PVC conduit and pipe uses up to 40 phr $CaCO_3$ usually 2–3 μ size. The properties of white rigid PVC siding compound without and with 10–40 phr 3 μ $CaCO_3$ are given in Tables 5-19–5-20 with tin mercaptide stabilizer and with lead stabilizer. Propylene-modified PVC polymer readily accepts $CaCO_3$ (Table 5-21).

Table 5-15. 90°C. PVC electrical wire insulation compound

Formulation (pbw)	
PVC Geon 102 EP-F1 BFG	100
TINTM triisononyltrimellitate, Exxon Chem. Co.	52
Epoxidized soybean oil, elec. grade	5
Pigment 33 calcined clay, Freeport Kaolin	10–15
$CaCO_3$, Atomite	0–10
Dythal XL dibasic lead phthalate NL	8
Stearic acid	0.25

Properties

Pigment portion of formulation P33	$CaCO_3$ Atomite	Electrical volume resistivity at 90°C ohm-cm $\times 10^{14}$
15 phr	0 phr	2.7
15	5	2.7
10	5	2.3
10	10	2.3

Table 5-16A. Vinyl Plastisol with 0.5% Ken-React TTMDTP-55 on Atomite.

Formulation:	
Geon 121 BFG	100 pbw
DOP	60
Epoxidized soybean oil	5
Bacd Zn stabilizer	2

ATOMITE, pbw	0	20	30	40	60	0	0	0	0
KENREACT 55 PRE-TREATED ATOMITE	0	0	0	0	0	20	30	40	60
BROOKFIELD LVF VISCOSITIES, POISE, 71°F									
Initial 6 RPM	120	197	223	280	505	182	200	265	460
12 "	105	163	185	222	375	147	175	210	345
30 "	88	132	150	169		118	146	158	
7 Day 6 RPM	160	235	273	395	630	223	265	348	516
12 "	137	197	227	313	462	185	220	278	395
30 "	117	160	180			140	175		
PHYSICAL PROPERTIES–FILMS CAST, CURED @350°F									
Tensile Strength, PSI									
10 Minutes Cure	2007	1913	1414	1360	780	1867	1455	1520	1286
20 Minutes Cure	2242	2242	2045	1958	1522	2261	2150	2000	1636
100% Modulus, PSI									
10 Minutes Cure	1129	1042	1000	955	–	982	978	1000	912
20 Minutes Cure	1119	1045	1217	1174	1045	1149	1250	1233	1109
Elongation, %									
10 Minutes Cure	300	260	190	185	75	250	200	200	260
20 Minutes Cure	325	305	250	240	200	290	275	225	210
Shore Hardness, A									
10 Minutes Cure	75	75	76	77	79	75	76	77	79
20 Minutes Cure	77	77	78	80	82	77	78	80	82
Spec. Gravity	1.198	1.274	1.310	1.343	1.405	1.274	1.309	1.342	1.404

Table 5-16B. Vinyl plastisols.

Formulation:	Geon 121 PVC BFG	100
	Dicapryl phthalate	85
	Paraplex G-62 epoxidized soybean oil	15
	Ba Cd Zn Stabilizer	2

	24 hour Viscosity	200% Modulus	Tensile Strength PSI	Elongation %	Shore A Hard	Sp. Gr.
3 μ $CaCO_3$, Atomite						
0 pbw	1300	780	1335	387	61	1.15
10	1600	795	1370	386	62	1.19
20	2400	800	1395	380	63	1.22
30	3000	800	1370	370	64	1.24
40	4000	795	1292	356	65	1.27
50	5000	782	1190	334	66	1.30
75	8000	732	925	292	69	1.36
100	12000	654	735	250	72	1.42
150	30000	515	518	207	75	1.52
200	65000	–	309	123	77	1.61
5 μ $CaCO_3$ Snowflake						
0 pbw	1300	780	1335	388	61	1.15
10	1500	780	1350	383	62	1.19
20	1800	778	1330	375	63	1.22
30	2400	772	1290	368	64	1.24
40	3000	767	1230	357	65	1.27
50	3600	758	1155	345	66	1.30
75	5200	730	970	310	69	1.36
100	8100	692	825	273	72	1.42
150	16800	540	603	222	75	1.52
200	33000	–	392	180	77	1.61
10 μ $CaCO_3$, Duramite						
0 pbw	1300	780	1335	388	61	1.15
10	1350	780	1342	376	62	1.19
20	1500	778	1318	383	63	1.22
30	1800	770	1270	376	64	1.24
40	2100	760	1218	367	65	1.27
50	2400	750	1154	355	66	1.30
75	3000	715	1020	328	69	1.36
100	3600	670	867	308	72	1.42
150	5500	570	700	284	75	1.52
200	9600	480	595	270	77	1.61
12 μ $CaCO_3$ No. 1 White						
0 pbw	1300	780	1335	387	61	1.15
10	1400	755	1260	373	62	1.19
20	1600	735	1202	360	63	1.22
30	2100	720	1148	348	64	1.24
40	2500	709	1090	337	65	1.27
50	3000	700	1038	323	66	1.30
75	4500	690	923	287	69	1.36
100	6000	668	770	260	72	1.42
150	10000	–	505	195	75	1.52
200	17400	–	224	125	77	1.61

Table 5-17. Mechanically frothed PVC plastisol foam under PVC floor tile.

Formulation (pbw):		
	Geon 135 PVC BFG	30
	Geon 120X241 PVC BFG	30
	Geon 106F2 PVC blending resin BFG	40
	DOP plasticizer	60
	Santicizer 160 butyl benzyl phthalate	25
	Gen. Elec. SS4255 (50% silicone)	4
	Ba Cd Zn phosphite phenate stabilizer	3
	$CaCO_3$	0, 25 or 50

$CaCO_3$		Brookfield LVF VISC., #3 Spindle, 60 RPM		Density, lb/ft^3		Tensile str. PSI	Elong. %	Taber abras., mg. loss	Bashore resil., %
μ	PHR	Initial	3 Days	Before curing	Cured 300° F 30 min				
		CPS	CPS						
0	0	1670	1920	30	31	131	110	2320	6
3	25	3052	2640	35	34	125	120	3410	7
3	50	6100	6250	43	37	107	100	3810	6
6	25	2280	2240	38	34	120	100	3740	7
6	50	4800	4500	47	40	134	120	2700	7
14	25	2300	1860	38	34	116	100	3870	5
14	50	2920	2520	47	37	111	100	3360	7

6.14 PVC Floor Tile. PVC floor tiles are sometimes called "asbestos floor tiles," because the darker colors may contain asbestos as one of the fillers. $CaCO_3$ loadings of about 80–400 phr small granular size are usual, with total filler loadings up to about 500 phr. Tables 5-22–5-24 contain formulations and properties.

Tile installation requires easy break of notched (scored) tile, and this is achieved by high loadings. Lighter colors use beneficiated $CaCO_3$.

6.2 Glass-Reinforced Polyesters—SMC + BMC. (SMC = sheet molding compounds, BMC = bulk molding compounds.)

6.2.1 Sheeting Molding Compounds.[29, 30, 31] SMC has found many specialized uses, including automotive panels, mower bodies, skimobiles, lawn tractors, air conditioner housings, appliance and electronic housings, etc. SMC sheets are made on SMC machines that spread two layers of $CaCO_3$-filled polyester resin, automotive low-shrink grade, onto release-coated paper, and chop and randomly distribute glass fibers between the polyester layers. The SMC is rolled up and aged for 1-3 days in a warm oven to increase the viscosity so it can be easily handled for compression molding. The SMC must have essentially no shrinkage when molded, and the finished part must have high microsmooth surface. There must be no sink marks opposite reinforcing ribs—this is critical property. The $CaCO_3$ filler, used at about 150 phr loading, must have size and distribution to meet the above requirements and provide a low controlled viscosity in the low-shrink polyester resin at specified loading, either without or with a release agent such as zinc stearate. $CaCO_3$ moisture must be less than 0.10% for correct thickening during aging. $CaCO_3$ median particle size is about 3–6 μ; up to 200 phr loading can be used with a low viscosity one component polyester resin. See Table 5-25.

6.2.2 Bulk Molding Compounds, Premix, Preform, and Mat. $CaCO_3$ in the 3-6 μ range is suited to these uses. Loading is usually about 250 phr $CaCO_3$ for bulk molding compounds and 100–150 phr for premix, preform, and mat, although in some cases, over 400 phr $CaCO_3$ is possible. The $CaCO_3$ must provide low, uniform viscosity in the polyester, by controlled $CaCO_3$ particle size, distribution, and purity.

Table 5-18. Rigid PVC pipe, ASTM D1784 class 12454 B or C.

Formulation (pbw)				
$CaCO_3$, Atomite	0	10	20	30
PVC Plaskon SR414-3 (I.V. = .94)	100	100	100	100
Aristowax 165	1	1	1	1
Acryloid K 120N, R and H	1.5	1.5	1.5	1.5
TiO_2 R900	1	1	1	1
Calcium stearate	0.8	0.8	0.8	0.8
Thermolite 31 Super, M and T	0.5	0.5	0.5	0.5
Milling and compression molding: 195°C (190 PSIG steam) in mill rolls and press patterns.				
Unaged properties measured at 23± 2°C.				
1ZOD impact, ft lb/in. notch	0.65	0.93	1.00	0.98
Tensile strength, psi (0.2 in./min)	8210	7765	6965	5325
Tensile modulus, psi (0.2 in./min)	416,000	448,000	496,000	554,000
Flex. strength unaged, PSI, 0.1 in./min	11,368	10,545	10,135	9,716
Deflection temp. °C, 264 psi	73	74	74	75
Chem. resistance when aged:				
93% H_2SO_4, 14 days 55±2°C				
Weight change, %	−0.052	−0.061	−0.070	−0.090
Flex. yield strength change, %	+3.3	+4.8	+4.7	+5.0
80% H_2S_4, 30 days 60±2°C				
Weight change, %	−0.026	−0.070	−0.087	−0.108
Flex. yield strength change, %	+11.4	+13.4	+14.6	+12.0
ASTM No. 3 Oil, 30 days 23±2°C				
Weight change, %	+0.029	+0.107	+0.088	+0.075

ASTM requirements	Min.	Max.
1ZOD, ft lb/in. notch	0.65	–
Tensile strength psi	7000	–
Mod. of elast. in tension psi	4,000,000	–
Deflection temp. °C, 264 psi	70	–
Chem. resistance	Decrease, max. %	Increase, max. %
93% H_2SO_4, wt. change %	0.1	5.0
flex. str. change %	25.0	5.0
80% H_2SO_4, wt. change %	5.0	5.0
flex. str. change %	15.0	15.0
ASTM No. 3 Oil, wt. change %	1.0	1.0

(See also Table 9-5 and Figs. 9-7 and 9-8.)

Table 5-19. White rigid vinyl siding compound.

Compounds (pbw)	Tin Mercaptide Heat Stabilizer 1	2	3	4	5
Geon 103 EP – F7 PVC resin BFG	92	92	92	92	92
CPE 3614 chlorinated polyethylene DOW	8	8	8	8	8
Acryloid K-120N acrylic resin, R and H	3	3	3	3	3
Titanox 2071 titanium dioxide NL	10	10	10	10	10
Calcium carbonate, Atomite	0	10	20	30	40
Thermolite 31 tin mercaptide stabilizer M and T	1.75	1.75	1.75	1.75	1.75
Wax S (Am. Hoechst) lubricant	0.8	0.8	0.8	0.8	0.8
Wax V (Am. Hoechst) lubricant	0.2	0.2	0.2	0.2	0.2
Premix: intensive mixer. Milling: 5 min 188°C. (160 psig steam rolls)					
Compression molding: 188°C (160 psig steam)					
Properties Before Aging					
Gardner impact 45 mils, in. lb	150	120	100	55	45
Gardner impact 90 mils, in. lb	160+	160+	140	120	90
Elec. vol. resistivity 10^{14} ohm-cm	40	20	12	14	12
Tensile at break, psi (20 in./min)	7676	7304	6382	6233	5478
Elongation, % (20 in./min)	30	30	30	30	30
Hardness, Shore D-2	83	83	83	83	83
Taber abrasion loss, H22, mg	340	360	390	607	622
Ross flex, injured, cycles	1977	1951	1826	2018	2026
Ross flex, uninjured, cycles	3466	3248	3229	3056	3228
Sp. G.	1.44	1.50	1.55	1.60	1.64
Color	white	white	white	white	white
Properties After Ultraviolet Radiation 72 hr 6 in. from G.E. bulb G15T8					
Gardner impact 45 mils, in. lb	115	95	70	47	37
Gardner impact 90 mils, in. lb	160+	160+	140	120	90
Elec. vol. resistivity 10^{14} ohm-cm	25	12	5	4	3
Tensile at break, psi	7927	7430	6933	6000	5524
Elongation, %	30	30	30	30	30
Hardness, Shore D-2	83	83	83	83	83
Taber abrasion loss H22, mg	520	550	640	680	820
Ross flex, injured, cycles	1775	1884	1669	2008	2072
Ross flex, uninjured, cycles	3049	3008	3003	2938	3189
Color	lt. yellow	off-white	off-white	off-white	off-white

(See also Table 9-5 and Figs. 9-7 and 9-8.)

Table 5-20. White rigid vinyl siding compound.

Compounds (pbw)	Lead Stabilizer and Lubricant System 1	2	3	4	5
Geon 102 EP – F7 PVC resin BFG	92	92	92	92	92
CPE 3614 chlorinated polyethylene DOW	8	8	8	8	8
Acryloid K-120N acrylic resin, R and H	3	3	3	3	3
Titanox 2101 titanium dioxide NL	10	10	10	10	10
Calcium carbonate, Atomite	0	10	20	30	40
Tribase AG tribasic lead sulfate NL	2	2	2	2	2
Dyphos XL dibasic lead phosphite NL	2	2	2	2	2
Plastiflow LPC lead-calcium soap NL	1	1	1	1	1
Premix: intensive mixer. Milling: 5 min 188°C (160 psig steam on rolls)					
Compression molding: 188°C (160 psig steam)					
Properties Before Aging					
Gardner impact, 45 mil in. lb	140	110	90	40	38
Gardner impact, 90 mil in. lb	160+	160+	140	90	80
Elec. vol. resistivity 10^{14} ohm-cm	55	25	21	19	15
Tensile at break, psi (20 in./min)	7904	7238	6367	5744	5488
Elongation, % (20 in./min)	30	30	30	30	30
Hardness, Shore D-2	83	83	83	83	83
Taber abrasion loss H22, mg	255	330	337	570	580
Ross flex, injured, cycles	2125	1931	1605	1607	1729
Ross flex, uninjured, cycles	4705	3490	3225	2826	3549
Sp. G.	1.49	1.54	1.59	1.65	1.68
Color	white	white	white	white	white
Properties After Ultraviolet Radiation 72 hr 6 in. from G.E. bulb G15T8					
Gardner impact, 45 mil	115	90	70	42	40
Gardner impact, 90 mil	160+	160+	140	90	80
Elec. vol. resistivity 10^{14} ohm-cm	39	14	7	3	2
Tensile at break, psi (20 in./min)	7927	7327	7142	6000	5273
Elongation, % (20 in./min)	30	30	30	30	30
Hardness, Shore D-2	83	84	85	85	86
Taber abrasion loss, H22, mg	480	560	580	790	805
Ross flex, injured, cycles	2562	1860	1626	1678	1822
Ross flex, uninjured, cycles	6582	4173	3326	2901	3281
Color	med. gray	off-white	off-white	off-white	off-white

(See also Table 9-5, and Figs. 9-7 and 9-8.)

Table 5-21. Propylene-modified rigid PVC.

Recipe: Propylene Modified PVC (pbw)		Recipe: PVC, Unmodified (pbw)	
A1RCO 401	100	GEON 103 EP F7	100
TR1BASE XL NL	5	TR1BASE XL NL	5
LEADSTAR* NL	0.4	LEADSTAR* NL	0.4
DS-207* NL	0.4	DS-207* NL	0.4
Properties:		Properties listed in left column.	
Tensile at break, psi (20 in./min)			
0 phr $CaCO_3$,	7680		8330
12.5 Atomite	7612		8290
25	7200		7360
50	5380		5240
100	3930		3980
Elongation at break, % (20 in./min)			
0 phr $CaCO_3$,	15		20
12.5 Atomite	20		20
25	20		20
50	20		15
100	15		15
Ross flex, cycles to break ASTM-D1052			
0 phr $CaCO_3$,	527		367
12.5 Atomite	498		320
25	310		232
50	84		39
100	1		1
Hardness, Shore D2			
0 phr $CaCO_3$,	81		84
12.5 Atomite	82		85
25	83		86
50	84		87
100	87		88
Sp.G.			
0 phr $CaCO_3$,	1.439		1.449
12.5 Atomite	1.514		1.524
25	1.581		1.591
50	1.694		1.704
100	1.864		1.873

*Lead stearate lubricants, NL Industries

Table 5-22. PVC asbestos tile.

Formulation (pbw)	
PVC/14% VAc VYHH UCC	50
PVC, inherent visc.* 0.70 Pliovic S-50 Goodyear	50
Butyl benzyl phthalate, Santicizer 160 Monsanto	10.6
DIDP diisodecyl phthalate	21.8
Epoxidized soybean oil	8.8
Heat stabilizer for asbestos–see References 26, 28, 29	5.9
Asbestos grade 7R Johns Manville	73.5
Talc, 18–20% on 325 mesh, International 6N, Vanderbilt	73.5
$CaCO_3$ 20–30 μ unpurified, 85 ± 3% through 325 mesh	294.1
Milling: 110°C.	
Properties	
Hardness, Shore D2	74
McBurney Indentation, 23.5°C, 1 min**	12.8 mils
McBurney Indentation, 23.5°C, 10 min	17 mils
McBurney Indentation, 46°C, 30 sec	21.2 mils

*Inherent viscosity, ASTM D1243.
**ASTM F142

Table 5-23. PVC homogenous flexible flooring.

5-23.1 PVC homopolymer.

Formulation (pbw)			
PVC QSQF UCC	100	100	100
$CaCO_3$, Atomite	85.8	118	166.2
Clay, hard kaolin, Snobrite	85.8	118	166.2
NBR Chemigum N6B Goodyear	16.7	10	0
DOP	30	36	45
Epoxidized soybean oil	8.3	10	12.5
Zinc soap, Metasap 635, Nopco	3.3	4	5
A-C617A* Allied Chem.	3.3	4	5
Properties			
Tensile str., psi	1265	1175	1050
Tens. mod., 10% elongation, psi	915	990	990
Elongation, %	90	40	30
McBurney indent., 24°C, 1 min, mils	23.5	22.2	19
McBurney indent., recovery, 1 min, %	70	68	67
Taber abras., H18 wheels, mg	370	510	1140
Ross flex, cycles to break	4135	491	2
Sp.G.	1.66	1.77	1.88

5-23.2 Vinyl chloride/vinyl acetate co-polymer.

Formulation as in 5-23. 1 except:

PVC/AC VYFG UCC	100 pbw	100 pbw	100 pbw
Properties			
Tensile strength, psi	1150	1010	930
Tensile modulus, 10% elongation, psi	650	785	685
Elongation, %	95	55	25
McBurney indent., mils	25	21.2	21
McBurney indent., recovery, %	67	63	60
Taber abras. H18, mg	290	580	1190
Ross flex, cycles to break	16,000	570	2
Sp.G.	1.66	1.77	1.88

*Polyethylene, mol. wt 1500, m.p. 102°C, density 0.91

Table 5-24. PVC homogenous flexible flooring.

5-24.1 Low cost formulations (pbw).

PVC QSQF UCC	100	100	100
$CaCO_3$, Atomite	221.5	166.2	111
Clay, hard kaolin, Snobrite	111	166.2	221.5
DOP	45	45	45
Epoxidized soybean oil	12.5	12.5	12.5
A-C617A* Allied Chem.	5	5	5
Zinc soap, Metasap 635 Nopco	5	5	5
Properties			
McBurney indent., 1 min, mils	21.9	19	17.8
McBurney indent., recovery, 1 min, %	68	67	66
Taber abras., H18, mg loss	970	1140	1120
Ross flex, cycles to break	98	2	1
Hardness, Shore A-2	98	99	100
Sp.G.	1.88	1.88	1.88

5-24.2 Medium cost formulations (pbw).

$CaCO_3$, Atomite	157.2	118	78.8
Clay, kaolin, Snobrite	78.8	118	157.2
PVC QSQF UCC	100	100	100
DOP	36	36	36
NBR Chemigum N6B Goodyear	10	10	10
Epoxidized soybean oil	10	10	10
A-C617A* Allied Chem.	4	4	4
Zinc soap, Metasap 635	4	4	4
Properties			
McBurney indent., mils	23	22.2	19
McBurney indent., recovery, %	69	68	68
Taber abras., H18, mg loss	490	510	540
Ross flex, cycles to break	1935	491	20
Hardness, Shore A-2	98	99	100
Sp.G.	1.77	1.77	1.77
Milling and compression molding: 177°C (120 psig steam)			

*Polyethylene, mol. wt 1500, m.p. 102°C density 0.91

6.2.3 Marine Polyester Water Absorption. Ground $CaCO_3$ of 3-5 μ size at a loading of about 175-200 phr in marine polyester passes ASTM D1201 requirement of 0.15% maximum water absorption during 24-hour immersion, while clays, anhydrous calcium sulfate, and some precipitated $CaCO_3$ absorb excessive amounts of water during immersion. See Table 5-26 for results and recipe.

6.3 Polyethylene

Low density polyethylene and high density polyethylene can be successfully loaded with ground calcite as shown in Table 5-27. Considerable effort has been devoted to finding ways to increase the physical properties of loaded polyethylene. Successful solutions to the problem are as follows:

- Co-polymerize polar groups in the polyethylene, such as vinyl acetate groups, ethyl acrylate groups (Table 5-28), or carboxylic acid groups (Table 5-29).
- Cross-link the filled polyethylene with an organic peroxide, that is, "cure" the polyethylene, making it a "thermoset" material. High density polyethylene can be cross-linked (Table 5-30) as well as low density polyethylene (Table 5-31). Blends of low density polyethylene with ethylene-

Table 5-25. 1. Low-shrinkage reinforced polyester compounds.

Materials	Premix, preform, or mat	Bulk molding compound	Sheet molding compound
	FORMULATIONS (PBW) AND PHYSICAL PROPERTIES		
Low shrinkage resin(s)	100	100	100
$CaCO_3$, 5 μ, P3587 T-W	125	250	150
Zinc stearate	3	3	3
T-Butyl perbenzoate	0.75	0.75	0.75
MgO thickener	–	–	3
Treated glass fiber, ¼ in.	60	90	–
Treated glass fiber strand, cut to 2 in.	–	–	125

ASTM	Properties at 23 + 2°C.	(Cured 7 min 155°C to 60–70 Barcol Hardness)		
D790	Flexural strength, psi	10,000–14,000	13,000–16,000	27,000–31,000
D790	Flexural modulus psi	1,400,000–1,700,000	1,700,000–2,200,000	1,500,000–2,100,000
D638	Tensile strength psi	5,000–7,000	5,000–7,000	12,000–14,000
D638	Tensile modulus psi	1,400,000–1,700,000	1,900,000–2,200,000	1,700,000–2,000,000
D256	1ZOD impact, notched, Ft lb/in.	5–7	4.5–6.5	13–24
D2583	Hardness, Barcol	60–70	60–70	60–70

Table 5-25. 2. Optimum calcium carbonate selection for reduced density in SMC/BMC.

2a. Viscosity without glass fibers

Brookfield viscosities in typical SMC paste, without glass fiber: Recipes Nos. 1–5 have equal volumes of filler and/or bubbles. The recipe consists of: Rohm & Haas P19D–100 pbw, fillers including 3*M* glass bubbles B37/2000, as below.

Recipe No.	1	2	3	4	5	6
Glass bubbles	0 phr	5.33	10.64	0	5.33	7.11 phr
T-W P-3587	156	117	78	–	–	–
Atomite	–	–	–	156	117	117
Visc., 90°F	20,500	8000	7700	28,000	12,000	16,500 CPS.
Rel. filler vol.	1.0	1.0	1.0	1.0	1.0	1.08

2b. Physical Properties with Glass Fibers

Physical property information with typical BMC recipes: Nos. 1–5 have equal filler volume. Recipe: Marco GR 13024–100 pbw, zinc stearate–3 phr, t-butyl perbenzoate–1 phr, JM glass fiber ¼ in. CS-308A, as shown, 3*M* glass bubbles B37/2000 as shown, $CaCO_3$ as shown.

Recipe No.	1	2	3	4	5	6
Glass bubbles	0 phr	7.5	15	0	7.5	10
T-W P-3587	220	165	110	–	–	–
Atomite	–	–	–	220	165	165
Glass fiber	80	80	80	80	80	85
Sp.G.	1.95	1.75	1.53	1.97	1.78	1.70
Flex str., psi	21,339	17,786	14,158	18,748	19,480	20,308
Tens. str., psi	6849	5617	4206	6363	6521	7060
Izod, notched	7.0	6.5	6.3	6.0	6.9	7.4
Barcol hardness	60	50	32	62	52	56
$/liter	$1.03	1.10	1.18	1.04	1.12	1.13
Rel. filler vol.	1.0	1.0	1.0	1.0	1.0	1.08

Table 5-26. 1. Water absorption of a marine-type polyester.

Formulation	Increase in wt during immersion 24 hr @ 23 ± 1°C	Soluble matter lost during immersion	Water absorbed (sum of wt increase plus solubles lost)
	TESTED BY ASTM METHOD D570[d]		
100 pbw Polylite 31-447[a] 1 phr BPO paste[b] Filler as below[c]			
Unfilled Polylite 31-447 no filler	0.398%	0.0%	0.398%
Low absorption W.W. clay			
88 phr Alsilate LO (Freeport)	0.183%	0.072%	0.255%
100 phr Alsilate LO	0.194%	0.060%	0.254%
Hi absorption air float Kaolin			
100 phr Snobrite Clay	0.229%	0.009%	0.238%
Low absorption ppt'd $CaCO_3$			
175 phr Suspenso (Diamond)	0.119%	0.022%	0.141%
Med. absorption ppt'd $CaCO_3$			
175 phr Surfex MM (Diamond) (surface treated)	0.107%	0.004%	0.111%
Anhydrous $CaSO_4$			
175 phr W-1320 (Pfizer)	0.024%	0.182%	0.206%
Ground natural $CaCO_3$			
50 phr Atomite	0.237%	0.021%	0.258%
175 phr Atomite	0.095%	0.006%	0.101%
200 phr Atomite	0.085%	0.006%	0.091%

[a]Reichhold Chemicals Inc. marine-type polyester resin.
[b]Reichhold Chemicals Inc. 50% benzoyl peroxide in mixed plasticizers.
[c]Quantity of filler was determined by the amount needed to achieve good moldability in the absence of glass fiber.
[d]Specimens compression molded at 162°C, 0.125 in. thick and cut to 3.0 in. × 1.0 in. size.

Table 5-26. 2. ASTM D1201 requirements for polyester compression-molded speciments for marine application.

	Max. water absorption
Type 1–General purpose with mineral fillers	0.15%
Type 2–General purpose with mineral and cellulose fillers	1.0%
Type 3–General purpose putty type with mineral fillers	0.5%
Type 4–Putty type with mineral fillers, superior electrical properties	0.15%
Type 5–High impact, glass fiber filled, mat form	0.15%
Type 6–High impact, glass fiber filled, putty form	0.25%

Table 5-27 A. Atomite-filled, low density polyethylene, Union Carbide Corp. DYNH-1, 0.075 in. thickness.

U.C.C. DYNH-1	Stearic Acid on Atomite	TENSILE STRENGTH At Yield psi	TENSILE STRENGTH At Break psi	300% Modulus psi	Elongation, %	Tear Strength lb/in.	Hardness, Shore D2
0 phr Atomite	–	1295	1722	1272	532	486	50
2.5 " "	0%	1350	1719	1333	490	501	50
2.5 " "	1%	1353	1640	1271	490	444	50
5 " "	0	1279	1765	1241	480	442	51
5 " "	1	1359	1810	1385	480	488	51
10 " "	0	1393	1483	1296	460	492	51
20 " "	0	1432	1449	1261	440	474	52
20 " "	1	1307	1479	1198	440	447	52
40 " "	0	1530	1512	–	230	348	53
40 " "	1	1350	1306	–	230	368	53
50 " "	0	1515	1298	–	170	356	54
60 " "	1	1355	1199	–	115	344	54
80 " "	1	1321	1192	–	55	312	55
100 " "	1	1343	1267	–	20	322	56

Above about 50 phr of Atomite, an external lubricant such as stearic acid is needed for proper release on mill and in molds.

Table 5-27 B. Film properties of Atomite-filled DYNH-1; compression-molded, 8-12 mil film thickness.

		Tensile at Yield psi	Tensile at Break psi	Elongation %	Tear Strength lb/in
DYNH-1 unfilled		1253	1627	500	405
10 phr Atomite		1267	1440	490	420
20 " "		1389	1373	280	365
20 " "	+1% stearic acid	1093	1133	278	367
40 " "		1323	1227	190	243
40 " "	+1% stearic acid	1150	1083	193	245
50 " "		1342	1147	155	215
60 " "	+1% stearic acid	1220	1109	130	209
80 " "	+1% stearic acid	1200	1100	80	201
100 " "	+1% stearic acid	1207	1067	40	187

Table 5-27 C. Atomite-filled, high density polyethylene, Phillips Marlex 5502.

	Stearic Acid, % on Atomite	TENSILE STRENGTH At Yield psi	TENSILE STRENGTH At Break psi	Elongation %	Film Tear lb/in	Hardness Shore D
0 phr Atomite	–	3830	2323	60	780	67
5 " "	0	3708	2210	33	768	68
10 " "	0	3528	2120	32	743	68
20 " "	0	3352	3352	30	792	69
20 " "	1	3356	2086	30	616	69
40 " "	0	3346	3346	20	607	69
40 " "	1	3105	2096	20	575	69
60 " "	0	2997	2997	12	500	69
60 " "	1	2901	2901	12	475	69
80 " "	1	2741	2213	12	453	69
100 " "	1	2596	2596	5	408	69
60 " "	0	–	–	–	500	69
80 " "	0	–	–	–	423	69
100 " "	0	–	–	–	420	69

Table 5-28. Low density polyethylene EVA and Ethylene-Methyl Acrylate Co-polymer (EMA).

5-28.1. Low density polyethylene (no vinyl acetate) Petrothene NA-280, USI.

Amount of Atomite	0 phr	25 phr	50 phr	75 phr	100 phr
Tensile strength, psi	1600	1614	1698	1716	1675
300% modulus, psi	1600	–	–	–	–
Elongation, %	600	200	50	20	10
Hardness, Shore D-2	52	55	57	59	61
Taber abras., H22, mg loss	60	120	160	205	260
Ross flex, injured, cycles	16356	1989	170	37	6
Elec. vol. res., ohm-cm $\times 10^{14}$	60+	60+	60+	15	12
Sp.G.	0.923	1.065	1.184	1.288	1.378

5-28.2. Petrothene NA-294, USI, 6% vinyl acetate co-polymer.

Tensile strength	2056	1370	1388	1544	1658
300% modulus, psi	1222	1304	–	–	–
Elongation, %	600	500	170	60	40
Hardness, Shore D-2	47	49	51	55	56
Taber abras., H22 mg loss	50	100	160	230	450
Ross flex, injured, cycles	200,000+	32,427	7501	1521	209
Elec. vol. res., ohm-cm $\times 10^{14}$	60+	60+	60+	40	24
Sp.G.	0.926	1.066	1.186	1.290	1.380

5.28.3. Ultrathene UE-637, USI, 10% vinyl acetate co-polymer.

Tensile strength	1893	1428	1253	1334	1480
300% modulus, psi	1040	1143	–	–	–
Elongation, %	700	500	250	100	50
Hardness, Shore D-2	45	46	50	52	55
Taber abras., #22, mg loss	90	170	240	365	510
Ross flex, injured, cycles	200,000+	45,040	10,629	3687	337
Elec. vol. res., ohm-cm $\times 10^{14}$	60+	60+	60+	49	39
Sp.G.	0.927	1.068	1.187	1.291	1.381

5-28.4. Ultrathene UE-630, USI, 15% vinyl acetate co-polymer.

Tensile strength, psi	1869	1639	1242	1115	1114
300% modulus, psi	913	928	1090	1110	–
Elongation, %	700	650	500	350	150
Hardness, Shore D-2	40	42	43	44	46
Taber abras., H22 mg loss	190	215	265	480	575
Ross flex, injured, cycles	200,000+	200,000+	200,000+	94,101	19,710
Elec. vol. res., ohm-cm $\times 10^{14}$	60+	60+	60+	60+	33
Sp.G.	0.937	1.078	1.198	1.302	1.392

5-28.5. Ethylene/methyl acrylate (85/15) Co-polymer, Poly-Eze 70053 Gulf Oil Co.

Formulation					
EMA Poly-Eze 70053	100 phr	100	100	100	100
$CaCO_3$ Atomite	0	25	50	75	100
Stearic acid	0.5	0.5	0.5	0.5	0.5
Properties					
Tensile strength, psi	2063	1614	1293	1029	939
300% modulus, psi	975	888	878	833	939
Elongation, %	700	700	600	600	600
Hardness, Shore D-2	35	38	40	42	44
Taber abras., H22, mg loss	105	250	430	785	1005
Ross flex, injured, cycles	200M+	200M+	81,280	33,022	8412
Elec. vol. res., ohm-cm $\times 10^{14}$	60+	60+	43	14	7
Sp.G.	0.938	1.078	1.198	1.301	1.392

Milling: 148°C.

Compression molding: 188°C.

Table 5-29. Carboxy-polyethylene coupling agent for $CaCO_3$ in low density polyethylene.

Formulation: Low density PE: Alathon PE.6647 DuPont 100 pbw
Coupling agent: Carboxy-PE: Surlyn A-1652 DuPont 0-2 phr

Amount of $CaCO_3$, Atomite:	0 phr	25 phr	50 phr	75 phr	100 phr
No Coupling Agent (No Surlyn)					
Properties					
Tensile strength, psi	1892	1591	1584	1547	1531
300% modulus, psi	1500	1591	—	—	—
Elongation, %	500	450	190	40	20
Hardness, Shore D-2	49	51	53	54	55
Taber abras., H22, mg loss	60	220	240	410	415
Ross flex, injured, cycles	128M	37M	6480	2175	232
Elec. vol. res., ohm-cm $\times 10^{14}$	60+	60+	60+	14	0.1
Sp.G.	0.919	1.059	1.179	1.282	1.372
0.5 phr Surlyn					
Properties					
Tensile strength, psi	1868	1548	1692	1866	1904
300% modulus, psi	1400	—	—	—	—
Elongation, %	530	100	50	40	20
Hardness, Shore D-2	50	52	54	56	58
Taber abras., H22, mg loss	60	140	185	260	335
Ross flex, injured, cycles	74M	21M	5650	540	43
Elec. vol. res., ohm-cm $\times 10^{14}$	60+	60+	60+	44	5
Sp.G.	0.919	1.058	1.178	1.281	1.371
1 phr Surlyn					
Properties					
Tensile strength, psi	1960	1555	1640	1880	1960
300% modulus, psi	1427	—	—	—	—
Elongation, %	510	260	80	40	30
Hardness, Shore D-2	50	52	54	56	58
Taber abras., H22 mg loss	80	120	180	260	320
Ross flex, injured, cycles	107M	20M	5520	490	111
Elec. vol. res., ohm-cm $\times 10^{14}$	60+	60+	60+	53	27
Sp.G.	0.919	1.057	1.177	1.279	1.369
2 phr Surlyn					
Properties					
Tensile strength, psi	1882	1616	1777	1947	2050
300% modulus, psi	1461	—	—	—	—
Elongation, %	510	270	90	50	40
Hardness, Shore D-2	50	52	54	56	58
Taber abras., H22, mg loss	75	130	215	225	290
Ross flex, injured, cycles	103M	15M	3127	1100	142
Elec. vol. res., ohm-cm $\times 10^{14}$	60+	60+	60+	39	30
Sp.G.	0.919	1.057	1.175	1.277	1.366

Milling: 110°C

Compression molding: 134°C

Table 5-30. Cross-linked high density polyethylene.

Formulation (pbw)					
HDPE Alathon 7516	100	100	100	100	100
Percadox 14/40*	2	2	2	2	2
$CaCO_3$ Atomite	0	25	50	75	100
Properties					
Tensile strength, psi	3018	2793	2654	2555	2480
300% modulus, psi	3018	2793	2654	—	—
Elongation, %	480	480	560	140	80
Hardness, Shore D-2	63	64	65	66	67
Taber abras., H22, mg	30	60	110	135	145
Ross flex, injured, cycles	200M+	200M+	200M+	200M+	30530
Elec. vol. res., ohm-cm $\times 10^{14}$	60+	60+	60+	55	7.5
Sp.G.	0.967	1.107	1.227	1.329	1.419

Milling: 50 PSIG steam (148°C)

Curing: 15 min 160 PSIG Steam (188°C) 30 tons hydraulic pressure on 28 in.^{2} circular plates (free-floating pistons).

*Percadox 14/40, 2 phr on "as received" basis, 60% calcined clay and 40% α,α' bis (t-butyl peroxy) diisopropyl benzene, (Noury Chem Corp.).

Table 5-31. Cross-linked low density polyethylene.

5-31.1. Cross-linked with PERCADOX 14/40, consisting of 60% calcined clay/40% α, α' bis (t-butyl peroxy) diisopropyl benzene, Noury Chem. Corp., Akzona

Formulation (pbw)	100	100	100	100	100
Alathon PE-6647 DuPont	2	2	2	2	2
$CaCO_3$ Atomite	0	25	50	75	100
Properties					
Tensile strength, psi	2979	2813	2413	2000	1803
300% modulus, psi	1379	1321	1296	1244	1215
Elongation, %	560	540	520	490	470
Hardness, Shore D-2	50	51	53	54	55
Taber abras., H22, mg loss	55	120	235	330	575
Ross flex, injured, cycles	200M+	200M+	200M+	200M+	200M+
Elec. vol. res., ohm-cm $\times 10^{14}$	60+	60+	60+	60+	31
Sp.G.	0.926	1.063	1.181	1.283	1.372

5-31.2. Cross-linked with dicumyl peroxide, Hercules Di-cup T, 2 phr, and 100 phr PE-6647.

$CaCO_3$ Atomite (pbw)	0	25	50	75	100
Properties					
Tensile strength, psi	2679	2705	2275	2039	1655
300% modulus, psi	1358	1263	1227	1221	1206
Elongation, %	550	560	520	500	430
Hardness, Shore D-2	50	51	53	54	55
Taber abras., H22, mg loss	105	130	250	490	700
Ross flex, injured, cycles	200M+	200M+	200M+	200M+	200M+
Elec. vol. res., ohm-cm $\times 10^{14}$	60+	60+	60+	60+	40
Sp.G.	0.920	1.057	1.175	1.277	1.366

Milling: 108°C 5 min

Curing: 188°C 15 min, 30 tons pressure on 6 in. diameter plates

Table 5-32. Low density polyethylene in ethylene-propylene-diene rubber (EPDM).

Formulation–DuPont PE and EPDM (pbw)						
Polyethylene PE 6647	0	0	20	20	40	40
EPDM, Nordel 1560	100	100	80	80	60	60
$CaCO_3$ Atomite	0	25	0	25	0	25
Percadox 14/40	5	5	5	5	5	5
Spider Sulfur	0.3	0.3	0.3	0.3	0.3	0.3
Properties Unaged						
Tensile strength, psi	2560	3100	2177	2345	2307	2124
300% modulus, psi	365	382	593	586	805	773
Elongation, %	670	650	665	700	670	560
Hardness, Shore A-2	77	80	88	90	94	96
Hardness, Shore D-2	18	20	25	26	29	30
Bashore resilience %	70	64	57	53	45	43
Taber abras., H22, mg loss	130	265	90	230	100	200
Elec. vol. res., ohm-cm × 10^{14}	60+	60+	60+	60+	60+	60+
Sp.G.	0.87	1.00	0.88	1.01	0.90	1.03
Aged 150°C 24 hr, forced air oven						
Tensile strength, psi	961	1855	2714	2104	3084	3026
300% modulus, psi	396	472	649	700	874	879
Elongation, %	560	620	610	610	590	630
Hardness, Shore A-2	80	83	90	93	98	100
Hardness, Shore D-2	20	22	27	28	31	32
Bashore resilience, %	72	69	57	53	49	47

Milling: 126°C 5min.

Curing: 160°C 15 min, 30 tons, 6 in. diameter plates, semi-permanent silicone mold releasant.

Table 5-33. Injection Molded Polypropylene homopolymer with 40% filler, ASTM D2146.

Recipe (pbw)	$CaCO_3$	Talc
PRO-FAX 6523 Hercules	100	100
Carbon Black, Raven Black 30	0.83	0.83
$CaCO_3$ Atomite	66.67	—
Talc, Desert 57 J-M	—	66.67
Properties		
Flow rate, 230°C., GM/10 min	3.3	2.6
Hardness, Shore D-2, initial	80	81
Hardness, Shore D-2, 15 sec	76	75
IZOD impact strength, unnotched, ft lb/in.	9.3	3.8
Tensile strength at yield, psi, 2 in./min	3590	4550
Tensile strength at break, psi, 2 in./min	2370	3320
Elongation, %	17	4
Flexural modulus, psi	399,000	617,000
Flexural strength, psi	6890 (5% offset)	8430 (yield)
Stiffness in flexure, psi	418,000	626,000
Deflection temp. at 66 psi, °C	118	137
Vicat softening point, °C	154	156
Density, g/cm^3	1.23	1.23

Speciments were injection molded to ASTM D638 Type I Tensile Bar 1/8 in. thickness, and flex. Bar 5 in. × 1/2 ft × 1/8 in., with 10 oz. Buhler model 175 screw-injection machine.

(See also Table 9-4 and Figs. 9-3–9-6.)

Table 5-34. Compression Molded Polypropylene homopolymer with 30–40% filler.

Recipe (pbw)	CALCIUM CARBONATE 30%	CALCIUM CARBONATE 35%	CALCIUM CARBONATE 40%	TALC 30%	TALC 40%
PROFAX 6523	100	100	100	100	100
Atomite	43	53.8	66.7	—	—
Talc	—	—	—	43	66.7
Properties at 23 ± 2°C					
Izod unnotched, ft lb/in.	13.7	13.6	9.9	5.4	3.4
Flexural modulus, psi	425,000	441,000	471,000	412,000	619,000
Flexural strength, psi	9,680	8,810	8,550	9,360	9,430
Tensile strength, psi	4,450	4,360	3,580	4,640	4,240
Elongation, %	70	11.7	17.2	10.5	4.0

(See also Table 9-4 and Figs. 9-3–9-6.)

Table 5-35. Polypropylene co-polymers and homopolymers.

	Co-polymer: Eastman Polyallomer 5021A						
$CaCO_3$, Atomite, phr:	0	25	50	75	100	125	150
Properties							
Tensile strength, psi	4483	3925	3343	3153	3017	2518	2400
Elongation, %	60	20	20	20	20	20	20
Hardness, Shore D-2	71	72	73	74	75	76	77
Taber abras., H22, mg loss	165	310	395	410	485	535	700
Ross flex, injured, cycles	200M+	200M+	200M+	200M+	200M+	200M+	200M+
Elec. vol. res., ohm-cm × 10^{14}	60+	60+	60+	60+	60+	60+	60+
Sp.G.	0.900	1.039	1.158	1.261	1.351	1.431	1.502
	Co-polymer: Hercules Pro-Fax 8523						
Properties							
Tensile strength, psi	3803	3416	3345	2632	2572	2244	2105
Elongation, %	50	20	20	20	20	10	10
Hardness, Shore D-2	72	73	73	74	75	76	76
Taber abras., H22, mg loss	65	145	230	310	350	470	480
Ross flex, injured, cycles	200M+	128M	32M	22M	725	44	1
Elec. vol. res., ohm-cm × 10^{14}	60+	60+	60+	60+	60+	25	24
Sp.G.	0.904	1.043	1.162	1.265	1.356	1.435	1.506
	Homopolymer: Eastman Tenite 4231A						
Properties							
Tensile strength, psi	5413	4301	3954	3519	3254	2960	2576
Elongation, %	40	40	30	30	20	10	10
Hardness, Shore D-2	77	78	79	80	81	82	83
Taber abras., H22, mg loss	190	230	285	320	390	495	525
Ross flex, injured, cycles	200M+	200M+	200M+	200M+	37M	14M	436
Elec. vol. res., ohm-cm × 10^{14}	60+	60+	60+	60+	60+	29	24
Sp.G.	0.902	1.041	1.160	1.263	1.354	1.433	1.504
	Homopolymer: Hercules 6523						
Properties							
Tensile strength, psi	5925	4597	3925	3471	3267	3244	2912
Elongation, %	30	20	20	10	10	10	10
Hardness, Shore D-2	79	80	81	82	83	84	85
Taber abras., H22, mg loss	150	240	310	390	580	690	715
Ross flex, injured, cycles	200M+	200M+	86M	45M	1	1	1

Table 5-35. *(Continued)*

Elec. vol. res., ohm-cm × 10^{14}	60+	60+	60+	60+	60+	46	28
Sp.G.	0.904	1.043	1.162	1.265	1.356	1.435	1.506

Heat Aging Seven Days 150°C. Forced Air Oven

Formulation: 100 pbw Resin, 50 phr $CaCO_3$, 0.4 phr DLTDP, 0.2 phr Naugard 432 Hindered Phenol

Aged Properties	Polyallomer	PROFAX 8523	Tenite 4231A	PROFAX 6523
Tensile strength, psi	3586	2947	4296	4275
Elongation, %	30	10	10	10
Hardness, Shore D-2	73	73	79	81

Formulation: 100 pbw Resin, 50 phr $CaCO_3$, 0.4 phr DLTDP, 0.3 phr Naugard 445 substituted diphenylamine.

Tensile strength, psi	3614	2912	4076	4093
Elongation, %	50	10	10	10
Hardness, Shore D-2	73	73	79	81

(See also Table 9-4 and Figs. 9-4–9-6.)

propylene-diene rubbers are readily cured with organic peroxides (table 5-32).

- Coat the calcium carbonate with a material having both polar groups and hydrocarbon chains, such as Ken-React TTS, or with a carboxylated polyolefin such as duPont Surlyn Ionomer or Exxon Dexon. These methods are in development (see Section 7).

6.4 Polypropylene

Both propylene homopolymer and copolymers readily accept $CaCO_3$ of 3 μ particle size at loadings of 30–40% of the total formulation (that is, about 43–67 phr $CaCO_3$). Comparative properties of talc and $CaCO_3$ filled polypropylene, tested in accordance with ASTM D2146 are listed in Table 5-33 and 5-34. Propylene homopolymer and copolymers with $CaCO_3$ loadings from 0 to 100 phr are characterized in Table 5-35. Automotive panels use 35–40% filler; about 30% $CaCO_3$ is used in polypropylene "paper" that absorbs and retains the correct amount of ink when printed.

6.5 ABS: Acrylonitrile-Butadiene-Styrene

Table 5-36 shows the loading of ABS resin with $CaCO_3$. Both ABS and glass-reinforced polyesters are examples of successful loading of resins containing substantial quantities of styrene.

6.6 Miscellaneous

$CaCO_3$ fillers are used to some extent in polyurethane foam, epoxy compounds, and phenolic resins. In polyurethane foam, the $CaCO_3$ particles serve as nuclei to produce a uniformly small cell size. Epoxy potting compounds with $CaCO_3$ filler are well known. Acid-catalyzed phenolic resins are not suited to $CaCO_3$ filler: only alkaline types of phenolics can be filled with $CaCO_3$.

7. NEW DEVELOPMENTS

There are three main methods of improving the properties of calcium carbonate filled plastics: (1) add polar copolymer during polymerization (2) add coupling agent to the filler or to the polymer or to the formulation, (3) cross link the formulation. Combinations of the above methods can sometimes be used.

Examples of adding copolymers:

Carboxylated polypropylene (propylene-acrylic acid copolymer):
Hercules Profax PC072
Exxon Dexon 1001, 1002, XPA-5A, XPA-6

Table 5-36. Acrylonitrile-butadiene-styrene (ABS) terpolymer, Borg-Warner Cycolac TD 1001, with chlorinated polyethylene, Dow Chem. CPE 3614 impact modifier

No impact modifier (No CPE)
Formulation: 100 pbw ABS, 0.4 phr Leadstar (normal lead stearate), 0.4 phr DS-207 (dibasic lead stearate); $CaCO_3$, Atomite

Atomite, pbw	0	12.5	25	50	100
Tensile strength, psi	5900	5493	4941	3990	3120
Elongation, %	20	20	15	10	5
Hardness, Shore D-2	76	76	77	78	80
Ross flex, injured, cycles	820	751	404	1	1
Sp.G.	1.043	1.119	1.188	1.310	1.504
	5 pbw CPE, 95 pbw ABS				
Tensile strength, psi	5880	5363	4951	4180	3070
Elongation, %	30	30	30	20	12
Hardness, Shore D-2	75	75	76	77	79
Ross Flex, injured, cycles	1881	1636	1531	47	1
Sp. G.	1.050	1.126	1.196	1.319	1.511
	10 pbw CPE, 90 pbw ABS				
Tensile strength, psi	5500	5086	4715	3760	2930
Elongation, %	40	30	30	25	15
Hardness, Shore D-2	74	74	75	76	78
Ross flex, injured, cycles	3783	3473	2525	74	1
Sp.G.	1.057	1.133	1.203	1.325	1.518
	15 pbw CPE, 85 pbw ABS				
Tensile strength, psi	5040	4941	4358	3690	2720
Elongation, %	40	30	30	30	20
Hardness, Shore D-2	73	73	74	75	77
Ross flex, injured, cycles	2230	2008	354	43	1
Sp.G.	1.065	1.141	1.210	1.333	1.526

Milling and molding: 177°C.

Table 5-37. Ken React TTS coupling agent.

High Density Polyethylene

Formulation (pbw)						
Gulf 901d HDPE	100	100	100	100	100	100
$CaCO_3$, Atomite	0	42.86	43.42	68.03	103.73	157.07
Ken React TTS	0	0	1.30	2.04	3.11	4.71
Properties						
Melt index, g/10 min	19.67	16.31	18.83	18.42	17.71	17.81
Tensile strength, psi	2050	1961	2334	1725	1765	1804
Tensile modulus, psi	98,000	217,900	180,600	163,400	150,900	130,800
Impact, ft lb/in. notch	0.926	0.531	0.572	0.785	0.967	1.014

Table 5-38. Specifications for $CaCO_3$ and plastics.

$CaCO_3$ Requirements:
- ASTM D1199, D34, D1366.
- Military spec. MIL-C-15198A
- Maritime administration spec. 52-MA-524 B
- Food Chemicals Codex: ground limestone and precipitated calcium carbonate
- U.S. Pharmacopeia precipitated calcium carbonate

Plastics Requirements, ASTM
- PVC flexible sheet D1593
- PVC insulation for wire and cable D1047, D2219, D2220, D2633
- PVC rigid compound D1784, D3010
- PVC rigid pipe D2836, D1785, D2241, D2665 etc.
- PVC foam D1565
- Crosslinkable polyethylene D2647, D2655, D2656
- Polyethylene sheeting D2103
- Polyester molding compounds D1201
- Polypropylene molding and extrusion materials D2146
- Ethylene propylene rubber D2000, D2770, D2802, D2768

Plastic and Elastomer Abbreviations, Definitions, ASTM
- Plastics D1600
- Elastomers D1418
- Plastics: test definitions E6
- Plastics: test temperature D618
- Conversion to metric units E380
- Elastomers: Automotive classification D2000

Processing
- Milling, compression molding, curing: D15, D3010
- Testing machine calibration E4, E74
- SPE Guide for Uniform Testing and Reporting Identification Properties of Plastic Materials, Society of Plastics Engineers, 1972.

Table 5-39. Testing of $CaCO_3$ by ASTM methods

Particle diameter distribution by sedimentation	D422
Particle diameter by screening	E161, D546
Bulk density	C128, D1895
Aqueous electrical resistivity	D2448
Hunter reflectance	E97, D2244
Oil Absorption	D281
Moisture	D280, Method A
Specific gravity	D153
pH in water (5% $CaCO_3$, 23°C)	D1208

Carboxylated polyethylene
- DuPont Ionomer Surlyn
- Exxon Dexon 2000, XEA-1, XEA-2, XEA-3, XEA-5.

Carboxylated ethylene-propylene polymer:
- Dexon XPA-4, XPA-8

Carboxylated ethylene-vinyl acetate polymer:
- Dexon XEA-4.

Table 5-40. Plastic and elastomer tests.

General Tests, ASTM	
Tensile strength, modulus, elongation	D638, D412
Tear strength	D1004, D624
Hardness, Shore A and D	D1706
Hardness, Barcol	D2583
Taber Abrasion	D1044
Ross flex	D1052
Electrical volume resistivity	D257
Specific gravity, density	D792, D1505
Flexural modulus and strength	D790
Stiffness in flexure	D747
Deflection temperature	D648
Vicat softening point	D1525
Izod impact strength	D256
Shrinkage in mold	D955
Melt flow rate	D1238
Water absorption	D570
Bashore resilience	D2632
Forced air oven aging polypropylene	D3012
Permanent set	D412
PVC Tests, ASTM	
Torque rheometer: mixing of ingredients	D2396
Torque rheometer: fusion rate	D2538
Plastisol melt viscosity, Brookfield	D1824, D2393
Oven aging	D2115
Pipe, falling weight impact	D2444, D3029
Flooring indentation, McBurney	F142
Rigid PVC milling and compression molding	D3010
Low temp. film impact	D1593

PVC Potable Water Pipe

- National Sanitary Foundation, Ann Arbor Mich., NSF Standard 14
- Plastic Pipe Institute, N.Y., N.Y.; Technical Reports TR2, TR3, TR4
- Pipe Requirements: ASTM D1785, D2241
- Pipe Tests: ASTM D1784, D2827, D2152, D1599, D2444

Table 5-41. Plastic and elastomer suppliers.

Ref.: Table 5-	For complete listing see *Modern Plastics Encyclopedia*
12, 14, 15, 16 17, 19, 20, 21	B. F. Goodrich Chemical Co., 6100 Oak Tree Road, Cleveland, Ohio 44131 (216) 524-0200
13, 22, 23, 24, 27	Union Carbide Corp., Chemicals & Plastics, 270 Park Ave., New York, N.Y. 10017 (212) 695-5054
18	Robintech, Inc., 3421 Vestal Road, Vestal, New York 13850 (607) 729-9381
18, 19, 20, 25	Rohm & Haas, Independence Mall W., Philadelphia, Pa. 19105 (215) 592-3000
19, 20, 36	Dow Chem. Co., 2020 Dow Center, Midland, Mich. 48640 (517) 636-1000
21	Air Products & Chemicals, Inc., Plastics Div., 5 Executive Mall, Swedesford Rd., Wayne, Pa. 19087 (215) 687-6150
22, 23, 24	Goodyear Tire & Rubber Co., Chemical Div., 1144 E. Market St., Akron, Ohio 44316 (216) 794-4012
23, 24	Allied Chem. Corp., Plastics Div., P.O. Box 2365R, Morristown, N.J. 07960 (201) 455-4064
25	W. R. Grace & Co., Marco Chem. Div., 1711 West Elizabeth Ave., Linden, N.J. 07036 (201) 862-5600
25, 26	Reichhold Chemicals, Inc., 525 N. Broadway, White Plains, N.Y. 10602 (914) 948-6200
27	Phillips Petroleum Co., Chemical Dept., Plastics Div., Bartlesville, Okla. 74003 (918) 661-6600
28	U.S. Industrial Chemicals Corp. div. National Distillers & Chem. Corp., 99 Park Ave., New York, N.Y. 10016 (212) 697-0700
28	Gulf Oil Chemicals Co., Plastics Div., Chemical Row, Orange, Texas 77630 (713) 886-7491
29, 30, 31, 32	E. I. duPont de Nemours & Co., 1007 Market St., Wilmington, Del. 19898 (302) 774-2421
33, 34, 35	Hercules, Inc., 910 Market St., Wilmington, Del. 19801 (302) 656-9811
35	Eastman Chemical Products, Inc., sub. Eastman Kodak Co., P.O. Box 431, Kingsport, Tenn. 37662 (800) 251-0351
36	Borg-Warner Chemicals, International Center, Parkersburg, W.Va. 26101 (304) 485-1771

Recently developed coupling agents,[32] Ken-React organic titanates, improve both the physical properties and rheological processing properties of $CaCO_3$-filled plastics (Tables 5-16A and 5-37). One possible disadvantage is that oxidation may cause the compound to become yellow or orange in color. Therefore oxidative conditions at high processing temperatures should be avoided where light colors are important.

During the years of rapid growth, the plastics industry has faced a multitude of problems. As these problems are solved, it is expected that more time will be available for cost-saving projects such as economizing on resin by adding filler. Steady, long term growth of $CaCO_3$ filler consumption in plastics has occurred quietly, with $CaCO_3$ proving its worth in an increasing number of applications.

8. ACKNOWLEDGMENTS

Thompson, Weinman and Co. laboratories furnished the technical data in Tables 5-12–5-36. Mr. Levente L. Koller and Mr. Robert Baker conducted the laboratory work.

Table 5-42. Plasticizer suppliers.

Ref.: Table 5-	Material	For complete listing see *Modern Plastics Encyclopedia* and "Materials and Compounding Ingredients for Rubber"	
12, 13, 16, 23, 24	DOP	Many suppliers, incl. Ashland, Exxon, FMC, Hatco, Harwick Std., Wallace & Tiernan, Monsanto, Witco, etc.	
12	P72	Hatco Chem. Div., W. R. Grace, King George Post Rd., Fords, N.J. 08863 (201) 738-1000	
12	TCP	Many suppliers, incl. Celanese, FMC, Harwick Std., Montrose, Monsanto, etc.	
12, 15, 17, 22, 23, 24	Epoxidized soybean oil (Soybean oil epoxide)	Many suppliers, including:	
		Paraplex G62	Rohm & Haas
		Drapex 6.8	Argus Chem.
		Estynox 140, 203	NL, Baker Castor Oil Div.
		Hatcol 16	Hatco
		PX-800	U.S. Steel, USS Chemicals Div.
		Flexol EPO	Union Carbide Corp.
		Truflex E-54	Thompson Chem.
13	Chlorinated paraffin, 40% Cl.	Chlorowax 40 Unichlor 40	Diamond Shamrock or, Neville Chem.
14, 16, 22	DIDP	Many suppliers. See DOP above for phthalate ester suppliers.	
15	TINTM	Exxon Chem. Co. USA, 60 W. 49th St., New York, N.Y. 10020 (212) 974-3000	
17, 22	Santicizer 160	Monsanto Co., 800 N. Lindburgh Blvd., St. Louis, Mo. 63166 (314) 694-1000	
17	DCP Dicapryl phthalate	Hatco Chem.; Wallace & Tiernan	

Table 5-43. Heat stabilizer suppliers.

Ref.: Table 5-			Ref.: Table 5-		
12, 13, 15 16, 17, 22	BaCd Zn phosphite phenate (liquid)	Argus; Ferro; Interstab; see References 26, 28, 29	20	Tribase AG	NL Industries
			20	Dyphos XL	NL Industries
			21	Tribase XL	NL Industries
14 15	Tribase E-XL Dithal XL	NL Industries, Industrial Chemicals Div., Wycoff Mills Rd., Hightstown, N.J. 08520 (609) 448-3200	22	Stabilizer for asbestos	Argus, Ferro, Cincinnati Milacron, 27, 29, 30
			23, 24	Metasap 635 Zinc soap	Nopco Chem. Div., Diamond Shamrock Chem. Co., 350 Mt. Kemble Ave., Morristown, N.J. 07960 (201) 267-1000
18, 19	Thermolite 31 Super	M & T Chemicals, 28			

Table 5-44. Lubricant suppliers.

12, 14, 15	Stearic acid	Many suppliers, incl. ADM; Armour; Emery Inds; Harwick Std.; Wallace & Tiernan; Humko; Merck; Swift; Wilson, etc.
18	Calcium stearate	Many suppliers, incl. C.P. Hall; Harshaw Chem.; Harwick Std.; Tenneco; WCD, Witco
18	Aristowax 165	Union Oil Co. of Calif., distributed by Sonneborn Div. Witco Chem. Co., New York, N.Y.
19	Wax S, Wax V	American Hoechst Corp., Wax Dept., Rt. 202-206N, Somerville, N.J. 08876 (201) 685-2423
20	Plastiflow LPC	NL Industries
21, 36	Leadstar	NL Industries
21, 36	DS-207	NL Industries
25	Zinc stearate	Many suppliers, incl. Amer. Cyanamid; C.P. Hall; Harshaw Chem.; Harwick Std.; Nopco; Smith Chem. & Color; Sherwin Chems.; Whittaker Clark & Daniels (WCD); Witco Chem.

Table 5-45. Supplier of miscellaneous materials used in tests, Tables 5-14–5-37.

14, 15	Pigment 33 calcined clay electrical grade	Freeport Kaolin Co., 733 Third Ave., New York, N.Y. 10017 (212) 687-8100
17	Silicone SS4255	General Electric Co. Silicone Products Div., Waterford-Mechanicville Rd., Waterford, NY 12188 (518) 237-3330
18	TiO_2 R900	duPont
19, 20	Titanox 2071, 2101	NL Industries
22	7R asbestos	Johns-Manville, Greenwood Plaza, Denver, Colo. 80217 (303) 770-1000
25	Glass fiber	
22	6N Talc	International Talc Co., 420 Lexington Ave., New York, N.Y. 10017 (212) 532-8533
24	Snobrite clay	Thompson-Weinman
25	MgO thickener	Merck & Co., 126 E. Lincoln Ave., Rahway, N.J. 07065 (201) 381-5000
25	Glass bubbles	3M Company 3M Center, St. Paul., Minn. 55101
25	t-butyl perbenzoate	Reichhold Chemicals Inc., see Table 5.41 for address
26	BPO Paste (50% benzoyl peroxide)	
26	Alsilate LO Clay	Freeport Kaolin Co. 733 Third Ave., New York, NY 10017 (212) 687-8100
	$CaSO_4$ W-1320	Pfizer (Table 5-7)
30, 32	Percadox 14/40	AKZONA, NOURY Chem. Corp., Burt, N.Y. 14028 (716) 778-8554
31	Di-cup T	Hercules Inc. see Table 5-41 for address
32	Spider Brand Sulfur	C.P. Hall Co., 73rd & Central Ave., Chicago, Ill. 60638 (312) 767-4600
33	Raven Black 30	Columbian Carbon Co., sub. Cities Service Co., 3200 W. Market St., Akron, Ohio 44313 (216) 867-5800
33, 34	Talc, Desert 57	Johns-Manville, Denver, Colo. 80217 (303) 770-1000
37	Ken-React TTS	Kenrich Petrochemicals, Foot of E. 22nd St., Bayonne, N.J. 07002 (201) 436-5610

References

1. Perry, J. H., *Chemical Engineers' Handbook*, McGraw-Hill, New York, 1975.
2. Riegel, E. R., *Industrial Chemistry*, Van Nostrand Reinhold, New York, 1975.
3. American Society for Testing and Materials, 1916 Race St., Philadephia, Pa. 19103. Revisions published annually.
4. Kirk, R. E. and Othmer, D. F., *Encyclopedia of Chemical Technology*, Interscience, New York, 1948.
5. Seymour, R. B., "Fillers for Plastics," *Modern Plastics Encyclopedia*, McGraw Hill, New York, 1974, p. 217–226.
6. *Food Chemicals Codex*, 2nd Edition, National Academy of Sciences, Washington, D.C., 1972.
7. Lange, N. A., *Handbook of Chemistry*, 10th Edition, McGraw Hill, New York, 1967.
8. Hodgman, C. D., *Handbook of Chemistry and Physics*, The Chemical Rubber Publishing Co., Cleveland, Ohio, 1975.
9. J. E. Hartitz and R. A. Yount, Society of Plastics Engineers, Palisades Section Minitec, Jan. 19, 1977.
10. Walker, R., *Materials and Compounding Ingredients for Rubber* (*Rubber Blue Book*), Bill Communications Inc., New York, 1975.
11. Marsich, Ellen E., *Raw Materials Index, Pigment Section*, May 1974, National Paint and Coatings Assoc., 1500 Rhode Island Ave. NW, Washington, D.C. 20005.
12. Karasek, F. W., "Micromeritics," *Research/Development Magazine*, Sept. 1970. The Sedigraph instrument is made by Micromeritics Instrument Corp., 800 Goshen Springs Rd., Norcross, Ga. 30071. (404) 448-8282.
13. Penn, W. S., *PVC Technology*, Maclauren and Sons, London, 3rd edition.
14. Thompson, Weinman and Co. published and unpublished technical data are listed in Tables 5-12–5-36, using materials listed in Tables 5-41–5-45. ASTM methods and standard calibration of testing instruments are used: see Tables 5-38–5-40.
15. *SPE Guide for Uniform Testing and Reporting Identification Properties of Plastic Materials*, updated in 1972 and issued by Engineering Properties and Structure Division, Society of Plastics Engineers. This guide states ASTM test methods, recommends sample thickness, and specifies standard units for reporting results. This guide was followed for most tests. Where exceptions occur, experimental conditions are stated.
16. Deanin, R. D. and Schott, N. R., *Fillers and Reinforcements for Plastics*, Advances in Chemistry Series 134, *Am. Chem Soc.*, Washington, D.C., 1974.
17. Lead and tin stabilizer technical data: N. L. Industries, Inc., Industrial Chemicals Div., Wycoff Mills Rd., Hightstown, N.J. 08520. (609) 448-3200.
18. *Thermolite Stabilizers for Rigid PVC*, M. and T. Chemicals Inc., sub. American Can Co., Rahway, N.J. 07065. (201) 499-0200. Also literature on various PVC stabilizers.
19. Struber, V. R., *Theory and Practice of Vinyl Compounding*, published by the Argus Chemical Corp., sub. Witco Chem. Corp., 633 Court St., Brooklyn, N. Y. 11231. (212) 858-5678. *PVC Pipe and Conduit* also from Argus.
20. PVC stabilizer literature: Ferro Corp., Chemical Div., 7050 Krick Rd., P.O. Box 46349, Bedford, Ohio 44146. (216) 232-4330.
21. PVC stabilizer literature: Interstab Chemicals Inc., 500 Jersey Ave., New Brunswick, N.J. 08903. (201) 247-2202.
22. Mentovay, L. W., *Torque Rheometer-Plastics Bibliography*, C. W. Brabender Instruments, Inc., 50 E. Wesley St., S. Hackensack, N.J. 07606. (201) 343-8425.
23. Kromer, H. M., *35th Anniversary Symposium, a Compendium of Technical Papers*, Haake Inc., 224 Saddle River Rd., Saddle Brook, N.J. 07662. (201) 843-2320.
24. Sarvetnick, H. A., *Plastisols and Organisols*, Van Nostrand Reinhold, New York, 1972.
25. Park, R. A., *Dispersion Resin Handbook*, Chemical Div., Firestone Plastics Co., Pottstown, Pa.
26. Park, R. A., *Viscometers*, see above address.
27. National Sanitation Foundation *Standard No. 14. Thermoplastic Materials, Pipe, Fittings, Valves, Joining Materials,* N.S.F. Testing Laboratory, Inc., P.O. Box 1468, Ann Arbor, Mich. 48106.
28. Plastic Pipe Institute, *Standards for Plastic Piping*, Technical Report PPI-TRS-March 1971. Plastic Pipe Inst., 250 Park Ave., N.Y., N.Y. 10017. (212) 687-2675.
29. Nussbaum, H. W. and Czarnomski, T. J., *Smooth Surface Premix and Sheet Molding Compound Technology*, publication of Marco Chem. Div. of W. R. Grace and Co., Research and Development Laboratory, Linden, N.J. 07036. (201) 925-6610.
30. *Paraplex P-19 Polyester Resins*, Publication No. PL-751a, Rohm and Haas, Independence Mall West, Philadelphia, Pa. 19105. (215) 592-3000.
31. Deis, W. H. and Ness, W. W., Merck and Co. Chem. Div., Rahway, N.J., "Magnesium oxide and hydroxide for SMC," *Modern Plastics:* 94–98 (Nov. 1974).
32. Monte, S. J. and Bruins, P. F., "New Coupling Agent for Filled Polyethylene," *Modern Plastics:* 68–72 (Dec. 1974). Ken-React TTS literature available from Kenrich Petrochemicals, Inc., Foot of East 22nd St., Bayonne, N.J. 07002. (201) 436-5610.

6

Kaolin, Hydrous and Anhydrous

Thomas H. Ferrigno and Michael Taranto

Improde
Trenton, New Jersey

Georgia Kaolin Co.
Elizabeth, New Jersey

CONTENTS

1. INTRODUCTION

Kaolin, which is commonly called clay, is a hydrous alumino silicate mineral. It is available in two primary grades: the natural hydrous form and a calcined, anhydrous form. The larger average particle sizes consist of mixtures of blocky and platy particles. The finer fractions are almost entirely thin plates.

The hydrous forms are non-abrasive, chemical resisting, and have relatively high surface area, which promotes high viscosities. Their naturally acid pH must be considered in many systems wherein reactivity may occur. They disperse quite readily in most plastics, especially with the aid of dispersants or surfactants. In reinforced and thermoset plastics, they control flow properties to provide more uniform composite properties.

Calcined kaolin is considerably harder, but provides improved electrical properties in thermosets and thermoplastics.

Many surface-treated types of each grade are available to provide outstanding water resistance, electrical properties, and reinforcement, in a variety of systems.

2. MANUFACTURE

2.1 Sources

Kaolin is the common term for the mineral kaolinite, one of a number of minerals called clays. Clay is a rock term applying to soft, earthy ores which are plastic when mixed with water. Vast deposits are mined extensively in the Piedmont Plateau area of Georgia and South Carolina and in Texas. The ores are generally associated with quartz, mica, carbon, iron, and titanium oxides and sometimes a wide variety of other clay minerals. Open-pit mining is usually employed.

2.2 Air-Floated Grades

These grades are produced from dried ores which are passed through roller mills with internal air-classifiers to remove coarse impurities and to delaminate the ore to its natural state of subdivision. Two grades are produced; the South Carolina type, which is extensively used as a rubber reinforcer, due to its high surface area and fine particle size; and the coarser, better crystallized, Georgia type.

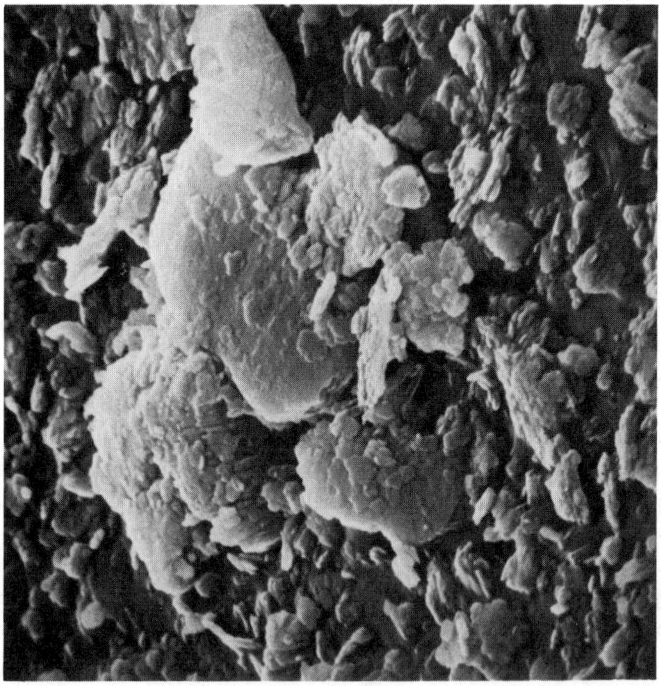

(A) ←— 10 microns —→

(B) ←— 10 microns —→

(C) ⟵ 2 microns ⟶

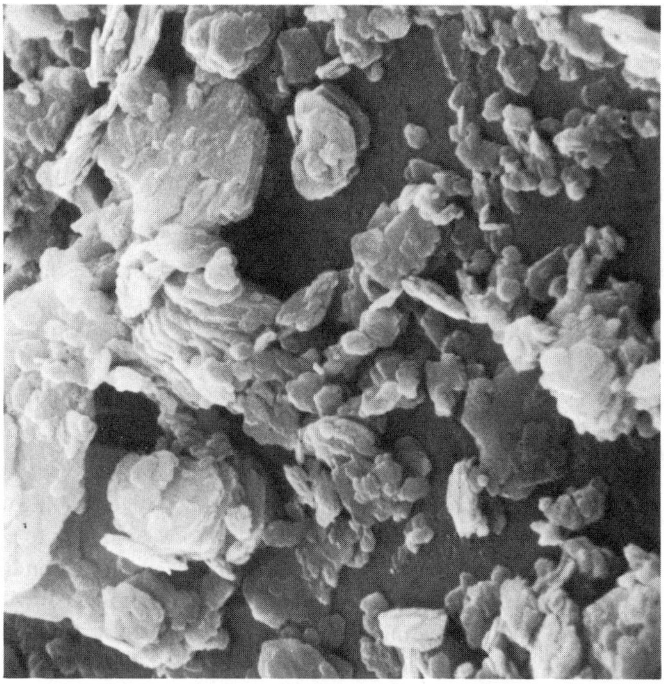

(D) ⟵ 2 microns ⟶

Fig. 6-1. Electron micrographs of kaolin particle of Georgia origin:
A – Coarse, air-classified.
B – "Hydrite Flat D," coarse, water-leached, 4.5 μ average particle size.
C – Fine, beneficiated; note absence of "stacks."
D – Water-leached, 1.5 μ average particle size.
(*Courtesy of Georgia Kaolin Co.*)

2.3 Water-Leached Grades

Water-leached grades are processed as water suspensions, treated with decolorizing agents, and fractionated with centrifuges to obtain specific particle size distributions, after which they are vacuum filtered to remove soluble impurities. The coarser grades are obtained by drying the filter cake, whereas the finer grades are re-slurried in water and spray-dried. In either case, the dried products are milled to reduce agglomeration for plastics uses.

2.4 Beneficiated Grades

These grades are made by a number of patented processes devised to either physically or chemically remove deleterious coloring matter and to obtain very pure and white (high brightness) fillers.

2.5 Calcined Grades

The water-leached grades are generally used as the base kaolin to be calcined at lower temperatures for superior electrical properties or at higher temperatures for exceptional whiteness. Both are milled to reduce the hard agglomerates which form on calcining.

2.6 Surface-Treated Grades

All of the above grades are treated with surface-modifying agents to obtain improved dispersibility, electrical properties, water resistance, and reinforcement in plastics systems.

3. COMMERCIAL PRODUCTS AND PROPERTIES

3.1 Physical

The single crystal of kaolin is a two-layer hydrous alumino silicate, consisting of chemically bonded layers of silica and Gibbsite (hydrous alumina), and has the form of a hexagonal platelet.. In the process of formation, some crystals were bonded together quite strongly to give rise to a range of natural sizes. Particles larger than about 10 μ esd (equivalent spherical diameter) are stacks of platelets with heights about equal to their average diameters. Those which are finer have different aspect ratios according to their esd and the method of delamination. Below about 2 μ esd, all particles are thin platelets with average platelet diameters of 4–10 times the thickness.

Properties for the various grades are reported in Table 6-1. Refined grades with hard impurities removed are non-abrasive, whereas calcined grades are quite hard and abrasive, as indicated by Mohs hardness values. Micrographs are shown in Fig. 6-1.

3.2 Chemical

Because the two sides of the kaolin particle are different and broken bonds at edges are highly active, kaolin exhibits a strong tendency to resist dispersion, by forming strongly bonded agglomerates. This is more pronounced with decreasing particle size. The active surfaces are, however, readily reactive with organo-silanes, various metallic salts such as vinyl stabilizers, polar polymers, lubricants and the like, to obtain a fairly complete state of dispersion. Surface treatment of hydrous or calcined kaolin is preferred to obtain the physical advantages along with improved bonding with polymers, ease of dispersion, better flow and non-reactivity in critical applications.

The naturally acid pH of the mineral causes undesirable reaction with epoxies and vinyls, unless the acid sites are inhibited by surface treatments; many such products are available for specific applications.

Both natural and calcined kaolin are more resistant to acids and alkalis than all but the most resistant polymers. Removal of the water of constitution requires temperatures in excess of 1000°F, much higher than plastics processing temperatures.

3.3 Electrical

Table 6-2 lists selected electrical properties of fillers and insulations. Particularly emphasized is the effect of hydrophobic surface treatment on the electrical properties of calcined kaolin.

Table 6-1. Properties of kaolin fillers.

Properties	AIR CLASSIFIED		WATER CLASSIFIED			BENEFICIATED			CALCINED		SURFACE MODIFIED		
	Coarse	Fine	Coarse	Med	Fine	Coarse	Med	Fine	Partial	Fully	Silane	Resin	Cationic
Specific gravity	2.58	2.58	2.58	2.58	2.58	2.58	2.58	2.58	2.50	2.63	2.58	2.58	2.58
pH (20% in water)	3.8–5.5	3.8–5.5	3.8–7.5	3.8–7.5	3.8–7.5	3.8–7.5	3.8–7.5	3.8–7.5	4.2–6.0	4.2–6.0	4.0–9.0	6.5–7.5	7.0–8.0
Hardness (Mohs)	2	2	2	2	2	2	2	2	4–6	6–8	2–8	2	2
Refractive index	1.56	1.56	1.56	1.56	1.56	1.56	1.56	1.56	1.62	1.62	1.56–1.62	1.56	1.56
BET surface area, m^2/g	10–12	20–24	6–10	14–16	18–22	7–8	10–14	18–22	5–12	5–12	8–24	14–16	14–16
Oil absorption, ASTM D-281	28–32	34–40	28–32	32–40	40–48	50–60	45–55	40–50	45–60	45–90	28–60	28–31	24–33
Average particle size (esd), μ	1.5–1.8	0.3	4–9	1.8–0.7	0.7–0.2	2.0–3.0	1.5–2.5	0.5–1.0	1.5–1.8	0.9–3.0	0.3–3.0	0.7	0.6–4.5
Free water content, %, max.	1.0	1.0	1.0	1.0	1.0	1.0	1.0	1.0	0.5	0.5	0.5–1.0	0.5–1.0	0.2–0.5
Brightness–G.E. %, MgO Std	74–84	74–80	79–85	85–87	87–90	85–86	86–88	89–91	85–90	90–96	74–92	85–87	80–87
Price range F.O.B. Plant, 50 lb bags, $/ton	30–40	25–35	70–75	75–80	80–175	90–95	100–105	105–120	140–235	120–240	110–310	200–300	115–125

Typical Chemical Analyses

Element	Reported as %	Hydrous Kaolin	Calcined Kaolin
Silicon	SiO_2	45.4	52.1–52.9
Aluminum	Al_2O_3	38.8	44.4–45.2
Iron	Fe_2O_3	0.3	trace
Titanium	TiO_2	1.5	0.8–2.0
Calcium	CaO	0.1	–
Sodium	Na_2O	0.1	–
Potassium	K_2O	trace	–
Loss on ignition	H_2O	13.8	0.5–0.9

The major constituents reported in these analyses are combined as complex aluminum silicate and do not exist as free oxides.

Typical Trace Metal Analysis of Water-Leached Kaolin

Metal	Parts per Million, by Weight*
Mercury	Less than 3
Arsenic	Less than 10
Barium	10–90
Copper	1–9
Lead	1–9
Zinc	10–90
Cadmium	Not detected
Bismuth	Not detected
Antimony	Not detected
Tin	Not detected
Nickel	Not detected
Cobalt	Not detected
Molybdenum	Not detected
Manganese	Not detected

*Kaolin is generally recognized as safe (GRAS) by FDA for use in food-contact applications.

Table 6-2. Electrical properties of selected insulator materials.

Insulator Material	Volume Resistivity, Ohm-cm: Dry	Volume Resistivity, Ohm-cm: Humid	Dielectric Strength, volts/mil	Dielectric Constant K_e @ 1 mc	Dielectric Loss
Kaolin, water-leached	10^{13}	10^{6}	70–120[a]	2.6	0.10
Kaolin, Calcined	10^{13}	10^{8}	60–100[a]	1.3	0.06
Kaolin, Calcined, hydrophobic surface	10^{13}	10^{12}	80–150[a]	1.3	0.003
Talc	10^{14}	10^{9}	200–400[a]	5.5–7.5	0.001
Mica	10^{17}	10^{10}	50–60[a]	2.0–2.6	0.10
Polyethylene	10^{13}	10^{13}	460[b]	2.3	0.0001
Polyester, thermoset	10^{14}	10^{12}	380–500[b]	2.8–4.1	0.01
PVC	10^{16}	10^{16}	700–1300[b]	2.9	0.01

[a]Tested as a powder.
[b]Tested as a film.

4. SUPPLIERS

Primary suppliers are listed in Table 6-3. Most have distributor or sales agents to represent them in many areas of the U.S. and in foreign countries.

5. COSTS AND AVAILABILITY

Representative costs are given in Table 6-1. Supplies are assured for all grades beyond the foreseeable future. Less truckload quantities are available from many agents and distributors in principal cities throughout the world at prices reflecting the costs of shipping and storing.

6. APPLICATIONS

6.1 Thermosets

Kaolin is principally used in reinforced plastics, where its platiness and its viscosity increasing effect prevents fiber "blooming" or the appearance of fibers at the molded surface. Grades for this use should have very low +44 μ (+325 mesh) contents since larger particles filter on fiberglass mats and cause a non-uniform distribution of filler, crazing, and poor physical properties. Figure 6-2 indicates the viscosity effects of a range of particle sizes.

Calcined kaolin also is used in electrical laminates to improve volume resistivity, and to reduce the concentration of titanium dioxide, due to the whiteness imparted.

In this and other applications, composite properties are generally influenced by the degree of dispersion of the filler and its packing (P_f), as detailed in Chapter 2. Because dispersion is so important to obtaining maximum filler effectiveness, surface-treated grades are becoming more generally used.

6.2 Thermoplastics

The finer grades of kaolin are particularly effective for increasing tensile strength and modulus of low T_g thermoplastics, especially surface-

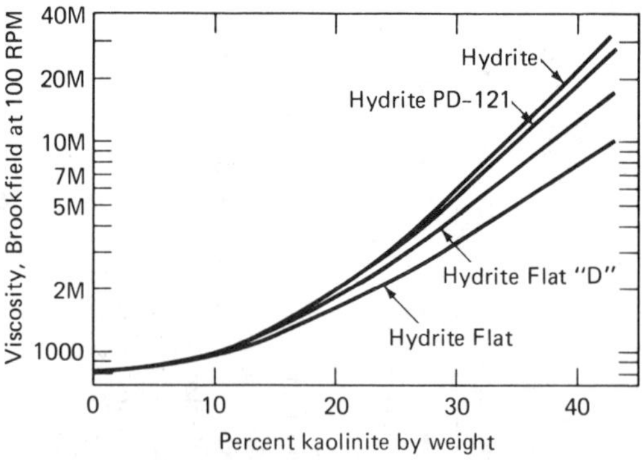

Fig. 6-2. The effect of kaolin particle size and concentration on thermosetting polyester viscosity. Average particle size, μ: Hydrite Flat, 7; Hydrite Flat "D," 4.5; Hydrite PD-121, 1.5; Hydrite, 0.77.

Table 6-3. Major kaolin suppliers in the United States.

	Shipping Point	Grades Marketed
Anglo-American Clays Corp., 52 Executive Park South, Atlanta, Ga. (404) 321-6304	2	A, B, C
Burgess Pigment Co., Kaolin Rd., Sandersville, Ga.	2	A, B, C, D, E
Dresser Minerals, Box 98, Kosse, Texas. (713) 781-5900	8	A, D
Freeport Kaolin Co., 733 3rd Ave., New York, N.Y. (212) 986-4030	4	A, B, C, D, E
Georgia Kaolin Co., 433 N. Broad St., Elizabeth, N.J. (201) 352-9800	1, 2	A, B, C, D, E
J. M. Huber Co., Clay Div., Twiggs County, Huber, Ga. 31040 (912) 745-4751	8, 9	A, B, C, D, E
Minerals & Chemicals Div., Engelhard Industries, Middlesex Essex Turnpike, Menlo Park, Edison, N.J. (201) 321-5000	3	A, B, C, D, E
National Kaolin Products Co., Box 431, Aiken, S. C.	6	A
Southeastern Clay Co., Box 1055, Aiken, S.C.	6	A
Thiele Kaolin Co., Box 337, Wrens, Ga. 30833 (404) 547-2593		
Thompson, Weinman & Co., Box 130, Cartersville, Ga. 30120 (404) 382-5353	3	A
United Sierra Div., Cyprus Mines Inc., 555 South Flower Street, Los Angeles, Ca. (213) 489-3700	2, 10	A, B

Key to shipping points

1. Dry Branch, Ga.
2. Sandersville, Ga.
3. McIntyre, Ga.
4. Gordon, Ga.
5. Huber, Ga.
6. Aiken, S.C.
7. Natka, S.C.
8. Kosse, Texas
9. Langley, S. C.
10. Jeffersonville, Ga.
11. Wrens, Ga.

Key to grades marketed

A. Air classified
B. Water leached
C. Beneficiated
D. Calcined
E. Organo-modified

Grades within classifications vary slightly in particle size and other properties.

treated types, without attendant severe loss of elongation and impact strength.

Vinyl wire insulation is substantially improved by calcined kaolin at up to 30 phr. Hydrophobic surface treatment further improves electrical properties under wet or humid-aged conditions.

7. FUTURE DEVELOPMENTS

The use of kaolin is expected to follow the trend of greater filler use due to the increasing cost of polymers.

Many surface-treated products have been developed in the past and it is expected that producers will utilize prior technology and further refine these products to offer a wider variety of more versatile kaolin fillers.

References

1. Lamb, R. E., *Rubber Age* **107, No. 4:** 42-51 (Apr. 1975).
2. Amond, Carl and Morris, Horton, H., "Kaolin Clays in Polyester Molding Compounds," Freeport Kaolin Company Publication.
3. "Fillers for Encapsulation," *Plastics Technology:* 41–49 (June 1962).
4. Hofferth, B. F. "Fillers for Vinyls"–presented February 6, 1969 at the SPE/Georgia Institute of Technology Seminar on Vinyl Plastics.
5. Frankenhoff, E. B., Maxel, J. M., and Miller, E. R., "Bulk Molding Compound" Society of Plastics Industry, Inc., Section 3-B, pg. 1–8, 1969.
6. Schwartz, E. W., *Plastics Technology:* 31-34 (July 1959).
7. Georgia Kaolin Co. Technical Service Bulletin, TSBT-84, pg. 1–10, Aug. 25, 1958.
8. Solomon, D. H., Dr., "Catalytic Properties of Pigments," *Tappi Monograph* (March 1969).
9. Hauser, E. A. and Leggett, M. B., *J. Amer. Chem. Soc.*: 62, 1811 (May 1940).

10. Seymour, R., *Chemtech:* 422–425 (July 1974).
11. Fallick, Bixler, Marsella *et al.* "Ceraplast, A New Family of Reinforced Plastics"–presented to 22nd Annual SPI Meeting, Reinforced Plastics Division, Washington, D.C., Jan. 31–Feb. 3, 1967.
12. Georgia Kaolin Co. Calcined Clays for the Elastomer Industry, TD-8.
13. Olyphant, Murray Jr., *Insulation:* 42–46 (Aug. 1966).
14. Larson, G. P., "Kaolinite Fractions, their effect on physical properties of reinforced plastics," *Modern Plastics* (May 1958).
15. Conley, R. F., "Properties of Electrical Grade Fillers," Georgia Kaolin Co.

7

Feldspar and Nepheline Fillers

Thomas H. Ferrigno
Improde
Trenton, New Jersey

CONTENTS

1. INTRODUCTION

Feldspar and nepheline are similar minerals and serve the same purposes. They are anhydrous, alkali, alumino silicates having excellent chemical resistance and are offered in coarse to medium particle sizes. As low surface area fillers, their uses are comparable to calcium carbonate and silica but they are more versatile. Costs are 2-3¢/lb FOB plant.

Advantages in plastics are:

Transparency to translucency in most polymers;
Ease of wetting, dispersion, and deairing;
Little to no effect on promoter, accelerator, or additive activity;
Low viscosity and high mobility at high concentrations;
Less color pigment required to obtain a given intensity;
Improved chemical, weather, and abrasion resistance;
Suitable for use in compositions contacting food;
Not a health hazard as a dust or when ingested.

Disadvantages are:

Coarser particle sizes are somewhat abrasive;
Settling in low-viscosity systems.

2. MANUFACTURE

2.1 Feldspar

Feldspar is obtained from white granite by quarrying the rock, and crushing and grinding to -20 mesh. The washed product is then subjected to froth-flotation to remove silica, mica, and garnet impurities. The mineral obtained is then dried and ground in air-swept, alumina-lined pebble mills in a closed-circuit with fractionating equipment to obtain the desired particle size distribution. Yields are in the order of 50% of the ore. Extensive deposits are mined in western North Carolina and comprise billions of tons.

2.2 Nepheline

Nepheline is the mineral derived from the syenite rock and is commonly referred to as nephelìne syenite to indicate its source. Workable deposits have little contamination but require selective mining techniques to choose high-grade ore since secondary refining processes are limited to magnetic separation or electrostatic separation to remove emery, a mixture of black iron oxide and alumina. Grinding and fractionating techniques are the same as for feldspar. High-grade deposits consisting of millions of tons are mined and processed in an area northeast of Toronto, Canada.

3. COMMERCIAL PRODUCTS AND PROPERTIES

3.1 Chemical

As a class, these minerals comprise the tectosilicates which constitute 75% of the earth's crust. Granite and associated formations are the most common rock-forming agents. Feldspar and nepheline are closely associated, the latter being classified as a feldspathoid because it was formed from a silica deficient magma or molten rock. The primary difference between the two materials is the ratio of silica to alumina in their analyses. This is of no consequence in their use as fillers.

As with all silicate minerals, the elements exist as atoms bonded with commonly shared oxygens in the crystal lattice. Except for isolated defects and at broken bonds, the elements and their ions are not available for chemical reaction unless the structure is dissolved or otherwise decomposed. Minerals of this class are highly resistant to chemical reaction under conditions much more severe than polymers are able to withstand. The pH of feldspar depends on the exposed surface since fracture occurs at the weakest bonds, the sodium, potassium, and calcium oxide bridges in the crystal lattice. The exposed ions bond with (chemisorb) hydronium ions of water to yield free hydroxyls and thus a basic pH which increases as the particle size decreases. Whereas feldspar cleaves along planes at nearly right angles to each other, nepheline cleaves predominantly along a crystal plane exposing the alkali oxide bonds. Nepheline then, has a more nearly constant effect on pH over a wide range of particle sizes. The potassium and sodium feldspars, as single species, are very resistant to acids whereas the calcium type is not. The commercial mineral is a solid solution called anorthoclase, predominating in the properties of the former.

Nepheline is derived from the Greek word meaning a cloud since the crystals become cloudy when immersed in 10% hydrochloric acid. The greater numbers of exposed cation sites make nepheline somewhat less acid resistant than feldspar. Both minerals have a very high order of resistance to alkali.

3.2 Physical

Neither commercial product is a single species of mineral, but are solid solutions. As such, they do not cleave regularly but fracture irregularly. The particles produced by grinding may have only one characteristic similarity to the pure mineral. For feldspar there is always one 90° angle giving rise to blocky or splintery particles. Nepheline syenite tends to cleave along parallel planes and the particles appear to be splintery or irregular with flat faces. Both product lines have a minimum of ~0.5 μ particles because of the expense of grinding these tough materials. Figure 7-1 shows photomicrographs of each.

As indicated in Tables 7-1 and 7-2, the Mohs hardness values are relatively high. As the particle size is reduced to below about 5 μ, both feldspar and nepheline syenite reveal a more regular, blocky shape with fewer sharp-edged particles and therefore considerably reduced abrasiveness. These minerals are among the lowest surface area types because of their particle shapes and low fines contents.

The oil absorption test measures the packed volume of filler particles, data for which are included in Tables 7-1 and 7-2. Since very low concentrations of fine particles are present, these fillers provide highly packed systems when combined with fine to very fine fillers and pigments as described in Chapter 2.

3.3 Electrical

Feldspar has somewhat better electrical properties than nepheline syenite, but poorer

Minex 4

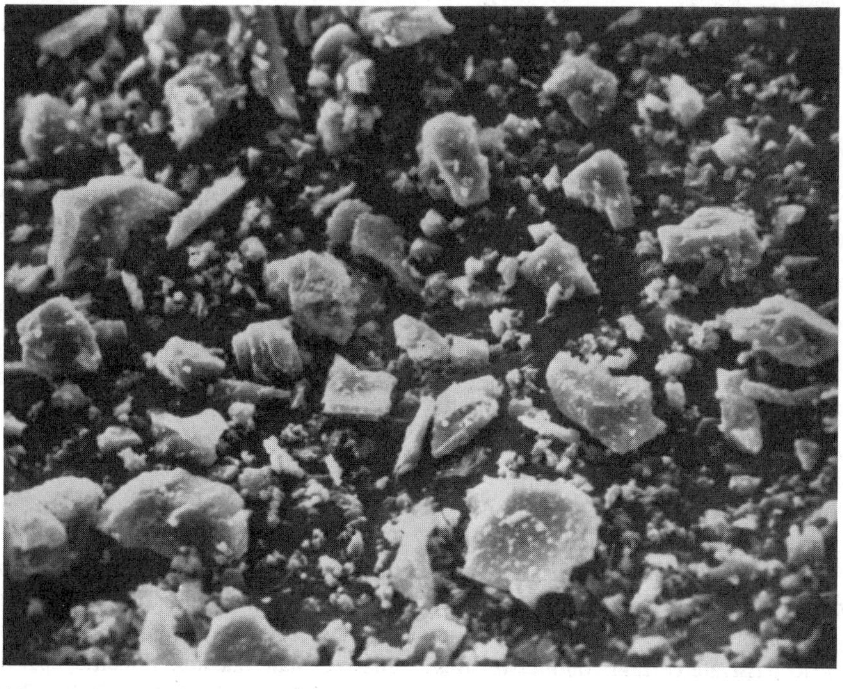

LU-330

Fig. 7-1. Photomicrographs of nepheline (Minex 4) and feldspar (LU-330) at 1300×. (Courtesy *Lawson-United*).

Table 7-1. Properties of Feldspar fillers.[a]

Trade Name	LU-390	LU-340	LU-330	LU-320
Coarseness parameter,				
99.9% finer than, microns	90	40	30	20
+325 mesh, % by wt	2.5	0.05 max.	0	0
Mean particle size, microns	12	9	5	4
Oil absorption, ASTM D 281	16–17	18–19	20–22	22–24
Packed volume, % (from oil absorption data)	68	66	63	61
Surface area, m^2/g	0.8–0.9	1–1.3	1.5–1.9	1.8–2.3
Hegman dispersion value, ASTM D 1210	2	4.5	5.5	6.5
pH, 10% in water	7–7.3	7.5–8	8–8.5	9
Specific resistance, ohm-cm	16,000	14,000	–	–
Properties common to all:				
Specific gravity	2.6			
Free water, 100°C, % max.	0.5			
Refractive index, mean	1.53			
Mohs hardness	6–6.5			
Chemical analysis, % by wt[b]				
Silicon dioxide	67.8			
Aluminum oxide	19.4			
Calcium oxide	1.7			
Sodium oxide	7.0			
Potassium oxide	3.8			
Iron oxide	0.08 max			
Titanium dioxide	tr			
Magnesium oxide	tr			
Manganese oxide	none			
Copper oxide	none			
Ignition loss	0.2			
Chemical formula 0.91 (Na, K)$(AlSi_3O_8)$ 0.09 $Ca(Al_2Si_2O_8)$				

[a]Lawson-United Feldspar and Mineral Co., P.O. Box 309, Spruce Pine, N.C. 28777 704-765-4383
[b]The oxides reported in the analysis are chemically combined and do not exist as free oxides.

than silica. Feldspar is especially suitable where high concentrations of filler are used to obtain surface resistivity. Dielectric strengths for feldspar in epoxy and polyester compounds are equal to or slightly lower than silica. Conductivity bridges on the surfaces of the particles due to sorbed water, crystal defects, or contaminants, such as black iron oxide in nepheline syenite, are responsible for electrical conductivity and are frequently grossly influenced by the polymer and its ability to bond to the filler surface or to passivate certain conductive sites. Thus, pure potash feldspar provides excellent electrical resistance in ceramic compositions where the matrix is a glass and no surface sites exist to provide electrical conductivity. Silicate minerals are non-conductors, so the degree of electrical resistance is largely a matter of purity rather than differences in mineral types.

3.4 Thermal

Thermal conductivity for these minerals is in the range of 5.6–8.2 g-cal./(sec. cm^2)°C/cm compared with values of 0.2–0.5 for plastics, and therefore they are considerably more conductive than plastics. Specific heat values are about 0.21 g-cal./g/°C for both fillers versus about 0.3–0.5 for most plastics. On a volume basis (specific heat × density), the fillers have higher values than plastics.

Table 7-2. Properties of Nepheline Syenite fillers[a]

Trade Name	Minex 2	Minex 3	Minex 4	Minex 7
Coarseness, microns	105	74	44	17
+325 mesh, % by wt	4.2	1.7	0.02	0
Mean particle size, microns	16	14	7.5	4.5
Oil absorption, ASTM D 281	21-22	21-22	22-23	28-29
Packed volume, % (from oil absorption data)	62	62	61	56
Surface area, m^2/g	0.5-0.6	0.8-0.9	0.9-1.0	1.4-1.5
Hegman dispersion value, ASTM D 1210	–	1.5-2	3.5-4	5.5-6
Specific resistance, ohm-cm	–	4,000	3,600	3,300
Properties common to all:				
Specific gravity	2.61			
Refractive index, mean	1.53			
Mohs hardness	5.5-6			
pH, 10% in water	9.9			
Chemical analysis, % by wt[b]				
Silicon dioxide	61.0			
Aluminum oxide	23.3			
Sodium oxide	9.8			
Potassium oxide	4.6			
Calcium oxide	0.7			
Iron oxide	tr			
Magnesium oxide	tr			
Titanium dioxide	none			
Manganese oxide	none			
Copper oxide	none			
Ignition loss	0.6			
Chemical formula (Nepheline) (Na, K)($AlSiO_4$)				

[a]Indusmin Limited, Commerce Court West, P.O. Box 40, Toronto, Canada. U.S. Distributor; American Nepheline Corp., P.O. Box 3766, Columbus, Ohio 43214.
[b]The oxides reported in the analysis are chemically combined and do not exist as free oxides.

3.5 Optical

One of the primary advantages of these fillers is their mean refractive index of 1.53. Since these minerals wet and disperse readily, their surfaces are quite well contacted by the polymer. Most polymers have refractive indexes near 1.53, so filled compounds may obtain clarity or translucency not provided by other fillers. Figure 7-2 illustrates the show-thru (lying on a surface) properties of filled plastisol films. No other fillers provide this degree of light transmittance at equivalent concentrations.

4. SUPPLIERS

Present suppliers are listed in the footnotes of Tables 7-1 and 7-2. A total of about 10,000 tons per year of all products are sold to the plastics trade.

5. COSTS AND AVAILABILITY

5.1 Prices

Current prices are \$35-\$55/ton for the Minex, LU-390, and LU-340 products in 50 lb multiwall bags, FOB plants. Bulk shipments cost \$7-\$8 less per ton.

5.2 Availability

At this writing, the Lawson-United products LU-320 and LU-330 are not available, but may be in the future. Both Indusmin and Lawson-United LU 390 and 340 are available in quantities well beyond present demands. Bulk shipments are made in 100 ton covered hopper cars and 20 ton bulk trucks. Bag shipments are made in 50-ton rail cars and 22-ton trucks directly from the plants with mixed shipments taking the carload prices.

6. APPLICATIONS

6.1 Thermoplastics[1,2]

Compounding. In both dry-blending and liquid mixing, it is preferable to add these fillers last. They do not agglomerate as fine fillers do and they wet and disperse very readily, so adding them last reduces abrasion or over-mixing in liquid dispersions which may result in excessive air entrainment.

Processing. These fillers have been used in all methods of liquid and melt processing. Depending on the section thickness or degree of work involved, all grades may be used. For high-shear melt processes, the finest grades are most suitable (as in injection molding) where minimum abrasiveness is required. Both filler types exert minimum effects on melt-flow or viscosity compared with other fillers having about the same particle size distributions. Table 7-3 presents representative data for a plastisol compound and indicates differences due to particle size distributions.

Physical properties. Feldspar and nepheline syenite produce the same effects on physical properties and are superior to equivalent particle size calcium carbonate fillers (Table 7-4). Abrasion resistance is materially improved by these fillers provided that the polymer bonds adequately to the filler surfaces and has a moderate to high tensile strength. Highly plasticized

Table 7-3. Rheological properties of plastisols.

Formulation:	phr
Tenneco 1755	100
DOP	60
V-1515	3
L-U filler	as indicated

Filler Parts by Weight	None –	LU-390 20	LU-390 50	LU-340 20	LU-340 50	LU-330 20	LU-330 50	LU-320 20	LU-320 50
Brookfield Rheology									
Viscosity @ 1 sec.$^{-1}$, poise;									
1 day	75	72	114	83	158	89	197	102	286
2 days	62	62	116	73	160	77	198	99	265
Yield, dynes/cm^2;									
1 day	0	1	0	1	19	3	44	6	98
2 days	1.4	1	0	2	14	1	27	6	98
Flow index, n;									
1 day	0.97	0.96	0.85	0.90	0.74	0.90	0.74	0.87	0.71
2 days	1.01	0.99	0.90	0.96	0.76	0.90	0.73	0.87	0.73
Thixotropic index B;									
1 day	7	3	3	3	3	4	6	5	9
2 days	6	4	2	3	4	3	4	5	5
Apparent viscosity, poise;									
2.5 rpm–1 day	85	85	153	106	290	117	407	149	677
2 days	72	70	158	89	290	100	391	142	628
20 rpm–1 day	72	71	111	80	145	84	184	98	270
2 days	61	63	112	72	146	75	179	95	256
Severs Viscosity									
100 psi, poise	69	190	573	291	2253	339	2503	496	3754
Shear rate, sec^{-1}	1650	598	103	391	50	336	45	223	30

Data Courtesy *Tenneco Chemicals, Intermediates Div.*

Table 7-4. Comparison of physical properties of vinyls containing equivalent particle size fillers.

Filler, phr	Tensile, psi	Elongation, %
	Semi-rigid[a]	
None	4050	305
LU-340, 50	2900	220
LU-340, 75	2550	190
Limestone, 50	2550	210
Limestone, 75	2180	170
	Flexible[a]	
None	2640	440
LU-340, 50	2200	325
LU-340, 100	1775	275
Limestone, 50	2120	320
Limestone, 100	1600	250

[a]Semi-rigid formulation contained 25 phr DOP, flexible had same formulation with 50 phr DOP.

or weak matrixes (polyethylene, for example) do not have sufficient bonding and tenacity to prevent the filler particles from being dislodged. As with all filled composites, stresses in the matrix are magnified, resulting in reduced physical properties and, particularly, impact resistance. Inherently tough or rigid matrixes are less affected, especially if the finest fillers are used (see Chapter 2). Strain whitening is also more noticeable with these fillers due to the clarity imparted.

Optical effects are illustrated by Fig. 7-2. Similar results are obtained in semi-rigid and rigid compounds. These fillers transmit radiation from 0.3–1.5 μ wavelength. A small proportion of ultraviolet is absorbed and the remainder transmitted, therefore UV absorbers are effective. Infrared is largely transmitted by the fillers so that heating for vacuum forming of filled sheets is more rapid than with the unfilled sheets.

Thermal stability is illustrated by the heat stability tests of Fig. 7-3 where the improvement due to feldspar addition may be seen. Although the improvement is partly due to dilution of the plastisol by the highly purified filler, bonding of the electronegative chlorine at cationic (positive) sites of the filler surface reduces the chlorine mobility and therefore polymer degradation.

Other thermoplastics. As with all fillers, polar polymers such as ABS, EVA, nylon, urethanes, etc., are most amenable to being

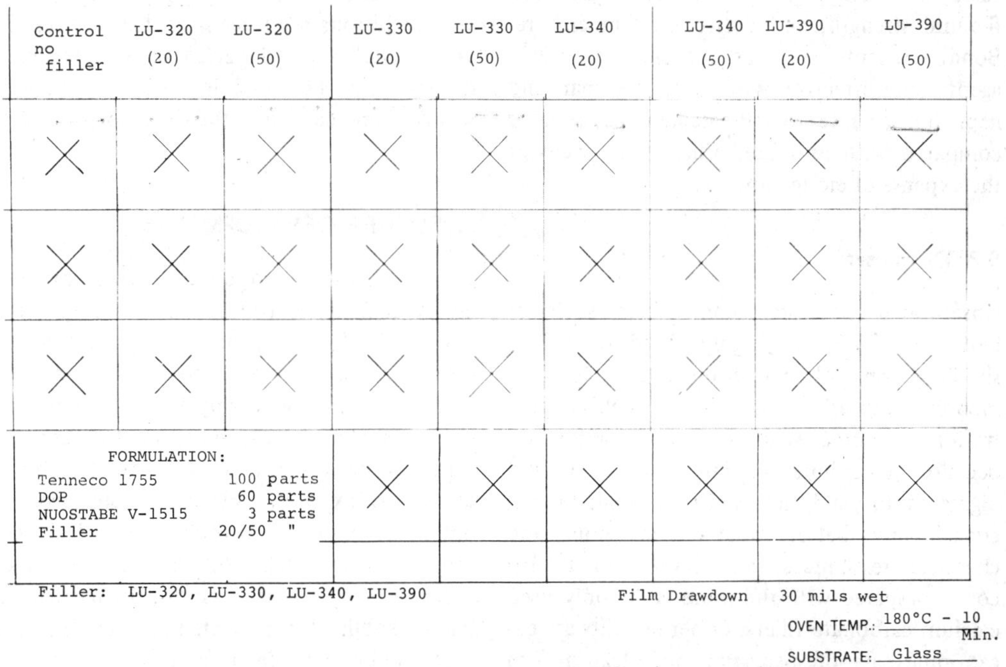

Fig. 7-2. Film clarity chart.

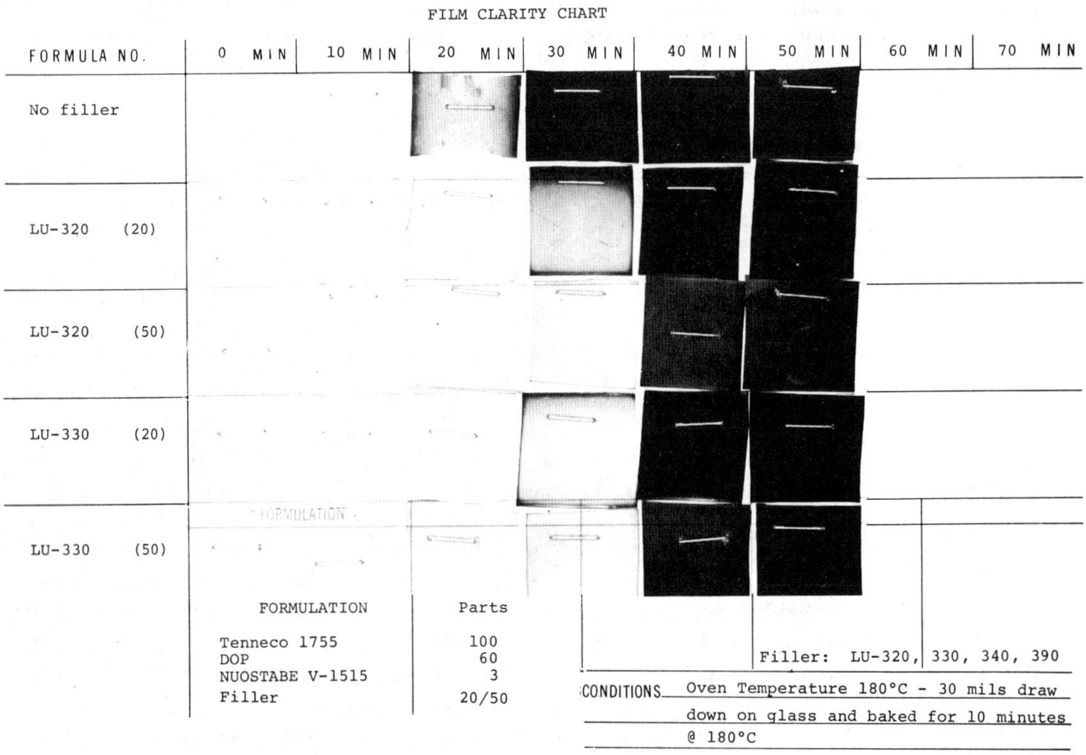

Fig. 7-3. Heat stability tests of filled plastisols.

filled, with resulting improvements in modulus, flexural strength, and deflection temperature. Bonding agents such as the silane coupling agents are effective with both feldspar and nepheline syenite to increase tensile strengths compared with untreated filler, but always at the expense of elongation.

6.2 Thermosets

Unsaturated polyester compounds such as SMC and BMC are highly filled to reduce shrinkage on polymerization and to increase moduli. Both fillers provide the high concentrations required with such compounds, especially in conjunction with finer fillers and pigments, to yield more uniform molded parts, considerably better weather, abrasion, and chemical resistances, and savings in tinting costs compared with the more commonly used calcium carbonate fillers. Other benefits are exceptional ease of dispersion and deairing, no effect on promoter or accelerator activity, excellent flow, molding and mold detail properties, and show-thru clarity where it is desired to see the fibrous reinforcement distribution. The same benefits are imparted to matched metal die molding, sheet molding, gel coats, filament winding, and casting of the filled compounds.

7. FUTURE DEVELOPMENTS

The particular limitation of current products lies with their relative coarseness compared to talc and kaolin. Much finer fillers, finer than 10 microns, will provide considerably improved thermal stability with vinyls, and higher tensile strengths, elongations, and reduced abrasiveness generally in all polymers. Also, these fillers require less expensive silane coupling agent to obtain dramatic improvements in all physical properties since their surface areas are in the order of $1/6$–$1/10$ that of equivalent particle size talc or kaolin. It is possible that such fine products may be available in the future.

Because of the superior properties imparted to composites by these fillers, their use is expected to increase more rapidly than indicated in a recent survey.[3]

References

Producers

Lawson-United Feldspar and Mineral Co., P. O. Box 309, Spruce Pine, N.C. 28777. (704-765-4383)

Indusmin Limited, Commerce Court West, P. O. Box 40, Toronto, Canada. (416-863-7088)

Literature

1. Lawson-United Fillers in Vinyls.
2. Ferrigno, T. H., "The Case of the Fugitive Filler," 22nd ANTEC, SPE, Montreal (March 1966).
3. *Modern Plastics*, p. 59 (October 1975).

8

Silica Fillers, Extenders, and Reinforcements

James E. Moreland
Malvern Minerals Company
Hot Springs, Arkansas

CONTENTS

1. INTRODUCTION

There are at this time as many as 22 distinct phases of silica and each one of the phases is chemically SiO_2. Currently, not all of the phases are useful in plastics but a number of them are. Interestingly, the various phases, and even phase variations, offer distinct advantages and disadvantages. For example, a synthetic colloidal form of silica under the amorphous phase offers great surface area, thixotropic effect, and distinct reinforcing properties. Polymers filled with this form have greatly increased viscosity which may result in processing difficulties. There are flocculated varieties of the colloidal form which give flatting characteristics and composite properties somewhat different from the particulate colloidal phase. Moreover, the synthetic forms just mentioned are so fine in particle size, they offer little in the way of hardness in finished composites. They lack the purity and "anhydrous" characteristics that some of the naturally occurring, low quartz phase varieties offer.

Natural silica products in the low quartz phase come from quartzite and the tripolitic forms, which would generally be termed natural, microcrystalline quartz. These forms of silica are usually less expensive than the synthetic, colloidal forms. They too, have advantages and disadvantages. These low quartz varieties offer much lower surface area, hence better rheological flow. Excellent electrical insulation, heavier loading potential, greater purity, and good mechanical properties are other benefits. The drawbacks would be increased wear of processing equipment due to hardness of this filler, and the lack of reinforcing properties because of lower surface area.

Diatomaceous silica is very high in SiO_2 content. Quantitatively, the silica content will range about 90%. Other constituents comprising the remaining 10% are found under sub-section 7.3. The uncalcined variety contains up to 4% H_2O while calcined varieties have about 0.5%. Uncalcined distomaceous silica has a rather high surface area of 15 m^2/g while the calcined grades measure 3.0 m^2/g. Oil adsorption is quite high, from 90-150%. The effect of diatomaceous silica on processing and composite properties is somewhere between synthetic amorphous silicas and the naturally occurring type. The diatomaceous silica particles have a highly porous structure, and the phase of silica is micro amorphous and opaline silica. (See SEM at end of chapter)

Fused silica is manufactured by subjecting high purity quartzite to temperatures in excess of 4,000°F. Finely divided fused silica has a coefficient of thermal expansion in the range of $0.54 \times 10^{-6}/°C$ from 0-1,000°C. This low thermal expansion is very desirable in composites subjected to wide variations in temperatures.

All of the silica materials discussed are in adequate supply based on current demand. The synthetic, amorphous silica material prices vary widely from 12¢-\$1.60 a pound. The reasons for difference in price range are degrees of fineness and purity. The naturally occurring quartz phase material prices range from 1¢-20¢ a pound. The diatomaceous silica sells at about 10¢ a pound. Fused silica is about 22¢ a pound.

Surface modification via the hydroxylated surface of silica is a subject of wide interest. Silanol activity on the various phases is believed to vary due to origins. However, surfaces can be modified to offer organofunctionality, hydrophobicity, and non-functionality. Commercial activity in surface modification is limited now, but the future will call more and more for functional particulate silica materials with the interface tailored to a specific end use.

A dramatic source of silica of the opaline variety will come from rice hulls. Rice hulls are about 15-25% SiO_2. The hulls have been an environmental burden and waste product over the years. This fascinating source of silica is under careful study and recent developments should provide a variety of silica based products. The lower index of refraction and other unique opaline silica properties from plantlife origin will be serving an unfilled need in the plastics field.

2. SILICA, AMORPHOUS–PYROGENIC OR FUMED

2.1 Abstract

This particulate amorphous silica is colloidal and has a surface area as much as 380 m^2/g.

These products provide thixotropy, antisettling, antiblocking, and true reinforcement. The rather negative results against such contributions are poor flow properties and increased viscosity. Pyrogenic silica is broadly used for specific results in both thermoplastic and thermosetting resins, and elastomers.

2.2 Manufacturing

Pyrogenic silica was developed in 1941 in Germany. It was developed by a Degussa chemist, Dr. Klöefper who was searching for a highly active, light colored reinforcing filler to replace conventional rubber carbon blacks. Pyrogenic silica is manufactured[1] by hydrolysis of silicon tetrachloride in an oxygen-hydrogen flame:

$$2\,H_2 + O_2 \longrightarrow 2\,H_2O$$
$$SiCl_4 + 2\,H_2O \longrightarrow SiO_2 + 4\,HCl$$
$$\overline{2\,H_2 + O_2 + SiCl_4 \longrightarrow SiO_2 + 4\,HCl}$$

A homogeneous mixture of $SiCl_4$ vapor, hydrogen, and air is burned in a cooled combustion chamber. Particle size and, thus, surface area and surface structure can be varied by the $SiCl_4$ concentration, flame temperature, duration of stay of the SiO_2 nuclei, and variations of the amount of inert gas in the flame. After leaving the reaction chamber, the combustion gases containing the entire amount of SiO_2 as an aerosol pass through a coagulation zone–under turbulance, and the SiO_2 flocculates. The *flocks* with a diameter of 1-200 μ, depending on the grade, are separated almost quantitatively from the HCl containing combustion gases by centrifugal force and filtration. In order to remove the residual HCl which adheres to the large surface as a result of the production process, the silica is aftertreated with moist hot air, thus lowering the HCl content to 0.025%. The bulk density of the freshly produced silica (15 g/l) is increased to that of the commercially available grades, i.e., to 60 g/l or 1250 g/l (compressed material) by means of special vacuum equipment.

2.3 Summary of Properties

Aerosil is X-ray amorphous, which is a decisive reason for the fact that it is safe to work with, i.e., the highly dispersed silica does not cause silicosis even under extreme conditions. Many special reports written in Germany, France and the U.S.A. support this statement.[1] In addition, it is not toxic. Despite the fact that pyrogenic silica is physiologically safe, excessive inhalation should be avoided by use of suitable ventilation of the work room or by wearing breathing masks.

The two broad areas of usage for pyrogenic silica are in rubber and plastics. The advantages offered by the use of pyrogenic silica are primarily the following:

- its thickening and thixotropic effect for liquids,
- its effect as an antisedimentation agent,
- its reinforcing effect in elastomers,
- as an antiblocking agent in plastics films,
- and, good dispersibility under many applications.

The silica particles have silanol groups on their surfaces while Si-O-Si bonds are located inside these particles (Fig. 8-1). Development of the silanol groups is attributed to incomplete condensation during the silica process. An SiOH density of 3 Si/OH/100 Å^2 is attained.[1] The SiOH groups on the surface of adjacent particles have a strong tendency to interact with others through hydrogen bonds. As a result, a three dimensional lattice structure composed of silica particles can develop, and this produces an increase in viscosity in liquid system. These relatively weak hydrogen bonds break down as even light shear forces are applied. Aside from increased viscosity and thixotropy, the silanol

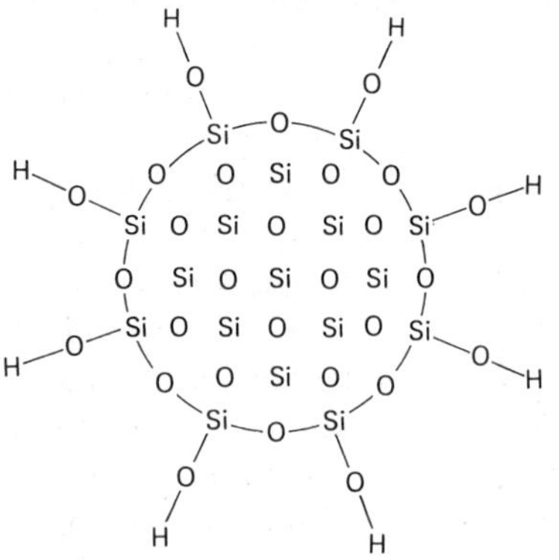

Fig. 8-1. Structure of an aerosil particle.

Table 8-1 Physical-chemical data of pyrogenic silicas.

		AEROSIL					
		130	200	300	380	R 972	MOX 170
Appearance		fluffy white powder					
BET surface area	m^2/g	130±25	200±25	300±30	380±30	120±30	170±30
Average primary particle size	milli-microns	16	12	7	7	16	15
Tamped volume (DIN 53 194)							
standard material	ml/100 g	approx. 1700	approx. 1700	approx. 1700	approx. 1700	approx. 2000	approx. 1700
densed material ("V")	ml/100 g	approx. 1000	approx. 1000	approx. 1000	approx. 1000	approx. 1000	approx. 1000
Moisture (DIN 53 198, method A) (2 hours at 105 °C/220 °F)	%	< 1.5	< 1.5	< 1.5	< 1.5	< 0.5	< 1.5
Ignition loss*) (DIN 52 911) (2 hours at 1000 °C/1800 °F)	%	< 1	< 1	< 2	< 2.5	< 2	< 1
pH-value (DIN 53 200) (4% aqueous suspension)		3.6–4.3	3.6–4.3	3.6–4.3	3.6–4.3	3.5–4.1**)	3.6–4.2
SiO_2*	%	>99.8	>99.8	>99.8	>99.8	>98.3	>98.3
Al_2O_3	%	< 0.05	< 0.05	< 0.05	< 0.05	< 0.05	0.3–1.3
Fe_2O_3	%	< 0.003	< 0.003	< 0.003	< 0.003	< 0.01	< 0.01
TiO_2	%	< 0.03	< 0.03	< 0.03	< 0.03	< 0.03	< 0.03
HCl	%	< 0.025	< 0.025	< 0.025	< 0.025	< 0.05	< 0.025
Sieve residue by Mocker (DIN 53 580)	%	< 0.05	< 0.05	< 0.05	< 0.05	–	< 0.1
X-ray structure		amorphous					
Packaging (net)							
normal material	kg	10	10	10	10	5	5
densed material	kg	20	20	20	20	10	12.5

* of material dried for 2 hours at 105 °C (220 °F)
** in 1 : 1 mixture of water and acetone or methanol

groups are reactive sites for surface treatment. Silanes are used to render silica hydrophobic, and are conceivably organofunctional via ambifunctional silane coupling agents.

Depending on the surface area, the average particle diameters of pyrogenic silica determined by electron micrographs and X-ray diffraction vary between 30 to 100 Å. Surface area ranges from 50 to 380 [m^2/g] BET surface.[1]

A typical line of pyrogenic silica products as furnished by Degussa is shown in Table 8-1.

2.4 Material Suppliers, Types, Costs and Availability

Table 8-2 shows producers of pyrogenic silica.

Table 8-2.

Manufacturer	Product Name/s	Particle Size Information	Remarks	Location of Source
Degussa	Aerosil® 130	7–16 Millimicron average particle size	Lower surface area	Frankfurt/Main West Germany and New York, New York, USA
X	X 200	X X X X X	Median X X	
X	X 300	X X X X X	Greater X X	
X	X 380	X X X X X	X X X	
X	X 972	X X X X X	Hydrophobic	
Cabot Corp.	Cab O Sil® LM5	X X X X X	Lower surface area	S. Boston, MA, USA
X X	X X X M-5	X X X X X	Greater X X	X X X X
X X	X X X EH-5	X X X X X	X X X	X X X X
Kalusch	Information and products similar to that listed above.			Soviet Union
Nippon Aerosil	X X X X X X X X			Japan

Pyrogenic silicas range in price from $1.13 to $1.60/lb. It is reported (Degussa) that availability is suitable for current demand.

2.5 Future Costs and Availability

At this time major producers are expanding production capabilities in the United States and Europe. Supply for normal growth should be adequate.

2.6 Applications

2.6.1 Rheology. In various systems thickened with pyrogenic silica, the addition of small amounts of certain additives produces a considerable increase in viscosity and thixotropy. The most pronounced viscosity increase is obtained with highly polar substances, such as ethylene glycol or glycerine (Fig. 8-2). In most cases, the rheological properties of systems containing pyrogenic silica and additives will remain unchanged even after prolonged storage.

2.6.2 Elastomers. Pyrogenic silica is used in hot-vulcanized silicone rubber primarily for reinforcing, i.e., for improving the mechanical properties. Contrary to carbon black and coarser white fillers, these silicas permit the manufacture of translucent and transparent silicone rubber. Excellent electrical properties are maintained, even at high temperatures. Pyrogenic silica is also highly effective as a reinforcing filler of natural and synthetic rubber. Its use can result in improved mechanical properties such as tensile and tear strength, translucency and transparency, and high abrasion resistance.

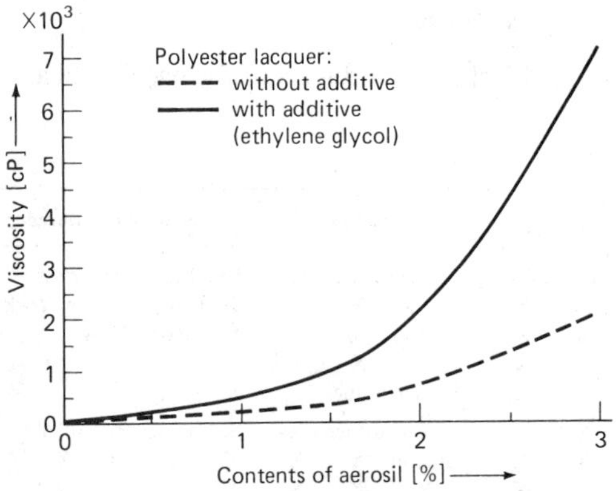

Fig. 8-2. Effect of glycol additive on viscosity of polyester aerosil system.

2.6.3 Epoxy. In liquid epoxy systems, the thickening effect of pyrogenic silica results from the interaction among silanol groups (SiOH) on the surface of adjoining particles, and between these silanol groups and the amino and imino groups of hardeners. The silica particles form a three-dimensional structure in such systems. Their use is mainly for control of rheological flow behavior while maintaining the excellent electrical and physical properties of the epoxy system.

2.6.4 PVC (Polyvinyl Chloride). In the PVC sector, pyrogenic silica is used to modify rheological properties, as in other systems. To lessen thickening behavior, treated hydrophobic pyrogenic silica has the function of reducing the rate of viscosity increase. Treatment involves changing the surface from the abundance of silanol groups (SiOH) to particles terminated in single methyl (CH_3) or other organic groups per site.

2.7 Future Uses

At this time it is difficult to foresee dramatic new fields of application. Cost demands performance and these materials perform as excellent thixotropic fillers in the plastics field. Further work may offer promise in broadening dry-flow behavior in fine particle technology.

2.8 Significant Literature

Significant articles regarding theoretical and applied technology are abundant. For example, the publication Aerosil® Ra 42-0-5-874K by Degussa contains 118 references to works covering pertinent phases of manufacturing, properties and application. Other pertinent bulletins by Degussa are the Technical Bulletins, *Pigments for Plastics* (#13), *Epoxy Resins* (#27), and *PVC* (#51).

3. SILICA, AMORPHOUS–PRECIPITATED (HYDRATED) AND SILICA GEL

3.1 Abstract

Precipitated silica is colloidal and produced by a chemical reaction in a water solution. The ultimate fineness is about .002 μ. Silica gel is a porous, particle type amorphous silica formed as a result of the chemical reaction of sodium silicate and a mineral acid. The particle size range is between 2.0 and 25.0 μ. The surface area of the precipitated silica is as much as 143 m^2/g. The silica gel from 175 to 800 m^2/g. The performance of these amorphous silicas is similar to pyrogenic silicas; however, the usage is controlled by price while the performance is controlled by purity, particle size, and the geometry of the particles.

3.2 Manufacturing

Precipitated amorphous silica is classed as wet-process, hydrated silica because it is produced by a chemical reaction in a water solution. From this solution it is precipitated as an ultra-fine, spherical particle. (Fig. 8-3.)

Silica gel is an amorphous silica formed as a result of the chemical reaction of sodium silicate and a mineral acid such as sulphuric. The colloidal silica sol is polymerized further to silica hydrogel. The hydrogel is washed free of reaction salts and controlled processing produces either an aerogel or xerogel. This gel can be reduced if desired to very small particle size by various means. The final product is an inert, high purity, porous material. Control is exercised over pH, pore volume, pore size and particle size. Surface treatments, both

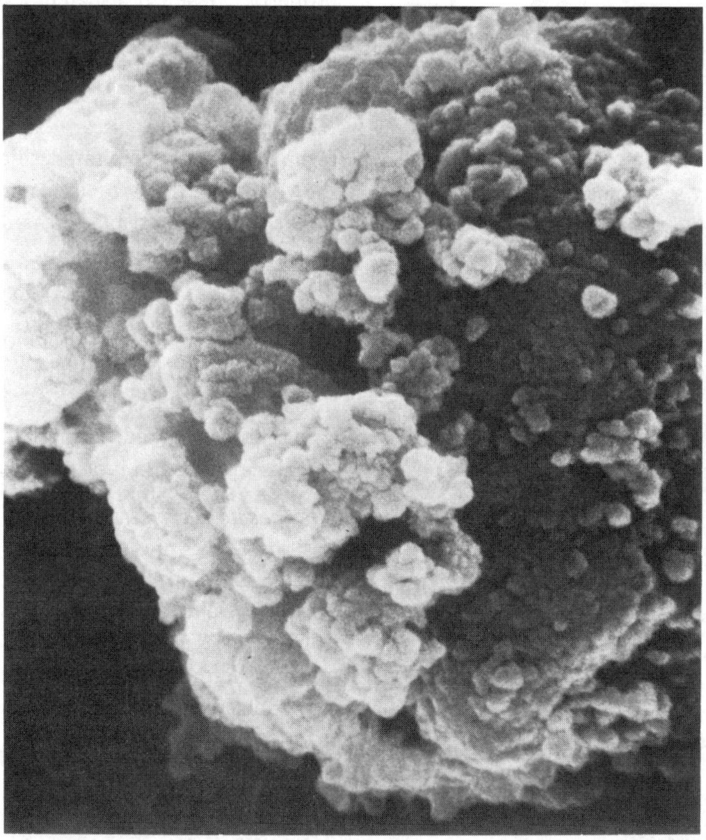

LOW DENSITY SYLOID 244 26,000 X 1 MICRON

Fig. 8-3.

internal and external, can be provided for special applications.

3.3 Summary of Properties

Table 8-3. Chemical and physical properties of precipitated and silica gel.

	Precipitated (HiSil® 200 Series)	Silica Gel (SYLOID®)
SiO_2	94.0%	99.5%
Al_2O_3		.06%
NaCl	1.7%	
CaO	.63%	.08%
Cu and Mn	.003%	
Moisture	5.3%	.01-.04%
TiO_2		.03%
Na_2O		.06%
MgO		.05%
Trace elements as oxides		.02%
Surface area m^2/g	143	175-800
Particle size	0.002 μ (ultimate	2.0-25.0 μ (average
Refractive index	1.46	1.46
Specific gravity	2.05	2.1
pH	7.0	2.0-8.0
Absorption g/100 g	164 (Dibutyl phthalate)	50-310 oil (same as g/100 g)
Description	Agglomerated loose structure	Finely divided granular powder
	Other Information	
Stability	Basically inert	Basically inert
Solubility		Insoluble except in HF and strong bases
Smell and taste		Odorless and tasteless

HiSil, product of PPG Industries.
Syloid, product of W. R. Grace & Co., Davidson Chemical Division.

3.4 Material Suppliers

See Table 8-4.

3.5 Cost and Availability

Price Structure–Precipitated Silica. The current price range of precipitated silica is 12¢–13.5¢ a pound in bulk. Add 3-4¢/lb for bagged material.

Price Structure-Silica Gel. Material is available in fiber drums or paper bags. Powders range in price from 60¢-$1.00 a pound. Granules range in price from 50¢-70¢ a pound.

Availability–Precipitated Silica. Availability of precipitated silica is described as good.

Availability–Silica Gel. Supply of silica gel worldwide is projected to be adequate.

3.6 Applications

Thermoplastics–Precipitated Silica. Precipitated silica is added to vinyl plastics as a release agent to smooth out calendered sheets. In plastisols used to coat textiles, these silica materials impart mar-resistance and flatting. Also, in PVC, hardness and stiffness are increased without detracting significantly from other physical properties. When transparency is desirable, precipitated silicas can be added successfully when other denser and coarser materials would interfere with clarity.

Table 8-5 below[2] is an example of the performance of a precipitated silica (HiSil® 233) showing the effects of different fillers on hot flow, abrasion resistance, hardness, and cold break in a PVC resin system.

Thermoplastics–Silica Gel. Contributions to thermoplastics include plate-out prevention, anti-blocking of film, pigment dispersion aid, flow control of resin powders, viscosity control agent, processing aid, intercoat adhesion, selective adsorption, and moisture removal.[3]

Thermosets–Precipitated Silica. Phenolic compositions that are used in the fabrication of brake linings utilize precipitated silica to improve heat and wear resistant properties. Some of the processing aid characteristics covered under Thermoplastics–Precipitated Silica are also applicable in thermoset usage.

Thermoset–Silica Gel. Same as Thermoplastics–Silica Gel.

3.7 Future Uses

Precipitated Silica. The price range and versatility of these silica materials qualifies them for

Table 8-4.

Manufacturer	Product Name/s	Particle Size Information	Remarks	Location of Source
		PRECIPITATED		
PPG Industries	Hi-Sil® 210	Surface Area 140 to 160 m^2/g	Pelletized	Pittsburgh, Pa., U.S.A.
" "	" 215	" " " " " "	Compressed powder	" " "
" "	" 233	" " " " " "	Powder	" " "
		SILICA GEL		
Davidson Division	Syloid® 72	Average Particle Size 4 microns	Density 11 lbs/ft^3	Baltimore, Md., U.S.A.
W. R. Grace Co.	X 244	X X X 4 microns	X 7 X	X X X
	(NOTE: There are a number of other Syloid® grades other than those listed above. Average particle size ranges from 2.0 to 25.0 microns).			
Glidden Durkee SCM	Information similar to that listed above.			Baltimore, Md., U.S.A.
J. M. Huber Company	X	X X X X X		Havre de Grace, Md., U.S.A.
Monsanto Corp.	X	X X X X X		St. Louis, Mo.

broader usage. The supply is seen as adequate and costs will increase with inflationary trends.[4] PPG Industries is looking forward to plastic alloy compounds such as HiSil®, Plastic Alloy Compounds designed as thermoset silica–reinforced blends of ethylene vinylacetate resins with modifying resins.

Silica Gel. Markets for these products will not change much in the future in the plastics arena. Supply is foreseen as adequate.

3.8 Significant Literature

Precipitated Silica

PPG Industries	HiSil® 200 Series Silicas	Bulletin 85A
PPG Industries	Silene Reinforcing Pigment	Bulletin 86A
PPG Industries	HiSil® Plastic Alloy Compounds	Bulletin 150-G-44
PPG Industries	HiSil® Reinforcing Fillers for PVC	Bulletin 150-G-47

Silica Gel

W. R. Grace & Co.	Anti-Block Agents for Plastic Films	Bulletin PA 90-F
W. R. Grace & Co.	Prevention of Plate-out in Plasticized Vinyls	Bulletin PA 87
W. R. Grace & Co.	Prevention of Plate-out in Rigid Vinyls	Bulletin PA 90-E

Table 8-5. Effects of filler type on PVC hot flow, abrasion resistance, hardness, and cold break.

Base Formula	
PVC Resin	100 parts
Primary Plasticizer (dioctyl phthalate)	80 parts
Secondary Plasticizer	11 parts
Hexanetriol Stabilizer	10 parts
Filler	60 parts

Identification	Hot Flow, % at 70°C	Relative Abrasion Resistance, %	Hardness, Shore D	Cold Break, °C
Blank	22	100	13	-60
Hi-Sil 233	3	41	34	-40
Hi-Sil EP	6	34	28	-30
Hi-Sil 422	8	38	24	-40
Ground Whiting	12	32	21	-50
Precipitated Calcium Carbonate	11	35	21	-45

Acker, E. G., "The Characterization of Acid-Set Silica Hydrosols, Hydrogels, and Dried Gel," *J. Colloid and Interface Science*, 32, (1): 41, Jan. 1970.

4. SILICA, LOW QUARTZ, PULVERIZED AND MICROCRYSTALLINE

4.1 Abstract

Pulverized Silica (Quartzite)–High purity and broadly classified ground or pulverized silica are available from several producers. These products are usually processed from pure quartz sands. The ore in some instances is a loose sand while other ore sources are quartzite sandstone wherein the rounded, spherical particles are held together with a siliceous cement. These materials are available in a very broad range of particles ranging from +150 μ to 5 μ and smaller.

4.2 Manufacturing

Ground quartzite is mined from very vast deposits which abound in many geographical areas worldwide. The portion that is used as fillers would be approximately 9.7% of the total production.[5] This 9.7% covers both plastics and coatings and was equal to 130,950 tons in 1968.[5] The mining operations are huge, involving railroads or rubber-tired trucks capable of carrying in excess of 100 tons a load. The ore is crushed by primary crushers, and taken to settling tanks to remove foreign matter. It is then dried, pulverized in tube mills, and either screened or air floated into the many grades.

4.3 Summary of Properties

The following data are quite typical of ground quartzite (Table 8-6). The table below is from published data.[6] The table covers only four commercial grades and in this instance these are finer grades. Coarser grades are available.

4.4 Material Suppliers and Products Information

See Table 8-7, page 147.

4.5 Costs and Availability

There are two size categories of ground quartzite. One group of products is 44 μ (325 mesh) and coarser. Most producers have these grades available and they range in price from 01¢–02¢/lb. The other group is 44 μ (325 mesh) and finer. These are more specialized products requiring sophistication in plant design plus special equipment to "super" classify. This entails closer quality control, additional personnel, and lower production. These requirements cause prices to range from 03¢–06¢/lb, generally.

The coarser varieties of ground quartzite are in abundant supply and should remain so. The finer varieties (especially in the 10 μ and finer grades) do get in short supply periodically. Overall, finely divided quartzite is plentiful and prices should not fluctuate broadly.

4.6 Applications

4.6.1 Thermoplastics. The hardness of ground quartzite (7 Mohs) limits its use in typical thermoplastics, typical meaning the lower priced, high volume types. The hardness of the quartzite makes these materials brittle, which is usually undesirable. In the engineering type of thermoplastic materials, ground quartzite can sometimes be used. These finely divided silicas along with ambifunctional coupling agents actually enhance some properties, for example in nylon. Hardness however, causes an additional concern in the wear of the mixing and extrusion equipment. This problem can be eliminated through the use of new, hardened material in processing equipment.

4.6.2 Thermosets. This area of plastics can utilize the good mechanical and electrical properties of ground quartzite. Thermosets are cured basically by chemical reaction. Elevated temperature and pressure are used to shorten cycle times. Utilization of this variety of silica in thermoset materials imparts dimensional stability, good electrical insulation properties, greater thermal conductivity, and reduction of cost in finished composites.

Table 8-6. Typical physical and specific properties on Min U Sil®/s.

	STANDARD GRADES			
Specific Properties	Min-U-Sil 5	Min-U-Sil 10	Min-U-Sil 15	Min-U-Sil 30
Surface area, m^2/g	2.06	1.1	0.84	0.54
Avg. particle size, μ	1.9	3.5	4.0	8.8
Fineness of grind–Hegman (ASTM D1210)	7½	7	6½	5–5½
Grade bulk density, lb/gal.	5.442	6.123	6.926	8.423
Apparent bulk density, lb/ft^3	40.7	44.8	51.8	63
Oil absorption, lb of oil/100 lb Min-U-Sil (Gardner-Coleman ASTM D1483)	39.5	36.2	31.5	26.9
Oil absorption, lbs of oil/100 lb Min-U-Sil (Spatula Rub-Out ASTM 281)	36.2	32	29.3	24.1
Water absorption, lb of water/100 lbs Min-U-Sil (Spatula Rub-Out Basis ASTM 281)	43.5	40	35.5	32.5
Dry color (green filter reflectance)	90	91.5	91	90.5
Moisture content	← 0.05% maximum →			
Packaging	← 50 lb 4-ply multiwall bags →			

Typical chemical analysis of Min-U-Sil 30

Silicon dioxide (SiO_2)	99.46%	Specific gravity	2.650
Iron oxide (Fe_2O_3)	0.048%	Refractive index	1.547
Aluminum oxide (Al_2O_3)	0.210%	pH	7.0
Titanium dioxide (TiO_2)	0.027%	Porosity	Non-porous
Calcium oxide (CaO)	0.021%	Bulking value	22.144 lb/gal.
Magnesium oxide (MgO)	Trace		0.04522 gal./lb
Loss on ignition at 1200°F	0.25		

4.6.3 Elastomers. The use in elastomers is restricted almost totally to silicone rubber products. In RTV and silicone gum stocks, the finest particle size grades are used as semi-reinforcing extender fillers. They reduce costs and preserve good electrical insulation values.

4.7 Future Uses

Ground or pulverized quartzite will continue to be used where its characteristics are desirable. The hardness and relatively coarse size can cause abrasion of processing equipment and limit its use. The advent of new and harder metal surfaces in mixing equipment could broaden the acceptance of this filler. The superfine grades (10 μ and smaller top size) will enjoy steady growth.

5. SILICA, LOW QUARTZ, NATURAL MICROCRYSTALLINE "AMORPHOUS"

5.1 Abstract

Natural Microcrystalline, "Amorphous"–The words "natural microcrystalline" define this

Table 8-7.

Manufacturer	Product Name/s	Particle Size Information	Remarks	Location of Source
Pennsylvania Glass Sand Corp.	30 μ Min-μ-Sil®	Typical maximum particle Size 30 μ	Super classified ground quartzite.	Pittsburgh, Pa., USA
X X X X	15 μ X	X X X X 15 μ	X X X X	X X X
X X X X	10 μ X	X X X X 10 μ	X X X X	X X X
X X X X	5 μ X	X X X X 5 μ	X X X X	X X X
X X X X	Supersil®	#140, #160, #200, and #325 Mesh	Standard ground, classified quartzite.	X X X
Ottawa Silica Co.	Microsil® Sil-co-Sil®	#285, #290, #295, #390, #395 and #398	X X X X	Ottawa, IL., USA
Quarzwerke GmBH	Sikron® H-100, H-200, H-300 and H-500. Quarzfeinsmehl–mostly 325 mesh fineness with sub-sieve variations.		Standard ground, air classified quartzite.	5 Cologne, W. Germany
X X –	Sikron® F-300, F-500 and F-600. Quarzfeinsmehl–typical 30 μ, 15 μ, and 10 μ maximum particle size.		Super classified ground quartzite.	X X X
X X	Quarzmehl w-1, w-3, w-4, w-6, w-8, w-10, and w-12–Seven typical maximum particle size 50 mesh to 100 mesh (U.S. Sieve)		Thought to be screened quartzite.	
Wedron Silica	–		–	400 W. Higgins Rd. Park Ridge, IL 60068

occurrence of natural silica, that is, the ore is a natural, sub-divided particulate mineral in the earth. The geologic term TRIPOLI covers these silicas as found in Southern Illinois, Southwest Missouri, Northeastern Oklahoma, and Arkansas. The minerals *do differ* in characteristics, but are broadly defined as tripolitic. Several producers have referred to the Southern Illinois silica as "amorphous." The term amorphous for this mineral is misleading and is a misnomer.[6] The term cryptocrystalline quartz would more accurately define this mineral. Cryptocrystalline refers to a crystalline structure so fine that no distinct particles are recognizable even under microscopic examination. A photomicrograph (1765X) will show hexagonal structures. These structures could be further described as a bound and free state, subdivided. The bound state of subdivision describes the condition of agglomerated clusters held together by cohesive forces but still made up of particulates. The free state describes the individual particulates free and clear of the agglomerates. Moreover, data show this tripolitic silica to have a specific gravity and index of refraction of low or alpha quartz.

5.2 Manufacturing

The tripolitic ore is mined by room and pillar method. The ore comes from the natural deposits in large chunks. It is then ground and pulverized. Both typical and sophisticated types of machinery are used to separate this material into many grades of fineness. The range varies from +74 μ (200 mesh) to 10 μ (1250 mesh). Total tripoli production in the United States is generally in a range from 70,000–100,000 tons per year. The Southern Illinois type of tripoli comprises in the neighborhood of 50% of the total.

5.3 Summary of Properties

Below are general characteristics of this type of silica.

Low cost	Lower abrasive coefficient	Good suspension
Neutral pH	Low moisture content	Good dispersion

Table 8-8. Typical specifications of different grades of Imsil®.

Chemical analyses and physical data contained herein are representative of average samples for each grade.

Physical Characteristics (Typical)	AIR FLOATED GRADES						MICRONIZED GRADES		
	200	250	54	1160	1240	0	Imsil® A-25	Imsil® A-15	Imsil® A-10
Oil absorption (Gardner-Coleman)	29–31	29–31	29–31	29–31	29–31	29–31	29–31	29–31	29–31
Specific gravity	2.65	2.65	2.65	2.65	2.65	2.65	2.65	2.65	2.65
Weight per solid gallon, lb.	22.07	22.07	22.07	22.07	22.07	22.07	22.07	22.07	22.07
Bulking value	0.04531	0.04531	0.04531	0.04531	0.04531	0.04531	0.04531	0.04531	0.04531
Apparent density (Scott volumeter) lb./cu ft	32–34	31–33	30–32	29–31	28–30	27–29	24–26	22–24	21–23
pH value	7	7	7	7	7	7	7	7	7
Moisture limit at 105°C	0.25%	0.25%	0.25%	0.25%	0.25%	0.25%	0.25%	0.25%	0.25%
Refractive index	1.54 to 1.55	1.54 to 1.55	1.54 to 1.55	1.54 to 1.55	1.54 to 1.55	1.54 to 1.55	1.54 to 1.55	1.54 to 1.55	1.54 to 1.55
Hardness (Moh's scale)	6.5	6.5	6.5	6.5	6.5	6.5	6.5	6.5	6.5
Specific resistance (ohms)	25,700	25,700	25,700	25,700	25,700	25,700	25,700	25,700	25,700
Molecular weight	60.09	60.09	60.09	60.09	60.09	60.09	60.09	60.09	60.09
Melting point	3100°F (1722°C)	3100°F (1722°C)	3100°F (1722°C)	3100°F (1722°C)	3100°F (1722°C)	3100°F (1722°C)	3100°F (1722°C)	3100°F (1722°C)	3100°F (1722°C)
Hegman gauge (fineness of grind)				3+	4+	4.5+	5+	6+	7+
200 mesh Tyler screen (thru)	90/95%	96/99%	99.58%	99.98%	99.99%	99.99%	100.0%	100.0%	100.0%
325 mesh Tyler screen (thru)	75/79%	80/87%	90/95%	96/98%	98/99.4%	99.5+%	99.99%	100.0%	100.0%
400 mesh Tyler screen (thru)			84.58%	92.65%	95.5%	98.92%	99.96%	99.99%	99.99%
Av. part. size (Fisher sub sieve)			3.00	2.95	2.82	2.35	1.97	1.82	1.55
Specific surface area, cm^2/g			7,547	7,673	8,029	9,635	11,493	12,440	14,607
G. E. Brightness									
Wavelength 457 mμ	85%	86%	86.5%	87.5%	86.0%	87.75%	90.5%	89.5%	88.5%
Particle Size Distribution*									
40 μ diameter, below				92%	94%	95.5%	99.0%	100.0%	100.0%
20 μ diameter, below				69%	74%	82.0%	96.0%	100.0%	100.0%
15 μ diameter, below				58%	63%	73.0%	90.0%	99.0%	100.0%
10 μ diameter, below				46%	50%	61.0%	77.0%	96.0%	99.0%
7.5 μ diameter, below				39%	42%	58.0%	66.0%	87.0%	91.0%
5.0 μ diameter, below				29%	31%	46.0%	51.0%	70.0%	76.0%
*Andreason Pipette ± 1% accuracy									
Specifications for shipment									
Screen (thru)	200 Mesh 90/95%	200 Mesh 96/99%	325 Mesh 90/95%	325 Mesh 97 ± 1%	325 Mesh 98/99.4%	325 Mesh 99.5 + %	400 Mesh 99.9 + %	<15 Microns 99.0%	<10 Microns 99.0%
Moisture	0.25% Max.	0.25% Max.	0.25% Max.	0.25% Max.	0.25% Max.	0.25% Max.	0.25% Max.	0.25% Max.	0.25% Max.

Low soluble salts	Stable	High loading
Purity	High GE brightness	

In Table 8-8 you can review the typical data published by Illinois Minerals Company of Cairo, Illinois.

5.4 Material Suppliers and Product Information

See Table 8-9, below.

5.5 Cost and Availability

Cost of the Southern Illinois silica products ranges from 1.5¢–5¢/lb. Fineness usually controls price. These tripolitic minerals are vast in their occurrence and reserves will be adequate for years to come.

5.6 Applications

This filler has been used extensively in adhesives and polyurethanes, epoxy, polyester resins, and silicone rubber. The applications of these products are primarily the same as for the ground quartzite products. The tripolitic, microcrystalline silicas from Southern Illinois are perhaps more versatile. If the particle size is comparable, the tripoli materials are lower in abrasive coefficient. Rheologically, they offer more in thixotropy than the pulverized types. The oil adsorption is quite high and resin demand for wet-out is higher. The purity is high and the electrical properties are generally equal to that of other microcrystalline quartz. Their small particle size makes them suitable as a semi-reinforcing material for elastomers.

There is a unique feature characteristic of tripolitic silica particulates that is very attractive to R&D. These materials are unique from the standpoint that they contain 30–40% of individual sub-micron non-agglomerated particles, which potentially could be separated as a commercial product. A low quartz phase product of this type, a free-state subdivision in the sub-micron range would be a truly reinforcing product due to its high surface area.

5.7 Future Uses

Microcrystalline silica from Southern Illinois will find greater demand. They are reasonably priced and prices should not fluctuate greatly. The advent of increased use of chemicals for improving the particle-resin interface should increase usage even more.

6. SILICA, LOW QUARTZ, NATURAL MICROCRYSTALLINE, NOVACULITE

6.1 Abstract

This microform of silica, or quartz, is also a tripoli but coming from the Novaculite Uplift in the West Central part of Arkansas. This is a

Table 8-9.

Manufacturer	Product Names	Particle Size Information	Remarks	Location of Source
Illinois Minerals Co.	#200 Insil®	90/95% Passing 200 mesh (74 μ)	Air Floated	Cairo, Il., U.S.A.
X X	#250 X	96/99% X X X	X X	X X X
X X	#1160 X	96/98% X 325 X (44 μ)	X X	X X X
X X	#1240 X	98/99% X X X X	X X	X X X
X X	#0 X	99.5% X X X X	X X	X X X
X X	A-25 X	99.9% X 400 X (37 μ)	Micronized	X X X
X X	A-15 X	99.0% X 15 μ	X X	X X X
X X	A-10 X	99.0% X 10 μ	X X	X X X
Tammsco, Inc.	Velveteen R	96/98% Passing 325 mesh (44 μ)		Tamms, Il., U.S.A.
	Gold Bond R	98/99% X X X		X X X
	00 Silica Smoke	99.5% X X X		X X X
	S Micron	99.99% X X X		X X X
	Neosil A	99.99% X 20 μ		X X X
	Neosil XV	99.5% X 15 μ		X X X

true mountain range topped off by three strata of novaculite. The microcrystalline novaculite is the upper stratum and in the case of the material which is used in the plastics industry, the stratum is actually overturned and on the bottom. The inverted formation has protected the purity over the eons of time. It is a true natural microform as the ore, to a large extent, is less than 10 μ. The purity, year in and year out, is superior to most other silicas. It is almost void of alkaline earths and is superior in electronics applications. Southern Illinois tripolis are made up of finer, basic particulates but are more agglomerated, while the novaculite ore is more of a free-state subdivided material.

6.2 Manufacturing

The novaculite ore, as mentioned, is a fine white powder upon reaching the mill. There are large chunks in the powder. These chunks give body and momentum in propelling the feed. If the ore feed were all powder, the mass would not go through the machinery. Hammer mills

Table 8-10. Typical properties and characteristics of NOVACITE®.

NOVACITE®
Physical and Chemical Data

Typical Chemical Composition

SiO_2	99.12%
Fe_2O_3	0.04%
TiO_2	0.015%
CaO	0.0%
MgO	0.0%
Loss	0.20%
Al_2O_3	0.61%

SPECIFIC GRAVITY
True specific gravity at 70°F 2.650

HARDNESS

Mohs Scale	7.0
Knoop Scale	820

PRESENCE OF H2O
3 hr. @ 110°C 0.0%
30 min. @ 1000°C 0.0%

INDEX OF REFRACTION
1.550

BULK DENSITY:

Loose Packed–lb./cu ft	52.97
Dense Packed–lb./cu ft	82.75
lb./gal.	11.090

TRUE DENSITY
22.07 lbs./gal.

MOLE SiO2–60 g

MOLE to VOLUME RATIO–22.6 ml.

Approximate Packing Density Functional Surface Sites
1.2×10^{19}/g NOVACITE®

Silane Coupling Receptability–Excellent at very low concentrations.

Transmissibility for Radiations–Very transparent to the ultraviolet and to the visible spectrum, but opaque for the infrared beyond 7.0 μ.

OIL ABSORPTION
a. 20 lb raw linseed oil to wet exactly 100 lb.
b. Spatula rub-out method 17 to 20% from coarse to finest respectively.

WATER ABSORPTION
100 g adsorb 8.7 g water

SURFACE
Highly polar

ACIDITY
ph in distilled water 6.0–6.3
Acid Number 0-(-0–18)

THERMAL STABILITY
Stable from absolute 0°–573°C.

HIGH-LOW INVERSION

On heating	573.7°C
On cooling	573.8°C

SPECIFIC HEAT
Mean specific heat between 0 and 200°C
192 cal/(g)/(C°)

HEGMAN FINENESS GAUGE
L-207A NOVACITE®–7½

SORPTION OF AQUEOUS PLUTONIUM
30%/100 ppm Novacite®

MICROSCOPIC DESCRIPTION–Small magnification–well rounded crystalline discs. Small proportion agglomerate. High Magnification–(2000) All shapes, finely divided quartz containing many minute cracks and fissures.

CHARACTERISTICS

1. Rheologically superior.
2. Low oil adsorption.
3. Defloculated dispersion in applied systems.
4. Unsurpassed electrical insulation properties.
5. Low abrasive coefficient compared to other naturally occurring silicas.
6. Highly reactive silanol surface.
7. Purity.
8. Chemically inert.
9. Low long-term water pick up.
10. Anhydrous.
11. Receptive to ambifunctional silane coupling.
12. No ionic contamination.

Table 8-10. *(Continued)*

"Per Cent Finer Than Sieve Opening" NOVACITE® GRADES

Micron Diameter	U.S. Sieve Series Number	200	325	625	1250	Daper	L 207A	L 337	L 338
74	200	98.12	99.97	99.04	100.00	100.00	100.00	100.00	100.00
53	270	92.14	99.50	99.88	100.00	100.00	100.00	100.00	100.00
44	325	82.82	97.98	99.08	99.99	100.00	100.00	100.00	100.00
37	400	76.07	93.07	98.88	99.92	100.00	100.00	100.00	100.00
30	475	66.90	82.20	92.80	99.10	100.00	100.00	100.00	100.00
20	625	56.20	67.10	81.10	92.00	97.70	100.00	100.00	100.00
15	950	48.30	57.50	74.00	86.30	95.80	100.00	100.00	100.00
10	1250	34.60	36.10	50.80	65.40	78.40	100.00	100.00	100.00
5	2500		16.91		19.84		50.00	50.00	67.00
3	4500		9.30		11.64		21.00	22.00	38.00
1	12500		5.79		3.45		2.0	3.00	5.00
Avg. Part Size*		11.5 μ	9.5 μ	8.5 μ	6.5 μ	5.5 μ	3.65 μ	3.45 μ	2.40 μ

*Fisher Sub Sieve Sizer.
Footnote: Sub sieve analysis under 10 micron determined by Andreason Pipette method and coulter counter.

deagglomerate the larger pieces. There is a rotary kiln in the system to drive off the free moisture.

Other than finer grade special processing, the manufacture of standard grades is 90% air floated with 10% regrind through a pebble mill. The finished novaculite grades are particularly suited to chemical treatment using coupling agents, both ambi and mono functional. These products can also be rendered hydrophobic.

Malvern Minerals, the manufacturer of microcrystalline novaculite materials, markets a complete line of treated materials. The treatment process is proprietary.

6.3 Summary of Properties

Table 8-10 lists the typical properties and analyses of microcrystalline novaculite, as marketed under the trade name NOVACITE® by Malvern Minerals Company of Hot Springs, Arkansas.

6.4 Material Suppliers

See Table 8-11, page 152.

6.5 Cost and Availability

Cost of microcrystalline novaculite ranges from 02¢–20¢/lb in carload quantities. The degree of fineness controls prices. These materials are regarded as plentiful in source based on present demand. The silane or chemically treated materials sell in a range from 09¢–26¢/lb in carload quantities.

6.6 Applications

This filler has been used in the following resin systems:

Epoxy	Silicone Rubber	Polyester
DAP	RTV Silicone	PBT
		PET

Table 8-11.

Manufacturer	Product Name/s	Particle Size Information	Remarks	Location of Source
Malvern Minerals Co.	#100 Novacite®	100% Passing 40 mesh (Fisher Average Particle Size)	Screened Product	Hot Springs, Ark., U.S.A.
X X X	#200 X	99% X 200 X (74 μ) 11.5 μ	Standard Air Classified	X X X
X X X	#325 X	98% X 325 X (44 μ) 8.5 μ	X X X	X X X
X X X	#625 X	99% X 325 X (44 μ) 8.0 μ	X X X	X X X
X X X	#1250 X	99.99% X 325 X (44 μ) 7.0 μ	X X X	X X X
X X		micro		
X X X	#550 X	100% X 30 μ mesh 6.0 μ	Super X X	X X X
X X X	Daper X	100% X 15 μ X X 5.5 μ	X X X	X X X
X X X	L-207A X	100% X 10 μ X X 4.0 μ	X X X	X X X
X X X	L-337 X	100% X 10 μ X X 3.0 μ	X X X	X X X
X X X	L-338 X	100% X 10 μ X X 2.5 μ	X X X	X X X
X X X	XL-396 X	100% X 6 μ X X 2.0 μ	X X X	X X X
X X X	Novakup®*			

*(All grades listed above, except #100, are available surface treated with silanes, titanates, glass resins, and other polymer coatings, under the tradename NOVAKUP®).

HDPE		
Polymide	Polyurethane	PPS
	Polypropylene	
Nylon	Phenolic	Polycarbonate

These materials are primarily used in thermoset plastics and engineering types of thermoplastics. There are new chemical coupling agents, which are not silanes, that appear to enhance the properties in some filled commodity thermoplastics such as polypropylene, polyethylene, and polystyrene. Also, these chemicals may broaden novaculite usage in many elastomers

Table 8-12. L-207A NOVACITE® in GE silicone rubber.

L-207A Novacite Filled General Electric Silicone Rubber		100 PARTS GENERAL ELECTRIC SE900 REINFORCED GUM 1.5 PARTS CADOX TS-SO WITH A. 25 Parts L-207A Novacite®	B. 85 Parts L-207A Novacite®
Press Cure			
Shore A		61	71
Ten. st.		995	640
Elongation		400	370
Sp. Grav.		1.29	1.55
1 hr/300°F			
Shore A		63	72
Ten. st.		1070	640
Elongation		430	340
Tear (Die B)		70	80
24 hr/480°F			
Shore A		62	73
Ten. st.		765	720
Elongation		260	180
96 hr/480°F			
Shore A		65	78
Ten. st.		715	695
Elongation		260	150
Diel. Const.	10^2	2.94	3.16
" "	10^3	2.93	3.13
" "	10^4	2.92	3.12
" "	10^5	2.91	3.11
Dis. Fact.	10^2	.006	.008
" "	10^3	.00186	.0036
" "	10^4	.0014	.0022
" "	10^5	.0006	.0011
Loss Index	10^2	.01764	.02528
" "	10^3	.00545	.01127
" "	10^4	.00409	.00686
" "	10^5	.00175	.00342
V.R. (ohm-cm)		7.8×10^{15}	6.9×10^{15}

Table 8-13. L-207A NOVACITE® in epoxy castings with and without coupling agents, wet and dry.

Flexural strengths and electrical properties of L-207A NOVACITE -filled epoxy

Ref	Coupling Agent	Con'c. %	As Received Flex, psi	2 Hr Boil Flex, psi	72 Hr Boil Flex, psi	DK 10^2	Dry* 10^5	DK 10^2	Wet** 10^5	DF 10^2	Dry 10^5	DF 10^2	Wet 10^5
1736-119-1	–	0	17,100	14,500	11,300	4.25	3.96	4.29	4.01	.0055	.028	.0057	.028
119-2	Z-6020	0.01	18,200	14,200	13,200	4.16	3.90	4.23	3.96	.0056	.028	.0054	.029
119-3	Z-6020	0.03	20,800	15,800	15,200	4.24	3.94	4.28	3.98	.0056	0.28	.0059	.029
119-4	Z-6020	0.05	20,300	18,400	16,700	4.09	3.84	4.14	3.88	.0054	.027	.0052	.029
119-5	Z-6020	0.10	19,900	20,500	18,500	4.22	3.96	4.32	4.02	.0054	.026	.0053	.029
119-6	Z-6040	0.01	19,400	17,100	14,000	4.17	3.89	4.24	3.94	.0056	.028	.0058	.029
119-8	Z-6040	0.05	20,200	17,000	15,500	4.13	3.86	4.21	3.91	.0060	.029	.0056	.030
119-10	Z-6040	0.10	18,700	14,900	14,000	4.00	3.77	4.10	3.85	.0048	.026	.0048	.027
129-1	Z-6040	0.50	18,800	15,900	16,600								
119-11	XZ-8-5106	0.01	19,100	15,700	15,400	4.07	3.82	4.14	3.89	.0052	.027	.0050	.028
119-12	XZ-8-5106	0.03	20,500	14,600	13,700	4.16	3.89	4.21	3.95	.0052	.027	.0050	.029
119-13	XZ-8-5106	0.05	20,900	16,000	15,300	4.16	3.90	4.19	3.94	.0047	.027	.0051	.028
119-14	XZ-8-5106	0.10	22,500	19,600	15,100	4.02	3.81	4.05	3.84	.0044	.024	.0042	.026

*Dry As Received
Wet – 24 hr in 23°C water.

Table 8-14. Unfilled, L-337 NOVACITE® filled, and L-337 NOVAKUP® 1100 1.0 filled Nylon 12. (Nylon 12 Mobay Chemical) (Silane Union Carbide A-1100)

	Unfilled	"Novacite L-337" No Silane	"Novakup L-337" A-1100
Flexural strength, psi			
Initial	8,400	10,000	11,300
After 16 hr in 50°C water	6,200	8,700	9,900
Standard deviation[a]	116	348	120
Flexural modulus, 10^5 psi			
Initial	1.8	3.1	2.8
After 16 hr in 50°C water	0.9	1.8	2.1
Tensile strength, psi			
Initial	5,800	6,000	6,500
After 16 hr in 50°C water	5,100	4,600	6,400
Standard deviation[a]	59	113	47
Ultimate elongation, %	7.5	3.5	24.6
Impact strength, Izod, ft-lb/in., unnotched	31.9	3.6	22.2
Heat deflection temperature, °C, 264 psi	52	68	70

[a]Based on 5 observations.

and rubbers, as well as the silicones. This technology is in its infancy; however, some positive results are apparent at this time.

Novaculite has several unique properties as compared with other silica products. Some are listed below.

Lower abrasivity—Reduced wear
Lower Specific Heat—Higher melt index and reduced cycle time (.19 cal/g/C° vs. other silica .32 cal/g/C°)
Platey particles—Interlamellar flow, lower viscosity
Low oil adsorption (17 to 20%)—Lower viscosity & better rheological flow, heavier loading.

The properties just listed qualify this type of silica for widespread usage including thermoplastic, thermoset, and elastomer composites.

Table 8-12 shows the use of L-207A in General Electric silicone rubber formulations. Table 8-13 shows the use of L-207A NOVACITE® in an expoxy system, plus the utilization of several coupling agents and the effect of coupling agents of dry and wet system physical and electrical properties.

Table 8-14 shows the dramatic improvement properties of a L-337 NOVAKUP® 1100 1.0 filled Nylon 12 (Mobay Chemical).

6.7 Future Uses

The unique properties of the novaculite material will provide diversification in plastics composites and should enjoy steady growth. The treated varieties of the novaculite materials will be exciting in development and should eventually grow rapidly. Particulate mineral treatment is a very young technology and at this time would be difficult to project precisely. There are the strong indications of acceptability and growth.

6.8 Significant Literature and Patents

"Low Quartz Microforms and Their Contribution to Plastics & Resinous Systems," *Advances in Chemistry Series 134 Fillers and Reinforcements in Plastics*.

Gillespie, Thomas and Wiley, Ralph M., "Determination of London-Van Der Wals Constants,"

Journal of Physical Chemistry, **66**: (1077), 1962.

Higuchi, W. I. and Stehle, R. G., "Rheometric Study of Particle-Particle Interactions in Suspensions," *Journal of Pharmaceutical Sciences*, **54**, (2) Feb. 1965.

Novel Coupling Agents for Mineral Reinforced Polyamides, John Newbould Polymers Dept., General Motors Research Labs, Warren, Michigan.

The Use of NOVACITE® (A Natural Microform of Low Quartz) in Plastics & Resinous Systems, Malvern Minerals Company.

Making a Good Product Better–NOVAKUP®, Malvern Minerals Company.

Kaplan, Stephen L., *Polyimide Resin for Fire-Resistant Moldings and Laminates*, Ciba-Geigy Corp. Ardsley, New York (Plastics Design & Processing).

U.S. Patent 3,745,140 Shapable Fiber–Reinforced Novaculite Filled Low Molecular Weight Polyethylene Terephthalate, Leon Segal, Allied Chemical Company, N.Y., N.Y.

7. SILICA, DIATOMACEOUS

7.1 Abstract

Diatomaceous silica is an naturally occurring opaline phase of silica. The ore is actually the skeletal remains of plankton. The variety of shapes is a world of fascination under the microscope. Diatomaceous comes from the word diatom, which is a single-celled aquatic plant.

7.2 Manufacturing

Natural diatomite is a soft, chalky rock of very light density. In modern operations, crude is selectively quarried and transported to primary crushers. The ore is then crushed, milled, and dried in heated air currents. Milling takes place as crude is conveyed through air blowers.

Following cyclone separation, the natural powder is subjected to calcination in rotary kilns. White products of varying particle size are produced by air classification systems. For grades designed for antiblock useage, particular attention must be given to assure that no oversize particles are present.

7.3 Summary of Properties

Table 8-15 shows the typical chemical analysis of flux calcined Celite® from Johns Manville Corporation.

Table 8-15. Typical chemical analysis % weight flux calcined celite® for anti-block.

H_2O	0.1%
Ignition loss	0.2
SiO_2	91.9
Al_2O_3	3.3
Fe_2O_3	1.2
CaO	0.5
MgO	0.5
Na_2O	1.8
K_2O	0.3
Other	less than 0.5

Table 8-16 provides typical physical properties of three grades of Celite®.

The shape of particles of the Johns Manville Super Floss product is shown in Fig. 8-4 at 300X, and Fig. 8-5 at 1000X by scanning electronmicrograph.

7.4 Material Suppliers and Product Information

See Table 8-17 on page 157.

7.5 Cost and Availability

Diatomaceous silica is readily available and occurs worldwide. Approximately 19% of the total usage is in plastics and the price range average is close to 03¢/lb according to current calculation.[7]

7.6 Applications

The major use for processed diatomaceous silica in plastics industry is for anti-blocking

Table 8-16. Typical physical properties.

	Super Floss	Silver Frost K-5	Celite White Mist
Appearance	Fine, fluffy	White powder	
Color	White	White	White
325 mesh residue %	0.1	trace	trace
Oil absorption (G-C)	120	115	160
Specific gravity	2.3	2.3	2.3
pH	9–10	9–10	9–10
Refractive index	1.48	1.48	1.48
Surface area, sq/m/g		– 0.7 to 3.5	

of low density polyethylene film. Most usage is in the form of master batches containing from 10–50% diatomaceous silica. Regulated amounts of master batch are added to film resin in a range generally to provide from 500–3,000 ppm of anti-block in the finished film.

7.7 Future Uses

This type of silica is readily available at a relatively low price. Diatomaceous silica has unique properties which should be more effectively utilized. This will occur as the demand for higher performance fillers develops. Its use in other fields, such as for filtration, is currently greater than in plastics. This may change as plastic filler usage increases.

8. SILICA, FUSED (VITREOUS)

This amorphous silica is considered to be a high performance filler in that it is specifically used in application of extreme temperature environ-

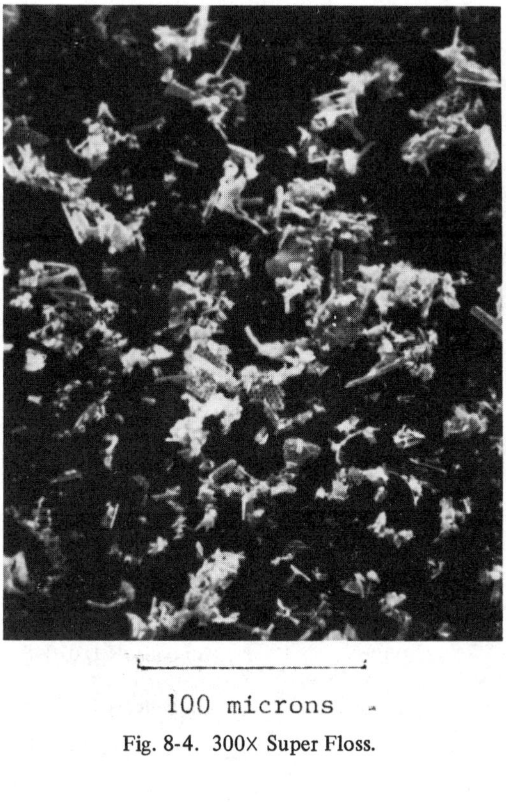

100 microns

Fig. 8-4. 300X Super Floss.

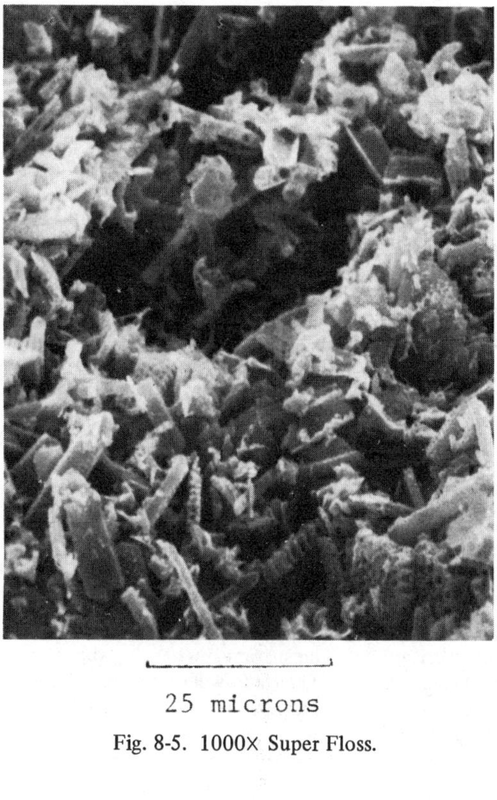

25 microns

Fig. 8-5. 1000X Super Floss.

Table 8-17.

Manufacturer	Product Name(s)	Particle Size Information	Remarks	Location of Source
Johns Manville Corp.	Celite® Super Foss	99.9% passing 325 mesh (44 μ)	Calcined-Cyclone Separation	Denver, Colorado, U.S.A.
	Celite® Silver Frost K-5	100.0% (trace) passing 325 mesh (44 μ)	Calcined-Cyclone Separation	Denver, Colorado, U.S.A.
	Celite® White Mist	100.0% (trace) passing 325 mesh (44 μ)	Calcined-Cyclone Separation	Denver, Colorado, U.S.A.
Eagle Picher, Ind.	Products similar to those above.			Cincinnati, Ohio., U.S.A.
Dicalite	Products similar to those above.			Avenel, New Jersey, U.S.A.

ments. It has almost negligible thermal expansion from near absolute zero (-459.6°F) to its melting point (3100°F). It is, therefore, an ideal filler for composites that are exposed to varying temperatures and require an extremely low thermal expansion and resistance to thermal shock. Other characteristics of fused vitreous silica are similar to those of crystalline silicas.

8.1 Manufacturing

Fused silica is manufactured from high purity quartz sand by fusing in an arc furnace using graphite electrodes. The resultant ingot is then reduced to various grain and powder fractions using standard ceramic comminution processes.

8.2 Summary of Properties

The standard fractions available range from coarse grains from -4 + 10 mesh to a fine -325 mesh powder with a surface area of 9 m^2/g. The lower specific gravity of 2.18 versus 2.65 for quartz sand aids in settling problems in certain systems. The thermal expansion of fused silica, 0.3×10^{-6} per °F, is significantly less than other common fillers and thus allows for substantial reduction of the overall thermal expansion of a resin system.

The typical chemistry of fused silica is shown below.

SiO_2	99.6
Al_2O_3	0.2
Fe_2O_3	0.04
$Na_2O + K_2O$	0.02

8.3 Material Suppliers and Product Information

See Table 8-18, below.

8.4 Cost and Availability

The current price of fused silica is around 22¢/lb. Variations in particle sized distribution controls price, as in most particulate materials. For current demand, fused silica supply is sufficient, barring some new and dramatic large

Table 8-18.

Manufacturer	Product Name/s	Particle Size Information	Remarks	Location of Source
Glassrock Products	Glasgrain®, GP Series Standard fractions available in coarse products -4 to +10 mesh, to fine 325 mesh powder.		Grains and powders come from reduced ingots.	Atlanta, Georgia, U.S.A.
Thermal American Fused Quartz	Transparent Fused Quartz Powder.	Approximately 300 mesh	$33.00/lb	Montville, N.J. 07045
Thermal American Fused Quartz	Spectrosil Powder	200 % 300 mesh.	Superpure Quartz $100.00/lb	Montville, N.J. 07045

Table 8-19. Physical and electrical properties of GP 31, 71 and 3T Quartz in epoxy systems.

Formulation	30 p* epoxy 70 p GP3I 5 p hardner	120 p* epoxy 75.2 p GP7I 9.6 p hardener	120 p* epoxy 37.6 p GP3T 9.6 p hardener
Volume % of powder	52.0%	24.6%	14.0%
Flexure strength, psi	12,700	12,000	11,300
Specific gravity	1.7	1.4	1.3
Dielectric constant			
Measured at 75°F			
Frequency 60	Not determined	4.07	4.80
10^3	4.09	3.56	3.86
10^6	3.90	3.57	3.59
Dissipation factor			
Measured at 75°F			
Frequency 60	Not determined	.007	.009
10^3	.007	.010	.009
10^6	.010	.008	.010

*p stands for parts.

use. The price levels at which fused silica sell demand specific performance. It would be considered a specialized material.

8.5 Applications

The Glasgrain GP Series includes several carefully controlled variations of particle size distributions and surface areas. Several are excellent reinforcing extenders in plastics. They are especially suited for use in electrical applications. Glasgrain fused silica displays more desirable physical characteristics in plastics than most mineral fillers. Fused silica allows for high loading, light weight, low shrinkage, low thermal expansion, excellent thermal shock resistance, and, of primary importance, good electrical properties. Further, the Glasgrain GP Series contains selected powders, which allow for variations in flow and cure rates and strength. They are used in silicone, epoxy, high molecular weight fluorocarbons, and other resins. Table 8-19 is GP 31, 71, and 3T in epoxy systems.

In epoxies, use of GP powders shows excellent strength per weight. The volume stability afforded by powders, as compared to other mineral filler materials, can be seen by comparison of the Thermal Expansion Coefficients shown by Curlee for fused silica and aluminum oxide filled epoxies, which are as follows in Table 8-20.

The low specific gravity of fused silica allows for weight reduction; it also allows for a pound of fused silica to go a lot further than a pound of other mineral materials. A comparison is shown in Table 8-21.

The oil absorption test (Garnder-Coleman method) for various GP Powders falls in a range

Table 8-20. Comparison of GP Quartz powders to other mineral filler materials.

Formula:	70 pbw Al_2O_3 30 pbw epoxy 6 pbw Shell Z	70 pbw (GP3I) 30 pbw epoxy 6 pbw Shell Z	75 pbw (GP3I) 30 pbw epoxy 5 pbw Shell Z
Coefficient of thermal expansion (−65° to +165°F); $\times 10^{-6}$ in./in./C	26	18.8	14.8
Specific gravity, g/cc:	2.37	1.72	1.78

Table 8-21. Volume comparisons of GP powder to other typical fillers.

	Specific Gravity	Number of grams needed to equal 1 gram volume of GP Powder
GP (I) Powders	2.15	1.0
Al_2O_3	3.97	1.85
SiO_2 (Quartz)	2.65	1.23
Glass	2.7	1.26
Calcium carbonate	2.7	1.26
Asbestos (chrysotile)	2.2	1.05
Kaolinite	2.6	1.21

between 22 and 30 lbs of oil per 100 lbs of powder.

8.6 Future Uses

Only nominal attention has been directed to the plastics field. Increased involvement should broaden use in plastics calling for greater thermal stability coupled with stringent electrical insulation requirements. Thermoset and engineering thermoplastics are good targets.

8.7 Significant Literature

Curlee, Richard M., "A Low Coefficient of Thermal Expansion Epoxy Encapsulating Compound," 1112 Sandia Laboratory, Albuquerque.

GLOSSARY

Aerogel—The material formed during the removal of free water without shrinkage of the gel.

Amorphous—Without crystalline lattice work.

Anti-blocking—Tendency of a film not to stick together.

Calcined—To heat to a high temperature without fusing. This is done to remove impurities or change the physical characteristics of a mineral or compound.

Colloidal—A state of existence of matter in a particle size range of 0.001–0.5 microns (although there is some disagreement in size range).

Flatting—Surface capable of dispersing light rays to give dried surface a nonglossy, matte finish.

Flux Calcined—Calcining utilizing a fluxing substance.

Fluxing Substance—As applied to the subject being discussed, a substance used to promote the calcining of minerals.

Hydrogel—The undried material formed at gelation.

Interfacial Crosslinking—The condition that results when one end of a coupling agent links with the filler and the other end of the coupling agent interacts with the polymer matrix.

Mar Resistance—Resistant to score, scratch, or blemish.

Microcrystalline—A crystalline substance which has been reduced to a powder with a portion of its particles less than one micron. It could be a microform either natural in its microcrystallinity, or reduced into this fine state mechanically. There would be evidence of the orderly arrangement of at least a large share, if not all, of its atoms.

Novaculite—A silicic sedimentary rock that is dense, hard, even textured, light colored, and characterized by dominance of microcrystalline quartz over chalcedony.

Pyrogenic Silica—Synthetic, amorphous silica manufactured by hydrolysis of silicon tetrachloride in an oxygen-rich flame.

Quartz (Alpha)—The very prevalent natural form of silicon dioxide, stable in its phase from absolute zero to 574 C° ± 2°.

Quartzite—High purity quartz rock either metaquartzite or orthoquartzite. The orthoquartzite is specifically granular while the metaquartzite is recrystallized sandstone. Also included are the high purity, deagglomerated, loose quartzite sands.

Rheology—The science of the flow of materials.

Silane Coupling Agent—A silicone chemical which is usually ambifunctional in that the chain has a hydrolyzable group (ethoxy, methoxy, e.g.) on one end, and an organo functional group (epoxy, amine, mercaptan, e.g.) on the other.

Silanol—The hydroxyl group present on the surface of the silica.

Silica Sol—The clear mixture of sodium silicate and acid prior to gelation.

Super Classification—Primary, secondary, or tertiary processing of a finely divided particulate material to create a state of unusually fine subdivision. This particle size range of subdivision would be narrow, with controlled top particle size measurement. This type of product is more difficult to manufacture and volume is somewhat limited. Special equipment (as opposed to ordinary separators) is required.

Thixotropy—The property of a film or liquid remaining thick and stable at rest, but shearing easily to a fluid state when force is applied, and returning to the thick and stable phase when force is removed.

Tripoli—A porous, friable silicic sedimentary rock that varies in color. Colors could be white, pink, red or yellow. The material is very finely divided, naturally.

Viscosity—Flow resistance.

Wet Strength—The physical or electrical values of a system (filled or unfilled) measured after the system has been subjected to moisture by one method or another.

Xerogel—The material formed during the removal of free water with controlled shrinkage of the gel.

References

1. *Aerosil® Manufacture, Properties and Applications*, Degussa, Frankfurt am Main, West Germany.
2. Service Bulletin 150-G-47, PPG Industries, Technical One Gateway Center, Pittsburgh, PA.
3. Marvel, Don, *Silica Gel*, W. R. Grace and Co., Baltimore, MD.
4. PPG Industries, One Gateway Center, Pittsburgh, PA, Donald D. Dunnom, Mgr. Transportation Industry Polymer Technical Service.
5. Patton, Temple, Editor, *Pigment Handbook*, I, John Wiley & Sons, 1973, p. 136.
6. Patton, Temple, Editor, *Pigment Handbook*, I, John Wiley & Sons, 1973, p. 129.
7. Patton, Temple, Editor, *Pigment Handbook*, I, John Wiley & Sons, 1973, p. 142.

9
Talc

Joseph A. Radosta and Nikhil C. Trivedi
Pfizer MPM Division
Easton, Pennsylvania

CONTENTS

1. INTRODUCTION

There are several forms of ore from which talc products can be made. However, the talc of major interest to plastics users is a finely ground product consisting of thin platelets and is preferably white in color. Due to the platy nature of this special form of talc, it is considered to be a *reinforcing filler* in many plastic applications. Its low cost qualifies the talc as an extender (lowers the cost of the compound and extends the resin with minimal sacrifice in physical properties) while its platyness, or high aspect ratio, further qualifies the talc as a reinforcing filler (enhances the performance properties of the compound). White, platy, fine ground talcs having a high, dry brightness are readily available from certain suppliers.

Composites filled with a platy talc always exhibit a higher stiffness and creep resistance, both at ambient and elevated temperatures, when compared with composites filled with a particulate filler such as calcium carbonate. For example, when polypropylene is filled with a 40% loading of talc, its stiffness is tripled from a value of about 200,000 psi for the unfilled pp to about 600,000 psi for the talc filled pp. A 40% loading of calcium carbonate only increases the stiffness up to about 400,000 psi. A trade-off that usually accompanies high loadings of filler is reduction of impact strength. This effect can be minimized through proper selection of talc particle size and size distribution, surface treatments on the talc, and resin formulation.

The mineral talc is a hydrated magnesium silicate with the theoretical formula: $3MgO \cdot 4SiO_2 \cdot H_2O$, or 31.7% MgO, 63.5% SiO_2, and 4.8% H_2O. Composition will vary according to the locality in which the talc is mined. For example, Montana talcs approximate the theoretical composition, while California talcs often contain calcite ($CaCO_3$) and dolomite ($CaCO_3 \cdot MgCO_3$). The current price range for talcs used for plastics applications is 2-6¢/lb, depending on the fineness of grind, color, and ore body.

2. PRODUCTION METHODS FOR TALC

The designation "talc" covers a wide range of natural products. For reporting purposes, the United States Bureau of Mines considers talc, soapstone, and pyrophyllite as one group.[1] It is important to know that commercially available talc may not always contain the mineral talc as a major constituent. Some of the major contaminants associated with talc are listed in Table 9-1.

Table 9-1. Minerals commonly found in a talc deposit.

Talc	$3MgO \cdot 4SiO_2 \cdot H_2O$
Anthophyllite	$7MgO \cdot 8SiO_2 \cdot H_2O$
Calcite	$CaCO_3$
Chlorite	$Al_2O_3 \cdot 5MgO \cdot 3SiO_2 \cdot 4H_2O$
Diopside	$CaO \cdot MgO \cdot 2SiO_2$
Dolomite	$CaCO_3 \cdot MgCO_3$
Magnesite	$MgCO_3$
Quartz	SiO_2
Serpentine	$3MgO \cdot 2SiO_2 \cdot 2H_2O$
Tremolite	$2CaO \cdot 5MgO \cdot 8SiO_2 \cdot H_2O$

The presence and the relative amounts of these various contaminants of talc are often indicative of the areas from which it is mined. The Montana talc deposits constitute the major known U.S. reserves of talc. They are by far the best and the purest natural talcs. New York talcs, on the other hand, are classed as low grade talcs containing, on the average, only 40-60% talc. Vermont talcs are high in their magnesite content, while some California talcs contain varying proportions of several hard impurities. Methods employed for the production of talcs are dependent on the amounts and characteristics of the associated impurities.

2.1 Mining

The mining of talc is carried out in the classical mining methods employed commonly by mining companies. The two types of mining are open pit and underground. Many of the California talc operations are open pit mining operations, whereby the ore body is mined subsequent to the removal of overlying waste material. When the ore body is located at greater depths and in mountainous terrain, the underground mining method is often utilized. The mining costs for the latter method tend to be higher than those for an open pit operation.

2.2 Dry Processing of Talc

The Montana talcs generally approach the theoretical composition of pure steatitic talc (see the section on chemical properties of talc). Production of talcs in Montana is therefore quite straightforward. The first of several beneficiation steps applied to the ore is known as Selective Mining. This consists of greater care at the mine site in both blasting as well as subsequent loading of ore from carefully selected areas in the mine. The talc ore is then subjected to various comminution and classification operations. Most of the equipment and procedures utilized for the processing of talc are described elsewhere in detail.[2,3,4]

The crude ore is "crushed" in jaw crushers or rolls where the large lumps (as large as 36 in.) are crushed to a size of 4-6 in. The crushed ore is then "washed" to remove adherent impurities and then "hand sorted" to further remove apparent contaminants. Attempts are now being made to mechanize the sorting operation by installing optical or electronic sorters that would reject any material with reflectance values different from a pre-set, programmed value.

Further size reduction of the hand-sorted talc is carried out in cone crushers and roller mills. Further "micronizing" of the ground ore is carried out in hammer mills, tube mills, pebble mills, or fluid energy mills. These fine grinding units are operated under carefully controlled parameters of air pressure, moisture, and heat content, and in a closed loop with pneumatic classifiers to obtain products with desired fine particle sizes and particle size distributions. The choice of fine grinding method, the type of equipment used for grinding, and the type of air separators used for classification of the product are very critical to the quality of the talc as a filler. These techniques are often closely guarded proprietary information. The grinding machines and their mode of operation determine the particle shape and size of the product. The surface-related properties of the ground product are often thought to be governed by the grinding technique used for its production. Commercial products of dry processed ground talc vary in size from 100%–70 micrometers (commonly referred to as "a top size of 70 μ") to 99%–10 μm (a top size of 10 μ).

To summarize, dry ground talcs are partially beneficiated, and have wide size distributions, color, and purity, which vary with the ore.

2.3 Beneficiated Ground Talc

New York and Vermont talcs are known to contain varying amounts of impurities that may be deleterious to the use of talc as a filler. Some of the Vermont talcs are therefore beneficiated prior to grinding. Froth flotation is commonly used for the beneficiation of talc. This is because the Eastern talcs are hydrophilic in nature and are thus "naturally flotable." Using various surfactants, the surface properties of the impurities and/or talc are modified to affect the separation of one from the other.

A typical flowsheet might consist of selective mining, crushing, wet screening, beneficiation by flotation, wet (or dry) grinding in pebble mills, and further micronizing. Air or water classification is widely practiced to achieve efficient size classification of the various products. Where color of the product is a critical parameter, bleaching with suitable bleaching agents is also commonly incorporated as a beneficiation step to attain a brighter product.

2.4 Calcined Talc

For special applications, including certain plastics applications, selected pure talc ore is calcined in rotary kilns at high temperatures. The calcined talc is then ground in Raymond mills to products with a median particle diameter of 25 μm. The calcined talc has low calcium content, low acid solubles, and low impurities. (See section 3.3.)

2.5 Other Processes for the Production of Talc

In addition to froth flotation and bleaching, as mentioned in section 2.3, wet magnetic separation, dry high intensity magnetic separation, high gradient magnetic separation, foam fractionation, and selective flocculation may be utilized at various stages of processing of talc. The choice of one or more of the operations depends on the end use of the product and the amounts and natures of impurities associated with talc.

3. PROPERTIES OF TALC

3.1 Chemical Properties

Theoretical chemical composition of talc is as follows:

Element	Wt %
Mg	19.2
Si	29.6
H	0.5
O	50.7

Compound	Wt %
MgO	31.7
SiO_2	63.5
H_2O	4.8

The chemical formula for pure talc is:

$$3MgO \cdot 4SiO_2 \cdot H_2O$$
$$Mg_3Si_4O_{10}(OH)_2$$
$$Mg_3Si_4H_2O_{12}$$

Talc is a hydrated magnesium silicate and is mineralogically classified within the sub-class "phyllosilicates" with trioctahedral three layer structures.[5] The structure of pure talc consists of a layer, or sheet, of brucite (magnesium hydroxide, $MgO \cdot H_2O$) sandwiched between two sheets of silica (SiO_2). These layers are superimposed upon each other. Adjacent talc layers are held together by weak van der Waal type forces. These layers slide over each other with considerable ease when shearing action is applied to the mineral.

Talc is inert to most chemical reagents. It remains undecomposed in contact with acids. Chemical analyses of various high grade talc products are shown in Table 9-2.

Table 9-2. Chemical analyses of selected commercial talc samples.

Location	CHEMICAL ANALYSES, %												Reference
	SiO_2	MgO	Ratio SiO_2/MgO	CaO	Na_2O	Al_2O_3	K_2O	MnO	FeO	Fe_2O_3	CO_2	LOI	
Pure Talc/ Theoretical Compn.	63.46	31.79	1.99	–	–	–	–	–	–	–	–	4.75	
Alabama	62.17	32.34	1.92	–	–	0.51	–	–	–	1.43	–	2.98	[11]Neathery, et al (1967)
Alabama	52.57	24.62	2.14	0.51	4.75	1.88	–	0.05	–	5.62	–	7.70	[12]WC&D, Inc. (1957)
California	60.20	27.98	2.15	2.60	–	1.25	–	–	–	2.50	–	5.70	[9]Ladoo, R.B. (1923)
California	59.61	30.01	1.99	0.84	0.26	1.65	–	–	0.92	–	–	5.94	[10]Engel & Wright (1960)
California	57.40	23.91	2.40	13.55	0.44	1.29	–	–	0.86	–	–	2.20	"
Georgia	41.02	28.60	1.43	4.76	–	4.23	–	–	–	5.85	–	15.51	"
Montana	62.65	30.23	2.07	–	0.20	0.31	–	–	–	1.51	0.27	4.95	"
Montana	62.50	30.60	2.04	0.25	0.08	0.50	0.04	0.09	–	0.70	0.60	4.64	"
New York	66.23	25.71	2.58	2.26	–	1.05	–	0.16	0.22	0.13	0.56	3.86	"
New York	60.59	34.72	1.75	–	–	0.13	–	1.16	0.21	–	–	3.77	[11]Neathery, et al (1967)
Vermont	60.48	28.52	2.12	0.02	0.03	0.82	–	0.09	4.59	0.10	–	4.94	[10]Engel & Wright (1960)
Vermont	42.73	33.16	1.28	0.10	–	1.17	–	–	–	4.93	–	17.69	[13]Ind. Minerals (1975)
Virginia	39.54	24.84	1.59	5.93	0.08	3.72	–	1.60	7.12	3.62	9.50	5.04	[10]Engel & Wright (1960)
Canada	47.55	32.21	1.48	6.65	–	2.37	–	–	–	0.80	–	5.87	[12]WC&D (1957)
China/Micronized MW4	59.40	31.40	1.89	0.66	–	0.68	–	–	–	0.18	–	7.10	[13]Ind. Minerals (1975)
India/Finex	62.20	32.20	1.93	0.10	–	0.30	–	–	–	0.30	–	4.90	"
Austria/Best White	61.54	30.09	2.04	1.81	–	1.74	–	–	–	0.76	–	3.65	"
France/Luzenac 005	57.52	31.60	1.82	0.53	–	3.08	–	–	–	0.83	–	6.48	"
Italy/ValChisone MT120	60.52	31.75	1.90	0.53	–	1.27	–	–	–	1.03	–	5.51	"
Norway/Ground FWXD	39.80	33.10	1.20	0.60	–	0.60	–	–	–	6.7	–	19.70	"
Switzerland	61.51	30.93	1.99	3.70	–	0.83	–	–	0.12	–	–	2.84	[11]Neathery, et al (1967)
UK/Shetland	34.80	34.50	1.00	0.4	–	1.3	–	–	–	6.2	–	23.60	[12]Ind. Minerals (1975)

3.2 Physical Properties

Scanning electron micrographs of some typical California and Montana talc products are shown in Figs. 9-1 and 9-2.

3.2.1 Hardness. Pure talc is the softest known mineral and has a slippery feel. On Moh's scale of hardness (1-10), talc is the standard for a hardness of 1. Calcite, $CaCO_3$, has a hardness of 3 and diamond has a hardness of 10 on this scale. Commercial talc products often tend to be harder than pure talc because of the presence of impurities, such as calcite and tremolite.

3.2.2 Color. The color of natural talc varies. It may be white, gray, yellow, pale blue, or pale green. Talc has a characteristic silvery or pearl-like luster. Upon grinding, however, all talcs yield a gray to white powder with varying degrees of dry brightness.

3.2.3 Particle Size and Shape. Most of the commercially available talc products used in filler applications are very finely ground. Suppliers of talc are generally able to alter the particle size distributions of their products for special needs. This is generally done by changing the operating conditions of the grinding and classification steps. The crystal form of talc may be platelike (lamellar), foliated (leafy), fibrous (acicular) or massive. The particle shapes of commercial talcs are characteristic of their origin. However, most of the commercial talcs that are sold for use as plastics fillers, have the platy configuration.

In the case of platy talcs, one property often mentioned in the literature is the aspect ratio. It is the ratio of the average diameter of a platelet particle to its thickness.

3.2.4 Surface Properties. Depending on the source from which it is mined, talc shows different surface characteristics. The Montana talcs are hydrophobic in nature while the California talcs are predominantly hydrophilic.

5,000 X   5μ

Fig. 9-1. California talc. Pfizer Talcron CP 38-33. Magnification 5000X.

5,000 X   5μ

Fig. 9-2. Montana talc. Pfizer Microtalc MP 15-38. Magnification 5000X.

Figs. 9-1 and 9-2. SEM micrographs of platy talc.

Surface area and oil absorption are governed, to a large extent, by the fineness of the product. Surface roughness, particle shape, and pore volume are other factors that influence these properties. In general, talc exhibits large surface area in relation to its mass because of its plate-like structure.

3.3 Electrical and Thermal Properties

Talc is a non-conductor of electricity. Therefore, it is used in the manufacture of high frequency electrical insulators.

Talc exhibits a stability towards heat up to temperatures as high as 1650°F (~900°C). It has low thermal conductivity and high resistance to heat shock. Upon heating, talc loses a part of its associated water without any change in its optical or crystalline properties. Water in the range of 1 molecule/mol talc is driven off at temperatures between 380–500°C. At temperatures in excess of 800°C, it loses its combined water of crystallization and dissociates into enstatite via an endothermic reaction. (See section on calcined talc). Calculated Gibbs energy of formation of talc at 25°C and 1 atm pressure is -1319.4 ± 1.5 kcal/mol.[6] Experimentally, the Gibbs energy of formation of talc at 25°C, 1 atm pressure, is determined to be -1319 kcal/mol.[7]

3.4 Optical Properties

Talc exhibits three indices of refraction. For pure talc, the indices are, $x = 1.539$, $y = 1.589$ and $z = 1.589$. It has a small and variable optic axial angle. Typical GE brightness values for micronized talc products range between 85 and 94 using MgO as 100 at a wavelength of 457 nm.

3.5 Miscellaneous Properties

The specific gravity of talc is 2.7–2.8 g/cm^3 (22.5–23.3 lb/gal). Its bulking value has a range of 0.043–0.045 gal/lb. The bulk density varies between 10–15 lbs/cu ft. for the finest micronized commercial products. For coarser products, the bulk density could go as high as 30 lbs/cu ft. The ambient pH value of a 5% solids suspension of most talc fillers is 9.0–9.5.

4. SUPPLIERS

Major suppliers of talc and a selection of their products are listed in Table 9-3. The list was compiled from information derived from National Paint and Coatings Association–Raw Materials Index (Pigment Section).[8] In addition to the suppliers listed here, there are other local producers with a relatively insignificant share of the market. Most of the products meet one or more of the following specifications–

ASTM Specification D605-66T
Federal Specification TT-P-403A
Maritime Administration
  Specification 52-MA-523b, Type I
Maritime Administration
  Specification 52-MA-523b, Type II
Military Specification MIL-P-15173A, Type A
Military Specification MIL-P-15173A, Type B

Most of the above specifications govern the particle properties–either the top particle size or the size distribution, and seem to be generally acceptable to the paint industry.

5. COST AND AVAILABILITY

5.1 Cost of Talc in 1975

Cost of talc has tended to be a dynamic situation. Costs of comparable products are quite competitive among the various suppliers. Purity, dry brightness, top particle size, and particle size distribution are prime factors in determining costs of various products. Currently, the price of talc ranges from 2-6¢/lb. This translates to a cost of 0.2-0.6¢/cu in. Future pricing of talc will be mainly dependent on (1) changes in energy costs for crushing, grinding, and drying the ore, (2) changes in transportation costs of the ore and of the final product, and (3) changes in state and federal tax laws and mining laws.

5.2 Availability

Supplies of talc are currently adequate. Major talc deposits have been located and mined in California, Montana, and New York. Good quality ores are also available in Vermont, North Carolina, Alabama, Texas, and Georgia.

Table 9-3. Major suppliers of talc.

Supplier	Product Name	Top Particle Size (μ)	Maximum +325 Mesh, %	Remarks
Cyprus Industrial Minerals Co.	Beaverwhite	75	3.5	Montana Talc, High Dry Brightness
555 South Flower St.	Glacier 325	45	0.3	Montana Talc, DB ≈ 80
Los Angeles, CA 90071	Mistron Frost	14	nil	California Talc, High Dry Brightness
(213) 489-3700	Mistron Vapor	12	nil	Montana Talc, "
(Formerly: United Sierra Corp.)	C-400	45	0.3	California Talc, "
Engelhard Minerals & Chemicals Corp.	Emtal 41	30	0 to 0.2 }	Vermont Talc, Platy, Low DB (<80)–
430 Mountain Ave.	Emtal 42	45	0 to 3.0 }	contains large amounts of Magnesite;
Murray Hill, N.J. 07974				contains ~5% Fe_2O_3
(201) 464-7000	Emtal 4190	15	nil	Waterfloated Vermont Talc, DB = 80
(Recently acquired Eastern	Emtal 500	40	0 to 0.2	" " " DB = 83
Magnesia Talc Co. from	Emtal 599	15	nil	" " " DB = 87
Johnson & Johnson)				
Johns-Manville Corp.	Desertalc 54	55	0.6	California Talc, High Dry Brightness
Greenwood Plaza	Desertalc 57	21	Trace	" "
Denver, Colorado 80217	Desertalc 23	85	1.0	" "
(303) 770-1000	Mikro 507	15	Trace	" "
(Parent Company of L. Grantham	Mikro 707	17	Trace	" "
Corp., Desert Minerals is a				
subsidiary of L. Grantham Corp.)				
Pfizer Inc., MPM Division	Microtalc CP 10-40	10	nil	California Talc, Platy, DB = 95
235 East 42nd St.	Microtalc CP 14-35	14	nil	" " DB = 95
New York, N.Y. 10017	Microtalc CP 20-30	20	nil	" " DB = 94
(212) 573-2323	Talcron CP 38-33	38	Trace	" " DB = 94
	Talcron MP 45-26	45	0.5	Montana Talc, Platy, DB = 88
	Talcron MP 40-27	40	Trace	" " DB = 91
	Microtalc MP 30-36	30	nil	" " DB = 88
	Microtalc MP 15-38	15	nil	" " DB = 91
	Microtalc MP 12-50	12	nil	" " DB = 92
	Talcron CP 46-27	46	1	California Talc, Platy, DB = 92
R. T. Vanderbilt Company	NYTAL 200	50	2	New York Talc, may be tremolitic,
33 Winfield Avenue				beneficiated, DB = 90
Norwalk, Connecticut 06855	NYTAL 300	25	0.06	" "
(203) 853-1400	NYTAL 400	15	Trace	" " DB = 91
(Recently acquired International	I.T. X[a]	NA	2	" may be tremolitic, DB = 90
Talc Co.)	I.T. 3X	NA	2	" "
	I.T. 5X	NA	0.75	" "
Whittaker, Clark & Daniels, Inc.	367 Georgia Talc	NA	1.5	Dry Brightness = 65
1000 Coolidge Street	399 LoMicron Talc	12	nil	" = 92
South Plainfield, N.J. 07080	1313 Canadian Talc	NA	0.4	" = 92.5
(201) 561-6100	2611 Talc	NA	nil	" = 92
	2794 Visco XX	NA	Trace	" = 92

[a]All "International" talcs were formerly known as "Asbestine".
NA = Information Not Available

Some of the major California talc deposits are located within the Death Valley National Monument. New legislation may retard mining activities in that area, but the overall availability of talc should not be seriously affected.

Beneficiation methods have been evolving and are being improved for the upgrading of talc ores. More efficient grinding and drying methods are being utilized. Grinding capacities at various operations are being expanded. Future developments in talc products will be towards meeting individual demands more closely. Products with specific top particle sizes and particle size distributions may be specially made for each end use.

6. PLASTIC APPLICATIONS

6.1 Talc as a Reinforcing Filler

The commercial grades of talc used for plastic applications are fine ground products consisting of thin platelets. Due to the platy nature of this special form of talc, it is considered to be a *reinforcing filler* to distinguish it from the other particulate mineral fillers. Its low cost qualifies the talc as an extender (lowers the cost of the compound and extends the resin with minimal sacrifice in physical properties), while its platyness or high aspect ratio further qualifies the talc as a reinforcing filler (enhances the performance properties of the compound).

6.2 Melt Compounding Techniques

Polymers filled with a platy talc always exhibit a higher stiffness and creep resistance, both at ambient and elevated temperatures, than polymers filled with a particulate filler such as calcium carbonate. However, in order to maximize these enhanced properties, it is essential that proper melt compounding techniques be used to incorporate high loadings of talc into the polymer. Proper compounding of the fine talc particles requires all the special melt compounding actions (smearing, folding, stretching, wiping, compressing, and shearing) to fully wet the talc particles with the molten polymer and achieve a high degree of dispersion resulting in a homogeneous composite. To accomplish this, a batch or continuous melt compounding system must be used to produce pellets of properly dispersed filler in the resin. These filled, precompounded pellets are then processed in injection molding or extrusion equipment to make the final product.

In general, all the major types of plastics compounding equipment in current use could do an adequate job. The list includes batch and continuous internal intensive mixers, two-roll mills, twin screw extruders, and a variety of single screw compounding extruders. The advantages and disadvantages of each type of compounding system are fully discussed elsewhere,[14] including Chapter 32 of this Handbook. The choice of the proper system will depend on factors such as the heat sensitivity of the resin, the volume of material to be compounded, and the available capital.

6.3 Typical Properties in Various Plastics

6.3.1 Polypropylene. The major applications for talc filled polypropylene are in the automotive and appliance industries because of the increased stiffness and high-temperature creep resistance imparted by the talc. Additional advantages imparted by the talc are good surface quality of the finished part, lower mold shrinkage, and easier processing, with the talc acting as a flow promoter. In the automotive industry, specific applications for talc in polypropylene include fan shrouds, heater housings and ducts, battery heat shields, and fluid pump parts. In the appliance industry, specific applications include refrigerator door liners, heater and vacuum pump housings, and washing machine agitators.

Typical physical properties of 20% and 40% talc filled polypropylene composites are shown in Table 9-4. The outstanding effectiveness of the platy talc in increasing the modulus of polypropylene, both at ambient and elevated temperatures, with minimum detriment in tensile strength reduction is shown in Fig. 9-3–9-6.

Table 9-4. Typical physical properties of 20% and 40% talc-filled polypropylene. (All properties determined at 23°C.)

Property	ASTM Procedure	Unfilled polypropylene	20% Talc filled polypropylene	40% Talc filled polypropylene
Flexural modulus (tangent)	D-790	230,000 PSI	390,000 PSI	600,000 PSI
Tensile yield strength	D-638	5,600 PSI	5,800 PSI	5,400 PSI
Tensile elongation at yield	D-638	8.5%	4.1%	3.1%
Notched Izod impact strength	D-256A	0.58 ft-lb/in.	0.47 ft-lb/in.	0.42 ft-lb/in.
Heat deflection @264 psi	D-648	62°C	72°C	88°C

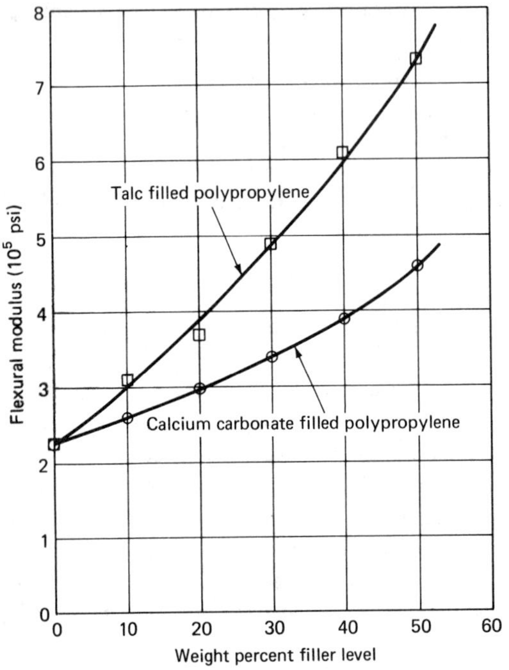

Fig. 9-3. Effect of filler level on flexural modulus of polypropylene at 23°C.

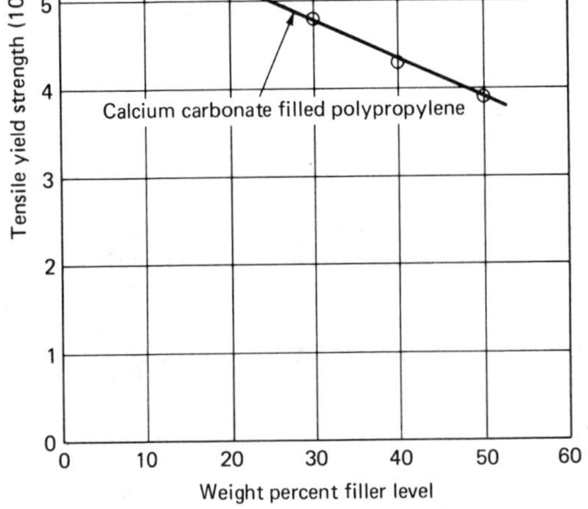

Fig. 9-4. Effect of filler level on tensile yield strength of polypropylene at 23°C.

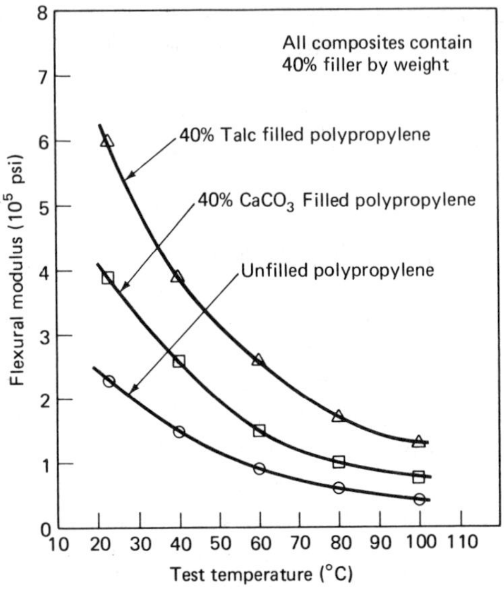

Fig. 9-5. Temperature dependence of flexural modulus for polypropylene materials.

While it has been generally assumed in the past that the increased rigidity and creep resistance of filled polymers are usually attained at the sacrifice of impact strength, this has been shown to be not true in all cases.[15] Certain grades of treated calcium carbonate have the

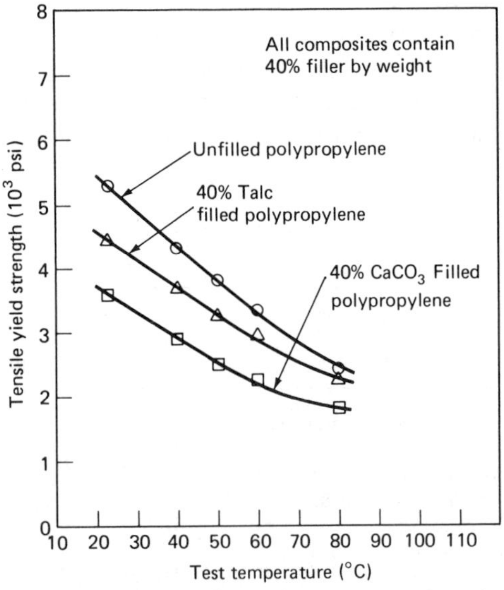

Fig. 9-6. Temperature dependence of tensile yield strength for polypropylene materials.

unique advantage of dramatically increasing the impact strength above that of the polypropylene homopolymer. Considerable effort is presently underway to develop a talc that does not cause a decrease in impact strength and still maintains its other outstanding properties in polypropylene. Another approach for certain critical applications requiring both high stiffness and impact strength is to use a proper mixture of talc and treated calcium carbonate to tailor the polypropylene composite to fit the required end use properties.

There are many polypropylene applications, particularly in the automotive industry, where long term thermal stability is an important consideration.[16] One standard test for thermal stability is the aging of test samples in an air oven at 150°C. Days to failure is determined using visual signs of degradation as the failure point. In general, any filler of fine particle size and high surface area will cause a significant decrease in the thermal stability of a nominally stabilized polypropylene resin filled at high loadings (typically 40% by weight). The higher the surface area of the filler, the lower the thermal stability of the composite. To counteract this effect, there are many commercial stabilizer additives on the market that when added in small amounts will increase the thermal stability of a talc filled polypropylene composite back to that of the original unfilled homopolymer. Use of talcs containing low amounts of metal impurities and having particle size distributions aimed at lowering the surface area will reduce the amount of stabilizer required for acceptable performance.

6.3.2 Other polymers. When incorporated in other polymers, platy talc will generally cause property improvements of the same order of magnitude as with polypropylene. Typical physical properties of talc filled rigid PVC, medium molecular weight polyethylene, and high impact polystrene are shown in Table 9-5.

In a recent paper[15], it has been shown that for high density polyethylene, a 30% loading of talc gives the optimum impact properties to the composite. Using rigid PVC as an example, Figs. 9-7 and 9-8 again show the outstanding ability of platy talc in increasing the modulus of a polymer, both at ambient and elevated temperatures, with negligible loss in tensile strength. Talc has also been used in various grades of nylon and phenolic molding compounds to achieve higher strength and stiffness.

6.3.3 Predictive equations for modulus of talc filled polymers. Using the Halpin-Tsai equations, one can obtain good predictions of the modulus for any filled polymer system at various loading levels and at various temperatures. The Halpin-Tsai equations are based on micromechanics theory and are fully discussed by others.[17, 18, 19] A paper describing the use of these equations to predict the modulus for talc filled polymers at any temperature and loading level was published in 1975.[20, 21]

Table 9-5. Typical physical properties of talc-filled polymers. (All properties determined at 23°C.)

			MEDIUM MOLECULAR WEIGHT POLYETHYLENE		RIGID PVC		HIGH IMPACT POLYSTYRENE	
Property	ASTM Procedure	Unit	Unfilled	40% Talc filled	Unfilled	30% Talc filled	Unfilled	40% Talc filled
Flexural modulus (tangent)	D-790	PSI	170,000	480,000	510,000	1,050,000	340,000	840,000
Tensile yield strength	D-638	PSI	7,600	8,900	8,200	7,900	4,800	5,500
Tensile elongation at yield	D-638	%	10.4	8.8	4.5	2.8	2.9	1.6
Notched Izod impact strength	D-256A	ft-lb/in.	3.67	3.72	0.59	0.23	2.75	0.45

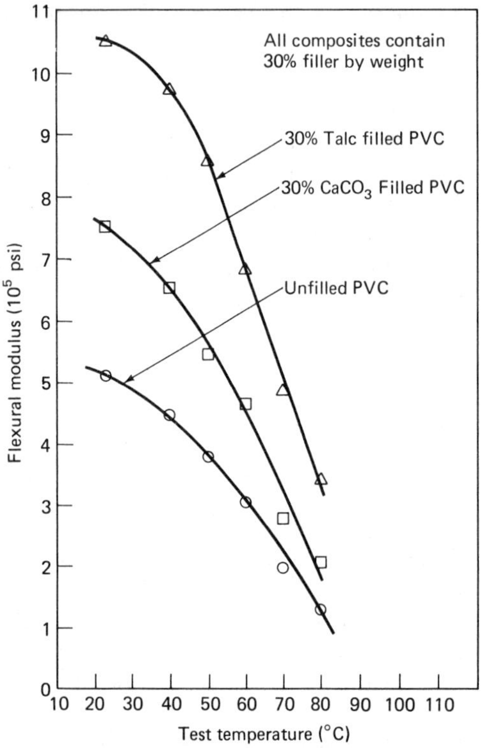

Fig. 9-7. Temperature dependence of flexural modulus for rigid PVC materials.

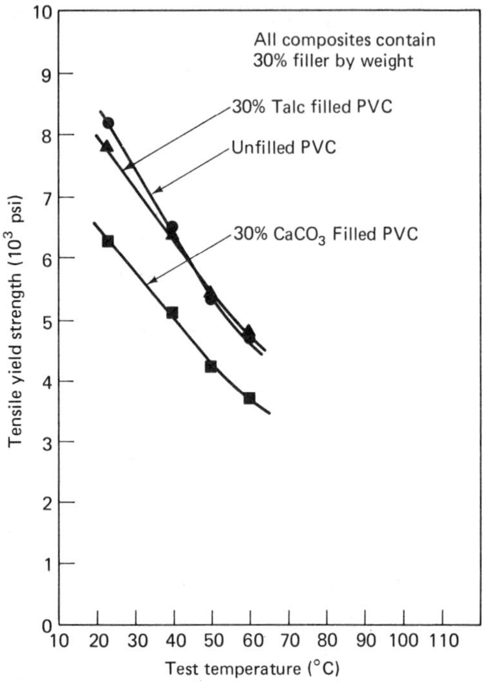

Fig. 9-8. Temperature dependence of tensile yield strength for rigid PVC materials.

7. NEW DEVELOPMENTS

New technology in the area of grinding and beneficiation is presently being applied to produce platy talcs having finer particle sizes, lower impurity levels, and higher aspect ratios than are presently commercially available. There is a major effort to understand the effects of talc particle size and size distribution on the major processing[22] and performance properties of filled plastics. With the full understanding of how the size distribution affects performance in different polymer systems, new products will be available that are tailored to meet specific end use applications. The development of a talc product that improves the impact strength of a polymer while still maintaining its other outstanding properties is a major goal for talc producers.

At the present time, there is extensive research on surface modifications of talc to produce benefits that include higher stiffness at elevated temperatures, increased thermal stability, and higher impact strength. Silane coupling agents are available which can produce a good bond onto the talc surface. Lower priced coating agents and reduced costs for application are needed to make these products a commercial success.

8. ACKNOWLEDGMENTS

The authors wish to express their appreciation to the Minerals, Pigments, and Metals Division of Pfizer Inc. for permission to publish this work. The plastics laboratory at the MPM division of Pfizer has furnished the technical data in Tables 9-4 and 9-5 and Figures 9-3 to 9-8.

References

1. U. S. Bureau of Mines, *Minerals Yearbook*, 1973.
2. Taggart, A. F., *Handbook of Mineral Dressing*, John Wiley, New York, 1945.
3. Pryor, E. J., *Mineral Processing*, 3rd ed., Applied Science Publishers, London, 1965.
4. Perry, J. H., *Chemical Engineers' Handbook*, McGraw-Hill, New York.
5. Mason, B. and Berry, L. G., *Elements of Mineralogy*, W. H. Freeman, San Francisco, California, 1968.

6. Bricker, O. P., Nesbitt, H. W., and Gunter, W. D., "The Stability of Talc", *Am. Mineralogist*, **58:** 64-72 (1973).
7. Hostetler, P. B., Hemley, J. J., Christ, C. L., and Mantoya, J. W., "Talc-Chrysotile Equilibrium in Aqueous Solutions," *Geol. Soc. of America: Abstr. Programs*, **3:** 605-606 (1971).
8. Marsick, E. E., "Raw Materials Index-Pigment Section," National Paint and Coatings Association, Washington, D. C., May, 1974.
9. Ladoo, R. B., "Talc and Soapstone: Their Mining, Milling, Products and Uses," U. S. Bureau of Mines Bulletin No. 213, p. 2 (1923).
10. Engel, A. E. J. and Wright, L. A., "Talc and Soapstone," in *Industrial Minerals and Rocks*, A.I.M.E., New York, N.Y., p. 835, 1960.
11. Neathery, T. L., et al., "Talc and Asbestos at Dadeville, Alabama," U. S. Bureau of Mines Report of Investigation No. 7045, p. 53 (1967).
12. Whittaker, Clark and Daniels, Inc., "Talcs, Magnesium Silicate, Pyrophyllite, Soapstone, Steatite," Bulletin No. 54-1, pp. 9, 51, 75, 101, 177, 191, 259 (1957).
13. *Industrial Minerals* pp. 15-31, London, April, 1975.
14. Schoengood, A. A., "Compounding", *SPE Journal*, **29:** (2), 21 (Feb. 1973).
15. Radosta, J. A., "Impact and Flexural Modulus Behavior of Calcium Carbonate and Talc Filled Polyolefins", 34th SPE-ANTEC, April 1976. Also Pfizer Report No. MPM-2-76.
16. Bragg, T. P. and Held, M. D., "Choosing the Right Talc for Reinforcing Polypropylene," *Plastics Engineering*, **30:** (9), 30 (Sept. 1974).
17. Ashton, J. E., Halpin, J. C., and Petit, P. H., "Primer on Composite Materials: Analysis," Technomic Publication, Stamford, Conn., 1969.
18. Kardos, J. L., "Structure Property Relations in Short-Fiber Reinforced Plastics," *Critical Reviews in Solid State Sciences*, **3:** Issue 4, 419-450, (Aug. 1973), CRC Press, Cleveland, Ohio.
19. Halpin, J. C. and Kardos, J. L., "Moduli of Crystalline Polymers Employing Composite Theory," *J. Appl. Phys.*, **43:** 2235 (1972).
20. Radosta, J. A., "Effect of Talc and Calcium Carbonate on the Elevated Temperature Properties of Polypropylene and PVC," 33rd SPE-ANTEC, May 1975.
21. PFIZER REPORT No. MPM-1-76 (Note correction in modified Halpin-Tsai equation for talc filled polymers.)
22. Chapman, F. M., and Lee, T. S., "Influence of Talc Filler on the Melt Rheology of Polypropylene," 27th SPE-ANTEC, 1969.

10
Miscellaneous Mineral Fillers

Benjamin M. Walker
Walker Engineering Associates
Madison, Connecticut

CONTENTS

1. INTRODUCTION

A number of mineral materials, although less commonly used as plastics fillers than others, are desirable because of the special effect they have upon a specific property for a particular application. This chapter describes several such mineral materials, indicating properties and sources of supply. It also gives an indication of the distinctive type of effect which may lead to its choice.

2. DESCRIPTION OF MISCELLANEOUS FILLERS

2.1 Carbide Fillers

2.1.1 Silicon carbide, SiC.

Trade Names:
"Carborundum"–Carborundum Co.
"Carbolon"–Exolon Co.
"Crystolon"–Norton Co.
"Carbonite"–General Abrasives Div.

Source: Produced by reaction of carbon and silica sand upon heating in a furnace.

Properties:
Bluish black iridescent crystals.
Specific gravity 3.217 (198 lbs/cu ft).
Hardness–Very hard.
–Mohs scale 9
–Knoop, 11 kg/min 2500
Decomposes at 2210°C; m.p. 2700°C.
Coefficient of thermal expansion,
per °C 4.7×10^{-6}
per °F 2.7×10^{-6}
Specific heat, 75°F 0.22
Heat conductivity, cal/g/sec/cm²/°C/cm, 20°C 0.49
, Btu/hr/sq ft/in./°F, 75°F 500.
Modulus of elasticity, 77°F 68×10^{-6} psi
Compressive strength, 77°F 150,000 psi. (500,000 psi per Norton Co.)
Electrical resistivity, ohm cm, 75°F 10.
Insoluble in water and alcohol
Soluble in fused alkali.

Special Filler Use: To increase abrasion resistance.

2.2 Metallic Oxides

2.2.1 Aluminum oxide, alumina, Al_2O_3.

Source: Obtained from bauxite ore by chemical treatment, including leaching with hot caustic soda, followed by precipitation of the hydrated aluminum oxide by hydrolysis. The hydrate is then washed, filtered, and calcined to obtain anhydrous aluminum oxide. It may also be obtained from coal mine waste water as aluminum sulfate, then reduced to alumina. It also occurs as a natural mineral called corundum. Impure natural varieties are emery, ruby, and sapphire.

Properties:

White powder, balls, or lumps of various mesh
Specific gravity 3.4-4.0
Melting point 2030°C
Tensile strength, 20°C $8\text{-}50 \times 10^3$ psi
Modulus of elasticity, 20°C $15\text{-}52 \times 10^6$ psi
Coefficient of thermal expansion, 100°C, °C–$5.5\text{-}8.0 \times 10^{-6}$
Specific heat, 20°C, cal/g/°C 0.21
Heat conductivity, 100°C, cal/g/sec/cm^2/°C/cm 0.069
Insoluble in water; difficultly soluble in mineral acids and strong alkalis

Grades: Technical; C.P.; fiber; high purity; calcined.

Filler Uses: Improves electrical resistivity and thermal conductivity of polymers. Main use has been in polyesters and epoxies. Also reduces coefficient of thermal expansion and improves chemical inertness; improves hardness and abrasion resistance.

2.2.2 Beryllium oxide, beryllia, BeO.

Source: Produced by heating beryllium nitrate or hydroxide.

Properties:

White amorphous powder (*Caution: Dust is poisonous!*)
Specific gravity 2.86-3.02
Melting point 2530-2570°C
Hardness, Mohs 9
Elastic modulus, 20°C 45×10^6 psi
Coefficient of thermal expansion, 100°C, °C $8.0\text{-}8.4 \times 10^{-6}$
Specific heat, 20°C, cal/g/°C 0.24
Heat conductivity, cal/g/sec/cm^2/°C/cm 0.50
Insoluble in water; soluble in acids and alkalis

Grades: Technical; C.P.; pure.

Special Filler Use: Improves thermal conductivity for "heat sinks" without loss of desirable electrical properties.

2.2.3 Iron oxide. A range of compounds from ferric oxide (red), ferric oxide hydrated (orange), ferroferric oxide (black), and blends of these to give brown and a wide range of intermediate colors.

A. Iron oxide, red, ferric oxide, Fe_2O_3.

Source: Natural hematite ore; by product in some industries.

Properties:

Dense, dark-red powder or lumps
Specific gravity 5.12-5.24
Melting point 1565°C
Hardness, Mohs 5-6
Insoluble in water; soluble in acids

Grades: Technical, 99.5%, pure (96% min.), electronic.

Uses: In small percentages as a pigment.

Special Filler Uses: To improve high temperature stability; the Gamma form of iron oxide is used with resins such as polypropylene to produce high specific gravity to yield a hard, high modulus composite.

B. Iron oxide, black, ferroferric oxide, ferrosoferric oxide, magnetic iron oxide, black rouge, Fe_3O_4

Source: Natural magnetite ore; action of air, steam, or carbon dioxide on iron.

Properties:

Reddish or bluish black powder
Specific gravity 5.18
Melting point 1538°C (decomposes)
Hardness, Mohs 5.5-6.5
Insoluble in water, alcohol, ether; soluble in acids

Grades: Technical, pure (96% min).

Special Filler Use: In magnetic inks and ferrites for electronic industry.

2.2.4 Lead oxide.

A. Lead oxide, yellow; litharge, plumbous oxide, lead monoxide, PbO.

Source: By oxidizing metallic lead in air; natural mineral litharge.

Properties:

Crystals varying from yellow to yellow-red depending upon purity
Specific gravity 9.53
Melting point 888°C
Hardness, Mohs 2
Insoluble in water; soluble in acids and alkalis

Grades: C.P., fused, powdered

Special Filler Use: For high specific gravity compounds.

B. Lead oxide, red; red lead, minium, plumboplumbic oxide, Pb_3O_4.

Source: By heating litharge in a furnace in air; natural mineral minium.

Properties:

Bright red powder

Specific gravity 8.32-9.16

Melting point–decomposes at 500-530°C.

Hardness, Mohs 2.5

Insoluble in water; partly soluble in acids

Grades: Technical, 95%, 97%, 98%.

Special Filler Uses: For opacity to radiation, as in shielding for X-ray equipment; for molding frangible bullets.

2.2.5 Magnesium oxide, magnesia, MgO.

Source: By calcining magnesium carbonate, basic magnesium carbonate, or magnesium hydroxide; by heating magnesium chloride with lime.

Properties:

White powder, either light or heavy depending upon whether prepared by oxidizing magnesium to a low bulk density material or heating basic magnesium carbonate to a high bulk density product.

Specific gravity 3.6 (varies)

Melting point 2800°C

Boiling point 3600°C

Elastic modulus, 20°C 12×10^6 psi

Coefficient of expansion, 100°C, °C 11.0×10^{-6}

Specific heat, 20°C, cal/g/°C 0.23

Heat conductivity, 100°C, $cal/g/sec/cm^2/°C/cm$ 0.082

Dielectric constant, 25°C, 60 cycles 9.7

Insoluble in water; soluble in acids and ammonium salt solutions

Grades: Technical; C.P.; USPXVII, F.C.C., 99.5%; fused; low-boron; rubber; semiconductor.

Special Filler Use: Gives increased stiffness, hardness, and creep resistance to composites, with resins such as polypropylene; hard calcined grade is useful for dissipation of heat from resin systems; increases the viscosity of polyester formulations in SMC and BMC compounds.

2.2.6 Titanium dioxide, titanium white, titania, titanic oxide, TiO_2.

Source: From ilmenite ore, treated with sulfuric acid and further processing to yield primarily *anatase.*

From rutile ore, by chlorination and oxidizing the chlorides.

(Note: As a pigment, the *rutile* form has greater opacity than the anatase form).

Properties:

White to black powder depending upon purity

Two crystalline forms: anatase and rutile

	anatase	*rutile*
Specific gravity	3.8	4.3
Melting point	1560°C	1640°C (decomposes)
Hardness, Mohs	5-6	6-6.5
Index of refraction	2.5	2.7

Both forms are insoluble in water and dilute acids; both forms are soluble in hot concentrated sulfuric acid and alkalis.

Elastic modulus, 20°C $12\text{-}15 \times 10^6$ psi

Water absorbtion 0.02%

Coefficient of expansion, 100°C, °C 7.5×10^{-6}

Specific heat, 20°C, cal/g/°C 0.17

Heat conductivity, 100°C, $cal/g/sec/cm^3/°C/cm$ 0.015

Grades: Technical; pure; USPXVII; single crystals, whiskers.

Special Filler Use: Pigment, opacifier (greatest hiding power of white pigments); used in amounts to 50 phr for vinyl compounds; rutile imparts weather resistance, heat resistance, and rigidity.

2.2.7 Zinc oxide, Chinese white, zinc white, ZnO.

Source: By oxidizing of pure zinc metal; by roasting zinc oxide ore (franklinite) with coal and subsequent air oxidation; by similar treatment of other ores.

Properties:

Amorphous, odorless, white or yellowish white powder.

Absorbs carbon dioxide from the air; has high UV absorbtion.

Specific gravity 5.47
Melting point 1975°C
Hardness, Mohs 4.5
Insoluble in water and alcohol; soluble in acids and alkalis.

Grades: American process, lead free; French process, lead free; green seal, red seal, white seal (according to fineness); leaded, USPXVII, single crystals.

Special Filler Use: Filler and accelerator-activator in rubber and plastics; improves resistance to weathering (using 10 phr with polypropylene); promotes hardness, heat stability, and electrical conductivity to polymers; has been used in silicones, polyesters, and polyolefins.

2.2.8 Zirconium oxide, zirconium dioxide, zirconia, zirconium anhydride, baddeleyite, ZrO_2.

Source: Natural, as baddeleyite; by heating zirconium hydroxide or zirconium carbonate.

Properties:
Heavy, white to yellow amorphous powder
Specific gravity 5.73
Melting point 2600°C
Hardness, Mohs 6.5
Refractive index 2.2
Most heat resistant of common refractories
Insoluble in water and most acids or alkalis at room temperature; soluble in nitric acid, hot concentrated hydrochloric, hyrofluoric, and sulfuric acids

Grades: Reagent; technical; crystals; fused; C.P. (99%); hydrous.

Special Filler Use: With resins such as polypropylene to produce a hard, high modulus composite; with polyester resins to promote a faster cure.

2.3 Sulfates

2.3.1 Barium sulfate, barite, barytes, $BaSO_4$.

Source: Natural as mineral barite; by treating a barium salt with sodium sulfate; as by-product of hydrogen peroxide manufacture.

Properties:
Fine, white, odorless, tasteless, bulky powder
Specific gravity 4.25-4.5
Melting point 1580°C
Hardness, Mohs 3-3.5
Insoluble in water, organic solvents, acid, and alkali solutions; soluble in concentrated sulfuric acid.

Grades: Technical, dry, pulp, bleached, natural; C.P.; USPXVII; X-ray.

Special Filler Use: As filler and white pigment; increases specific gravity, frictional resistance, chemical resistance, and X-ray opaqueness; promotes a faster cure as a filler in polyester resins; used in polyurethane foams to provide high density, higher load bearing properties, and improved processing.

2.4 Sulfides

2.4.1 Molybdenum disulfide, molybdenum sulfide, molybdic sulfide, molybdenite, MoS_2.

Source: Purification of natural molybdenum ore, molybdenite; reaction of sulfur or hydrogen sulfide on molybdenum trioxide.

Properties:
Black, lustrous powder
Specific gravity 4.80
Melting point 1185°C
Hardness, Mohs 1-1.5
Coefficient of friction 0.02-0.06
Magnetic permeability—Diamagnetic
Insoluble in water; soluble in aqua regia and concentrated sulfuric acid

Grades: Technical, Technical Fine, suspension grades depending upon particle size.

Special Filler Use: As an internal lubricant, reduces wear, reduces friction, increases resistance to high temperature, increases dielectric strength, improves processability. Also increases elastic modulus, tensile strength, and flexural strength.

2.4.2 Zinc sulfide, ZnS.

Source: Natural, in the ores sphalerite and wurtzite; by passing hydrogen sulfide into a solution of a zinc salt.

Properties:
Yellowish—white powder; stable if kept dry
Specific gravity—alpha (from wurtzite) 3.98
—beta (from sphalerite) 4.10
Melting point—beta form changes to alpha at 1020°C, sublimes at 1180°C.

Hardness, Mohs 3.5
Insoluble in water; soluble in acids

Grades: Technical; C.P.; fluorescent or luminous; single crystals.

Special Filler Use: In production of colored molding powder.

2.5 Silicates

2.5.1 Aluminum silicate (varying proportions of Al_2O_3 and SiO_2).

Sources: Occur naturally in clays, such as kaolinite (China clay), halloysite, and many other minerals, usually with other minerals and impurities.

Mullite, $3Al_2O_3 \cdot 2SiO_2$, is made by heating other aluminum silicates, such as cyanite, sillimanite, and andalusite, to high temperatures.

Properties:
Fine, white powder, colorless crystals
Specific gravity 3.15
Melting point 1810°C
Insoluble in water

Sillimanite, fibrolite, Al_2OSiO_4, occurs naturally as fine, fibrous masses.

Properties:
White, gray, brown, or greenish; luster vitreous.
Specific gravity 3.23
Hardness, Mohs 6-7

Cyanite, kyanite, Al_2OSiO_4, is a natural mineral; same composition as sillimanite and andalusite.

Properties:
Blue, green, white, gray; luster-vitreous.
Specific gravity 3.56-3.66
Hardness, Mohs 5-7

Andalusite, Al_2OSiO_4, is a natural mineral; same composition as sillimanite and cyanite.

Properties:
Gray, greenish, reddish, or bluish
Specific gravity 3.1-3.2
Hardness, Mohs 7-7.5

Special Filler Use: Of these aluminum silicates, mullite is the form of particular interest; although it has no properties that are significantly different from other aluminum silicates, its usage has declined; it has slip characteristics that permit reduction in the amount of lubricant required in the molding composition.

2.5.2 Zirconium silicate, zircon, $ZrSiO_4$ or $ZrO_2 \cdot SiO_2$.

Source: Naturally occurring mineral.

Properties:
Brown, gray, red, colorless; luster adamantine
Specific gravity 4.5-4.68
Melting point dissociates at 3,200°F
Hardness, Mohs 7.5-8.0
Specific heat, Btu/hr/°F 0.132
Thermal conductivity, Btu/hr/sq ft/°F/in. 14.5
Coefficient of thermal expansion, °F 4.2×10^{-6}
Dielectric constant, avg. 12.7
Refractive index, avg. 1.95
Insoluble in water, dilute acids, cold, dilute alkalis and hot, concentrated sulfuric acid; slightly soluble in hydrofluoric acid; soluble in hot, concentrated alkalis
Hygroscopicity–Nil
Toxicity–nontoxic

Special Filler Use: Improves electrical insulation properties; promotes faster cure in polyester resins; increases heat stability; greatly reduces absorbtion of water, ethyl alcohol and benzene; reduces transmission of vapors of water and solvents; reduces shrinkage and cracking; increases thermal shock resistance; reduces aging and color degradation; serves as a primary pigment and pigment extender.

Grades: Zircon Granular "TAM"–a pure grade, Zircon Granular G–99.8%

2.6 Titanates

2.6.1 Barium titanate, $BaTiO_3$, $2BaO \cdot 3TiO_2$

Source: Fusion of titanic anhydride and barium carbonate.

Properties:
Light gray-buff powder, or yellow cubical and octahedral crystals.
Specific gravity 5.91-5.95
Melting point 1400-1430°C
Dielectric constant, 25°C 1,000-10,000
Dissipation factor, 25°C, 60 cycles 0.056
Resistivity, ohm-cm, 25°C 10^{11}-10^{13}
Insoluble in water and alcohol; slightly solu-

ble in dilute acids; soluble in concentrated sulfuric and hydrofluoric acids

Special Filler Use: Used with such resins as epoxies to produce castings with low loss and stable dielectric properties. Also used to modify the dielectric constant of a compound.

3. SUMMARY OF SPECIAL USES OF THESE MINERAL FILLERS

The miscellaneous mineral fillers described in this chapter, like fillers and reinforcements generally, will affect many of the properties of the compounds into which they are incorporated. These effects are discussed at length in section II Chapter 2, under "Guide to Selection and Usage of Fillers and Reinforcements," and include a wide range of factors, such as physical, chemical, and electrical properties, processability, and cost. However, in addition to these general effects, the fillers listed here have some unusual effects upon certain specific properties. These special effects may determine their choice for an unusual modification of properties which is especially desired for a particular application. The Table lists these miscellaneous fillers and the effect each is likely to have upon the properties listed. The *special effect* on properties for which a particular filler might be chosen is indicated in the table. The extent of the effect will depend not only upon the type of filler used, but also on many other factors, including the amount of filler used, the type of polymer used, and details of compounding and processing.

4. MATERIAL SUPPLIERS

4.1 List of Materials and Their Suppliers

Note: This is a partial list of suppliers. Others may be found by referring to references 11, 12, and 13.

4.1.1 Carbide Fillers.

Silicon carbide:

Carborundum Co.	"Carborundum"
Norton Co.	"Crystolon"
Exolon Co.	"Carbolon"
General Abrasives Co., Div. U.S.I.	"Carbonite"

4.1.2 Metallic oxides.

Aluminum oxide:
Aluminum Co. of America
Degussa, Inc.
General Abrasive Co. "Lionite"
Hammill & Gillespie, Corp.
Kaiser Aluminum & Chemical Corp., Kaiser Chemicals Div.
H. Kohnstamm & Co.
Mallincrodt Chemical Works
Reynolds Metals Co.
Whittaker, Clark & Daniels, Inc.

Beryllium oxide:
Kawecki Berylco Industries Inc.

Iron oxide:
United Mineral & Chemical Corp.
Pfizer Minerals, Pigments and Metals Div.

Lead oxide:
American Hoechst Corp., Chemicals and Plastics Div.
Eagle-Picher Industries, Inc.
NL Industries, Inc., Industrial Chemicals Div.
United Mineral & Chemical Corp.

Magnesium oxide:
Basic Chemicals, Div. Basic, Inc.
C.P. Hall Co.
Dow Chemical Co.
General Electric Co.
Kraft Chemical Co.
Michigan Chemical Corp.
Morton Chemical Co.
Pigment and Chemical Co.
Frank Samuel & Co.
Standard Lime and Refractories Co.
United Mineral & Chemical Corp.
Whittaker, Clark & Daniels, Inc.

Titanium dioxide:
H. Kohnstamm & Co.
New Jersey Zinc Co.
Titanium Pigment Corp.

Table 10-1. Special uses of miscellaneous mineral fillers

	Density	Cost	Tensile Strength	Elastic Modulus	Impact Strength	Hardness	Abrasion Resistance	Lubricity	Thermal Expansion	Thermal Conductivity	Thermal Endurance	Crazing, Cracking	Shrinkage
Carbides:													
Silicon Carbide	I	I		Ⓘ	D	Ⓘ	Ⓘ	D	D	Ⓘ			D
Metallic Oxides													
Aluminum oxide	I	Ⓓ		I	D	I	I	D	D	Ⓘ		D	D
Beryllium oxide	I	I		I	D	I	I	D	D	Ⓘ		D	D
Iron oxide	Ⓘ			Ⓘ		Ⓘ			D		Ⓘ (red)		D
Lead oxide	Ⓘ (yellow)		D	I	D	Ⓓ (red)							D
Magnesium oxide	I	I			D	I			D	Ⓘ			
Titanium oxide	I	I		I		I			D		Ⓘ		D
Zinc oxide	I				D	Ⓘ					Ⓘ		D
Zirconium oxide	I			Ⓘ	D	Ⓘ							D
Sulfates													
Barium sulfate	Ⓘ	D		I	D			D					D
Sulfides													
Molybdenum disulfide	I	I	I	I		D		Ⓘ			I		D
Zinc sulfide	I	I			D				D				D
Silicates													
Aluminum silicate	I												
Calcium silicate	I	Ⓓ	Ⓘ	I		I	I		D	I			D
Zirconium silicate	I					I			D	I	Ⓘ		D
Other Fillers													
Barium Titanate	I	I											

KEY: I—Increases property value of polymer.
D—Decreases property value of polymer.
○—Indicates *Special Effect* upon polymer property which would lead to its use.

Zinc oxide:
Harshaw Chemical Co., Div. Kewanee Oil Co.
New Jersey Zinc Co.

Zirconium oxide:
NL Industries Inc., TAM Div.

4.1.3 Sulfates.

Barium sulfate:
Pfizer Minerals, Pigments and Metals Div.
Hammill & Gillespie, Inc.
Industrial Minerals, Inc.
Kraft Chemical Co.
Landers-Segal Color Co.
NL Industries Inc., DeLore Div.
H.M. Royal, Inc.
Smith Chemical & Color Co.
Thompson, Weinman & Co.
Whittaker, Clark & Daniels, Inc.

Table 10-1 (*Continued*)

	Dimensional Stability	Electrical Insulation	Electrical Conductivity	Creep	Radiation Absorption (X-Ray opaque)	Viscosity	Speed cure or handling	Chemical Resistance	Weather Resistance	Moisture Absorption	Processibility	Fire Retardancy	Magnetic Properties
Carbides													
Silicon Carbide	I			D							D		
Metallic Oxides													
Aluminum oxide	I	(I)		D		I		I	I		D	I	
Beryllium oxide	I			D		I					D		
Iron oxide	I			D		I					D		(I) (yellow)
Lead oxide				I	(I) (red)								
Magnesium oxide	I			D		I					D		
Titanium oxide	I					I			(I)		D		
Zinc oxide	I		(I)	D		I	(I)		(I)				
Zirconium oxide	I			D		I	(I)						
Sulfates													
Barium sulfate	I				(I)		(I)	(I)			I		
Sulfides													
Molybdenum disulfide		I									I		
Zinc sulfide	I					I							
Silicates													
Aluminum silicate											(I)		
Calcium silicate	I			D		I					D		
Zirconium silicate	I	(I)		D		I	(I)			(D)			
Other Fillers													
Barium Titanate		(I)											

4.1.4 Sulfides.

Molybdenum sulfide:

Climax Molybdenum Co.
Acheson Colloids Co.
Bel-Ray Co.
CERAC, Inc.
Dow Corning Corp.
Electronics Space Products Inc.
Graphite Products Corp.
McGee Industries Inc., McLube Div.
Meta-Therm Corp.
Pesses Co., The
Rocky Mt. Research, Inc.
VAR-LAC-OID Chemical Co.
White & Bagley Co.

Zinc sulfide:

American Hoechst Corp., Chemicals & Plastics Div.

Aceto Chemical Company, Inc.
Atomergic Chemtals Co., Div. Gallard-Schlesinger Chem. Mfg. Co.
CERAC, Inc.
Eagle-Picher Industries, Inc.
Electronic Space Products, Inc.
Research Organic/Inorganic Chemical Corp.
Rocky Mountain Research, Inc.

4.1.5 Silicates.

Aluminum silicate:
Georgia Kaolin Co.
R.T. Vanderbilt Co.

Zirconium silicate:
NL Industries Inc., TAM Div.
Frank Samuel and Co., Inc.

4.1.6 Other Fillers.

Barium titanate:
CERAC, Inc.
Electronic Space Industries, Inc.
NL Industries Inc., TAM Div.
United Minerals & Chemical Corp.

4.2 Addresses of Suppliers

Aceto Chemical Co., Inc., 126-02 Northern Blvd., Flushing, N.Y. 11368
Acheson Colloids Co., 1635 Washington Ave., Port Huron, Mich. 48060
Aluminum Co. of America, 1501 Alcoa Bldg., Pittsburgh, Pa. 15219
American Hoechst Corp., Chems & Plastics Div., Route 202-206, Somerville, N.J. 08876
Atomergic Chemetals Co., Div. Gallard-Schlesinger Chem. Mfg. Corp., 584 Mineola Ave., Carle Place, N.Y. 11514
Basic Chems, Div. Basic Inc., 845 Hanna Bldg., Cleveland, Ohio 44115
Bel-Ray Co., Inc., P.O. Box 526, Farmingdale, N.J. 07727
Carborundum Co., Abrasives Marketing Div., P.O. Box 337, Niagara Falls, N.Y. 14302
CERAC, Inc., 13460 W. Silver Springs Rd., Menemenee Falls, Wis. 53051
Climax Molybdenum Co., One Greenwich Plaza, Greenwich, Conn. 06830
Degussa Inc., 2-T Penn. Plaza, New York, N.Y. 10001
Dow Chemical Co., Midland, Mich. 48640
Dow Corning Corp., Savage Rd. #154, Midland, Mich. 48640
Eagle-Picher Industries, Inc., 580 Walnut St., Cincinnati, Ohio 45202
Electronic Space Products, Inc., 854 S. Robertson Blvd., Los Angeles, Calif. 90035
Exolon Co., 1000 E. Niagara St., Tonawanda, N.Y. 14150
General Abrasive Co., Div. USI, Inc., 2100 College Ave., Niagara Falls, N.Y. 14305
General Electric Co., Plastics Business Dept., 1 Plastics Ave., Pittsfield, Mass. 01201
Georgia Kaolin Co., 433 N. Broad St., Elizabeth, N.J. 07202
Graphite Products Corp., P.O. Box 207, Brookfield, Ohio 44403
Hall, C.P. Co., 7300 S. Central Ave., Chicago, Ill. 60638
Hammill and Gillespie, Inc., P.O. Box 104, Livingston, N.J. 07039
Harshaw Chem Co., Div. Kewanee Oil Co., 1945 E. 97 St., Cleveland, Ohio 44106
Industrial Minerals, Inc., 3 S. Congress St., York, S.C. 29745
Kawecki Berylco Industries, Inc., 220 E. 42nd St., New York, N.Y. 10017
Kohnstamm, H. and Co., 161 Ave. of the Americas, New York, N.Y. 10013
Kraft Chem. Co., 917 W. 18 St., Chicago, Ill. 60608
Landers-Segal Color Co., 78 Delavan St., Brooklyn, N.Y. 11231
Mallincrodt, Inc., P.O. Box 5439, St. Louis, Mo. 63160
Meta-Therm Corp., P.O. Box 1494 FDR Station, New York, N.Y. 10022
McGee Industries, Inc., McLube Div., 9 Crozerville Rd., Aston, Pa. 19014
Michigan Chemical Corp., 351 E. Ohio St., Chicago, Ill. 60611
Morton Chem. Co., Div. of Morton-Norwich Products, 110 N. Wacker Drive, Chicago, Ill. 60606
NL Industries Inc., TAM Div., P.O. Box C Bridge Sta., Niagara Falls, N.Y. 14305

Norton Co., 1 New Bond St., Worcester, Mass. 01606

New Jersey Zinc Co., A Gulf Western Co., 2045 City Line Rd., Bethlehem, Pa., 18017

Pesses Co., P.O. Box 24226, No. 8 Pepper Creek Dr., Pepper Pike, Ohio 44124

Pfizer Minerals, Pigments & Metals Div., 235 E. 42 St., New York, N.Y. 10017

Pigment & Chem. Co., Ltd., 3300 Cavendish Blvd., Montreal, H4B 2M8, Quebec, Canada

Pilot Engineering Co., Div. of Reagent Chem. Inc., Jeffersonville Industrial Park, Jeffersonville, Ind. 47130

PPG Industries, Inc., Chem. Div., One Gateway Ctr., Pittsburgh, Pa., 15222

Research Organic/Inorganic Chemical Corp., 507-519 Main St., Belleville, N.J. 07109

Reynolds Metals Co., 6601 W. Broad St., Richmond, Va. 23218

Rocky Mountain Research, Inc., 17400 Hwy 72, Golden, Colo. 80401

Royal, H.M. Co., Inc., 690 Pennington Ave., Trenton, N.J. 08601

Samuel, Frank & Co., C.E. Minerals Div. of Combustion Engineering, Inc., 443 South Gulph Road, King of Prussia, Pa., 19406

Smith Chemical & Color Co., Inc., 104-20 Dunkirk St., Jamaica, L.I., N.Y. 11412

Standard Lime and Refractories Company, Executive Plaza II, Hunt Valley, Md. 21030

Thompson, Weinman and Co., P.O. Box 130, Cartersville, Ga. 30120

Titanium Pigment Div., NL Industries, Inc., 100 Chevalier Ave., S. Amboy, N.J. 08879

United Mineral & Chem. Corp., 129 Hudson St., New York, N.Y. 10013

Vanderbilt, R.T. Co., Inc., 30 Winfield St., Norwalk, Conn. 06855

VAR-LAC-OID Chem. Co., 666 S. Front St., Elizabeth, N.J. 07202

White & Bagley of Mich., Inc., 7131 Westfield, P.O. Box 04067 NW Station, Detroit, Mich. 48204

Whittaker, Clark & Daniels, Inc., 1000 Coolidge St., South Plainfield, N.J. 07080

5. SUMMARY OF COSTS OF MINERAL FILLERS

The following table lists prices as of October, 1975 of these miscellaneous fillers as used as plastics additives. The actual prices of these materials are subject to substantial and rapid changes, mostly upward at the present time. However, the prices listed will give a good indication of the relative prices for the various fillers, and will be of value in showing the order of magnitude of price for a particular filler. Note that the prices listed are quoted for *large lots*, of carloads or 20-ton lots. Prices will be correspondingly higher for smaller lot purchases. Other quality and purity levels of these fillers are also available and prices will deviate from those shown, depending upon the grade chosen. However, the grades listed are ones commonly used as plastics fillers.

Material	Price per lb, carload or 20 ton lots, as of 9/29/75
Carbides	
Silicon carbide	
8–10 mesh	$0.58
600 mesh	1.85
1000 mesh	4.71
Metallic Oxides	
Aluminum oxide	
alumina, activated, granular	0.1365
Beryllium oxide	
U-15 or U-18 size	45.00 (10–50 lb lot)
Iron oxide	
red pure	0.20–0.23
black, syn.	0.31–0.38

Material	Price per lb, carload or 20 ton lots, as of 9/29/75
Lead oxide	
yellow, *Litharge*, commercial, powder	0.245
red, Lead, red, 96–98%	0.28
Magnesium oxide	
U.S.P., light	0.74
Magnesia, technical, chemical grade	0.070–0.075
Titanium dioxide	
anatase	0.435
rutile, regular grade	0.435
Zinc oxide	
U.S.P.	0.44
pigment, French process	0.425
Zirconium oxide	
C.P. white	2.22 (1.50 in 12/73)
electric fused, lump	No prices (0.505 in 12/73)
milled	No prices (0.64 in 12/73)
Sulfates	
Barium sulfate technical (Barytes, dry ground, southern, off color)	0.0235
water ground, white	0.0554
unbleached, extra fine, pigment grade	0.54
Sulfides	
Molybdenum disulfide	
technical	1.90
technical fine	2.04
suspension grade	2.12
Zinc sulfide	
standard, "L" grade	0.65
Silicates	
Zirconium silicate	
Zircopax	0.29
Zircopax A	0.29
Superpax	0.34
Superpax A	0.45
Excelopax	0.53
Other compounds	
Barium titanate	
"P" or "C" grades	0.615

References

1. Seymour, Raymond B., "Fillers for Plastics," *Modern Plastics Encyclopedia*, 1974–75, pp. 213–226.
2. Fisher, H. L., "Fillers for Molding Compounds," *Modern Plastics Encyclopedia*, 1949, pp. 312–313.
3. Mascia, L., "The Role of Additives in Plastics," Soc. Plastics Engineers, Publication #590, 1974.
4. Simonds and Church, *Encyclopedia of Basic Materials*, 1967.
5. Koenig, J. H., "New Developments in Ceramics," *Materials in Design Eng.* (March 1958).
6. Javitz, A. E., "The Epoxy-Resin System for Embedded Circuits and Components," *Elec. Mfg.*, 74: (April 1955).
7. Clark, F. M., *Insulating Materials for Design and Engineering Practice*, John Wiley & Sons, 1962.
8. Lee and Neville, "Epoxy Resins," McGraw-Hill, 1957, pp. 146–154.
9. "*The Vanderbilt Rubber Handbook*," R. T. Vanderbilt Co., Inc., 1968.
10. "Condensed Chemical Dictionary," Van Nostrand Reinhold, 1972.
11. *Chemical Week 1975 Buyer's Guide*, McGraw-Hill.
12. *Rubber Red Book*, Palmerton Publ. Co., 1969.
13. *Ceramics Data Book*, Cahner Publ. Co., 1969.
14. Deanin and Scott, "Fillers & Reinforcements for Plastics," *Adv. in Chem. Series*, (Plueddemam, E. P.) American Chemical Society, p. 89.

Section IV
Metallic, Conductive, and Magnetic Fillers

11
Metal Fillers

John Delmonte
Glendale, California

CONTENTS

1. INTRODUCTION

Metallic fillers add special qualities to plastic composites, enhancing properties such as thermal conductivity, electrical conductivity, response to magnetic fields, heat capacity, and radiation shielding. Metal-powder-filled resin composites find important applications as repair and maintenance products, adhesives, magnetic tapes, corrosion resistant coatings, and in specialized uses for electronic industries. Metallic fillers are available as a consequence of the considerable activity in powder metallurgy, which has yielded good quality steel, iron, copper, and copper alloys. There is also a growing interest in aluminum, zinc, silver, nickel, and lead powders for specialty plastics formulations.

2. METHODS FOR PRODUCING PRINCIPAL METALLIC POWDERS

Metallic powders are produced in substantial tonnage for powder metallurgical purposes. These powders are directed principally toward fulfilling requirements which emphasize the compaction and molding of metal powders, followed by elevated temperature sintering processes. This chapter is concerned primarily with the utilization of metallic powders in combination with organic resin matrices. While the objectives of these combinations are completely different than those for the fabrication of metal parts, one must recognize that the origins of metal powders stem principally from the powder metallurgy activity.

Virtually all pure metals, and most alloys, can be obtained in powder form. Iron powder is extensively used in powder metallurgy because it is economical, and the properties of finished parts are adequate for many structural purposes. Iron powders for powder metallurgy have been upgraded in recent years for improved compressibility and dimensional stability. Most fabricated parts, such as gears and pinions, weigh less than 5 lbs. However, newer techniques, such as isostatic pressing, permit the manufacture of considerably larger components. Significant inroads have been made in automotive equipment, farm machinery components, appliances, office machines, power tools, and hardware. Principal advantages in the use of powdered metals (P/M) for fabricating parts lie in low scrap losses, low labor input (dies shape the parts instead of machining), high volume production, and improved strength factors by alloying during fabrication. The use of premixed powders is also on the increase.[1]

2.1 Iron

Iron, chiefly sponge iron, is the most common base metal for P/M parts. Other common compositions include electrolytic iron, silicon steels, specialty iron compositions such as sulfurized iron, iron-carbon, iron-copper-carbon, nickel steels, stainless and nickel-molybdenum-steels. Identities of various grades have been established by the Powder Metallurgy Parts Association. Electrolytic iron powder and iron powder made by thermal decomposition of iron carbonyl are used for some specialty items such as magnetic tapes, inks, and pharmaceutical products, where the higher costs are justified. Iron powder made primarily by the reduction of high purity, finely ground Swedish iron ore, has been used widely in the United States. Mill scale has been another important source of iron oxide for making iron powder. Other methods have been the atomization of high-carbon iron melted in a cupola, followed by decarbonization or atomization of molten, low-carbon steel with high pressure water jets.

2.2 Copper

Copper and its alloys are used extensively as powders. They are prepared by several methods, including air atomization of molten copper, electrolysis of copper solutions, and hydrogen reduction of copper salts in solution. The latter process requires elevated temperatures and pressures. A similar process has been used for obtaining nickel and cobalt powders. Measures are taken to inhibit the effect of atmospheric oxygen on some grades of copper powders. Typical applications are motor brushes which are prepared from alloyed copper (65%) and graphite (35%). Brasses, bronzes, and nickel-silvers are processed as alloys into powdered form. These products have fulfilled major roles in porous filters and bearings. The latter are impregnated with lubricants for long-time serviceability.

2.3 Aluminum

Aluminum and its alloys may be comminuted in a hammer mill and particle size further reduced to a desired size by ball milling. The molten metal is also atomized by high velocity air into small spherical particles. These particles are then further reduced in a ball mill until the desired fineness is obtained. The powders are graded by air flotation separation. The grades intended for pigment coatings are flat and are further polished by friction during a polishing operation with rotating brushes. Small amounts of organic lubricants may be added to impart a floating quality, so that the flake will tend to leaf into a laminar orientation during application. "Bronzing" powders, copper, and gold colored particles are prepared in a similar manner, producing flat, polished, thin flakes which will achieve a plated metallic effect when formulated with surface coatings. During late 1976—aluminum powdered metal components began to make their appearance from Alcoa in Pittsburgh.

2.4 Lead and Zinc

Softer, easily melted metals, such as lead and zinc, may be atomized into fine particles. Lead has specialized functions in radiation shielding. Zinc is used as an electrolytic and sacrificial element in external coatings. These powders are identified with applications entailing the use of organic resin matrices. Significant trends in the development of powder metallurgy technology are shown in references 1 and 2. In the latter reference, further data are presented on the isostatic compression of P/M parts. The processing limits of P/M parts production have recently been reviewed.[3] Attention is called to raw material variables, listing apparent density, particle shape, particle size and distribution, oxides on the surface of particles, and other contaminants. Basic processing differences in handling powdered metals are shown in the schematic diagrams of Fig. 11-1. The use of *metal fiber* rein-

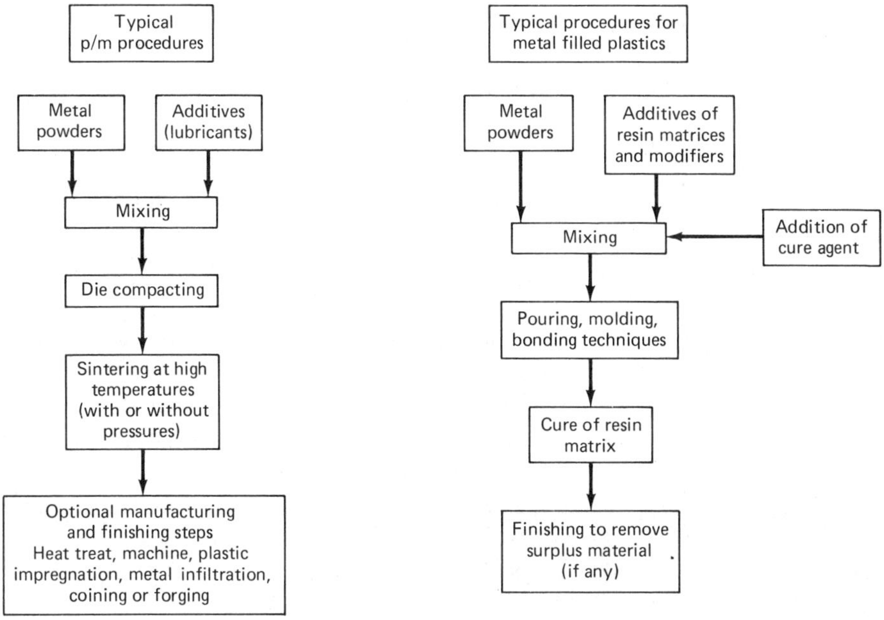

Fig. 11-1. Procedures for powder metallurgy compared with metal filled plastics.

forcements is discussed in Chapter 31 of this Handbook.[4,5]

3. PROPERTIES OF TYPICAL COMMERCIAL METAL POWDERS

Powder metallurgical activity has fostered interests in the manufacture of the more common types of powdered metals, iron, copper, aluminum, and their alloys. Specialty manufacturers of magnetic devices and electronic components have stimulated interest in nickel, cobalt, tungsten, manganese, silver, beryllium, bismuth, tin, and others. Manufacturers of protective coatings and radiation shields have been interested in finely divided lead, cadmium, zinc, indium, bronze, brass, zirconium, and molybdenum. Some of the powdered metals are offered in an alloyed form such as ferro-zirconium, ferro-boron, ferro-chrome, ferro-manganese, ferro-silicon, ferro-titanium, ferro-vanadium, and ferro-columbium. Examination of manufacturers' literature on powdered metals demonstrates emphasis on the needs of the powder metallurgical industry, where quantities utilized far exceed the requirements of the plastics industry. Tabulated properties on powders reflect the stronger commercial interests for P/M parts, sintered bearings, welding rods, brazing, motor brushes, friction discs, etc. Only in a few instances have manufacturers designated metal powders for plastics fillers and for coatings. The following is a list of commercially available metal powders and the companies that supply them.

Iron Powder

Hoeganeas Corp.
Riverton, N.J.

(Plant capacities of iron, steel, and high alloy steel powders (1973)–150,000 tons/year

Ancor C-2024 Sponge Iron Powder

Chemical	%
Total Fe	97.00
H_2 Loss	1.15
C	0.06
S	0.03
P	0.01

Physical	
Bulk density	2.4 g/cc
Sieve analysis, U.S. Standard	
+100	0.5
−100 + 140	5.5
−140 + 200	12.0
−200 + 230	5.0
−230 + 325	18.0
−325	59.0

Glidden Metals
SCM Corporation
900 Union Commerce Bldg.
Cleveland, Ohio 44115

Powders for Molded Ferro-magnetic Cores

Electrolytic grades A-233, A-210, A-230 for oscillator cores, choke coils, and transformer cores

Reduced iron powder—B-214 for toroids and threaded cores

Soft carbonyl iron powder K-291 for tuning cores and I.F. transformers. Useful in the frequency range of 200–1500 KHz.

Electrolytic iron flake A-270 for high permeability, low frequency applications to 20 KHz.

Electrolytic iron for small magnetic field clutches for aircraft servomechanisms—A-290.

Aluminum Powders

Alcan Metal Powders
Div. of Alcan Aluminum Corp.
Box 290
Elizabeth, N.J. 07207

Aluminum Powders for Rubber Compounds, Chemical Formulation, Plastics Tooling and Fillers for Plastics (99.3% Al, min.)

Typical Analysis

Grade	Control Screen	%	Screen	Bulk Density g/cc
MD-13	20	30 minus	325 mesh	1.3
MD-101	100	80 minus	325 mesh	1.1
MD-201	200	92 minus	325 mesh	1.0

Aluminum Powders and Needles

Aluminum Metallurgical Co.
424 Nichols Road
P. O. Box 11430
Kansas City, Mo. 64112

Almeg 100P Al Alloy Powder—
Aver. Bulk Density 1.04 g/cc

Mesh	%
−90 + 100	10–20
−100 + 140	20–30
−140 + 200	20–30
−200 + 325	20–30
−325	15–25

Copper Powders

Alcan Metal Powders
Div. of Alcan Aluminum Corp.
Box 290
Elizabeth, N.J. 07207

Copper Powders for Rubber Compounds, Conductive Cements, Fillers for Plastics (99.4% min. Cu)

Typical Analysis

Grade	Control Screen	%	Screen	Bulk Density g/cc
MD-41	40	92	+100 mesh	1.8
MD-151	150	58	minus 325 mesh	2.5
MD-180	100	50	minus 325 mesh	2.7
MD-201	200	94	minus 325 mesh	2.6

Glidden Metals
SCM Corporation
900 Union Commerce Bldg.
Cleveland, Ohio 44115

Copper Powders

Powders for bearings and structural parts	100–150 mesh
Powders for adjusting dimensional changes in P/M parts	200–325 mesh
Powders for formulation with asbestos and other non-metallics for clutch facings	150–325 mesh
Powders used in copper/graphite motor and generator brushes	100–150 mesh

Crescent Bronze Powder Co.
116–122 W. Illinois St.
Chicago, Ill. 60610

Metallic Pigments for Coatings

Gold bronze powders are alloyed of copper and zinc (for interior use).

Copper bronze powders are from pure electrolytic copper (for interior use).

Aluminum powder from pure primary aluminum ingot (for exterior use).

Mesh Sizes (All Grades)

Extrabrilliant grades	150
Brilliant grades	200
Fine grades	275
Extra fine grades	325
Super fine grades	400

U. S. Bronze Powders, Inc.
P. O. Box 31, Rte. 202
Flemington, N.J. 08822

Copper Powders

Flake and dendrite grades for manufacture of carbon brushes used in electrical equipment.

Spherical grades for chemical and plastic filler applications.

Mesh Sizes (Nominal)

Spherical Copper Powder	–100 to –200
Bronze Powders (90 Cu-10 Sn)	–100 to –200
Brass Powders (90 Cu-10 Zn)	–60 to –100
Nickel Silver Powders (64 Cu/18 Zn/18 Ni)	–100

Greenback Industries, Inc.
29501 Greenfield Rd.
Southfield, Mich. 48076

Non-Ferrous Powders

Pure tin powders, 75 to 95% (–325 mesh)

Molding grade (99.25% min. copper)

260NT30 Bulk density–2.65–2.75 g/cc
Particle size–16–19 microns
Screen–5–12% +150 mesh
18–28% +200 mesh
20–30% +250 mesh
15–25% +325 mesh
25–35% –325 mesh

Friction and brake lining grade (99.00% min. copper)

240NT95 Bulk density – 2.30–2.80 g/cc
Particle size–8–11 microns
Screen–10% max. +325 mesh
85% min. –325 mesh

Alcan Metal Powders
Div. of Alcan Aluminum Corp.
Box 290
Elizabeth, N.J. 07207

Miscellaneous Metal Powders–Selected Examples

Metal/Grade	Control Screen	Typical Analysis %	Typical Analysis Screen	Bulk Density g/cc
Lead, MD-101 (99.8 Pb)	100	95 minus	325	4.4
Lead, MD-104 (99.8 Pb)	100	36 minus	325	5.4
Lead, MD-201 (99.7 Pb)	200	98 minus	325	4.5
Manganese, MD-101 (99.8 Mn)	100	75 minus	325	3.1
Manganese, MD-201 (99.8 Mn)	200	93 minus	325	2.8
Silicon, MD-101 (98.7 Si)	100	70 minus	325	0.8
Silicon, MD-201 (98.6 Si)	200	90 minus	325	0.7
Silver, MD-101D (99.9 Ag)	100	15 minus	325	3.6
Silver, MD-101S (99.9 Ag)	100	20 minus	325	6.4
Silver, 750 Flake (98.4 Ag)	200	99 minus	325	1.1
Tin, MD-101 (99.8 Sn)	100	96 minus	325	3.2
Tin, MD-201 (99.8 Sn)	200	98 minus	325	3.2

Chemalloy Co. Inc.
950 County Line Rd.
Bryn Mawr, Pa. 19010

Miscellaneous Metal Powders

Special alloys and powders for the welding rod and steel industries. Literature lists iron powder, magnesium powder, manganese powder, nickel powder, and molybdenum powder.

Babcock & Wilcox
Nuclear Materials Div.
P. O. Box 1260
Old Forest Rd.
Lynchburg, Va. 24505

Nuclear fuels and materials including metallurgical products such as zirconium, hafnium, and titanium powders. Manufactured from the hydrides.

Particle size: −40 mesh down to 5 microns average diameter.

Brazing powders down to −325 mesh of 48 Ti/48 Zr/4 Be and 70 Ti/15 Cu/15 Ni.

Stellite Division
Cabot Corporation
1020 W. Park Ave.
Kokomo, Ind. 46901

Powders for hard facing tools, several cobalt base alloys, nickel base alloys, nickel base alloys, and iron base alloys.

Belmont Metals Inc.
330 Belmont Ave.
Brooklyn, N.Y. 11207

Non-Ferrous Powders

Aluminum, cadmium, copper, lead, tin, bismuth powders, and shot are available.

Silver Powder

Handy & Harmon
850 Third Ave.
New York, N.Y. 10022

Silver Powders (98.0 Ag min.) (Cu–0.02 max. Fe–0.05 max.)

	Screen Mesh
Silflake 131	−325 mesh–90%
Silflake 135	−325 mesh–90%

Zinc Dust

American Smelting & Refining Co.

Zinc Dusts for Corrosion Resistive Coating

	Screen Mesh
Grade No. 1 – 97.0% Zn 0.12% Pb 0.04% Cd 2.8% ZnO	96% through 325 mesh
Grade L-15 – 96.0% Zn	(min. 98% less than 15 microns)

4. TYPICAL MATERIAL SUPPLIERS AND TYPICAL CURRENT PRICES

The manufacturers and prices shown below are representative of key manufacturers. Most of these prices were effective in late 1975.

Supplier		Per Pound Price
Iron Powders		
Hoeganaes Corp. Subsidiary of Interlake Inc. Riverton, N.J. 08077 (609) 828-2220	Truckload prices, in 100-lb containers, FOB Riverton, N.J.	\$0.175–0.273 \$0.222–0.370
Copper Powders		
Glidden Metals 900 Union Commerce Bldg. Cleveland, Ohio 44115 (216) 771-5121	Truckload prices, 1 to 599 lb, FOB Hammond, Ind.	\$0.82–0.96 \$1.08–1.22
Specialty Copper Powders for Coatings		
Crescent Bronze Powder Co. 116–122 W. Illinois St. Chicago, Ill. 60610 (312) 644–8124	100 lb containers, 1-lb containers,	\$4.15–11.90 \$5.15–12.90
Copper and Copper Alloy Powders		
U.S. Bronze Powders Inc. P. O. Box 31, Route 202 Flemington, N.J. 08822 (201) 782-5454	Copper powders Bronze powders Brass powders Nickel silver powders	\$0.91–0.99 \$1.25–1.45 \$0.86 \$1.33
Aluminum Powder		
Alcan Metal Powders Box 290 Elizabeth, N.J. 07207 (201) 353-4600	Truckload prices 600-lb containers	\$0.53–0.60 \$0.68–0.75
Copper Powder		
Alcan Metal Powders Box 290 Elizabeth, N.J. 07207 (201) 353–4600	30,000 lb and over Less than 600 lb	\$0.25–0.54 \$0.45–0.74
Zinc Powder		
New Jersey Zinc Co. 2045 City Line Rd. Bethlehem, Pa. 18017 (215) 691-5000	Truckload zinc dust, Under 5000 lb	\$0.48 \$0.52

5. HISTORY OF COSTS AND AVAILABILITY

The history and future costs of powdered metals would have to be considered on an individual basis, since each metal has factors that have controlled its rate of growth. We have seen that most commercial metals have been available in a finely divided form, and while their usage is possible, the reasons for their selection do not always have compelling economic justifications. However, the examples which will be cited in the following section on applications, will be made on the basis of unique features of certain metal powders. One must always weigh such material selections on the basis of lower cost, viable alternatives.

The end of December, 1975, reveals the following basic costs of metals which must, necessarily, influence the price of metal powders prepared from them:

Aluminum (99.5%)	\$0.41/lb
Copper–wirebar producers	\$0.63625-0.6500/lb
Nickel–electrolytic cathodes	\$2.20/lb
Steel–cold finish carbon	\$16.65-\$17.10/cwt
Silver–Handy and Harman	\$4.18/troy oz

During 1973 there were shortages in the availability of Fe and Fe/C powders, due to be alleviated by more capacity.[6] Kempton[7] reports that in 1973, shipments of Fe powder exceeded 193,000 tons; Cu powder, 31,000 tons; and nickel powder, 7,600 tons. He estimated that the total value of all metal powders sold exceeded \$350 million in 1973. Better than 50% of the market lies in automotive parts–principally from Fe powder.

6. APPLICATIONS

Finely divided metallic particles have found numerous industrial applications when dispersed in organic resin matrices. There are specific and unique advantages to be gained through the utilization of metallic powders in organic binders. The choice of a metal filler may be influenced by functional, economic, or esthetic considerations. Uninformed individuals may anticipate that the presence of metal powders, such as steel powders, will impart stiffness and strength to comparatively weak organic polymers. Unfortunately, this is not necessarily true and, quite the contrary, a loss of overall tensile strength may occur. As will be seen from the review of applications, other factors, such as density, appearance, or response to magnetic fields may prompt the use of steel powders in organic polymers.[8]

6.1 General Influence on Properties of Plastics

A discussion of the utilization of finely divided metallic particles should be prefaced by the general observation that commercially available grades are usually used. This acknowledges the fact that the more common metallic particles, like copper, iron, and aluminum are encased in oxide films due to exposure to air and/or lubricants present during grinding processes arising during their preparation. In many instances, this surface contamination is of little concern, though in some of the more sophisticated electrical and magnetic applications, the manufacturer may be prompted to employ preparatory measures to remove contamination on powders to optimize their performance. The presence of absorbed moisture on the surfaces of metallic particles and their oxidized surfaces will also have an influence on the final properties of the composite structure to which they have been added. During formulation, predrying of fillers will frequently enhance properties of the final product.

Not all surface contaminants are disadvantageous, as the formulator of plastics materials may find advantages in introducing trace amounts (less than 1/2 of 1% for example) of silane coupling agents to aid properties. Many types have been used, such as gamma glycidoxypropyltrimethoxy silane and N-(trimethoxysilylpropyl)-ethylenediamine, though final selection should be influenced by the organic resin binder. In general, adhesion qualities are enhanced between the metal filler and binder and the adverse effects of moisture are minimized.

There are unexpected results which arise from combinations of metals and organic resin plastics. Metal-organic complexes may form at the interfaces which may function as an accelerator to enhance the cure. On the other hand, they may facilitate degradation as, for example, when small amounts of iron or zinc are present in formulations of polyvinyl halides.

The influence of volumetric loading has a marked effect on electrical, thermal and physical properties. Changes which occur may be anticipated, as suggested in Table 11-1, as long as the resin matrices surround the metal particles. However, when the volumetric loading of filler increases to the point of metal particle to metal particle contact, the properties may be expected to change quite abruptly. This is particularly evident in measurements of electrical conductivity of silver filled epoxy, where a loading

Table 11-1.

Property	Effect of increasing volumetric concentration of metallic particles in resin matrix
Tensile and flexural strength	Modest decrease
Compressive strength	Increase
Flexural and compressive moduli of elasticity	Increase
Shrinkage during cure	Decreased
Density	Function of volume loading and relative densities increase in most cases.
Volumetric resistance and dielectric strength	Decrease
Dielectric constant	Increase
Dissipation factor	Increase
Thermal conductivity	Increase
Thermal expansion coefficient	Usually decrease
Glass transition temperature	Little or no change

is finally reached where small additions in silver concentration achieve a marked decrease in electrical resistivity.

The decrease in physical properties shown in Table 11-1 may be attributable in part to processing variables which entrap air and moisture during blending. Ultimate tensile strength and flexural strength, particularly, will reflect these variables. On the other hand, the denser packing of particles under compressive loads usually leads to higher ultimate strengths. Impact strength was not included in the previous comparisons because the geometry of the particle, such as a fine whisker or fiber, has more influence than the type of metal. Also reported in the Russian literature[9] is the ability of colloidal copper and colloidal lead to enhance the cure and strength of epoxy systems.

There are limits on volumetric loading of metallic particles into resin compositions, dictated principally by the processing and blending operations. A point is reached wherein further additions of filler drastically alter the rheology of the system, as evidenced by a marked increase in viscosity. The end products are then principally pastes, the thixotropy of which will also depend on particle sizes and shapes, and other additives which enhance thixotropy and insure particle suspension. Higher filler loadings are also attained in solvent-polymer systems which anticipate solvent evaporation, and the deposition of films which have a high percentage of metal.

Particle sizes and shapes have decided influences on the properties of plastics composites prepared with metallic particles. There is a wide selection possible. Flat, planar particles of aluminum and copper alloys offer pleasing surface finishes to coating compounds. Filamentary whiskers of ferrous alloys will enhance the impact strength of resin compositions. The formulator must be aware of the more dense, compacted powders formed during atomization, as contrasted to the more dendritic, less dense powders characterizing electrolytic processes. There are unique chemical processes, such as the decomposition of carbonyls which yield pure spherical particles. Metallic coatings have also been deposited on finely divided inorganic or glass spheres, in order to attain certain economies, as in the case of silver. The uses of metallic coated glass beads have been limited, up to the present time, because they have only recently been gaining acceptance in this type of application.

Recently, metallic particles have been characterized more precisely in size and shapes. This is one of the factors that has led formulators to copy the success of portland cement concrete by trying to attain better packing and, thus, better properties. The efficiency of "packing" to achieve physical, electrical, or thermal properties may be improved by experimenting with various proportions of different particle sizes and shapes. Much more work has been conducted in this manner with inorganic fillers, and lessons may be learned from these experiences to guide the use of metallic powders. See Chapters 2 and 4 for an insight into methods for efficient packing of filler particles.

6.2 Specific Properties

6.2.1 Thermal Conductivity. Improved thermal conductivity and electrical conductivity are obtained with the aid of metallic fillers, derived from metals with high heat conductivity and low specific resistance to the flow of electrical currents. Finely divided fillers which are substantially free of oxide and other insulative surface layers are effective. Of available metallic fillers, copper and aluminum are used principally for improved thermal conductivity, and while gold and silver are good candidates, their costs limit them to more exotic electrical applications. Particle size, shape, distribution, and concentration are important considerations to enhancing thermal conductivity, and maximum loading is generally sought after. A typical example, in Fig. 11-2, shows aluminum nodules or needles with smaller aluminum particles dispersed in an epoxy matrix with one principal objective—improved heat transfer in epoxy tooling used for shaping preheated thermoplastics. Further enhancement of heat conductivity occurs when aluminum or copper fibers or gauze are interspersed and the overall composite densified under pressure. Limiting factors are the fabricating costs which must always be weighed against machining of tools from a solid block of metal. Elongated metal particles are more effective than spherical or irregular particles. Length/diameter ratios of 50 are reported quite effective in improving heat conductivity (10). While there are modest improvements of heat conductivity with the addition of metallic fillers, loading to the point where the composite barely pours results in conductivity with a three- or four-fold improvement over the base resin matrix, but still very much lower than the solid metal structure. Resin matrices of cast thermoset resins have thermal conductivities averaging $3\text{–}5 \times 10^{-4}$ calories/sq cm/cm/sec.

6.2.2 Heat Capacity. Heat capacity of a composite may be considered as a performance criterion depending upon specific heat, thermal conductivity, and volumetric loading of the metallic fillers. The thermal conductivity of the organic matrix is low and its specific heat approximately 0.3. Hence, a substantial gain in heat capacity, functioning as a heat sink for an electromechanical device, is achieved through the juxtaposition of a mass of solid metal or a poured-in-place composite of metallic filled resin system. Iron powders, as well as copper and aluminum, are useful in enhancing the heat capacity of a pourable, thermosetting resin system. Non-metallic fillers used as "heat sinks" are silica and aluminum oxide as well as the more expensive beryllium oxide. Product designers may find that a two-layer application will serve their purpose. The first resin layer is used to coat and to electrically insulate an electrical component, followed by a composite mass heavily loaded with metallic filler to enhance the heat capacity and to function as a heat sink.

6.2.3 Glass Transition Temperature. As noted previously, the physical presence of metallic powders in a resin matrix does not necessarily improve the glass transition temperature of the

Fig. 11-2. Aluminum-needle-filled epoxy tooling.

resin or its ability to operate at higher temperatures. Empirical thermal test methods which reflect creep characteristics or penetration tests may show slight improvements in heat distortion due to presence of metallic fillers, but not necessarily improved Tg values.

6.2.4 Thermal Expansion Coefficient. At low filler loadings, where the dominating structure or continuum is the resin matrix, the matrix will have the principal influence upon the thermal expansion coefficient of the overall composite. However, when volumetric loading of metallic fillers is sufficiently high, the filler will have a pronounced influence. This has already been demonstrated for certain inorganic mineral fillers, the selection and concentration of which may be used to lower the value of the expansion coefficient. As temperatures exceed the glass transition temperature of the mixture, thermal expansion coefficient of the system is markedly increased.

6.2.5 Electrical Properties. The addition of metal fillers decreases insulation properties but beneficial electrical effects may be obtained when superior electrical insulation is not required. In this case, the metal particles will improve thermal conductivity and therefore reduce operating temperature and hot spots. This can lead to longer service life. Also, metallic particles in an insulative sheath will decrease voltage gradients across a dielectric. Other considerations regarding specific electrical properties are listed below.

1. Dielectric strength. A marked decrease in dielectric strength occurs as metal filler loading is increased.

2. Volume Resistivity and Surface Resistivity. Reduced values of electric resistivity would be a consequence of metallic particle loading, particle shape, and size distribution. Higher loadings decrease resistivity, high aspect ratio metal particles reduce resistivity more as compared to equal loadings of lower aspect ratio particle.

3. Dielectric Constant. The introduction of metal powders into a dielectric medium will increase values of dielectric constant (see examples of aluminum particle filled epoxies in Table 11-2). However, the fall-off in resistivity may be sufficiently pronounced to negate any advantages gained from an increased dielectric constant. Metallic fillers and their controlled distribution in a dielectric medium offer a method of developing composites with a variable dielectric constant.

4. Dielectric Loss Factors at various electrical frequencies would become greater as metallic fillers offset the low loss factors of the dielectric.

5. Magnetic Compounds. The sophisticated industry of magnetic tapes, inks, discs, and cards reflects the many considerable skills which

Table 11-2. Properties of metal-filled liquid epoxy resins.

Composition of mixture −100 pbw of resin plus indicated pbw of aluminum filler	Density (g/cc)	Volume of aluminum (% of total)	Modulus of elasticity in flexure (psi × 10^5)	Thermal conductivity (cal/cm/sec) $cm^2/°C$)	Volume resistivity R.T. (ohm-cm × 10^{11})	Volume resistivity 200° F (ohm-cm × 10^9)	Dielectric constant at 1 kc and R.T.
no filler	1.16	0	4.5	0.00054	1.0	6.3	4.4
+10 pbw Al #1[a]	1.18	4.1	4.6	0.00059	1.1	6.4	4.7
+30 pbw Al #1	1.26	11.4	5.7	0.00075	0.89	6.1	6.5
+50 pbw Al #1	1.35	17.7	6.4	0.00100	0.81	5.6	8.7
+100 pbw Al #1	1.54	29.8	9.5	0.00150	0.45	5.3	10.0
+200 pbw Al #1	1.77	46.0	11.0	0.00220	.05	–	–
+10 pbw Al #2[b]	1.20	4.1	5.1	0.00065	1.2	6.0	4.9
+30 pbw Al #2	1.28	11.4	5.2	0.00076	0.62	5.4	5.7
+50 pbw Al #2	1.38	17.7	5.8	0.00085	0.16	3.8	8.8
+100 pbw Al #2	1.48	29.8	6.7	0.00120	–	–	10.0

[a] Alcoa No. 123 finely divided aluminum powder.
[b] Fine aluminum needle-like particles 1/3 in. long.

led to the utilization of magnetizable particles contiguous with organic binders. Functional requirements, some of which have been outlined in patent literature, have resulted in advanced products serving the communications industry. The ability of nickel, cobalt, and iron alloys to be readily magnetizable, even as finely divided particles in an organic matrix, has been the basis of many industrial applications. The high speed recording of electronic impulses is a trademark of satellite communications to computer storage discs or tapes. Magnetic particles in various physical states are contributory to their success.

For larger magnetic components, as distinct from tapes, the ability to shape complex pieces by pressing of powdered magnetic materials has distinct advantages. The harder magnetic materials are difficult to machine and to grind. For optimum magnetic properties related to high concentrations of flux density and largest response to electrical fields, powdered magnetic materials which are pressed and heat sintered together, fulfill important industrial roles. Tool costs and heat treating costs are among factors to be considered. Where a lesser degree of magnetic response is acceptable, the distribution of magnetic particles in a fluid resin base, such as an epoxy, and subsequent addition of a curing agent, yields an easily castable compound, capable of being shaped into a magnetizable component. By this technique, costly grinding operations are obviated. The distribution of magnetic fillers and application of magnetic compounds is such an important subject that this subject is treated in a separate section in this Handbook. The reader is urged to refer to Chapter 12 for more specific details.

6. Electrical Conductivity. Contributions of metallic particles to resin matrices to enhance electrical conductivity are numerous, and the subject of numerous patents and publications. There are many commercial conductive cements and solders for electrical and electronic manufacturers. Principal methods for the addition of good (silver, copper, gold) electrically conductive fillers into resin matrices include:

(a) Large concentrations of finely divided metal into liquid resin base by planetary mixing, accompanied by three-roll milling. When conductive lacquers are desired, the presence of solvents in the resin facilitates addition of the metallic powder, but settling of filler becomes a problem.

(b) Similar introduction of metal coated spherical particles (glass beads or spheres are favored) requires less drastic mixing to avoid cracking of spheres. Cost savings in reduced amounts of expensive metallic filler are apparent.

(c) Nickel, copper, and aluminum particles (5–10%) are loaded with thermoplastic particles such as polyvinyl chloride. Fusion and high pressure lead to a reticulated metal structure with minimum addition of costly fillers.[11]

(d) Blends of graphite or conductive carbon particles or fibers with metallic silver or copper, for example, will lower cost but suffer a loss in metallic appearance.

Due to the influence of silver particle shape and concentration in the plastic matrix, volume resistivity versus concentration of silver must be determined for a particular processing method. The amount of silver in the resin matrix is critical (usually in excess of 70% by weight of silver), as well as the method by which they are blended together. In two-component adhesive systems, the addition of a small amount of curing agent may adversely affect this balance, and sometimes silver filler must also be added to the liquid curing agent to maintain an adequate portion of silver to effect good conductivity. Palladium-silver pastes are reported to offer good adhesion and solderability, minimizing silver migration on certain semiconductor devices.

Metal-filled molding compounds have also been prepared for improved electrical conductivity. A German and a Japanese patent describe copper filled resin systems for molding applications.[12]

6.2.6 Physical Properties. Reference should be made to Table 11-1 which summarizes the general influences of metallic fillers upon the physical properties of resin matrices. As noted, there are no significant improvements in most physical properties. However, a large portion of the reduction of tensile and flexural strength can be attributed to the blending processes, which involve large amounts of filler entrapped air and thus contribute to discontinuities in the composite. Properties are improved by vacuum mixing of the resin-filler-cure agent system, and the behavior of cured composites which contain small amounts of silane coupling agents are markedly improved during humid aging. Recent published experiments upon iron and aluminum filled epoxies bear out the complex relationships which exist under imperfect interfacial bonding conditions and physical properties.[13]

There are instances where improvements in certain physical properties of metal powder filled epoxies have been observed. Filler compaction in a cured composite under direct compressive loading may occur during compression deformation and creep processes, leading to higher ultimate compressive strength and compressive modulus. Undoubtedly, filler particle size and distribution will have a bearing upon the results and there are indications that creep rates under compressive stress are reduced when metal particles contact one another.

As may be expected, fibrous metallic fillers will improve impact strength and this class of metallic fillers has been used for this purpose, particularly in applications such as non-metallic tooling. The choice of the metallic fibers over glass fibers has usually been prompted by the desire for improved heat transfer in the tool or fixture.

Discussions of metal-powder-filled plastics should once again remind the reader that when high physical properties are required, the processes of powder metallurgy, briefly reviewed in this chapter, are the directions to follow. The strengths accruing from metal-to-metal contacts in the absence of organic binders are substantially greater than those expected from resin binders, one is left with a metal porous structure that may be useful in applications such as impregnated metal bearings. In the examples that follow, the continuing presence of resin binders is essential to the application and functioning of the composite.

6.3 Typical Applications

Applications of metal filled organic polymers are numerous and from the review of typical examples in the paragraphs which follow, we will be able to recognize those qualities, including property enhancement, which motivated their selection. In the examples to be examined, one is not necessarily limited to the metal powders cited, such as classic iron, copper, or aluminum, which are the choice in many cases because of lower cost and availability.

6.3.1 Reduction of Friction. Contributions of metallic powders to the reduction of friction on non-metallic bearing, journal, or sliding surfaces have been observed. Non-metallic bearing surfaces have advantages of quietness of operation, ability to alleviate stresses of impact loading and, as in the case of nylon bearings and laminated phenolic journals, utilize small amounts of water for lubrication. "Teflon" has also been used on bearing surfaces because of its chemical inertness and low coefficient of friction. The inclusion of fine small particles of lead, zinc, bronze, indium, graphite, or molybdenum disulfide in such compositions will contribute to reduced friction.[19] Also, the metallic filler can reduce the wear rate of a Teflon bearing by a factor of 1000. The role of powdered metals in fabrication of pressed and sintered bearings, coupled with entrained lubricating oils in a semi-porous structure, is a major activity of P/M technology. Powdered metal fillers are frequently used in braking devices and clutches because they enhance performance. A particular example is the use of powdered copper in formulations with asbestos in clutch facings. The presence of copper will assist dissipation of heat and prolong the life of the organic binders, such as phenolic resins.

6.3.2 Reduction of Moisture Transmission. Thin laminar grades or leafing grades of aluminum powder will increase the resistance to moisture vapor transmission of thin organic protective coatings. The amount of metallic pigmentation required for this purpose is not large (usually less than 10%), but improvements are noteworthy. The same benefits are not as evident in thicker structures, where moisture absorption will depend upon the innate qualities of the resin binder.

6.3.3 Electrically Conductive Adhesives. As already discussed under electrical properties,

Table 11-3

Percentage powdered silver* (by weight)	Volume resistivity cured adhesive, ohm-cm	Al/Al Lap shear, psi
75	0.023	1,480
72.5	0.36	1,540
71.2	6.8	1,560
70.0	640.0	1,700
0	–	2,900

*Silflake 135, Handy & Hardman

finely divided silver and copper are found in numerous commercial adhesives where good electrical conductivity is required. These adhesives are not inexpensive, usually priced at about $80.00/lb for the silver filled variety. They find application to electronic components and devices where the heat of brazing or silver soldering is prohibitive or bonding is better accomplished with room temperature setting, silver filled epoxy cement. There is a delicate balance between adhesive strength, as measured by ASTM lap shear methods, volume resistivity, and percentage weight of silver, as shown in Table 11-3. The unfilled epoxy resin adhesive, used in this example, has a lap shear of 2,800 to 3,000 psi.

Good lap shear strength is important for small electronic components, as well as good stability of properties (good electrical conductivity and good lap shear) during service and exposure to humid conditions. Aging data is important and details must be carefully studied before selection is made. One-component (heat cured) and two-component (room temperature set) epoxy silver filled adhesives are sold in tube kits, 4 oz, and 1 lb packages. Powdered copper, silver coated copper flakes, and powdered gold filled epoxy adhesives are also sold commercially to electronics industries for similar purposes.

6.3.4 Solid Rocket Propellants. Finely divided aluminum powder is used in some solid rocket propellant systems of resin matrix and oxidizer fuel to enhance the thrust and specific impulse of the rockets. Performance control made possible with finely divided metallic powders opens new possibilities of optimizing performance. One should also bear in mind the considerable background of powdered metals such as magnesium in pyrotechnic and incendiary devices.

6.3.5 Anti-Fouling Compounds. A classic example of the use of copper and copper oxide in organic matrices is their application as anti-fouling compounds in "bottom" paints on ocean going vessels. These have proven to be the most effective deterrents to marine growths on the bottom of ships. Marine growth, particularly in warm and polluted waters, has been known to accumulate on naval vessels to such an extent that the speed of the ship has been greatly reduced. The copper ions will tend to inhibit the growth. Replenishment is required. In practice, this is accomplished by a controlled rate of erosion of the coating in sea water per year, exposing fresh copper and its salts. Many organic coatings have been prepared in this manner, and while other salts, particularly those of tin, have proven effective, copper and copper oxides dominate the formulation of anti-fouling compounds.

6.3.6 Corrosion Resistant Paints. The success of hot, galvanized steel in withstanding corrosive effects of industrial and commercial atmospheres has led to the utilization of powdered zinc in the preparation of certain types of surface coatings. Zinc, as sacrificial element in contact with steel, has prolonged the serviceability of industrial steel structures. While some coatings have been prepared for sales with large amounts of powdered zinc, the tendency of gas formation within the containers due to reaction of zinc with small amounts of water, has led to the separate addition of powdered zinc to the resin matrix prior to its application as a surface coating. While patent literature discloses chemical modifications to permit prior addition of zinc, most "zinc-rich" coatings offer zinc separately, as two or three-component system. For example, 90–95% zinc dust is used for maximum corrosion resistance in epoxy/polyamide coatings.

6.3.7 Maintenance and Repair Products. Probably the most publicized groups of powdered metal/resin matrix products have been those which have found application to maintenance and repair. Expressions such as "steel filled pastes," coatings, and castings, as well as aluminum filled and bronze filled have become well known in maintenance and repair activities, with applications proliferating into the many thousands. Much of the early enthusiasm was generated by companies such as Devcon Corporation which have been successful in marketing and merchandizing these materials. Metal oriented industries took notice when promotional efforts directed attention, for example, to the "cold-poured steel." This segment of the industry has matured and, while no real synergism of metals and resin binders resulted in startling, new physical properties, useful compounds based on epoxy resins in combination with available powdered metals are now widely accepted. Steel and aluminum powders are extensively used and, to a lesser degree, stainless steel powders, zinc, lead, brass, and bronze powders. Applications are diverse, usually entailing the mixing of a liquid metal-filled resin base and a liquid hardener, with fast setting times (for small parts) of 10-20 minutes, or longer working times of 40-120 minutes (for larger parts being prepared or repaired). Cure takes place at room temperatures, while materials are set overnight to become hard enough for filing or machining. The formulator is not limited to rigid plastics because elastomeric types, such as silicone and liquid urethane resins, may be used, and "flexibilized" steel and aluminum filled compounds are perfectly feasible, as well as with other powdered metals. Typical applications include:

1. Steel-filled pastes trowelled over rivet heads on welded tanks and hulls.

2. Bronze filled repair compounds to build up surfaces of eroded propellers.

3. Aluminum filled compounds to repair aluminum alloy structures and enhance bonding strength. Benefits are also found in aluminum filled adhesives for cellulosic structures.[14]

4. Cast aluminum filled thermosets for model and pattern making (Fig. 11-3).

Fig. 11-3. Aluminum-filled plastics models.

5. Attachments of metal hardware to portland cement concrete are best accomplished by aluminum (gray colored) filled epoxy cements or iron (black colored) filled epoxy cements (Fig. 11-4).

6. Flanges, valves, shafts, radiators, machinery, pumps, etc., are among hundreds of components and equipments repaired by metal filled epoxies, usually under field conditions.

7. Foundry uses range from pattern making, duplicate models, match plates, core boxes, molds, etc.

8. "Epoxy-Metal Repair and Hull Smoothing," Military Specification MIL-C-24176, U.S. Dept. of the Navy (1966).

Typical of the repair and maintenance metal filled products are those shown in Table 11-4, illustrating properties of cured, steel filled epoxy pastes and liquids, and aluminum filled epoxy pastes and liquids. The differences in cured physical properties are not large.

6.3.8 Tooling. Certain types of non-metallic tooling have effectively utilized metal-filled plastic composites. The methods for fabricating such tooling are well established, utilizing techniques of casting of liquid resin systems about models or patterns, or laminating of large structures employing liquid resin systems and glass cloth reinforcements. Tooling which

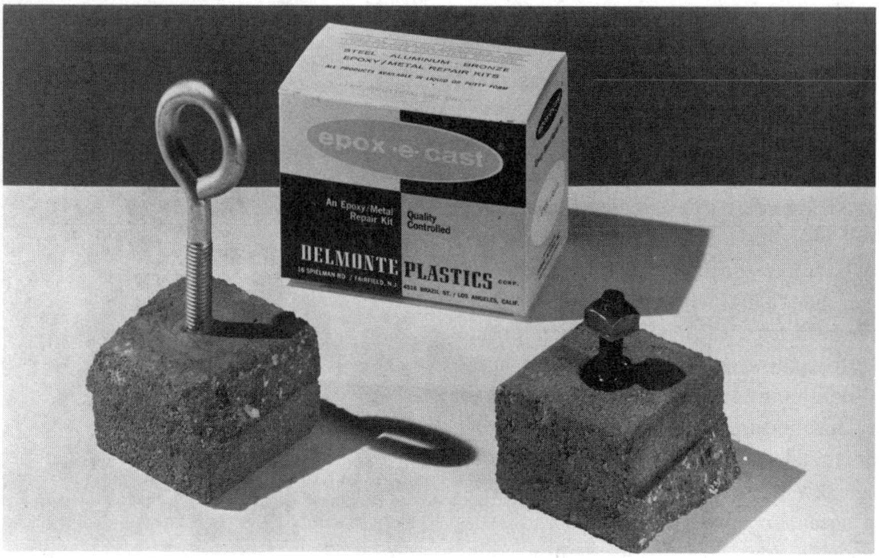

Fig. 11-4. Metal-filled epoxy for anchoring bolts.

operates at high temperatures usually requires metal filled bonding formulations to improve heat transfer. These include the following:

1. Short run, high temperature plastic injection molds.
2. Vacuum forming dies for thermoplastics, sometimes with controlled porosity to aid application of vacuum.
3. Metal bonding fixtures and assembly tools used in aerospace industries.
4. Holding fixtures for welding.
5. Molds for the foaming of urethane parts.
6. Core driers for foundry components.

Metal filled non-metallic tooling has, in some instances, utilized embedded electrical heating elements to aid the fabrication operation. In the formulation of high temperature tooling, special high temperature epoxy resin systems are usually selected as the resin matrix. See reference 15 for a review of metal filled epoxy tooling.

6.3.9 Application Requirements for Radiation Shielding. Non-metallics and metal powders combinations achieve effective radiation shielding. With the aid of a liquid resin matrix, it is possible to combine metallic elements which are incompatible metallurgically.[16] Among the better known radiation shields encountered by private citizens are the flexible, lead filled elas-

Table 11-4. Physical properties of cured Al and Fe filled epoxies.*

Product	Density, g/cc	Durometer hardness	Ultimate compressive strength, psi	Ultimate tensile strength, psi	Ultimate flexural strength, psi	Modulus of elasticity in compression, psi	Thermal expansion coefficient in/in/°F.
Epoxy-Steel paste	3.0	93 (D)	20,000	8,000	12,000	1×10^6	3×10^{-5}
Epoxy-Steel liquid	3.0	93 (D)	20,000	10,000	13,000	1×10^6	3×10^{-5}
Epoxy-Aluminum paste	1.8	92 (D)	20,000	8,000	10,000	0.9×10^6	2.8×10^{-5}
Epoxy-Aluminum liquid	1.8	92 (D)	20,000	10,000	11,000	0.9×10^6	3×10^{-5}

*By applicable ASTM procedures.
EPOX-E-CAST data sheet, Furane Plastics, Inc.

tomers used as drapable sheets, offering personal protection to X-rays in medical and dental procedures. The more sophisticated aspects of radiation shielding are detailed by Henry in his book on radiation protection,[17] and the latest edition of radiological health published by the U.S. Department of Health, Education, and Welfare.[18] Considerable data have been assembled on the absorption coefficients of the more common materials of construction and metals which enables a designer to select the proper mass of material to reduce radiation. H. Henry has arranged a practical table of half-life layers. These are the thicknesses of a specified substance which, when introduced into the path of a given beam of radiation, reduces the exposure rate by one-half. (See Table 11-5). The penetration ability of beta particles as a function of energy up to 4 MeV is shown in Fig. 11-5, as taken, in part, from Radiological Health Handbook referenced earlier. The advantages of the more dense fillers are apparent. Alpha particles are not shown since they have low penetration ranges. Shielding against thermal or low energy neutrons associated with nuclear reactors entails complex empirical studies which have been conducted on the capture cross-sections of various elements. Mass is not a criterion. Effective neutron shielding, as a matter of fact, is added by hydrogen atoms which afford a good neutron shield. Among elements prominent for controlling this radiation are indium, cadmium, and boron, with their relatively high neutron capture cross-section. As more attention is directed to shielding of gamma, beta, and neutron radiation, metals in resin matrices will begin playing more prominent roles in the nuclear age.

6.3.10 Body Solders. While, in principle, an auto body solder should be viewed as a form of repair compound, increasing interest is evinced by automobile manufacturers in body solders for the assembly of exterior shells and body components and warrants separate attention. This promises to be a large and growing market for metal powder filled epoxy and polyester resin systems. The following features influence automotive body solder specifications:

1. Gap filling qualities of metal filled epoxy pastes, bridging tolerances and gaps which may occur during assembly.
2. Good sandability, grindability, and compatibility so that sheet metal parts and the filled gaps may be coated over with other coatings, and the presence of the gap not be noticed.
3. One-component pastes capable of curing rapidly in a matter of 1-2 minutes at 150°C.
4. Ability to exhibit acceptable performance under extremes of processing during manufacturing, and weather exposure extremes during service.

6.3.11 Density Controls. While seemingly an innocuous requirement, requirements for adjustments to density appear on moving components ranging from missile warheads to flywheels. Designs call for weight concentrations in a specific volume, necessitating the addition of a filled composition, usually a pourable,

Table 11-5. Half-Life thicknesses for various materials (Reference 17).

Gamma Energy (Mev)	0.1	0.2	0.5	1.0	2.0	5.0
Material			Half-Life thickness (cm)			
Aluminum	1.60	2.14	3.05	4.17	5.92	9.11
Iron	0.26	0.64	1.07	1.49	2.09	2.84
Copper	0.18	0.53	0.95	1.33	1.86	2.47
Lead	0.012	0.068	0.42	0.90	1.34	1.44
Water	4.14	5.10	7.17	9.82	14.05	23.02
Air	35.5	43.6	61.9	84.5	120.5	195.8
Concrete	1.75	2.38	3.40	4.65	6.60	10.28

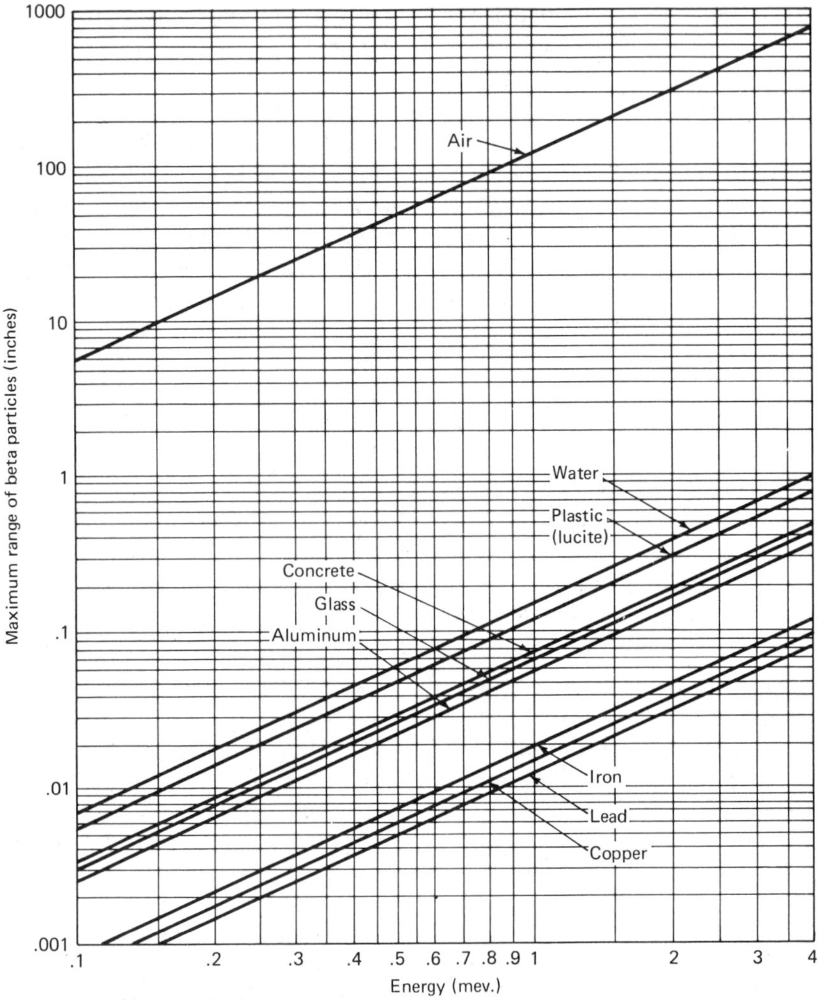

Fig. 11-5. The maximum range of beta particles as a function of energy in the various materials indicated. (From SRI Report No. 361, "The Industrial Uses of Radioactive Fission Products". With permission of the Stanford Research Institute and the U.S. Atomic Energy Commission.)

dense metal-filled resin base. The available densities are large, ranging to low values for aluminum powder-filled systems to lead or tungsten powder-filled systems. Inasmuch as the metal particles are insulated from one another by the resin matrix, unusual combinations may be effected.

6.3.12 Metal Filled Resilient Pads. The most likely method of obtaining a truly flexible metal body is to prepare formulations with metallic powders in a resilient resin base. Among the liquid blends of flexible resin matrices into which metallic powders have been incorporated are silicones, polyurethanes, and plasticized polyvinyl resins. Milling of synthetic rubbers and metal powders is also feasible in preparation of molded resilient pads of metal filled rubbers. The applications for these systems generally fit into the manufacture of sound dampening pads and radiation shielding pads and clothing. Lead filled aprons and coverings are, for example, common apparel in the laboratories of X-ray technicians.

Gasket seals have been prepared from metal-filled resilient compounds, and qualities such as improved heat transfer, improved resistance to compression set, and some lubricity are among qualities imparted by the metal powders to the gasketing composition. However, the lower cost and analogous qualities of many other gasketing compounds account for their preferred use. A particular advantage of metal-filled resilient gasket exists in its fabrication advantages when gasketing for large gap differences at different planes. For these non-laminar situations, a case may be developed for special cast, metal-filled gasketing compounds.

6.3.13 Magnetic Clutches. An example of the versatility of iron-filled organic liquid or semi-liquid systems is the development of special forms of magnetic clutches. As a magnetic field is applied to the iron-filled fluid with a non-magnetic encasement, the material will stiffen and, in effect, arrest the movement or rotation of parts in the fluid. Servomechanisms on some aircraft components are controlled in this manner.

Metal powders contribute in many ways to the efficacy of electrical and mechanical designs. In many instances, the choice of powdered metals is based upon the qualities of particulate metal powder rather than the behavior of the solid metal. Thoughtful utilization based on specific property enhancement will encourage the continued use of metal powders in a field replete with low cost inorganic fillers.

7. TRENDS IN METAL FILLED RESIN COMPOSITES

The increasing availability of more grades of metallic powder will alert resin formulators to prospects of the enhancement of properties of composites. The P/M industry is openly "bullish" on the prospects of larger volume applications. Each year sees further instances of savings effected by eliminating scrap and reducing labor costs in designs of machine components involving metals. Newer techniques of isostatic pressing, gleaned in part from aerospace industries, permit the fabrication of larger size powdered metal parts. Because of these growing involvements, it is likely that larger quantities of metal powders for other uses, such as in resin composites, will be available.

In highlighting trends which offer large potential growths, we can draw upon examples and applications noted in Section 7:

Body Solders. The filling of seams and cracks in automobiles with metal filled plastics has been receiving considerable attention from the major manufacturers of automobiles. The market would also include repair and maintenance activities. Estimates of potential body solders metal/resin base for the automobile industry are placed at well over 2 million lb/yr.

Electronic Industry. Though quantities may not be large, the dollar volume of metal filled resin systems for the electronic industries is growing substantially. Copper, silver, palladium, and gold are the principal materials involved. Electrically conductive solders, semiconductor devices, special storage batteries, capacitors, heat sinks, printed circuit boards, etc., all represent components requiring powdered metals in a large growth industry.

Radiation Shielding. We are embarked upon a nuclear age. Radiation controls and shielding are becoming of increasing interest to all nations. The unique shielding qualities of certain metals to beta, gamma, and neutron radiation offer promise of increasing applications of such metals, with planned distribution in resin matrices. Power plants and personnel protection will require these protective measures.

Bibliography

1. Vaccari, J. A., P/M Parts, Materials and Processes, Manual 242, *Materials Eng.*, p. 85 (July 1967).
2. McIntire, H., "Trends in Powder Metallurgy," *Batelle Institute J.*, p. 16 (Jan. 1968).
3. Feir, M., "Processing Limits of P/M Parts Production," MF74-146, Society of Manufacturing Engineers, Dearborn, Michigan.
4. Hamling, B., to Union Carbide Corp., U.S. Patent 3,406,025, Oct. 15, 1968.
5. Scheffler, L., "Why Use Whiskers In Plastics," Reinforced Plastics Div., S.P.I., Washington, D.C., Feb. 1967.
6. DuMond, T. C., *Iron Age*, **212**(5), 30, 31 (Aug. 2 1973).
7. Kempton, H., "Progress in Powder Metallurgy," Vol. XXX, 194, *National Powder Metal. Conf. Proceedings*, 207-213, Met. A-7505-72 (1974).
8. Delmonte, J., *Metal Filled Plastics*, Van Nostrand Reinhold Co., 1961.
9. Chemical Abstracts 68-5-40404v and 64-11387j, U.S.S.R.
10. Hansen, D. and Tomkiewicz, R., "Heat Conduc-

tion in Metal Filled Polymers," *Polymer Eng. and Science*, **15**, 353 (May 1975).
11. Turner, D., U.S. Patent 3,708,387, Jan. 2, 1973.
12. Chemical Abstract 74-2-13734, Jan. 25, 1971 and Germ. Off. 2,014,433, Oct. 22, 1970.
13. Chemical Abstract 83-180116v, "Dynamic Properties of Metal Filled Epoxy Polymers," (Dec. 1975) and Paipetis, S. *et al. Fiber Sci. & Tech.* 8(3) 221–42 (1975).
14. Tiedeman, G., and Gillern, M., (Weyerhaeuser), U.S. Patent 3,872,051, Mar. 18, 1975.
15. Delmonte, J., "Metal Filled Epoxies," S.P.E. RETEC (July 14 1965).
16. Delmonte, J., "Radiation Resistant Plastics," S.P.I. Reinf. Plastics Div., Chicago, Feb. 1965.
17. Henry, H., *Radiation Protection*, p. 485 Wiley-Interscience Publ. Co., 1969.
18. U.S. Department of Health, Education and Welfare, *Radiological Health Handbook*, Jan. 1970.
19. Bearing Materials, *Materials Eng.* p. 53 (Jan. 1970).

12
Magnetic Fillers

Zoltan E. Kerekes
Ceremet Engineering Associates
Carteret, New Jersey

CONTENTS

1. INTRODUCTION

The magnetic phenomena found in many metallic and ceramic materials can be utilized to produce useful plastic magnets, when fine magnetic particles are dispersed in polymer matrices. At the current time, most of the magnetic composites utilize a flexible polymer. The characteristics of the flexible magnet can make it useful for many engineering applications. Unique processes are employed for orientation of the hard, platelet-shaped magnetic particles, which are embedded in the plastic matrix, to obtain optimum magnetic characteristics.

The ceramic magnetic powders called ferrites are the major ingredient in plastic magnets. Its low cost qualifies the ferrite for use in two major markets, automotive and appliances. The loading of ferrite in the plastic matrix ranges from 88 to 92 weight percent, and the maximum energy product of these magnets can be enriched to 1.1×10^6 Gauss Oersteds in oriented form. This compares with a typical ceramic magnet with a range of 1 to 3.5×10^6 and an alnico magnet at 1.3 to 9×10^6. The plastic magnets are nonconductive, light weight, readily molded, and commercial. A glossary of terms related to magnetic phenomena is given at the end of this chapter.

2. METHODS FOR PRODUCING FERRITE POWDER AND FLEXIBLE MAGNETS

2.1 Ferrite Formulation and Production Methods

2.1.1 Formulation. The major material used for soft and hard ferrite production is iron oxide (Fe_2O_3). The second components added to the formulation are oxides or carbonates of zinc, manganese, barium, strontium, lead, etc., depending on whether soft or hard ferrite production is desired (see Tables 12-1 and 12-2).

2.1.2 Production. Both types of ferrites are produced in similar fashion. The raw materials are mixed together and milled either wet or dry. After the mill is discharged and the material separated from the grinding media, it is then dried in a glass baking dish at 150°C for 24 hr. The dried, caked powder should be granulated through a 20 mesh screen and placed on a firing setter, then calcined at about 1100°C for 2 hr to form the ferrite by solid state diffusion in

Table 12-1. Soft ferrite compositions.

	COMPOSITION WT %			
Comp. Code	Fe_2O_3	NiO	ZnO	MnO_2
1 (Ni–Zn)	63.0	12.5	24.5	–
2 (Zn)	66.0	–	34.0	–
3 (Mn–Zn)	67.0	–	13.0	20.0
4 (Mn–Zn–Ni)	66.5	3.0	23.7	7.0

the oxides system. The calcined material is ball-milled again in order to produce a particle size of approximately 1 micron or less, to obtain the optimum magnetic properties of the ferrite item. The resultant dry powder can be processed to final form by any standard ceramic method such as dry pressing, wet pressing, or extrusion. The hard ferrites' maximum energy product and final magnetic properties can be greatly improved by orienting the ferrite crystals in a magnetic field during the formation of the ferrite particles. This orientation is best conducted in wet pressing a slurry, because the crystal can be arranged more easily.

After the mixing is completed, the roller spacing is reduced to achieve a thin sheet with maximum orientation of the ferrite particles. After this final rolling, to about 0.010 in., the sheet is ready for punching out layers for the die pressing process. The pressing die is pre-heated to 320°F and the pre-weighed laminations are inserted into the die, which is then closed by applying 50 psi pressure. This pressure is held until the laminations have attained die temperature. The pressure is then increased uniformly to 16,000 psi and held until the interfaces fuse together. The pressed pieces are

Table 12-2. Hard ferrite compositions.

	COMPOSITION WT %		
Comp. Code	Fe_2O_3	$BaCO_3$	$SrCO_3$
5 ($BaFeO_{19}$)	82.0	18.0	–
6 ($SrFeO_{19}$)	85.0	–	15.0

removed from the die and submitted for firing, which is a two stage operation:

1. *Binder Bakeout* at a rate of 30-35°F hr from 100° to 550°F.
2. *Sintering* at a rate of 400-425°F/hr up to a temperature of 1800°F.

The time and temperature are carefully monitored to control shrinkage. The green density of the laminate increases from 3.7 g/cm^3 to 5.3 g/cm^3 during sintering. The energy product of the sintered pieces that can be achieved is more than 3.5×10^6 Gauss-Oersteds.

2.2 Flexible Magnet Production Methods

2.2.1 Rolling Process. A flexible matrix process was utilized in the middle of 1960 to successfully produce mechanically oriented hard ferrite permanent magnets. The process was developed by General Motors Research Group, and was called "Ferriroll Process." The process uses a multilayered stack of flexible sheets of platelet-shaped magnetic particles to form highly oriented, ferrite, permanent magnets.[1,2] Lead ferrites are especially suited to the process, because of the plate-like shape of the particles which are susceptible to the mechanical orientation which takes place during the milling process. Essentially, the process consists of mixing of ferrite powder and the binder (low molecular weight polyethylene) in a rubber mill until a homogeneous flexible sheet is formed.

2.2.2 Tape Casting Process. The plastic magnet can be made in a form of flexible permanent strip, which offers the product designer a uniquely desirable combination of physical and magnetic properties at lower cost than other magnetic materials. The magnetic strip is made by a special process that orients the hard ferrites (Ba, Sr, Pb) during manufacturing and prior to magnetization under the influence of an external magnetic field. The magnetic strip can be bent or twisted without loss of its magnetic properties. The flexible magnetic strip can be used in a wide range of applications.

Tape casting or doctor blading has long been used as a fabrication technique for producing

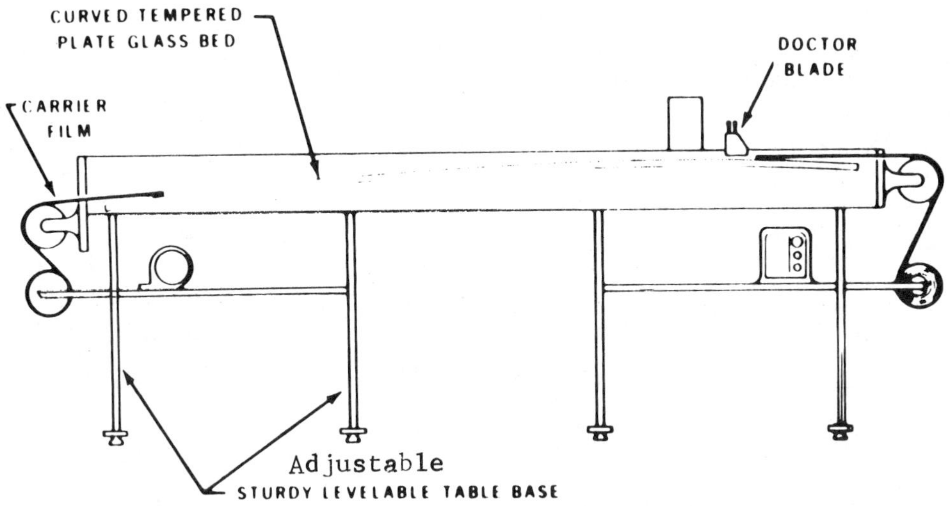

Fig. 12-1. Precision tape casting machine.

flexible ceramic sheet product. Basically, the technique comprises the suspension of ceramic powder (ferrite) in a liquid and the flow of this suspension under a blade onto a flat surface. Most often, the suspending medium includes solvent, deflocculent, binder, and plasticizer. The ingredients are thoroughly mixed with the ceramic (ferrite) powder to yield a suspension or slip with a paint-like consistency. In this process, the slip is poured onto a moving plastic tape, under a blade called the doctor blade, to produce a thin layer which, after partial drying, can be removed from the carrier film. Shapes can be cut or punched from the dry tape, and subsequent curing or firing forms flexible or hard, dense magnetic pieces. Typical tape casting machine and two-stage doctor blade construction are shown in Fig. 12-1 and 12-2.[3]

2.2.3 Extrusion Process. The Plastiform^{R} magnetic strip, manufactured by Bunting Magnetic Company, consisting of barium ferrite #5 bonded in Natsyn by a matrix process, can be manufactured with conventional magnetization when two pole magnets have equal strength poles on each side of the strip, or can be furnished with "multipole" polarity. The multipoles can run longitudinally and parallel along one face or both faces of the strip. The

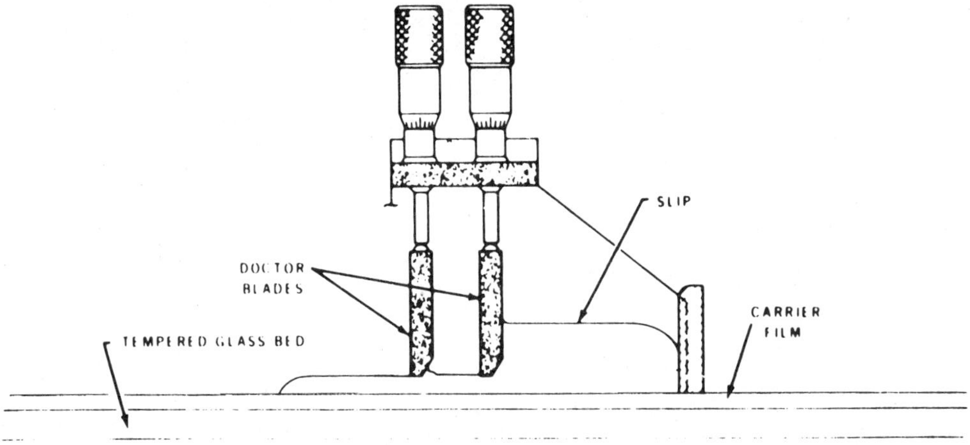

Fig. 12-2. Doctor blade showing two-blade construction and optimized flow conditions.

Koroseal[R] is an extruded flexible magnetic strip which consists of a flexible binder impregnated with barium ferrite.[4] Figure 12-3 shows the different pole arrangement in the flexible magnetic strips. The strips are produced in various widths ranging from 0.25–10. Various strips and products are shown in Fig. 12-4. The standard size strip dimensions are tabulated below:

	DIMENSION		
Type	T	W	L
W(Plastiform)	0.030–0.312 in.	3.0 in.	up to 100 ft
N(Koreseal)	0.060–0.250 in.	0.25–1.0 in.	up to 100 ft

A specially prepared high energy powder, highly suitable for embedding into a plastic or rubber matrix, may then be extruded into a magnetic strip or bonded magnet. Flexible and plastic magnets are compounded with RM 200 Barium, manufactured by Greenback Industries, Inc., a hard ferrite powder which has a highly intrinsic coercive force and is highly resistant to self-demagnetization. The RM allows ease in processing and permits the user to produce variable magnetic characteristics simply by changing the loading ratio of the magnetic powder to the compound.[5]

2.2.4 Injection and Compression Molding. The barium ferrite particles can be encapsulated

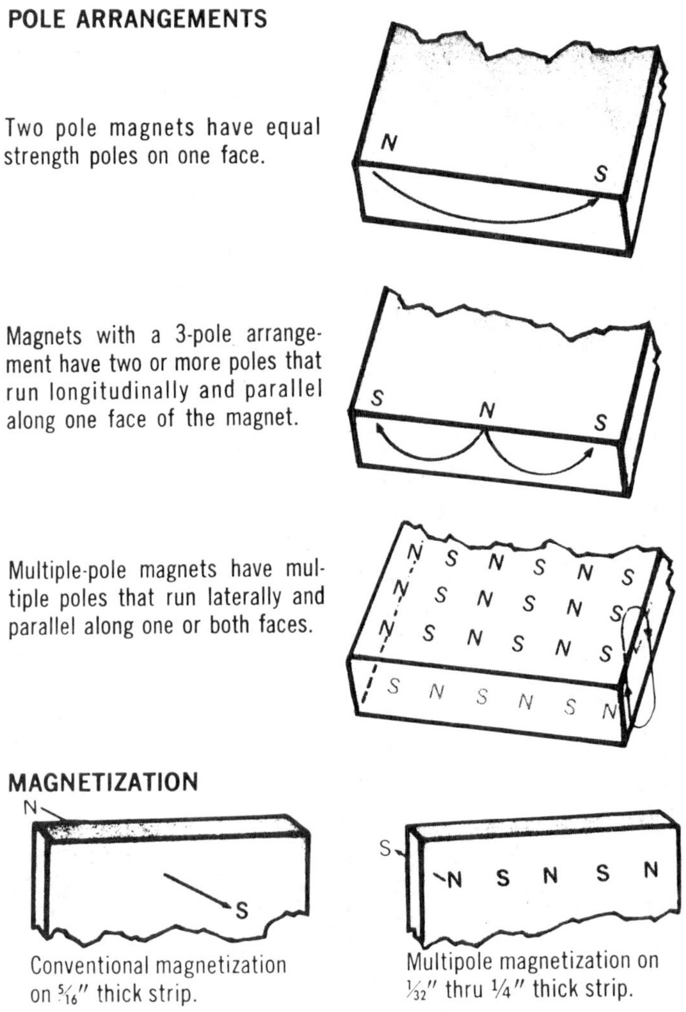

Fig. 12-3.[4]

Fig. 12-4.[4]

with a thin film of nylon-12 to produce structural ferrite magnets used for such applications as low-torque fractional–hp motors, switches, TV beam bending magnets, toys, etc. The nylon encapsulated compound, available in plastic and powder processing form, can be molded on a standard injection machine or by conventional compacting.[6] The plastic compacting grade compositions, manufactured by Rilsam Corporation are given in Table 12-3.

The ferrite powder can be mixed with many polymers or waxes from 87-94% by wt (60-70% by vol.). A few of the binders are listed below:

Polypropylene
Ethylene-ethyl acrylate copolymer
Polyisoprene
Polyvinyl chloride
Styrene
Carnauba Wax
Baker's Castorwax

Table 12-3. Composition of nylon matrix magnets.

		COMPOSITION	
Comp. Code	Process	$BaFe_{12}O_{19}$	Nylon-12
FPC 80.20	Injection molded	80	20
FEC 88/12	Injection molded	88	12
FPC 90/10	Powder compacted	90	10
FPC 95/5	" "	95	5
FPC 97/3	" "	97	3

2.3 Key Advantages of Flexible Permanent Magnets (FPM)[7]

2.3.1 Multiple Poles. The flexible strip can be magnetized with multiple-poles in one piece, either before or after installation, into electrical equipment as shown in Fig. 12-5.

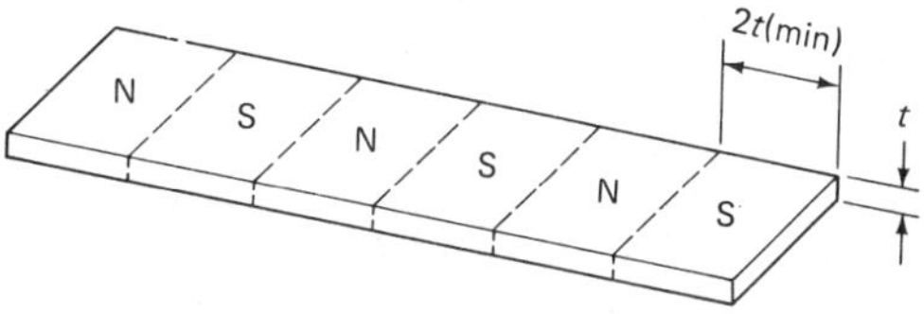

Fig. 12-5.[7]

2.3.2 No Stray Flux Field. Lines of magnetic force enter and leave each pole at the surface of the strip, not at the edges. The result of the field envelopes is illustrated in Fig. 12-6.

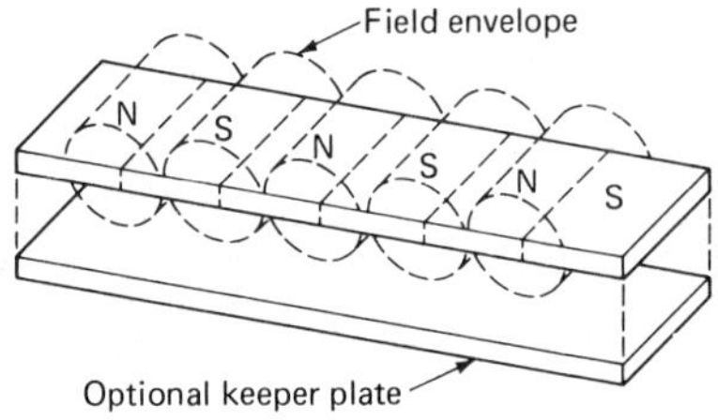

Fig. 12-6.[7]

2.3.3 Tailored Flux Field. Field pattern between adjacent poles can have sharp or tapered transitions, as shown in Fig. 12-7.

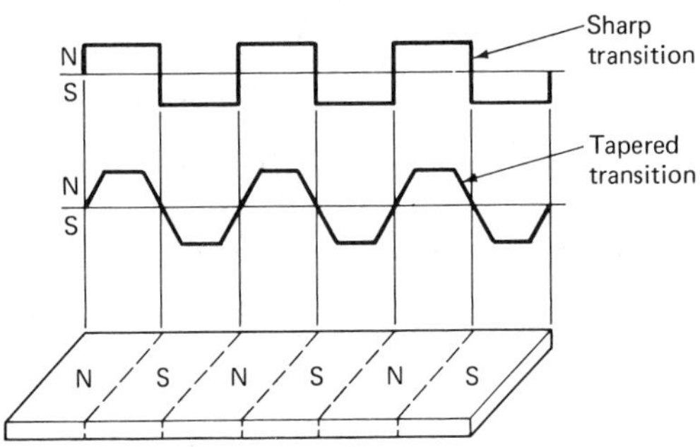

Fig. 12-7.[7]

2.3.4 Customized Pole Arrays. Multipole arrangement can be arrayed in any configuration compatible with magnetic principles, as presented in two cases shown in Fig. 12-8.

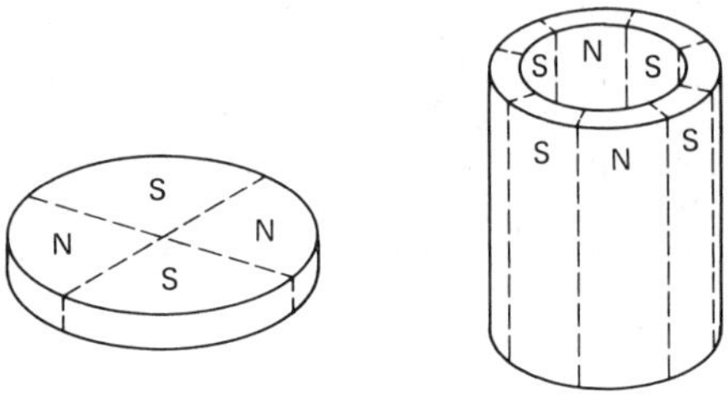

Fig. 12-8.[7]

2.3.5 Easy Fabrication. The hard rubber matrix FPM materials can be worked by ordinary shop tools for property evaluation in prototype work without costly tooling. The cutting steps are shown in Fig. 12-9.

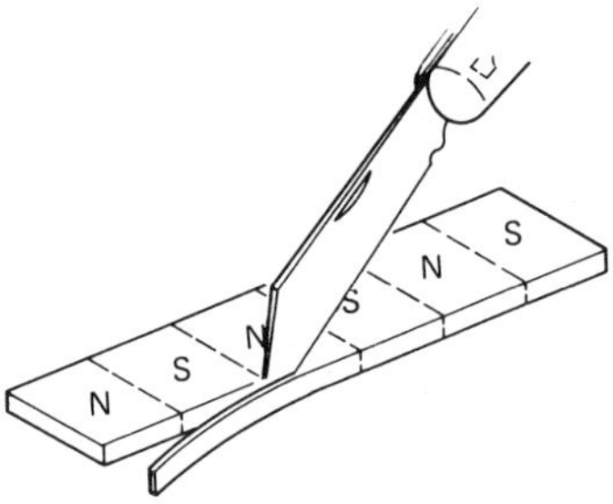

Fig. 12-9.[7]

3. PROPERTIES OF FERRITE POWDER AND FLEXIBLE MAGNET

Table 12-4. Magnetic characteristics and properties of hard ferrites.[8]

MATERIAL DESIGNATION #	Orientation	Br (Gauss)	H_c (Oe)	Intrinsic coercive force, IH_c (Oe)	BH max. (MGO_e)	Comp.	Manufacturers' Code
FB_1	Unoriented	2200	1900	3500	1.1	Ba	1
FB_2A	Oriented	3850	1800	1850	3.4	Ba	1
FB_3A	Oriented	3600	2400	2500	2.9	Ba	1
FB_3B	Oriented	3400	3000	3400	2.6	Sr	1
FB_4A	Oriented	4150	1800	1850	3.8	Ba	1
FB_4B	Oriented	3800	3200	3500	3.2	Sr	1

Table 12-5. Magnetic characteristics and properties of plastic magnets.[8]

Material Designation	Br (Gauss)	H_c (Oe)	Intrinsic coercive force, IH_c (Oe)	BH max. (MGO_e)	Comp.	Manufacturer's Code
MBP-1	1000	900	3000	0.2	$SrFe_{12}O_{19}$	1
MBS-3	1600	1350	3000	0.55	$BaFe_{12}O_{19}$	1
MBS-6	2200	1900	3000	1.1	$BaFe_{12}O_{19}$	1

Note: MBP-1 Consists of ferrite magnet powder and thermoplastic resin, shaped into magnets by injection molding.
MBS-3 Same as MBD-1 but shaped into magnets by compression molding.
MBS-6 Same as MBS-3 in the presence of magnetic field orientated.

Table 12-6. Magnetic properties of flexible magnetic strips.[8]

	W		N
Type Material Designation	Plastiform 1	Plastiform 1-H	Koroseal 22-980
Br (Gauss)	2,200	2,140	2,015
H_c (Oe)	1,480	1,940	1,600
$(BH)_{max}$ (MGOe)	1.08	1.04	0.93
Manufacturer's code	2	2	2

Table 12-7. Physical characteristics of flexible magnetic strips.

Properties	W	N(22-980)
Electrical		
Insulation resistance × 10^6 ohms	7.55	–
Volume resistance × 10^{10} ohm-cm	3.54	–
Arc resistance sec.	102	–
Mechanical		
Ultimate tensile strength, psi	640	1300
Elongation, %	3.25	150
Hardness (Durometer "D")	65	92
Thermal		
Coefficient of linear thermal expansion × $10^{-5}/°F$	2.61	3.0
Thermal conductivity, Btu/hr/ft^2/°F/in.	4.28	–
Maximum operating temperature, °F	200	175
Specific gravity, gr/cm^3	–	3.72
Manufacturer	Bunting Magnetic Company	

Table 12-8. Magnetic properties of flexible magnets prepared from RM 200 powder.[5]

Magnetic Powder	Composition	Br (Gauss)	H_c (Oe)	IHC (Oe)	$(BH)_{max}$ Wt %	Load Wt %
RM 200		1600	1300	2500	0.65	89.5
RM 200	$BaFe_{12}O_{19}$	1700	1400	2450	0.75	90.5
		1800	1450	2400	0.85	91.5

Ferro barium ferrite 300 Series embedding powder is also available for flexible magnet application. These powders are a low lead free product which pulverizes to fine particle size for easy dispersion in a polymeric matrix.

Table 12-9. Magnetic properties of 300 series embedding powders.[9]

MAGNETIC POWDER									
Type	Size*	Price/Lb	Composition	Binder	Br Gauss	H_c (Oe)	IH_c (Oe)	$(BH)_{max}$ (MGOe)	Loading Wt %
303	1.6	0.215–0.235	$BaFe_{12}O_{19}$	A	1800	1350	1850	0.74	90.0
				B	1700	1300	1850	0.665	90.0
304	1.8	0.25–0.27	$BaFe_{12}O_{19}$	A	1800	1550	2500	0.76	90.0
				B	1700	1500	1600	0.685	90.0
334	1.7	0.235–0.255	$BaFe_{12}O_{19}$	A	1800	1450	2200	0.75	90.0
				B	1700	1400	2225	0.675	90.0

*Microns

Table 12-10. Composition of binders.[9]

Type Binder	Parts/Weight	%	Material	Manufacturer	Grade	Specific Gravity
	800	73.87	Chlorinated Polyethylene	Dow Chemical	M × 2243.25	
	200	18.47	Polymeric Plasticizer	Rohm & Haas	Paraplex G-50	
A	40	3.69	Epoxidized Soyabean Oil	Archer, Daniels Midland	Admex 710	1.18
	28	2.58	Ra-Cd Stabilizer	Ferro Corp.	#1827	
	15	1.39	Lubricant	–	Stearic Acid	
	680	63.20	Chlorosulfonated Polyethylene	Dupont	Hypalon 45	
B	320	29.74	Polyisobutylene	Exxon Chem. Co.	Vistanex 140	1.03
	54	5.02	Resin-Derived Alkyd Resin	Hercules	Neolyn 20	
	22	2.04	Lubricant	–	Stearic Acid	

Table 12-11. Physical characteristics of injection molded nylon, encapsulated barium ferrite compound and process parameters.[6]

Properties	COMPOSITION CODE FPC 80/20 Rilsan Corp.	FPC 88/12 Rilsan Corp.
Bulk density, gr/cc	1.42	1.59
Tensile stress, ksi	6.8	5.9
Flexural modulus, ksi	810	820
IZOD impact strength, ft-lb/in	1.7	1.2
Melting point, °F	482	482
Heat deflection temperature, °F (at 66 psi)	410	320
Hardness rockwell (at 75°F)	83	93
Mold temperature, °F	150	
Injection temperature, °F	580	
Injection pressure, ksi	20	
Screw speed (rpm)	300	

4. MATERIAL SUPPLIERS

Code	Name	Location
1.	TKF Electronics Co.	931 South Douglas St. El Segundo, Cal. 90245 Tel. (312) 593-2060
2.	Bunting Magnetic Co.	2100 Estes Ave. Elk Grove Village, Ill. 60007 Tel. 312-593-2060
3.	Geral Magnetic	1001 Erwin Ave. Detroit, Mich. 48234 Tel. 313-921-7558
4.	Ferro Ottawa Chemical	700 N. Wheeling St. Toledo, Ohio 43605 Tel. 419-691-3507
5.	Rilsan Co.	139-T Harristown Rd. Glen Rock, N.J. 07452 Tel. 201-•47-3300
6.	Indian General	Crow's Hill Rd. Keasby, N.J. 08832 Tel. 201-826-5100
7.	Ampex Corporation	465 W. Maude Ave. Sunnyvale, Cal. 94086 Tel. 408-245-5242
8.	Ceramic Magnetic Inc.	87-T Fairfield Rd. Fairfield, N.J. 07006 Tel. 201-227-4222
9.	FerroxCube Co.	Mt. Marion Rd. Saugerties, N.Y. 12477 Tel. 914-246-2811
10.	Allen-Bradley Co.	1201 S. 2nd St. Milwaukee, Wis. 53204 Tel. 414-671-2000
11.	Stackpole Carbon Co.	1942 Stackpole Street Saint Marys, Pa. 15857 Tel. 814-781-1234
12.	Arnold Engineering Co.	Box G Marengo, Ill. 60152 Tel. 815-568-2000
13.	Greenback Industries Inc.	Greenback, Tenn. 37742 Tel. 615-856-3021

5. AVAILABILITY AND COST

5.1 Availability

In 1970 United States manufacturers had produced about 60 million lb of ferrites, broken down by:[10]

Hard ferrite	35×10^6 lb
Soft ferrite	25×10^6 lb
Total	60×10^6 lb

These ferrites required 33,250 tons of raw materials valued at over $10 million. The hard ferrite expected to enjoy substantial growth in 1975, the forecast being broken down into:

Hard ferrite	60×10^6 lb
Soft ferrite	25×10^6 lb
	85×10^6 lb

The dollar values of the two types of ferrites are broken down into:

	1970	1975
Hard ferrite	$ 61 million	$105 million
Soft ferrite	77 "	77 "
	$138 million	$182 million

Most of the growth in hard ferrites came from their increased use in automobiles, up to 35 million lb in 1975.

5.2 Cost

Table 12-12. Cost of barium ferrite.

MAGNETIC POWDER				
Powder designation	Composition	Particle size	\$ Price/lb	Supplier Code No.
Permacal W-7	Barium Ferrite ($BaFe_{12}O_{19}$)	1.8 μ	0.495-0.285*	3
303	($BaFe_{12}O_{19}$)	1.6 μ	0.215-0.235	4
304	($BaFe_{12}O_{19}$)	1.8 μ	0.25-0.27	4
334	($BaFe_{12}O_{19}$)	1.7 μ	0.235-0.255	4

*Less than 1000 lb and over 40,000 lb.

Table 12-13. Cost of magnetic strip.

	DIMENSION				
Type	Thickness (in.)	W (in.)	L	Price \$/ft	Supplier Code No.
Plastiform W	0.030-0.312	3.0	up to 100′	0.63-16.24**	2
Koroseal N	0.060-0.250	0.25-1.0	up to 100′	0.21-3.63	2

**Less than 10 ft and over 1000 ft.

6. APPLICATIONS AND END USES OF PLASTIC MAGNETS

Soft ferrites are used in three major areas.[11]

- Television receivers
- Communication equipment
- Computer memories

The largest tonnage of soft ferrites is used for television applications such as horizontal deflectors, transformers, and deflection yokes. Each black and white TV set contains approximately 1 lb of ferrite, whereas the color sets utilize more than 2 lb. A significant amount of the soft ferrites used in communication is primarily for telephone equipment. The soft ferrite is superior to metal cores, especially in the low megacycle region, because of the low eddy current loss properties. The greatest amount of soft ferrites was used in the early seventies in the computer industry for the memory core application. The type of ferrite used is a square loop ferrite named after the shape of its hysteresis loop. This type of ferrite has a sharp switching characteristic when being magnetized from one direction to the opposite direction. Therefore, a bit of information is stored by the memory core ferrite depending on its direction of magnetization. Each memory system in a computer contains many millions of ferrites. The size of each ferrite core is approximately 20 mils or smaller in diameter. The small size core is necessary in order to occupy the smallest space to obtain the highest core density for the maximum capacity of memory storage. Also, the switching time of a ferrite is dependent on its mass, the smaller the mass, the faster the switching time. A 50 mil diameter core requires about 1 microsecond switching time, while a 20 mil diameter core only requires 0.2 microseconds. The soft ferrites are used in microwave application, such as AM radio antenna cores and tape recorder heads.

Hard ferrites are used for DC motors in two major areas, automotive and appliances.[12] In the automotive industry, the flexible permanent magnet (FPM) is used in windshield wipers, window lifts, seat adjustors, and air conditioner blowers. In the appliance industry, the FPM is used in electric toothbrushes, electric shavers, portable electric tools, etc. Other special applications for the FPM are: toys, door gaskets, Hall-effect-switches, Reed switches, AC timers, Stepper motors, and sensors (for engine ignition and antiskid systems).

Plastic magnets have an advantage over metal and ceramic magnets in their shape versatility, in conjunction with dimensional accuracy, and low cost. Therefore, they find widespread usage in applications such as:

1. Rotor magnets in synchronous motors.
2. Ring magnets in induction motors.
3. Multi-pole disc magnets in rotary speed detectors.
4. Color adjustment magnets in color TV receivers.
5. Beam centering magnets in TV receivers.

Glossary of Magnetic Terms[13,14,15]

Coercive Force (H_c)—The magnetizing force that must be applied to a magnetic material in a direction opposite to the residual induction to reduce the induction to zero.

Curie Temperature—Temperature in degrees centigrade at which a magnetized sample is completely demagnetized due to thermal agitation.

Demagnetization Curve—That portion of the normal hysteresis loop in the second quadrant showing the induction in a magnetic material as related to the magnetizing force.

Energy Product Curve—The curve obtained by plotting the product of the coordinates of the demagnetization curve (B_dH_d) as abscissa against the induction (B_d). The maximum energy product expressed $(B_dH_d)_{max}$ corresponds to the maximum value of the external energy.

Ferromagnetic Material—A material which exhibits hysteresis phenomena and whose permeability is dependent upon the magnetizing force.

Gauss—The c.g.s. unit of magnetic induction.

Hysteresis Loop—A normal hysteresis loop is a graphical representation of the relationship between the magnetizing force and the resultant induced magnetization of a ferromagnetic material, when the magnetic force is cyclically varied.

Induction (Magnetic Flux Density)—(B)—The magnetic flux per unit area of a section normal to the direction of flux. The unit of measurement for flux density in the c.g.s. system is the Gauss.

Magnetizing Force—(H)—The magnetomotive force per unit length at any given point in a magnetic circuit. In the c.g.s. system the unit is the Oersted.

Oersted—The c.g.s. unit of magnetizing force.

Permeability—(μ)—The ratio of the magnetic induction in a given medium to the induction which would be produced in a vacuum with the same magnetizing force.

$$\mu = \frac{B}{H} \text{ Gauss/Oersted}$$

Permeance Coefficient—(P)—Ratio of the total external permeance to that of permeance of the space occupied by the magnet.

$$(P = B_d/H_d)$$

Reluctance—(R)—The reciprocal of permeance.

Remanence—(B_d)—The flux density which remains in a magnetic circuit after the removal of an applied magnetomotive force.

Residual Induction—(Br)—The magnetic flux density corresponding to zero magnetic force in magnetic material.

Soft Magnetic Material—A magnetic material, metallic or ceramic (ferrite), which once magnetized is very easily demagnetized, i.e., requires a slight coercive force to remove the residual magnetism. (The hard ferrite behaves reciprocally.)

Saturation—The condition under which all elementary magnetic domains have become oriented in one direction. In this condition, the applied external field produces no increase in intrinsic induction.

References

1. Bergstrom, J. W., "General Motors Research Publication GMR-636, Feb. 17, 1967.
2. Bergstrom, J. W., Sutherland, G., and Tealer, D. H.,

"Ferrirol Process," *Ceramic Age* pp. 54–58, (July 1967).

3. Runk, R. B. and Andrejco, M. J., "A Precision Tape Casting Machine for Fabricating Thin Ceramic Tape," *Ceramic Bulletin*, **54**, (2) 199–200 (Feb. 1975).
4. Bunting Magnetic Co. Bulletin, "Flexible Permanent Magnetic Strip," Catalog No. 270-2.
5. Greenback Industries Inc., personal communication, 1975.
6. New Materials–Significant Development, "Magnetic Nylon Compounds," *Modern Plastics*, pp. 16–17 (Nov. 1975).
7. Hinderaker, P. D., "Flexible Magnets Shed Hardware Image," **48**(3) 94–98 (Feb. 12, 1976).
8. TDK Ferrite Magnet Bulletin, BME 37-003A.
9. Ferro Ottawa Chemical Corp. Bulletin, "Specifications of Ferro Barium Ferrite (300 Series) Powder," July, 1974.
10. Charles H. Kline Co. Publication Bulletin–Fairfield, N.J. "Ceramic Raw Materials"–Product–Market–New Development.
11. Buessem, W. R., Gigliotti, O. V., Linwood, S. H., and Proske, J. W., "Soft Ferrite–A Review," *Powder Metallurgy International*, **6**(3) 120–123 (1974).
12. Delmonte, J. *Metal-Filled Plastic, Section 9–Magnetic Materials*, pp. 147–168, Van Nostrand Reinhold, New York, 1961.
13. General Electric Co. Bulletin, "Permanent Magnet Manual," 1264-10 M, 1963.
14. Indiana General Co. Bulletin, "Design and Application of Permanent Magnets," Manual 7.
15. Stackpole Co. Bulletin, "Magnetic Theory," No. 61B.
16. Tokar, M., and Koenig, J. H., "High Lead Ferrite," *J. Applied Physics*, **39**(6) 2977 (May 1968).
17. Kerekes, Z. E., "Effects of Samarium Oxide on the Microstructure and Magnetic Properties of Lead Ferrite," Ph.D. Thesis, Rutgers University, 1973.
18. Taeler, D. H., "The Future of Hard Ferrites," *Ceramic Age*, p. 28 (Aug. 1968).
19. Sixtus K. J., Kronenberg, K. J., and Tenzer, R. K., "Investigation on Barium Ferrite Magnets," *J. Applied Physics*, **27**, (69), 1051–1057, (1956).
20. Collins, J. R., "Advances in Magnetic Materials," *Electronic World*, pp. 49–52 (Dec. 1967).

Section V
Fire Retardants

13

Antimony Oxide

H. H. Waddell and Irving Touval
M&T Chemicals, Inc.
Rahway, New Jersey

CONTENTS

1. INTRODUCTION

Antimony has been used by mankind since antiquity. Its name is derived from the ancient Greek, meaning "seldom found alone." This name aptly describes the modern use of antimony oxide to reduce the flammability of plastics, since antimony oxide alone is practically useless as a flame retarder. However, it is extremely effective when used in combination with selected halogen compounds.

Antimony oxide is an opaque white mineral powder which has been used as a pigment because of its good color and high hiding power. However, it is no longer competitive with more economical white pigments, such as titanium dioxide, and its high cost and density also make it unattractive as an ordinary filler or extender. The major use of antimony oxide now is as a flame retarder, reducing the flammability of combustible materials.

Synergism between antimony oxide and halogens accounts for the widespread use of antimony oxide in a variety of plastics. Halogen compounds can be used alone as flame retarders, but adding a relatively small amount of antimony oxide improves performance sharply or achieves a given level of performance with a lower flame retarder content. Since flame retarder additives generally reduce physical properties of plastics and cost more than base resins, it is desirable to minimize the amount used. Antimony oxide is usually added at 1-15 wt %, and affects physical properties like other fine particle size minerals. The halogen compound usually has a much greater effect on plastic properties than antimony oxide because larger quantities are added. The halogen compound is often an organic liquid or low-melting solid that acts as a plasticizer which lowers modulus and strength. An additive system must be selected for the fabrication method as well as to meet the flammability and physical requirements of the intended end use.

The use of flammable materials, such as plastics, is restricted by a great and growing number of codes, regulations, and standards. Standard flammability test methods required for compliance with regulations governing flammability of materials used in aircraft, appliances, automobiles, buildings, fabrics, electrical products, etc., are listed and described.

The United States consumes about 40% of the world production of primary antimony, but domestic mines produce only about 15% of our needs. Most of the antimony oxide used domestically is produced from imported ores or crude oxide, but about one-third is imported from producers in Belgium, England, France, and Japan. This heavy dependence on foreign sources subjects domestic antimony supply to worldwide economic and political influences. As a result, the price of antimony oxide has fluctuated widely and will probably continue to do so. During recent shortages in 1970 and 1974, the price of antimony oxide exceeded $2/lb, but at the close of 1975 supplies were ample, and the price was reduced to $1.65/lb.

2. METHODS OF PRODUCING ANTIMONY OXIDE

2.1 Natural Sources

Antimony oxide occurs naturally as the minerals cervantite, senarmontite, and valentinite, which are constituents of some ores and oxidation products of other antimony minerals. Oxide ores from Algeria and Mexico are smelted to recover antimony metal. However, none of the natural oxides are recovered and used in their native form. All of the antimony oxide used in plastics is manufactured to obtain products with consistent purity, color, and particle size.

2.2 Commercial Manufactured Grades

The antimony oxide of commerce is antimony trioxide, Sb_2O_3, which is a stable, white, crystalline mineral powder. Theoretically, it contains 83.53% antimony in the three valence state, and the balance is oxygen. Actually, small amounts of higher antimony oxides and other elements are also present. Since antimony oxide is produced by direct oxidation of ores or metal, impurities present in the ore or metal are found in the oxide in amounts varying from barely detectable traces to about 0.5 wt %. Such contaminants include arsenic, copper, iron, lead, nickel, sulfur, tin, and many other elements. None of these trace impurities appear to affect the efficiency of the antimony oxide as a flame retarder, but they may affect oxide color and chemical activity. Domestic manufacturers control the purity and color of their products, so that trace elements generally have no significance to plastic formulators. However, the chemical composition of antimony oxide from unfamiliar sources, and particularly from the People's Republic of China, should be checked carefully before use to be sure that it does not contain excessive amounts of impurities such as arsenic or iron.

Details of commercial manufacturing processes are proprietary, but the general methods are described in the technical literature.[1] It is general knowledge that the Harshaw Chemical Company and M&T Chemicals Inc., produce oxide directly from high-quality imported antimony sulfide ores in their plants at Cleveland, Ohio, and Baltimore, Maryland, respectively. Chemetron Corporation and McGean Chemical recently began to supply oxide produced from crude imported oxide in a Cleveland plant. Chemetron Corporation has an interest in the plant which produces the crude oxide in South Africa. NL Industries operates a large smelter in Laredo, Texas, where both metal and oxide are produced.

Antimony trioxide exists in two crystal forms, and commercial products contain mixtures of the two. The cubic form, senarmontite, predominates in commercial products and is the stable, low-temperature form. The rhombic form, called valentinite, is formed at temperatures over 570°C and converts to the cubic form at temperatures below 570°C, but the rate of conversion is undetectable at room temperature, so commercial grades contain varying amounts of rhombic crystals. Properties of the two crystal forms are shown in Table 13.1, and photomicrographs in Figs. 13-1 and 13-2 illustrate the difference in crystal structure. Antimony oxide is produced by vaporizing antimony sulfide or metal, oxidizing the vapors, and crystallizing antimony oxide from the vapor

Table 13-1. The two crystal forms of antimony trioxide.[2]

Crystal form	Cubic	Rhombic
Name	Senarmontite	Valentinite
Specific gravity	5.2	5.67
Melting point, °C	656	656
Refractive index	2.087	2.18, 2.35, 2.35

phase. Temperatures of 600-800°C are employed in the process. When sulfide ore is used, gaseous sulfur oxides and solid slag are formed as waste products. No waste products are formed when antimony metal is burned. Some antimony oxide is produced by a "double fuming" process, in which impure oxide is revaporized and recrystallized as a commercially pure product. Most of the oxide manufactured in the United States is produced from high-

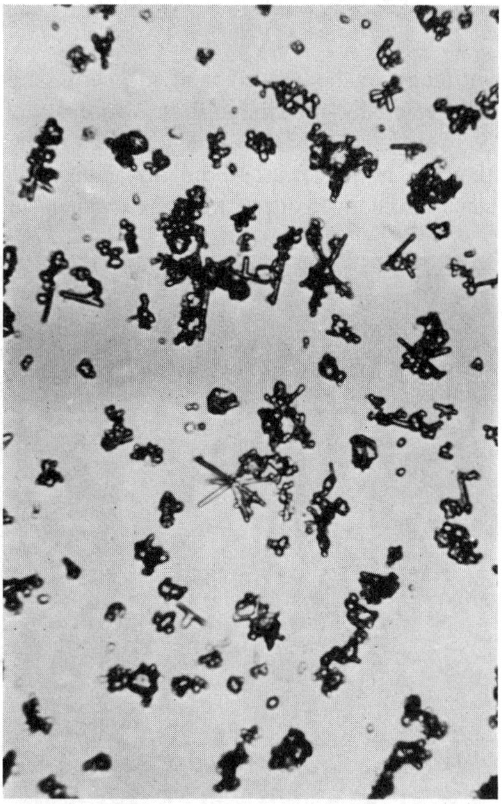

Fig. 13-1*. Antimony trioxide containing needle-like rhombic valentinite crystals mixed with cubic senarmontite (magnified 650 times, average particle size 1.53 microns).

*Photographs supplied by M&T Chemicals Inc.

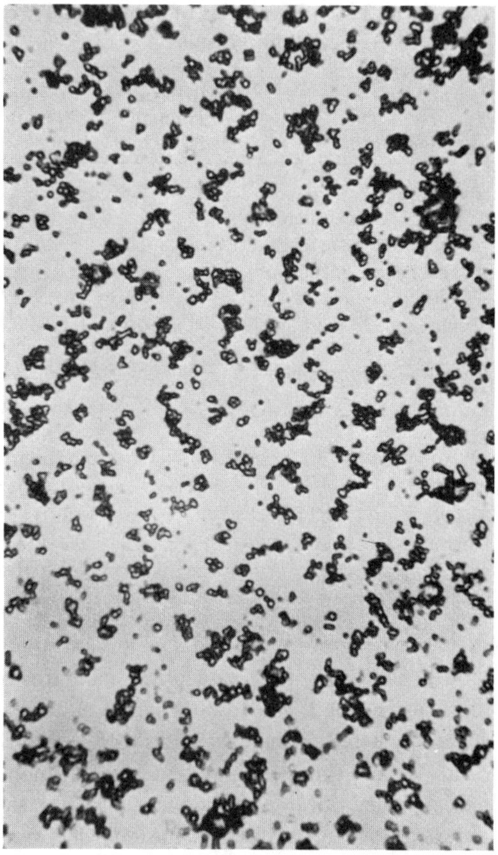

Fig. 13-2*. Typical commercial antimony trioxide, consisting primarily of cubic senarmontite crystals (magnified 650 times, average particle size 0.95 micron).

grade stibnite (antimony sulfide) ore from South Africa or South America. Much of the oxide imported from Europe and Japan is made by oxidizing metal. Antimony metal in turn is obtained either from smelting oxide or sulfide ores or as a co-product of copper, gold, lead, and silver recovery processes.

3. PROPERTIES OF COMMERCIAL GRADES

3.1 General Physical and Chemical Properties

Domestic grades of antimony oxide are equivalent and completely interchangeable as flame retarders. However, differences in particle size do cause significant differences in tint strength

*Photographs supplied by M&T Chemicals Inc.

Table 13-2. Properties of commercial antimony oxide.

Physical properties	High tint strength grade	Low tint strength grade
Specific gravity	5.3–5.8	5.3–5.8
Avg. particle size, microns (Fisher Sub-sieve method)	0.8–1.3	1.6–2.5
Oil absorption, g/100 g of Sb_2O_3	9–12	9–12
Residue on 325 mesh, wt %	0.001–0.1	0.001–0.8

Chemical composition, wt %	All grades
Antimony oxides, as Sb_2O_3	99.0 min.
Arsenic	0.1–0.4
Lead	0.05–0.30
Iron	0.001–0.04
Nickel	0–0.004
Moisture (2 hr at 105°C)	0.1–0.2

Fig. 13-3*. Dustless antimony oxide.

or pigmenting power. Generally, the antimony oxides fall into two categories—standard or high tint strength and low tint strength. Other antimony-based flame retarders offer significantly lower tint strength than any antimony oxide grade, but these products also are significantly less effective as flame retarders than antimony oxide. The color of antimony oxide is affected by trace impurities. Manufacturers control the color of their products, even though they are no longer used primarily as pigments, since antimony oxide is a strong pigment and

Table 13-3. Toxicity of antimony oxide.

Test		Results
Acute oral toxicity study[a]—albino rats.		$LD_{50} > 34.6$ g/kg
Acute dermal toxicity study—albino rabbits.		$LD_{50} > 2.0$ g/kg
Eye irritation tests—albino rabbits.	Unwashed eyes	Extremely irritating
	Washed eyes, 4 sec. contact	Severely irritating
Primary skin irritation—albino rabbits.		Minimally irritating
Skin sensitization—albino guinea pigs.		Not a sensitizer
Primary skin irritation—Human.		Non-irritating
Skin sensitization—Human.		Not a sensitizer
Evaluation employing animal and human test subjects was conducted by a repeated insult patch test.		

[a]In another study, single dose quantities of 20 g/kg failed to kill any of the six test rats.[3]

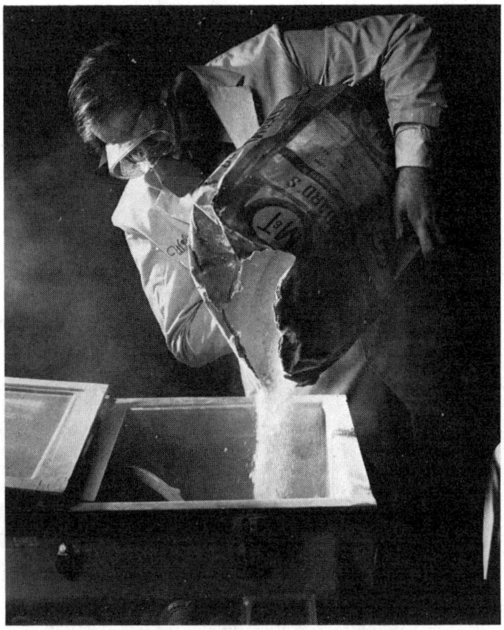

Fig. 13-4*. Regular antimony oxide.

*Photographs provided by M&T Chemicals Inc.

will affect the color of plastics in which it is used. Properties of commercial grades vary somewhat, and manufacturers' specifications should be referred to for exact details. However, properties generally fall into the range of values shown in Table 13-2.

3.2 Safe Handling Procedures

Suppliers are required to provide U.S. Department of Labor Material Safety Data Sheets and to disclose precautions for handling antimony oxide safely. Their instructions should be followed carefully. Instructions generally include providing protection against inhaling dust, avoiding ingestion, and prompt removal from skin and clothing by washing with soap and water. Industrial experience with antimony oxide has been good under normal conditions of usage. To reduce the dust formed when handling antimony oxide, suppliers offer "dustless" grades which have been treated with small amounts of ethylene glycol, mineral oil, or plasticizer for various end uses. Reducing dust creates healthier working conditions, improves housekeeping, and reduces material losses. The effectiveness of the treatment is clearly shown by photographs in Figs. 13-3 and 13-4.

3.3 Toxicity

Studies of the toxicity of antimony oxide performed recently by Industrial Bio-Test Laboratories Inc., under the sponsorship of the Manufacturing Chemists Association produced the results given in Table 13-3.

4. COMMERCIAL GRADES OF ANTIMONY OXIDE

4.1 Domestic Suppliers and Their Products

	BRAND NAME DESIGNATIONS		
Names and addresses of domestic suppliers	High tint strength grades	Low tint strength grades	Other antimony flame retarders
Chemetron Corporation 111 East Wacker Drive Chicago, Ill. 60601	H-Grade	L-Grade	
The Harshaw Chemical Co. Div., Kewanee Oil Company 1945 East 97th Street Cleveland, Ohio 44106	KR Grade White Star Grade White Star "M" Grade	KR-LTS Grade	
M&T Chemicals Inc. Post Office Box 1104 Rahway, N.J. 07065	Thermoguard[a] S Antimony Oxide Thermoguard[a] S-800 Dustless Antimony Oxide		Thermoguard[a] FR Flame Retarder Thermoguard[a] S-711 Flame Retarder
McGean Chemical Co., Inc. 1250 Terminal Tower Cleveland, Ohio 44113	H	L	Sodium Antimonate
NL Industries Inc. Industrial Chemicals Div. Post Office Box 700 Hightstown, N.J. 08520	Regular Grade Red Star Grade White Star M Grade	Grade 10	Oncor[b] 23A Oncor[b] 75 RA

[a]Trademark, M&T Chemicals Inc.
[b]Trademark, NL Industries.

5. FACTORS AFFECTING SUPPLY AND PRICE

5.1 Antimony Uses and Trends in Usage

Each year about 40% of the world's production of primary antimony is consumed in the United States.[4] However, mines in the U.S. produce only about 15% of our needs. Since we are a major consuming nation, but have little domestic production, the price and supply of antimony in the United States reflect worldwide political and economic influences. Only a few countries are major ore producers, so supplies can be seriously disrupted by the action of any one of them. The leading producing nations, in order of their proven reserves of ore, are the People's Republic of China, Bolivia, Republic of South Africa, U.S.S.R., and Mexico.

Antimony is widely scattered around the world, and proven reserves are 80-100 times annual consumption. However, rich deposits are small and often uneconomical to develop. Most of the world's supply results from recovery of other minerals, notably copper, gold, lead, silver, and tungsten. Historically, most antimony consumed was used as the metal to make alloys used for ammunition, printing type, battery plates, bearings, pewter, and cable sheaths. These uses are now obsolete or declining; even consumption in auto batteries declined when high prices in 1970 prompted a reduction in antimony content of battery plates. Nonmetal uses included chemicals for matches, ammunition primers, tracer bullets, rubber vulcanization and use of antimony oxide as a pigment, glass fining agent, and ceramic enamel opacifier. These uses also have declined or been discontinued as antimony prices made other materials more competitive. However, during World War II, antimony oxide was used with chlorinated wax to "flameproof" canvas, and since then its use as a flame retarder has grown. Now, the single largest use of antimony oxide is as a flame retarder.

In 1974, 24.2 million lb (11,000 metric tons) of antimony oxide were used as a flame retarder for plastics.[5] During that year, domestic production was 25.0 million lb (11,320 metric tons)[6], and imports totaled 12.5 million lb (5,660 metric tons), so 65% of the total new supply was consumed as a flame retarder for plastics. A thorough discussion of antimony supplies and uses and a good bibliography of antimony references is found in *Trends in Usage of Antimony, Publication NMAB-274* (1970), which is available from the National Technical Information Center. The most complete statistics on production, consumption, pricing, etc., are published by the Bureau of Mines. Their *Antimony Quarterly–Mineral Industry Surveys* and their annual *Minerals Yearbook* are supplemented by reviews at about five-year intervals in *Mineral Facts and Problems* bulletins.

5.2 Price and Availability

Price and availability can change rapidly, so users are advised to keep in close contact with suppliers and to watch trade publications for market information. *American Metal Market*, *Chemical Marketing Reporter*, and *Metals Week* are 3 trade publications which provide prices and news of significant developments in antimony markets. Prices for commercial grades from all suppliers are generally the same, and all sell products f.o.b. plant with freight allowed. However, when ore supplies are restricted, it is not unusual to find prices significantly different from various suppliers as a result of their different raw material costs. Imported oxide tends to be priced below domestic products as an incentive when supplies are adequate because most users prefer dealing directly with a domestic manufacturer to obtain normal credit terms, familiar quality, technical service, and other usual advantages of buying from American sources. When supplies are restricted, the price of imported oxide generally exceeds that of domestic products, and supply is apt to be interrupted. Historical price information for antimony metal can be obtained from Bureau of Mines publications, but until recently the data on antimony oxide prices have been limited. However, metal and oxide both are produced from the same basic ores, and their prices rise and fall together. Average annual metal prices shown in Table 13-4 generally indi-

Table 13-4. Price history of antimony metal.

Year	Antimony metal price, annual average, $/lb
1960	.3130
1961	.3389
1962	.3475
1963	.3475
1964	.4222
1965	.4575
1966	.4575
1967	.4575
1968	.4575
1969	.5757
1970	1.4419
1971	.7118
1972	.5900
1973	.6845
1974	1.8176
1975 (9 months)	1.823[a]

[a]The antimony oxide price for truckload quantities was $1.65/lb at the end of 1975.

cate the way antimony oxide prices have fluctuated.

6. THE USE OF ANTIMONY OXIDE AS A FLAME RETARDER

6.1 Introduction

Plastics decompose when they are exposed to a flame. The temperature at which they decompose and the products which are formed determine whether or not they will ignite and burn. Plastics which decompose at relatively low temperatures and form combustible hydrocarbon gases are more easily ignited than plastics which decompose at high temperatures and form either chars or noncombustible gases. Ignition and the ensuing consumption of plastics by flame can be retarded by adding antimony oxide and a halogen compound. They are used together because antimony oxide by itself has almost no flame retardant effect, and halogens are only effective at relatively high loadings. However, together they constitute an extremely effective flame retarder system.[7] The optimum combination of these materials is one which contains 3 moles of halogen for each mole of antimony. Since more of the halogen compound is needed, it exerts the greatest effect on the physical properties of the plastic. The kind of halogen compound used also affects the flame retarder efficacy of the total flame retarder system. Therefore, halogen compounds must be included in any discussion about the uses of antimony oxide as a flame retarder.

6.2 Halogen Compounds

6.2.1 Types Used. A halogen compound is selected on the basis of economy, stability, and general properties as well as its efficacy as a flame retarder. Organic bromine and chlorine compounds are the most commonly used halogen sources. Of these, bromine compounds are more effective,[8] but chlorine compounds are more widely used because they are less expensive and more stable. The halogen compound must be stable during processing and during the subsequent useful life of the plastic it is added into. Poor stability can result in corrosion of processing equipment and discoloration or other degradation of the plastic.

Inorganic halogen compounds are seldom used because most of them are ineffective, unstable and water-soluble. No inorganic halogen compounds are used commercially with antimony oxide on a large scale, so this chapter will consider only organic halogen compounds.

6.2.2 Efficacy of Various Types of Organic Halogen Compounds. The efficacy of the organic halogen compounds is related to the ease with which they decompose and release reactive halogen.[9] Long chain halogenated aliphatic compounds, which have relatively low halogen-to-carbon bond strength, are the most effective flame retardants. They usually melt and decompose at relatively low temperatures, generating either elemental halogen or the halogen acid. However, the very property which makes them so effective also limits their use to plastics which can be processed at temperatures below which the halogenated compound decomposes. Halogenated cyclo-aliphatic compounds have stronger halogen-to-carbon bonds than halogenated linear aliphatics. They melt and decompose at higher temperatures but are less effective than the linear aliphatic compounds. The halogenated cyclo-aliphatics are usually stable up to temperatures of 350°C or more. They are used primarily in plastics which require high processing temperatures. Halogenated aromatics are the least effective class of organic halogen compounds. Their high halogen-to-carbon bond strength makes them so stable that in many cases they release their halogen after the plastic has already been decomposed.

6.3 Interaction of Halogen Compounds with Antimony Oxide

As stated previously, antimony oxide is an ineffectual flame retarder by itself. It must be activated to be effective, and a halogen compound is used as the activator. The halogen compound decomposes upon exposure to heat, liberating halogen acid and/or halogen. These decomposition products then react with antimony oxide to produce either antimony trihalide or antimony oxyhalide, which are the flame retarding antimony species. The release of halogen and its reaction with antimony oxide are illustrated by the following formulas, in which R is an organic radical and X is a halogen, such as bromine or chlorine.

$$2RHX \longrightarrow 2RR + 2HX$$
$$2RHX \longrightarrow 2RH + X_2$$
$$6HX + Sb_2O_3 \longrightarrow 2SbX_3 + 3H_2O$$
$$X_2 + Sb_2O_3 \longrightarrow SbOX + SbO_2X$$
$$3SbOX \longrightarrow SbX_3 + Sb_2O_3$$

At least 2 hypotheses[10,11,12] have been proposed to describe the flame retardant mechanism of antimony trihalides. One theory proposes a reaction between the polymer and antimony trihalide which results in the formation of a char instead of volatile, flammable gases. The char further reduces volatile gas formation by functioning as a thermal barrier which protects the underlying polymer from further thermal degradation. The flame retardant also forms an inert atmosphere which suppresses combustion by reducing the oxygen supply to the flame. The other hypothesis contends that antimony trihalide volatilizes into the flame, where it decomposes into various antimony compounds and halogen free radicals. These decomposition products function as flame inhibitors by dissipating the energy of the flame and altering the flame chemistry. The antimony compounds function as energy dissipators, and the halogen free radicals alter the flame chemistry.

The flame inhibiting efficiency of antimony oxide/halogen combinations is governed by the halogen to antimony ratio, the availability of the halogen compound, and the decomposition products generated. Mixtures which contain a 3:1 halogen/antimony mole ratio and a halogen compound which readily generates the halogen acid will form antimony trihalide as the exclusive product. If the ratio is less than 3:1, or if elemental halogen is the main decomposition product, a mixture of the antimony trihalide and antimony oxyhalide is formed.

The following results of thermogravimetric analysis (TGA) of such combinations illustrates these points.[13] A 50/50 weight mixture of a chlorinated paraffin (70% chlorine) and antimony oxide has a 3 to 1 mole ratio of chlorine to antimony. As shown in Fig. 13-5, this mixture has a 72% weight loss between 300-375°C when subjected to TGA. The loss represents 94% of the potential amount of antimony trichloride the mixture can form. The TGA pattern of another mixture which also has a 3:1 mole ratio but contains a cycloaliphatic halogen source which does not readily form HCl, is quite different. As shown in Fig. 13-6, this mixture decomposes in 2 stages. In the first stage between 325 and 410°C, a 42% weight loss occurs. This represents only 58% of the antimony trichloride the system is capable of forming. The remaining trichloride is formed

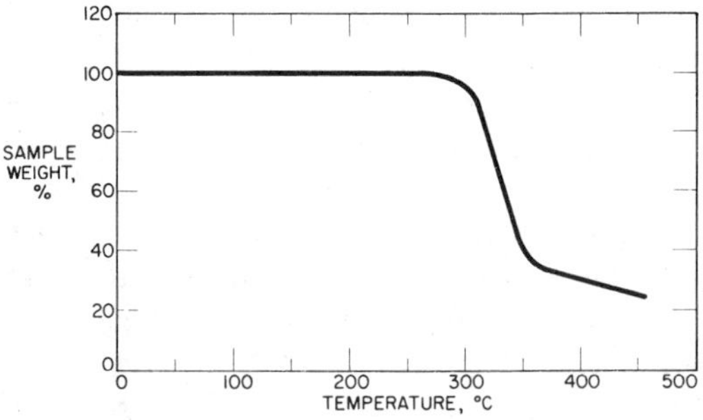

Fig. 13-5. TGA of a chlorinated paraffin and antimony oxide mixture.

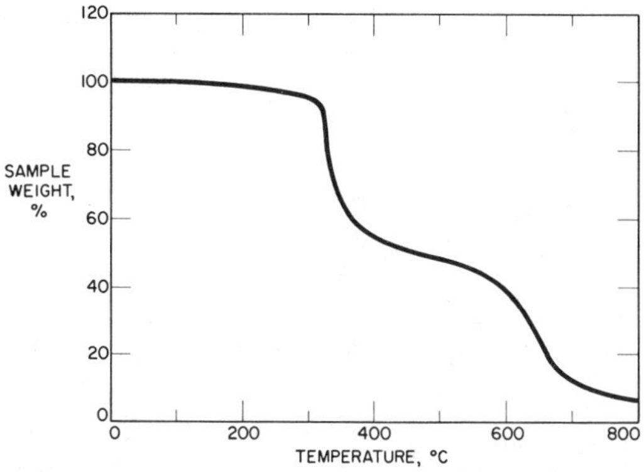

Fig. 13-6. TGA of a cycloaliphatic chlorine and antimony oxide mixture with a 3:1 Cl:Sb mole ratio.

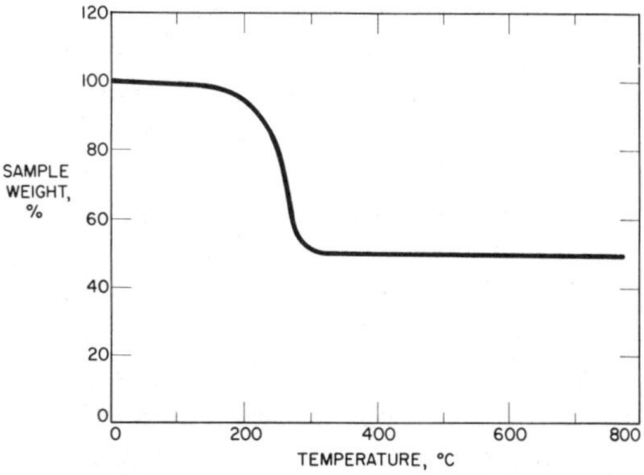

Fig. 13-8. TGA of a 50:50 weight % mixture of tetrachloro phthalic anhydride and antimony oxide.

between 575 and 675°C as the antimony oxychloride decomposes to antimony trichloride and antimony oxide.[14] However, if the mole ratio of the combination is increased to 8:1, more antimony trichloride is formed during the initial decomposition, and it is a more effective flame retardant system. This is shown in Fig. 13-7.

It is essential that the halogen compound does not volatilize before it decomposes and reacts with the antimony oxide. Tetrachloro phthalic anhydride, an aromatic chlorinated compound, sublimes below its decomposition temperature of 255°C. It does not form the hydrogen chloride or chlorine necessary to react with antimony oxide. Therefore, when these materials are mixed and subjected to thermogravimetric analysis, all of the weight loss can be attributed to the chlorine source alone. Fig. 13-8 is a thermogram of a 50:50 weight percent mixture of tetrachloro phthalic anhydride and antimony oxide. It shows a 50% loss in weight between 175 and 275°C, indicating that no reaction between the two ingredients has occurred. The most efficatious combinations of chlorine sources with antimony oxide are those that produce antimony chloride at or just above the decomposition temperature of the plastic.

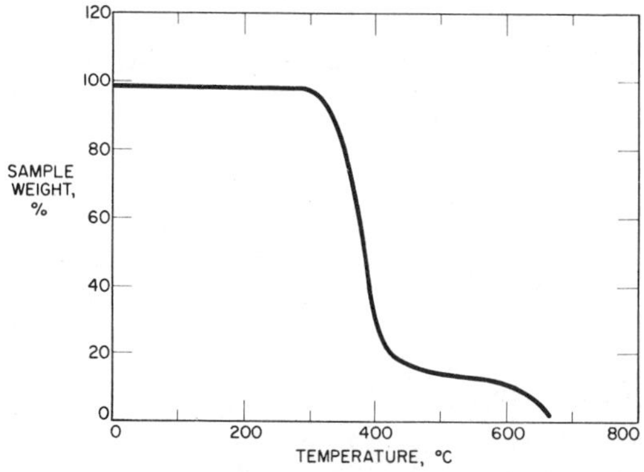

Fig. 13-7. TGA of a cycloaliphatic chlorine and antimony oxide mixture with an 8:1 Cl:Sb mole ratio.

6.4 Formulations for Thermoplastics

6.4.1 Polyethylene. Polyethylene is processed between 275 and 350°F. Therefore, it can be flame retarded with antimony oxide and either low-melting halogenated paraffins or the more stable halogenated cyclo-aliphatics or aromatics. Typical formulations and their relative flame retardency are shown in Table 13-5.

Table 13-5. Flame retardant polyethylene formulations.

Component	WT % A[15]	WT % B[16]	WT % C[17]
Polyethylene	85	60	82
Antimony oxide	7.5	13	6
Chlorinated paraffin (70% Cl_2)	7.5	–	–
Dechlorane Plus[a]	–	27	–
Decabromobiphenyl oxide	–	–	12
Results			
Oxygen index	24.0	24.0	27.0
UL94 class	V-2	V-1	V-0

[a]Trademark, Hooker Chemical Co.

Formulation A contains a mixture of a chlorinated paraffin and antimony oxide in a 3:1 mole ratio. It has an oxygen index of 24.0 and a V-2 UL94 rating.* It is the least flame retardant of the three formulations. However, increasing the loading of flame retardant will increase its UL94 rating to V-1 and possible V-0. Formulation B contains 40% of the flame retardant mixture in a 7:1 mole ratio. It has a 24.0 oxygen index and a V-1 UL94 rating. Although it is more flame resistant than Formulation A, it is also more costly. Reduction in the flame retardant content by as little as 10% will seriously affect the flame resistance. Formulation C contains bromine and antimony in a 3:1 mole ratio. It is the most flame resistant of the three formulations. It has a 27.0 oxygen index and a V-0 UL94 rating. However, it suffers from poor light stability and will rapidly discolor.

6.4.2 Polypropylene. The 340–450°F temperature required to process polypropylene excludes the use of chlorinated paraffin because of its low melting point and poor heat stability. Only high melting halogen sources are used. Both brominated and chlorinated compounds are very effective in combination with antimony oxide, as shown in Table 13-6.

Formulation B, which contains bromine and antimony oxide, is more effectively flame retarded than Formulation A, which contains chlorine and antimony oxide. Both formulations have an oxygen index of 27, but the bromine formulation rates better in the UL94 test. Formulation A is often preferred over Formulation B, although it is slightly more expensive, because Formulation B has poor light stability. Its light stability can be improved by the incorporation of UV stabilizer at increased costs.

*See Glossary of Test Methods for description of tests.

Table 13-6. Flame retardant polypropylene formulations.

Component	WT %	
	A[16]	B[17]
Polypropylene	60	80
Antimony oxide	10	5
Dechlorane Plus	30	–
Decabromobiphenyl oxide	–	10
Results		
Oxygen index	27.0	27.0
UL94 Class	V-1	V-0

Table 13-7. Flame retardant polystyrene.

Component	WT %		
	A[7]	B[18]	C[17]
Polystyrene	75	80	85
Antimony oxide	15	5	5
PVC	15	–	–
Dechlorane 604[a]	–	15	–
Decabromobiphenyl oxide	–	–	10
Results			
Oxygen index	NA	24.5	24.5
UL94 Class	NA	V-1	V-0

[a]Trademark, Hooker Chemical Company.

6.4.3 Styrenics. Polystyrene and ABS will be discussed together, since they both have approximately the same flame retardant requirements. Typical formulations are shown in Table 13-7.

Polystyrene can be flame retarded with aliphatic, cyclo-aliphatic, or aromatic halogen donors and antimony oxide. All three halogen sources will increase the oxygen index to approximately 24.5 and impart V-1 to V-0 ratings in the UL94 test. The flame retardant requirements for ABS are almost identical to polystyrene, as shown in Table 13-8.

Table 13-8. Flame retardant ABS

Component	WT %		
	A[7]	B[18]	C[17]
ABS	85	83	85
Antimony oxide	5	4	5
PVC	10	–	–
Dechlorane 604	–	13	–
Decabromobiphenyl oxide	–	–	10
Results			
Oxygen index	NA	25.0	25.0
UL94 Class	NA	V-1	V-0

Table 13-9. Effect of plasticizer on flammability of PVC.

	OXYGEN INDEX	
phr Plasticizer	No Sb_2O_3	3 phr Sb_2O_3
0	42–44	55–60
45	21.5	27.9
90	19.2	23.6

Formulation, parts per hundred:
PVC–100
Stabilizer–2.5 (Ba/Cd type)
Plasticizer, DOP–As shown.

6.4.4 Polyvinyl Chloride. Rigid PVC is an inherently flame retardant polymer which does not require additional flame retardants. Flexible PVC, on the other hand, is combustible and does require the addition of flame retardants. Since flexible PVC already contains chlorine, the addition of antimony oxide alone will markedly increase its flame retardancy. The quantity required will depend upon the amount and type of plasticizer used. This is indicated in Table 13-9.

No discussion of flame retarder uses of antimony oxide would be complete without some mention of proprietary antimony-based products. Such products contain less antimony than antimony oxide, and their flame retarder efficacy is generally directly related to their antimony content. However, one such product, Thermoguard* S-711 Flame Retarder, is reported to be more effective than antimony oxide for reducing the flame spread rating of plasticized PVC, despite its lower antimony content.[19]

6.5 Formulations for Thermosets

6.5.1 Polyesters. The flame resistance of polyesters can be enhanced by using halogenated dicarboxylic acids or polyhydric alcohols as monomers. Approximately 25% chlorine or 15% bromine is required to obtain a 30.0 oxygen index. The same degree of flame retardancy can be obtained with lesser amounts of halogen, or higher oxygen indices can be obtained with the same amount of halogen, by adding antimony oxide to the formulations as shown in Table 13-10.

Table 13-10. Effect of antimony oxide and chlorine on flammability of polyesters.

Wt % Chlorine	Wt % Antimony oxide	Oxygen index
25	0	30.2
25	1	33.4
25	3	38.7
20	1	30.2
12.5	2.5	30.0

Formulators can obtain a desired combination of cost, flame retardance, and physical properties by blending regular and halogenated resins and adding antimony oxide.[21] Nonreactive halogenated organics and antimony oxide can also be used to reduce the flammability of polyesters. Since the resins are liquid, low melting halogenated paraffin as well as the higher melting halogenated cyclo-aliphatic or aromatic compounds can be used. Typical formulations are given in Table 13-11.

Therefore, 17 to 23 wt % of a 9:1 or a 3:1 chlorinated paraffin/antimony oxide mixture will impart a 9-11 second flame out time to a polyester which would otherwise be completely consumed in the ASTM D635 test.

6.5.2 Polyurethanes. The use of antimony oxide and halogen in flexible urethane foams is one of the more recent advances in flame retardancy. In 1971 a patent[22] describing the use of antimony oxide, polyvinyl chloride, and zinc oxide in flexible urethane foams was issued to Harrington of General Tire Corporation.

Table 13-11. Flame retardant polyesters.[20]

	Wt %	
Non-halogenated polyester	83.3	77.0
Chlorinated paraffin (70% Cl_2)	12.5	11.5
Antimony oxide	4.2	11.5
Mole ratio Cl:Sb	9:1	3:1
ASTM D635 Flame Out (Sec.)	11	9
Oxygen index	27.0	27.0

*Trademark, M&T Chemicals Inc.

Table 13-12. Flexible urethane foam formulations. (Parts by Weight)

Component	A	B	C
3000 M.W. Polyoxy-propylene triol	100.0	100.0	100.0
Surfactant	1.0	1.0	1.0
Amine blowing agent	0.15	0.15	0.15
Stannous octoate	0.40	0.40	0.40
Deionized water	4.0	4.0	4.0
Toluene diisocyanate	52.0	52.0	52.0
Polyvinyl chloride resin	0	22.0	22.0
Antimony oxide	0	7.0	14.0
Zinc oxide (French process)	0	3.0	0
Self extinguishing time, sec.	Totally consumed	25	70
Distance burned, in.	6.0	1.6	2.4

He found that this combination of flame retardants reduced the self-extinguishing time of the foam to 25 seconds in the ASTM D1692, less than half that obtained when zinc oxide was not used.

6.6 Effect of Flame Retarders on Physical Properties

It would be of academic interest to show the effect of antimony oxide alone on the physical properties of plastic, especially since the halogenated organic is the major component of the flame retardant mixtures. Therefore, any consideration of changes in physical properties must include the entire combination. The changes in physical properties depend mostly on the characteristics of the halogen compound

Table 13-13. Effect of flame retarders on mechanical, thermal and electrical properties.[24]

Property	ASTM Test method	Polypropylene		Polystyrene		ABS	
Additive level, wt %[a]		None	40	None	30	None	38.1
Density	D792	0.905	1.188	1.050	1.240	1.035	1.310
Tensile strength	D638						
Yield, psi		5,400	4,000	–	–	–	–
Ultimate, psi		3,600	3,300	6,800	3,200	6,400	4,200
Ultimate elongation, %	D638	30	20	2.2	1.3	4.1	2.1
Flexural strength	D790						
Yield, psi		8,300	7,500	15,800	8,600	11,700	8,700
Modulus, psi		2.8×10^5	3.9×10^5	5.0×10^5	4.5×10^5	3.6×10^5	4.9×10^5
Impact strength, notched Izod, ft/lb/in. notch	D256	0.9–1.0	0.4–0.7	0.4–1.5	0.2–0.4	3.4–5.0	0.4
Hardness							
Rockwell	D785	95	96	122	121	102	101
Shore D	D1706	76	79	87	86	76	82
Heat distortion temperature, °C	D648						
66 psi		102	117	83	86	92	89
264 psi		63	74	78	82	82	82
Resistivity	D257						
Volume, ohm-cm		2.5×10^{14}	2.5×10^{14}	$>2.5 \times 10^{14}$	$>2.5 \times 10^{14}$	$>2.5 \times 10^{14}$	2.3×10^{14}
Surface, ohm-cm		2.5×10^{14}	2.5×10^{14}	$>2.5 \times 10^{14}$	$>2.5 \times 10^{14}$	$>2.5 \times 10^{14}$	2.5×10^{14}
Dielectric constant, 10^6 cps	D150	2.3	2.5	2.6	2.7	2.8	2.9
Dissipation factor, 10^6 cps	D150	3.2×10^{-4}	3.5×10^{-4}	2.4×10^{-4}	4.4×10^{-4}	6.2×10^{-3}	5.5×10^{-3}
Dielectric strength, S/TV/Mil	D149	500	540	510	430	410	460

[a]66% Dechlorane.
33% Antimony oxide.

and the loading required to obtain the necessary flame retardancy.[23] Plasticizing additives such as the halogenated paraffins tend to reduce tensile strength and increase elongation. Semi-plasticizing compounds, some of the halogenated paraffins, and cycloaliphatics have a variable effect. Nonplasticizing compounds, the high melting cycloaliphatics and aromatics, behave as non-reinforcing fillers. They reduce ultimate tensile strength, elongation, and impact strength while raising the modulus and tensile yields. The effect of antimony oxide and Dechlorane mixtures on the physical properties of some common thermoplastics is shown in Table 13-13.

References

1. Wang, C. Y., *Antimony–Its Geology, Metallurgy, Industrial Uses and Economics*, London, Charles Griffin & Co., 1952.
2. *CRC Handbook of Chemistry and Physics*, 56th Ed., p. B-73.
3. Smyth, H. F. and Carpenter, C. P., *J. Industrial Hygiene and Toxicology*, **30**, 6368.
4. Miller, M. H., *United States Mineral Resources–Antimony*. U.S. Geological Survey Professional Paper 820.
5. *Modern Plastics*, p. 42–44 (Sept. 1975).
6. *Antimony Quarterly–Mineral Industry Surveys*, U.S. Department of Interior, Bureau of Mines, Dec. 18, 1975.
7. Lyons, J. W., *The Chemistry and Uses of Fire Retardants*, John Wiley & Sons, New York, 1970.
8. Nametz, R., Polymer Conference Series. The Flammability Characteristics of Polymer Materials, Wayne State University, 1966. Paper.
9. Delman, A., Polymer Conference Series, Flammability Characteristics of Polymeric Materials, University of Detroit, 1969, Paper.
10. Read, N. J. Heigaway-Bury eg., *J. Soc. Dyers and Colourists*, **74**, 823 (1958).
11. Schmidt, W. G., *Trans. J. Plastic Inst.* 247 (1965).
12. Lindermann, R. F., *Ing. Eng. Chem.* **61**, 70 (1969).
13. Touval, I., "The Use of Stannic Oxide Hydrate," *J. Fire and Flammability* (1972).
14. Pitts, J. J., "Antimony-Halogen Synergistic Reactions in Flame Retardants," *J. Fire and Flammability*, **3**, 51 (1972).
15. Touval, I. and Waddell, H. H., "What You Should Know About Flame Retardants," *Plastics Technol.* (July 1971).
16. Hooker Chemicals Data Sheet #345.
17. Tabor, T. E., FRCA Meeting, New Orleans, Louisiana, 1974. Decabromo Biphenyl Oxide–A New Flame Retardant Additive for Plastics.
18. Hooker Chemicals Data Sheet #349.
19. M&T Chemicals Data Sheet #307.
20. Colelli, C. C., Paper presented at First National Flame Retardancy of Plastic Conference. N.Y.U. Continuing Education Series, Miami, Fla., Oct. 9, 1973.
21. M&T Antimony Based Flame Retarders for Unsaturated Polyester Resins, Quick Facts Sheet No. 15, M&T Chemicals Inc.
22. U.S. Patent 3,574,149.
23. Howarth, J. T., Lindstrom, R. S., Sheth, S., and Sidman, K. R., "Flame Retardant Additive," *Plastic Technol.* (Mar. 1973).
24. Dechlorane Additives for Fire-Retardant Polymers, Bulletin No. 66, Hooker Chemical Corporation.

Glossary of Test Methods

There are hundreds of test methods used to determine the flammability of plastics. These are designed to measure at least one flammability parameter, such as ignition, surface flammability, heat contribution, smoke generation, fire gases, and fire endurance. The results from one test cannot be used to predict the results from another or how the plastic will behave under actual fire conditions. Many of these tests are used to establish whether materials comply with various governmental regulations and trade association standards. Some of the most important tests used today are cataloged and listed below.

Test Method	Materials Tested	Ignition Source	Property Measured	Description
Underwriters Laboratories Standard for Safety UL-94	Plastics used in electrical applications	Bunsen burner flame	Duration of self-sustained combustion	A vertically oriented sample is exposed to a bunsen flame twice for 10-second ignition intervals. The material is classified according to the duration of flaming and nonflaming (afterglow) combustion and the capability of any dripping material to ig-

Test Method	Materials Tested	Ignition Source	Property Measured	Description
				nite cotton placed 12-in. below the sample (flaming drip). *Class* / *Flame* / *Afterglow* / *Flaming Drip* V-0 / 0–5 sec. / 0–25 sec. / No V-1 / 0–25 sec. / 26–60 sec. / No V-2 / 0–25 sec. / 0–60 sec. / Yes
FAA Regulation for Compartment Interiors Par. 25.853	Aircraft interior materials	Bunsen burner flame	Flame spread, char length, self-extinguishment time	1. A sample held in a vertical configuration as shown in Fig. 13-9 is ignited with a bunsen flame for 60 seconds and then removed. The specifications of the test are: (a) a maximum burn length of 6 in., (b) a maximum self-extinguishment time of 15 seconds and (c) any material which drips away from the specimen must extinguish itself in a maximum of 3 seconds. 2. A sample held in a horizontal configuration is ignited with a bunsen flame for 60 seconds. Maximum allowable flame propagation rates of 2.5 to 4 in./min are specified for materials intended for various ultimate uses.
ASTM D 2862 Oxygen Index Test	Paper, plastics, textiles and all organic materials	Burner flame	Minimum oxygen concentration necessary to sustain combustion	A sample in a holder illustrated in Fig. 13-10 held in a vertical position is burned in a mixture of oxygen and nitrogen. The oxygen is gradually adjusted to a concentration at which combustion can be sustained for 3 minutes. The apparatus is shown in Fig. 13-11.
Underwriters Laboratories Hot Wire Ignition	Plastics used in electrical applications	Electrically heated wire	Ignition time in minutes	Electricity is passed through a resistance wire wrapped around the sample. The time for sample ignition is noted.
Underwriters Laboratories Hot Current Arc Ignition	Plastics used in electrical applications	High voltage electrical arcs	Ignition	Electrodes resting on a sample are moved together until they arc. They are then moved apart until the arc is ruptured. The number of arcs required to ignite the sample is recorded.
ASTM D 635	Rigid plastics	Bunsen burner	Flame spread rate	A sample in a horizontal configuration as illustrated in Fig. 13-12 is exposed to a bunsen burner flame for 30 seconds and allowed to burn until the flame goes out or the sample is consumed. The flame spread rate is reported in inches/minute.

Test Method	Materials Tested	Ignition Source	Property Measured	Description
ASTM D 1692	Foams	Bunsen burner with wing tip	Flame spread rate	A sample held in a horizontal configuration is exposed to the bunsen flame for 60 seconds and is allowed to burn until the flame goes out or the sample is consumed. The flame spread rate is reported in inches/minute.
DOC FF-1-70 and DOC-2-70 Methenamine Pill Test ASTM D 2859	Carpets and floor coverings	Methenamine pill	Area of flame, spread	A standard sized methenamine pill placed in the center of a sample is ignited and allowed to burn out. The maximum flame spread allowed is 7 in. from the ignition source.
MVSS 302	Plastics and materials used as automobile interiors	Bunsen burner flame	Flame spread rate	A horizontal specimen is ignited with a bunsen flame. Flame spread rate must not exceed 4 in./min. to pass.
ASTM D 1433	Thin plastic sheets	Butane burner	Flame spread rate	A sample held at a 45° angle is ignited with a butane flame. Flame spread rate is calculated in inches per minute.
ASTM E 119 Fire Endurance Test	Construction materials	Gas flame (2000°F)	Heat penetration	One side of a full sized wall is exposed to a gas flame (2000°F). The temperature on the other side of the wall is monitored with thermocouples. The wall is rated by the amount of time it takes to reach 250°F on the cool side.
Bureau of Mines Flame Penetration	Plastic foams used in mines	Propane torch	Flame penetration resistance	A 1-in. thick specimen in the vertical position is exposed to a pencil flame of a propane torch. The time for the flame to penetrate the specimen is recorded.
ASTM E 84, UL 723, Steiner Tunnel	All construction materials	Two gas burners	Flame spread, heat contribution, smoke	A 25-ft sample mounted along the ceiling of a tunnel is exposed to the flames of two gas burners. The distance burned, the flame spread rate, the heat contribution, and smoke are compared to those obtained from cement board (rated zero) and red oak (rated 100). Materials are classified as follows: Class A–0–25 ft flame spread Class B–26–75 ft flame spread Class C–76–225 ft flame spread Class D–over 225 ft flame spread
ASTM E 162 Radiant Panel	Construction materials	Gas fired ceramic plate	Flame spread index is defined as product of the heat contribution and the flame spread $I_s = Q \cdot F_s$	A specimen 30° to the horizontal is exposed to a radiant panel at 1258°F. The rate of flame spread and heat contribution are recorded.

Test Method	Materials Tested	Ignition Source	Property Measured	Description
National Bureau of Standards	Plastics, fabrics, foams, etc.	Electric furnace, flame	Smoke density by light obscuration	1. A sample is placed vertically to an electric furnace in a sealed chamber. The amount of light obscuration through the chamber is recorded. 2. A sample is placed vertically to an electric furnace and ignited with a propane burner in a sealed chamber. The amount of light obscuration is recorded. A smoke density is illustrated in Fig. 13-13.
ASTM D 1929	Plastics, fabrics foams, paper, etc.	Hot air, flame	Self-ignition temperature, flash ignition temperature	1. A sample is placed in a cup which is gradually heated. The temperature at which self-ignition occurs is recorded. 2. A sample is placed in a cup which is gradually heated. A pilot flame is placed above the cup. The temperature at which ignition is attained is recorded.

Fig. 13-9. Vertical burn test housing showing mounted specimen and burner. (*Courtesy: FMC Corp., Industrial Div.*)

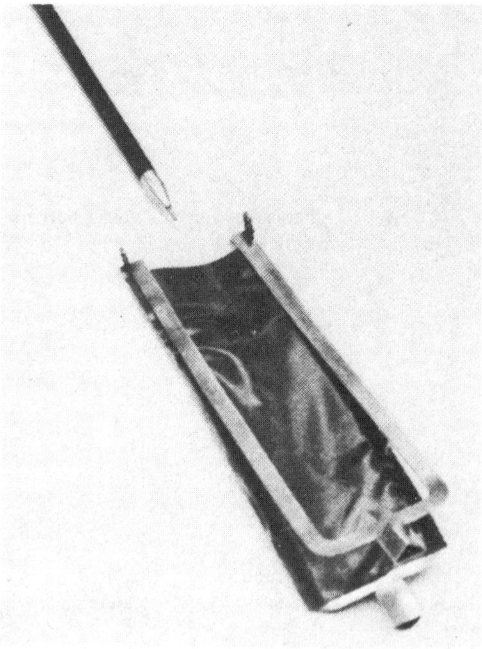

Fig. 13-10. Special holder for film specimen in G.E. Oxygen Index Tester. Flame is applied from top as indicated by pencil. (*Courtesy FMC Corp., Industrial Chemicals Div.*)

Fig. 13-11. Oxygen index apparatus. (*Courtesy FMC Corp., Industrial Chemicals Div.*)

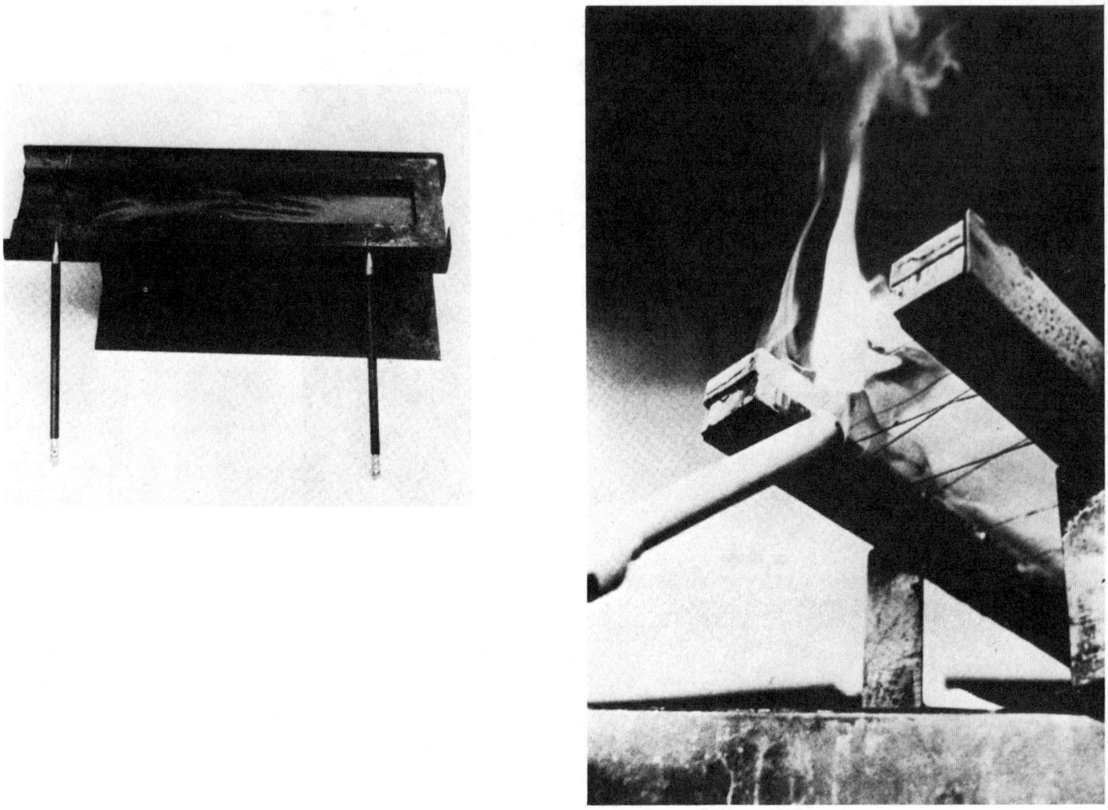

Fig. 13-12. Horizontal burn test. (*Courtesy FMC Corp., Industrial Chemicals Div.*)

Fig. 13-13. Smoke density chamber. Sample mounted and burned in holder at bottom. Smoke density measured by photo electric cell mounted in center. (*Courtesy FMC Corp. Industrial Chemicals Div.*)

14

Alumina Trihydrate

E. A. Woycheshin and I. Sobolev
Kaiser Aluminum and Chemicals Corp.
Pleasanton, California

CONTENTS

1. INTRODUCTION

Alumina trihydrate (gibbsite) is a refined mineral filler which offers a bonus of flame retardance and smoke suppression in most plastics applications. An intermediate in the manufacture of aluminum metal, it is produced by the Bayer process from bauxite ore. It is available in a number of grades, which contain varying amounts of minor impurities and range in particle size from coarse materials almost entirely +325 mesh (>45 microns) to very fine products less than one micron in mean particle diameter. Minor impurities in the commercial product include soda, iron oxide, and silica. The price range of standard grades in 1975 was 5–12¢/lb, depending on brightness and particle size. Surface-modified products were somewhat higher.

Sometimes called hydrated alumina or simply alumina hydrate, alumina trihydrate is nontoxic and relatively inert chemically. With a Mohs hardness of about three, it exhibits only low-to-moderate abrasiveness. As a filler in plastics exposed to electrical arcing, it provides the additional benefit of arc and track resistance. Since its surface has a low sorptive capacity, trihydrate has low resin demand in liquid polyester, epoxy, and acrylic systems. Like other non-reinforcing fillers, however, it generally lowers strength properties. It also carries a slight cost premium over unrefined fillers. Because it undergoes thermal decomposition, it is not suitable for processing at temperatures above about 220°C (428°F). In its composition, alumina trihydrate is actually not a hydrate, but rather crystalline aluminum hydroxide, $Al(OH)_3$. The bonding between the aluminum and hydroxide ions is predominantly ionic, and all of the hydroxide ions are equivalent. On heating to temperatures above about 220°C (428°F), alumina trihydrate decomposes endothermically to alumina, releasing 34.6% by weight water between about 220 and 600°C. This ability to act as a heat sink is the principal reason for its utility as a flame retardant in polymers.

Estimated U.S. consumption in rubber and plastics was 250,000 tons in 1974. The principal applications are in rubber carpet backings and adhesives, glass-reinforced, unsaturated polyester sanitary ware, appliance housings, and a variety of other molded electrical products, acrylic vanities, countertops, and panels. Epoxies, crosslinked polyolefins, and flexible vinyl represent other, relatively low volume, end uses. Application of alumina hydrate in several other polymeric substrates is still under development. As in the past, further growth of alumina hydrate as a flame-retardant filler will be fostered by new flammability regulations.

2. METHODS FOR PRODUCING ALUMINA TRIHYDRATE

Almost all the Western world's alumina is produced from bauxite by the Bayer* process.[1,2] Most bauxite ore, actually an impure form of alumina hydrate, comes from open-pit mines. The largest deposits are found in Jamaica, Australia, Surinam, USSR, and Guyana. The composition of the ore varies, but the alumina is present chiefly as alpha alumina trihydrate (gibbsite), along with smaller amounts of monohydrate (boehmite and diaspore). In the following, the term trihydrate, or simply hydrate, will be used because it is generally accepted by workers in the field, even though the compound is crystalline aluminum hydroxide.

In the Bayer process, the bauxite is digested with hot caustic soda under pressure to solubilize the alumina as sodium aluminate:

$$\text{Gibbsite: } \text{“}Al_2O_3 \cdot 3H_2O\text{”} + 2NaOH \longrightarrow 2NaAlO_2 + 4H_2O$$

$$\text{Boehmite: } \text{“}Al_2O_3 \cdot H_2O\text{”} + 2NaOH \longrightarrow 2NaAlO_2 + 2H_2O$$

Iron oxide and titania are not dissolved and settle out as a "red mud." Complex sodium aluminum silicates are also precipitated and are removed with the red mud. The sodium aluminate solution is separated from the red mud residue by gravity sedimentation and/or filtration. A flocculant is added to achieve practical settling rates. High caustic levels must be maintained to prevent premature crystallization of the alumina trihydrate. The sodium aluminate solution is cooled and agitated in the presence of alumina trihydrate seed crystals, whereupon it hydrolyzes on the surface of the seed to form crystalline alumina trihydrate:

$$2NaAlO_2 + 4H_2O \longrightarrow \text{“}Al_2O_3 \cdot 3H_2O\downarrow\text{”} + 2NaOH$$

The coarse crystal clusters, with a mean particle diameter of 60–80 microns, are separated and washed, while fine crystals are recycled as seed. Most of the product is calcined to aluminum oxide for use in the Hall-Heroult process for making aluminum. Some undergoes further processing and is used in the manufacture of refractories, dessicants, catalysts, and chemicals. A portion is dried and sold as Bayer hydrate, usually after grinding to specified particle sizes. The same types of commercial equipment are used for grinding alumina hydrate as for calcium carbonate. The Bayer product contains small amounts of organic impurities. While most of the organic matter in bauxite is decomposed during the digestion process, some humate residues survive and remain on the trihydrate crystals. Except for a reduction in brightness, no other effect relating to performance in plastics has been observed.

Another type of hydrate is produced by the "sinter" process,[1,2] which was developed to reclaim alumina from the sodium aluminum silicates in red mud waste. Red mud is mixed with limestone and sodium carbonate and calcined in rotary kilns to form a sinter containing dicalcium silicate and sodium aluminate. The sodium aluminate is dissolved, separated from the insoluble dicalcium silicate and other residue, and treated as in the Bayer process to precipitate alumina trihydrate. Since organic matter is burned in the rotary kilns, the sinter product has less organics and is brighter than Bayer hydrate. It is ground to specified particle sizes in the manufacture of a range of commercial grades. A brighter hydrate with a lower content of organics can also be produced by precipitation from redissolved Bayer hydrate. All of these processes yield hydrates with average particle diameters in the 30–80 micron range. Submicron size hydrates are produced by other processes. These bright, precipitated products have very narrow particle size distributions. They find use as paper pigments and as additives in paints, rubber, and plastics.

*Named after Karl Josef Bayer, who invented the process in Austria and patented it in 1888.

3. PROPERTIES OF ALUMINA TRIHYDRATE

3.1 Chemical Properties

Alumina trihydrate is not truly a hydrate but a crystalline aluminum hydroxide, $Al(OH)_3$.[3] The bonds between the aluminum and hydroxyl ions are predominantly ionic, and all the hydroxyl groups are essentially equivalent. The product from the Bayer process has the crystal lattice structure of the mineral gibbsite.[4] The major impurity is soda, with the sodium ions dispersed throughout the crystal lattice. Other minor impurities include silica, iron, zinc, calcium, and gallium. Trace amounts of titanium, manganese, and magnesium may also be present. Some organic impurities carry over from the bauxite ore in the Bayer process and affect the brightness of the product. Typical chemical compositions of the two major types of commercial hydrate are shown in Table 14-1.

Alumina trihydrate is insoluble in water and therefore nonleachable. However, it does dissolve in strong mineral acids and alkalies. It is non-toxic; in fact, amorphous forms of aluminum hydroxide are used in gastric antacid preparations. It is not considered to be a skin irritant or sensitizer. Hydrate dust should be treated like other inert particulate matter, and standard industrial hygiene procedures should be followed in handling it.

3.2 Physical Properties

Alumina trihydrate (gibbsite) has a density of 2.42 g/cc. Its average refractive index, determined with white light at 25°C by matching to oils of known refractive index, is 1.579. It has a Mohs hardness of 2½–3½, comparable to that of $CaCO_3$, and hence is only moderately abrasive. Brightness, or blue-light reflectance, determined according to TAPPI standard T646, varies from 86–92 for Bayer hydrates to 96–100 for sinter and submicron grades. Particle sizes of commercial products range from coarse grades which are almost all +325 mesh (>45 microns) to very fine grades less than one micron in average particle diameter. The particle size distribution of ground materials is generally broad, while that of coarse and fine precipitated grades is relatively narrow. Except for the submicron grades, surface areas and oil absorption values are quite low.

Table 14-1. Chemical composition of commercial alumina trihydrates.

	TYPICAL CHEMICAL ANALYSES, %	
	Bayer hydrates	Sinter hydrates
Aluminum as Al_2O_3	64.9–65.0	65.0–65.2
Iron as Fe_2O_3	0.007–0.02	0.003–0.006
Silica, SiO_2	0.01–0.02	0.007–0.01
Soda as Na_2O	0.25–0.4	0.15–0.2
Free moisture	0.02–0.3	0.02–0.3
Loss on ignition (H_2O)	34.4–34.7	34.4–34.7

3.3 Thermal Properties

The chemical thermodynamic properties of alpha alumina trihydrate (gibbsite) at 25°C are shown in Table 14-2.[5] Its specific heat varies nearly linearly from 0.282 cal/g at 20°C to 0.363 cal/g at 176.4°C.[6] On heating to temperatures above about 220°C (428°F) alumina trihydrate decomposes endothermically to alumina and water:

$$2Al(OH)_3 \longrightarrow Al_2O_3 + 3H_2O$$

The weight loss on ignition of commercial grades is close to the theoretical value of 34.6%. The measured heat of dehydration is 0.337 kcal/g at 210°C.[7]

TGA and DTA curves for a typical Bayer hydrate at a heating rate of 5°C/min. are shown in Figs. 14-1 and 14-2. At more rapid heating rates, both curves are displaced toward higher temperatures. The major endothermic peak at 300°C corresponds to the decomposition of alpha alumina trihydrate (gibbsite) to transition (chi) alumina. The smaller peak at 230°C is due

Table 14-2. Thermodynamic properties of α-$Al_2O_3 \cdot 3H_2O(c)$ at 25°C.

Formula weight	156.0072
Heat of formation, kcal/mol	−612.5
Gibbs free energy of formation, kcal/mol	−546.7
Entropy, cal/deg-mol	33.51
Heat capacity, cal/deg-mol	44.49

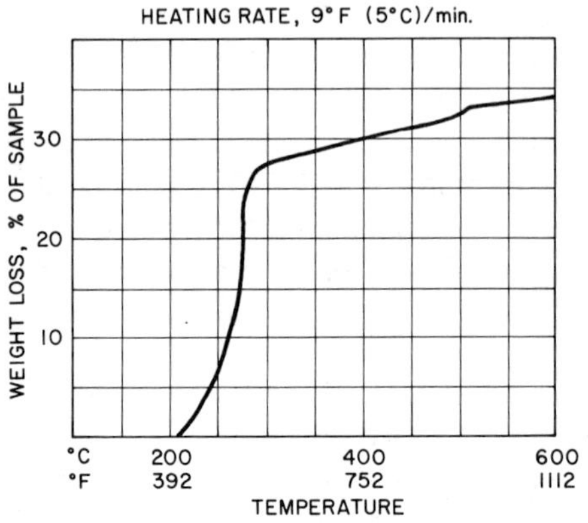

Fig. 14-1. Thermogravimetric analysis of H-46 alumina hydrate. (From Sobolev, I., and Woycheshin, E. A., "Alumina Hydrate as a Flame-Retardant Filler for Thermoplastics," *J. Fire Flamm./Fire Retardant Chem.*, **1**, 13–25, February 1974. Technomic Publishing Company, Inc., 265 West State Street, Westport, Connecticut 06880.)

to hydrothermal conversion of a portion of the alpha trihydrate to alpha-monohydrate (boehmite):

$$Al(OH)_3 \longrightarrow AlOOH + H_2O$$

The monohydrate subsequently decomposes to transition (gamma) alumina as indicated by the

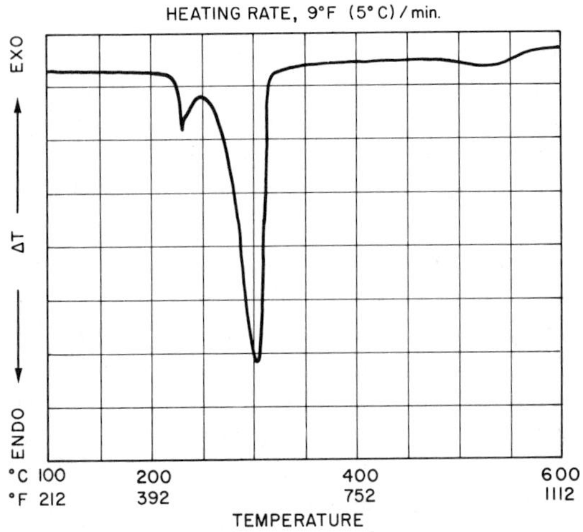

Fig. 14-2. Differential thermal analysis of H-46 alumina hydrate. (From Sobolev, I., and Woycheshin, E. A., "Alumina Hydrate as a Flame-Retardant Filler for Thermoplastics," *J. Fire Flamm./Fire Retardant Chem.*, **1**, 13–25, February 1974. Technomic Publishing Company, Inc., 265 West State Street, Westport, Connecticut 06880.)

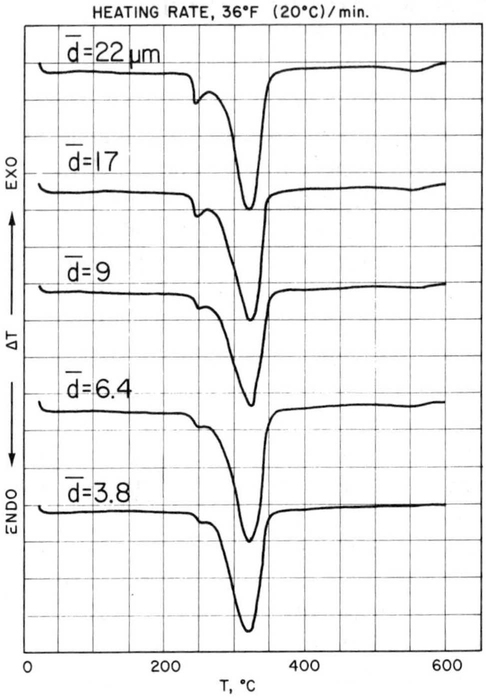

Fig. 14-3. Effect of particle size on the DTA of alumina hydrate. (From Woycheshin, E. A., and Sobolev, I., "Effect of Particle Size on the Performance of Alumina Hydrate in Glass-Reinforced Polyesters," *J. Fire Flamm./Fire Retardant Chem.*, **2**, 224–241, November 1975. Technomic Publishing Company, Inc., 265 West State Street, Westport, Connecticut 06880.)

broad endothermic peak at about 530°C:

$$2AlOOH \longrightarrow Al_2O_3 + H_2O$$

These peaks due to boehmite formation and decomposition are more pronounced for coarse trihydrate and diminish as the mean particle diameter decreases (Fig. 14-3). At temperatures below 205°C (400°F) alumina trihydrate is quite stable at atmospheric pressure. Between 205° and 220°C (428°F) slow decomposition occurs. During processing of hydrate fillers in polymers, therefore, residence time in this temperature range should be limited. Above 220°C decomposition becomes rapid.

3.4 Flame Retardance and Smoke Suppression

It is now generally recognized that the main mechanism by which alumina hydrate retards the burning of polymers is endothermic dehydration. In the burning process, volatile polymer fragments are produced through an endothermic cracking reaction in the pyrolysis zone by a portion of the heat of combustion transferred back to the polymer. These volatile products travel to the combustion zone, where they react exothermically with oxygen. By acting as a heat sink and sacrificially absorbing a portion of the heat of combustion transmitted back to the condensed phase, alumina trihydrate retards pyrolysis and hence reduces the burning rate. Dilution of combustion gases by water vapor and participation by water in condensed phase reactions are considered of secondary importance, but may be significant in certain cases. Absence of free-radical quenching effects in the vapor phase during combustion of alumina trihydrate-filled epoxy resins has been demonstrated.[8]

The smoke inhibiting activity of alumina trihydrate filler in SBR foams has been attributed to its promotion of solid-phase charring in place of soot formation.[9] Smoke suppression in these systems, therefore, is also probably directly related to endothermic dehydration, since dissipa-

Table 14-3. Principal suppliers and types of alumina trihydrate.

Supplier	Hydrate type	Particle size range	Price range, ¢/lb[a]
Primary producers:			
Alcan Aluminum Corp. 111 W. 50th St. New York, N.Y. 10020 tel. (212) 582-2070	Bayer	15–97% +325 mesh	4.7–5.2
Aluminum Company of America 1501 Alcoa Building Pittsburgh, Pa. 15219	Bayer Sinter Sub-micron	6.5 microns–98% +325 mesh 6.5 microns–65% +325 mesh 0.4–0.8 microns	4.7–6.85 5.4–7.85 12.0[b]–17.5[b]
Kaiser Chemicals Div. Kaiser Alum. & Chemical Corp. P.O. Box 1031 Baton Rouge, La. 70821 tel. (504) 293-9761	Bayer	6 microns–97% +325 mesh	4.7–6.85
Reynolds Chemicals Div. Reynolds Metals Co. Richmond, Va. 23261	Bayer Sinter	98% +325 mesh 0.1–60% +325 mesh	4.7 5.4–12.0[b]
Other suppliers			
Great Lakes Foundry Sand Co. Mineral Products Div. 2855 Coolidge Highway Troy, Michigan 48084 tel. (313) 649-3700	Bayer Sinter[c]	1.3–15.5 microns 1.3–15.5 microns	7.9[b]–9.55[b] 8.6[b]–11.75[b]
Industrial Minerals & Chemicals, Inc. P.O. Box 2204 Dalton, Ga. 30720 tel. (404) 278-8318	Bayer	4–20 microns	6.5–11.0
Solem Industries, Inc. 3550 Broad Street Atlanta, Ga. 30341 tel. (404) 457-1722	Bayer[c] Sinter[c]	1.5–25 microns 1.5–25 microns	d d

[a]Mid-1975 price, 40,000 lb minimum, in bulk unless otherwise indicated.
[b]Bags.
[c]Silane-treated grades also available; surface-modified products range in price from 14 to 22.5¢/lb.
[d]Prices range from 5 to 17¢/lb, depending on particle size, brightness, and presence of additives.

tion of heat in the condensed phase will reduce pyrolysis and favor competing crosslinking reactions.

4. COMMERCIAL GRADES

The principal suppliers of alumina trihydrate and the general types and average particle sizes of the grades they offer are listed in Table 14-3. The list includes primary producers who manufacture trihydrate from bauxite ore by the Bayer and other processes and other suppliers who grind and surface-treat the product to prepare a variety of grades. Many of these products are also sold by distributors of plastic fillers and additives.

5. ALUMINA TRIHYDRATE AVAILABILITY–PRESENT AND FUTURE

As mentioned earlier, almost all the world's aluminum is produced with alumina hydrate as an intermediate. In the U.S., estimated production of this intermediate in Bayer-process plants is 8 million metric tons annually. In the growing market for flame retardants, alumina hydrate has shown a market penetration which, in terms of tonnage, is quite impressive. The current estimated U.S. consumption of 250,000 tons/year in flame retardants (including ground and otherwise refined material) thus represents only 2.8% of this pool of raw material. Virtually all of the material consumed in the U.S. is manufactured domestically. Although because of energy requirements there may be a trend away from hydrate-based processes for aluminum in the future, economics will favor the continued utilization of existing Bayer plants, even if raw materials other than bauxite were to be used. The supply of alumina hydrate for flame-retardant applications thus appears assured over the long term and, with the construction of additional drying facilities, should sustain a many-fold increase in present consumption.

6. APPLICATIONS

Although the flame-retardant properties of alumina hydrate were certainly recognized by the late 1950's,[10] the material did not come into large volume use in polymers until the late 1960's. At that time, new consumer safety-oriented legislation was passed requiring carpets to have a certain degree of flame retardance. Alumina hydrate then began to replace calcium carbonate as a filler in synthetic rubber carpet backings. The next large market was in glass-reinforced unsaturated polyesters, and here again these developed in response to new flammability regulations. All of these applications are discussed in more detail below. Uses of alumina hydrate will continue to grow for the same reasons as the usage of other flame retardants, namely as a result of legislative action in response to consumer safety-oriented public demand.

The reasons for the growing popularity of alumina hydrate are its cost-performance advantages. The product is relatively low in cost compared to other flame retardants. It is easy to handle and is non-toxic, is free of halogens, and does not produce corrosive combustion products. Unlike many organic additives, it is not volatile, does not exude from polymers, and is resistant to leaching by all liquids except strong acids or bases. Grades are available displaying high whiteness, which is resistant to aging. While offering these many desirable features, however, alumina hydrate has two properties that impose limitations on its use: like many other fillers it is non-reinforcing, and it undergoes dehydration at temperatures above about 205°C. Thus it is understandable that this filler has gained acceptance in SBR latex in carpet backings, in crosslinked polyethylene insulation, and in glass-reinforced polyesters. All of these are chemically crosslinked systems that are processed and used at temperatures below 205°C. The chemical crosslinking compensates for whatever strength losses would be caused by a non-reinforcing filler. In the case of polyesters, glass fiber reinforcement provides additional strengthening. In thermoplastics, alumina hydrate, in common with other non-reinforcing fillers, generally reduces strength. Applications in such polymers are therefore still rather limited. There have been efforts to improve strength properties by use of coupling agents, and although silane-treated grades of alumina hydrate are now being offered commercially, consumption in thermoplastics has not been significantly increased. End uses of alumina hydrate in thermosets and thermoplastics are covered in more detail in the following sections.

6.1 Thermosets

Because non-reinforcing fillers generally reduce the strength of crosslinked polymers less than that of thermoplastics, thermosets today account for over 95% of total alumina hydrate consumption by polymers.

6.1.1 Polyesters. The effectiveness of alumina trihydrate as a flame-retardant filler in glass-reinforced unsaturated polyesters has been recognized for some time, and its economic and performance advantages in these systems have been repeatedly documented.[11-22] As a substitute for other fillers in premix-type systems (e.g., SMC, BMC), it imparts flame retardance, smoke inhibition, and arc tracking resistance. Halogenated polyesters, both chlorinated and brominated, when filled with alumina trihydrate are upgraded in fire resistance and smoke suppression. More recently, trihydrate has found widespread commercial application in spray-up systems for FRP bathroom components.

To impart significant flame retardant properties to polyesters, relatively high loadings of alumina trihydrate must be used. In order to achieve self-extinguishing characteristics in the back-side torch test for FRP tubs and showers,[23,24] for example, a 45-50% by wt concentration of alumina trihydrate in the polyester resin is needed. This test has been employed since 1972 because of the fire hazard presented by the use of plumbers' torches during installation. To accommodate such high filler loadings in spray-up systems, special low-viscosity resins have been developed. Reactivity was raised to compensate for the absorption of exothermic heat due to the heat capacity of the filler. Moreover, the chopped glass content had to be reduced to 18-20% to provide adequate glass wet-out. These modifications resulted in practical filled systems for spray-up application.[18]

Generally, the alumina trihydrate grades which have proved most useful in spray-up polyesters are those with an average particle size in the range of 6-14 microns. If the average particle size is finer than about 6 microns, the viscosity of the resin-hydrate mix will be too high for convenient handling in conventional spray-up equipment. If the average particle size is much coarser than about 14 microns, settling in the tanks and material supply lines may become a problem. Wet-out and thixotropic properties are also optimum for trihydrate in the 6-14 micron range. Viscosity depressants such as triethyl phosphate or dimethyl methyl phosphonate are sometimes added to improve processability and provide additional flame resistance.

Strength properties of the laminate are affected very little by the particle size of the alumina trihydrate.[22] Flame retardance is also relatively independent of average particle size, except for very fine grades (i.e., less than about 4 microns average diameter) which are slightly more efficient at equivalent loadings.[21,22] Use of such fine grades in the necessary concentrations, however, is usually precluded by the unmanageably high viscosities of the suspensions which result. Because of the lower glass content of filled systems, there is a slight reduction in tensile strength, impact strength, and flexural strength, compared to general-purpose unfilled systems containing 25-30% glass.[18] Flexural modulus (stiffness), on the other hand, is increased. Filled systems also require a somewhat greater effort to roll out because of the slower wet-out of the reinforcement. There is an increase in the weight of the laminate, filled systems having a specific gravity of about 1.6 compared to about 1.3 for general purpose systems.

In spite of the above disadvantages, alumina hydrate filled systems have found widespread use in FRP bathtubs, tub-shower units, sinks, and vanities because they meet flammability requirements, yet cost less than standard non-flame retardant systems.

6.1.2 Latex Foam and Adhesives. Latex foam rubber and carboxylated latex adhesives for carpet backing represented the first large market for alumina trihydrate in polymers and still consume more than any other application. Alumina trihydrate is used as a flame-retardant filler in precoat, adhesive, and foam compounds, having replaced calcium carbonate or clay fillers, which had no significant effect on flammability.

The foam compounds normally contain 20-150 parts hydrate per 100 parts latex, while the adhesive compounds contain 75-250 parts.

Typical foam backing formulations contain an SBR latex, natural rubber latex, alumina trihydrate, KOH, NH_4OH, feldspar, zinc oxide, and sulfur accelerators, an emulsifier such as potassium oleate, and frequently antioxidants and other proprietary additives. This mixture is mechanically frothed, ammonium acetate gelling agent is added at an in-line mixer, and the frothed compound is applied continuously to the back of a pre-coated carpet. Cure is completed in an oven at 120-150°C. Specific examples of foam backing formulations can be found in the patent literature.[25,26]

Typical carboxylated latex adhesive formulations contain alumina trihydrate, water, calcium carbonate, and a thickener. They are normally applied to the back of the carpet with a roller applicator. A recently developed foamed adhesive is prepared by mechanically frothing a surfactant-containing latex compound. With alumina hydrate rather than calcium carbonate or clay as filler, latex-backed carpets can pass the so-called "pill test" (DOC FF 1-70). The test is designed to measure ease of ignition from a small source such as a burning cigarette. In the test, a methenamine (hexamethylenetetramine) pill is ignited under standardized conditions and the carpet area charred by the flame is measured.

Although additional types of carpet backings, such as polyurethane and PVC foams are coming into use, latex backings are expected to represent the largest share of polymer end uses for alumina hydrate for several more years.

6.1.3 Acrylics. A relatively new but rapidly growing application for alumina hydrate is cast, crosslinked acrylics. It is used in sheets and fabricated shapes in bathroom vanities, wall panels, kitchen counters, and similar products. Here the alumina hydrate is employed not only as a flame retardant filler, but because its optical properties impart to the cast acrylic a marble-like translucency. Most of these products are marketed as synthetic marble.[27]

Sheets or shaped products are generally cast from a mixture containing approximately 58% by wt alumina hydrate, 30% methyl methacrylate monomer, 10% poly(methyl methacrylate) prepolyer, 1% ethylene dimethacrylate crosslinking agent, and less than 1% of a thermal initiator such as lauroyl peroxide or a redox initiator system that can initiate polymerization at ambient temperature. Curing generally takes place at elevated temperatures.[28]

6.1.4 Epoxies. Alumina trihydrate has been shown to be especially effective as a flame retardant filler in epoxy resins.[8] Used at loadings of 40-60% wt, it results in a sharp increase in oxygen index,* up to twice that of the unfilled resin. Flammability measurements in O_2/N_2 and N_2O/N_2 atmospheres have demonstrated that it acts primarily in the condensed (molten or solid) phase. Alumina trihydrate filler also improves the arc and arc-track resistance of epoxy resins.[29,30] This property, perhaps more than its flame retardant effect, accounts for its use in electrical applications. In general, cyclo-aliphatic epoxy resins have better arc and arc-track resistance than bisphenol resins. However, even the bisphenol resins are upgraded in arc resistance by addition of trihydrate. Best performance properties result through the use of a fine alumina trihydrate with low soda content. The soluble soda (i.e., soluble in the epoxy resin) is apparently more significant than total soda.[30] Cyclo-aliphatic epoxy resins filled with alumina trihydrate have shown considerable promise for use in such applications as transformers, insulators, and switchgear.

6.1.5 Polyurethanes. The growing emphasis on flame retardance in polyurethanes—particularly foams—has encouraged efforts to utilize alumina hydrate in these polymers.[31-34] Prospects for alumina hydrate appear better in flexible rather than in rigid foams, where several technical problems must first be overcome. Even with flexible foams, further development is still needed to boost the cost efficiency of alumina hydrate. Although there is considerable research

*The minimum volume-percent oxygen in a slowly rising mixture of oxygen and nitrogen which is just able to support combustion of a material burning in a candle-like manner (ASTM D2863).

activity, commercial utilization of alumina hydrate in polyurethanes is still rather limited.

6.1.6 Ethylene Copolymers. Alumina hydrate offers low-cost flammability reduction for crosslinked polyethylene[35,36] and crosslinked ethylene-vinyl acetate copolymers.[37] Filled polymers of this type are generally used because of their good electrical insulating properties. Although market data for crosslinked polyethylene and other polymers in electrical insulation have been reported,[38] no estimates have been published for the amounts of alumina hydrate consumed for this end use.

6.1.7 Other Thermosets. Alumina hydrate alone or in combination with other additives has been claimed as an effective flame retardant for diallylic phthalate molding compounds,[39] melamineformaldehyde molding compounds,[40] and intumescent coatings.[41] An interesting application is molded panels in which alumina hydrate is the principal constituent, the only other component being 2-20% of a urea or melamine-formaldehyde resin binder.[42,43] None of these end uses consumes significant quantities of alumina hydrate as yet.

6.2 Thermoplastics

As indicated earlier, alumina hydrate behaves as a nonreinforcing filler, and tends to dehydrate at temperatures above 205-220°C. Its application in thermoplastics is attractive in those cases where non-reinforcing fillers have been or can be used, and where processing temperatures are relatively low.

6.2.1 PVC. In view of the above limitations, it is not surprising that plasticized PVC has been the subject of more literature reports than any other thermoplastic as a substrate for alumina hydrate. Vinyl-coated fabrics now appear to represent the only commercial application of alumina hydrate in a thermoplastic polymer. Incorporation of alumina hydrate in plasticized PVC can be accomplished quite readily where the hydrate replaces another filler such as calcium carbonate. For a high level of flame retardance, all the filler should be replaced by alumina hydrate, and a phosphate plasticizer should be employed.[44,45] For example, replacement of 30 phr (based on polymer) calcium carbonate by alumina hydrate in a phthalate-plasticized PVC reduced the flame spread from 14 to 7 in. and the smoke rating from 215 to 175 in the 2-ft Monsanto tunnel test. Substitution of nearly one-third of the phthalate by phosphates further reduced the flame spread to 5-6 in., but raised the smoke rating to 190-200.[44] Addition of zinc borate reportedly offers another way of enhancing the flame retardance of alumina hydrate in phosphate-free PVC.[46] Alumina hydrate has also been claimed to fireproof certain nonwoven textile fabrics when added to PVC, acrylic, or poly (vinyl acetate) latex binders.[47]

In rigid PVC, which has inherently very low flammability, the principal effect of alumina hydrate is to reduce smoke. For example, replacement of 40% wt calcium carbonate by alumina hydrate in such a polymer reduced the smoke density rating in the Rohm & Haas XP2 chamber from 71 to 50%.[48] These and other results for rigid PVC are shown in Table 14-4. Being non-reinforcing fillers, both alumina hydrate and calcium carbonate lowered the tensile strength but, interestingly, not the impact strength of the polymer.[48]

6.2.2 Polyolefins. Alumina hydrate has been reported to reduce the flammability of polyethylene,[48] polypropylene,[48] and polybutene.[49] The results that can be expected with a filler loading of 40% for polyethylene and polypropylene are illustrated by the comparison with calcium carbonate, an inert filler, in Table 14-4. Both fillers reduce the strength of these polymers appreciably, however.

6.2.3 Polystyrene. The effectiveness of alumina hydrate as a flame retardant for polystyrene appears to depend on the flammability test method at least as much as on the concentration. As shown by the results in Table 14-4, 40% alumina hydrate yields a moderately higher oxygen index than an equivalent concentration of calcium carbonate, but has virtually no effect on the burn rate in the horizontal UL 94 test or on smoke generation. The apparent ineffective-

Table 14-4. Effect of alumina hydrate on the flammability and strength of thermoplastics[a]

Filled polymer[b] (Filler content 40%)	Oxygen index	UL 94 Horiz. burn rate, in./min	Smoke density rating, %	Tensile strength lb/in.^{2}	Impact[c] strength ft-lb/in.
Polyethylene, unfilled		0.79	4	3250	0.41
Polyethylene + $CaCO_3$		0.95	4	2940	0.31
Polyethylene + H-46[e]		SE[d]	2	2440	0.35
Polypropylene, unfilled	17.4	0.90	15	5280	0.44
Polypropylene + $CaCO_3$	19.0	0.93	5	3210	0.39
Polypropylene + H-46	22.1	0.52	6	3310	0.37
ABS, unfilled		1.64	89	5760	4.30
ABS + $CaCO_3$		1.31	82	3300	0.52
ABS + H-46		0.78	57	3280	0.50
Rigid PVC, unfilled		SE	89	7420	0.46
Rigid PVC + $CaCO_3$		SE	71	4340	0.60
Rigid PVC + H-46		SE	50	4180	0.50
Polystyrene, unfilled	17.9	1.43	84	3820	0.21
Polystyrene + $CaCO_3$	19.6	1.19	80	2840	0.19
Polystyrene + H-46	22.6	1.23	78	2910	0.21

[a]Data from ref. 48, which should be consulted for further details on materials and methods.
[b]The alumina hydrate was Kaiser Aluminum's H-46 with a median particle diameter of 20 microns; the calcium carbonate was "Atomite" from Thompson, Weinman & Co.
[c]Notched Izod impact.
[d]Self-extinguishing.
[e]Grade of alumina hydrate.

ness of alumina hydrate in the horizontal burn test has been ascribed to the low pyrolysis temperature of polystyrene, as a result of which cracking of the polymer to volatile, flammable products (mostly monomer) may proceed faster than the endothermic dehydration of the alumina.[48] In the oxygen index test, the flame appears to be more effectively cooled by the water of dehydration, imparting some degree of flame retardance to the alumina. Impact polystyrene containing 66% alumina hydrate has been reported to have an oxygen index of 29.[50] No data on its performance in other flammability tests are available. Foams produced from impact polystyrene loaded with 50% alumina hydrate have also been claimed to be fire-resistant.[51] Obviously, more data are needed before any generalizations can be made regarding the flame retardance of alumina hydrate in polystyrene that is not crosslinked.

6.2.4 Other Thermoplastics. In ABS, alumina hydrate is moderately effective as a flame retardant and smoke suppressant but, being a non-reinforcing filler, it reduces impact strength to the same extent (which is quite significant) as calcium carbonate.[48] Results for a 40% loading are presented in Table 14-4.

There appear to be no reports in the literature on the use of alumina hydrate in engineering and other high-performance thermoplastics, undoubtedly because all of these polymers are processed at temperatures above 220°C. Attempts to mold nylon 66 with alumina hydrate at normal processing temperatures, for example, lead to dehydration and the formation of voids and surface imperfections in the polymer. Whether alumina hydrate could serve as a foaming agent for such polymers has not been discussed in the literature. One way of circumventing the need for high processing temperatures is through monomer casting. This approach, however, is obviously applicable to very few high-melting thermoplastics. Only one example appears to have been reported so far in the case of polylactam. Foamed and unfoamed polymers, prepared from E-caprolactam and an anionic polymerization initiator in the presence of 20–80% alumina hydrate, have been claimed to have enhanced flame retardance.[52]

6.3 Additives and Synergists for Alumina Hydrate

With an increasing awareness of the many benefits offered by alumina hydrate, researchers have made efforts to boost its effectiveness and to improve its compatibility with polymers. The effort has so far been aimed towards two different objectives: to enhance flame retardance with synergists, and to minimize strength losses through the use of coupling agents. Both approaches have shown promise, and considerable progress can be expected in the future.

6.3.1 Synergists. Alumina hydrate has been occasionally used in combination with other flame retardants, as for example with phosphate esters in PVC. In nearly all such cases, the flame-retardant effects are additive in a predictable way. In two instances, however, unexpectedly greater than additive effects have been claimed for combinations with zinc borate in halogenated polyesters,[53] and with hexabromobiphenyl in diallylic phthalate molding compounds.[39]

6.3.2 Coupling Agents. Silanes, which have enjoyed growing use as coupling agents for silica and other inorganic fillers, have also been found to have utility when applied to alumina hydrate. Various silanes have thus been reported to improve the flexural strength of polyester resins filled with alumina hydrate[15] and the tensile strength of epoxy resins, containing this filler.[15,54] Crosslinkable ethylene-vinyl acetate copolymer compounds containing vinyl silane-treated alumina trihydrate reportedly exhibit improved moisture and heat resistance and flame retardancy.[55,56] These silane-treated products are now commercially available.[57] Two other interesting coupling agents reported for alumina hydrate are alkylketene dimers for filled polypropylene[58] and isocyanates for filled ethylene-propylene rubber.[59]

7. CURRENT AND PROJECTED UTILIZATION

Estimated U.S. consumption of alumina trihydrate in rubber and plastics was 500 million lb in 1974.[60] Currently, the major market for alumina trihydrate filler is in latex foams and adhesives for carpet backing. Approximately 390 million lb was used for this application in 1974. About 110 million pounds of trihydrate was used in plastics,[61] most of which went into glass-reinforced polyesters for spray-up systems as well as BMC and SMC compounds. Vinyls, epoxies, and polyolefins consumed smaller amounts. Usage of alumina trihydrate in plastics is expected to grow to about 440 million lb by 1980 and almost 2 billion lb by 1990.[62] As the cost of organic polymers rises, emphasis on the use of fillers as extenders for plastics will increase. Fillers which have other performance advantages (e.g., flame retardance) will have a decided edge. Usage in SBR rubber latex, unsaturated polyesters, vinyls, and epoxies will, of course, continue to grow. However, additional new markets are expected to develop in crosslinked polyethylene electrical insulation compounds and in flexible polyurethane foams.

References

1. Pearson, T. G., *The Chemical Background of the Aluminum Industry*, Royal Institute of Chemistry, London, 1957.
2. Gerard, G. and Stroup, P. T., *Extractive Metallurgy of Aluminum, Vol. 1, Alumina*, John Wiley, New York, 1963.
3. Wefers, K. and Bell, G. M., *Oxides and Hydroxides of Aluminum*, Technical Paper No. 19, Alcoa Research Laboratories, 1972.
4. Schoen, R. and Roberson, C. E., "Structures of Aluminum Hydroxide and Geochemical Implications," *The American Mineralogist*, **55**, (Jan.–Feb. 1970).
5. *Selected Values of Chemical Thermodynamic Properties*, NBS Technical Note 270-3, Department of Commerce, Jan. 1968.
6. Takahashi, Y., Yamada, K., Fukunaga, T., and Mukaibo, T., "Heat Capacity of Hydrated and Crystalline Aluminas," *Denki Kagaku*, **41**, No. 4 (1973).
7. Yamada, K., Fukunaga, T., Takahashi, Y., and Mukaibo, T., "Heat of Dehydration of Hydrated Aluminas," *Denki Kagaku*, **41**, No. 4 (1973).
8. Martin, F. J. and Price, K. R., "Flammability of Epoxy Resins," *J. Applied Polymer Sci.*, **12**, 143 (1968).
9. Lawson, D. F., Kay, E. L., and Roberts, D. T., "Mechanism of Smoke Inhibition by Hydrated Fillers," *Rubber Chem. Technol.*, **48**, No. 1 (1975).

10. Thompson, D. C., Hagman, J. F., and Mueller, N. N., "Flame Resistance of Neoprene," *Rubber Age* (Aug. 1958).
11. Connolly, W. J. and Thornton, A. M., "Aluminum Hydrate Filler in Polyester Systems," *Modern Plastics* (Oct. 1965).
12. Ampthor, F. J. and Kroekel, C. H., Developments in Low-Profile SMC for Flame Retardant and Electrical Applications, SPI-RP/C Institute Proceedings, Feb. 1972.
13. Goins, O. K. and Chae, Y. C., Tetrachlorophthalic Anhydride Polyesters–Flame Retardancy, UV Stability, and Reduced Smoke, SPI-RP/C Institute Proceedings, Feb. 1972.
14. Rizzi, M. A., Total System Approach to Mixing and Molding of Glass Reinforced Polyester Compounds, SPI-RP/C Institute Proceedings, Feb. 1972.
15. Ranney, M. W., Berger, S. E., and Marsden, J. G., Silane Coupling Agents in Mineral-Filled Composites, SPI-RP/C Institute Proceedings, Feb. 1972.
16. Evans, T. R., Barrentine, E. M., Goold, R. D., and Willard, B. S., Formulated Fire Retardant Polyester Systems for Press Molded Laminates, SPI-RP/C Institute Proceedings, Feb. 1972.
17. McNally, J. S., Dzik, C. J., and Williams, D. R., Low Shrinkage Plus Flame Retardance, SPI-RP/C Institute Proceedings, Feb. 1972.
18. Sprow, T. K., Connolly, W. J., and Kerle, E. J., Filled Polyester Spray-Up Systems Offering Improved Fire Hazard Classifications, SPI-RP/C Institute Proceedings, Feb. 1973.
19. Trampenau, R. H. and Evans, T. R., Fire Retardant FRP in Construction, SPI-RP/C Institute Proceedings, Feb. 1973.
20. Lundberg, C. V., Hydrated Alumina as a Fire-Retardant Filler in Styrene-Polyester Casting Compounds, 166th A.C.S. Meeting, Chicago, Aug. 1973.
21. Bonsignore, P. V. and Manhart, J. H., Alumina Trihydrate as a Flame Retardant and Smoke Suppressive Filler in Reinforced Polyester Plastics, SPI-RP/C Institute Proceedings, Feb. 1974.
22. Woycheshin, E. A. and Sobolev, I., "Effect of Particle Size on the Performance of Alumina Hydrate in Glass-Reinforced Polyesters," *J. Fire Flamm./Fire Retardant Chem.*, **2**, 224 (Nov. 1975).
23. IAPMO Standard PS 11-72, International Association of Plumbing and Mechanical Officials, 1972.
24. ANSI Standard Z124.1-1974, American National Standards Institute, 1974.
25. Sanderford, F. M. and Ray, J. C., U.S. Patent 3,699,041 to Kaiser Aluminum & Chemical Corp., 1972.
26. Geppert, G. A. and Woosley, R. D., U.S. Patent 3,874,889 to Aluminum Co. of America, 1975.
27. "The Bathroom Also-Ran Now Sets the Pace," *Modern Plastics* (Mar. 1975).
28. Duggins, R. B., U.S. Patent 3,847,865 to du Pont, 1974.
29. Lee, H. and Neville, K., *Handbook of Epoxy Resins*, McGraw-Hill Book Co., New York, 1967.
30. Bruins, P. F., *Epoxy Resin Technology*, John Wiley, New York, 1968.
31. Kumasaka, S., *et al.*, U.S. Patent 3,737,400 to Toyo Rubber Chem. Ind. Corp., 1973.
32. Norman, A. J. and Cobbledick, D. S., U.S. Patent 3,810,851 to General Tire & Rubber Co., 1974.
33. Taylor, R. C., *et al.*, U.S. Patent 3,821,067 to Atlantic Richfield Co., 1974.
34. Kehr, C. L., *et al.*, U.S. Patent 3,897,372 to W. R. Grace, 1975.
35. Burton, P., U.S. Patent 3,741,929 to ITT Corp., 1973.
36. Aishima, I., *et al.*, Japan, Kokai 74 59,841 to Asahi Chem. Ind. Co., 1974; *C.A.* **81,** 153769e (1974).
37. Walters, R. B., Ger. Offen. 2,439,490 to General Electric Co., 1975.
38. "No Slack for Wire and Cable," *Modern Plastics* (Nov. 1974).
39. Thomas, J. L., U.S. Patent 3,826,777 to FMC Corp., 1974.
40. Talsma, H., U.S. Patent 3,786,041 to Standard Oil Co., 1974.
41. Saheki, T., and Ueda, M., Japan, Kokai 74 92,139 to Okura Industrial Co., 1974; *C.A.* **83,** 12379g (1975).
42. Kato, Ch., *et al.*, Japan, Kokai 74 120,978 to Kitajima, 1974; *C.A.* **82,** 172199f (1975).
43. Kato, Ch., Japan, Kokai 74 131,244 to Dainippon Printing Co. (1974); *C.A.* **83,** 29123s (1975).
44. Morgan, A. W., *et al.*, The Use of Phosphate/Alumina Trihydrate Systems in Flame Retarding Vinyl Compositions, SPE 30th ANTEC Technical Papers, Vol. XVIII., p. 475, Chicago, May 15–18, 1972.
45. Mathis, T. C. and Hinchen, J. D., The Effect of Phosphate Ester/Filler Interactions on the Flammability and Smoke Generation of PVC, SPE 31st ANTEC Technical Papers, Vol. XIX, p. 343, Montreal, May 7–10, 1973.
46. Kawawada, S., *et al.*, Japan, Kokai 74 113,845 and 74 114,657 (1974) to Hitachi Cable, Ltd.; *C.A.* **82,** 112769g and 172027y (1975).
47. Okuda, I., Japan, Kokai 75 48,272 to Kuraray Co., 1975; *C.A.* **83,** 116864c (1975).
48. Sobolev, I. and Woycheshin, E. A., "Alumina Hydrate as a Flame-Retardant Filler for Thermoplastics," *J. Fire Flamm./Fire Retardant Chem.*, **1,** 13 (Feb. 1974).
49. Ohtsuka, K., *et al.*, Japan, Kokai 74 125,453 to Chisso Corp., 1974; *C.A.* **82,** 112755z (1975).
50. Sasano, T. and Hirate, H., Japan, Kokai 73 76,985 to Teijin Ltd. (1973); *C.A.* **80,** 109231d (1974).
51. Harada, Y., *et al.*, Japan, Kokai 74 103,957 to Showa Denko (1974); *C.A.* **82,** 99399q (1975).

52. Duggins, R. B., U.S. Patent 3,733,283 to du Pont (1973).
53. Draganov, S. M., Ger. Offen. 2,105,443 to U.S. Borax & Chemical Corp., 1972.
54. Stevens, J. J., U.S. Patent 3,647,742 to Union Carbide Corp., 1972.
55. North, J. A. and Kuckro, G. W., U.S. Patent 3,832,326 to National Distillers and Chemical Corp., 1974.
56. North, J. A. and Kuckro, G. W., U.S. Patent 3,922,442 to National Distillers and Chemical Corp., 1975.
57. "There are New Ways to Tame the Flame," *Modern Plastics* (May 1975).
58. Sobolev, I. and Woycheshin, E. A., U.S. Patent 3,826,775 to Kaiser Aluminum & Chemical Corp., 1974.
59. Yamamoto, K., *et al.*, Japan, Kokai 74 130,440 to Sumitomo Chemical Co., 1974; *C.A.* **83**, 29555j (1975).
60. Robinson, A. L., "Flame Retarding Materials," *Rubber World* (Oct. 1975).
61. "Flame Retardants," *Modern Plastics* (Sept. 1975).
62. "Plastics–The Next 25 Years," *Modern Plastics* (Oct. 1975).

15
Miscellaneous Flame Retardants

John V. Milewski and Harry S. Katz

Los Alamos Scientific Lab.
Los Alamos, New Mexico

Utility Research Co.
Montclair, New Jersey

CONTENTS

1. INTRODUCTION

As stated in the preceding chapters, antimony oxide and alumina trihydrate are used extensively as flame retardant fillers. In addition, a large number of different compounds have been marketed for this purpose. This chapter describes many alternate flame retardant fillers. Some are liquids and these have been included, even though they are exceptions to this Handbook's scope of only solid fillers. The field is growing rapidly, with new compounds appearing frequently. The range of composition is extremely broad, and some arbitrary choices of grouping have been made. In each of these classifications, typical materials that are suitable for use in plastics have been listed, even though some of these have been used primarily in wood, paper, textiles, or rubber.

Because of the diversity of the field of flame retardants, it is difficult for most compounders to keep abreast of new developments. Also, modified mixing procedures or special equipment may be required for the proper utilization of the flame retardants. As a result of this, there has been an emergence of small processors who specialize in custom formulation and manufacture of flame retardant concentrates. This chapter identifies many of these processors as well as the trade names of the most common flame retardants, and most of the companies dealing with flame retardant chemicals. References relating to each type of flame retardant are given for each material. A number of general references are given at the end of the chapter.

Synergism plays an important part in the performance of various flame retardants. It is probable that the optimum combinations of the wide variety of flame retardants have not yet been discovered. This point is illustrated by the rapid growth in the whole or partial replacement of antimony oxide or alumina trihydrate by molybdates or borates, as is discussed in parts of this chapter. Undoubtedly, the field of synergistic fillers will exhibit extensive development and rapid growth.

2. TRADE NAMES AND COMMON NAMES OF FLAME RETARDANT CHEMICALS

Aerotex	American Cyanamid
Aflamman	Quehl and Company GMBH
Aflammit	Quehl and Company GMBH
Agent CD-202	GAF Corporation
Albert-Schaumschutz	Chemische Werke Albert
Amgard	Albright and Wilson
Antoxal	Eronel Industries
Arko	Arkansas Company
Aroclor	Monsanto
Atlac 711-OSA	Atlas Chemical
Bonrex	Saki Chemical Ind. Co., Ltd.
Brominex	Great Lake Chemical Company
Bromkal	Chemische Fabrik Kalk GMBH
Busan 11-M1	Buckman Laboratories
Caradol 385 R	Shell International
Carapor Paste R	Shell International
Cereclor	ICI U.S.
Cereclor S 52, 42	Kraft Chemical
Chlorez	Dover Chemical
Chlorowax	Diamond Shamrock
Citex	Cities Service
Clarechem	Claremont Poly Chemical Corporation
Clophen	Naftone
C.M.	DuPont
Conco Finish	Continental Chemicals
Cyagard	American Cyanamid
Dai-Guard	Daihachi Chemical Ind. Co., Ltd.
Daran	W. R. Grace and Company
Dawsonite	Alcoa
Dechlorane	Hooker
Delvet	Diamond Shamrock
Disflamoll DPK	Naftone
E-XXX	Stauffer Chemicals
Emkaprof	Emkay Chemical
Empee	Monmouth Plastics Inc.
Escoflex CDP	East Coast Chemicals
Ferrocene	Arapaho Chemicals
Fire Brake ZB	U.S. Borax
Fire Guard	Polymer Research Corporation
Firemaster TF-1	Michigan Chemicals
Fire Retardant C	Monsanto
Fire Shield	Chemtron
Fi-Retard	Arkansas Company
Fire Trol	Arizona Argochemical Corporation
Flamarest	Avco Systems Division
Flame/Chek	Chemtech Corporation
Flame Proof	Osmose Wood Preserving Company of America
Flame Proof XXX	Apex Chemical Company, Inc.
Flame Proof DCA	Tanatex Chemical Corporation
Flame Retardant A,B	BASF Corporation
Flame Retardant H	Clarkson Laboratories Inc.
Flame Retardant P, MT, TX	Tanatex Chemical Corporation
Flame Snub-Polymix, Intermix CA, Acetate	Arkansas Chemical
Flamithane	Raychem Corporation
Flammentin	Quehl and Company, GMBH
Flammex AP	Berk Ltd.
Flammschutz Albert DS	Chemische Werke Albert
Flamolin	U.S. Industrial Chemicals
Flamort	Flamort Chemicals
FRB 1,2	Daihachi Chem. Ind. Co., Ltd.
FR Texcryl	Poly-Acryls
Fyrol	Stauffer Chemical Corporation
Fyrolflex	Stauffer Chemical Corporation
Gaftex	GAF Corporation
GHA	Great Lakes Foundry Sand Company
Harshaw KR	Harshaw Chemical Company
HB	Teijin Chemicals Ltd.
Hydrafil	Great Lakes Foundry Sand Company
Isobor	Isochem Resin Company
Isochem	Isochem Resin Company
Isonol	Upjohn
Imco	IND (Industrial Mineral and Chemicals)
Kloro 3000	Keil Chemical
Korsite	World Chemical Industries Inc.
Kronitex WT	FMC Corporation
MCC-100, 200, 300	Monsanto
Microfil	Great Lakes Foundry Sand Company
Molybdic Oxide	Climax Corporation
Molywhite 101	Sherwin Williams
Niax RO-350	Union Carbide
OBC, OCC, ODM, OMS	Basic Chemicals
Oncor 55	NL Industries
ORCO-NRC	Organic Chemical Corporation
Paroil	Dover Chemical
PC-45	Upjohn
Pearagel (C1)	Osmose Wood Preserving Company of America
Phos-Chek	Monsanto
Phosclere P.134	Noury and Van Der Lande N.V.
Phosflex	Stauffer Chemical
Phos Gard	Monsanto

Plastector	Saki Chemical Industries
Plastriox	Societe Nouvelle Des Mines
Polyflam	Schulman, A Inc.
Proban	Albright and Wilson
Proban 210, 420	Quehl and Company
Propylan	Lankro Chemical Ltd.
PU-72	Daihachi Chemical Industries
Purex	Saki Chemicals
Pyrofix	Uetikon
Pyrosan Ca, B	Laurel Products Corporation
Pyrotard B	Scientific Chemical
Pyrothan	Uetikon
Rez-o-sperse	Dover Chemicals
Rulan	E. I. DuPont
Salxex	Mitusbishi Petroleum
Santicizer	Monsanto
Sumiguard	Sumitomo Chemical Company Ltd.
Synachlor	Pearsall Chemical
Tanotard	(C.S.) Tanner
TCPP	British Colanese
Tego-Tex	(Th) Goldschmidt AG
Thermoguard	M&T Chemical Inc.
Thermolin	Olin Chemicals
Ucet A-10	Union Carbide
Unichlor	Neville Chemical
Vircol	Mobil Chemical
VC-611	Mobil Chemical
White Star	Harshaw Chemical
X-Fire System NA, A	Pearsall Chemical Corporation
X-12	E. I. DuPont
2155-A Fr	Inmont Corporation

3. CONCENTRATORS AND CUSTOM FORMULATORS OF FLAME RETARDANT CHEMICALS

Adell Plastics, Inc.
Ampacet Corporation
Chemetron Corporation–Chemical Product Division
Customcolor Inc.
Coz Chemical Company–Division of Allied Products Corporation
Crompton and Knowles Corporation–Plastics Color Division
Ferro Corporation
Marine Plastics Division–Northern Petrochemical Company
Merex, Inc., Division of RPC Industries
Monmouth Plastics, Inc.
Podell Industries Inc.
Polychrome Dispersions, Inc.
Reed Plastics
Service Color Corporation
A. Schulman Inc.

4. SUPPLIERS OF FLAME RETARDANTS

Adell Plastics, Inc.
4530 Annapolis Road
Baltimore, Md. 21227
Phone: (301) 789-7780

Albright and Wilson, Ltd.
Oldbury Division
1 Knightsbridge Green
London SW1
United Kingdom

Alcoa
Aluminum Co. of America
1501 Alcoa Building
Pittsburgh, Pa. 15219
Phone: (412) 555-2354

Aldrich Chemical Co., Inc.
940 West St. Paul Avenue
Milwaukee, Wis. 53233

Allied Chemical Corp.
Specialty Chemicals Division
P. O. Box 1087R
Morristown, N.J. 07960
(Phone) (201) 455-2000

American Cyanamid Company
Bound Brook, N.J. 08805
Phone: (201) 356-2000

Amoco Chemicals Corp.
Mail Code 4102
200 E. Randolph Street
Chicago, Il. 60601
Phone: (312) 856-3200

Ampacet Corp.
250 S. Terrace Avenue
Mt. Vernon, N.Y. 10550
Phone: (914) 699-9100

Apex Chemical Co., Inc.
200 South First Street
Elizabethport, N.J. 07206
Phone: (201) 354-5420

Arapahoe Chemicals
2855 Walnut Street
Boulder, Colo. 80302
Phone: (303) 442-1926

Arizona Agrochemical Corp.
P. O. Box 2191
Phoenix, Ariz. 85001

Arkansas Company, Inc.
P. O. Box 210
Newark, N.J. 07101
Phone: (201) 589-0516

Atlas Chemical Industries, Inc.
Chemicals Division
Wilmington, Del. 19899

BASF Corp.
P. O. Box 8467
Charlotte, N.C. 28208

Basic Chemicals
845 Hanna Building
Cleveland, Ohio 44115
Phone: (216) 241-5000

Berk Ltd.
Chemicals Manufacturing Division
Statford, London E163NX
United Kingdom

British Celanese, Ltd.
Chemical Sales
Foleshill Road
Coventry, 88771 STDOC03
United Kingdom

Buckman Laboratories, Inc.
1256 N. McLean Blvd.
Memphis, Tenn. 38108
Phone: (901) 278-0330

Chemische Werke-Albert
6202 Wiesbaden-Biebrich
Postfach 9101
West Germany

Chemtech Corp.
7882 Folk Avenue
St. Louis, Mo. 63143

Cities Service Company
Drawer 4
Cranbury, N.J. 08512
Phone: (609) 655-7280

Claremont Polychemical Corp.
501 Winding Road
Old Bethpage, N.Y. 11804
(516) 420-1500

Clarkson Laboratories, Inc.
1450 Ferry Avenue
Camden, N.J. 08104

Climax Molybdenum Company
1600 Huron Parkway
Ann Arbor, Mich. 48106
Phone: (313) 761-2300

Continental Chemical Corp.
270 Clifton Blvd.
Clifton, N.J. 07015

Coz Chemical Company
Providence Road
Northbridge, Mass. 01534
Phone: (617) 234-7438

Crompton and Knowles Corp.
Plastics Color Division
Campus Drive
Somerset Valley Industrial Campus
Somerset, N.J. 08873
Phone: (201) 469-5600

Dart Industries, Inc.
Fiberfill Division
1701 N. Heidelbach Avenue
Evansville, Ind. 47717
Phone: (812) 424-3831

Diahachi Chemical Industry, Co., Ltd.
54, 3-Chome Chodo
Higashiosaka Osaka-Pref.
Japan

Diamond Shamrock Corporation
1100 Superior Avenue
Cleveland, Ohio 44114
Phone: (216) 694-5000

Dover Chemical Corporation
15 and Davis
Dover, Ohio 44622
Phone: (216) 343-7711

Dow Chemical Co.
Abbott Road. Center
Midland, Mich. 48640
Phone: (517) 636-1000

E. I. duPont de Nemours & Co., Inc.
1007 Market Street
Wilmington, Del. 19898
Phone: (302) 774-2421

East Coast Chemicals Co.
417 Main Street
Little Falls, N.J. 07424
Phone: (201) 256-7576

Eastman Chemical Products, Inc.
Kingsport, Tenn. 37662
Phone: (800) 251-0351

Emkay Chemical Company
319–325 Second Street
Elizabeth, N.J. 07206
Phone: (201) 352-7053

Eronel Industries
12607 Cerise Avenue
Hawthorne, Calif.

FMC Corporation
200 Market Street
Philadelphia, Pa. 19103
Phone: (215) 299-6000

Ferro Corporation
7050 Krick Road
Bedford, Ohio 44146
Phone: (216) 232-4330

Fine Organics
205 Main Street
Lodi, N.J. 07644
Phone: (201) 472-6800

Flamort Chemical Company
746 Natoma Street
San Francisco, Calif. 94103
Phone: (415) 621-7825

Formulated Resins, Inc.
P. O. Box 508
Greenville, R.I. 02828
Phone: (401) 949-2060

GAF Corp.
140 West 51 Street
New York, N.Y. 10020
Phone: (212) 582-7600

(tH.) Goldschmidt AG
Chemische Farbiken
43 Essen 1
Postfach 17
West Germany

(W. R.) Grace and Co.
Polymers & Chemicals Division
62 Whittemore Avenue
Cambridge, Mass. 02140

Great Lakes Chemical Corporation
P. O. Box 2200
W. Lafayette, Ind. 47906
Phone: (317) 463-2511

Great Lakes Foundry & Sand Company
Mineral Products Division
2855 Coolidge
Troy, Mich. 48084
Phone: (313) 649-3700

Harshaw Chemical Company
1945 East 97th Street
Cleveland, Ohio 44106
Phone: (216) 721-8300

Hexcel Corporation
11711 Dublin Boulevard
Dublin, Calif. 94566
Phone: (415) 828-4200

Hooker Chemical Corporation
47th Street and Buffalo Avenue
Niagara Falls, N.Y. 14302
Phone: (716) 285-6655

Humphrey Chemical Corporation
P. O. Box 2
Edgewood Arsenal, Md. 21010
Phone: (301) 679-1700

ICI United Stated, Inc.
New Murphy Road
Wilmington, Del. 19897
Phone: (302) 575-3000

IMC Chemical Group, Inc.
52 Sobin Park
Boston, Mass. 02210
Phone: (617) 268-5100

Industrial Minerals and Chemical, Inc.
P. O. Box 2204
Dalton, Ga. 30720
Phone: (404) 278-8318

Inmont Corp.
150 Wagaraw Road
Hawthorne, N.J. 07507
Phone: (201) 427-6700

Isochem Resins Co.
99 Cook Street
Lincoln, R.I. 02865
Phone: (401) 723-2100

Kaiser Aluminum and Chemical Corp.
Pleasanton, Calif. 94566
Phone: (415) 462-1122

Kaiser Chemicals
300 Lakeside Drive
Oakland, Calif. 94604
Phone: (415) 271-2211

Keil Chemical Division
3000 Sheffield Avenue
Hammond, Ind. 46320
Phone: (219) 931-2630

H. Kohnstamm and Co., Inc.
161 Avenue of the Americas
New York, N.Y. 10013
Phone: (212) 929-7000

Kraft Chemical Company, Inc.
917 West 18th Street
Chicago, Ill. 60608

Lankro Chemicals, Ltd.
P. O. Box 1
Eccles, Manchester
England M300BH

Laurel Product Corporation
2600 E. Tioga Street
Philadelphia, Pa. 19134

M&T Chemicals, Inc.
P. O. Box 1104
Rahway, N.J. 07065
(201) 499-0200

Merex Inc.
14110 S. Broadway
Los Angeles, Calif. 90061
Phone: (213) 532-0337

Michigan Chemical Corp.
2 North Riverside Plaza
Chicago, Ill. 60606
Phone: (312) 454-7900

Mobil Chemical Company
P. O. Box 26683
Richmond, Va. 23261
Phone: (804) 798-4291

Monmouth Plastics, Inc.
814 Asbury Park
Asbury Park, N.J. 07712
Phone: (201) 775-5100

Monsanto Company
800 N. Lindbergh Blvd.
St. Louis, Mo. 63166
Phone: (314) 694-1000

Morton Chemical Company
110 N. Wacker Drive
Chicago, Il. 60606
Phone: (312) 621-5565

Naftone, Inc.
425 Park Avenue
New York, N.Y. 10022

National Lead Company
P. O. Box 420
Hightstown, N.J. 08520

Neville Chemical Company
Neville Island
Pittsburgh, Pa. 15225
Phone: (412) 331-4200

Northern Petrochemical Company
2350 E. Devon Avenue
Des Plains, Ill. 60018
Phone: (312) 295-2400

Noury and Van Der Lande N.V.
P. O. Box 10
Brink, 13, Deventer, Holland

Nyacol, Inc.
Megunco Road
Ashland, Mass. 01721
Phone: (617) 881-2220

Olin Corp.
120 Long Ridge Road
Stamford, Conn. 06904
Phone: (203) 356-2000

Organic Chemical Corporation
P. O. Box 4258
East Providence, R.I. 02914

Osmose Wood Preserving Company of America, Inc.
980 Ellicott Street
Buffalo, N.Y. 14209

Pearsall Chemical Corporation
P. O. Box 437
Houston, Tex. 77001
Phone: (713) 682-6331

Philadelphia Quartz Company
P. O. Box 840
Valley Forge, Pa. 19482
Phone: (215) 293-7200

Podell Industries, Inc.
3 Entin Road
Clifton, N.J. 07014
Phone: (201) 773-3700

Polychrome Dispersions, Inc.
13429 S. Western Avenue
Los Angeles, Calif. 90249
Phone: (213) 532-1480

Polymer Research Corporation
2186 Mill Avenue
Brooklyn, N.Y. 11234
Phone: (212) 444-4300

Quehl & Co., GMBH
D-672 Speyer
Postfach 186
West Germany

Raychem Corporation
300 Constitution Drive
Menlo Park, Calif. 94025

Reed Plastics Corporation
Holden Industrial Park
Holden, Mass. 01520
Phone: (617) 829-6321

Saki Chemical Industrial Co., Ltd.
Ebisujima-Cho, Sakai City
Osaka, Japan

A. Schulman, Incorporated
3550 W. Market Street
Akron, Ohio 44313
Phone: (216) 666-3751

Scientific Chemicals, Inc.
3800 S. Racine Avenue
Chicago, Ill. 60607

Service Color Corporation
11524 W. Addison Street
Franklin Park, Ill. 60131
Phone: (312) 455-9100

Shell International Chemical Co., Ltd.
Shell Center
London SE1
United Kingdom

Sherwin Williams Co.
P. O. Box 6520
Cleveland, Ohio 44101
Phone: (216) 566-2960

Societe Nouval Des Mines De La Lucette
Siege Social
4, Rue De Rome
Paris 8e
France

Solem Industries, Inc.
3400 Oakcliff Road
Atlanta, Ga. 30340
Phone: (404) 458-9107

Stauffer Chemical Company
Westport, Conn. 06880
Phone: (203) 222-3000

Strem Chemicals, Inc.
150 Andover Street
Danvers, Mass. 01923

Sumitomo Chemical Company, Ltd.
15 5-Chome, Kitahama
Higashi-Ku, Osaka
Japan

Swift Chemical Company
1211 West 22nd Street
Oak Brook, Ill. 60521

Tanatex Chemical Corporation
Page and Schuyler Avenues
P. O. Box 388
Lyndhurst, N.J. 07071

(C.S.) Tanner Company
450 Furman Hall Road
P. O. Box 3867 Park Place
Greenville, S.C. 29608

Teijin Chemicals Ltd.
1-1, 2-Chome,
Uchisaiwai-Cho, Chiyoda-Ku
Tokyo, Japan

Tenneco Chemicals, Inc.
P. O. Box 365
Piscataway, N.J. 08854
Phone: (201) 981-5000

U.S.I. Chemicals Company
99 Park Avenue
New York, N.Y. 10016
Phone: (212) 949-5700

Uetikon
Chemische Fabrik Uetikon
8707 Uetikon (Schweiz)
West Germany

Union Carbide Corporation
Chemicals and Plastics
270 Park Avenue
New York, N.Y. 10017
Phone: (212) 551-2345

U.S. Borax and Chemical Corp.
3075 Wilshire Blvd.
Los Angeles, Calif. 90010
Phone: (213) 381-5311

Upjohn Company
Polymer Chemicals
P. O. Box 685
LaPorte, Tex. 77571
Phone: (713) 479-1541

Velsicol Chemical Corporation
341 E. Ohio Street
Chicago, Ill. 60611
(312) 670-4610

Ware Chemical Corporation
P. O. Box 3375
Bridgeport, Conn. 06005
Phone: (203) 366-4941

Weston Chemical, Inc.
103 Spring Valley Road
Montvale, N.J. 07645

White Chemical Corporation
P. O. Box 278
Bayonne, N.J. 07002
Phone: (201) 437-0050

5. DISCUSSIONS OF VARIOUS FLAME RETARDANT CHEMICALS

5.1 Ammines

The metal-ammine complexes are an emerging class of flame retarding agents. $Ni(NH_3)_6(BF_4)_2$ and $Zn(NH_3)_4(BF_4)_2$ may be used to suppress flame in polyethylene and polypropylene. Some of these metal-ammine complexes, however, decompose when exposed to moist air. The zinc and cadmium ammine halides are not thermally stable at room temperature. The cobaltic ammine complexes, however, are stable at room temperature and when exposed to moist air, and are thus superior flame retardants for flexible urethane foam. $Co(NH_3)_6Cl_3$ is a more effective flame retardant than either $Ni(NH_3)_6(BF_4)_2$ or $Zn(NH_3)_4(BF_4)_2$, which contains less NH_3 by weight.

References for Ammines

1. Little, R. W. "Flame Proofing–Textile Fabrics," American Chemical Society, Monograph Series No. 104. Reinhold, New York, 1947.
2. Pitts, T. T., *J. Cell Plastics* 7 (4) 202 (1971).
3. Wendlandt, W. W. and Smith, T. P., "The Thermal Properties of Transition Metal Ammine Complexes," Elsevier, Amsterdam, 1967.

5.2 Borates

Zinc borate is the borate most widely used as a flame retardant for plastics. It is also used extensively in paint formulations, including intumescent coating systems. Zinc borate can be used as an economical partial replacement and sometimes complete replacement for antimony oxide in polyvinyl chloride, polyesters, halogenated polyesters, and, to some degree, polyolefins.

Barium metaborate, when used in combination with phosphorus compounds or materials containing a halogen donor, can effectively replace 40-90% of the antimony trioxide normally used to impart flame resistance to paints, plastics, textiles, rubber, and adhesives. In some instances, all of the antimony trioxide can be replaced.

Sodium, calcium, lead borates and ammonium, potassium and tetraamine zinc tetrafluoroborates are also used as flame retardants for specialized applications.

5.2.1. Physical and Chemical Properties of Barium Metaborate:

Busan®* 11-MI has the following composition	Minimum 90% barium metaborate, calculated as $BaB_2O_4 \cdot H_2O$
Specific gravity	3.25 to 3.35
Weight per solid U.S. gallon	27.5 lb.
Bulking	0.036 U.S. gal. per lb.
Refractive index	1.55 to 1.60
Oil absorption, spatula rub-out method	30
pH of saturated solution at 21°C (70°F)	9.8 to 10.3

Solubility in water at 21°C (70°F). Typical 0.3%; max. 0.4% (obtained by adding 10 g of Busan 11-MI to 200 ml of distilled water, agitating 2 hr at room temperature, filtering, and determining the BaO and B_2O_3 in the filtrate)

Fusion temperature	900–1050°C (1650–1920°F)
Thermal gravimetric test	Initial weight loss at about 200°C (392°F)

*Busan® is the trade name for Buckman Laboratories, Inc.

5.2.2. Physical and Chemical Properties of Zinc Borate. Firebrake®† ZB zinc borate is specially manufactured for use in fire retardant polymer systems. The water of hydration is retained even at temperatures above 500°F., thereby allowing Firebrake ZB to be used in systems requiring high temperature processing.

Formula: $2ZnO \cdot 3B_2O_3 \cdot 3 \cdot 5H_2O$

% ZnO (theo.)	37.45
% B_2O_3 (theo.)	48.05
% H_2O (theo.)	14.50

†U.S. Borax and Chemical Corporation.

Formula weight	434.66
Color: White	
Refractive index	1.58

Particle size: Nominally–325 mesh. Mean particle size is in range of 2–10 microns.

Crystal density	2.8 g/cm^3
Bulk density	lb/cu ft, typical
tight pack	30–50

Solubility: Virtually insoluble in water below 100°F. Sparingly soluble at higher temperatures. Can be hydrolyzed with acids and bases.

Oil absorption	45 g oil/100 g

5.2.3 Products, Suppliers and Prices. Zinc borate is available from United States Borax and Chemical Corporation as Firebrake ZB and from Humphrey Chemical Company. Price per pound in 250-lb drums in quantities of 20,000 lb and up is:

	East Coast	West Coast
	58¢	55¢
100 to 19,500 lb	74¢	73¢

Barium meta borate is available under the trade name of Busan 11-MI from Buckman Laboratories, Inc. and is also available from Ware Chemical Corporation. Price per pound of Busan 11-MI f.o.b. Memphis, Tenn. is:

Less than 1 ton	1–4 tons	18 tons or more
53¢	46¢	40¢

Net weight per bag is 50 lb.

5.2.4 General Applications of Borates. See Table 15-1, p. 258.

5.2.5 Complete Replacement of Antimony Oxide with Zinc Borate. Several high performance resins containing zinc borate (Firebrake ZB) as the only inorganic fire retardant have been tested for flame spread and oxygen index. The data obtained are given in Table 15-2. The data indicate that in all the resin systems evaluated except for Atlac A711-050, 10 phr of zinc borate will upgrade the resin system to a class 1 resistant material. In the case of Atlac A711-050, the results were slightly below the class 1 rating.

5.2.6 Complete Replacement of Antimony Oxide with Zinc Borate–Alumina Hydrate Mixtures. Zinc borate and alumina trihydrate

Table 15-1. General applications of borates.

Compound	Formula	Application
Zinc borate	$2ZnO \cdot 3B_2O_3 \cdot 3 \cdot 5H_2O$	Paint coatings, PVC, polyesters, halogenated polyesters, polyolefins, high and low density polyethylene, polypropylene, polyurethane, polystyrene, and neoprene.
Sodium metaborate	$Na_2O \cdot B_2O_3 \cdot 4H_2O$	Paint coating.
Sodium borate FR28	$Na_2O \cdot 3 \cdot 6B_2O_3 \cdot 3 \cdot 7H_2O$	Paint coating.
Colemanite	$2CaO \cdot 3B_2O_3 \cdot 5H_2O$	Epoxies.
Lead tetraborate	$PbO \cdot 2B_2O_3$	Paint coating.
Ammonium tetrafluoroborate	NH_4BF_4	Polypropylene, polyethylene.
Potassium tetrafluoroborate	KBF_4	Rigid urethane foam.
Tetraamine zinc tetrafluoroborate	$Zn(NH_3)_4(BF_4)_2$	Polypropylene.
Barium meta borate	$BaB_2O_4 \cdot H_2O$	PVC, ABS, polyethylene, polypropylene, paint coatings.

have been found to be synergistic when used in the halogenated polyester laminates shown in Table 15-3. All of the compositions with this combination of additives achieved ratings of 15 or lower on the radiant panel test, and had oxygen indices of 34.5% or higher which corresponds to Class 1 performance. These findings have great value for fabricators who normally use alumina hydrate as a filler in their polyester products, because in these systems zinc borate is essentially as effective as antimony oxide on a pound-for-pound basis. For example, 4 phr of zinc borate in a Dion 6692 laminate filled with 10 phr of alumina hydrate gives better performance than 6 phr of antimony oxide, which is well within Class 1 performance. For Atlac A711-050, Stepan CX 717, and the Roblen Resin, 5 phr of zinc borate + 10 phr of alumina hydrate compares favorably with or exceeds 5 phr of antimony oxide. For Hetron 92, 5 phr of zinc borate + 15 phr of alumina hydrate gives predicted Class 1 performance, although

Table 15-2. Comparative fire tests for laminates (26% fiberglass) prepared from various halogenated polyesters containing zinc borate.

Resin	Zinc borate (Firebrake ZB) conc. (phr)	Radiant panel flame spread I_s	Oxygen index (%)
Hetron 92	0	23	30.4
Hetron 92	5	15	34.1
Hetron 92	15	14	35.7
Dion 6692	0	28	32.0
Dion 6692	10	19	34.2
Atlac A711-050	0	29	28.1
Atlac A711-050	5	32	32.9
Atlac A711-050	15	34	35.0
Stepan CX 717	0	12.5	34.1
Stepan CX 717	5	7.2	34.6
Stepan CX 717	15	8.3	38.6
Roblen Resin	0	28	31.1
Roblen Resin	5	27	40.8
Roblen Resin	15	21	38.0

Courtesy of U.S. Borax and Chemical Corporation.

Table 15-3. Comparative fire tests for laminates (26% fiberglass) prepared from various halogenated polyesters containing zinc borate-alumina hydrate mixtures.

Resin	Zinc borate Firebrake ZB* conc. (phr)		Alumina hydrate conc. (phr)	Radiant panel flame spread, I_s	Oxygen index (%)
Hetron 92	–	No additives	–	23	30.4
Hetron 92	5		15	10	38.5
Dion 6692	–	No additives	–	28	32.0
Dion 6692	4		15	8.6	36.3
Dion 6692	10		15	14	39.1
Dion 6692	10		10	11	38.0
Atlac A711-050	–	No additives	–	29	28.1
Atlac A711-050	5		5	24	34.5
Atlac A711-050	5		10	10	38.3
Stepan CX 717	–	No additives	–	12.5	34.1
Stepan CX 717	5		15	2.5	37.0
Stepan CX 717	10		15	3.5	39.0
Roblen Resin	–	No additives	–	28	31.1
Roblen Resin	5		15	13	49.6
Roblen Resin	5		10	18	49.6

*Trade Name U.S. Borax. (*Courtesy of U.S. Borax and Chemical Corp.*)

antimony oxide in this chlorinated resin gives even better results. Thus, for manufacturers already using fillers, addition of zinc borate plus alumina hydrate makes good sense economically and provides outstanding fire performance. In fact, this approach applies equally well to any system where up to 20-25 phr of solids can be tolerated.

5.2.7. Partial Replacement of Antimony Oxide with Zinc Borate. Partial replacement of antimony oxide presents an effective alternative to complete replacement, particularly in systems where economy is being sought with minimum change from standard formulations. The results shown in Table 15-4 were obtained with laminates based on a chlorinated resin prepared by styrenating Hetron 19 base, and with Dion 6692. For the chlorinated resin, potential Class 1 performance can be maintained with three-quarters of the antimony oxide replaced by zinc borate. Since the flame spread is well below 15 and the oxygen index above 35%, the measurements shown with the brominated Dion resin also indicate Class 1 performance. The partial replacement approach has been used ex-

Table 15-4. Comparative fire tests for laminates (26% fiberglass) prepared from two halogenated polyesters containing zinc borate antimony oxide mixtures.

Resin	Antimony oxide conc. (phr)		Zinc borate Firebrake ZB* conc. (phr)	Flame spread, I_s	Oxygen index (%)
Hetron Resin	–	No additives	–	15	28.0
Hetron Resin	2		0	9.1	37.3
Hetron Resin	2		6	3.7	40.2
Hetron Resin	1		6	12.5	36.8
Hetron Resin	0.5		6.5	5.4	34.1
Dion 6692	–	No additives	–	28	32
Dion 6692	1.5		1.5	3.0	38.3

*Trade name U.S. Borax. (*Courtesy of U.S. Borax and Chemical Corp.*)

tensively in polyvinyl chloride formulations and the results have been outstanding. Partial replacement or use of zinc borate as an extender for antimony oxide is strongly recommended for all systems in which complete replacement is either inconvenient or not fully effective.

5.2.8 Non-Halogenated Systems. In many end uses, less stringent fire resistance than that corresponding to Class 1 is adequate to meet existing codes. Such applications can often be served by inexpensive general purpose polyesters containing chlorinated or brominated additives along with zinc borate or antimony oxide. A number of firms using zinc borate have developed self-extinguishing formulations using this approach. In systems of this type, partial replacement of antimony oxide at a 1:1 ratio is used as the point of departure. When this proves effective, further development involves replacement of increasing proportions of antimony oxide. Partial replacement invariably provides economy and quite often gives higher performance as well as the side benefits mentioned in the previous pages.

5.2.9 Usage Procedure. Zinc borate requires no special handling. It is a free-flowing, finely divided (−325 mesh) white powder which easily disperses in resins. In general, it is used in the same way as any of the conventional, dry powder fillers. Recommendations for concentration levels depend on a number of factors including desired performance, glass content, filler content, type of resin, etc. General guidelines recommended by U.S. Borax for using their product, Firebrake ZB, in order to achieve various levels of fire retardancy, are given below.

1. Class 1 Performance:
   Use high halogen resins (Atlac A711-050, Dion 6692, Hetron 92, Stepan CX 717, etc.) and
   a. Add 10–15 phr of zinc borate.
   b. Add 4–5 phr of zinc borate plus 10–15 phr of alumina hydrate or
   c. Add 1.5–5 phr of zinc borate plus 1.5 phr of antimony oxide.
2. Class II Performance:
   Use resin systems containing 15-20% chlorine or 10–15% bromine, and follow above dosage recommendations.
3. Self Extinguishing Performance:
   Use approach No. 2 above or use a general purpose polyester with a chlorinated or brominated additive plus 10–20 phr of a mixture of zinc borate and antimony oxide.

5.2.10 Toxicity of Zinc Borate. U.S. Borax reports that tests performed by an independent laboratory resulted in classification of their brand of zinc borate, Firebrake ZB as nontoxic by ingestion or dermal application. It is not a primary skin irritatnt or corrosive material, and it is not an eye irritant. All of these findings were evaluated in accordance with Regulations (Federal Register, August 12, 1961, et. seq.) under the Federal Hazardous Substance Act.

References for Borates

1. Woods, W. G., *Boron Compounds as Flame Retardants in Polymers*, U.S. Borax Research Corp., Anaheim, Calif., Dec. 31, 1967.
2. Anon, *Paint Varnish Prod.*, **51**, 9 (1961).
3. Connell, G. A. and Holmes, G. D., *Forestry*, **36**, 91 (1963).
4. Ling, T. H., to Anaconda Wire and Cable Co, U.S. Patent 3,287,312 (1966).
5. McCluer, J. D., to Thermoid Co., U.S. Patent 2,948,641 (1960).
6. Parker, J. A., Riccitiello, S. R., Gilwee, W. J., and Fish, R., *SAMPE J.*, **41** (Apr./May, 1969).
7. U.S. Borax Service Bulletins IC 66, IC67, IC 67B.

5.3 Carbonates and Bicarbonates

Although sodium and potassium bicarbonates and ammonium carbonates are contained in dry powder fire extinguishers, they cannot be used in various polymeric matrices because they all decompose at very low temperatures. Thus, they do not survive the temperatures required for most processing procedures. However, many paint formulations are cured at room or low temperature, and therefore carbonates and bicarbonates are used extensively as paint flame retardants. Calcium carbonate, however, is used as a cheap, inert filler in many polymer applications. $CaCO_3$ shows significant flame retardancy predominantly through its diluent effect. Suppliers of sodium and ammonium bicarbonates are Allied Chemical Corporation, and Diamond Shamrock Chemical Company.

References relating to the use of carbonates and bicarbonates as flame retardants are:

1. Vleek, L., Australian Patent 19,455/34 (1934).
2. Eickner, H. W., *Forest Prod. J.*, **12** (4), 194 (1962).
3. Tank, W. K. and Neill, W. K., *J. Polymer Sci.*, **C-6**, 65 (1964).

5.4 Inorganic Halide

Most of the inorganic halides that are used as flame retardants are used in wood and paper. However, ammonium chloride is used in polyethylene and polyester, while ammonium bromide has been used in melamine-formaldehyde and styrene resins. Thermal decomposition studies on NH_4Cl and NH_4Br show that both halides first sublime (at 200 and 250°C, respectively), and then decompose into ammonia and the corresponding hydrogen halide. In addition to the flame-retardancy effect of the NH_3 and HCl or HBr in the gas phase, the solid NH_4Cl and NH_4Br particles reduce the energy of the flame by the energy requirement for sublimation. Another chloride used as a flame retardant for paper is $MgCl_2 \cdot 6H_2O$, while the following three chlorides are extensively used as flame retardants in wood: $AlCl_3 \cdot 6H_2O$, $CaCl_2 \cdot H_2O$ and $ZnCl_2$. Both the chloride ion and water of hydration take part in the flame retardancy mechanism. Suppliers of inorganic halides are Allied Chemicals Corp., E. I. DuPont, Boston Chem. Co., and Kaiser Chemicals.

Some references relating to the use of inorganic halides as flame retardants are as follows:

1. Tang, W. K. and Neill, W. K, *J. Polymer Sci.*, **C-6**, 65 (1964).
2. Berk, Ltd., by D. H. Derbyshire and B. J. Bailey, British Patent 1,068,040 (1967).
3. Singleton, F. G. and Kennan, P. J., to H. H. Robertson Co., U.S. Patent 3,061,492 (1962).
4. McMaster, E. L., Eichhorn, J., and Nagle, F. G., to Dow Chemical Co., U.S. Patent 3,058,927 (1962).
5. Nagle, F. B. and McMaster, E. L., to Dow Chemical Co., U.S. Patent 3,133,037 (1964).
6. Dorset, B. C. M., *Text. Mfr.*, 87, 243 (1961).
7. Bonotto, S. and Loni, M. C., *Energ. Nucl. (Milan)*, **13** (11), 631 (1966).
8. Angell, H. W., *Proc. Forest Prod. Res. Soc.*, **5**, 107 (1951).

5.5 Molybdenum Compounds

Molybdenum compounds have been found to be effective flame retardants and smoke suppressants for plasticized polyvinyl chloride and halogen-containing thermoset polyesters. In some formulations, it is possible to replace all of the antimony oxide normally used with molybdenum trioxide and maintain the same level of flame retardancy. The effectiveness of molybdenum compounds as a co-additive with antimony oxide, alumina trihydrate, zinc borate, or organic phosphates has been demonstrated. Molybdenum compounds have been shown to reduce the smoke generation from halogen-containing thermoset polyesters. Molybdenum compounds as flame-retardant chemicals have, in many cases, been found to be cost effective. Also, much of the world's molybdenum compounds are mined and produced in the United States from large reserves. This will assure the user a stable, long term, domestic supply. Physical property data on molybdenum fire retardants are presented in Table 15-5.

5.5.1 Products, Price and Availability of Molybdenum Compounds. The following compounds are available from the Climax Molybdenum Company:

Commercial Products

- Undensified Molybdenum Trioxide (MoO_3) —Commercial price @ \$2.42/lb, Grade L.
- Densified Molybdenum Trioxide (MoO_3)— Commercial price @ \$2.42/lb, Grade L.
- Ammonium Di-Molybdate $(NH_4)_2Mo_2O_7$— Commercial price @ \$3.22/lb.

Development Products

- Ammonium Octamolybdate $(NH_4)_4Mo_8O_{26}$
- Moly FR-201 (20% Ammonium dimolybdate and 80% Aluminum tryhydrate).
- Moly FR-10–Zinc moloybdate based com-

Table 15-5. Physical properties of molybdenum containing fire retardents.

Property	Pure molybdic oxide Grade L-undensified (MoO_3)	Screened technical[a] Grade molybdic oxide hammer milled twice	Moly FR-10[a]	Moly FR-21[a]	Ammonium octamolybdate[a] hammer milled twice
Specific gravity	4.7	4.7	5	3	3.1
Particle size					
Average particle size (microns) (Fisher Sub-Sieve Sizer)	0.4	5.0	0.5	2	3.0
Range (microns)	0.2–15.0	0.5–15.0	–	–	0.5–30
Bulk density	15 lbs/ft^3	90 lbs/ft^3	–	–	–
Color	light blue-gray	light blue-gray	white	white	white

[a]Developmental products of the Climax Molybdenum Company. Screened Technical Grade Molybdic Oxide, sized to 20 mesh and down, available commercially.

pound–estimated commercial price–\$.90/lb.

- Moly FR-21–Calcium zinc molybdate based compound–estimated commercial price–\$.60/lb.

5.5.2 Discussion and Application of Molybdenum Compounds. A study determined the effectiveness of molybdenum compounds as flame retardants and smoke suppressants in various polymer systems. The results are given in Table 15-6. The bulk of the work was concentrated on flexible PVC and halogenated rigid polyesters, since molybdenum compounds function most effectively in these polymers.

5.5.3 Polyvinyl Chloride: Effect of Plasticizer Level. The results obtained with flexible PVC are illustrated in Fig. 15-1 which compares the oxygen indexes of PVC samples containing 6 phr of antimony oxide with the oxygen indexes of similar samples containing 6 phr of molybdenum trioxide (MoO_3), undensified grade (Climax), or a mixture of 3 phr of Sb_2O_3 plus 3 phr of MoO_3. At low plasticizer levels, molybdenum trioxide is about as effective as Sb_2O_3. The MoO_3-Sb_2O_3 mixture is more effective or just as effective as Sb_2O_3 at all of the plasticizer levels examined. These results show

Table 15-6. The use of molybdenum compounds in various polymers.

Polymer	Flame retardant	Smoke suppressant
Polyolefins	?	?
High impact polystyrene	?	No
ABS	Yes	Yes
Rigid PVC	Yes	Yes
Flexible PVC	Yes	Yes
Nylon	Yes	?
Polyurethanes	Yes	?
Epoxies	Yes	?
Rigid polyesters	Yes	Yes

Courtesy Climax Molybdenum Company.

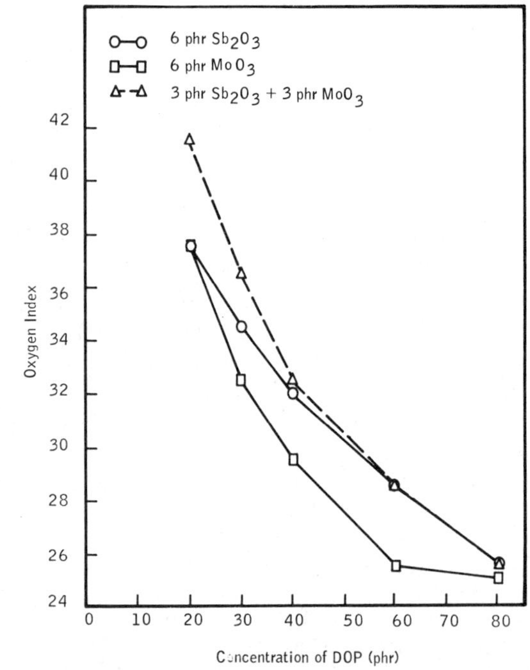

Fig. 15-1. Oxygen index of DOP plasticized PVC (formulation: 100 pts PVC, 3 pts epoxy soya, 26 pts $CaCO_3$, 2 pts stabilizer, plus DOP). (*Courtesy of Climax Molybdenum Company*)

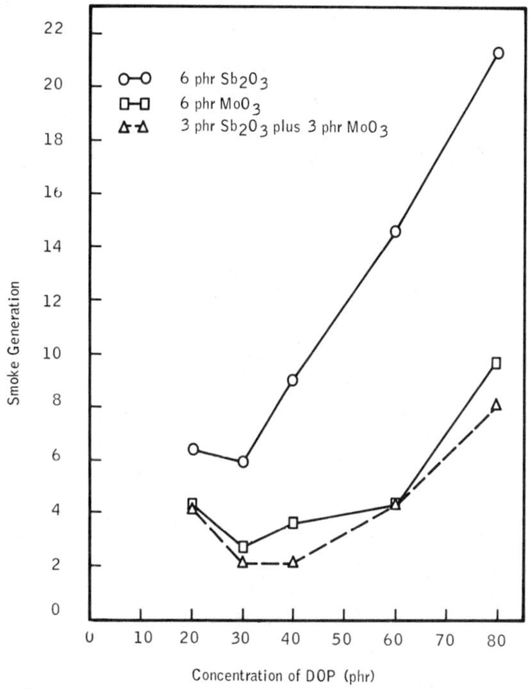

Fig. 15-2. Smoke generation of DOP plasticized PVC (formulation: 100 pts PVC, 3 pts epoxy soya, 26 pts $CaCO_3$, 2 pts stabilizer, plus DOP). (*Courtesy of Climax Molybdenum Company*)

that MoO_3 is an effective flame retardant either by itself or when used with Sb_2O_3. The amount of smoke formed by these formulations was measured using the Michigan Chemical Smoke Densitometer[1,2] and these data are shown in Fig. 15-2. The addition of antimony oxide produces much more smoke than MoO_3 or MoO_3 plus Sb_2O_3 in plasticized PVC. Thus, MoO_3 acts as an effective smoke suppressant for plasticized PVC.

5.5.4 Typical Wire Covering Formulations. Molybdenum compounds were also found to be useful when added to a typical wire insulation formulation plasticized with 30 phr of diisodecyl phthalate (Table 15-7). The 1 phr Sb_2O_3 plus 1 phr MoO_3 mixture is as effective a flame retardant as 2 phr Sb_2O_3 and it produces much less smoke. Both of the ammonium polymolybdates are quite effective additives. The molybdates of calcium and zinc are modest flame retardants and good smoke suppressants. Moly FR-21 is a white material, which contains calcium, zinc, and molybdenum, and this material is a good smoke suppressant in this formulation.

Table 15-7. Properties of typical wire insulation formulation[a].

		SMOKE DATA	
Flame Retardant	Oxygen Index[b]	Smoke Generation	Percent Reduction[c]
None	27.5	23.2	–
1 phr Sb_2O_3	31.5	28.2	(22)
2 phr Sb_2O_3	32.5	26.7	(15)
1 phr MoO_3	29.0	12.2	47
2 phr MoO_3	30.5	4.8	79
1 phr Sb_2O_3 plus 1 phr MoO_3	32.5	6.1	74
2 phr ammonium octamolybdate	30.5	7.2	69
2 phr ammonium decamolybdate	31.5	4.3	81
2 phr calcium molybdate	30.0	11.5	50
2 phr zinc molybdate	30.0	8.4	64
2 phr Moly FR-21	29.0	8.3	64
1 phr Moly FR-10 plus Moly FR-21	30.5	11.3	51
1 phr Sb_2O_3 plus 1 phr Moly FR-21	31.5	6.8	71

[a]Formulation: 100 pts PVC, 30 pts diisodecyl phthalate, 7 pts tribasic lead sulfate stabilizer, 0.4 pts lubricant, 0.4 pts lead stearate, plus flame retardant.
[b]75 mil (1.9 mm) thick samples.
[c]Compared to control. (–) means increase in smoke.

(*Courtesy of Climax Molybdenum Company.*)

Table 15-8. Properties of phosphate ester plasticized PVC samples[a].

Flame Retardant	Oxygen Index[b]	Smoke Generation (SMOKE DATA)	Percent Reduction[c] (SMOKE DATA)
Tricresyl phosphate			
None	27.0	11.5	–
3 phr Sb_2O_3	29.5	9.0	22
3 phr MoO_3	29.5	7.8	32
1.5 phr Sb_2O_3 plus 1.5 phr MoO_3	31.0	3.2	72
3 phr Ammonium Octamolybdate	31.5	5.8	50
Tri(isopropylphenyl)phosphate			
None	26.5	12.6	–
3 phr Sb_2O_3	28.0	9.1	28
3 phr MoO_3	29.0	2.9	77
1.5 phr Sb_2O_3 plus 1.5 phr MoO_3	29.5	3.4	73
Tri(dichloropropyl)phosphate			
None	28.0	6.0	–
3 phr Sb_2O_3	30.5	6.2	(3)
3 phr MoO_3	32.0	2.8	53
1.5 phr Sb_2O_3 plus 1.5 phr MoO_3	33.0	3.3	45

[a]Formulation: 100 pts PVC, 30 pts DOP, 15 pts phosphate ester, 5 pts epoxy soya, 35 pts $CaCO_3$, 1 pt stabilizer, plus flame retardants.
[b]75 mil (1.9 mm) thick samples.
[c]Compared to control. (–) means increase in smoke.

(Courtesy Climax Molybdenum Company.)

5.5.5 Phosphate Ester Plasticized PVC. The results obtained with three representative phosphates are shown in Table 15-8. In these formulations, MoO_3 is equivalent or superior to Sb_2O_3, and the 1:1 mixture of Sb_2O_3–MoO_3 is also more effective than Sb_2O_3 alone. Ammonium octamolybdate is a better flame retardant than MoO_3 or Sb_2O_3 in the tricresyl phosphate formulation. All of the molybdenum containing formulations produced much less smoke than the comparable Sb_2O_3-containing formulations. Molybdenum compounds are very promising additives for phosphate ester plasticizer PVC formulations.

5.5.6 Rigid Polyesters. Atlac 711-05A is a brominated (20% Br) corrosion resistant resin manufactured by ICI United States, and the data obtained with this resin show molybdenum compounds to be useful flame retardants and smoke suppressants for halogenated rigid polyesters. This data is summarized in Table 15-9. Mixtures of MoO_3 with Sb_2O_3 are very effective. The MoO_3 containing formulations also produce less smoke than the Sb_2O_3 containing formulations. The ammonium molybdates alone are not as effective flame retardants as antimony oxide, but they are good smoke suppressants whereas antimony oxide increased smoke. The mixtures of alumina trihydrate with MoO_3 or $(NH_4)_2Mo_2O_7$ are excellent flame retardants and produce extremely low levels of smoke.

References for Molybdates

1. Michigan Chemical Co., *Oxygen Index-Smoke Densitometer, Instruction Manual*, Chicago, Illinois.
2. DiPietro, J. and Stepniczka, H., *J. Fire and Flammability*, **2**, 36 (1971).
3. Weilin, E. D., *Flame Retardancy of Polymeric Materials*, Vol. **3**, p. 203, edited by W. C. Kuryla and A. J. Papa, Marcel Dekker, New York, 1975.
4. Pitts, J. J., *J. Fire and Flammability*, **3**, 51 (1972).
5. Brauman, S. K., Fishman, N., Brolly, A. S., and Chamberlain, D. L., *J. Fire and Flammability*, **6**, 41 (1976).
6. Church, D. A. and Moore, F. W., Molybdenum-based Flame Retardants are now in the Plastics Ball Game, *Plastics Eng.*, **31** No. 12 (Dec. 1975).

Table 15-9. Properties of halogenated polyesters.

		SMOKE DATA	
Flame retardant[b]	Oxygen index	Smoke generation	Percent reduction[c]
Atlac 711-05A Resin Samples[a]			
None	33.0	8.4	–
1% Sb_2O_3	34.5	14.0	(67)
3% Sb_2O_3	42.0	16.2	(93)
5% MoO_3	36.0	5.4	36
7% MoO_3	39.0	4.3	49
1% Sb_2O_3 plus 1% MoO_3	39.5	3.9	54
1% Sb_2O_3 plus 2% MoO_3	40.0	7.1	16
5% Ammonium octamolybdate	36.5	2.7	68
5% Ammonium dimolybdate	34.0	2.6	69
15% Alumina trihydrate	35.0	4.3	49
8% Alumina trihydrate plus 2% MoO_3	40.0	1.2	86
8% Alumina trihydrate plus 2% ammonium dimolybdate	38.5	1.0	88
Hetron 92-C Samples			
None	32.5	4.8	–
5% Sb_2O_3	41.0	4.8	0
5% MoO_3	37.5	2.1	56
2.5% Sb_2O_3 plus 2.5% MoO_3	42.0	2.2	54

[a]Samples contain 25% fiberglass and are 1/8 in. (3.2 mm) thick.
[b]Percent is calculated on resin portion only.
[c]Compared to control. (–) means increase in smoke.
(*Courtesy Climax Molybdenum Company.*)

7. Moore, F. W. and Church, D. A., *Molybdenum Compounds as Flame Retardants and Smoke Suppressants for Polymers*, presented at International Symposium on Flammability and Flame Retardants, Toronto, Ontario, May 7, 1976.

5.6 Inorganic Phosphates

In the class of the inorganic flame retardants, ammonium phosphates have widespread use. Diammonium phosphate (DAP) and monobasic ammonium phosphate (MAP), however, have limited usage as flame retardants because they are very soluble in water. Also, the thermal stability of diammonium phosphate is low and ammonia is lost when heat is applied. The polymeric ammonium polyphosphates are relatively water insoluble and exhibit greater thermal stability than the simple ammonium orthophosphates. The phosphates of the alkaline earth elements offer an advantage, in certain applications, over the ammonium phosphates, in that they are more insoluble. They are often used in fairly low concentration with other flame retarding compounds.

Although it has a low thermal stability similar to DAP, magnesium ammonium phosphate is highly water insoluble, and thus less susceptible to leaching. When $Mg(NH_4)PO_4$ is used in flexible urethane foam as an additive filler, little flame retardancy occurs. In a post-treatment, however, when the insoluble phosphate is coated to fill the surface of the open foam matrix, the resultant product is extremly flame resistant, with remarkably little loss in physical properties. General application information on inorganic phosphate as flame retardant is given in Table 15-10.

References related to inorganic phosphates as flame retardants are:

1. Pruitt, R. M., *J. Cell Plastics*, **6**, 262 (1970).
2. Little, R. W., *Flameproofing Textile Fabrics*, American Chemical Society Monograph Series No. 104, Reinhold, New York, 1947.
3. Miles, C. E. and Lyons, J. W., *J. Cell. Plastics*, **3**, 539 (1967).
4. Liberti, F. P., *Paint and Varnish Prod.*, **51**, 57 (Nov., 1961).
5. Eickner, H. W. and Schaffer, E. L., *Fire Technol.*, **3**, 90 (1967).

Table 15-10. Application of inorganic phosphate flame retardants.

Compound	Formula	Application
Diammonium phosphate (DAP)	$(NH_4)_2HPO_4$	Rigid and flexible urethane foam
Monobasic ammonium phosphate (MAP)	$NH_4H_2PO_4$	Formaldehyde polymers, rubber, rigid urethane foam, paint coatings
Triammonium pyrophosphate	$(NH_4)_3HP_2O_7$	Rigid urethane foam
Ammonium polyphosphate	$(NH_4PO_3)_n$	Rigid urethane foam, polyesters, epoxies, paint coatings
Ammonium hexafluorophosphate	NH_4PF_6	Polystyrene
Magnesium ammonium phosphate	$Mg(NH_4)PO_4 \cdot 6H_2O$	Flexible urethane foam, paint coatings
Lithium orthophosphate	$Li_3PO_4 \cdot 1/2\ H_2O$	Flexible polyester, urethane foam
Sodium orthophosphate	$Na_3PO_4 \cdot 12H_2O$	Rigid urethane foam
Magnesium orthophosphate	$Mg_3(PO_4)_2 \cdot 5H_2O$	Paint coatings

6. *Celanese Resins*, Tech. Bull. 0667, Division of Celanese Coatings Co., Louisville, Ky.
7. Vandersall, H. L., *Phos-Chek P/30 Brand Fire Retardant–its Use in Intumescent Organic Solvent Paints*, Rept. No. 7088, Inorganic Chemical Division, Monsanto Co., St. Louis, Mo., Apr. 23, 1968.
8. Monsanto Co., *Phos-Chek P/30 Phosphorus-Based Fire Retardant*, Tech. Bull. I-270, Inorganic Chemical Division, St. Louis, Mo.
9. Shen, C. Y., Stahlheber, N. E., and Dyroff, D. R., *J. Amer. Chem. Soc.*, **91**, 62 (1969).

5.7 Organic Phosphates

Most flame retardant compounds are solids and share a common disadvantage of reducing many physical properties of plastics. In PVC resin systems, where clarity is a requirement, liquid flame retardants must be chosen. The liquid organic phosphates offer a viable solution to this problem by bringing both flame retardancy as well as excellent plasticizing effect in PVC formulations.

Phosphorous compounds are effective flame retardants because they reduce the percentage contribution of hydrocarbon fuel and produce a more carbonaceous char. As a general rule, plasticizers are more expensive than the standard phthalate type such as DOP (dioctyle phthalate). Therefore, it is important to study the economics and the effect on physical properties of the vinyl compound, when considering the use of phosphate plasticizers. Additional considerations in choosing phosphate plasticizers for PVC are those associated with processing such as volatility, stability to processing heat, solvating effectiveness, and additional effects on physical properties such as low-temperature flexibility and resistance to long term aging in the finished product.

TCP or Tricresyl Phosphate was one of the first and most popular organic phosphates to be used for its flame retardant properties. Its major application was as a partial or complete replacement for DOP or DOA where flame retardancy was required. TCP is marketed under the trade names of Phosflex 179, Kronitex, and Lindol. TCP has low air, oil and water loss and therefore is one of the most permanent monomeric plasticizers available. TCP is preferred for imparting non-burning characterization to colored sheeting or to transparent films where pigmenting or opacifying flame retardants are undesirable. Physical test data of a stabilized plastisol formulation are shown in Table 15-11. A successful replacement for TCP is Kronitex 100 (triaryl phosphate) TAP. TAP is made from 100% synthetic feed stocks and is not dependent on naturally derived cresylic acids, which directly affect the availability of TCP. In some cases, TCP and TAP can be used interchangeably, and TAP is priced somewhat lower than TCP. As measured by oxygen index tests, the flame retardancy of TAP is identical to that of TCP. The relative flame retradancy of PVC compounds containing several phosphate esters is given in Table 15-12. TAP is particularly recommended for use in plastics not only for its flame retardancy characteristics and economic advantages but also because of its unique solvating characteristics. Phosphates have long

Table 15-11. Test data TCP in stabilized PVC.

The physical data listed below were obtained with milled and molded plastic containing Kronitex TCP plasticizer, a polyvinyl chloride resin and 1.5 phr of a barium-cadmium stabilizer.	63 phr
Modulus @ 100% elongation	
@ 25°C and 50% RH, psi	1400
Ultimate tensile strength	
@ 25°C and 50% RH, psi	2800
Ultimate elongation	
@ 25°C and 50% RH, %	350
Hardness, Shore Dur. A	
@ 25°C and 50% RH	70
Loss in water	
10 days @ 25°C, %	0.1
Loss in oil	
10 days @ 25°C, %	3.5
Loss in gasoline	
10 days @ 25°C, %	25.7
Loss, carbon volatility	
24 hr @ 70°C, %	0.5
Oxygen index, 10 mil film @ 61 phr	29.4
Low temperature flexibility	
ASTM D746 procedure, @ 63 phr, Tb (°C)	-7
ASTM D1043–Clash and Berg, Tf°C	6.4
T4°C	6.9
Fluxing rate	
@ 310°F, min.	0.9
Heat stability	
Pt-Co Color 50, min.	15
Pt-Co Color 500, min.	20
Black, min.	30
Ultraviolet light stability	
Color change, Moderate	
Bleed, None, but tacky	
Migration	
Nitrocellulose film, Moderate	
Nitrocellulose lacquer, Slight	
Varnish, Slight	
Compatibility	
Room temperature, No Bleed	
24 hr @ 105°C, No Bleed; moderate discoloration & odor	

(*Courtesy of FMC Corporation, Industrial Chemical Division.*)

been known for their high fusion rates in plastisols, but their high solvating rates may result in relatively high viscosity increases. TAP is unusual in that it is slower solvating than TCP, yielding a more stable viscosity over extended periods.

FMC has developed a phosphate ester which provides a low specific gravity and high viscosity. It is called Kronitex TXP and is trixylenyl

Table 15-12. Flame retardance of phosphate esters in PVC.

	OXYGEN INDEX	
	61 phr	88 phr
Kronitex 100 (Triaryl phosphate) (TAP)	29.4	27.7
Tricresyl phosphate (TCP)	29.4	27.9
Isodecyl diphenyl phosphate (IDDP)	26.4	25.4
Trioctyl phosphate (TOF)	24.8	23.5
Cresyl Di phenyl phosphate (CDD)	31.0*	
Kronitex 50	30.5*	

*60 phr.
(*Courtesy of FMC Corporation, Industrial Chemical Div.*)

phosphate. TXP is recommended specifically for electrical grade applications and for the heavily filled compounds where its high viscosity induces better milling action. Physical test data of a stabilized plastisol formulation are listed in Table 15-13.

Cresyl disphenyl phosphate (CDP) is the least costly of the natural derived phosphate esters and also has the highest specific gravity and lowest viscosity. These are characteristics which are particularly recommended for plastistols. CDP has a greater compatibility with cellulose acetate than TAP and is slightly better than TCP in low temperature flexibility. It also has the highest phosphorous content of the triaryl phosphates.

FMC has recently introduced a new synthetic triaryl phosphate flame retardant plasticizer designated Kronitex 50. Kronitex 50 is a low molecular weight analog to the current synthetic triaryl phosphate industry standard Kronitex 100. Like Kronitex 100, Kronitex 50 is produced from synthetic raw materials which are not subject to the erratic supply situations of the natural based cresylic phosphate esters. Kronitex 50 is designed to be a more economical replacement for the naturally derived cresyl diphenyl phosphate (CDP) for use in flame retarding various polymer and elastomer systems. Kronitex 50 is notably similar to CDP in physical properties as illustrated in Table 15-14. However, because of its lower specific

Table 15-13. Test data TXP in stabilized PVC.

The physical data listed below were obtained with milled and molded plastic containing Kronitex TXP plasticizer, a polyvinyl chloride resin and 1.5 phr of a barium-cadmium stabilizer.	63 phr
Modulus @ 100% elongation @ 25°C and 50% RH, psi	1500
Ultimate tensile strength @ 25°C and 50% RH, psi	2800
Ultimate Elongation @ 25°C and 50% RH, %	320
Hardness, Shore Dur. A @ 25°C and 50% RH	72
Loss in water 10 days @ 25°C, %	0.15
Loss in oil 10 days @ 25°C, %	2.0
Loss in gasoline 10 days @ 25°C, %	27.0
Loss, carbon volatility 24 hr @ 70°C, %	0.5
Oxygen Index, 10 mil film @ 61 phr	29
Low temperature flexibility	
ASTM D746 procedure @ 65 phr Tb (°C)	3
ASTM D1043-Clash and Berg. Tf., °C	-2.7
T_4. °C	14.0
Fluxing rate @ 310°F, min.	1.1
Heat stability	
Pt-Co Color 50, min.	15
Pt-Co Color 500, min.	20
Black, min.	30
Ultraviolet light stability	
Color change, moderate	
Bleed, None, but tacky	
Migration	
Nitrocellulose film, Moderate	
Nitrocellulose lacquer, Very slight	
Varnish, Very slight	
Compatibility	
Room temperature, No bleed	
24 hr @ 105°C, No Bleed; moderate discoloration and odor	

(Courtesy of FMC Corporation, Industrial Chemical Div.)

gravity, Kronitex 50 offers additional economic advantages.

When used in PVC compounds Kronitex 50 is comparable to CDP in flame retardance, low temperature performance, volatility, compatibility, extraction, and heat and light stability.

When utilized as the sole plasticizer for PVC, cresyl diphenyl phosphate demonstrates an

Table 15-14. Typical physical properties of K-50 VS CDP.

	Kronitex® 50 Triaryl phosphate	CDP
Appearance	Clear liquid	Clear liquid
Color, APHA	30	30
Odor	Essentially odorless	Very slight
Specific gravity, 25°/25°C	1.168	1.188
Viscosity, cps		
20°C	70	51
25°C	55	42
38°C	29	23
Acidity, % as acetic acid	.01	.01
Refractive index, H_D, 25°C	1.553	1.559
Moisture, %	< .1	< .1
Weight, pounds/gallon, 25°C	9.8	9.9

approximate 15% greater plasticizing efficiency than Kronitex 50 as illustrated (Table 15-15) by the 100% modulus and Shore "A" Hardness values. However, when incorporated at the more common use level (25% replacement for the primary plasticizer system i.e. DOP) the differences in plasticizing efficiency become negligible. Kronitex® 50 triaryl phosphate is recommended for use in flame retarding PVC, flexible polyurethane foam, cellulose-acetate/butyrate, nitrocellulose, and various elastomer and engineering resin systems.

A schedule of typical prices for various organic phosphates is given in Table 15-16.

Tris (2-ethylhexyl) phosphate or trioctyl phosphate is produced by Tenneco. It is one of the best low temperature plasticizers for PVC resins. It contributes not only to low temperature flexibility but fungus resistance and flame retardance. TOF is compatible with most resin systems. When used in plastisols as the sole plasticizer, it provides a very low initial viscosity. Viscosity increase over 30 days is slight. The plastisols are essentially non-dilatent under high or low shear. TOF has been widely used in vinyl cable coatings, synthetic rubber compositions, and coated military tarpanlins. Physical test data of a stabilized formulation of PVC and TOF are given in Table 15-17. Chemical and physical property data on these organic phos-

Table 15-15. Typical performance data of K-50 VS CDP.

Plasticizer	DOP	Kronitex® 50 Triaryl phosphate	CDP	DOP/Kronitex® 50 Triaryl phosphate	DOP/CDP
phr	60	60	60	45/15	45/15
Tensile properties					
100% Modulus, psi	1025	1400	1175	1060	1000
Tensile strength, psi	2290	2710	2820	2340	2400
% Elongation	420	340	360	400	430
Oxygen index					
L. O. I.	22.5	30.5	31.0	25.0	25.0
Low temperature properties					
T_F, °C	-30	-8	-9	-25	-24
T_b, °C	-31	-7	-7	-29	-28
Volatility-activated carbon, % wt. loss					
24 hr @ 70°C	1.4	1.3	1.3	1.4	1.3
24 hr @ 90°C	10.3	7.2	6.9	8.5	7.3
Extraction–% wt. loss					
1% Ivory flake, 24 hrs. @ 70°C	2.8	12.3	12.8	7.5	6.8
Hexane–3 hr @ 23°C	30.0	7.4	4.5	26.5	25.9
Mineral oil 24 hr @ 23°C	2.6	1.0	1.2	2.9	3.0
Compatibility–bent loop					
10 days, 50°C 100% R.H.	←		No exudation		→
Oven heat stability					
Time to initial discoloration 15 min. intervals @ 350°F	45	30	30	45	45
Light stability, fadeometer					
120 hr	←		Slight yellowing		→
Shore "A" Hardness					
Initial/15 seconds	85/74	88/77	88/75	85/75	83/74

Formulation	phr
PVC	100
Ba/Cd stabilizer	2.5
Plasticizer	As indicated

Date *Courtesy FMC Corp. Industrial Chemical Div.*

Table 15-16. Organic phosphorous compounds, cents per pound*

	Tank car/ tank truck 40,000 lb min.	Truckload drums 24,000 lb min.	Less than truckload 10 drum min.	Drums net contents
Kronitex 50	$0.64	$0.69	$0.73	530 lbs
Kronitex 100	$0.62	$0.67	$0.71	520 lbs
Kronitex CDP	$0.69	$0.74	$0.78	545 lbs
Kronitex TCP	$0.74	$0.79	$0.83	530 lbs
Kronitex TXP	$0.74	$0.79	$0.83	520 lbs

Delivery: Prices are F.O.B. Nitro, West Virginia. Tank cars trip leased with 7 days free detention time. $25 per day thereafter.

Terms: Net 30 days.

*FMC Corporation, Industrial Chemical Division, 150 White Plains Road, Tarrytown New York 10591
(914) 631 7900
(212) 829 7558 New York City

Table 15-17. Test data TOF in stabilized PVC.

The physical data listed below were obtained with milled and molded plastic containing Kronitex TOF plasticizer, a polyvinyl chloride resin and 1.5 phr of a barium-cadmium stabilizer.	54 phr
Modulus @ 100% elongation @ 25°C and 50% RH, psi	1500
Ultimate tensile strength @ 25°C and 50% RH, psi	2780
Ultimate elongation @ 25°C and 50% RH, %	310
Hardness, Shore Dur. A @ 25°C and 50% RH	77
Loss in water 10 days @ 25°C, %	0.6
Loss in oil 10 days @ 25°C, %	23.5
Loss in gasoline 10 days @ 25°C, %	22.5
Loss, carbon volatility 24 hr @ 70°C, %	4.5
Oxygen Index, 10 mil film @ 61 phr	24.8
Low temperature flexibility	
ASTM D746 procedure @ 54 phr Tb (°C)	-65
ASTM D1043-Clash and Berg. Tf. °C	-55.5
T4., °C	-14.5
Fluxing rate @ 310°F, min.	1.2
Heat stability	
Pt-Co Color 50, min.	5
Pt-Co Color 500, min.	10
Black, min.	20
Ultraviolet light stability	
Color change, None	
Bleed, None	
Migration	
Nitrocellulose, Severe	
Lacquer, Considerable	
Varnish, Considerable	
Compatibility	
Room temperature, No bleed	all
24 hr @ 105°C, No bleed	all

(Courtesy of FMC Corporation, Industrial Chemical Div.)

phate flame retardants are compared in Table 15-18.

There are several large volume compounds that are halogenated phosphates and could be listed either in this section or the next. These are products such as tris (2-chloroethyl) phosphate, tris (2, 3-dibromopropyl) phosphate, and tris (dichloropropyl) phosphate. They are effective in reducing the flammability of polystyrene.

Suppliers of Organic Phosphates

Aldrich Chemical	Monsanto Company
Arapahoe Chemicals	Stauffer Chemical Company
FMC Corporation	Strem
Hooker Chemical Corporation	Weston Chemical Inc.

5.8 Organic Halogens

Numerous organic halogens have been developed as flame retardants for use in a variety of thermoplastics that include polyethylene, polypropylene, ABS, polyester, and high-impact polystyrene. They generally are used in combination with antimony oxide and other synergists. Bromine and chlorine-containing compounds are the primary types of organic halogen flame retardants. It is generally believed that bromine is a more effective flame retardant than chlorine, but the type of chemical structure to which the halogen is attached is also a major factor. This structure affects the strength with which the halogen is bonded to the molecule, and hence the relative ease with which it is released during exposure to a flame.

Flame retardants in comercial use include compounds with aliphatic chlorine and/or bromine, aromatic bromine and alicyclic chlorine. The aliphatic chlorine types are sold as chlorinated paraffins and are used primarily in lower melting resins such as polyethylene, because of their limited heat stability. For higher melting resins, such as polypropylene and polystyrene, alicyclic chlorine and aromatic bromine chemicals are used. One of the largest volume alicyclic chlorine compounds is Dechlorane Plus made by Hooker Chemical. An aromatic bromine compound, decabromobiphenyl ether (DBBPE) is also widely used as an effective flame retardant in high-impact polystyrene, low density polyethylene, and thermoplastic polyester. DBBPE is normally used with antimony trioxide synergist. The same compound, some-

Table 15-18. Physical and chemical properties of organic phosphate flame retardants.

Kronitex	TCP	CDP	TXP	TOF	Kronitex 100
Formula	$(CH_3C_6H_4O)_3PO$	Type of $(RO)_3PO$	Type of $(RO)_3PO$	$(CH_3CH_2CH_2CH_2CH(CH_2CH_3)CH_2O)_3PO$	
Molecular weight (Avg.)	368	340	400	434	400
Specifications					
Specific gravity @ 20°C (ASTM D268-69)	1.160–1.175	1.195 ± 0.010	1.142–1.152	0.924–0.927	1.150–1.165
Moisture ASTM D1364-64	Max. 0.1%	Max. 0.1%	Max. 0.1%	Max. 0.1%	Max. 0.1%
Color Pt-Co ASTM D1209-69	Max. No. 75	Max. No. 75	Max. No. 100	Max. No. 100	Max. No. 75
Odor	Very slight	Very slight	Very slight	Mild	Very slight
Acidity ASTM D1613-66	Max. 0.01%	Max. 0.01%	Max. 0.01%	Max. 0.02%	Max. 0.01%
Average properties					
Boiling range @ 4 mm, °C	241 to 255	235 to 255	248 to 265	–	220 to 270
Mid-boiling point @ 4 mm, °C	249	242	255	216	250
Freezing point, °C	–33 (stiff gel)	–38 (stiff gel)	–25 (stiff gel)	–90	–
Pour point, °C	–28	–36	–20	–74	–25
Flash point, °F	490	470	495	420	470
Fire point, °F	550	550	550	–	550
Viscosity @ 20°C, cps	80	60–70	190	14.1	90
Vapor pressure @ 150°C, mm Hg	0.02	0.02	0.02	0.1	–
Vapor pressure @ 200°C, mm Hg	0.29	0.70	0.33	2.0	–
Surface tension @ 20°C, dynes/cm	39	42	38	–	–
Refractive index @ 25°C	1.55 ± 0.002	1.561	1.553	1.441	1.552
Specific heat @ 50 to 150°C	Avg. 0.47	Avg. 0.50	Avg. 0.50	–	–
Solubility in water @ 25°C, %	0.01%	0.01%	0.01%	0.01	–
Solubility, water in @ 25°C, %	0.4%	0.4%	0.03%	Appx. 1.4%	–
Solubility in mineral oil @ 25°C, %	6% by volume	3.5% by volume	8% by volume	Complete	–
Solubility in gasoline @ 25°C, %	Complete	15.5 by volume	Complete	Complete	–
Thermal expansion @ 10 to 40°C, per °C	0.00067	0.00067	0.00067	0.008	–
Weight per gallon @ 20°C, lb	9.70	9.96	9.58	7.70	9.6
Ortho content, %	1	1	1	–	–

(Courtesy of FMC Corporation, Industrial Chemical Div.)

times called decabromodiphenyl oxide (DBDPO) is made by Fine Organics, Dow, White Chem, and Great Lake Chemical. Cities Service Co., Citex BC-26 is described as an organic compound containing 29% Bromine and 40% chlorine and having a low melting point (392°F) and a high decomposition temperature (above 500°F). Both of these compounds, which are priced about the same at $1.75/lb are used in about 10% concentration along with 3-5% antimony oxide or other synergist type chemicals. Each has its strong points: BC-26 dissolves into the resin, offering good molding properties. Decabromodiphenyl oxide is somewhat more difficult to load due to its higher melting point but it imparts better resistance to heat distortion than does BC-26. BC-26 has good UV resistance, making it a better choice for outdoor applications.

Michigan Chemicals Firemaster 680 is a brominated aromatic additive (non-biphenyl) with good stability to withstand processing temperatures in the 500°F range. This is a 70% brominated compound which also provides good UV stability. Its price is about $1.50/lb in development quantities.

Cities Services Citex BT-93, a 67% bromine aromatic compound, possessing excellent UV stability. This compound has a 450°C decomposition temperature, thus offering exceptional thermal stability. It is recommended especially for high impact polystyrene which at a formulation level of 12% BT-93 and 3% antimony oxide, reportedly has better UV stability than an ABS with equivalent flame resistance. BT-93 is also recommended for high density polyethylene, low density polyethylene, and thermoplastic polyesters. Citex BT-93 is priced at about $1.75/lb.

Cities Service's brominated aliphatic compound BCL-462 is used mostly in polystyrene foam and is moderately priced at $1.30/lb. It is also effective in polystyrene applications and can attain UL94-V2 rating of 5% levels without a synergist.

Great Lakes Chemicals has two tetrabromobisphenol A compounds, BA-59P and heat stabilized FR-378. These compounds were initially used in epoxies and polycarbonates but now are being proposed as additives or partial replacements of the more expensive decabromo compounds—especially for high impact polystyrene applications. The tetrabromo compounds are used in a ratio of about 1.4 parts per part of decabromo replaced, but costs only about one half as much ($0.87 vs. $1.82). Other advan-

Table 15-19. Properties of halogenated organic products.

Product name	DBBPE	Citex BC-26	Dechlorane Plus 25	HFO-104	HFO-105
Thermal stability °C	330	260	>350	290	>300
Melting range °C	304	200	>350	180–185	136–138
Appearance	Creamy white free flowing powder	White powder	White powder	White powder	White powder
Bromine %	83	29	–	67	79
Chlorine %	–	40	65	–	–
Volatility °C for 10% wt loss TGA	–	–	–	290°C	231°C
V-O Use conc. %	10–13%	10–15%	10–15%	10–15%	10–15%
Synergist Sb_2O_3	3–5%	3–5%	3–6%	3–5%	3–5%
Solubility*					
Water	Insoluble	Insoluble	–	0.1	0.1
Acetone	Insoluble	Insoluble	–	1.1	3.6
Benzene	–	–	2.0	–	–
Density g/cc	3.0	1.84	1.8	–	–

*Solubility for all products expressed in grams per 100 ml of solvent.

tages include easier dispersability, lower bulk density, high melt flow, and improved Gardner impact properties.

Great Lakes has a new compound, DE-71, which is a pentabromo diphenyl ether. It is a liquid with 100% active bromine content. DE-71 was developed as an additive for epoxies and adhesives. This material is claimed to have good thermal stability to exotherm and soldering heat with no effect on electrical and physical properties. Another possible application for this material is flexible urethane. The development price is $2.00/lb.

Other companies that market organic halogens are GAF, Aldrich, Monsanto, Hooker, Polyscience, FMC, and Berk.

Physical and Chemical Properties of Organic Halogens and their Cost. Physical and chemical properties of a few typical organic halogen flame retardants are given in Table 15-19, along with two recently developed compounds. These compounds are: DBBPE–decabromobiphenyl ether, aromatic bromine; Citex-BC-26, 29% bromine + 40% chlorine, aliphatic Dechlorane Plus 25–65% alicyclic chlorine; HFO-104–pentabromophenylbenzoate, developmental compound by Hexcel; HFO-105–pentabromoethyl benzene, developmental compound by Hexcel.

5.9 Sulfates

The sulfates are utilized mainly as flame retardants in the cellulosic field of wood and paper. Ammonium sulfate, $(NH_4)_2SO_4$, is of the most use in this area. Ammonium alum, potassium alum, gypsum, and barium sulfate, along with ammonium sulfate, are involved as flame retardants in certain specific applications in the polymer field, as shown in Table 15-20. Suppliers of the sulfate flame retardants are Allied Chemical Corporation, American Cyanimid Co., and Diamond Shamrock Corporation.

References relating to the use of sulfates as flame retardants:

1. Herting, O., to Sani Paper Products Co., U.S. Patent 1,804, 417, (1931).
2. Lichtenstein, I., to Structural Paper Co., U.S. Patent 2,710,264, (1955).
3. William Old (Resiform) Ltd., by H. Butcher and M. J. Double, British Patent 1,085,773 (1967).
4. U.S. Rubber Co., British Patent 870,331 (1961).
5. Celcure & Chemical Co., Ltd., by E. M. Wallace, German Patent 1,056,356 (1959).
6. Bannister, J. W. and Rittel, C. A., British Patent 1,744 (1915).

Table 15-20. Application of sulfate flame retardants.

Sulfate	Formula	Resin
Ammonium alum	$NH_4AL(SO_4)_2 \cdot 12H_2O$	Polyesters
Potassium alum	$KAl(SO_4)_2 \cdot 12H_2O$	Polyesters
Gypsum	$CaSO_4 \cdot 2H_2O$	Polyesters
Barium sulfate	$BaSO_4$	Paint coatings
Ammonium sulfate–antimony trifluoride	$(NH_4)_2SO_4 \cdot 2SbF_3$	Polyethylene

Table 15-21. Projected use of flame retardants.

Flame Retardants used in Plastics (in thousand metric tons)

	1975	1980
Additives		
Alumina trihydrate	50.0	200.0
Antimony trioxide	9.0	20.0
Boron compounds	4.0	6.0
Bromine compounds	6.0	13.0
Chlorinated paraffins and cycloaliphatics	32.0	65.0
Phosphate esters, nonhalogenated	32.5	60.0
Others	7.5	16.0
TOTAL	153.0	405.0
Reactive systems		
Epoxy intermediates	5.0	11.0
Polycarbonate intermediates	1.3	2.0
Polyester intermediates	7.5	16.0
Styrenic intermediates	2.0	5.0
Urethane intermediates (flexible foams)	13.8	29.0
Urethane intermediates (rigid foams)	7.2	15.0
Others	4.7	10.0
Total	41.5	83.0

Source: *Modern Plastics*, May, 1976

7. Craig, R. W., to Diamond Alkali Co., U.S. Patent 2,627,475 (1953).
8. Rapp, H. C., to Raychem Corp., U.S. Patent 3,239,482 (1966).

5.10 Others

Some less popular flame retardant chemicals and their applications are:

Name	Formula	Application
Sodium metavanadate	$NaVO_3$	Cellulose
Lead arsenate	$PbHA_sO_4$	Polypropylene
Lead metaarsenate	$Pb_2As_2O_4$	Polypropylene
Sodium metasilicate	$Na_2SiO_3 \cdot 5H_2O$	Polystyrene coatings

6. FUTURE GROWTH

With the continued impetus caused by government regulations, rapid growth is assured for the flame retardant business as a whole, and an important segment will be plastics applications. The estimate shown in Table 15-21 projects a two- to three-fold increase in sales by 1980, as compared with 1975 levels. There will be continued research and development in reducing production costs of new materials, materials formulations, and products that have improved resistance to fire and smoke generation.

General References

1. Flame Retardants, *Plastic Technology Manufacturing Handbook* and *Buyers Guide*, Apr. 1976.
2. There are New Ways to Tame the Flame, *Modern Plastics*, May, 1975.
3. Flame Retardants Additive, '76, *Plastic World* (July 19, 1976).
4. "Flame Retardants R&D Phase Quickens," *Plastic Technol.* (July 1976).
5. Hilado, C. J., *Flammability Handbook for Plastics*, Technomic Publishing Co., Inc., Stamford, Conn., 1969.
6. Ranney, M. W., *Flame Retardant Polymers*, Noyes Data Corp., Park Ridge, N.J., 1970.
7. Lyons, J. W., *The Chemistry and Uses of Fire Retardants*, John Wiley, New York, 1970.
8. Kurylan, W. C. and Papa, A. J., *Flame Retardancy of Polymeric Materials, Vols.* 1 and 2, Marcel Dekker, New York, 1973.

Section VI
Organic and Special Purpose Fillers

16
Carbon Black

Errol J. Queen
Columbian Division of Cities Service Co.
Akron, Ohio

CONTENTS

1. INTRODUCTION

Carbon black is usually thought of as a reinforcer for elastomers, particularly rubber, in automotive tires. Although the greatest volume of carbon black does find its way into transportation products, millions of pounds of black are consumed annually by the inks, coatings, and plastics industries.

Generally, carbon black does not provide the reinforcing characteristics associated with rubber in most plastics polymers. It is used, however, as a filler for economy. At the same time, carbon black provides desirable end-performance properties, the most important being ultraviolet and thermal protection, control of electrical conductivity, and improved compound processing. Perhaps the most significant plastics application for carbon black is in wire and cable jacketing where cross-linked polyethylene is preferred. Carbon black, of course, is used in a wide range of thermoplastic and thermoset resins for coloring, providing a low-cost means of opacifying polymers, sometimes at very low concentrations.

Commercial carbon blacks range in arithmetic mean particle diameter from 10 to 300 mμ, vary considerably in their degree of agglomeration, and exhibit pH's between 3 and 9. The coarsest products provide the least blackness and the least ability to provide U.V. protection; they are, however, the least costly and because of low surface area, allow for the greatest loadings. Coarse blacks also tend to disperse the fastest. Consequently, the carbon blacks most often selected chiefly as fillers are coarse and low in cost (often \$.12–\$.25/lb). A number of specialty grade blacks, for example for conductive applications, are priced in the range of \$.30–\$.40/lb, while extremely fine particle size, high color products are priced as high as \$4.00/lb. Numerous intermediately-priced products are available.

Food and Drug Administration regulations limit the use of carbon blacks for products coming into contact with food to those produced by the channel or impingement process. The ultimate proof of safety rests with the final product manufacturer who is expected to make extraction tests. Manufacturers who intend to use carbon blacks (or any other fillers) in FDA applications should, therefore, fully acquaint themselves with all applicable requirements.

2. CARBON BLACK PRODUCTION METHODS

Carbon blacks are produced by a number of methods which differ significantly in technology and in the quality of the final products. The most important methods are the furnace, channel, and thermal processes, although lampblacks and acetylene blacks must also be mentioned. Furnace, channel, thermal, and acetylene blacks are classified as Color Index Pigment Black 6 (77266) while lampblack is categorized as Pigment Black 7 (77266).

2.1 Furnace Blacks

Well over 90% of the carbon black currently produced and consumed is made by the furnace process. Furnaces can be operated as all-gas or all-oil processes, but today's blacks are often produced by using a combination of the two. An oil feedstock, highly aromatic in nature, is introduced into a heated chamber which contains the combustion products of an oil or gas flame. The injected oil is thermally decomposed, forming carbon black particles. Water quenching, separation of the particulate component from the gases, and collection follows. The physical and chemical properties of the black thus produced by this continuous process depend on the selection of raw materials, velocity and configuration of ingredient addition, temperature, and furnace design. It is possible to carefully control particle size and degree of agglomeration ("structure") by varying these parameters.

Furnace blacks are usually characterized by relatively neutral pH surfaces, although additional treatment may be afforded to oxidize the black and alter its properties. Most blacks are beaded, either by tumbling in drums or by stirring in pin machines, in order to effect less dusting in end-use. Beaded blacks do, however, require additional shear or energy to process and disperse; under conditions of low energy mixing, therefore, powdered blacks may be preferred. Most manufacturers make their non-rubber grade blacks available in both beads (sometimes called "pellets") and powdered form.

2.2 Channel Blacks

Carbon blacks produced by the channel or impingement process accounted for the bulk of production prior to World War II. Long rows of small natural gas flames are allowed to impinge on channel irons upon which the carbon is formed. The black must be mechanically scraped from the channels, removed, and bagged. The gas feedstock is often enriched with oil in current manufacturing methods.

Channel blacks are generally characterized by low pH because of the chemisorption of oxygen-containing groups on the surface. While pigment acidity is not nearly as critical in filler selection for plastics as it might be for inks or coatings, the selection of a black with optimum surface chemistry should still not be overlooked. Because channel blacks do not require quench cooling, there is no introduction of ash from a water supply as is often true in the furnace process. The variables which affect carbon quality in the channel process are burner tip size and type, temperature, flame-to-channel distance, and the concentration of oxygen present.

Channel black production and consumption has decreased dramatically for two reasons. First, furnace blacks, as they are manufactured today, can be used in most applications where channel blacks used to be employed. Even in those polymers which require great acidity (high volatile content on the black), furnace blacks which have been oxidized in a post-treatment step can usually be employed. Secondly, the furnace process is a more efficient, flexible technique for carbon black production. It is, environmentally, a much cleaner process. Channel method is also more highly gas-dependent and recent trends seem to indicate that gas costs may rise more rapidly than the costs of petroleum oil fuels used in the furnace process. Impingement blacks, nonetheless, are still recommended for FDA applications, where the final plastics product comes into contact with food or drugs.

2.3 Thermal Blacks

Unlike channel or furnace blacks, thermal blacks are produced by the thermal decomposition of hydrocarbon gases in an oxygen-deprived environment. Gas is forced into a pre-heated refractory where it is decomposed, forming carbon black and hydrogen. Since the thermal method is a batch rather than continuous process (again, contrasting it with the channel or furnace processes), two generators are used alternately to effect continuous operation. Carbon is formed in one while the other is being heated. The reaction by which carbon is produced is endothermic, requiring the continued periodic application of external heat. Thermal blacks are very coarse in particle size, show relatively little tendency to aggregate into chain or grape-like "structures" (as do furnace and impingement blacks), and contain virtually no chemisorbed oxygen. Thermal blacks have historically been among the lowest in cost. This combination of properties and cost has led to its wide use as a filler for plastics and elastomers.

As in the channel process, thermal blacks are highly gas-dependent. Considerable research is being conducted in the industry to find alternative methods of producing coarse, low-structured thermal-like products.

2.4 Lampblacks

The ancient Egyptians and Chinese knew about the lampblack process by which oils or fats were burned and the soot deposited on cooled porcelain. In the 1800's, considerable lampblack was produced by this means, in large quantities, using more efficient collection chambers. The burning of petroleum products in restricted air supplies, while producing interesting products of large particle size and high structure, is often difficult to do with proper consideration for impact on the environment. Lampblack replacements, made by the cleaner furnace process, have provided much relief in recent years. There is little true lampblack production today. Lampblacks are known for their clean blue tone and their permanence to light, acid, and alkalis. They are of greater significance to the coatings industry than to plastics formulators.

2.5 Acetylene Blacks

Acetylene black, like thermal black, is formed from thermal decomposition. But unlike the thermal process, acetylene's decomposition liberates heat, maintaining the reaction without the application of additional heat.

Extremely high structure has led to the use of acetylene blacks in dry cell battery applications. High structure and low volatile content also results in exceptional electrical conductivity when compounded in plastics. The same kind of conductivity, however, has been achieved in specially designed furnace blacks for the plastics industry.

Table 16-1 illustrates the typical property ranges of carbon blacks produced by these five methods.

3. PROPERTIES OF CARBON BLACKS

The wide diversity of products commercially available today is illustrated in Table 16-2. Table 16-3 lists the industry equivalents for many industrial carbon blacks.

Unlike rubber-grade carbon blacks, industrial types are not generally produced to ASTM or industry specifications; comparisons shown in Table 16-3, therefore, are rough estimates of

Table 16-1. Chemical and physical properties of carbon blacks by production method.

Property	Furnace black	Channel (impingement) black	Thermal black	Lampblack	Acetylene black
Mean particle diam. (mμ)	13–70	10–30	150–500	50–100	350–500
Surface area (N_2) (m^2/g)	20–950	100–1125	6–15	20–100	60–70
Stiff paste oil absorption (cc/g)	0.65–2.0	1.0–5.7	0.3–0.5	1.0–1.7	3.0–3.5
pH	3.0–9.5	3.0–6.0	7.0–8.0	3.0–7.0	5.0–7.0
% Volatiles	0.3–4.0	3.0–17.0	0.1–0.5	0.3–10.0	0.3–0.4

(mμ) millimicron = 10 Å.

Table 16-2. Typical chemical properties of industrial (non-rubber grade) channel, furnace, and thermal blacks.*

			Surface area	Oil absorption									
				Venuto method fluid paste		Stiff paste endpoint		DBP					
	Blackness index	Particle diameter, mμ (arithmetic mean)	Sq. meters per gram	cc/gram	Gal./100 lb	cc/gram	Gal./100 lb	cc/100 grams	Tinting strength index	DPG adsorption index	pH	Fixed carbon %	Volatile matter %
Channel blacks													
Royal Spectra	253	10	1125	27.5	330	5.73	47.8	229	88	100	4.0	83.5	16.5
Neo Spectra Mark I	240	11	1065	25.8	310	3.66	43.9	219	89	100	4.0	86.2	13.8
Neo Spectra Mark II†	220	13	906	22.8	274	2.79	33.4	190	93	98	3.0	88.0	12.0
Neo Spectra AG†	220	13	900	18.2	218	2.75	33.0	169	93	95	3.0	86.0	14.0
Super Spectra	218	13	742	17.8	214	2.40	28.8	185	95	66	4.0	93.2	6.8
Neo Spectra Mark III†	190	14	605	15.4	185	1.54	18.5	145	102	84	3.2	88.5	11.5
Neo Spectra Mark IV†	179	15	335	13.3	160	1.62	19.4	127	101	32	5.0	95.8	4.2
Furnace blacks													
Raven 8000†	187	13	935	9.9	120	1.73	20.6	149	103	–	2.4	90.4	9.6
Raven 7000†	163	14	543	10.3	124	1.43	17.2	131	102	–	2.6	92.0	8.0
Raven 3500†	176	16	319	11.0	132	1.51	18.1	119	100	49	3.0	96.2	3.8
Raven 2000†	174	17	235	9.9	119	.88	10.5	60	112	19.8	9.0	97.8	2.2
Raven 1800†	167	18	214	13.7	164	1.04	12.5	96	107	19.0	9.0	98.5	1.5
Raven 1500†	170	18	187	7.9	95	.72	8.6	59	98	14	6.0	98.2	1.8
Raven 1250†	166	21	124	7.3	88	.68	8.2	54	97	7	7.0	98.7	1.3
Raven 1200†	160	24	123	9.4	113	.86	10.3	72	105	6	8.0	97.7	2.3
Raven 1170†	160	24	102	7.2	86	.69	8.3	58	102	6	7.0	99.0	1.0
Raven 1255†	160	22	130	5.4	65	.80	9.6	55	103	30	3.0	96.1	3.9
Raven 1035†	151	27	87	5.8	70	.78	9.3	55	102	9	5.0	98.3	1.7
Raven 1020†	151	27	82	7.1	85	.75	9.0	65	102	6	7.0	99.0	1.0
Raven 1000†	151	29	96	8.8	105	.75	9.0	61	96	5.4	7.0	98.0	2.0
Raven 890	130	29	65	8.5	102	1.21	14.5	102	93	4	8.0	98.8	1.2
Raven 850†	126	33	58	6.2	74	.88	10.5	78	85	3	7.0	98.6	1.4
Raven 500†	95	54	33	5.4	65	.88	10.6	75	68	4	7.0	98.8	1.2
Raven 450†	93	60	33	4.0	48	.68	8.1	65	60	2.4	8.0	99.3	0.7
Raven 420†	77	70	28	4.3	52	.78	9.4	77	52	2.0	9.0	99.6	0.4

Raven 410†	58	62	23	3.7	44	.69	8.3	70	48	2.0	9.0	99.7	0.3
Conductex 950†	116	21	245	14.5	173	2.18	26.3	175	94	25.4	6.5	99.2	0.8
Conductex SC†	155	17	200	11.7	140	1.30	15.6	110	101	17	8.7	97.8	2.2
Raven C Beads	138	19	155	11.1	133	1.37	16.4	120	98	11.7	7.0	98.0	1.6
Raven 1040†	135	29	85	5.2	62.4	1.17	14.0	105	95	21.6	3.0	96.5	3.5
Lampblack replacement													
Raven 14	92	60	36	–	–	1.00	12.0	–	69	16.4	2.7	97.8	2.2
Raven 16	77	62	23	–	–	1.08	13.0	–	48	3.3	5.1	99.4	0.6
Raven 22	65	70	17	–	–	1.17	14.0	–	40	2.2	6.3	99.5	0.5
Thermal black													
Statex MT Bds.	25	300	10	1.9	23	.34	4.1	35	18	0.4	9.0	99.5	0.5

**Courtesy Cities Service Co., Columbian Division.*
†Available as beads.
All products show tradenames registered Cities Service Company, formerly Columbian Carbon Co.
For practical purposes in formulating, specific gravities, are suggested as follows:
Channel 1.75
Furnace and Lampblack Replacement 1.80
Thermal 1.87

equivalency based primarily on blackness level. Manufacturers of carbon black for industrial applications, including plastic, control their products for several properties. The most significant ones are blackness, oil absorption, tint strength, and pH.

Blackness or jetness is a measure of particle size. The more jet a carbon black, the finer (and higher surface area) the product, all other things being equal. A secondary determinant of blackness is structure, i.e., the degree to which chain- or grape-like aggregates form; large clumps of particles in effect decrease surface area, lowering the color or blackness.

Both particle size and structure are even more significant in determining the rheological properties of a polymer filled with the pigment. As particle size decreases or structure increases, more polymer is needed to wet or coat the carbon particles or group. This gives rise to higher viscosity or stiffer compounds. For this reason, carbon blacks used strictly as fillers are coarse and often low-structured. Thermal black, which is substantially coarser than the largest particle sized furnace grade, has virtually no structure; particles tend to be separate entities. These coarse products, of course, provide the least blackness.

Furnace blacks such as those in the range of 60–70 mμ arithmetic mean particle diameter, are also often used because of relatively low polymer demand and economy. Coarse blacks disperse the fastest, too, because there is lower interparticle attraction. Higher structure, however, surprisingly affords better dispersion. This can be attributed to the greater shear generated (often at higher viscosity) in the mixing or dispersion phase.

A number of methods are used to report indices of particle size and structure. These include blackness ratings, electron microscope analyses, a variety of oil absorption tests, and tinting with white bases. Surface area, which is inversely related to particle size, is measured by either an iodine adsorption or a nitrogen adsorption method.

Besides particle size and structure, surface chemistry can be critical in the selection of a black. Surface chemistry, i.e., the degree to which oxygen-containing groups are attached to the carbon surfaces, can be measured in several

Table 16-3. Industrial carbon blacks industry comparators.

COLUMBIAN		CABOT		DEGUSSA	
Powder	Beads	Powder	Beads	Powder	Beads
Royal Spectra				Regent Super	
Neo Spectra Mark I				Regent	
Neo Spectra Mark II	Neo Spectra Beads			FW-2	FW-2 Beads
Neo Spectra AG	Neo Spectra AG Beads			FW-200	FW-200 Beads
Super Spectra				FW-1	FW-1 Beads
Neo Spectra Mark III	Neo Spectra Mark III Beads			FW-18	FW-18 Beads
Neo Spectra Mark IV	Neo Spectra Mark IV Beads			S-170	S-170 Beads
Raven 8000	Raven 8000 Beads		Black Pearls 1300		
Raven 7000	Raven 7000 Beads				
Raven 5250	Raven 5250 Beads				
		Monarch 1100	Black Pearls 1100		
Raven 3500	Raven 3500 Beads				
		Monarch 900	Black Pearls 900		
Raven 2000	Raven 2000 Beads	Monarch 880/800	Black Pearls 880/800		
Raven 1800	Raven 1800 Beads	Monarch 700	Black Pearls 700		
Raven 1500	Raven 1500 Beads				
Raven 1250	Raven 1250 Beads				
Raven 1200	Raven 1200 Beads	Regal 660R	Regal 660		
Raven 1170	Raven 1170 Beads			Printex 400	Printex 400 Beads
Raven 1255	Raven 1255 Beads	Mogul L	Black Pearls L	Special Black 15	
Raven 1035	Raven 1035 Beads	Regal 400 R	Regal 400	Printex 400	Printex 400 Beads
Raven 1020	Raven 1020 Beads	Regal 330R/300R	Regal 330/300	Printex 300	Printex 300 Beads
Raven 1000	Raven 1000 Beads	Regal 500R			
Raven 890/Raven 890 H	Raven 890 Beads	Elftex 12, 8, 5	Elftex Pellets/Elftex Pellets 130	Printex 30 (U, V)	Printex 30 Beads
Raven 850	Raven 850 Beads	Regal 99R	Regal 99I	Printex A	Printex A Beads
Raven 500	Raven 500 Beads				
Raven 450	Raven 450 Beads	Regal SR	Regal SRF-S	Printex 200	
Raven 420	Raven 420 Beads			Printex G	Printex G Beads
Raven 410	Raven 410 Beads	Sterling R	Sterling NS, NS-1	Lampblack 101	Lampblack 101 Beads
Conductex 950	Conductex 950 Beads	Vulcan XC-72R	Vulcan XC-72	Corax L	Corax L Beads
Conductex SC	Conductex SC Beads		Vulcan SC, C		
	Raven C		Vulcan 9, 9A50, 9A32		Corax 9
Raven 1040	Raven 1040 Beads				
Raven 14					
Raven 16					
Raven 22					

*Blackness comparison based on results obtained with pure white paraffinic oil represents true blackness and is not influenced by surface activity of the black or the flow and dispersing factors that make some vehicles excellent dispersing, wetting, and blackness developers.

ways. Surface acidity, i.e., the pH of a water slurry of the pigment, is an indirect measure of volatile content. Lower pH results from large concentrations of oxygen-containing volatiles. Another means of reporting surface activity in the Diphenyl Guanidine (DPG) index. The higher the number, the more surface activity. The most direct method consists of a weight-loss determination carried out at 950 ± 20°C. Highly acidic pigments are usually associated with faster dispersion, lower viscosity in some formulations, and higher color, although these effects are more pronounced in non-plastics applications.

One property which is directly related to particle size, structure, and volatile content is electrical conductivity. There are numerous plastics applications where carbon blacks are used to bleed off static charges which might tend to accumulate on extruded or molded parts. Conductivity achieved by the use of carbon black is dependent on the formation of reticulate chains of black particles through which electrons can flow. As a result, the black should be relatively fine in particle size and high in structure. Since oxygen-containing surface groups tend to act as insulators, relatively "clean," i.e., volatile-free carbons are most effective.

Commercial furnace blacks (such as Columbian's Conductex 950 or SC and Cabot's XC-72) have been specially manufactured to meet these requirements. Optimum conductivity depends on the appropriate selection of a black, the right concentration (usually over 25%), and good dispersion. Over-dispersion can result in a dimin-

Table 16-4. Electrical resistance of dry (non-dispersed) carbon blacks.

Carbon black	Dry resistivity (ohm-cm)	Carbon black	Dry resistivity (ohm-cm)
Royal Spectra	10.8	Raven 1170 Beads	113.0
Neo Spectra Mark I	8.4	Raven 1040	20.1
Neo Spectra Mark II	6.8	Raven 1040 Beads	74.0
Neo Spectra Beads	16.4	Raven 1035	11.6
Neo Spectra Mark III	11.0	Raven 1035 Beads	51.5
Neo Spectra Mark III Beads	32.7	Raven 1020	6.4
Neo Spectra Mark IV	6.1	Raven 1020 Beads	11.3
Neo Spectra Mark IV Beads	19.4	Raven 1000	3.1
Neo Spectra AG	3.8	Raven 1000 Beads	7.1
Neo Spectra AG Beads	16.1	Raven 890	2.2
Super Spectra	4.8	Raven 850	20.8
Raven 8000	2.5	Raven 850 Beads	52.2
Raven 8000 Beads	4.1	Raven 500	4.1
Raven 7000	10.2	Raven 500 Beads	12.6
Raven 7000 Beads	4.7	Raven 450	14.9
Raven 3500	3.6	Raven 450 Beads	104.3
Raven 3500 Beads	4.3	Raven 420	13.7
Raven 2000	2.1	Raven 420 Beads	178.8
Raven 2000 Beads	8.8	Raven 410	25.0
Raven 1800	1.6	Raven 410 Beads	99.8
Raven 1800 Beads	10.2	Raven C Beads	7.6
Raven 1500	2.3	Conductex SC	2.0
Raven 1500 Beads	6.0	Conductex SC Beads	6.1
Raven 1255	46.8	Conductex 950	0.8
Raven 1255 Beads	76.9	Conductex 950 Beads	1.3
Raven 1250	3.6	Statex MT Beads	800.1
Raven 1250 Beads	6.1	Raven 22	5.3
Raven 1200	2.3	Raven 16	5.4
Raven 1200 Beads	7.2	Raven 14	18.6
Raven 1170	10.7		

The reader should refer to Table 16-3 for approximate industry equivalents.

ished conductivity by shearing and destroying structured chains through which electrons can travel. By the same token, carbon users who require high electrical resistance in their products can achieve this through selection of blacks with large particle sizes, low structure, and higher volatiles.

Dry resistivity tests can be run on carbon blacks—usually in a one square centimeter chamber. These are not nearly as useful as laboratory formulations in which the black is dispersed in a polymer and conductivity determined on the final recipe. This is true because conductivity depends on degree of dispersion, loading, and electrical properties of the resin itself, as well as on the choice of black. Table 16-4 depicts representative, typical dry resistivity data on industrial carbon blacks. Figs. 16-1 through 16-4 show representative electron micrographs of different types of carbon black.

4. MATERIAL SUPPLIERS

The major domestic suppliers of carbon black specific for non-rubber applications are as follows:

Cabot Corporation
125 High Street
Boston, Mass. 02110
(617) 423-6000

Cities Service Company
Columbian Division
3200 West Market Street
Akron, Ohio 44313
(216) 867-5800

Degussa, Inc.
Pigments Division
2 Pennsylvania Plaza
New York, N.Y. 10001
(212) 675-5930

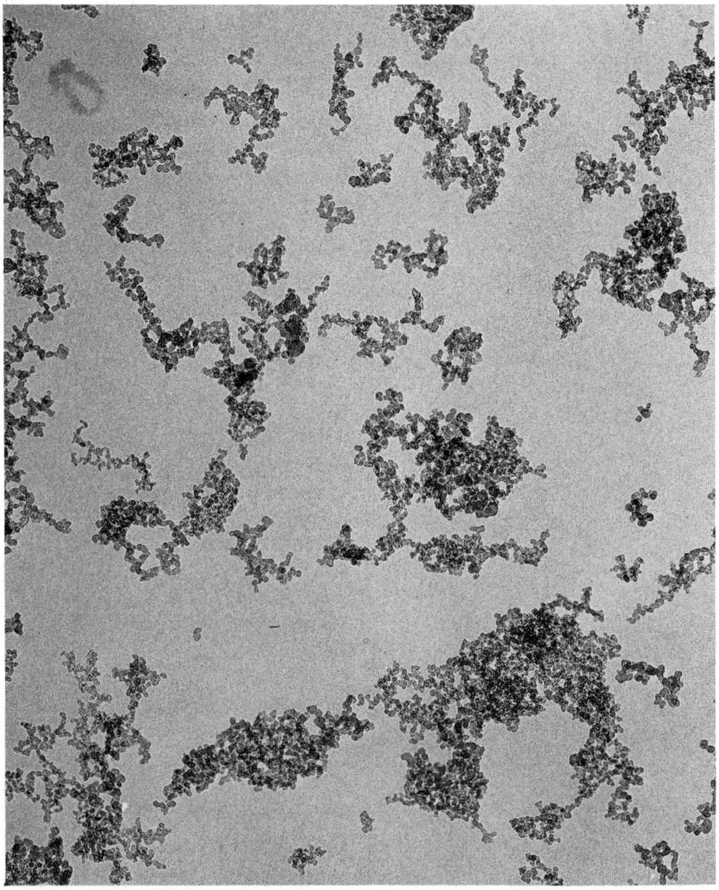

Fig. 16-1. Fine particle size channel black, magnification 75,000X.

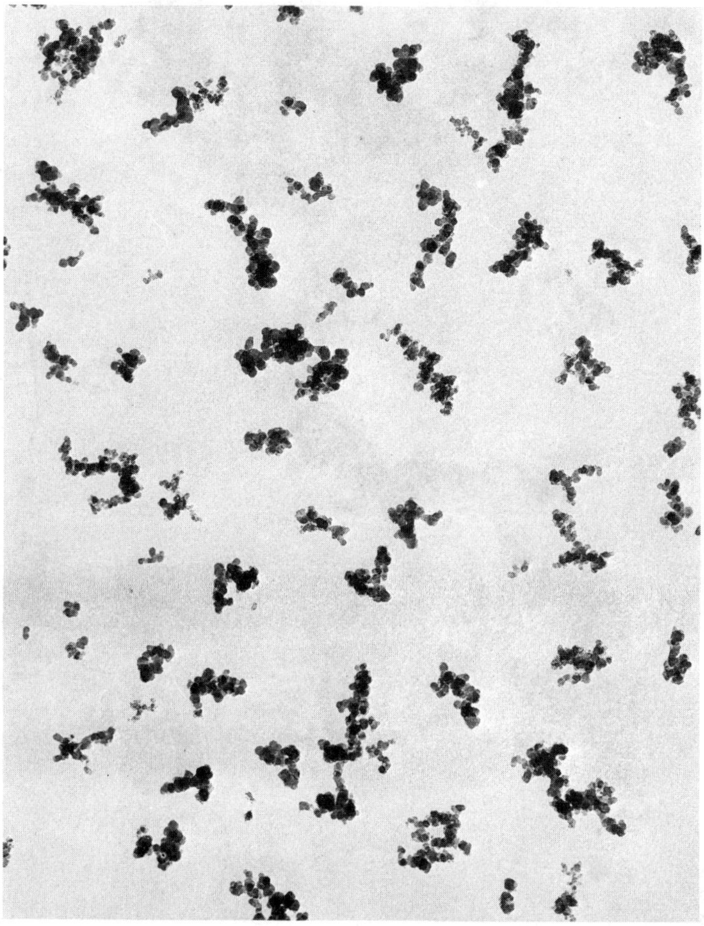

Fig. 16-2. Medium particle size furnace black, magnification 75,000×.

Other carbon black producers include the following:

Ashland Chemical Company
Carbon Black and Synthetic Rubber Division
P.O. Box 2219
Columbus, Ohio 43216
(614) 889-3333

J. M. Huber Corporation
Thornall Street
Edison, N.J. 08817
(201) 549-8600

Phillips Petroleum Company
1501 Commerce Drive
Stow, Ohio 44224
(216) 688-5562

R. T. Vanderbilt Company, Inc.
230 Park Avenue
New York, N.Y. 10017
(212) 686-6864

Witco Chemical Corporation
277 Park Avenue
New York, N.Y. 10017
(212) 826-1000

5. AVAILABILITY AND COST

Carbon black production and consumption have been steadily increasing, the most recent annual figures near 3 billion lb. This compares with a production rate of 1 billion lb at the end of World War II and 400 million lb in 1930. While during the 1940's, channel black produc-

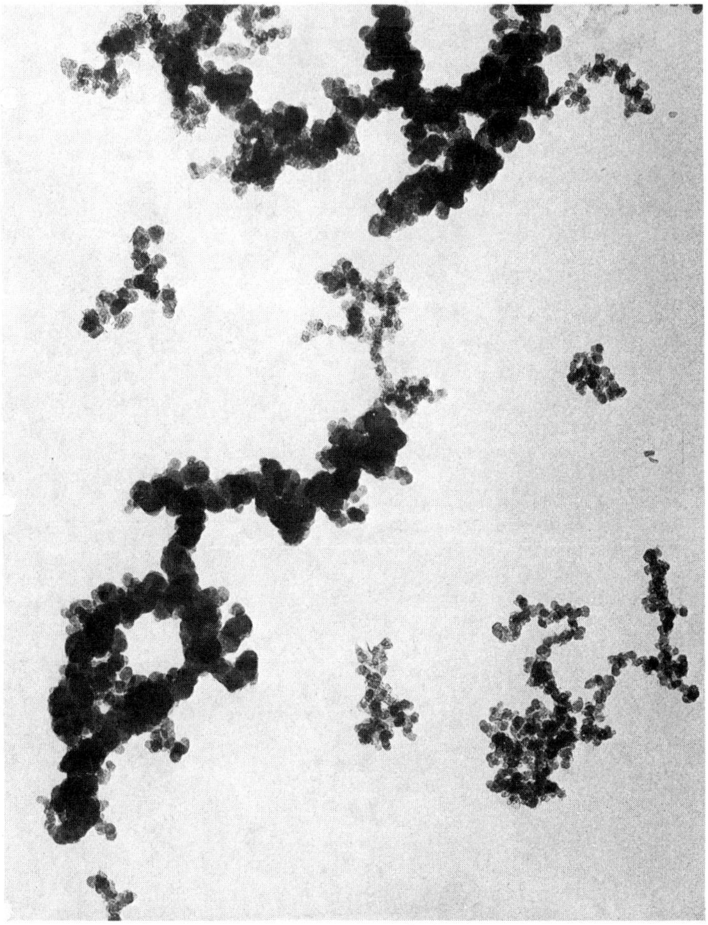

Fig. 16-3. Acetylene black, magnification 75,000X.

tion equalled furnace black production, channels represent a relatively insignificant portion of total production today. Channel blacks, which are highly gas dependent, have risen more sharply in price over the years than furnace blacks. This, coupled with environmental problems in producing channel grades, has led to channel blacks' virtual demise.

About 95 percent of all carbon black produced finds its way into rubber applications, including tires and mechanical rubber goods (hoses, belts, etc.). The remaining black is earmarked for industrial applications, the largest single area being printing ink, followed by plastics and paints.

Although carbon blacks manufactured for rubber and non-rubber applications are often similar in some chemical and physical properties, their quality control programs are inherently different. Rubber users are concerned with treadwear, reinforcing properties, modulus, and a host of other variables. Industrial users are more concerned with color, rheology, dispersion, and the factors which affect U.V., thermal, or electrical performance.

It would not be feasible to include, in this chapter, any meaningful tabulation of all available grades or prices since there are literally hundreds of grades of carbon black currently manufactured and marketed ranging in price from roughly $0.12–$4.00/lb. Unquestionably, as the costs for raw materials and energy, specifically oil feedstocks and natural gas increase, carbon black prices will rise accordingly.

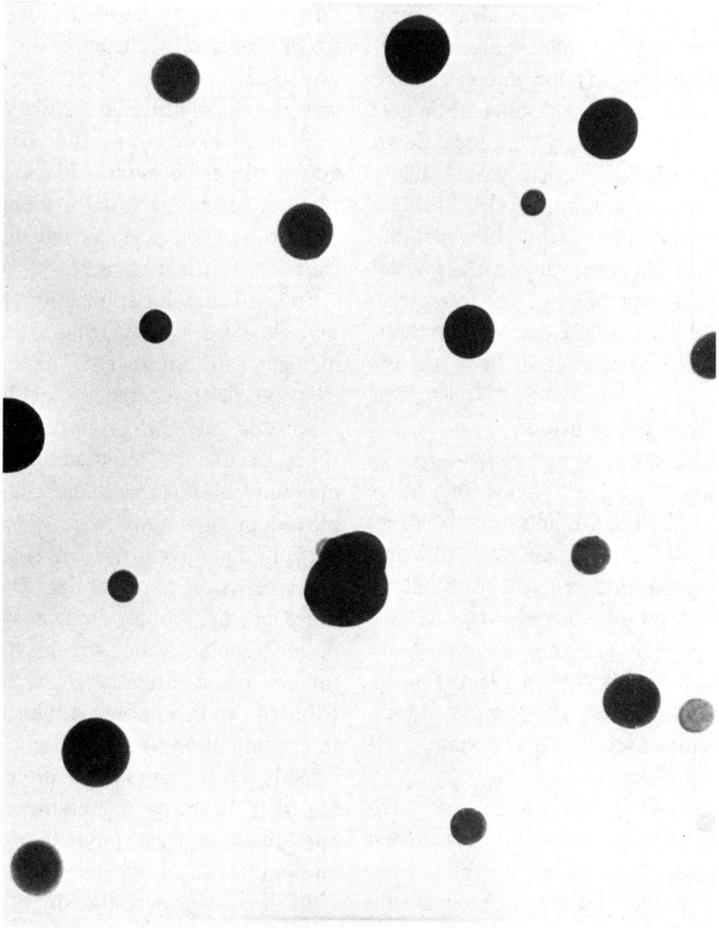

Fig. 16-4. Medium thermal black, magnification 75,000X.

6. APPLICATIONS—CURRENT AND FUTURE

Carbon blacks are most often used in plastics applications for one or more of the following reasons:

1. U.V. or thermal protection of the polymer against sunlight.
2. Control of electrical conductivity.
3. Improvement in polymer processing.
4. Cost reduction, as a filler.
5. Coloring.

The polymers are varied, but they include polyethylene and cross-linked polyethylene, styrene, ABS, polypropylene, homopolymer and copolymer vinyls, and many other thermoplastics. Carbon blacks are also being used in bulk and sheet molded polyesters, primarily for color, but also for the control of gel and exotherm times.

One of the largest single applications of carbon black is in wire and cable insulations. Products like Columbian's Raven C or Cabot's Vulcan 9A32 (in rubber terminology, super abrasion furnace of SAF grades) were designed to meet the specifications of Western Electric and others. The combination of relative fineness, high structure, and moderate volatility provides the needed loading capabilities, U.V. protection of the polymer, controlled conductivity, and ease of processing.

Carbon black prevents ultraviolet degradation of many polymers by two mechanisms. First, the carbon, because of its blackness, absorbs the damaging radiation. Second, the carbon acts as a free-radical trap, "catching" the breakdown products of degradation which would likely catalyze additional degradation. U.V. protection is best achieved when carbon blacks in the 16–20 mμ particle diameter range are dispersed well at concentrations between 2–3% by weight. Polyethylenes which would otherwise discolor, crack, or lose dielectric strength in 6 months weathering can be protected for 20 years or more with carbon black addition.

Studies have also shown that thermal degradation in low density polyethylenes and other polymers can be significantly inhibited by using the same kinds of black. This gives rise to many non-sunlight applications where, nevertheless, elevated temperatures may be encountered.

ABS polymers (for use in rigid pipes) are often compounded with carbon black for the same reasons. Similar U.V. and thermal stabilization effects are achievable in polyvinyl chloride and polypropylene.

Thermal blacks which, of course, are much coarser and lower in structure, are also often compounded into wire and cable applications. They are lower in cost and can be used at substantially higher concentrations in the polymer without material loss of physical properties. Thermal black imparts processing advantages, particularly in extrusions, is inert and easy to disperse because of its coarseness, and does not promote conductivity. Generally, 20 to 40 parts of medium thermal in 100 parts of polyethylene is used. The mechanical properties of a polymer so pigmented are essentially the same as those found in the unpigmented plastic. Improvement in U.V. protection and thermal stability, however, is attributed to the black.

Polyethylene which has been cross-linked by radiation or peroxide (such as dicumyl peroxide) can often be loaded with excessive amounts of medium thermal black, upwards of 75%, without dramatically destroying the polymer's physical properties. The cross-linking, in effect, adds strength and durability to the polymer.

In most plastic applications, it is important to appropriately disperse the black. Poorly-dispersed products, including thermal black, give rise to diminished tensile strength, impact resistance, elongation, and dielectric strength. Poorly-dispersed carbon black in polyethylene tape extrusions is often evidenced by pimpled or porous tape surfaces which would provide sites for electrical breakdown.

For critical color applications in plastics, finer particle sized blacks, often 15–20 mμ particle diameter are employed. These blacks also impart excellent resistance to U.V. Deterioration of certain physical properties, however, sometimes occurs, as depicted in Table 16-5. The medium color furnace and channel blacks employed in this study were of essentially equal particle diameter (16–17 mμ) and their impact characteristics were similar. One should note, however, the poorer impact strength in compounds which were not master-batched but, rather, loaded directly at 2%. This can be attributed to the poorer dispersion achieved by this formulating route.

Table 16-6 shows that when carbon blacks are well dispersed in polymers like polyethylene, the *long-term* physical properties of the compound, after weathering, can be substantially improved over the virgin polymer. In this study, artificial weathering was accomplished by means of an Atlas Twin Arc Weatherometer.

Table 16-5. Effect of carbon black on ABS IZOD impact strengths.

Carbon black	Incorporation technique	Izod impact strength (in.-lb)
Virgin polymer–No carbon black	–	5.02
Medium color channel	2% Direct loading	1.91
Medium color channel	20% Concentrate–2% Letdown	2.73
Medium color furnace	2% Direct loading	1.96
Medium color furnace	20% Concentrate–2% Letdown	2.78

Note: Cylolac TD (Marbon) was the polymer employed with compounding by Banbury mixing.

Table 16-6. Weathering studies—carbon black in low density polyethylene.

Formulation	Hours Atlas Weatherometer	Tensile @ break	Elongation @ break
Virgin resin	0	1880	132
(no carbon black)	1500	1300	41
Med. color channel)	0	1730	89
(2% direct loading)	1100	1760	87
	3000	1857	102
Med. color channel	0	1920	85
(20% conc.-2% letdown)	1100	1980	83
	3000	2019	101
Med. color furnace	0	1760	97
(2% direct loading)	1100	1680	94
	3000	1817	104
Med. color furnace	0	1920	94
(20% conc.-2% letdown)	1100	1980	81
	3000	2056	97

Note: NA310-06 (USI) was the polymer employed with compounding by Banbury mixing.

Carbon blacks are usually incorporated into plastics resins by Banbury or continuous mixer techniques. It is possible to attain relatively good dispersion, particularly with coarse blacks, in extrusion-type equipment, provided the residence time and shear are sufficiently high. Batch-processing on heated two-roll mills is also often used. One cannot overemphasize the importance of attaining good pigmentary dispersion.

There has been tremendous growth in recent years in thermosetting systems and particularly bulk and sheet molded polyesters. The resin is usually catalyzed by a peroxide, such as benzoyl peroxide. Additional ingredients often include glass reinforcers, promoters or accelerators, and inhibitors to prevent the premature polymerization of the polyester.

Carbon black has been employed in these bulk or sheet molded compounds to provide color, to reduce cost, and to control the reaction. The slowing down of gel and exotherm times has been attributed to surface chemistry of the black (the presence of oxygen and, to some degree, sulfur) and also to the physical adsorption of the peroxide catalyst onto the black's surface. By the same token, carbon blacks may tend to absorb inhibitors, antioxidants, or other trace ingredients in many recipes. The formulator should always consider this potential, and make necessary concentration adjustments, particularly when fine particle size blacks are used. At this time, it appears that coarse blacks in the range of 60–70 mμ particle diameter have gained the grestest inroads in molded polyester formulations.

There are many other unique applications for carbon black, such as phonograph record compounding. Here, carbons not only add color, but provide better wear properties and bleed off static. Phonograph records generally contain low levels of black and require excellence of dispersion. The pigments are high-color furnace or channel types, since these tend to induce the least background noise.

Carbon black consumption is also increasing in molded industrial parts such as automotive distributor caps made from phenolic compounds. In these applications, electrical properties rather than color, are most critical.

Carbon black may be used to intentionally impart high electrical conductivity, on the one hand, or extreme resistivity on the other. Conductivity is achieved when electrons can flow through the compound containing the dispersed black. This occurs when blacks of relatively fine particle size, high structure, and low surface volatility are dispersed well at adequate concentrations. The high structured black provides reticulate chains of intimate particle contact, thus establishing paths for electron flow. As a result, it is possible to over-shear such a dispersion, destroying these chains, resulting in lower conductivity. Small particle size provides greater numbers of particle contact points. Surface volatiles tend to act as an insulator and generally inhibit conductivity. Traditionally, highly resistive compounds were formulated with channel blacks which inherently contained large amounts of surface oxygen. With the gradual demise of channel blacks, furnace carbons of low structure and high volatile content are being used in these applications.

Table 16-7 illustrates applications of specific conductive carbon blacks in ethylene vinyl acetate and PVC applications. An increase in carbon loading from 20 to 30% dramatically increases conductivity. The reader should also note the detrimental effects of over-dispersing.

Because conductivity depends on the described reticulation effect, little conductivity can be expected at low loadings, usually below

Table 16-7. Comparative volume resistivity (ASTM D-991) (OHM-CM).

	Time in mill (min.)	Dispersion*	RESISTANCE (OHM-CM) Room temp.	Heated	Cooled to room temp.
Ethylene vinyl acetate copolymer					
30% Carbon black				@120°C	
Conductex 950	2	5.9	58	252	243
	5	3.7	36	128	196
	10	3.9	176	416	523
Conductex SC	2	4.7	128	906	699
	5	4.7	61	1088	832
	10	4.2	99	391	917
Polyvinyl chloride compound					
20% Carbon black				@90°C	
Conductex 950	2	–	433	527	320
	5	–	380	407	380
	10	–	6182	12806	3533
Conductex SC	2	–	1716	1288	2575
	5	–	4082	4173	4536
	10	–	2091	1523	3544
Polyvinyl chloride compound					
30% Carbon black				@90°C	
Conductex 950	2	–	49	127	11
	5	–	20	27	15
	10	–	345	399	388
Conductex SC	2	–	14	38	41
	5	–	24	40	32
	10	–	43	55	47

Ethylene vinyl acetate 30% Carbon black		Polyvinyl chloride 20% Carbon black		Polyvinyl chloride 30% Carbon black	
Polymer	210	Polymer	190	Polymer	164
Carbon black	90	Carbon black	63	Carbon black	93.5
Antioxidant	1.0	DOP	40.5	DOP	35.0
	301.0 parts	Soybean oil Epoxide	13.5	Soybean oil Epoxide	12.0
		Lead stearate stabilizer	1.0	Lead stearate stabilizer	1.0
			308.0 parts		305.5 parts

*Light microscope examination; 2% letdown; relative rating 1 to 10, Best to Worst.

10 or 15%, but the degree of dispersion and the polymer chosen will also affect conductivity.

The non-rubber use of carbon black is expected to increase significantly in the years ahead as the Age of Plastics continues to expand.

References

1. Boonstra, B. B., "Reinforcement of Polyethylene," *Reinforcement of Elastomers*, p. 529, Wiley-Interscience, New York, 1965.
2. Hawkins, W. L. and Winslow, "Antioxidant Properties of Carbon Black," *Reinforcement of Elastomers*, p. 563, Wiley-Interscience, New York, 1965.
3. Hawkins, W. L., "Protecting Polyolefins Against Photo- and Thermal Oxidation," *Rubber Plastics Weekly* (Feb. 24, 1962).
4. Garret, M. D., "Carbon Black, Bone Black, Lampblack," *Pigment Handbook*, Vol. I, pp. 709–743, Wiley & Sons, New York, 1973.
5. Schubert, B., Ford, and Lyon, "Analysis of Carbon Black," *Encyclopedia of Industrial Chemical Analysis*, Vol. 8, pp. 179–243, Wiley-Interscience, New York, 1969.

GLOSSARY

Acetylene Black—A carbon black manufactured by the thermal decomposition of acetylene, often used in battery applications.

Channel Black—A carbon black manufactured by impinging natural gas flames on channel irons, scraping, and collecting the product.

Impingement Black—See Channel Black.

Furnace Black—A carbon black manufactured by the thermal decomposition of oil feedstocks in oil and/or gas heated furnaces.

Lampblack—A carbon black manufactured by the burning of oils or petroleum products in restricted air supplies.

Structure—The property of carbon black which relates to the formation of chain or grape-like reticulate chains.

Volatile—The chemisorbed oxygen-containing surface groups found on carbon blacks. See ASTM D-1620-60 (1968).

17
Organic Fillers

Benjamin M. Walker
Walker Engineering Associates
Madison, Connecticut

CONTENTS

1. INTRODUCTION

This chapter describes organic materials which are presently of commercial importance as fillers and reinforcements for plastics. It also lists many of the other organic materials which have been used throughout the history of the plastics industry.

Initially, these additives were of natural origin, and were abundantly available at low cost. Most of them, like wood flour, cotton, and vegetable fibers, are fibrous in structure, and are composed largely of cellulose, with lesser amounts of lignin and other compounds. However, some of them are not fibrous in nature, and some are of animal origin. At present, synthetic organic materials, chiefly fibers, are also being used as plastics additives, principally for reinforcement of physical properties.

Like fillers and reinforcements generally, the organic material additives affect many of the properties of the compounds into which they are incorporated. The natural materials are generally less expensive than the resins with which they are compounded, and this is a most important reason for their use. However, both natural and synthetic organic additives have important functional effects, including improvement in processability and desirable modification of properties of the final compound. This chapter lists many of the organic materials that have been used as plastics additives. It also describes those which are of commercial importance at the present time, indicating the reasons for their use and the sources of supply.

2. DESCRIPTION OF ORGANIC FILLERS

2.1 General Discussion

Organic materials have been useful as additives for plastics throughout the history of the industry. When the earliest phenolic resins made by Dr. Leo H. Baekeland were first modified for molding in 1907, they were combined with wood flour as one of the additives. A great variety of organic materials have been used throughout the succeeding years, and several are still being used in substantial quantities for this purpose.

The reasons for using such organic materials, as with other types of additives, are varied.

These include reduction of cost of the molding compound, improved processability, and improved physical, chemical, and electrical properties of the final compound. As the field of plastics developed and expanded a variety of naturally occurring materials, readily available in quantity and at a low cost, were explored. Section 2.2 includes some of the materials that were tried and reported, and it appears that almost every conceivable natural material was considered at one time or another. The categories include wood flour, shell fibers, cotton fibers, and other vegetable fibers consisting mostly of cellulose, with lignin present in some of them. Also included are other starch and protein-containing natural materials of vegetable or animal origin. Not all of these materials are successful, but a few of them are effective in yielding desired properties, and are used extensively.

More recently, various synthetic organic materials have been used as plastics additives. Most of these are in the form of fibers, and are used for the reinforcement of the physical properties that they provide. These are also included in Section 2.2.

2.2 Organic Materials that Have Been Used as Fillers for Plastics. (*Note:* Not all of these have been successful)

NATURAL MATERIALS

- Wood Flour
  - Soft woods
    - pine, Ponderosa
    - pine, white
    - Norway fir
    - Douglas fir
    - spruce
    - redwood
  - Hard Woods
    - ash
    - beech
    - birch
    - maple
    - oak
  - Bark, fir
  - Sawdust, shavings, chips
  - Sawdust, rubber coated
- Shell Fibers
  - almond shell
  - apricot pit shell
  - Brazil nut shell
  - candle nut shell
  - coconut shell
  - filbert shell
  - peanut shell
  - pecan shell
  - walnut shell
- Cotton Fibers
  - cotton flock
  - cotton fabric, chopped
  - cotton cord, chopped
  - cotton linters
  - cotton seed hulls
  - cotton seed fiber
  - cellulose fibers
- Vegetable Fibers
  - hemp
  - coconut (coir fiber)
  - jute (Ref. 3)
  - ramie
  - sisal
- Miscellaneous Materials
  - beer hops, spent
  - coal, Anthracite fines, ground
  - coca bean material
  - coke
  - cork, granulated
  - corn cobs
  - corn stovers
  - citrus pulp, dried
  - flax shives
  - furfural process residues
  - hemp hurds
  - keratin
    - chicken feathers
    - cattle hoofs
    - hog bristles
  - leather
  - milkweed products
  - peat moss
  - soybean meal
  - starch, corn, potato, rice, tapioca, and wheat
  - wheat straw
  - wool fibers

SYNTHETIC MATERIALS

- Regenerated cellulose
  - rayon fabric
  - Cellophane, finely cut

Polyacrylonitrile fibers (Orlon*, Dynel**)
Polyamides, aliphatic (nylon) fibers
Polyamide, aromatic (aramid) fibers
Polyester fibers (Dacron*)
Polytetrafluoroethylene fibers (DuPont Teflon)*
Polyvinyl alcohol finers (Ref. 6)
Rubber dust, hard
Rubber dust, vulcanized

*DuPont.
**Union Carbide.

2.3 Wood Flour

Wood flour is a finely ground, dried wood product fibrous in structure, that acts as a reinforcing material for plastics. It is made mostly from softwoods, chiefly pine and spruce, but hardwood maple or ash flours are used where no wood resin content is desired. Woods containing essential oils, such as cedar, are not suitable. Wood flour is made from sawdust, chips, and shavings by grinding in a buhrstone mill, and has the appearance of wheat flour. Particle sizes commonly used for plastics fillers are 40, 60, and 80 mesh, with sizes as fine as 140 mesh available. Wood flour may absorb some resin or gums when mixed with some plastics, and mineral fillers may also be combined to vary the properties of the molded product.

2.4 Cotton

Cotton is a plant fiber which is useful as a reinforcing filler for plastics, and is especially used for thermosetting molding compounds. Cotton is a tropical plant, which grows in many parts of the world. However, it grows well in the southeastern United States as far north as Virginia. There are many commercial varieties, but the finest and longest fibers include Sea Island cotton, grown in the West Indies, and Egyptian or Peruvian cottons, grown in Egypt. The short-staple upland cotton is the most widely grown of cottons in the United States, but is spun only into coarse and medium yarns.

In plastics, the fibers of cotton are prepared and used in several forms, including flock, cordage, woven fabrics, nonwoven fabrics or felts, and chopped fabric. These fibers may improve moldability, reduce the cost of parts significantly, increase impact strength and other mechanical properties, and improve the appearance of the finished molded parts. For example, an unfilled phenolic resin is very difficult to mold. The resin alone is fluid and must be B-staged to a high viscosity in order to obtain a suitable molded part. Addition of an organic fibrous filler, such as cotton, tends to control this viscosity and enhances moldability.

2.5 Cellulose Fibers

These are derived from plant fibers by mechanical and chemical treatment to break up fiber bundles and remove some of the lignin or other compounds which may be present. All plant fibers are composed principally of cellulose, but the other materials present, and the structure and properties of the purified fibers, will depend upon the vegetable source used. Commonly used fibers are produced from wood, cotton, or other plant sources.

The term "cellulose fiber" is generally used to refer to vegetable fibers that have been treated to yield a residual product consisting of as high as 99% or more of pure cellulose. Cotton fibers have a high cellulose content initially, and were formerly the principal source of purified cellulose fibers. At present, wood flour is the principal source of cellulose fibers. Cellulose fibers of high purity are readily made from cotton fibers by boiling with sodium hydroxide solution, neutralization with acetic acid, and prolonged washing with distilled water. The product is regarded as a standard cellulose product with about 99.8% cellulose and 0.05% of ash.

Wood fibers normally contain 40–60% cellulose and 17–39% lignin. After purification by a chemical process, such as a sulfite or soda process, high purities of cellulose may be obtained. For example, Brown Company's "Solka Floc" may, when dry, contain over 99.5% cellulose. This material is white in color, has a brightness value to 92, a particle size of 40–165 μ, very low ash content, and a specific gravity of 1.58. Its apparent density is 9–34 lb/cu ft.

2.6 Sisal

Sisal, a plant having hard, strong fibers, is obtained mostly from Haiti, East Africa, and

Indonesia. Sisal fibers are not as strong as hemp, and are not as resistant to moisture. However, they are used largely for rope, cordage, and sacking, and they are also useful in laminated plastics.

2.7 Jute

Jute is another fiber used for making burlap, sacks, cordage, and upholstery fabrics. It has also been considered as a filler for plastics. Most of the commercial jute fiber comes from India, growing in tall, slender stalks like hemp. It is not as strong as hemp, but is widely used because of its low cost.

A recent study by Winfield and Winfield[3] reports that with proper compounding, treatment, and design, jute-reinforced polyester compositions show promise for use in constructing low-cost housing for developing countries. They reported that jute had been little used in plastics, "except on a decorative basis," and "provided little strength or stiffness in tensile and flexural tests." However, they were able to develop acceptable housing structures utilizing these low cost and readily available fibers.

2.8 Hemp

Hemp fiber is obtained from the stalk of the plant *Cannabis sativa*, which is grown principally in Southern Russia, Central Europe, Mediterranean countries, Asia, South America and the United States. The stalk is treated by a process called "retting," and yields fibers up to 75 in. long. They are coarse, strong, and more durable than cotton. These fibers are composed of as much as 78% alpha cellulose. Hemp fibers are used chiefly for cordage, sacking, and packings. They are also used as reinforcing fillers for plastics.

2.9 Starch

Starches from corn, potatoes, rice, tapioca, and wheat have recently been proposed as a filler for rendering plastics biodegradable (see G. Griffin[14]). The starches are readily obtained from these plant sources as a fine powder consisting of spherical or ellipsoidal grains ranging in particle size from 3–100 millimicrons. In general, they are insoluble in cold water, alcohol, and ether, and form a jelly with hot water. Specific gravity is 1.499–1.513. It does not melt, but decomposes or burns when heated.

A property that makes starches of interest as fillers for formulating biodegradable compounds is that they are subject to biological and oxidative attack. The action of enzymes and oxygen brings about their decomposition when buried in soil. They are inert to water at room temperature but are rapidly digested by the amylases; the porosity thus generated offers the most favorable circumstances for degradative attack.

Combining starches with polymers requires techniques adapted to avoiding degradation during processing. However, they resist dry heat in the normal plastics processing operations, and have been successfully compounded with low density polyethylene, polypropylene, and polystyrene. The processing of these compounds and the properties of the resulting products have been quite satisfactory.

2.10 Prices of Organic Fillers

The prices of the many organic fillers available cover a broad range, depending upon type, source, grade, and specifications. As many of them are of natural origin, prices will also vary widely with the availability of the material at a particular time and place, and upon the amount of processing necessary to make it suitable as a plastic filler. Thus, an attempt will not be made to list exact prices for these materials. Users should refer to the material suppliers for current information.

However, the possible range of prices that may be expected for some of the more common fillers is as follows:

	Price/lb
Wood flour	$.03–.06
Shell flour	.04–.06
Cotton flock, darker, general purpose	.17–.20
Cellulose fibers	.22–.35
Cellulose fibers, short, treated	.50–.62
Sisal	.24–.50

3. SPECIFICATIONS FOR COMMON ORGANIC FILLERS[1]

It is rather difficult to give a standard list of chemical and physical properties, as each producer of molding compound issues his own materials specifications and the test method to be used. However, the more important specifications can be generalized as follows:

	Wood flour	Cotton flock, Chopped cotton fabric	Cellulose flock	Sisal flock
Ash content	0.3%	0.5–1.5%	0.15%	0.3% max.
Acetone sol.	8.5% max.	2.0–3.0%	–	–
Oil absorption cc/10 grams	15–30	20–30	–	–
Ether extract	5.0% max	2.0% max	0.24%	0.5–1.5%
% Insolubles	0.0%	1.5–2.5%	0.0%	0.0%
Color	Softwood–Light cream Hardwood–Light tan	Bleached–white Colored	Bleached–white Unbleached–tan	Light tan
Appearance	Fibrous	Fibrous/macerated	Fibrous	Fibrous
Moisture	5.0–8.0%	5.0–8.0%	5.0–8.0%	12.0% max.
Bulk density ml/50 grams	Softwood–190–260 Hardwood–160–200	180–300	135–400	140–330
Weight/cu ft	Softwood–6–10 lb Hardwood–11–16 lb	7–11 lb	5–18 lb	4–16 lb
Particle size	5–140 mesh	10–200 mesh	10–200 mesh	Random cut–10–100 mesh Precision cut–1/32–1 in.

4. SUPPLIERS OF ORGANIC FILLERS

Anthracite
- M. E. Wallace Co.
- Shamokin Filler Co. (Carb-O-Fil)

Cellulose Fibers
- Brown Co. (Solka-Floc)
- Composition Materials Co.
- ITT Rayonier, Inc.
- Monsanto Co. (Santoweb)

Coal Fines
- Keystone Filler & Mfg. Co.
- Shamokin Filler Co.
- Slab Fork Coal Co., Chemical Products

Coke
- Asbury Graphite Mills, Inc.

Cork, Granulated
- Delgrosso Brothers, Inc. (Delcork)
- Maryland Cork Co.
- Southern Cork Co.

Cotton, Chopped
- Claremont Flock Corp.
- Danberg Chemical Co.
- Mundane Corp., Div. International Filler Corp.
- Plymouth Fibers, Inc.

Flock, Cellulose
- Brown Co.
- Danberg Co.
- Georgia-Pacific Corp.
- International Filler Corp.
- Stoners' Ink Co.

Flock, Cotton
- Basic Plastics Materials, Inc.
- Cellusuede Products, Inc.
- Claremont Flock Corp.
- Composition Materials Co.
- Danberg Chemical Co.
- Harwick Standard Chemical Co.
- International Filler Corp.
- Microfibers, Inc.
- Plymouth Fibers, Inc.

Ray-Cot Fibre Co.
Wood Flour, Inc.
Flock, Sisal
Basic Plastics Materials, Inc.
Danberg Chemical Co.
Flock, Others
Basic Plastics Materials, Inc. (Dacron, Rayon, nylon, Orlon, paper, wool)
Cellusuede Flock Corp. (Dacron, nylon, Dynel, Orlon, rayon)
Claremont Flock Corp. (acrylic, nylon, polyester, rayon, wool)
Danberg Chemical Co. (Dacron, nylon, Orlon, wool)
Harwick Standard Chemical Co. (rayon, silk)
Ray-Cot Fibre Co. (rayon)
Rubber Dust, Hard
Atlas Rubber, Inc.
Composition Materials Co.
Desser Tire and Rubber Co.
Sidney Freedman, Inc.
Laurie Rubber Reclaiming Co.
Lotz & Co.
H. Muehlstein & Co.
Herbert J. Muehlstein Co.
Richardson Co.
A. Schulman, Inc.
Leather Dust and Fibers
Composition Materials Co.
Rubber Dust, Vulcanized
H. Muehlstein
Starch, Corn and Potato
Stein Hall & Co.
Walnut Shell Flour
Agrashell, Inc.
Composition Materials Co.
Harwick Standard Chemical Co.
Wood Flour
Abco, Inc. (Wonderwood)
Acme Sawdust Co.
American Colloid Co.
Basic Plastics Materials Co., Div. Wood Flour, Inc.
Cellulo Co.
Composition Materials Co., Inc.
Connor Forest Industries
International Filler Corp.
Kraft Chemical Co.
Lignum Chemical Works
Love Wood Products Co.
Menasha Corp.
Frank Miller & Sons
Wilner Wood Flour Co.
Wood Flour, Inc.
Miscellaneous Fillers
Carbon Flours and Ground Anthracite Coal
Asbury Graphite Mills, Inc.
Pecan Shell Flour
Harwick Standard Chemical Co.

5. NAMES AND ADDRESSES OF ORGANIC FILLER SUPPLIERS

Abco, Inc., P.O. Box 99, Abbotsford, Wis. 54405
Acme Sawdust Co., Box 9282, San Antonio, Tex. 78204
Agrashell, Inc., 4560 East 26th St., Los Angeles, Calif. 90022
American Colloid Co., 5100 Suffield Court, Skokie, Ill. 60076
Asbury Graphite Mills, Inc., Warren County, N.J. 08802
Atlas Rubber, Inc., 1533 Fishburn Ave., Los Angeles, Calif. 90063
Basic Plastics Materials Co., Div. Wood Flour Inc., Howard St., Winchester, N.H. 03470
Brown Co., 251 South Lake Avenue, Pasadena, Calif. 91101
Cellulo Co., 27 North Ave., Hoboken, N.J. 07016
Cellusuede Products, Inc., 500 North Madison St., Rockford, Ill. 61101
Claremont Flock Corp., 1935 Elm St., Claremont, N.H. 03743
Composition Materials Co., Inc., 25 W. 43rd St., New York, N.Y. 10036
Connor Forest Industries, Laono, Wis. 54541
Danberg Chemical Co., 100 Pent Highway, Wallingford, Conn. 06493
Delgrosso Brothers Inc., 127 Main St., Franklin, N.J. 07416
Desser Tire and Rubber Co., 6208 South Alameda St., Los Angeles, Calif. 90001
Sidney Freedman, Inc., 410 W. 24th St., New York, N.Y. 10011
Harwick Standard Chemical Co., 60 S. Sieberling St., Akron, Ohio 44305
Georgia-Pacific Corp., P.O. Box 1236, Bellingham, Wash. 98225
International Filler Corp., 50 Bridge St., North Tonawanda, N.Y. 14120
ITT Rayonier Inc., 161 E. 42nd St., New York, N.Y. 10017
Kraft Chemical Co., 917 W. 18th St., Chicago, Ill. 60608
Keystone Filler & Mfg. Co., 214 Railroad St., Muncy, Pa. 17756
Laurie Rubber Reclaiming Co., Canal St., East Millstone, N.J. 08873

Lignum Chemical Works, foot of Jersey Ave., Jersey City, N.J., 07302

Lotz & Co., 80 Wall St., New York, N.Y. 10005

Love Wood Products Co., P.O. Drawer "O", Diholl, Tex. 75941

Maryland Cork Co., Suite 200, 5 Charles Plaza, Baltimore, Md. 21201

M. E. Wallace & Co., P.O. Box 581, Sunbury, Pa. 17801

Menasha Corp., 1680 Willeameete, Eugene, Oreg. 97405

Microfibres, Inc., 1 Moshassuck St., Pawtucket, R.I., 02860

Frank Miller & Sons, 13833 South Emerald, Chicago, Ill. 60627

Monsanto Co., Rubber Chemicals Div., 260 Springside Drive, Akron, Ohio 44313

H. Muehlstein & Co., 521 Fifth Ave., New York, N.Y. 10017

Herbert J. Muehlstein Co., 866 United Nations Plaza, New York, N.Y. 10017

Mundane Corp., Div. of International Filler Corp., 50 Bridge St., North Tonawanda, N.Y. 14120

Plymouth Fibers, Inc., Traffic and Palmetto Sts., Brooklyn, N.Y. 11227

Processed Fibers, Inc., Mill & Center St., Lockport, N.Y. 14094

Ray-Cot Fibre Co., 405 Water St., Warren, R.I. 10017

Richardson Co., 27th Ave. and Lake St., Melrose Park, Ill. 60160

A. Schulman, Inc., 790 East Talmadge Ave., Akron, Ohio 44309

Shamokin Filler Co., P.O. Box 272, Shamokin, Pa. 17872

Slab Fork Coal Co., Chemical Products Div., Slab Fork, W. Va. 25920

Southland Cork Co., P.O. Box 868, Norfolk, Va. 23501

Stein, Hall & Co., 605 Third Ave., New York, N.Y. 10016

Stoner's Ink Co., Quarryville, Pa. 17566

Wilner Wood Products Co., 300 Alpine St., Norway, Maine 04268

Wood Flour, Inc., 1 Howard St., Winchester, N.H. 03470

References

1. Killian, H. A., Private Communication, International Filler Corp., Sept. 1975.
2. Seymour, R. B., "The Role of Fillers and Reinforcements in Plastics Chemistry," *Advances in Chem. Series*, #134, edited by Deanin and Schott, American Chemical Society, p. 3, 1974.
3. Winfield, A. and Winfield, B., "Reinforced Plastics in Low Cost Housing," *Advances in Chem. Series*, #134, edited by Deanin and Schott, American Chemical Society, pp. 207–218, 1974.
4. Seymour, R. B., "Fillers for Plastics," *Modern Plastics Encyclopedia*, p. 222, 1974–75.
5. Seymour, R. B., "Fibrous Reinforcements," *Modern Plastics Encyclopedia*, p. 187, 1975–76.
6. Kawai, K., "Polyvinyl Alcohol Fibers," *Modern Plastics Encyclopedia*, pp. 376, 378, 1972–73.
7. "Fillers for Molding Compounds," *Modern Plastics Encyclopedia*, pp. 311–314, 1949.
8. Boustamy, K. and Hamed, P., "Short Cellulosic Fibers for Rubber," *Rubber World*, pp. 39–42, Nov. 1974.
9. Boustamy, K. and Arnold, R. L., "Short Fiber Composites: The Comparative Properties of Treated and Discontinuous Cellulose Fibers," paper presented at the American Chemical Society Rubber Division, Cleveland, May 8, 1975.
10. *Chemical Week 1976 Buyers Guide*, McGraw-Hill, New York.
11. *Rubber Red Book*, Palmerton Publishing Co.
12. *Materials Handbook*, Tenth Ed., McGraw-Hill, New York.
13. *The Vanderbilt Rubber Handbook*, R. T. Vanderbilt Co., New York, 1968.
14. Griffin, G. J. L., "Biodegradable Fillers in Thermoplastics," *Advances in Chem. Series*, #134, Edited by Deanin and Schott, American Chemical Society, pp. 159–170, 1974.

Section VII
Spherical Fillers

18
Solid Spherical Fillers

Bob Ryan and Barbara Tardy
Cataphote Division, Ferro Corp.
Jackson, Mississippi

CONTENTS

1. INTRODUCTION

The uniqueness of solid glass microspheres as a filler is primarily due to their spheroidal shape. As manufactured, the particles are perfect spheres with a smooth surface, resulting in many characteristics of a theoretically ideal filler. Among these characteristics are the minimum effect on viscosity and flow of the matrix, and the absence of the undesirable, high stress concentrations associated with irregularly shaped and sharp edged particulate fillers that result in severely reduced mechanical properties.

These particles are supplied in a large size range (5-700 microns) and with a wide variety of surface treatments. In some resin systems, the glass spheres provide reinforcement, but generally they are used as a cost reducing filler or for some special effect. The cost of silane-treated, closely screened glass spheres is about \$.15/lb in 1976. On a volume basis, this would be equivalent to a \$.37/lb resin.

There are other solid spherical fillers available besides glass spheres. These include organic polymeric spheres, zinc dust, metallic spheres, carbon spheres, spheroidal sand, and spheres from fly ash, all of which produce specific advantages.

The principal advantages of spheres in plastics are:

1. *Low surface area to volume ratio* which means low resin absorption. The oil absorption of $-44\ \mu$ cut of beads is 8 lb of oil/100 lb of bead as compared to $CaCO_3$ at 12-18. This allows for high loading up to 400% by weight with little effect on viscosity at 25% loadings.
2. *Regularity of shape* gives ease of wetting and uniform stress distribution with predictable physical properties.

3. *Non-scoring properties on metals* means less wear on fabrication and mixing equipment as well as molds. This is a consequence of the regular sphericity of the bead combined with their intermediate hardness and high crush strength. Injection mold gates show about the same wear with filled as with unfilled resin.
4. *Control of surface properties* by the use of coupling agents and other coatings allows the microspheres to become an integral part of the polymer system. This leads to higher bond strengths, and improved resistance to moisture and weathering.
5. *High crush strength* contributes high compressive strength to parts and resistance to breakage in molding or mixing equipment.
6. *High thermal stability and inertness* translates into flame retardance and insensitivity to the highest molding temperatures or pressures. Microspheres will not react and degrade the base resin.
7. *Closely controlled particle size* assures uniformity and is essential for flow through mold gates.
8. *Transparency* is a unique property which permits parts to be translucent. In addition, the spheres absorb U.V. radiation, which reduces degradation of the base resin.

The disadvantage of glass microspheres is their relatively high cost compared to nonspherical fillers.

2. METHODS OF PRODUCTION

Basically, there are two methods of producing solid spherical particles. The fire polishing of crushed particles accounts for the largest tonnage. The atomization of molten materials is a far second. There are variations of these two processes, such as the calcining of crushed particles and the formation of fibers from a glass melt with subsequent flame formation of spheres, but these special processes account for only a few percent of total production.

2.1 Firing Crushed Particles

The method of fire polishing crushed particles is used to produce approximately 250,000,000 lb of solid spheres each year in the United States. It consists of crushing and sizing scrap glass or cullet, then passing it thru an enclosed flame with subsequent collecting and cooling of the spheroidal product. The material is usually further screened to usable sizes. The raw materials for this process are, for the most part, plate glass, cullet from glass furnaces, and some bottle glass. The composition is essentially silicon dioxide 72%, sodium oxide 15%, and calcium and magnesium oxide 13% with minor variations; this is designated as "A" glass. Cullets are also made for specialty application by melting special glass compositions and dumping them into water to form frit.

Unless the cullet is unusually clean, it is water washed and air dried before crushing. Large hammer mills are used to crush the cullet. This produces particles small enough for fire polishing, 0.05-.001 in. Particles outside this range present many difficulties; the larger particles require too long a residence time in the flame and the smaller particles have such a large surface area that they tend to smother a flame. The crushed particles are classified by screen size.

The crushed and sized cullet is then passed thru a tube which encloses a flame or some other hot gas. Natural gas is the preferred medium. Each manufacturer's process differs as to tube length and diameter but most are 10-20 ft in length and 8-30 in. in diameter. A mixture of air and natural gas are introduced and ignited at one end. The cullet is usually introduced at this same end and carried thru the tube by the hot gasses, but it can also be introduced at the opposite end and allowed to fall thru the flame. The cullet remains in the hot medium for less than a second but the surface is heated to 1400-1600°F, which is enough to liquify it, and surface tension forms it into a sphere. The inner portion of the particles is heated to less than 500°F, which is below the softening point of "A" glass and is, therefore unaffected. "A" glass has a softening point of about 800°F but doesn't really melt until 1200-1300°F.

Since only a very thin area has been heated sufficently to become liquid, the sphere cools quickly below its softening point once it leaves the flame area and even though the temperature of a particle may be 600°F to 700°F when it is collected, the spheres will not deform or fuse together since they are well below their softening point at that temperature. The spheres are then screened and mixed to many specifications. If a surface treatment is required, it is applied as a final step.

2.2 Melting Glass Batch With Subsequent Atomization

The second method of production requires that a glass batch be formulated. This can be "A" glass but usually it is a formulation with a very low melt viscosity and high surface tension.

The batch ingredients, i.e., sand, soda ash, dolomite, feldspar, etc., are dry-blended and fed to a melting tank. The tank is held at a temperature above 2400°F and the batch is melted. The molten glass flows through an orifice in the form of a free falling molten stream. This stream is broken up to form small droplets which are cooled in air to form spheres. The stream can be broken up by any number of methods, such as impinging with air pressure, paddles, or splattering against a surface. The particles are formed into spheres by surface tension and harden as they fall toward a collector. These spheres are screened to size.

2.3 Calcining of Crushed Particles

In the calcining process, the sized cullet is introduced into a rotating heated tube along with a carrier medium such as carbon. The cullet is heated to its softening point and forms spheres. This process is used to form intermediate size spheres, i.e., 10–20 mesh.

2.4 Cenospheres from Fly Ash

The word *cenosphere* was first used as a name for the floaters on fly ash ponds in England. These floaters are light weight, hollow spheres composed of a glass composition formed from the fly ash. More detail on the floaters is given in the chapter on hollow spheres. More recently, the word cenosphere has taken on a much broader meaning and now encompasses any sphere formed from fly ash.

Close inspection of fly ash from coal burning power plants reveals that a major portion of what is known as fly ash is spherical glass-like particles. Recently, Ceno-Science Research, Inc. formed a joint venture with Northern States Power Company to develop separation and classification techniques for removal of cenospheres from the remaining fly ash residue. These glass spheres average about one half the diameter of the smallest glass beads that are currently being produced commercially by the more conventional methods. Most of these beads have a small gas bubble in the center and, as a result, they have an average specific gravity of 2.3 g/cc rather than 2.6 to 2.7 g/cc that would be expected from their high alumina glass composition.

2.5 Formation of Large Spheres

Large spheres, such as marbles, are made by depositing a blob of molten glass onto a set of metal screws. The glass is kept in continuous motion until it cools and a single marble is formed. This process produces spheres $1/4$ in. in diameter and larger.

2.6 Production of Other Spheres

Organic polymeric spheres are produced by the suspension polymerization process. In this method, organic monomers, i.e., styrene, polyester, acrylics, etc., are introduced into water along with suspension agents, soaps, and polymerization initiators. The monomers are agitated with the water to form droplets of monomer suspended in water. This suspension is maintained and the temperature of the water is increased to initiate the polymerization. The reaction is exothermic and the temperature is regulated by the water. When the reaction is complete, the liquid monomer has reacted and formed solid polymer beads which are separated from the water and dried.

Plastic spheres and carbon spheres from $1/6$ in. up to several inches in diameter are made by molding or machining individual parts.

3. PROPERTIES

3.1 Chemical Analysis

The compositions of two types of glass are given in Table 18-1.

Table 18-1. The chemical analysis of typical soda lime and cenosphere glass.

Glass (soda lime) (% by wt.)		Cenosphere (from fly ash) (% by wt.)	
SiO_2	72	SiO_2	54-55.15
CaO	8	Al_2O_3	25-30.90
MgO	6	CaO	7.87-10.0
Na_2O	14	MgO	2.0-3.11
		Na_2O	0-1.38
		K_2O	1.20-2.81
		TiO_2	0.78-1.15
		Fe_2O_3	0.46-1.74
		SO_3	1.24-2.00

3.2 Chemical Properties of Glass Surface

3.2.1. Untreated. The surface of newly formed soda lime glass is composed of a network of silicon and oxygen with ions of sodium, calcium, and magnesium interspersed in the network. Other elements are found in commercial glasses; some participate in the network and the rest appear in the interstices. A new glass surface is quite reactive but it quickly loses its reactivity on exposure to air. An exposed glass surface is inert due to the strong Si-O three dimensional bonding, but the ions held in this network can be leached out, exchanged, or even attract water to the surface of the glass. Soda lime glass is quite alkaline because of its surface sodium, calcium, and magnesium. When soda lime glass is washed, it gives a high pH because of the leaching of the alkali; this effect becomes less with successive washings. The surface can be attacked by hydrofluoric acid, which is one of the few substances which will break the Si-O bonds and dissolve glass.

3.2.2. Chemical Properties of Surface Coated Glass Microspheres. Glass spheres are coated with a large variety of materials to give specific properties. The most common coatings are silicones, silanes, metals, fluorocarbons, and polymers.

Glass microspheres are hygroscopic. Upon adsorption of moisture, they tend to agglomerate and present a dispersion problem. To counteract this tendency, they are coated with water repellants. Silicones are the most effective water proofing agents. Spheres treated with 3 $g/1000m^2$ of surface area are so resistant that they cannot be wetted even when immersed in water.

Although silicone coatings are used in some applications, silane coatings are preferred for composite materials since they create an improved bond between the glass sphere and the polymer matrix. Silanes are long molecules with two different types of terminal groups. One end contains the silane structure, which reacts and bonds readily with the glass surface. The other end is an organic functional group that is chosen for interaction with a specific polymer matrix. The commercially available organic functional groups include amino, epoxy, methacryloxy, vinyl, chloro, and mercapto silanes. These coatings are applied to the glass microspheres from solution, and reacted during the elevated temperature drying stage. The silane group reacts with the glass surface, leaving the organo-functional groups available for future interaction with the polymer matrix.

The improved bonding due to silane treatment may be illustrated by the case of the cured vinyl plastisol containing glass spheres. When coated spheres are forcibly removed, bits of adhered vinyl will be noted over most of the surface; uncoated spheres can be removed with less force and exhibit a clean surface.

No single silane serves best for all systems. Each polymer must be matched with a specific silane. Some of these matches are listed below. However, different formulations will often call for a different silane and only experimentation can determine the concentration and type of silane to be used. For the polymers listed in Table 18-2, the silanes with the designated organo-functional groups have been shown to work well.

Not all silanes give a water repellant surface but all lead to improved water resistance.

Microspheres are coated with metals, such as silver, to produce electrically conductive materials or for opaqueness. They are coated with polymers to improve bonding or specific wetting properties. Some silicone coatings are used to enhance electrical insulation properties. Fluorocarbons enhance non-wetting properties. There is a coating for almost any desired surface property and many types of these coatings are commercially available.

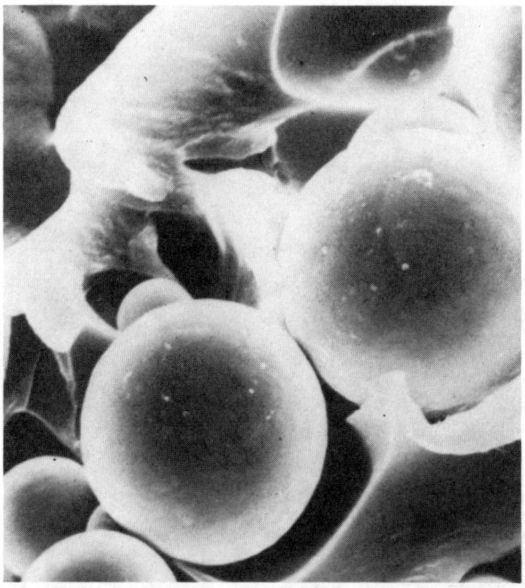

Fig. 18-1. Uncoated spheres–illustrates poor wetting or no bonding by the resin matrix.

Typical results on bonding of uncoated and coated glass spheres are shown in a PPO resin matrix in the photomicrograph in Figs. 18-1 and 18-2.

3.3 Physical Properties

3.3.1. Particle Shape and Size. Glass microspheres are supplied as at least 90% true spheres

TABLE 18-2. Recommended Silane Coatings

Polymer	Silane Functional Type(s)
Polyester	Epoxy, Methacryl
Nylon	Amino
Polyethylene	Amino, Vinyl
Polypropylene	Amino, Epoxy
Epoxy	Amino, Epoxy
ABS	Vinyl, Epoxy
Polystyrene	Epoxy, Methacryl
Polycarbonate	Amino
Melamine	Amino
Nitrile	Mercapto
Polysulfone	Amino
PVC	Amino

in sizes ranging from -325 to 20 mesh. The larger sizes contain a greater proportion of slightly ovoid particles. Figure 18-3 illustrates glass spheres in the typical size ranges that are available. These range from about 30 μ in diameter (-325 mesh) to 750 μ in diameter (20 mesh). Figure 18-4 is a photo of the small size grade of commercial cenospheres, which are available in a range from 0.3-30 μ in diameter.

3.3.2. Surface Area. The surface area of microspheres depends upon the size graduation. The surface area formula is $\frac{1099 \times 10^5}{D}$ = surface

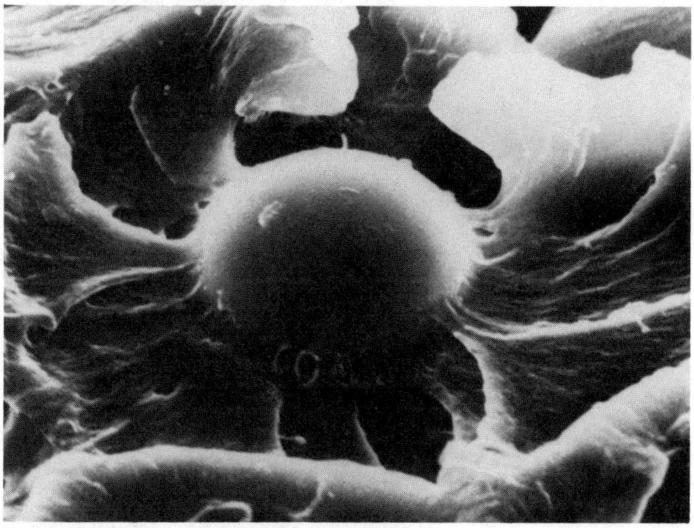

Fig. 18-2. Coated glass spheres (A100 silane)–illustrates good wetting and strong bonding by the resin matrix. *(Figs. 18-1 and 18-2 courtesy of A. Wambach, K. Trachte and A. DiBenedetto, "Fracture Properties of Glass Filled Polyphenylene Oxide Composites," Journal of Composite Materials; July 1968 Volume 2 Number 3. Reprinted with permission of Technomic Publishing Company, Inc., Stamford, Conn.)*

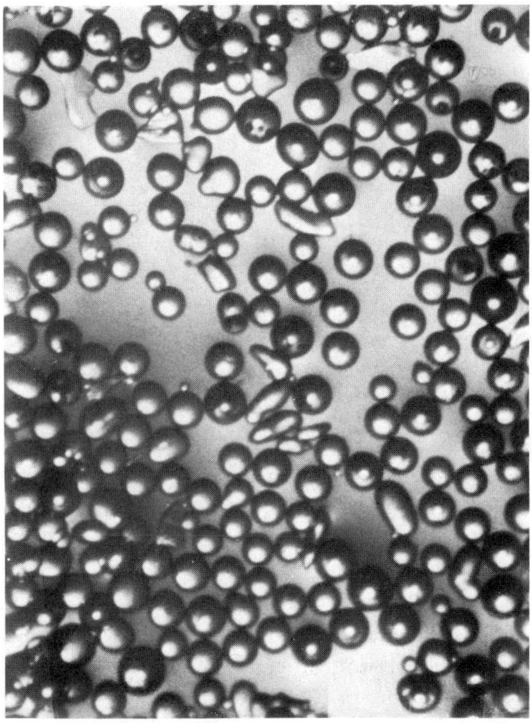

(Potter's 2427) −140 +270 mesh

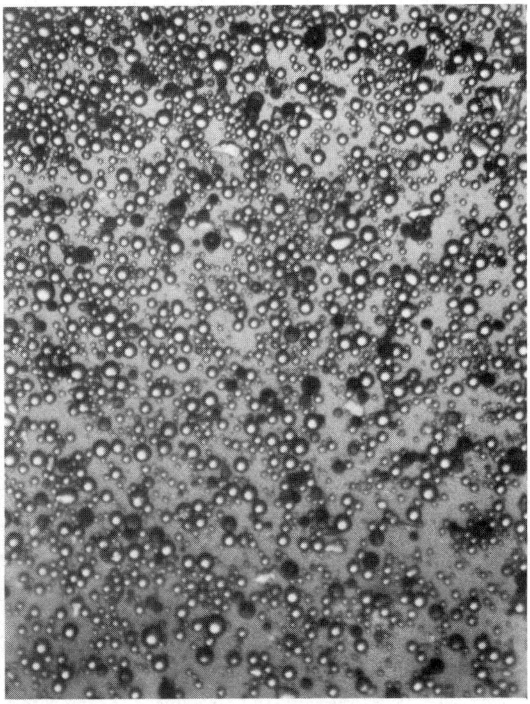

(Potter's 3000) −325 mesh

(Potter's 1721) –45 +80 mesh

(Potter's 1221) –20 +80 mesh

Fig. 18-3. Various sizes of solid glass beads–all photos at 50x magnification.

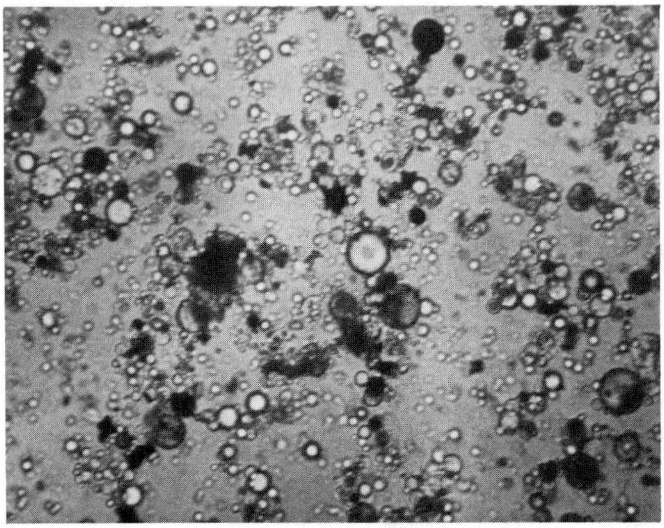

Fig. 18-4. Solid fly ash sphere–325 mesh–400x.

area in cm^2/lb, where D is measured in microns. This gives $2.5 \times 10^6 cm^2/lb$, of area for spheres with an average diameter of 44 μ and density of 2.48 g/cc.

3.3.3. Typical properties of soda lime glass microspheres are shown in Table 18-3.

Table 18-3. Properties of Soda Lime Glass Spheres

Refractive Index: Nd = 1.51
Density–2.48 g/cc
Bulk density–approximately 90 lb/cu ft
Hardness–5.5 Mohs' scale
Crush strength–40,000 psi
Coefficient of thermal expansion–$8.47 \times 10^{-6}/°C$ from 20–300°C
Thermal conductivity–.0025 cal/sec/cm^2
Specific Heat–0.27 cal/gm/°C average from 40–800°C
Sound transmission–5.3 km/sec @ R.T.
Dielectric constant–12.1 @ 1 KHz and 23°C
Volume resistivity–2×10^{13} ohm per cm @ 17°C

4. MATERIAL SUPPLIERS

4.1 Suppliers of Solid Glass Microspheres

Cataphote Division of Ferro Corporation
1001 Underwood Drive
Flowood, Miss. 39208
(601) 939-4612

Prod: Glass spheres
Price range $.14–.20/lb

Product Designation Size	U.S. Sieve Size	Size Diam(in.)
PF-1	−25 + 45	.0276–.0138
PF-2	−30 + 50	.0232–.0116
PF-3	−35 + 60	.0195–.0097
PF-4	−40 + 70	.0164–.0082
PF-5	−50 + 80	.0116–.0069
PF-6	−60 + 100	.0097–.0058
PF-7	−70 + 140	.0082–.0041
PF-8	−100 + 200	.0058–.0029
PF-9	−140 + 270	.0041–.0021
PF-10	−200 + 325	.0029–.0017
PF-11	−270 + 1000	.0021–.0005
PF-12	−325 and finer	.0017–.0002

Coatings	*Suggested plastic system*
R	Epoxy, phenolic, melamine, PVC, polycarbonate, polyethylene, nylon, polypropylene, polyurethane, phenol-formaldehyde
S	Polyester, polystyrene, SAN, ABS, acetal, epoxy
T	Polysulfone

Potters Industries, Inc.
377 Rt. 17
Hasbrouck Heights, N. J. 07604
(201) 288-4700
Prod: Solid glass spheres
Price range $.14–.20/lb.

Product Designation	Mesh Size
1221	−20 + 80
1721	−45 + 80
1922	−60 + 100
2024	−70 + 140
2429	−140 + 270
2900	−270 and finer
3000	−325 and finer (suggested for use in the plastics industry)

Product designation 3000

Coatings	*Suggested plastic systems*
CP01	Alkyd, polyester, ABS, polyethylene, polystyrene, polymethylmethacrylate
CP02	Epoxy, acetal, cellulosic, SAN Melamine, phenol-formaldehyde, phenolic, urethane, nylon, PPO, polycarbonate, polypropylene, polysulfone, PVC
CP03	Melamine, phenol-formaldehyde, phenolic, urethane, nylon, PPO, polycarbonate, polypropylene, polysulfone, PVC

Northern States Power Company
414 Nicollet Mall
Minneapolis, Minn. 55401
(612) 330-6067
Prod: Cenospheres. Sp. Gr. −2.35
Sizes available: 1/3–30 μ
Price range: $.06–.10/lb

Penn Virginia Materials Corporation
200 Erie Road
Eastlake, Ohio 44094
(216) 951-7711
Prod: Spherefil
Sizes available: (−)325 mesh and (−) 10 μ
Price range: $.05–.10/lb

Flex-O-Lite Divisions,
General Steel Industries, Inc.
P. O. Box 4366
St. Louis, Mo. 63123
(314) 351-4450
Prod: Solid Glass Spheres
Size range from 20 U.S. mesh and finer
Sieve separations minimum of 30 points
Price range: $.14–.20/lb

4.2 Suppliers of Metal Spheres

ASARCO
120 Broadway
New York, N.Y. 10005
(212) 732-9500
Prod: Zinc dust

Designation	Average Size Microns	Sp. gr.
No. 1	8	7.11
No. 50	7	7.10
No. 111	6	7.10
No. 555	6	7.09
HE	4	7.06
No. 280	2.7	7.00

4.3 Suppliers of Solid Plastic Spheres

Rohm & Haas
Independence Mall West
Philadelphia, PA 19105
(215) 592-3000
Prod: Acrylic and copolymer series

Dow Chemical Company
2020 Dow Center
Midland, Mich. 48640
(517) 636-1000
Prod: Styrene, divinyl benzene spheres (50–100 mesh and 100–200 mesh)

Diamond Shamrock Chemical Co., Plastics Div.
1100 Superior Ave.
Cleveland, Ohio 44114
(216) 694-5000
Prod: Polyvinyl chloride spheres (PVC-40)

E. I. Dupont
1007 Market St.
Wilmington, Del. 19898
(302) 774-2421
Prod: Polyvinyl acetate, acrylic, and acrylic copolymer spheres

Kureka Chemical Industry Co., Ltd.
1–8 Horidome-cho, Nihonbushi, Chuo-Ku Tokyo 103, Japan
New York Office: 420 Lexington Ave., New York City, N.Y.
(212) 683-1999
Prod: Solid carbon/graphite beads, 100–250 μ.

4.4 Suppliers of Large Plastic Spheres

Product Components Corp.
30 Lorraine Ave.
Mt. Vernon, N.Y. 10553
(914) 699-8640
Prod: Thermoplastic spheres $^1/_{16}$ in. to 6 in. made from nylon, teflon, polyethylene, etc.

Ace Plastic Co., Inc.
91-30 Van Wyck Expressway
Jamaica, N.Y. 11435
(212) 523-5500
Prod: Acrylic, acetate, nylon, teflon, cast phenolic, polycarbonate, delvin in $^1/_{16}$ to 4 in. range

Orange Products, Inc.
Passaic Ave.
Chatham, N.J. 07928
(201) 635-6666
Prod: Polystyrene, polypropylene, polyethylene, nylon, teflon spheres from 3/32 in to 1 3/4 in. and precision ground to any specification

Plastech Alloys, Inc.
1981 Placentia Ave.
Costa Mesa, Calif. 92675
(714) 642-1593
Prod: Glass bead-filled styrenics, polysulfone, acetals, olefins, modified PPO, nylon, urethane and polyesters

4.5 Suppliers of Sphere-filled Thermoplastics

LNP Corporation
412 King Street
Malvern, PA 19355
(215) 644-5200
Supplier of glass bead-filled ABS, SAN, nylon, etc.

Fiber Resin Corporation
27877 Groesbeck Hwy
Roseville, Mich. 48066
(313) 777-3131
Prod: Sastcast a liquid urethane glass sphere-filled casting material

Wellman, Inc., Plastics Division
75 Federal Street
Boston, Mass. 02110
(617) 482-0102
Prod: Nylon filled with 5–30% glass spheres, also combination of spheres and fibers

Fiberfil Division of Dart Industries, Inc.
P. O. Box 3333
Evansville, Ind. 47732
(812) 424-3831
Prod: Glass bead-filled nylon

5. HISTORY, FUTURE, COSTS AND AVAILABILITY

Glass spheres have been produced for more than 40 years. Eighty percent of current production is for retroreflective products such as signs and reflective traffic paint. Spheres as a filler for plastics were introducted in the mid sixties. During the 1974 resin shortage, the use of spheres gathered momentum. Since resin prices have stayed high and there has been little change in the price of spheres, their use is increasing. The availability of equipment for easy incorporation of spheres and experience showing that spheres do not cause excessive wear on processing equipment, are main factors leading to increased use by extruders and molders. The largest growth areas have been plastic furniture, housewares, and machine parts.

The price of glass spheres is approximately \$.15/lb (1975 price) and the density is 2.48 g/cc. The cost/volume ratio is therefore \$0.14/

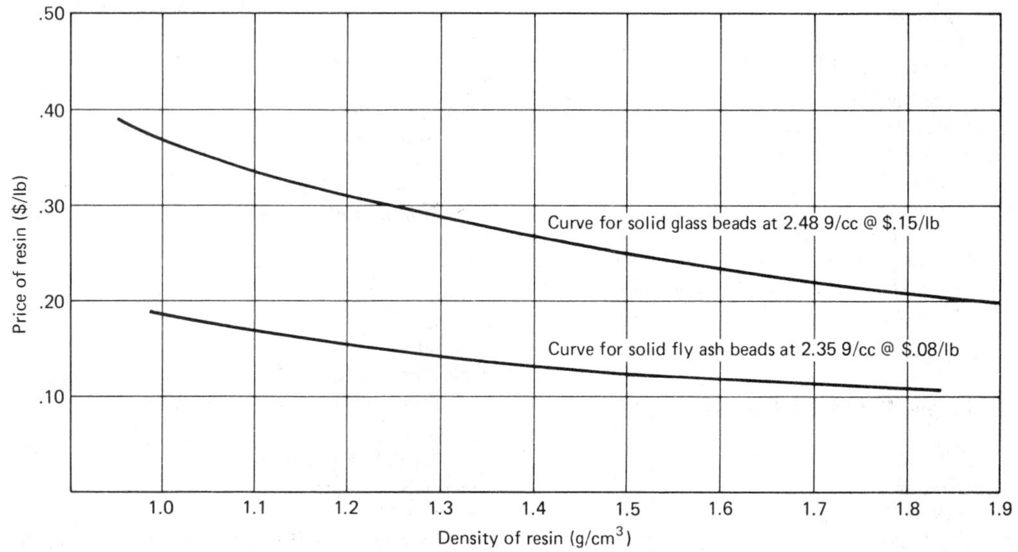

Fig. 18-5. Raw material cost of bead plastic system.

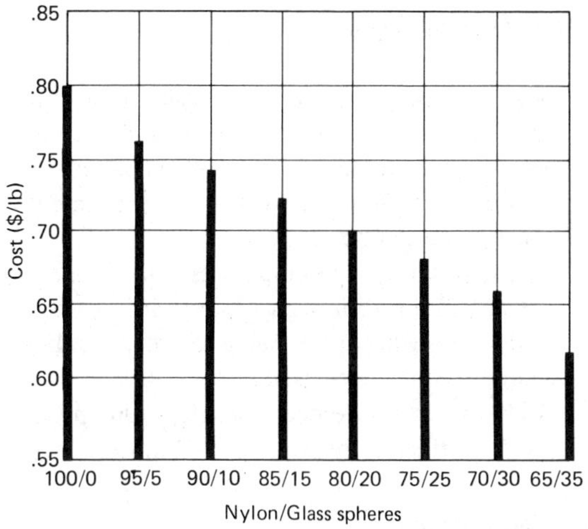

Fig. 18-6. Economics of glass sphere-filled nylon cost ($/lb) vs. nylon/glass spheres composite (% Vol).

cu in. This is equivalent to \$.37/lb plastic with a density of 1 g/cc. For solid fly ash beads at \$.08/lb and the density of 2.35 g/cc, the equivalent cost for a plastic with a density of 1 g/cc is \$.188/lb. Figure 18-5 depicts the economics of sphere filled resins. Figures 18-6 and 18-7 illustrate the economic effect (raw material cost) of the addition of glass spheres to nylon and silicone resin respectively. It is evident that when a lower cost filler is used, the total composite materials cost goes down rapidly.

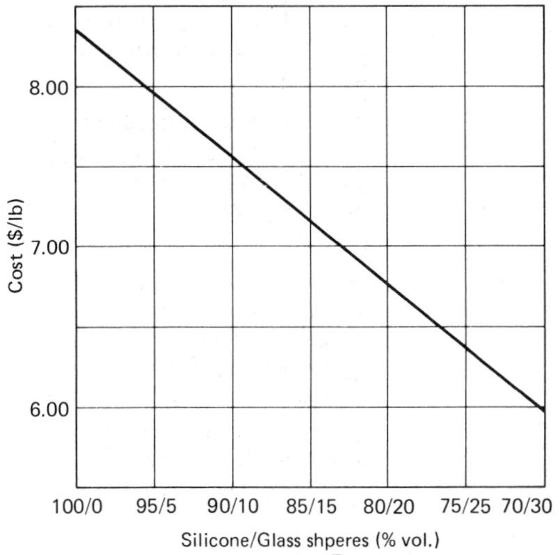

Fig. 18-7. Cost saving of glass sphere-filled sylgard 184 liquid resin (Dow Corning).

6. APPLICATION

6.1 Discussion

Glass spheres find application as a filler in any plastic system. Their low surface to volume ratio and uniform particle shape and surface make them an ideal filler. Through specific surface treatment, they can be made compatible with any polymer. They not only contribute bulk but they confer specific physical properties and improved processing.

Glass spheres are used to improve:

Tensile strength
Flexural modulus
Hardness
Abrasion resistance
Compressive strength
Heat deflection temperature
Water resistance
Corrosion resistance
Flame retardance
Improved dispersion of other ingredients
Electrical properties
Fracture toughness

The use of spheres in conjunction with fibers produces dramatic results. When spheres replace a portion of the fibers in FRP, they lower the melt viscosity of the system and improve fiber distribution and mold flow out. While improving processability, they act synergistically to improve impact resistance, stiffness, toughness and decrease shrinkage and surface imperfection, as well as lower cost by replacing high priced fibers.

It has been found that microspheres improve processability and lessen rejects of injection molded parts by controlling shrinkage and warpage, while also improving mold flow out and shortening cycle times. Spheres contribute little to expansion and contraction of parts, thereby controlling shrinkage. In one application (see abstract 8), shrinkage was reduced so much in one production item, that the screw holes in the glass sphere-filled part no longer fit the screw holes in another part injection molded without glass spheres. In a 3-ft long part, visual comparison of identical injection molded items showed the bead-filled parts were greater in length than unfilled parts. Their high heat capacity and uniform shape contribute to uniform cooling and also lessen warpage. This uniform cooling rate allows for shortened cycle time since parts can be ejected hotter without warping.

Glass spheres have been utilized as a clean-up aid in purging extruders and injection molding machines. Addition of the spheres to traditional polymers used for purging this equipment results in a more efficient clean-up operation and a faster changeover from one color or polymer to another.

6.2 Polymer Systems

Glass microspheres have been tested in a number of polymeric systems. Table 18-4 presents thermoplastic materials in which glass spheres have been shown to offer distinct improvements in manufactured items.

In addition to thermoplastics, glass spheres have also been utilized as a filler/reinforcement for thermosetting resins. Table 18-5 depicts their use in different thermosets.

Glass spheres have been tested by Union Carbide as a filler in high pressure foam injection

Table 18-4. Thermoplastics.

Polymer	Part(s)	Average Sphere Size (μ)	Surface Treatment	Filler Use Level (% by wt.)	Advantages
Polystyrene	Mirror frames, T.V. and stereo cabinets, housewares	30	Epoxy, vinyl silanes	20–35	Increased stiffness, improved heat deflection temp., increased compressive strength, shortened cycle time, improved paintability of part.
Nylon	Bearings, cameras, housewares	30	Amino silane	up to 40	Increased tensile and compressive strength flexural modulus, improved heat deflection temp., increased dimension stability, lowers shrinkage and water absorption.
San	Chevrolet glove compartment door	30	Proprietary (Potters Industries)	20	Smooth surface, improved resistance to solvent crazing when painted, reduced mold in stress, physical properties comparable to glass fiber-filled material except for impact strength.
ABS	Automobile trimming, T.V. and stereo cabinets, others	30	Vinyl	20–30	Increases flexural modulus, compressive strength and modulus, improved dimensional stability, lowers shrinkage, increases resistance to abrasion.
Polycarbonate	Injection molded physical testing, specimens	30	Amino silane	35	Improved moldability, 5% increase in tensile strength, 47% increase in tensile modulus, 8% increase in compression strength, 35% increase in compression modulus, 13% increase in flexural strength, 65% increase in flexural modulus, over a 60% decrease in impact strength.*

*Just as in nylon-6/6 (see abstract 9), it is highly probable that a glass bead/glass fiber combination in polycarbonate can serve to improve the overall physical characteristics, including impact strength, of the base resin better than either of the two alone.

Table 18-5. Glass sphere-filled thermosets.

Polymer	Parts	Spheres Range	Silane Surface Treatment	Advantages
Epoxies			Epoxy, amino	Permits high loadings with less viscosity build-up, physical properties comparable to or better than base resin, reduces shrinkage, increases resistance to abrasion, helps eliminate air pockets in laminates and coatings, reduces water absorption.
Polyesters	Grinding wheels, tools	Large spheres small spheres	Vinyl	
Polyurethanes	Boats, display items	(710–5 M)	Amino	
Silicones	Molds	710–5 M	Untreated	Improvement in physical properties permits higher loadings than irregular shaped fillers, definite cost saver to user.

molding of high impact polystyrene. In the foaming process, the glass spheres act as a nucleating agent for cell formations. Experimental results show a number of advantages achieved by using glass spheres. Among these are:

1. The mold filled faster and more completely.
2. Four or five times more injection pressure was available at the point of filling the mold.
3. There was less injection pressure drop filling the mold, 58 vs. 98% drop for the control.
4. Sets up faster; therefore, cycle time could be reduced.
5. The molded slabs were more rigid when taken from the mold.
6. A finer, more uniform cell structure was produced.
7. Superior rod drop impact resulted, 3 ft-lb vs. 2 ft-lb of the unfilled material.

Under the same processing conditions, the molded slabs of the glass sphere-filled material had a thicker skin and naturally a higher molded density.

Utilizing glass spheres in the foaming process produces swirl-free parts and parts with a more uniform cellular structure. Glass spheres are currently being tested in foamed polycarbonate with encouraging results.

6.3 Methods of Incorporation

Incorporation of microspheres into liquid thermosets is simply a matter of stirring in, in contrast to the special methods that must be used to uniformly disperse spheres into high viscosity thermoplastics. Microspheres are master batched by several suppliers, e.g., Wellman Corporation, LPN-Penn, Calif., Fiber Resins, Mich., Fiberfil, Mich., Plastic Alloys, Calif. These companies mix the spheres into plastic by the use of screw extruders, then pelletize the product. This method is also used by individual molders to incorporate fibers and spheres into the same batch. Some fabricators have used intensive type mixers such as a Banbury for preparing extrusion compounds.

There are proportioning machines made for extruders and injection molding machines. A machine of this type, manufactured by Whitlock Company, can be attached to the hopper of the extruder or injection molder and is equipped with a 0–40 rpm variable speed drive. The weight of filler added to the polymer can be varied by changing the rpm setting to a higher or lower number. In the extrusion process, glass spheres are continually added to the throat of the extruder and are mixed with the melted polymer in the barrel of the machine by the action of the screw. The feeder is set at the proper rpm to add a measured weight or volume of glass spheres to that particular plastic.

In the injection molding process, the feeder is electrically interconnected to the injection machine's screw control circuit and adds a predetermined weight of glass spheres to each shot of resin. The feeder is in operation only when the screw motor is running. Mixing takes place in the barrel of the machine.

7. CURRENT AND FUTURE UTILIZATION

In addition to increased raw material costs of petrochemical-based feedstock, which readily translates into higher resin prices, factors such as increased overhead costs, operating costs, and investment costs contribute to higher resin prices. To offset some of these costs, processors will take advantage of new technology that improves production efficiency, and also new resin-saving measures such as fillers.

Market analysts have predicted that consumption of fillers in the plastics industry will more than double by 1980, increase six-fold by 1990 and reach 15,000 metric tons (a 1,300% increase from 1975) by the end of the century.[1] Glass spheres are expected to take their share of the filler market. Estimated use level of glass spheres, flakes and bubbles in 1975 was 11 million lb, with an anticipated growth to 441 million lb by 2000. This projected phenomenal growth is attributable to a number of factors. With higher resin prices, fillers become an attractive cost saving device. Also, new technology has made fillers more than mere resin extenders. For example, when glass spheres are treated with various organo-functional silanes or organic titanates, there is an improvement in glass sphere/plastic adhesion and bonding. This results in improvement in mechanical properties over the untreated material and in some instances, an improvement over the virgin resin.[2]

Another key to the projected acceptance of glass spheres as a major filler for the plastic industry is the introduction of new machinery which solves the traditional problems associated with molding and extruding filled polymers. Since fillers used in the injection molding and extrusion process generally have higher densities and smaller particle sizes than the resin, filler/plastic separation in the hopper occurs. Three Japanese companies have developed equipment designed to solve this problem.[3] It can also be eliminated by using a Whitlock Model 640 Volumetric Feeder, manufactured by Whitlock Company, Farmington, Mich. (Such a model was described in the previous section, Methods of Incorporation.)

With the emerging technology in equipment, processors are no longer solely dependent on compounders to supply glass bead filled-resins. With the new machinery, in-house molding and extrusion of glass sphere-filled parts specifically designed to meet varying criteria have become a reality.

A real sleeper just emerging in the bead filler industry is fly ash spheres (cenospheres). The potential of large volumes of solid spheres at very low prices appears very realistic in the near future. If their prices can be dropped to below \$.05/lb, fly ash spheres could compete with the large volume fillers such as calcium carbonate and asbestos.

8. ABSTRACTS

1. Gerhard Illing, Epoxy-Coated Glass Spheres for Reinforcement of Nylon and Polyester Resin, *Proc. 29 SPI RP/Composite Conference*, 1974.

   Nylon and polyethyleneterephthalate filled with epoxy-coated glass spheres had superior mechanical properties and moldability characteristics than the virgin resin. Less machine and tooling wear were encountered with the sphere-filled system than with fiberglass. The material could be filled with as much as 70% glass spheres.

2. Breakthrough in Fillers, *Industry Week*, p. 18, April 2, 1973.

   The use of microscopic glass beads in combination with other fillers such as cellulose fibers, reduced die wear, increased molding characteristics, and improved dimensional stability and the resilience of the plastic in which it was used.

3. Otto R. Strauch and James E. Ritter, Use of Solid Glass Spheres in Tooling Resins, *Proc. 28, SPI RP/Composite Conference*, Section 3-G, p. 1, (1973).

Above 40 phr loading, glass sphere-filled epoxy was superior to the unfilled resin, calcium carbonate, and clay in tensile strength, flexural strength, flexural modulus, and water absorption.

4. Tiny Glass Beads Beef Up Plastic Tooling Compounds, *Product Engineering*, Mar. 1974.

   Advantages obtained from incorporating higher loadings of glass spheres in epoxy, polyester, and polyurethane were low shrinkage, gains in some physical properties, and improved dimensional stability.

5. Transfer-Moldable Foam Reduces Process Cycle, *Plastics World*, Jan. 21, 1975.

   Addition of glass microspheres to Eccomold LTM 1090 results in faster processing speeds, lower processing pressures and temperatures, less shrinkage, and lower cost than the base resin.

6. Barr, T. J., Glass Beads in GRP, Technical Center, Ferro Corporation, Independence, Ohio, unpublished paper.

   Glass beads were shown to have definite utility as a resin extender in polyester/fiberglass laminates. They even functioned as a secondary reinforcement when used in systems with less than an optimum level of a primary reinforcement, e.g., glass fiber. Employing only 15% fiberglass and 50% glass beads, flexural strength increased 50%, flexural modulus doubled, compressive strength and modulus increased over the 15% glass fiber-filled resin. No adverse effect on tensile strength was noted.

7. Solid Glass Spheres Increase ABS Abrasion Resistance—750%, *Plastic Design and Processing*, Sept. 1965.

   Glass spheres were employed to overcome a problem of insufficient abrasion resistance in plastic watering cups made from ABS and used in giant "egg factories." The life of the chicken's drinking cup was increased from 7 months to at least 5 years by the addition of tiny solid glass spheres at no increase in cost.

8. Tardy, Barbara A. and Albert Krock, Glass Microspheres: A Reinforcement-Extender For Plastic Furniture Products, *Proc. 30 SPI RP/Composite Conference*, Section 7-E, 1975.

   There are inherent advantages derived when glass microspheres-filled polystyrene is used in the manufacture of large furniture parts. Parts weighing up to 4 lb were molded. Molds filled quicker, easier, more uniformly, sink marks were eliminated and cycle time for molding the large items was reduced up to 16%. The sphere-filled material showed a decrease in shrinkage and warpage which resulted in less rejects, and increased production with a corresponding decrease in production cost.

9. Solid Glass Spheres Reinforcement-Filler For Nylon Compounds, Potters Industries Inc., *Technical Information Bulletin T 24-2.*

   The use of glass spheres in nylon can improve physical properties, molding conditions, and reduce cost. Addition of 40% glass spheres to nylon 6/6 raised tensile strength 19%, compressive strength 8 times, flexural modulus 41%, improved abrasion resistance, and resulted in better flow characteristics over the virgin resin. Impact strength was somewhat reduced. A combination of glass spheres and glass fibers was shown to be superior in many respects than either spheres or fibers alone.

References

1. "Even If Resin Is Plentiful, Filler Will Play A Big Role," *Modern Plastics*, p. 6 (Oct. 1975).
2. Strauch, Otto R. and Ritter, James E. "Use of Solid Glass Spheres In Tooling Resins, *Proc. 28.* SPI RP/Composite Conference, Section 3-G, p. 1 (1973).
3. *Modern Plastics*, p. 44 (June 1975).

Patents

1. U.S. Patent 2,460,977 (Feb. 8, 1949) E. P. Davis, et al. to 3M.

   The carbon powder method of making glass beads is described. The method employs a procedure of transforming irregularly shaped glass cullet particles into spheres by combining the cullet particles with powdered carbon.

2. U.S. Patent 3,190,737 (June 22, 1965) E. L. Schmidt to Flex-O-Lite Manufacturing Corp.

   The invention relates to improvements in apparatus and method for forming glass beads, and, in particular, is concerned with a method and apparatus for forming glass bead from ground glass, which is subjected to high temperatures and a vertical furnace tube.

3. U.S. Patent 3,138,444 (June 23, 1964) C. E. Searight, et al. to Cataphote Corp.
   A method of forming glass beads wherein the glass beads are formed directly from a molten stream of glass is described. Apparatus for fabricating small glass beads efficiently and economically is detailed.
4. U.S. Patent 3,150,947 (Sept. 29, 1964) C. C. Bland to Flex-O-Lite Mfg. Corp.
   The invention is concerned with the method for producing small glass beads by dispersing molten glass of low viscosity by the use of compressed fluid completely enclosed by a gas flame.
5. U.S. Patent 2,730,841 (Jan. 17, 1956) C. E. Searight.
   According to the patent, an improved method for producing silicone-coated glass beads is provided. Silicone-coated glass beads are produced by introducing a fog or mist of siloxane material into an enclosed zone, while passing glass beads at a temperature from about 300–500°F through such a zone.

Glossary

Atomization—To reduce to minute particles or to a fine spray.

Calcining—To heat to a high temperature but without fusing.

Cullet—Broken or refuse glass usually added to a new material to facilitate melting.

Exothermic—Characterized by or formed with the evolution of heat.

Frit—The calcined or partly fused materials of which glass is made.

Hygroscopic—Descriptive of a substance which has the property of adsorbing moisture from the air.

Inertness—The state of being chemically inactive.

Impinging—To strike or dash, especially with a sharp collision.

Leaching—The process of extraction of a soluble component from a substance by percolation of the substance with a solvent, usually water.

Mohs' Hardness Scale—A scale of hardness of minerals, running from 1 to 10, with talc the softest and diamond the hardest. Each mineral in the scale will scratch all those below it. The steps are not of equal value; the difference in hardness between 9 and 10 is much greater than between 1 and 2:

Diamond	10	Apatite	5
Corundum	9	Fluorite	4
Topaz	8	Calcite	3
Quartz	7	Gypsum	2
Orthoclase (Feldspar)	6	Talc	1

Molten—Fused or liquefied by heat.

Opaqueness—Descriptive of a substance that is neither transparent nor translucent.

Organic monomer—A molecule or compound containing carbon and of relatively low molecular weight and simple structure, which is capable of conversion to polymers, synthetic resins or elastomers by combination with itself or other similar molecules or compounds.

Orifice—An opening through which the molten glass is forced.

Polymer—A chemical compound or mixture of compounds consisting essentially of repeating units, formed by chemical reaction in which two or more small molecules combine into large molecules.

Synergistically—Acting together in such a manner that the total effect is greater than the sum of the effects taken independently.

Translucent—Admitting and diffusing light so that objects beyond cannot be clearly distinguished.

Transparency—Having the property of transmitting light without appreciable scattering so that bodies lying beyond are entirely visible. Pervious to any specified form of radiation.

19

Hollow Spherical Fillers

Bob Ryan and Barbara Tardy
Cataphote Division, Ferro Corp.
Jackson, Mississippi

CONTENTS

1. INTRODUCTION

Hollow spheres are one of the most useful fillers available to the plastics industry. Their sphericity, controlled particle size, and low density present the plastics product manufacturer with some profitable opportunities to modify his product. The replacement of high priced and scarce resin is the most obvious advantage; but lower and controlled density, uniform shrinkage, increased impact resistance, and mold fill-out are among other benefits. Hollow spheres are available in sizes from 0.001-2 in. and with densities from 0.1-0.7 g/cc, but the largest current usage is hollow glass spheres with an average diameter of about 3 mils (75 μ) and a density of 0.3 g/cc. Most of the current production is used in liquid casting systems with polyester, epoxies, and plastisols, where their properties of relatively low oil absorption, ease of dispersion, and inertness are unique. The apparent disadvantage of low crush resistance is not as severe as might be anticipated, since the spherical configuration withstands hydrostatic pressures in excess of 1500 psi. The price of hollow glass spheres in 1976 was approximately \$0.77/lb.

Hollow spheres are also made from materials other than glass, such as ceramics, carbon, and organic polymers. All are produced by a basic process where a blowing agent is added to a base material, which is crushed and subsequently heated for final expansion. The prices of these products range from about \$.20/lb for ceramic spheres from fly ash to \$35.00/lb for carbon spheres.

2. METHODS OF PRODUCTION

2.1 Inorganic

2.1.1. Firing of Crushed Particles Containing a Blowing Agent. Most hollow glass spheres are produced by passing small particles, containing a blowing agent, through a high temperature zone. The particle fuses or softens in the hot zone and gas is formed within the soft glass. The gas expands the particle to form a hollow sphere. The sphere is cooled in air and the walls solidify before the internal gas pressure is reduced. The hollow spheres are then collected by cyclone or bag collectors. Some of the principal types of inorganic hollow microspheres are described below.

3M Company "Glass Bubbles" are produced by passing glass particles through a flame. The process is described in U.S. Patent 3,365,315. Emerson and Cuming produces one of their "Microballon" products by passing spray dried sodium borosilicate solution through a flame, U.S. Patent 3,699,050. Philadelphia Quartz "Q-Cell" product is formed simply by spray drying special sodium silicate and ammonium pentaborate solutions, as described in U.S. Patent 3,796,777. "Fly Ash" hollow spheres are reclaimed from fly ash waste. The hollow fraction, or "floats," represent only 1-3% of the incineration product from coal burning power plants. They may be purchased damp with a moisture content of 20-25%. Several products, dried and graded, are available on the European market under such names as Cenospheres, Globalite, Lowkalite, Spherulite, Fillite, Armospheres, Microshells, Microballoons, and Floaters. Another method is to fire naturally occurring materials, such as "Kanium," to form hollow ceramic spheres as described in U.S. Patent 2,676,892.

2.1.2. Heating and Expanding Modified Pitch Particles. Hollow carbon spheres are produced by softening pitch material at a temperature of 300-450°F. The pitch is mixed with modifiers and dispersed to form fine particles ranging in diameter from 10-150 μ. The pitch particles are expanded by heating to form the hollow pitch microspheres ranging in diameter from 40-400 μ. The spheres are then carbonized at 1600-2000°F in an inert atmosphere to give hollow carbon microspheres.[1]

2.2 Organic

Organic low density spheres are produced by suspension polymerization of a thermoplastic bead with a blowing agent added. The blowing agent may be a low boiling hydrocarbon, an inert gas, or a latent gas material. (The latter term refers to any material, whether solid, liquid, or gaseous, which can be incorporated in the thermoplastic and converted to a gaseous state at an elevated temperature). When the thermoplastic bead is heated, it softens and the blowing agent vaporizes to form low density monocells or multicell spheres. The expanded particle size of monocells, such as Saran microspheres, ranges from 4-300 μ, and the size of multicell products, such as Dylite polystyrene, is usually about 500-5000 μ. Examples of current, low density organic spheres on the market are Saran Microspheres (vinylidene chloride/acrylonitrile spheres), phenolic microballons, urea-formaldehyde spheres, and polystyrene spheres.

3. PROPERTIES OF INORGANICS

3.1 Chemical Properties

The chemical properties of the spheres are determined by the chemical composition of the surface. Since this surface is glass, plastic, or carbon, they are generally inert and have no chemical effect upon the plastic matrix.

3.1.1. Chemical Analysis. A typical chemical composition of hollow glass spheres consist of the following:

Ingredients	W, %
SiO_2	60-80
Na_2O	5-16
CaO	5-25
$K_2O + Li_2O$	5-16
MgO	0-15
MnO_2	0-10
B_2O_3	0-20
P_2O_5	0-5

The range in chemical composition (% by wt.) of flyash cenospheres is:

Silica (SiO_2) 55-61
Alumina (Al_2O_3) 26-30
Iron oxides (as Fe_2O_3) 4-10
Calcium (as CaO) 2-6
Magnesium (as MgO) 1-2
Alkalis (as Na_2O, K_2O) 0.5-4.0
Carbon (loss on ignition) 0.01-2.0 (normally)

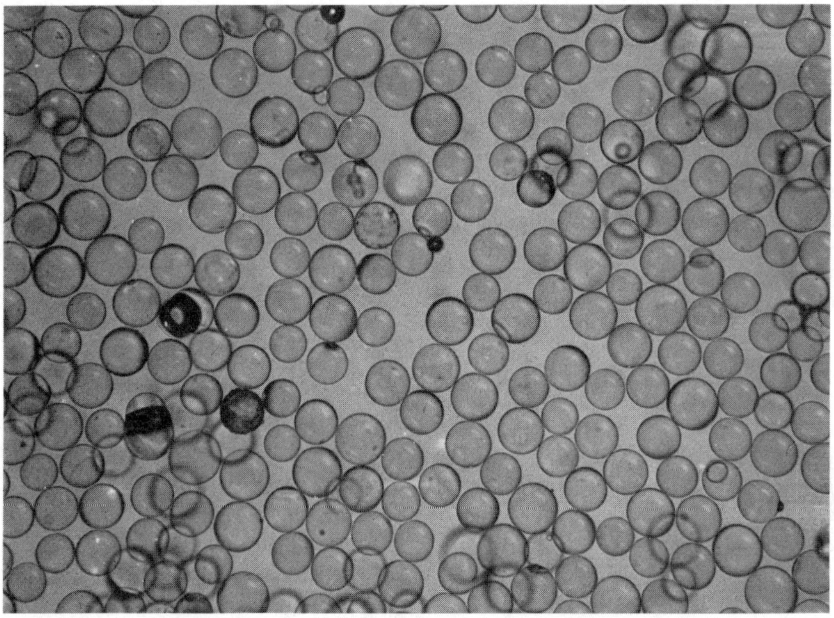

Fig. 19-1. Hollow glass microspheres, 3M type BB (63×). (*Photo by J. Varon, courtesy of Exxon Research and Engineering Company*)

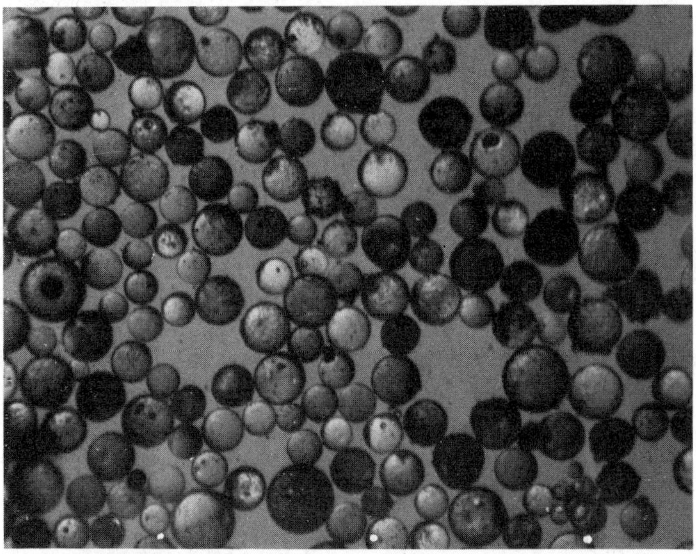

Fig. 19-2. Hollow glass microspheres from fly ash (63×).

Table 19-1. Typical physical properties of hollow glass spheres.

Size range	44–125 μ
Approximate diameter	75 μ
Wall thickness	1–2 μ
Density (particle)	0.08–0.64 g/cc
Bulk density	5–25 lb/ft^3 depending on density and size
Thermal conductivity	0.2–0.8 Btu/hr/ft^2/°F/in.
Thermal expansion	8.8 × 10^{-6}/°C from 60–440°C
Specific heat	0.27 cal/g°C approximate
Softening point	615°C, maximum safe use temperature for a prolonged period of time is approximately 550°C
Oil absorption	Approximately 36 g oil/100cc
Crush resistance	150–1500 psi hydrostatic pressure depending on specific product
Dielectric constant	1.2 @ MHZ to 8.6 GHZ
Dissipation factor	.002 @ 1 MHZ to 8.6 GHZ

3.2 Physical Properties

3.2.1. Photomicrographs. Figure 19-1 illustrates the uniform spheres and thin walls of commercial hollow glass microspheres. Figure 19-2 illustrates the thicker walls and wider size distribution of the hollow glass microspheres obtained from fly ash.

3.2.2. Particle Shape and Size. Hollow spheres are supplied in diameters from 20 μ to 1 in. but most products range from 20–130 μ and contain at least 80% true spheres. The wall thickness is 0.5–10 μ and is the determining factor of density and crush strength.

Table 19-2. Typical physical properties of fly ash hollow spheres.

Color	Off white
Size range	5 μ to 300 μ
Approximate mean diameter	50 μ
Wall thickness	2–5 μ
Density (particle)	0.35–0.7 g/cc
Bulk density	0.20–0.40 g/cc, 11–25 lb/ft^3
Thermal conductivity	0.69 Btu/in./ft^2/hr/°F
Specific heat	0.28 Cal/g°C approximate
Melting point	1200°C
Packing concentration	60–68% solids
Crush strength	500–5000 psi hydrostatic pressure

Table 19-3. Typical physical properties of hollow carbon/graphite microspheres.

		KUREKA KRECASPHERE	
	Versar Carbosphere	A50	A300
Size range	5–150	45–75	250–420
Approximate mean diameter	40	60	330
Wall thickness	1–2	1–2	6–12
Density (particle)	.20–.22	.15–.40	.10–.30
Bulk density	.12–.14	.10–.25	.05–.20

3.2.3. Appearance. The various products have different colors, such as white, light green, dark grey, and black.

3.2.4. Physical Properties. Tables 19-1–19-3 give the physical properties of hollow glass spheres, fly ash hollow spheres, and hollow carbon/graphite microspheres.

4. PROPERTIES OF ORGANIC SPHERES

4.1 Chemical Properties

Chemical properties of organic hollow spheres, and their resistance to chemical reagents, are determined by the chemical nature of the resin used in forming the spheres. Some, such as Eccospheres UF-O, which is a thermoset urea-formaldehyde hollow sphere, are relatively inert and are solvent and water resistant, while others, such as multicellular Dylite polystyrene are resistant to fungi and bacterial growth but have poor solvent resistance. As supplied, most organic spheres are stable at room temperature and can be readily stored without special handling precautions.

4.2 Physical Properties

Physical features are governed by the plastic surface, the thickness of the cell wall, and the size of the sphere, as well as by the nature of the encapsulated gas. Sphere size ranges from 4 μ to 1 in. Density is largely determined by size and wall thickness. The organic plastic spheres range in densities from 1 lb/ft^3-20 $lb/ft.^3$

5. MATERIALS SUPPLIERS

5.1 Suppliers of Inorganic Hollow Spheres

3M Company
3M Center
Saint Paul, Minn. 55101
(612) 733-1110
Product: Glass Bubbles
Price range from \$.70–\$1.00/lb

Series	Type	Avg. Particle Density (g/cc)	Size	Composition
B–B	B25B	0.23	90% by volume between 20 and 130 μ	Glass
	B30B	0.28		
	B40B	0.35		
	B15BX			
B–A & B35D	B18A	0.16		
	B22A	0.20		
	B35A	0.325		
	B40A	0.375		
	B35D	0.325		

Philadelphia Quartz Company
P. O. Box 840
Valley Forge, Pa. 19482
(215) 293-7200
Product: Q-CEL Inorganic Microspheres
Price range from \$.50–\$.75/lb

Product	Density	Avg. Size	Composition
Q-CEL	13 lb/ft^3 0.22 g/cc	75 μ	Sodium silicate

Pittsburgh Corning Corp.
800 Presque Isle Drive
Pittsburgh, Pa. 15239
Product: Celramic Glass Nodules
Price range from \$3.50–\$5.00/cu ft

Type	Density (±10%)	Size (In.)	Composition
8/12 mesh	8 lb/cu ft	0.66–0.094	Glass
12/20 mesh	12 lb/cu ft	0.33–0.066	Glass
20/40 mesh	20 lb/cu ft	0.016–0.033	Glass

Emerson & Cuming, Inc.
Dielectric Materials Division
Cauton, Mass. 02021
(617) 828-3300
Product: Microballoons
Product designation: Eccospheres
Price range form $.60–$10.00 lb

Type	Density, g/cc	Size (μ)	Composition
FTD-202, high strength	0.238	10–100	Insoluble glass
R, electronic grade	0.38	44–125	Sodium borosilicate glass
FTD-606, extra high strength	0.24–0.28	44–100	Insoluble glass
VT, surface treated	0.25–0.28	44–125	Silica
FT-102, hydrospace grade	0.25	10–250	Insoluble glass
FTL-200, low density hydrospace grade	0.209	44–125	Insoluble glass
IGD101, high strength	0.28–0.32	44–125	Sodium borosilicate glass
SI, hollow silica	0.18	30–125	Silica
IG 25, Low Density industrial grade	0.237	44–125	Sodium borosilicate glass
FTF15	0.28–0.32	2–44	Insoluble glass
FA-A, hollow ceramic	0.6	60–325	Ceramic (fly ash)
IG-101, industrial grade	0.34	89–124	Sodium borosilicate

Hargreaves PFA Marketing Ltd.
Queen Street
Woodesford, Nr. Leeds,
Yorkshire
Telephone: Garforth 87121
Product: Fly Ash
Price range from $.18–$.32/lb

Product Designation	Density, g/cc	Size (μ)	Composition
PFA Cenospheres	0.4–0.6	20–100	Ceramic fly ash

Fillite U.S.A., Inc.
801 30th Street
Huntington, W. Va. 25702
(304) 529-2436
Product: Fillite
Price range from $.18–$.30/lb

Product Designation	Density, g/cc	Size (μ)	Composition
Grade 52/7 "S"	0.4–0.7	300	Ceramic fly ash

Versar Inc.
General Technologies Division
6621 Electronic Drive
Springfield, Va. 22151
(703) 354-3350
Product: Hollow Carbon Spheres
Price range from $25 to $50/lb

Product Designation	Density, g/cc	Size (μ)	Composition
Carbo-Spheres	0.12–0.14	5–150	Carbon

Mitsubishi International Corp.
Petroleum Department
277 Park Avenue
New York, N.Y. 10017
(212) 922-3761

Product Designation	Density g/cc	Size (μ)	Composition
Krecasphere A-100	0.15–0.40	75–150	Carbon

Zircoa
Division of Corhart Refractories Co.
31501 Solon Road
Box 39217
Cleveland, Ohio
(216) 248-0500

Product Designation	Density	Size	Composition
Hollow Zirconium Spheres	–	–	Zirconium oxide

Norton Company
One New Bond Street
Worcester, Mass. 01602
(617) 853-1000

Product Designation	Density	Size	Composition
Alumina Bubble 8/72	–	72 mesh	Aluminum oxide

5.2 Suppliers of Organic Hollow Spheres

Emerson & Cuming, Inc.
Dielectric Materials Division
Canton, Mass. 02021
(617) 828-3300
Product: Eccospheres EP (Hollow Epoxy Spheres)

Type	Avg. Density, g/cc	Size, (In.)	Price ($/gal)
EP-100	0.21	0.07–0.15	9.25
EP-250	0.45	0.13–0.18	9.25
EP-300	0.30	0.19–0.25	9.25
EP-1500	0.33	0.25–0.50	9.25

Product: Eccospheres HS (Hollow Glass-Reinforced Thermoplastic Spheres)

Type	Density, g/cc	Approximate Size (In.)	Price ($/1000 pieces)
HS-30	0.13	2	210.00
HS-60	0.16	2	270.00
HS-90	0.20	2	315.00
HS-120	0.31	2	360.00

Union Carbide Corporation
Thermosetting Resins and Compounds
270 Park Avenue
New York, N.Y. 10017
(212) 551-4831
Product: Hollow Phenolic Microballons
Price range from $1.00-$1.50/lb

Product Designation	Density, g/cc	Size (In.)
BJO-0840	0.25-0.35	0.0002-0.005
BJO-0930	0.21-0.25	0.0002-0.005
BJO-0931	0.15-0.20	0.0002-0.005

Arco Polymers Inc.
1500 Market St.
Philadelphia, Pa.
(215) 557-2000
Product: Dylite Expandable Polystyrene
Price range from $.40-$.45/lb

Product Designation	Density	Size, Unexpanded (In.)
F-40/F-64-A	Approximately 38 pcf	0.0469-0.0234
F-40-B/F-64-B	Expansion yields	0.0331-0.0165
F-40-C/F-64-C	densities ranging	0.0234-0.0116
SE-57-A	from 0.8 to 20 pcf	0.0787-0.0165
SE-57-B	Bead size increases	0.0469-0.0165
SE-57-C	when expanded	0.0232-0.0117

Dow Chemical Company
2020 Dow Center
Midland, Mich. 48640
(517) 636-1000
Price, expanded $5/lb, unexpanded $3/lb

Product	Density, Unexpanded (lbs/ft^3)	Density, Expanded (lb/ft^3)	Size, Unexpanded (μ)	Size, Expanded (μ)
Saran Microspheres	82-83	1.8-2.2	4-20	10-100

BASF Wyandotte Corp.
1532 Biddle Ave.
Wyandotte, Mich. 48192
(313) 282-3300
Product: Expandable polystyrene beads.
Price range $.40-$.45/lb

Product	Bulk Density	Size
Styro por BR31	39 lbs/ft^3	20-50 mesh
Styro por BF55	39 lbs/ft^3	12-40 mesh
Styro por BR26	39 lbs/ft^3	14-40 mesh

6. HISTORY AND FUTURE COST AND AVAILABILITY

The technical development of hollow, unicellular, glass microspheres began in the early nineteen fifties with the Kanamite process as outlined in the McLaughlin Patent (U.S. Patents 2,676,892, April, 1954). The development of hollow microballoons from alkali metal silicate solution was documented by Veatch et al., U.S. patents 2,797,201, 1957, assigned to Standard Oil of Ohio, and in numerous subsequent patents. Emerson & Cuming Inc. licensed and later bought all rights to the Sohio patented process. E&C now manufactures a line of products that includes glass, silica, and urea formaldehyde microspheres and larger diameter epoxy hollow spheres. E&C has licensed Union Carbide under the Sohio patent to produce phenolic microspheres. 3M has developed their own process for the production of glass microspheres. Philadelphia Quartz has recently entered this market with a product which is produced by simply spray drying special alkali silicate solutions. Recently, there has been increased activity and interest in the utilization of fly ash "floats" as a low cost source for ceramic hollow microspheres.

Hollow microspheres range in diameter from 10-300 μ, and have a wall thickness of 1-4 μ and the visual appearance of a free flowing powder. In current technology, they are usually used as a filler in liquid thermosetting resins, but also in thermoplastics such as vinyl plastisols.

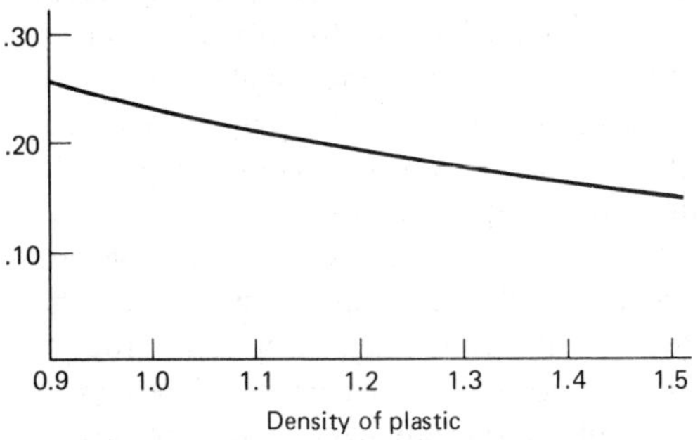

Fig. 19-3. Raw material cost of hollow glass microspheres in plastics of various densities. (Density of glass microspheres: 0.3 g/cc @ $.77/lb.) Raw material cost savings result if the price of the resin falls above the curve.

Figure 19-3 depicts the cost savings realized from utilizing hollow glass spheres in resins of various densities.

Glass and ceramic spheres are being marketed in the multi-million pounds per year range in the price range of $.50-1.00/lb for glass and $.18-.25 for ceramic (fly ash) spheres. About 500 lb of carbon microspheres were sold during 1975 at an average price of $35./lb. Hollow spheres will continue to show a strong growth trend and will be available in sufficient quantities to meet the needs of an expanding industry.

7. APPLICATION

7.1 Discussion

Hollow spheres are utilized in numerous areas to enhance and/or modify the properties of various resinous materials. Property improvements gained from using hollow glass beads are: water resistance, compressive strength, shrinkage and impact strength. Other significant advantages derived from utilizing hollow spheres are:

1. *Low density*—lighter finished parts can be made.
2. *Uniformity of size*—translates into consistency in batch to batch formulation of hollow spheres and resin.
3. *Low porosity and surface-to-volume ratio*—little resin is absorbed by the spheres, so low viscosity levels can be achieved at high filler loadings.
4. *Chemical stability and inertness*—of hollow glass spheres allows premixes of resin and spheres to be stored for later use. The glass spheres are inorganic and noncombustible, so they present no fire hazard.
5. *Crack arresting characteristics*—result in easy machining, sanding, nailing, and finishing when the matrix is cured.

Hollow spheres are used effectively in liquid resins such as epoxies and polyesters. Their use is limited to parts manufactured by low pressure processes. They can not be used in high pressure injection molding and extrusion because of their limited crush strength.

7.2 Application of Inorganic Hollow Spheres

7.2.1. Use in Thermosetting Compounds. In the recreation area, items fabricated from fiberglass reinforced polyester with hollow spheres added are strong and lightweight. Fiberglass reinforced polyester is widely used as the backup material for rigidizing thermoformed parts. The rigidizing process is needed because the thermoformed skin is too thin and does not have the desired stiffness and impact properties. When hollow glass spheres are added to the glass fiber, reinforced polyester, improved stiffness and drop impact resistance is noted. (Abstract 1)

Lightweight parts made by this process are being used in the recreation area for camper and pickup truck tops, boat hulls, and snowmobile shrouds.

Sporting goods manufacturers use hollow spheres in skis and tennis rackets. Some bowling ball cores are composed of hollow glass sphere composites. The bowling ball size can be maintained constant, while the weight of the ball can be varied by varying the content of hollow glass spheres in the core. (Abstract 2)

Hollow glass spheres are gaining use in the sanitaryware market. Bathroom items are being manufactured using hollow sphere-filled, fiber-glass reinforced polyester laminated to thermoformed acrylic sheet. (Abstract 2) In one application, a bathroom scale platform formerly produced from urethane foam is currently being manufactured utilizing hollow sphere-filled polyester. (2)

Cultured marble makers utilize the spheres to aid in weight reduction and improve crack resistance of parts. (3)

The hollow spheres are used in making syntactic foam cores for boat hulls and decks. (4) Unlike conventional foams, made by adding a blowing agent to the resin to form a cellular structure, syntactic foams are made by incorporating hollow spheres in liquid resins such as polyester, epoxies and polyurethanes. Syntactic foams are much stronger than conventional foams and, as a result, they are finding use in the hydrospace industry where their greater strength allows them to withstand high hydrostatic pressures up to 10,000 psi or more. (5)

The use of hollow spheres for making synthetic wood gives the manufacturers unusual benefits. Made from liquid resins (usually polyester), the synthetic wood product has the feel, look, and texture of real wood. A typical formulation recommended by Emerson & Cuming for synthetic wood consists of 100 parts of catalyzed polyester resin, 40 parts of wood flour, and 8 parts of IG101 microballoons. This casting has a specific gravity of 0.9 g/cc, flexural strength of 5000 psi, tensile strength of 3500 psi and compressive strength of 15,000 psi.

Hollow glass spheres find additional use in patch compounds. Patch compounds are commonly used in the polyester gel coat industry. When defects in finished parts occur, patch compounds are used to repair the part which is then sanded to shape.

Hollow glass spheres added to the polyester patch compound ease the sanding and finishing operation in direct proportion to the amount of spheres added. However, when the spheres are added, high shear mixing should be avoided due to the friable nature of the spheres. (Abstract 2)

Hollow glass spheres have been utilized in items manufactured for teaching aids. One such item formed from polyester and hollow spheres is a replica of a human torso including a detailed cranial section. This anatomy model was rationally cast in rubber molds. The addition of the hollow spheres reduced the item weight by 27% and material cost by 33%. (2)

Hollow microspheres have been useful as a filler for many ablative components, because of their contribution to properties such as light weight, insulation, and retention of compressive strength. The Apollo heat shield that protects the astronauts upon reentry into the earth's atmospheres consists of a microballoon filled epoxy novolac in a honeycomb structure. Other aerospace components have included insulation slivers for solid propellant rockets and rocket exit cones. More common forms of transportation use hollow spheres, too. In a commercial aircraft, hollow spheres helped solve the problem of producing a light-weight, high strength, low shrinkage part. Glass hollow spheres-resin mixtures provided greater strength

Table 19-4. Hollow fly ash microspheres (Fillite) in epoxy matrix.

Properties	35% Loading by weight	50% Loading by weight
Density lb/ft^3	53	47
Specific gravity	0.85	0.75
Tensile strength lb/in.^{2}	1,920	–
Shear strength lb/in.^{2}	4,115	1,817
Bulk modulus lb/in.^{2}	580,000	580,000
Co-efficient of linear expansion °F	40×10^{-6}	40×10^{-6}
Thermal conductivity BTU/hr/ft/ft^2 °F	0.1	0.1
Water absorption after 160 hours at 4000 lb/in.^{2}	0.55	1.83
Compressive modulus lb/in.^{2}	395,000	367,00
Compressive strength lb/in.^{2}	8,300	5,820
Volume resistivity	10^{12} ohm/cm^3	10^{12} ohm/cm^3
Surface resistivity	10^{12} ohm/cm^2	10^{12} ohm/cm^2
Insulation resistance at 18°C	11×10^6 to 60×10^6 ohms	11×10^6 to 60×10^6 ohms
Resistance to tracking	No tracking occurred	No tracking occurred

Source: Fillite U.S.A.

than plastic hollow sphere with a minimum increase in weight. (3)

The electronics industry has utilized the hollow fillers for adjusted-dielectric foams, potting and encapsulation systems, and low cost/volume molding compounds. The syntactic foams have low dielectric constants and low dissipation factors. Tables 19-4 through 19-6 present typical properties of syntactic foams made by use of hollow microsphere fillers. Table 19-4 gives the properties of a syntactic foam based on 35% by weight and 50% by weight of hollow fly ash microspheres in an epoxy matrix.

Table 19-5 compares some properties of syntactic foams made from glass microballoons and hollow carbon microspheres.[1]

Table 19-6 shows greater hydrolytic stability of hollow carbon microsphere syntactic foams as compared with glass microballons.[7]

7.2.2. Use in Thermoplastic Compounds. Hollow glass spheres work well as a bulk filler for some thermoplastic resins. Their use in a PVC plastisol results in a part with high strength and low density.[6] They find additional use in low density adhesives and sealants.

Table 19-5. Properties of syntactic epoxy matrix foams based on glass and carbon microspheres.

Property \ Microsphere		A-50	A-100	A-200	Glass Micro-Balloons*
Density	g/cc	0.66	0.68	0.68	0.65
Compressive strength	psi	12800	12100	10200	9420
modulus	10^3 psi	305	303	278	266
Hydrostatic crush point	psi	20000	18800	15900	14000
Bulk modulus	10^3 psi	480	456	444	336

*FTD-202.

Table 19-6. Hydrolytic stability of syntactic foams.

Filler	COMPRESSIVE STRENGTH (psi) Before	After Boil	Moisture Pick Up %	Aft.-Exposure Strength Density Ratio (psi)
Carbon microspheres	13,800	13,900	0.4	14,300
Glass microballoons	11,000	8,900	1.0	10,500
Silane-treated glass microballoons	11,400	10,500	0.5	12,400

7.3 Application of Organic Hollow Spheres

7.3.1 Use in Plastics Applications. Hollow organic spheres have found application in a wide range of products made by open mold processes using themoset resins such as polyesters, epoxies, polyurethane, phenolics, etc. They are used in synthetic marble and synthetic wood products and in other applications where reduced weight is a major requirement.

Hollow organic spheres are currently being utilized in a number of areas. Hollow spheres made of epoxy resin are used as a low density bulk filler for plastics and ceramics. Syntactic foams made by incorporating hollow organic spheres into liquid resins (epoxies, polyesters, urethanes, etc.) find use in the furniture market, in the marine industry as buoyancy materials for flotation, in the aerospace industry in ablative heat shields, and in adhesives and coatings for low weight. Prosthetic devices effectively use foams based on the plastic microballoon.

Saran microspheres, made by Dow Chemical Company, consist of vinylidene chloride/acrylonitrile. They are being utilized in syntactic foam preparations. When compared with glass and phenolic spheres, they were seen to exhibit similar physical properties at less weight per cent of spheres in syntactic foams of equivalent densities. In synthetic marble, the spheres improve impact resistance. Phenolic microballoons also find use in areas where reduced weight or controlled weight is essential, such as in bowling ball cores.

Syntactic foams are also made of hollow expandable polystyrene spheres. For special applications, high strength fiber glass-reinforced thermoplastic spheres such as Eccospheres HS, made by Emerson & Cuming Inc., are utilized in syntactic foams to create stronger and more efficient composite systems.

7.3.2. Other Usage. Hollow organic spheres have been incorporated in Portland cement and plaster of Paris. They can be mixed into explosives compositions to adjust density and detonation properties. In synthetic marble, the spheres improve impact and crack resistance.

8. CURRENT AND PROJECTED UTILIZATION

Manufacturers are continually seeking ways to improve, modify, and produce quality items to meet the ever increasing needs. Hollow spheres have become a useful material in areas where minimal weight is a major requirement.

The incorporation of hollow microspheres in liquid thermosetting resins results in a syntactic foam which has low density, high strength and good modulus-to-weight ratios, good machining properties, and excellent thermal and electrical characteristics. An important application for syntactic foams is for buoyancy end products, especially for deep submerging components, while the lower cost and lower density chemically foamed plastics may be suitable for surface flotation requirements. The chemically foamed plastics of higher density, 15 lb/cu ft and above, may be suitable for operating depths to 1800 ft, as in the case of offshore petroleum systems. However, for applications to depths of 10,000 ft, which will be required by submarines and rescue vehicles, only the glass syntactic foams at densities of 30–40 lb/cu ft have shown satisfactory results. The hydrostatic collapse strength of chemically foamed plastic with a density of 30 lb/cu ft is about 3,000 psi, whereas syntactic foams at the same density have a strength of 7,000–11,000 psi.

Emerson & Cuming and 3M supply formulated syntactic foam systems that can be used for casting and void filling in electronics or buoyancy applications. These are usually based on an epoxy matrix system. However, other matrices such as low density castable rubber materials are also available. Emerson & Cuming also sells molded syntactic foam in block form. Their Eccofloat 31 for deep water submergence has a density of 31 lb/cu ft and is available in blocks 12 in × 12 in. × 6 in.

The lower density Saran and the low cost ceramic products from fly ash may grow more rapidly than the current market leaders, glass and phenolic. Ceramic hollow microspheres have been used as an additive to achieve strong, heat insulating, and low absorption concrete. In this application, the microspheres maintain compressive strength while imparting unusually high insulation qualities, both heat and sound. They reduce shrinkage of the finished product, thereby reducing cracking. The water resistance, workability, and pumpability are improved. Multicellular glass modules are also available.

Hollow microspheres will be used increasingly in the following types of products:

Structural cores for wall panels
Furniture cast parts and decorative plaques
Flotation or buoyancy applications in boats and submarines
Ablative composites, such as heat shields and solid propellant-case insulations
Electronic applications, such as potting and encapsulation compounds
Artificial limb fabrication
Lightweight concrete

9. ABSTRACTS AND REFERENCES TO LITERATURE AND PATENTS

9.1 Abstracts

1. David I. Netting and Christopher G. Smith, Use of Hollow Silicate Spheres in Fiber Reinforced Plastics and Syntactic Foam Cores, *Proc. 30th SPI RP/Composite Conference*, (1975), Section 7-A.

   The addition of hollow glass spheres to fiberglass reinforced polyester laminates resulted in reduced raw material cost, improved impact resistance, reduced weight of the laminate, and increased stiffness of the part. Laminates composed of this material have been used in the furniture, recreation, marine, sanitaryware, and sporting goods industry.

2. Warren Beck, et al., Improved RP/C Marine Constructions Using Hollow Glass Microspheres, *Proc. 29th. SPI RP/Composite Conference*, (1974), Section 10-D.

   Hollow spheres were added to a gel coat back-up. Addition of the spheres resulted in reduced heat line cracking of the gel coat, reduced blistering and improved water resistance. The possibility of utilizing hollow spheres in laminate cores and as mold and patch compounds was discussed. The feasibility of utilizing hollow spheres in conventional spray-up equipment was demonstrated.

References

1. Amagi, Y. et al., "Hollow Carbon Microspheres From Pitch Material and Their Applications," Kureha Chemical Industry Co., Ltd., Tokyo Research Laboratory, Japan.
2. "Glass Bubbles Adapt Designs to the Resin-Cost Crunch," *Plastics World*, 24, (Apr. 21, 1975).
3. "New Era in Fillers," *Plastics World:* 54, (April 15, 1974).
4. "Yacht Design Waives The Rules To Rule The Waves," *Plastics World*, (July 23, 1973).
5. Emerson & Cuming Inc., Technical Bulletin 826 E&C 20M 4/69.
6. Alford H. E. and Veatch, F. "Glass Microballoon Particles, A Low-Density Filler," Reprinted from *Modern Plastics*, (Nov. 1961), by Emerson & Cuming, Inc.
7. Ferguson O.L. and Sharer R. G., "Syntactic Foams of Hollow Carbon Microspheres in Resin Matrix," Annual S.P.I. RP/CD-1970.

Patents

1. U.S. Patent 2,797,201 (June 25, 1957) F. Veatch, et al. to The Standard Oil Company, Cleveland, Ohio.

   The invention describes a method by which hollow spherical particles, which are substantially free from holes, are formed from film forming solutions to which a latent gas is added.

2. U.S. Patent 3,365,315 (June 23, 1968) Warren R. Beck, et al. to Minnesota Mining and Manufacturing Company.

This invention relates to minute glass bubbles formed by a process involving direct conversion of glass cullet particles into glass bubbles by heating.

3. U.S. Patent 3,699,050 (Oct. 17, 1972) Craig Henderson to Emerson & Cuming, Inc.
   This application relates to a product formed by the spray drying of an alkali borosilicate solution, which product is subsequently utilized as feed to a spheridizing furnace which produces hollow glass spheres of a diameter in a range from about 5-5000 μ.

4. U.S. Patent 3,796,777 (Mar. 12, 1974) David I. Netting to Philadelphia Quartz Company.
   Solutions of sodium silicate and ammonium pentaborate are mixed, spray dried, and the product from the spray drier is then further dried to a water content below about 7%.

10. GLOSSARY

Blowing Agent—A substance incorporated in a material for the purpose of producing a gas, which results in a foam product.

Encapsulation—The process in which a material, or an assembly of small, discrete units, is coated with, or imbedded in liquid resin which is then cured to a rubber or hard plastic.

Friable—A substance that is fragile and easily crumbled or pulverized.

Laminate—A composite made of any one of several types of thermosetting plastic (phenolic, polyester, epoxy, or silicone) reinforced by planar arrays of paper, cloth, asbestos, wood, or glass fiber.

Micron—A unit of length equal to one thousandth of a millimeter.

Noncombustible—A state of being incapable of ignition.

Syntactic Foam—A foam in which the gas is incorporated in the resin by the addition of hollow spheres as opposed to adding a chemical blowing agent to the resin.

Unicellular—A material consisting of a single cell.

Section VIII
Flake Reinforcements

20
High Aspect Ratio Mica and Other Flake Reinforcements

R. T. Woodhams and M. Xanthos
University of Toronto
Toronto, Canada

CONTENTS

INTRODUCTION

Flakes or platelets represent a special class of reinforcing fillers for thermoplastics and thermosets. Mica is one such type of filler and is a particularly abundant mineral. By virtue of its crystal habit, it may be easily cleaved into thin flakes by ordinary grinding methods and with special care, may be delaminated into very thin flakes. Such flakes, when suitably aligned in a plastic matrix, may impart a relatively high degree of reinforcement in terms of strength and stiffness. Such ultra-thin flakes are characterized by a rather large ratio of the average flake diameter to thickness, such a ratio being defined as the "flake aspect ratio." Conventionally ground micas usually possess rather small average aspect ratios and, therefore, cannot impart a high level of reinforcement. Furthermore, special processing techniques are required to minimize damage to high aspect ratio flakes and to permit maximum alignment of the flakes in the molded part.

This chapter describes methods for producing and characterizing high aspect ratio (HAR) mica and its use as a reinforcing agent in thermoplastics and thermosets. At high volume fractions in thermosets, modulus values equal to that of aluminum are possible. Strength is usually high when care is taken to eliminate voids and maintain good alignment of the flakes. Impact strength is generally low. The most satisfactory methods of fabrication involve modifications of techniques commonly used for glass fibers, particularly the use of spray-up and sheet moulding methods.

When reinforcing thermoplastics with high aspect ratio mica, great care must be exercised in order to minimize flake breakdown during normal mixing, compounding, extrusion, or injection molding operations. The use of hot runners, large gates, and low stresses permit better retention of strength and modulus during injection molding, for example. A distinct advantage of flakes compared to fibers is the greater ease of processing at high volume loadings so that highly filled composites are feasible with mica, usually up to 50% by volume or 75% by weight. Excellent flake alignment can often be achieved during extrusion or injection molding operations since the flakes tend to orient uniformly in response to flow shear stresses in a manner which enhances the planar reinforcement.

Silane coupling agents appear to function equally well with mica as with glass fibers and are recommended when long term retention of mechanical and physical properties in the presence of moisture is desired. In addition to relatively low cost, mica has several other advantages when used as a reinforcing agent in plastics, such as, particularly excellent dimensional stability of molded parts, low warp, planar reinforcement, reduced creep, excellent electrical properties, reduced wear of processing equipment, increased heat distortion temperature, reduced flammability, low permeability, plus excellent resistance to weathering and corrosive attack by acids or alkalis.

The major disadvantages are lower strength and impact toughness compared to glass fiber reinforced plastics. However, flow orientation is advantageous with mica but less desirable in the case of glass fibers. A judicious blend of mica and glass fibers may often yield a better balance of strength, stiffness, and impact toughness than is obtainable by the use of just mica or glass fibers alone.

While the greater part of this chapter is devoted to mica, there are several other classes of materials which can exist in flake form. Many inorganic compounds can be crystallized in the form of thin hexagonal flakes, particularly alumina, aluminum boride, and silicon carbide. Such extraordinary stiff and strong crystals exhibit properties comparable to the "whisker" or needle-shaped crystals developed in the 1960's. Amorphous materials such as glass, and polycrystalline materials such as the metals and ceramics, may also be produced in flake form. Synthetically produced flakes are too expensive for most domestic applications and their use is restricted to applications involving corrosion resistance, impermeability, scratch resistance, decorative coatings, electronics, and extraordinary mechanical properties. A brief description of these other flake materials is included for completeness.

Mica represents the cheapest and most abundant flake material with prices ranging from \$.01-.02 (U.S.)/lb to \$.16/lb for specially treated high aspect ratio grades. Glass flakes (non-reinforcing types) are normally priced at \$.23/lb. Most of the other flake materials discussed in this chapter have only been produced on a research scale and are not available commercially.

2. MICA

2.1 Introduction

Mica is a well known class of minerals noted for its ease of cleavage into thin, highly flexible, resilient sheets. The most common commercial varieties are muscovite, phlogopite and biotite. Most micas are predominantly aluminum silicates of potassium which may also contain various proportions of magnesium, iron, lithium, or fluorine, depending upon the particular source and variety. Since the chemical compositions are quite variable, micas obtained from different sources may range from colorless and transparent (mostly muscovite and synthetic fluorophlogopite) to highly colored, opaque varieties which may include shades of pink, amber, yellow, brown, green, violet, and black. Although the Suzorite variety of phlogopite has a lustrous, golden-brown sheen, other phlogopite varieties may be only slightly colored.

At the present time, the use of mica in the United States plastics industry is quite small and represents less than 1% of the total annual consumption. The largest annual consumption of mica is in gypsum plasterboard cement (61 million lb), followed by asphalt roofing compositions (57 million lb), paint additives (48 million lb), electrical and electronic components (26 million lb), molded rubber products (14 million lb), and plastics (1.8 million lb). Numerous miscellaneous uses make up the remaining 39 million lbs. Although mica has been used for many years as a cheap extending filler for plastics, particularly when high stiffness combined with good dielectric properties were desired, its use as a primary reinforcing agent for plastics has only recently been developed.[1,2,3,4,5,6] As normally wet or dry ground, mica has negligible reinforcing ability since the flakes do not possess the required shape for good reinforcement. However, when the flakes are suitably delaminated to form very thin flakes such as normally used for the manufacture of reconstituted mica papers, a high level of reinforcement becomes possible in most thermosetting and thermoplastic resins. These specially delaminated flakes have been designated as "high aspect ratio" mica,[7] where high aspect ratio (HAR) refers to the ratio of flake diameter to its thickness. In general, flake aspect ratios above 100 are needed for good reinforcement and care must be taken to preserve this aspect ratio in the final fabricated part. The high degree of reinforcement is made possible by virtue of the overlapping character of high aspect ratio flakes, similar to that of fish scales. At the highest volume loadings, mica reinforced composites have a modulus equal to that of aluminum, and similarly give a characteristic metallic sound when struck sharply or dropped on a hard surface. Strength values are also quite high (200–300 MN/m^2) so that HAR mica may be employed as a partial or total replacement for more expensive reinforcing agents such as chopped glass fibers or asbestos.

2.2 Delamination of Mica

As a result of the peculiar crystalline structure of the micas, the individual silicate layers are easily cleaved or split into thin flakes by various physical or chemical treatments. Each silicate layer is approximately 10 Angstroms thick, so that in theory, at least, the limiting thickness of the flakes is also 10 Angstroms. Since it is seldom possible to achieve a fully delaminated state, most flakes of mica will consist of multiple layers. In practice it is difficult to achieve flake thicknesses much below 1 μ, or the equivalent of 1000 silicate layers. In this respect, mica differs from the single crystal flakes derived from aluminum diboride, silicon carbide, and alumina, or from amorphous materials such as glass, or from polycrystalline materials such as steel. Each flake of mica is a miniature laminate, a feature which contributes a degree of inherent toughness and resistance to fracture perpendicular to the basal plane. The commercial delamination of mica has been the subject of numerous patents. These methods may be characterized as "wet" or "dry" according to whether the delamination was carried out in a dry state or in the presence of a fluid medium such as water. Since wet grinding involves the use of a fluid which must ultimately be removed by filtration and evaporation, wet ground products are considerably more expensive. However, wet grinding preserves the natural lustre and sheen of mica so that it commands a premium for special applications where this characteristic is desirable (such as in the manufacture of paint, wall paper, rubber, plastics, welding rod, textiles, etc.). Wet ground mica is normally characterized by clean-cut edges, higher aspect ratios, smooth surfaces, and the ability to disperse easily in many liquid vehicles. Although dry ground mica was formerly characterized by rough, abraded surfaces, hackled edges, and poor mixing qualities, recent improvements in dry grinding techniques now permit the production of very uniform, high quality flakes without resorting to wet grinding.[8]

Mechanical grinding invariably produces a broad distribution of shapes and sizes which must usually be fractionated in order to isolate the high aspect ratio component. After elutriation and sieving the HAR mica appears as shown in Fig. 20-1.

Dry grinding of mica is conventionally carried out by high speed hammer mills of the cage-disintegrator type equipped with screens and air separators. From the mill screens, the mica is further graded according to mesh size on multiple deck vibrating screens. Some firms have successfully used rod mills. Such dry ground mica is used principally in the manufacture of composition roofing, oil well drilling, etc., where extreme fineness, good color, and high sheen are not important. For ultrafine grinding, less than 325 mesh, fluid energy attrition mills are preferred which employ a high pressure jet of air, steam, or inert gas to effect the delamination.

In the time honored method of wet mica grinding, muller mills or chaser mills about 10 ft in diameter by 3½ ft deep are charged with a ton or more of scrap mica together with half a ton of water and ground with four heavy rollers mounted on horizontal arms and driven by a central shaft. The bottom of the pan and the rollers are usually faced with end grain hardwood blocks. The rollers or mullers require 40–50 hp to rotate them at 30–40 rpm. After a period of approximately 8 hr, the pulverized material is flushed into troughs where the heavier unground particles and grit settle out and the overflow goes to large settling tanks. After 24 hr, the supernatant water is drawn off and the recovered mica centrifuged, dried on a steam table or in a rotary drier, and finally screened to size.

The grinding of mica for the purpose of manufacturing reconstituted sheets of mica for electrical applications is a highly specialized technique. Large mica splittings, such as normally imported from India, are quite expensive since they must be hand cobbed and involve much specialized labor. Hence, it has been found economically attractive to process scrap mica into thin flakes and subsequently readhere them together again into continuous sheets. In order to prepare reconstituted sheets of highest strengths and optimum dielectric properties, it is necessary to delaminate mica into very thin flakes with high aspect ratios. Great care is needed to eliminate trace impurities that would impair perfect realignment and cohesion of the flakes after delamination. The patent literature reveals numerous techniques for producing such ultra thin flakes for the purpose of making reconstituted mica papers, binders, pigments, or fillers.[9,10,11,12,13,14] More efficient delamination is often obtained when muscovite mica is first calcined at 800°C.[15,16,17] At these temperatures, bound water is liberated and partially opens the crystal "books" so that only mild mechanical action is needed to complete the cleavage. Very intense mechanical action is made possible in fluid energy mills wherein the flakes are suspended in a fluid medium (usually an inert gas) and subjected to impingement by high velocity gas jets at elevated temperatures. "Micronized" mica is produced in a special type of dry grinding machine called a Micronizer. This ultrafine material in the 5–20 μ size range

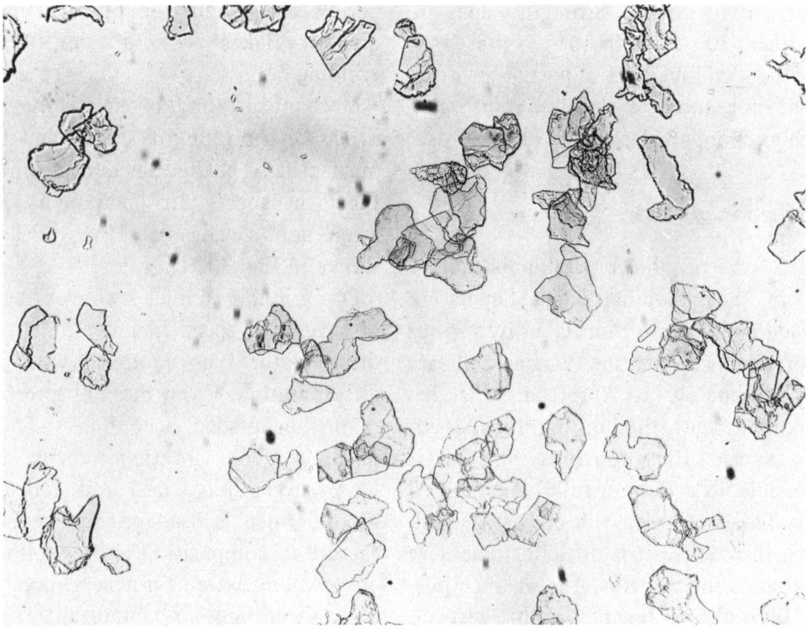

Fig. 20-1. A photograph of HAR mica flakes showing the highly irregular shapes and thicknesses of the individual flakes. Magnification 13.6×.

is produced by a disintegrator that has no moving parts but uses jets of high pressure, superheated steam. The Majac mill is another type of fluid energy mill which consists of a chamber with two horizontally opposed jets. A screw conveyor feeds the raw mica continuously and the particle size of the recovered product can be controlled over a broad range down to micron sizes. In the method of Ruzicka,[17] controlled delamination is claimed down to colloidal dimensions with aspect ratios as high as 1000/1 and surface areas of 2500 m^2/g. High rates of delamination, complete conversion and application to different types of mica, especially muscovite and phlogopite, are also claimed. The Ruzicka method involves shearing of the mica flakes in a swirling liquid or gaseous fluidized medium which is subjected to rapid temperature changes. The water of hydration, which is released at the elevated temperatures, causes a successive peeling of the flakes which can be controlled to yield flakes within a particular thickness and diameter range. Mica flakes have also been successfully delaminated into thin, high aspect ratio flakes using ultrasonics.[7,18,19] In these treatments, mica flakes were usually suspended in an aqueous medium and subjected to ultrasonic cavitation for a short period of time. The addition of -325 mesh silica sand to the suspension greatly increases the efficiency of delamination.[7,19] Stavely[21] has designed an apparatus for delaminating mica by passing a slurry through a progressively restricted annular space formed by a cylinder and an inner conical shaped rotor. The aspect ratios of the flakes are between 100 and 10,000 with diameters between 4 and 400 μ. Chemical treatments with strong acids,[22] wetting agents,[23] and soluble salts[24] have also been employed to permit a faster and more efficient delamination process. In the manufacture of reconstituted mica paper, care must be taken not to introduce ionic impurities which could contaminate the newly formed surfaces and thereby prevent readherence at a later stage. In order to prevent readhesion, a minor percentage of potassium ion, wetting agent, silane coupling agent, or colloidal additive may be added to the suspension prior to recovery. Such surface treatment of mica is usually desirable when the flakes are to be subsequently employed in resin systems.

2.3 Aspect Ratio Classification

Most delamination methods produce a very broad distribution of sizes and aspect ratios. In order to isolate those flakes with a particular diameter and thickness, the mixture must be classified both by sieving and elutriation. Sieving permits the isolation of any selected diameter range by choice of the appropriate sieves. Thus the designation -35 +50 U.S. mesh size implies that the diameters of the flakes were small enough to pass through 35 mesh (0.500 mm) but too large to pass through 50 mesh (0.297 mm). In most cases, however, such sieve fractions contain variable quantities of "fines" due to imperfect separation. Prolonged or repeated wet sieving helps to reduce the level of such fines. In dry sieving, static electricity can impair efficient separation by causing particles to adhere to each other or to the surfaces of the sieves. Flake materials also tend to "blind" sieves. That is, the larger flakes form an overlapping barrier which prevents the passage of smaller flakes which would otherwise pass through the openings. A vibrating Sweco Separator has been found to operate satisfactorily with aqueous mica dispersions.[25] Haller[9] has employed an oscillating screen in the form of a trough which operates continuously without clogging.

While sieving is a relatively simple mechanical operation which permits the isolation of any desired flake diameter range, such flakes will normally have different thicknesses or aspect ratios. Wide variations in flake thickness are undesirable for reinforcement purposes since the flakes do not efficiently align, create uneven stress distributions, and reduce strength and modulus properties in molded composites. Ideally, therefore, it would be desirable to isolate those flakes which have uniform thicknesses and diameters.

Elutriation provides a simple technique for isolating flakes with similar diameter and thickness dimensions. According to Heywood,[26] the terminal settling velocity of a non-spherical particle such as a mica flake is governed by the equations[27]

$$\frac{PR^2}{\rho u^2} = \frac{tD^2(\rho_m - \rho)\rho g}{\mu^2} \qquad (1)$$

and

$$R = \frac{uD\rho}{\mu} \qquad (2)$$

where $PR^2/\rho u^2$ is a dimensionless function involving the Reynolds Number R, and P, which is the force per unit area exerted on the face of the mica particle when oriented perpendicular to the motion of the fluid medium. The remaining symbols are

t = flake thickness in cm
μ = viscosity of the fluid medium in poises
g = gravitational constant in cm/sec^2
u = sedimentation velocity in cm/sec
ρ = density of fluid medium in g/cm^3
ρ_m = density of mica in g/cm^3
D = effective flake diameter in cm

The process of elutriation normally involves passing a current of air or water upward through a column in which the mica particles are suspended. By carefully adjusting the upward velocity, the thinner flakes whose sedimentation velocities are less than the upward velocity of air or water, will be recovered at the top of the column. If the particle is situated at the axis of a vertical tube in which the fluid is flowing in streamline motion, the velocity near the center of the tube will be twice the mean velocity due to the parabolic velocity profile in the fluid. Thermal gradients, suspension instability, and adsorbed air bubbles may interfere with efficient separation in water.

The classification of mica according to aspect ratio as proposed by Haller[9] is summarized in Fig. 20-2. Assuming that the entire area of the graph represents a complete distribution of flake geometries to be classified, the area *under* any of the hyperbolic curves indicates the particular flake sizes that will pass upwardly in a moving column of water and thus be separated from those flakes having geometries represented by the area *above* each respective hyperbolic curve, the latter flakes settling to the bottom of the column due to a greater sedimentation velocity.

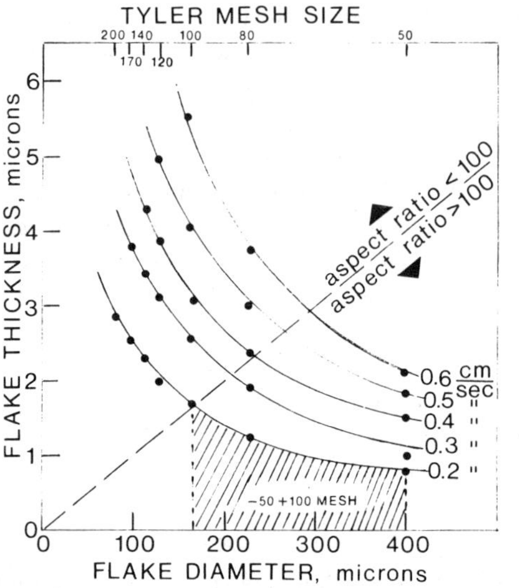

Fig. 20-2. The cross-hatched region represents the dimensions of all mica flakes (thicknesses and diameters) that would be isolated from a mixture using a water elutriation rate of 0.2 cm/sec and a combination of a 50 mesh and a 100 mesh Tyler screen. This mica fraction would have an average aspect ratio above 100.

Referring to Fig. 20-2, it is seen that the flake geometries elutriated at a flow rate of 0.2 cm/sec will range from extremely thick flakes with small surface areas to very thin flakes with large surface areas. By now combining the steps of elutriation and sieving, any particular flake aspect ratio may be isolated. For example, the use of a 100 mesh Tyler screen (147 μ openings) permits the retention of the flakes represented by the cross-hatched region in Fig. 20-2. It is evident that in order to isolate a particularly narrow aspect ratio distribution from any given sample, the proportion of recovered flakes will be correspondingly small.

Xanthos[25] successfully employed aqueous elutriation using a cylindrical glass column 15 cm (inside diameter) by 2 meters in length, followed by wet sieving using a Model H-2D Sweco Separator.[28] The addition of a silane coupling agent is often advantageous when improved long term moisture resistance is desired in molded composites. Moreover, it also prevents the freshly delaminated flakes from readhering to one another during the subsequent drying stage. Such adhering flakes may be difficult to redisperse in a resin and may create weak planes in the resulting composite.

The "Infrasizer" designed by Haultain[29] for the classification of fine powders by air elutration has been used in many laboratories throughout the world. The apparatus, as shown in Fig. 20-3, consists of seven vertical cones having different diameters and lengths. Air is metered at a constant flow rate into the longest cone, which contains a 50-g sample of the powdered material for classification. The smallest particles, which are entrained in the first cone, are carried over into the bottom of the second cone, and so on until the finest particles are finally trapped at the outlet of the remaining cone by means of filter paper. The air turbulence is considerable in the lower region of each cone, but becomes a gradual quiet upward movement devoid of turbulence in the upper regions. The flat top of each cone enables the particles in the center, where the velocity gradient is a maximum, to be carried over into each succeeding cone. After 30 min to 1 hr, the air flow is discontinued and the fractions collected from each cone. Using this technique, Lusis[30] was able to fractionate commercial muscovite mica samples and subsequently sieve each fraction so as to isolate 4 flake aspect ratios from 15 to 60. The use of conical shaped vessels apparently permits a more efficient classification than the use of cylindrical vessels.

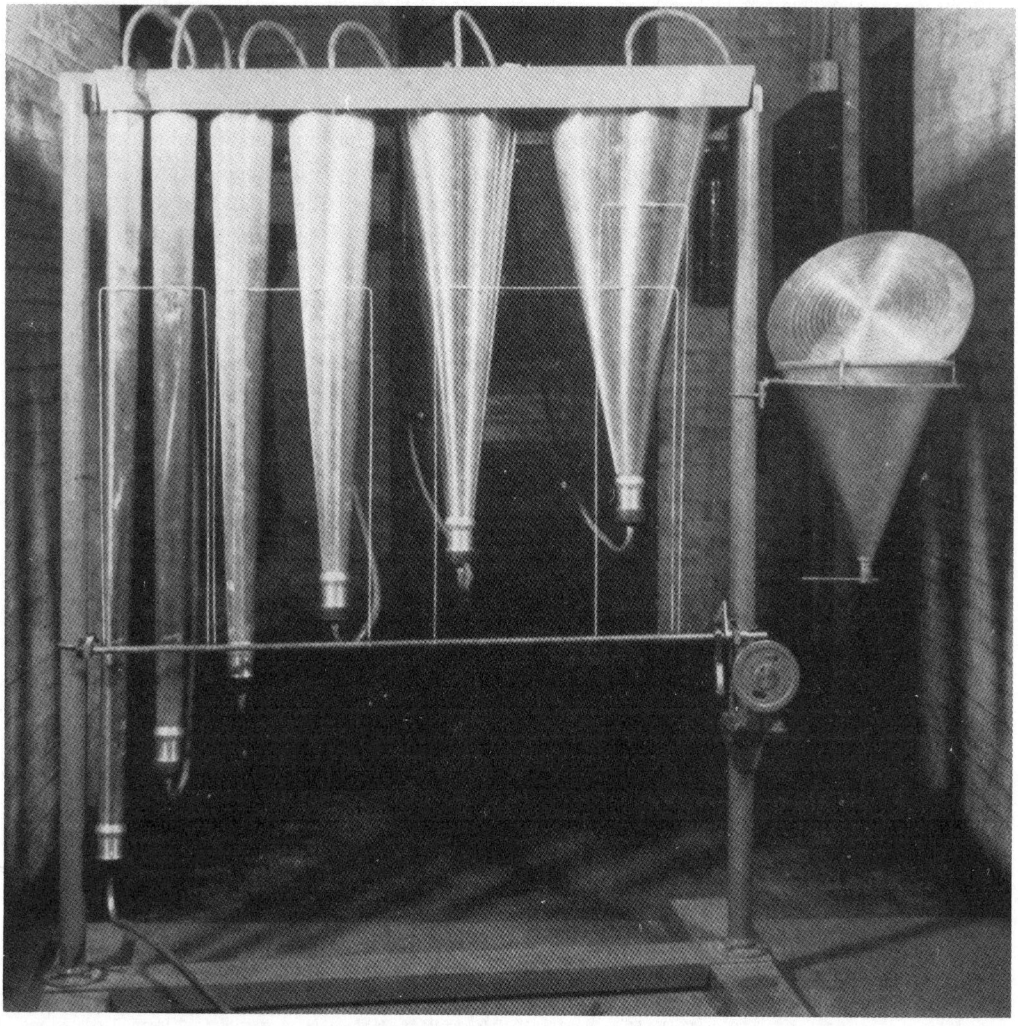

Fig. 20-3. The Haultain Infrasizer showing the seven cones with tapping wires. Air elutriation separates a 50 gram sample into six fractions in about 30 minutes.

Kaye and Jackson[31] have developed a method which combines the most advantageous aspects of both elutriation and sieving. The method, dubbed felvation, is quite rapid and permits the fractionation of small samples in minutes. The method of separation is similar to a distillation in which the fluidized particles are allowed to pass up a conical shaped column through a series of sieves placed across the column at intervals. Although the method was originally developed as a rapid analytical tool, the technique may also have merit for particle size control on a commercial scale.

Although the above classification methods are widely employed for research and characterization purposes, the use of elutriation on a commercial scale appears to be non-existent.

2.4 Aspect Ratio Measurement

The determination of the aspect ratio of a mica flake is made difficult not only by the non-uniform diameter, but also in many cases, by a non-uniform thickness. In order to obtain statistically significant values of the average diameter and average thickness, a sampling of at least 100 flakes is usually required. Since it is difficult to measure both the diameter and the thickness of each individual flake, the average diameter and the average thickness of, say, 100 flakes are measured independently so that the number average aspect ratio is given by

$$\bar{\alpha} = \frac{\bar{D}}{\bar{t}} \tag{3}$$

where $\bar{\alpha}$ is the average aspect ratio, $\bar{D}$ is the average diameter, and $\bar{t}$ is the average thickness. For broad distributions of aspect ratios, the *weight* average aspect ratio appears to correlate better with flake composite mechanical properties.[30] This is simply a reflection of the proportionately greater contribution of larger flakes to the composite mechanical properties. For well fractionated samples, however, the number average and the weight average aspect ratios are similar. For relatively large flakes, direct microscopic determination of the average diameters can be visually estimated using a calibrated reticule comprising circles of different diameters (Ferets statistical diameters). In other cases, the longest dimension of the flake may be measured[32] or a point count method using a grid may be employed to determine the total area covered by a known number of flakes.[7] In each case, an average diameter, equivalent diameter, or effective diameter is established for each sampling of flakes.

The average thickness may also be directly measured with a microscope. In one method, the flakes are molded into a flat specimen using a suitable resin (or using several plies of cellophane tape) and then the edge is polished for observation. Unfortunately, this method is often subject to rather large errors due to the limited resolving power of light microscopes. For flakes which are 2–3 μ thick and using a conventional microscope with a numerical aperture of 0.8, the error is between 12 and 18%. Maine and Shepherd[7] used a microbalance to weigh a sample of flakes which were deposited on a microporous filter grid. Using the point count method to determine the total area covered by all the flakes, the thickness could then be calculated from the known density of mica.

Leidner[7] used carbon black to estimate the thickness of mica flakes, which were deposited on a strip of cellophane adhesive tape. Carbon black was lightly sprinkled beneath and on top of the flakes adhering to the cellophane tape. A calibrated micrometer screw attached to the microscope objective was then used to measure the flake thickness by focussing alternately on the underlying carbon black particles and then on the surface particles. Locke[33] was able to accurately determine the thickness of mica flakes from microscopic examination of the interference fringes produced by mica flakes exposed to monochromatic light. From the known refractive index of mica, the thickness could be calculated from the distance between fringes.

As can be readily visualized, the determination of an average aspect ratio by any of the above techniques is extremely tedious and time consuming. An automatic particle size counter would considerably reduce the amount of labour and help to minimize differences in human judgement.

When the sizes of the mica flakes become

very small, below 40 μ, the problems of aspect ratio determination become considerably more difficult. Capes and Coleman[34] reemployed a classical technique to determine flake thickness and total area using a monolayer of mica (-140 + 200 mesh) on a Langmuir film balance. The flakes were pretreated with oleic acid in order to render them hydrophobic so they would readily float on the water surface. An accurately weighed sample was then placed upon the surface of the water, and the film compressed until a sharp increase in pressure was recorded. In measuring the thickness of the mica flakes, the standard deviation was about 15%, although accuracy could have been improved by using more data points. The flake diameter must be separately determined in order to establish an average aspect ratio.

It was found by Xanthos[25] that the bulk volume of a mica sample having a relatively narrow diameter distribution correlated reasonably well with the measured aspect ratio. Hence, by sieving any sample into several different diameter ranges, the aspect ratio and diameter distributions can be qualitatively estimated. The relationship between bulk volume and aspect ratio is given by

$$\alpha = 40V - 100$$

where α is the aspect ratio and V is the bulk volume in cm^3/g. V is determined by allowing a weighed quantity of mica (say 10 g) to settle after shaking in a 100 ml graduate cylinder filled with water. The volume occupied at equilibrium is easily read from the graduate cylinder. For broad diameter distributions, the flake packing geometry becomes non-linear and the correlation is poor. The felvation technique developed by Kaye and Jackson[31] may permit a more rapid characterization of such broad distributions.

3. PROPERTIES

3.1 General

Mica is the group name for a broad class of aluminum silicate minerals of the phyllosilicate family. Most of the micas crystallize in the monoclinic system and are characterized by nearly perfect basal cleavage which permits the larger crystals of "books" to be split into numerous extremely thin, highly flexible, and tough elastic flakes. These minerals can be classified into four principal types, named after the most common mineral in each group—muscovite, phlogopite, biotite, and lepidolite. Only muscovite and phlogopite have found appreciable commercial usage.

Although the precise chemical compositions may vary depending upon the origin of the sample, the approximate formulas of several common varieties are shown below:

Muscovite	$K_2Al_4(Al_2Si_6O_{20})(OH)_4$
Phlogopite	$K_2(MgFe^{+2})_6(Al_2Si_6O_{20})(OH, F)_4$
Biotite	$K_2(Mg, Fe^{+2})_6(Al_2Si_6O_{20})(OH)_4$
Lepidolite	$K_2Li_4Al_2(Si_6O_{20})(F, OH)_4$
Zinnwaldite	$K_2Li_2Fe_2Al_2(Al_2Si_6O_{20})(F, OH)_4$
Paragonite	$Na_2Al_4(Al_2Si_6O_{20})(OH)_4$
Fluorophlogopite (synthetic)	$K_2Mg_6Al_2Si_6O_{20}F_4$

All micas contain a small percentage of bound water, usually less than 5%. Vermiculite (meaning worm-like) is a hydrated form of biotite and contains a relatively large amount of bound water (11-21%). When strongly heated, vermiculite undergoes an extraordinary expansion into elongated worm-like structures which are several times greater than its original volume.

Referring to Fig. 20-4, mica is seen to consist of a sheet structure of silicon-oxygen tetrahedra (Si_2O_5 siloxane layers). In addition, there are layers of aluminum and hydroxyl arranged in octahedral form, one octahedral layer sandwiched between two siloxane layers. These three-layer units are loosely joined to similar layers above and below by cations such as potassium having 12-fold coordination with the alumino-siloxane layers. In other mica varieties, the potassium ions may be replaced with lithium, sodium, or calcium. The octahedral layers may consist primarily of gibbsite ($Al[OH]_3$)

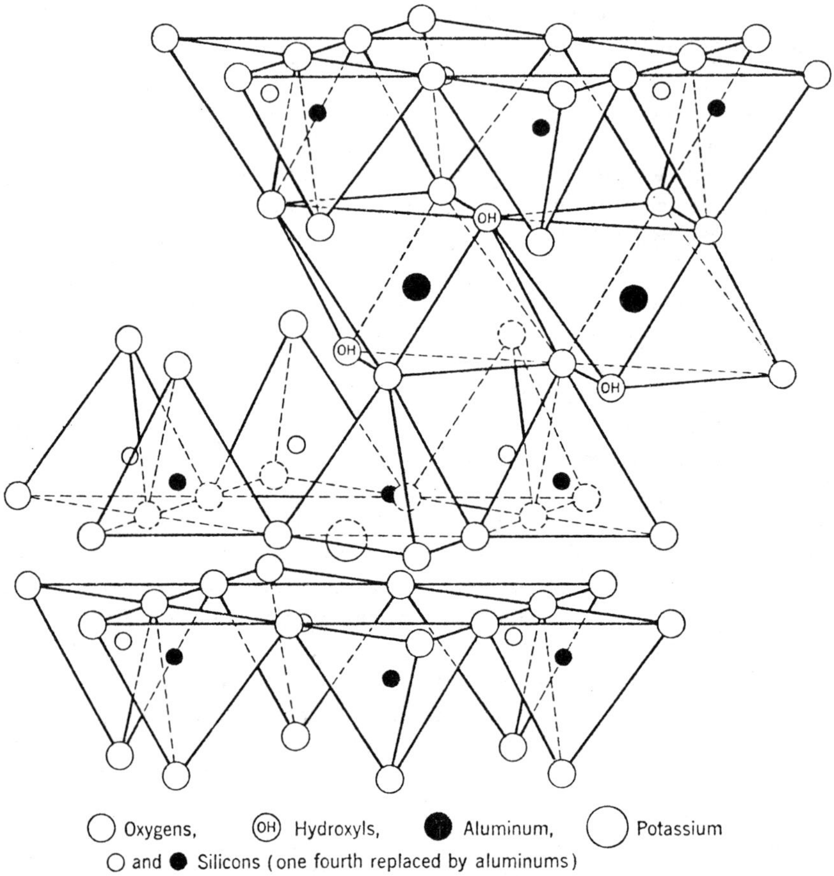

Fig. 20-4. A diagrammatic sketch of the crystal structure of muscovite shows a portion of the planar layers held together with potassium ions, after Jackson and West (1930).

as in muscovite, or brucite ($Mg[OH]_2$) as in phlogopite. The presence of iron or other impurities in the structure is often responsible for the color. Each triple layer is approximately 10 Angstroms thick.

The chemical compositions of three commercial micas are given in Table 20-1. Elements such as barium, chromium, magnesium, titanium, or vanadium may be present in minor amounts. Most micas contain 2.5–4.5% combined water. Suzorite mica is unusual in that the combined water content is quite low, viz., 1%, and is retained up to a temperature of 440°C.[35,36,37] Other varieties of mica begin to dehydrate at temperatures nearer 200°C.

The hardness of mica is relatively low, 2.5–4.0 on the Moh scale, so that abrasion of steel processing equipment is minor compared to glass filled plastics. The specific gravity may vary between 2.75 and 3.2 depending upon the particular variety. Phlogopite, which contains magnesium ions, tends to have a lower density than muscovite, and is only slightly higher in density than glass. Most micas are insoluble in strong acids or bases, but biotite and phlogopite may be etched by hot concentrated sulfuric or phosphoric acids, leaving a highly porous skeleton of pure silica. All micas are readily attacked by hydrofluoric acid and molten alkali metals.

The excellent dielectric qualities of muscovite and phlogopite account for their widespread usage in the electrical industry. In fact, films of muscovite and phlogopite one-thousandths of an inch thick are capable of withstanding 5000 volts without puncture.[38] The fact that mica is

Table 20-1. Chemical analysis of commercial phlogopite and muscovite micas.

	Phlogopite[a] [wt %]	Muscovite[b] [wt %]	Muscovite[c] [wt %]
SiO_2	40.74	47.90	48.30
Al_2O_3	15.76	33.13	31.55
MgO	20.56	0.69	1.99
FeO	7.83	–	–
Fe_2O_3	1.21	2.04	2.20
K_2O	9.96	9.80	7.86
Na_2O	0.46	0.80	1.20
BaO	0.45	–	–
CaO	<0.01	0.50	1.25
TiO_2	0.42	0.65	1.04
Cr_2O_3	0.14	–	–
MnO	0.07	0.01	0.01
F	2.16	–	–
P	–	.03	–
S	–	.01	–
H_2O^+			
[combined]	1.01	4.30	4.20
H_2O^- [free]	0.01	0.10	0.10

[a]Suzorite variety, Marietta Resources International Ltd.
[b]The English Mica Co.
[c]Whittaker, Clark and Daniels, Inc.

not affected by oil or water also enhances its desirability as an electric insulator. Micas are generally characterized by excellent weather resistance, good chemical resistance, low thermal conductivity, and prolonged stability at high temperatures. Muscovite is stable without loss of integrity up to 500-530°C, whereas phlogopite is stable to 1000°C.

Biotite is found as embedded crystals, disseminated scales, and in massive lamellar aggregations. The basal cleavage is distinct, but does not always carry across the width of the crystal. Thus, splitting properties are poor compared to muscovite and phlogopite and the splittings are less flexible and more brittle. The hardness ranges from 2.5-3 with a specific gravity from 2.75-3.1. As in the case of phlogopite, biotite is also attacked by sulfuric acid. Biotite, because of its high iron content, is a poorer dielectric material than muscovite or phlogopite. The index of refraction varies from 1.54-1.63.

Table 20-2 summarizes the known physical and chemical properties of muscovite and phlogopite.

3.2 Toxicity

Finely divided mica (<10 μ diameter) which is likely to become airborne and inhaled, should always be handled in a well ventilated area. The same precaution applies to all finely divided silicate minerals. Larger particles, 100 mesh or greater, are much less likely to be inhaled and thus may be considered relatively non-toxic. The accepted Threshold Limit Value (TLV) for mica dust is equal to that for talc, that is, 20 million particles/ft^3.[39] Table 20-3 summarizes the toxic hazard ratings for mica compared to talc, glass fibers, and glass flakes.

4. MATERIAL SUPPLIER GUIDE

The cost and selling price of mica depend on purity, method of grinding, aspect ratio, appearance, and end use. Grades designed for use as pigments are generally more expensive.

The selling prices of dry-ground micas range from \$.04-.08/lb (typically \$.04-.06/lb). Water ground and micromicas are sold from \$.08-.20/lb (typically \$.12/lb). The aspect ratio of most commercially available muscovites is generally below 30. Marietta Resources International Ltd. now markets high aspect ratio (α = 50-100) dryground Suzorite micas at development prices (\$185-\$260/ton, FOB Montreal). These high aspect ratio grades are particularly suitable for plastics reinforcement.

Table 20-4 contains a partial list of representative mica types presently available from North American suppliers. A current list of the available grades and prices may be obtained directly from the suppliers listed in Table 20-5.

5. ECONOMIC OUTLOOK

The availability of high quality sheet mica is restricted to only a few areas in the world, particularly India, Brazil, Malagasy Republic, and the Urals of the U.S.S.R. Fortunately, small particle size mica flakes occur in great abundance throughout the world since they are a common constituent of igneous, sedimentary, and metamorphic rocks. Muscovite and biotite, the

Table 20-2. Properties of muscovite and phlogopite micas.[40]

Property	Natural Muscovite	Natural Phlogopite
Density, g/cc	2.76–3.2	2.74–2.95
Hardness, Mohs	3–4	2.5–3.0
Transparency	Ruby [varies]	Amber-Brown [varies]
Shape	Roughly hexagonal or irregular flakes	Roughly hexagonal or irregular flakes
Cleavage	Basal, eminent	Basal, eminent
Crystal structure	Monoclinic	Monoclinic
Optical axial angle, 2V, degrees	38–47	30–10
Refractive index	1.552–1.611	1.54–1.69
pH	6.5–8.5	7.0–8.0
Melting point, °C	Decomposes	Decomposes–[>1300][1]
Water of constitution, %	4.5	3.2–[1.01][1]
Maximum temperature with little or no decomposition, °C	500–530	850–1000
Tensile modulus, GN/m^2 [psi]	172 [25×10^6]	172 [25×10^6]
Tensile strength, MN/m^2 [psi]	255–296 [37×10^3–43×10^3][a]	255–296 [37×10^3–43×10^3][a]
	3100 [450×10^3][b]	–
	690–900 [100×10^3–130×10^3][c]	690–900 [100×10^3–130×10^3][c]
Thermal conductivity [⊥ to cleavage], $cal/cm^2/sec/cm/°C$	16×10^{-4}	16×10^{-4}
Linear coefficient of thermal expansion per °C [⊥ to clevage]		
20°C to 100°C	15×10^{-6}–25×10^{-6}	1×10^{-6}–1×10^{-3}
100° to 300°C	16×10^{-6}–25×10^{-6}	2×10^{-5}–2×10^{-3}
Linear coefficient of thermal expansion per °C [‖ to cleavage]		
0° to 200°C	8×10^{-6}–9×10^{-6}	13×10^{-6}–14.5×10^{-6}
Specific heat [25°C]	0.206–0.209	0.206–0.209
Radiation resistance: Max. use temp. with 5×10^{14} thermal neutrons/cm^2, °C	500	550
Specific resistivity ohm cm [25°C]	10^{12}–10^{16}	10^{10}–10^{13}
Dielectric strength, V/mil [1.3 mils thick]	6,000–3,000	4,200–2,100
Dielectric constant, ϵ [1 MC]	6.5–9.0	5.0–6.0
Dissipation factor, tan δ [25°C]		
60 cycles	0.0008–0.0009	–
1 MC	0.0002	0.004–0.070
Loss factor, ϵ tan δ [1 MC]	0.0013	0.020–0.042
Chemical resistance	Very good[d]	Good[e]
Water solubles	Trace	Trace
Coefficient of static friction	0.2–0.4	0.2–0.4

[1]Suzorite variety [Marietta Resources International Ltd.].
[a]Measured on sheets [edges stressed,[40]].
[b]Measured on sheets [edges unstressed,[20]].
[c]Calculated value for the effective strength of high aspect ratio flakes in plastics.
[d]Attacked by HF and molten alkalis.
[e]Attacked by HF, H_2SO_4, H_3PO_4 and molten alkalis.

most common of the micas, occur abundantly in many granites, gneisses, schists, and pegmatites, these sometimes being entirely composed of mica. Large formations of phlogopite may be found in serpentized limestones, dolomite, and pyroxenites. Consequently, resources of mica are extremely large if not virtually unlimited. One of the largest phlogopite deposits in the world is located near Quebec in Canada with a proven reserve of nearly 30 million tons,

Table 20-3. Toxicity of mica dust, talc, glass fibers and glass flakes.[39]

	Mica dust [<5% free silica]	Talc	Glass fibers or flakes
Acute local	irritant 2 inhalation 2	inhalation 1	irritant 2 inhalation 1
Acute systemic	0	U	0
Chronic local	inhalation 2	inhalation 2	irritation 2
Chronic systemic	0	inhalation 3	U
Threshold limit value [TLV]	20 million particles/ft^3	20 million particles/ft^3	5 mg/m^3

Ratings: No toxicity = 0
Slight toxicity = 1
Moderate toxicity = 2
Severe toxicity = 3
unknown = U

which is sufficient to supply current world needs for over 100 years! In addition to its size, the deposit is unique for its high purity, having an average content of over 90% mica. Marietta International has begun construction of a processing plant near Montreal, Canada, for the purpose of marketing high aspect ratio mica in a wide range of particle sizes suitable for use in the plastics industry and as a reinforcing agent for cement, gypsum, and asphalt. This particular phlogopite variety is called Suzorite, after Suzor Township where the deposit was discovered. Other major deposits in Canada are located in the provinces of Quebec, Ontario, and British Columbia. Large deposits have been located in the Front Range of the Rocky Mountains extending from Colorado to Wyoming in the United States.[41] Other major deposits have been found in Mexico, South Africa, Korea, India, Brazil, and the Malagasy Republic. The long term outlook for the price of flake mica is predicted to remain constant until the year 2000 at approximately $30 U.S. per short ton (1968 constant dollars). During the 20 year period from 1949 to 1969 the price of scrap and flake mica in the United States fluctuated between $25 and $38 per ton (1968 constant dollars). With further improvements in low cost methods of mining and processing, ground mica can be expected to remain in the quoted range, rising only to offset increased production costs or taxation.[42] Slightly more than half of the supply of scrap and flake mica in the United States was produced in North Carolina, with nine other states accounting for the remainder of the output during 1971. Dry ground mica accounted for 86% of the total sales. Sixteen companies operating a total of 20 plants processed scrap and flake mica to small particle sizes. Of these plants, 15 produced dry-ground mica, 3 wet-ground, and 2 both wet- and dry-ground.[43] Development of a new muscovite mica deposit with an estimated reserve of 35 million tons began near Lancaster, Pennsylvania in 1971. The mica will be wet-ground for marketing.[42]

Dry and wet ground micas range in price from an average low of 1.67¢/lb for roofing applications to a high of 9.7¢/lb for use in plastics. Table 20-14 presents an interesting comparison of the average prices paid for ground mica in different use areas. The higher prices paid for mica in the rubber, plastics, and paint industries probably reflect the higher cost of wet ground mica compared to dry ground mica.

The world production of mica in 1971 was 183 thousand short tons, with 133 thousand tons being produced in the United States. The annual growth rate in the consumption of all forms of mica in the United States over the 20 year period from 1949 to 1969 varied from 1.5-2.9% per year. The recent availability of high aspect ratio mica as a primary reinforcing agent for plastics (as well as other materials), could represent a breakthrough in technology which may lead to a significant penetration of existing markets for reinforced plastics now using more expensive reinforcing agents such as glass fibers, or hazardous materials such as asbestos. In addition, new techniques, such as the spray application of powdered mica and polyester resins onto a thermoformed acrylic veneer, offers a relatively low cost method of fabricating large components for use in transportation, appliances, furniture, and housing.

The designer is frequently faced with a choice of materials for any given application. In the

Table 20-4. Commercial grades of mica—1975.

Supplier	Grade	Type	Nominal Mesh size	Suggested uses
Marietta Resources Inter. Ltd.	40S	Suzorite D.G.	-20 + 40	Plastics
”	40H	” ” HAR	-20 + 40	Plastics—corrosion resistant coatings
”	60S	” ”	-40 + 100	Plastics—rigidizing
”	60H	” ” HAR	-40 + 100	Plastics—corrosion resistant coatings
”	50S	” ”	-20 + 100	Plastics—rigidizing
”	50H	” ” HAR	-20 + 100	Plastics
”	200S	” ”	-100	Fillers
”	200H	” ” HAR	-100	Plastics
”	325H	” ” HAR	-325	Phenolics, polypropylene, paints, fillers
”	300WG	” W.G.	-200 + 325	Wall paper, decorative
”	325WG	” W.G.	-325	Rubber, mold release
Hayden Mica Co.	P-88 Davenite	Muscovite, W.G.	-100	Primers, sealers, paints, latex, PVA,
”	P-4 Davenite	” ”	-160	Acrylics—molding, plotting and
”	P-12 Davenite	” ”	-325	encapsulating compounds
The English Mica Co.	160 Mesh	Muscovite, W.G.	-160	Paints, rubber,
”	325 Mesh	” ”	-325	Plastics, wall paper
U.S. Mica Co., Inc.	50 Mesh	Muscovite, D.G.	-50	Welding electrodes, cements, adhesives,
”	100 Mesh	” ”	-100	asphalt, paints, fillers, wallboard joint cement
”	160 Mesh	” W.G.	-160	rubber
”	325	” ”	-325	Paints, plastics
Thompson, Weinman & Co.	Triple A	Sericite	96% thru 325 mesh	Joint cements, plastics,
”	Double A	”	90% thru 325 mesh	Paints, rubber, etc.
”	Standard A	”	80% thru 325 mesh	
”	W.O. Mica	”	70% thru 325 mesh	
Harwick Chemical Corp.	160-D	Muscovite, D.G.	160	
”	325-D	” ”	325	Rubber
”	325 Mesh Dry-Ground	” ”	325	
Diamond Mica Co.	160	Muscovite, W.G.	160	
”	325	” ”	325	Paints, plastics
”	Silversheen 160	” ”	160	
”	Silversheen 325	” ”	325	
Western Mica Co.	P-80-H	Muscovite, Fluid Energy Ground	-100	
”	P-80-F	”	-100	Paints, plastics
”	P-85	”	-100	
”	325	”	-100	

Table 20-5. North American suppliers of mica.

Supplier	Address
Construction Products Div., W. R. Grace Co.	Cambridge, Mass.
C. P. Hall Co.,	Chicago, Ill.
Diamond Mica Co.	Spruce Pine, N.C.
Engelhard Minerals and Chemicals Corp., Minerals and Chemicals Div.	Edison, N.J.
English Mica Co. (U.S. Mica Co. Inc.)	Stamford, Conn.
Franklin Minerals Products Co.	Wilmington, Mass.
Harris Mining Co.	Spruce Pine, N.C.
Harwick Chemical Corp.	Akron, Ohio
Hayden Mica Co. (Div. of Franklin Mineral Products Co.)	Wilmington, Mass.
Marietta Resources International Ltd.	Montreal, Quebec
Micalith Mining Co., Inc.	Phoenix, Ariz.
Mineral Mining Corp.	Mineola, N.Y.
3M Company	St. Paul, Minn.
Smith Color and Chemical Co.	Jamaica, N.Y.
Tamms Industries	Itasca, Ill.
Thompson, Weinman and Co.	Cartersville, Ga.
United Mineal and Chemical Corp.	New York, N.Y.
United States Gypsum Co.	Chicago, Ill.
Whittaker, Clark and Daniels	S. Plainfield, N.J.
Windsor Minerals, Inc.	Windsor, Vt.

future, as energy becomes more expensive, and hydrocarbon reserves decline, the incentives for incorporating cheap fillers and reinforcements into hydrocarbon polymers may become increasingly attractive. As long as the cost and quality of the reinforced plastics maintain their competitive edge with respect to metals, ceramics, and wood, the current growth rate of 8% per year should be maintained over the next decade. Longer range forecasts are uncertain due to the unpredictable effects of a rapidly expanding world population and limited resources.[44]

6. APPLICATIONS

6.1 Mica Reinforced Thermoplastics

High aspect ratio mica is an outstanding reinforcing agent for all thermoplastic materials. However, it is difficult to utilize HAR mica in thermoplastics, since normal compounding methods usually lead to breakdown of the flakes and attendant loss of aspect ratio. Since the degree of reinforcement depends largely upon the average aspect ratio in the finished product, the strength and modulus are far from optimum. The same difficulty is encountered in the processing of glass fiber reinforced thermoplastics, where care must be taken to preserve the aspect ratios of the glass fibers during the mixing and molding operations. When the same techniques used for glass fiber reinforced thermoplastics (RTP) are applied to mica reinforced thermoplastics (MRP), some attractive properties may be retained, particularly modulus and strength. These processing methods for glass fibers are well known in the industry.

In order to minimize breakdown of the mica flakes during processing, techniques of powder blending, latex blending, and encapsulation have been utilized, followed by compression molding of the resulting dispersion. In these cases, flake damage is negligible and good mechanical properties may be realized, as shown in Table 20-6.

Powder molding permits the incorporation of very high levels of mica, but the mechanical properties begin to decline above 60 volume percent (80% by weight), as shown in Fig. 20-5. The very attractive properties that can be attained at the highest volume loadings may warrant using such techniques for special purposes. Unfortunately, compression molding of thermoplastics (unlike that of thermosets) is a slow, expensive process, since the sample must be cooled before ejecting it from the mold. On the other hand, injection molding is a faster, more economical process, although it inevitably leads to particle breakdown of the mica to lower aspect ratios. Figure 20-6 shows the changes in flake diameter which occur in a compounding extruder during mixing with a commercial polypropylene resin (Exxon's Dexon E-117 containing 40% by weight Suzorite mica). Since the average flake thickness remains relatively constant, the changes in diameter reflect a proportionate change in aspect ratio. Further reduction in aspect ratio would also be expected during subsequent injection molding operations. Just how these changes in aspect ratio

Table 20.6. Mechanical properties of compression moulded mica reinforced thermoplastics

Polymer	Volume Fraction	FLEXURAL STRENGTH MN/m²	FLEXURAL STRENGTH (10^3 psi)	FLEXURAL MODULUS GN/m²	FLEXURAL MODULUS (10^6 psi)	System Used (Reference)
Poly(ethylene)	0.5	124	(18.0)	31.1	(4.51)	powder
Poly(propylene)	0.5	173	(25.1)	38.0	(5.15)	powder
Poly(styrene)	0.5	166	(24.1)	44.9	(6.51)	powder
Poly(styrene)	0.5	172	(24.9)	48.3	(7.00)	encapsulation[56]
Poly(styrene-co-acrylic acid)	0.5	165	(23.9)	44.0	(6.38)	latex[30]
Poly(styrene-co-acrylonitrile)	0.5	207	(30.0)	53.1	(7.70)	powder
Nylon-6,6	0.5	186	(27.0)	44.9	(6.51)	powder
Nylon-11	0.5	173	(25.1)	37.3	(5.41)	powder
Poly(butylene-terephthalate)	0.5	186	(27.0)	47.6	(6.90)	powder

affect the mechanical properties is shown in Table 20-7. Here it is shown that the process of mixing in a compounding extruder, followed by injection molding, can reduce the strength and modulus considerably (between 22 and 44%). The use of hot runners, larger sprues, and gates, plus the streamlining of melt flow, reduces attrition during such processing.[86] The results shown here were not conducted under such ideal conditions.

The rheological behavior of mica-poly(propylene) mixtures has been studied by Okuno and Woodhams (45). As expected, the addition

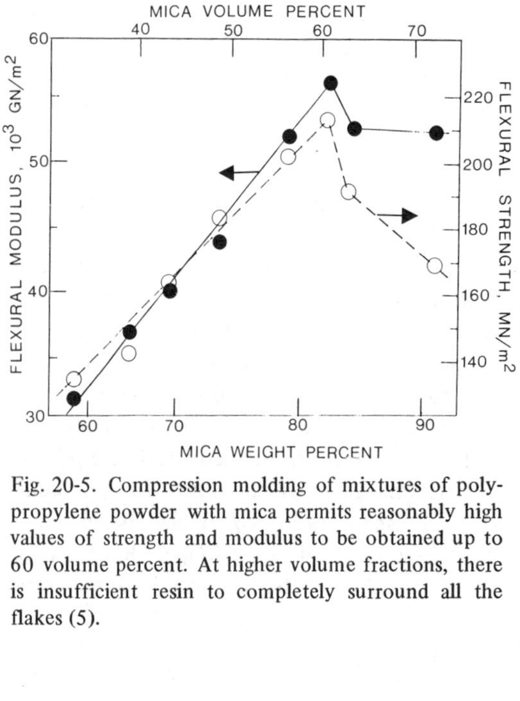

Fig. 20-5. Compression molding of mixtures of polypropylene powder with mica permits reasonably high values of strength and modulus to be obtained up to 60 volume percent. At higher volume fractions, there is insufficient resin to completely surround all the flakes (5).

of mica to a thermoplastic melt increases the viscosity so that higher processing temperatures usually become necessary unless a processing aid is employed. Figure 20-7 illustrates the dependence of extrusion temperature for poly(propylene) (Dexon E-117) upon both the concentration of mica and the average diameter of the flakes. It is seen that higher concentrations of mica as well as higher flake diameters (which are assumed to be proportional to the aspect ratio) both contribute to an increase in the melt viscosity and, therefore, higher processing temperatures. The viscosity behavior at a given temperature can be represented by the usual power law relationship

$$\tau = \eta \dot{\gamma}^n \tag{4}$$

where τ is the shear stress, η the apparent viscosity, $\dot{\gamma}$ the shear rate, and n is the flow index.

From experimental observation, it is seen that in isothermal flow, flakes tend to orient with their faces tangent to the velocity profile, as shown in Fig. 20-8. The photograph shows a longitudinal section of the extrudate with the flakes at the outer edge oriented parallel to the wall of the orifice and the direction of extrusion, whereas at the center of the extrudate the mica flakes are oriented perpendicular to the flow direction. In the case of injection molded parts, the flow is non-isothermal in the mold cavity and parallel orientation of the flakes is aided by the temperature gradient. Figure 20-9 shows the cross section of an ASTM tensile specimen which is cut parallel to

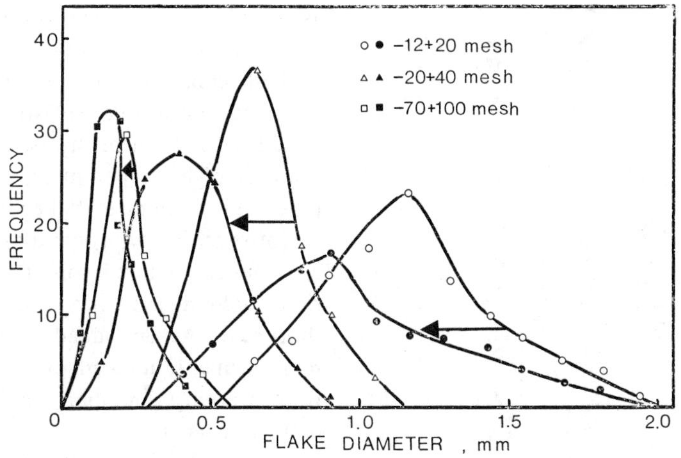

Fig. 20-6. The flake diameter distributions are shown for three sieve fractions before extrusion and after one pass in a two-stage mixing extruder. In each example, the mica was mixed with 60% by weight polypropylene (Dexon E-117). The decrease in the flake diameters after extrusion represents a loss of aspect ratio and hence the level of reinforcement attainable.

the flow direction. Except for the middle section, flake orientation is particularly good.

In the case of very thin injection molded parts, e.g., panels, housings, extruded sheets, etc., the mica flakes will predominantly orient parallel to the surface, giving good planar reinforcement. In such cases, the inherent properties of flakes are used to great advantage. In the case of thicker molded sections, parallel flake orientation is more difficult to control and the advantages of flake reinforcement are not so apparent, especially in terms of strength. The preferential alignment of flakes parallel to the molded surface, and the irregular alignment in the center of a molded part, imparts a sandwich type construction to the section. Consequently, differences are often observed between the flexural properties and the tensile properties. The self alignment of the flakes at the surface of a molded part produces a skin which is less susceptible to crack initiation than the interior section. Large differences have been observed between notched and unnotched Izod impact strengths as a result of this skin effect.

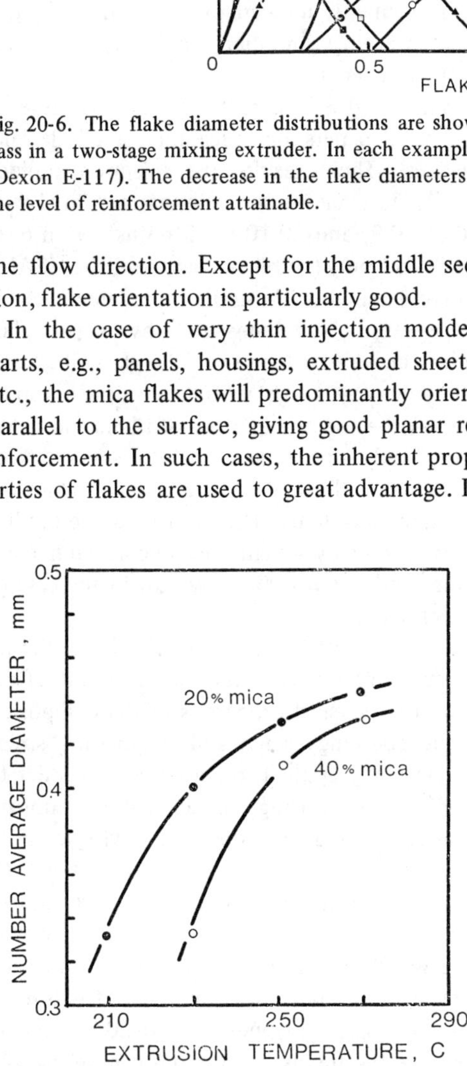

Fig. 20-7. The addition of mica to polypropylene (Dexon E-117) produces an increase in melt viscosity and processing temperature. These curves indicate the relative influence of flake diameter and flake concentration (20 and 40 weight per cent shown) on the optimum extrusion temperature.

The viscosity of mica filled thermoplastics is highly shear dependent, similar to the behavior of fiber filled plastics. However, unlike fiber filled plastics, much higher mica concentrations may be employed before the mixture becomes unprocessable. In general, a 50% by weight mica filled material processes under much the same conditions of temperature and pressure as a 20% by weight glass filled thermoplastic. This feature is a distinct advantage when high filler loadings are desired for overall reduced material costs and high stiffness. At such high volume loadings, the mica composites are extremely rigid, having modulus values two or three times higher than those of glass fiber filled polymers.

Table 20-7. Loss of mechanical properties due to flake degradation.[5]

Composite	Volume Fraction Mica	COMPRESSION MOLDING: FLEXURAL MODULUS (GN/m^2)	$(10^6$ psi)	COMPRESSION MOLDING: FLEXURAL STRENGTH (MN/m^2)	$(10^3$ psi)	INJECTION MOLDING: FLEXURAL MODULUS (GN/m^2)	$(10^6$ psi)	INJECTION MOLDING: FLEXURAL STRENGTH (MN/m^2)	$(10^3$ psi)	EXTRUSION FOLLOWED BY INJECTION MOLDING (GN/m^2)	$(10^6$ psi)	(MN/m^2)	$(10^3$ psi)
ABS/mica	0.50	41.6	(6.03)	154	(22.3)	36.9	(5.35)	105	(15.2)	32.5	(4.71)	99.9	(14.4)
Poly (styrene)/ mica	0.50	41.3	(5.99)	123	(17.8)	36.9	(5.35)	114	(16.5)	25.4	(3.68)	67.7	(9.82)
Poly (pro- pylene)/mica	0.50	37.9	(5.50)	172	(24.9)	26.9	(3.90)	86.1	(12.5)	—	—	—	—

using hot runner systems, up to 70% by weight mica may be satisfactorily injection molded.[86]

These considerations of melt rheology indicate that mica filled plastics should be processed at low shear stresses (large gates and runners, high temperatures, low viscosities) in order to minimize flake degradation. The thixotropic behavior of the filled melts also contributes to rapid solidification after injection molding, so that cycle times may be shortened. As an added advantage of planar reinforcement, molded parts are virtually free from warping and more faithfully reproduce mold dimensions.

Maine and Shepherd[2,5,46] have published a series of papers on mica reinforced thermoplastics. Their results for injection molded ASTM test pieces are summarized in Tables 20-8, 20-9, and 20-10 for 3 resins, nylon 6, 6, ABS, and poly (styrene-co-acrylonitrile) (SAN), and compared with similar measurements for these resins containing 20% by weight glass fibers, and also a blend of 20% glass fibers and 30% mica (by weight). Some advantages can be seen in the use of a mixture of fibers and flakes rather only one type of reinforcement. Note particularly the higher Izod impact values for the mixed systems. The limit of processability is about 65% by weight mica in cold runner systems (and about 40% by weight in the case of glass fibers).

As with most heterogeneous systems, the interfacial effects can have a profound effect upon the physical properties of the composite. Silane coupling agents and polymeric "sizes" are normally applied to glass fibers in order to modify the wetting characteristics, improve interfacial bonding, reduce sensitivity to moisture, enhance the mechanical properties, and reduce abrasion. Mica responds to the use of silane coupling agents in much the same way as glass fibers, probably because the outer silicate layers of mica are similar in structure to silica or glass. In general, a 10-20% improvement in the flexural properties can be realized by treatment of the mica with 0.25% by weight silane coupling agent. Wet strength retention is also improved. The amino-functional silanes have been found particularly effective for mica.[47] Table 20-11 lists some additional me-

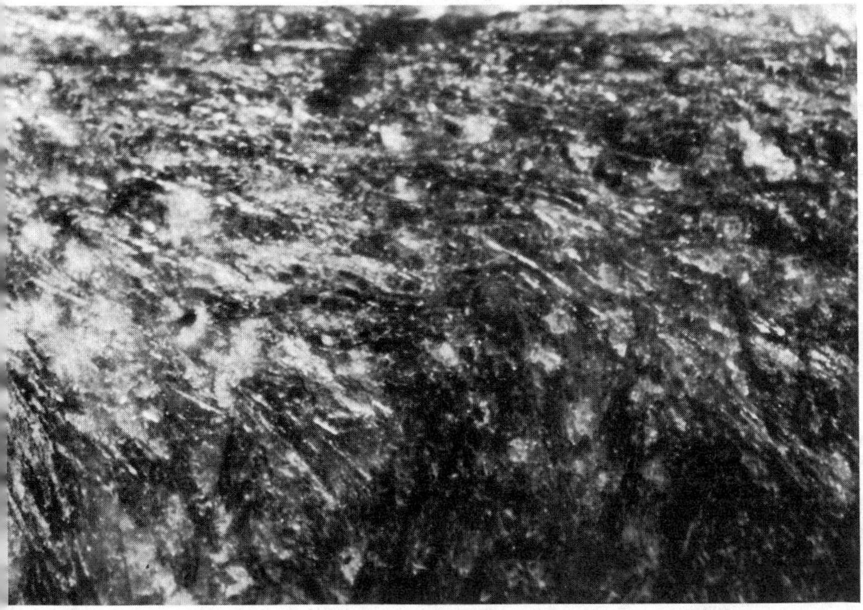

Fig. 20-8. Cross section of an extruded sample of mica reinforced polypropylene parallel to the direction of flow (40% by weight mica). Orientation of the flakes is predominantly parallel to the direction of flow near the wall of the orifice and perpendicular to the flow in the center of the orifice (extrusion temperature was 250°C at a shear rate of 13 sec^{-1}).

Table 20-8. Properties of Nylon 6/6 compounds.[2]

	20% GF	RFM 2030	RM 5000
Chopped glass fibers, wt %	20	20	–
Mica, wt %	–	30	50
Specific gravity	1.24	1.60	1.55
Mold shrinkage–6.2 mm section (mm/mm)	.005	.0038	.0054
Tensile strength MN/m^2 (10^3 psi)	107 (15.5)	124 (18.0)	94.8 (13.8)
Tensile modulus GN/m^2 (10^6 psi)	5.30 (0.769)	16.1 (2.34)	17.7 (2.57)
Flexural strength MN/m^2 (10^3 psi)	161 (23.4)	180 (26.1)	124 (18.0)
Flexural modulus GN/m^2 (10^6 psi)	6.47 (0.939)	14.1 (2.05)	13.3 (1.93)
Shear strength MN/m^2 (10^3 psi)	71.1 (10.3)	76.1 (11.0)	65.5 (9.51)
Compressive strength MN/m^2 (10^3 psi)	130 (18.9)	131 (19.0)	110 (16.0)
Izod impact strength			
–notched J/m (ft-lb/in.)	48 (0.90)	75 (1.4)	48 (0.90)
–unnotched J/m (ft-lb/in.)	410 (7.68)	400 (7.49)	220 (4.12)
Heat distortion temperature			
–at 66 psi (°C)	251	288	253
–at 264 psi (°C)	233	245	229

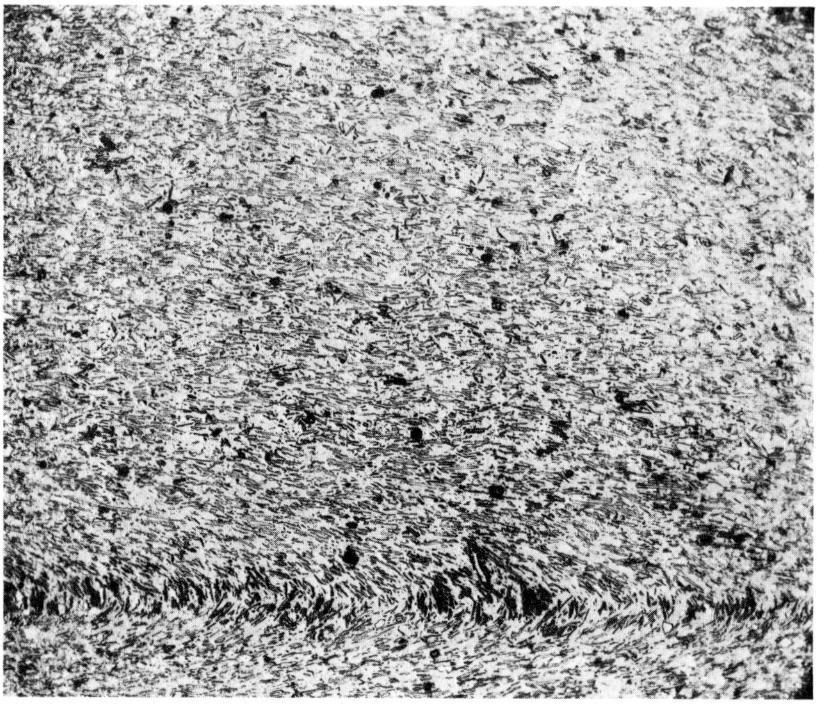

Fig. 20-9. The photograph shows a polished section cut parallel to the direction of injection of a tensile specimen comprising 40% by weight silane treated mica in poly(butylene-terephthalate). The flake orientation is good except for a thin region near the center of the specimen. The dark spots are largely voids.

Table 20-9. Properties of ABS compounds.[2]

	20% GF	AFM 2030	AM 5000
Chopped glass fibers, wt %	20	20	–
Mica, wt %	–	30	50
Specific gravity	1.19	1.52	1.52
Mold shrinkage–6.2 mm section (mm/mm)	.0006	.0008	.0014
Tensile strength MN/m^2 (10^3 psi)	73.7 (10.7)	80.0 (11.6)	65.2 (9.47)
Tensile modulus GN/m^2 (10^6 psi)	5.88 (0.81)	14.8 (2.15)	14.6 (2.12)
Flexural strength MN/m^2 (10^3 psi)	102 (14.8)	123 (17.9)	96.5 (14.0)
Flexural modulus GN/m^2 (10^6 psi)	5.58 (0.81)	14.0 (2.04)	14.7 (2.13)
Shear strength MN/m^2 (10^3 psi)	42.9 (6.22)	57.9 (8.41)	51.1 (7.41)
Compressive strength MN/m^2 (10^3 psi)	80.6 (11.7)	120 (17.5)	113 (16.4)
Izod impact strength			
–notched J/m (ft-lb/in.)	69 (1.3)	43 (0.8)	32 (0.6)
–unnotched J/m (ft-lb/in.)	230 (4.3)	150 (2.8)	130 (2.4)
Heat distortion temperature			
–at 0.45 MN/m^2 (°C)	100	110	109
–at 1.88 MN/m^2 (°C)	96	106	103

Table 20-10. Properties of SAN compounds.[2]

	20% GF	BFM 2030	BM 5000
Chopped glass fibers, wt %	20	20	–
Mica, wt %	–	30	50
Specific gravity	1.24	1.53	1.55
Mold shrinkage–6.2 mm section (mm/mm)	.0005	.0014	.0015
Tensile strength MN/m² (10³ psi)	102 (14.7)	82.7 (12.0)	68.2 (9.90)
Tensile modulus MN/m² (10⁶ psi)	9.09 (1.32)	16.1 (2.34)	17.6 (2.55)
Flexural strength MN/m² (10³ psi)	139 (20.1)	132 (19.2)	99.9 (14.5)
Flexural modulus MN/m² (10⁶ psi)	8.81 (1.28)	16.7 (2.43)	18.5 (2.68)
Shear strength MN/m² (10³ psi)	61.1 (8.87)	59.3 (8.60)	48.9 (7.10)
Compressive strength MN/m (10 psi)	131 (19.0)	140 (20.3)	107 (15.6)
Izod impact strength			
–notched J/m (ft-lb/in.)	48 (0.9)	43 (0.8)	27 (0.5)
–unnotched J/m (ft-lb/in.)	180 (3.4)	120 (2.3)	120 (2.3)
Heat distortion temperature			
–at 0.45 MN/m² (°C)	99	113	109
–at 1.88 MN/m² (°C)	97	110	108

chanical properties for silane treated mica in nylon and polyester thermoplastic resins.

One of the more attractive applications for the planar reinforcing properties of mica is in thermoplastic sheets for thermoforming or stamp forming.[48,49,68] The orientation of the flakes parallel to the surface during extrusion into sheets is further enhanced by subsequent drawing or thermoforming operations. In the case of fiber reinforced plastics, drawing leads to preferential orientation of the glass fibers, which produces regions of weakness and warping, whereas with mica reinforced sheets, the properties remain essentially uniform in all planar directions after thermoforming. As in the case of injection molded materials, best results are often achieved with a mixture of glass fibers and mica, with the glass fibers contributing strength and impact toughness and the mica contributing high rigidity and dimensional stability.

Mica filled thermoplastics such as nylon 6,6, poly(acetal), polypropylene, polyethylene, or polyesters may be hydrostatically extruded below their melting or softening points to produce various profiles with highly oriented flakes and polymer chains.[50] The "extrusion ratio" is the ratio of the original cross-sectional area to that of the extrudate. At extrusion ratios of 2:1 to 6:1, the tensile modulus, tensile strength, yield stress, and fracture strain were increased more than 100%. This technique is useful in the production of various profiles such as bars, strips, rods, and tubes.

Mica filled plastics are also characterized by good dielectric properties, reduced flammability, low abrasion (mica has a hardness of 2.5–3 Mohs), and low permeability.[51]

Table 20-11. Mechanical properties of mica filled thermoplastics.[47,5]

	Nylon 6 35 wt % mica	Nylon-6/Nylon-66 25/75 wt ratio 40 wt % mica	Polyester 35 wt % mica
Flexural strength MN/m² (10³ psi)	127 (18.4)	136 (19.7)	91 (13.2)
Flexural modulus GN/m² (10⁶ psi)	9.7 (1.41)	12.1 (1.75)	9.7 (1.41)
Tensile strength MN/m² (10³ psi)	72 (10.4)	–	48 (6.96)
Heat distortion temperature (°C at 1.88 MN/m²)	–	208	–

Mica has also been added to ABS to improve the adhesion of an electroplated metal coating.[52] Several advantages were observed, particularly a 4-fold increase in the rate of electroplating, good adhesion, and no observable cracking or blistering after repeated thermal cycling. The maximum peel strength was 15 kg/cm at 30% by weight mica. Figure 20-10 shows the appearance of the surface of the ABS test specimen after the copper plate had been peeled away. The long strands of polymer which have been pulled from the surface are characteristic of samples that show good adhesion.

Mica is compatible with a wide range of other fillers, particularly talc, bentonite, glass beads, microballoons, aluminum trihydrate, silica, perlite, sand, and calcium carbonate. A study of the packing efficiencies of spheres and flakes suggests that further improvements in the level of reinforcement can be gained by the judicious choice of filler and flake sizes in order to maximize the bulk density.[53] This behavior is similar to that found by Milewski, who studied the packing efficiencies of spheres and rods.[54] Such investigations help to establish the optimum packing density and the most favorable filler ratios for good reinforcement.

Mica may be used as a cheaper substitute for glass fibers in most reinforced thermoplastics when strength and impact resistance are not critical. Mica reinforced nylon 6/nylon 6,6 blends have been evaluated in injection molded automotive parts such as front ends, headlight and tail-light assemblies, louvres, and fender liners. Warping was negligible and the parts could be successfully painted and baked without distortion. Extruded pipe is a particularly attractive application since the flakes impart longitudinal as well as circumferential reinforcement by virtue of their preferential alignment parallel to the surface. A 2 in. extruded ABS pipe with 30% by weight Suzorite mica could not be easily fractured by repeated heavy impacts.

Toelcke has reported the use of mica as a reinforcing agent for large thermoplastic components made by a proprietary process.[55]

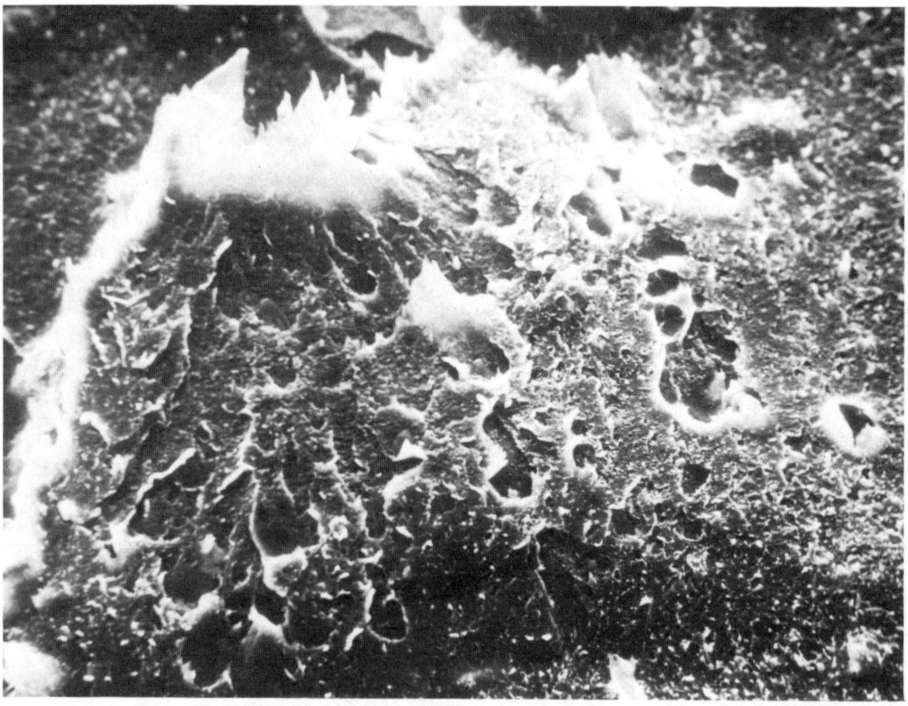

Fig. 20-10. Mica reinforced ABS plastics may be readily electroplated. This scanning electron microscope picture of the ABS surface after peeling away the electroplated film shows the plastic tendrils characteristic of good adhesion. Failures due to thermal cycling are also eliminated. Magnification 260X.

One shot moldings weighing nearly 300 lbs were routinely applied to the fabrication of rotary drum vacuum filters, cyclone cleaners, centrifugal pump components, valve bodies, plating barrel components, and gratings. Whereas glass fibers impart the highest level of physical property improvement, high aspect ratio mica was most effective for increasing rigidity. Combining both types of reinforcement provides a more attractive balance of properties.

In summary, mica reinforced thermoplastics offer to the product designer and engineer several outstanding advantages including:

- high rigidity
- high volume ratio of filler
- excellent chemical and corrosion resistance
- low permeability
- good dielectric properties
- low warpage
- low coefficient of thermal expansion
- reduced flammability
- faster molding cycles
- planar reinforcement
- increased heat distortion temperature
- good wet strength retention
- low abrasion and wear of processing equipment
- relatively low cost
- good part reproduction
- good paint adhesion

Further increases in the level of reinforcement will be possible as improved grades of high aspect ratio mica become available. Uniform flake thickness (which is obtainable by careful elutriation) and higher aspect ratios (preferably above 100) would further enhance the levels of reinforcement now obtainable. More effort will also be required to develop improved processing techniques which will minimize flake degradation and permit optimal flake alignment in the finished product.

6.2 Mica Reinforced Thermosets

Due to its outstanding dielectric properties, mica is often used as a special filler for thermosetting resins which are used in the electronics or electrical appliance industries. Thus, mica has been used to modify the electrical properties of phenolic, epoxy, silicone, polyester, or polyurethane resins. In such applications, mica also confers improved dimensional stability, high stiffness, increased heat distortion temperature, reduced permeability, and lower cost.

Phenolic resins consume the largest share of mica in plastics at the present time (excluding paints), where it is used to enhance dielectric properties, improve moisture resistance, increase dimensional stability, and reduce the linear coefficient of thermal expansion. Typical molding compositions employ about 150 pph iron-free mica (particle size range 10-80 μ diameter) in 100 parts of a novolac resin containing hexamethylenetetramine as a curing agent. When cured, these composites meet the requirements for automobile ignition equipment as well as numerous other electronic and electrical appliance applications. The consumption of mica filled phenolics has grown steadily at about 10% per year in the United States.

Finely divided mica is also employed as a filler for epoxy resins in concentrations up to 100 phr where it confers greater arc resistance, a higher dielectric constant, high resistivity, improved chemical and moisture resistance, reduced coefficient of thermal expansion, reduced shrinkage after curing, and increased modulus. The dielectric strength is also slightly increased. Mica may retard the cure of epoxy resins cured with aliphatic amines and accelerate those cured with aromatic amines.

The addition of finely divided mica to adhesive formulations has been observed to increase the lap shear strength.[57] For example, a general-purpose epoxy resin adhesive containing 14 phr fine mica gave a tensile lap shear strength of 16.3 MN/m^2 (2360 psi) with an aluminum to aluminum bond. A talc filled epoxy resin yielded the same strength at a much higher loading, 60 phr. The unfilled resin gave a lap shear strength of 7.1 MN/m^2 (1030 psi).

In these commercial applications, either wet or dry ground mica has been employed with an average flake aspect ratio below 30. Such low aspect ratios are not able to impart a substantial proportion of the inherently high modulus and

Table 20-12. Flexural properties of HAR mica reinforced thermosets (50 vol.% mica)[3,5]

	FLEXURAL STRENGTH		FLEXURAL MODULUS	
	MN/m²	(10^3 psi)	GN/m²	(10^6 psi)
Polyester resin	159	(23)	47	(6.8)
Epoxy resin	166	(24)	44	(6.4)
Phenolic resin	145	(21)	52	(7.5)
Commercial phenolic resin (low aspect ratio mica)	62	(9)	21	(3.0)
Commercial glass fiber reinforced bulk moulding compound (BMC)	100	(15)	13	(1.9)

strength of mica to the resultant composite. Hence, mica has been employed primarily as a special filler and has not been accepted as a particularly outstanding reinforcing agent, since the resulting composites prepared from such low aspect ratio micas are usually characterized by poor tensile and impact strengths. However, when care is taken to specially delaminate and classify the mica into thin flakes of uniform thickness, and subsequently incorporate these flakes into a thermosetting resin, at the same time retaining the high aspect ratio, then high levels of modulus and strength may be realized.[1,3] Table 20-12 compares the level of properties attainable with polyester, epoxy, and phenolic resins containing HAR mica. These experimental values are contrasted with a commercial mica-filled phenolic resin and a glass fiber reinforced polyester resin. It is apparent that a much higher level of reinforcement is achieved with HAR mica. This special form of mica has been called high aspect ratio (HAR) mica in order to distinguish it from non reinforcing grades. The following section summarizes the degree of reinforcement obtainable using HAR mica in thermosetting resins.

6.3 HAR Mica Reinforced Thermosets

The influence of the flake aspect ratio on the flexural properties of a vinyl ester resin (Derakane 411-45) is shown in Figs. 20-11 and 20-12. It is apparent that the aspect ratio has a strong influence on both the flexural strength and the flexural modulus. In general, the flexural properties show a rapid increase with

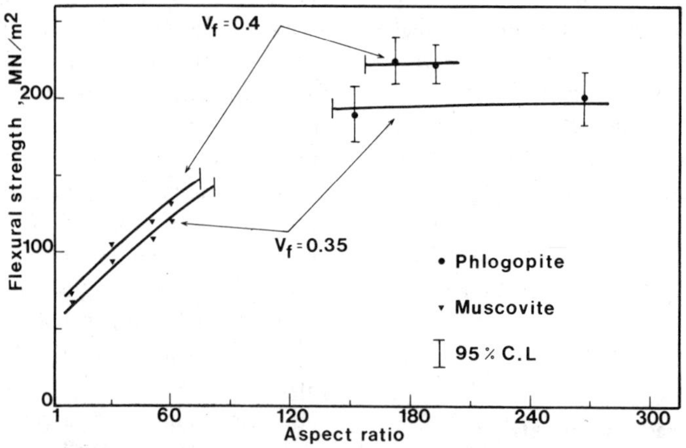

Fig. 20-11. The influence of flake aspect ratio on the flexural strength of a thermoplastic (▾) and a thermoset (•) resin containing two volume fractions of mica (0.35 and 0.40) are summarized. The thermoplastic resin was a polystyrene copolymer containing a minor proportion of acrylic acid. The thermoset was a commercial polyester resin (Derakane 411–45).

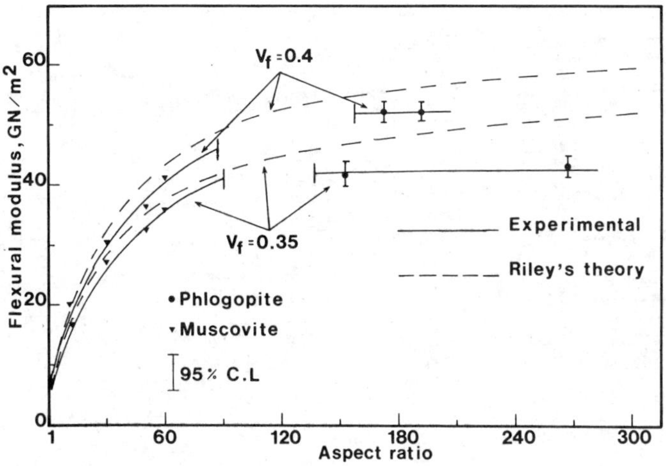

Fig. 20-12. The experimental data is compared with the theoretical predictions of Riley (78). The dependence of flexural modulus on aspect ratio is plotted for two mica volume fractions (0.35 and 0.40), and two different resin systems, using both muscovite and phlogopite micas.

aspect ratio up to an aspect ratio of 100 and thereafter approach a limiting value. The flexural strength is increased 4-fold and the flexural modulus 10-fold at the maximum level of reinforcement.

These high levels of reinforcement are frequently difficult to obtain unless care is taken to minimize voids, maintain good alignment, achieve uniform flake distribution, and preserve the original aspect ratio in the composite after fabrication.

Various techniques have been developed to permit a rapid and economical means of flake incorporation in liquid or solid resins derived from polyesters, epoxies, or phenolics. At high flake concentrations, usually above 50% by volume (75% by weight), the conventional methods of direct mixing (sigma blade, screw, ribbon) cause significant flake degradation with concomitant loss of mechanical properties in the resulting cured composite. The rate of decrease of the flake aspect ratio is particularly severe at high shear stresses.[45]

The impregnation of mica papers with liquid resins, often assisted by vacuum or pressure, has been widely employed in the manufacture of reconstituted or integrated mica papers. Reconstituted mica sheets or mica paper may be laminated under low pressure with epoxy or silicone resins in order to fabricate components for the electrical trade.[58,65] Other sheet materials, such as cellulose, glass fibers, mat, or film, may be interposed between the sheets of mica if desired. Such techniques are usually too slow or expensive for the majority of applications. Latex or solution blending techniques have been employed but the use of water or a volatile organic solvent requires a subsequent solvent removal stage. On the other hand, powder blending or dry blending can be easily accomplished in a few minutes using a tumbler, so that subsequent compression molding may be performed without any significant loss of aspect ratio. In a typical example, a polyester resin mixed with 50% by weight HAR mica in a cone vertical ribbon mixer gave, after compression molding, a flexural modulus of 31 GN/m^2 (4.5 $\times$ 10^6 psi) and a flexural strength of 93 MN/m^2 (13.5 $\times$ 10^3 psi). The same composition prepared by dry blending methods yielded a flexural modulus of 45 GN/m^2 (6.5 $\times$ 10^6 psi) and a flexural strength of 160 MN/m^2 (23 $\times$ 10^3 psi). Thus, high levels of mica may be compounded by dry blending techniques with only a slight loss of mechanical properties when compared to vacuum impregnation methods. Vacuum impregnation or the lamination of thin sheets which have been pressed between flat surfaces exhibits good orientation of the mica flakes, as illustrated in Fig. 20-13. A moderate pressure helps to keep the void content low.

Fig. 20-13. A polished cross section of a HAR mica reinforced polyester resin (Dow Derakane 411–45) showing excellent alignment and low void content (0.5%). Magnification 189×.

Mica flakes may also be applied by means of a spray gun using compressed air to fluidize and meter the mica directly into the vortex of the resin spray.[59,60,61] The spray-up gun also permits the application of chopped roving so that various combinations of flakes and fibers may be applied simultaneously or alternately.[85] Three currently used spray-up units are shown in Figs. 20-14, 20-15, and 20-16. Spray application minimizes loss of flake aspect ratio and helps to align the flakes parallel to the surface. The addition of a minor quantity of glass beads

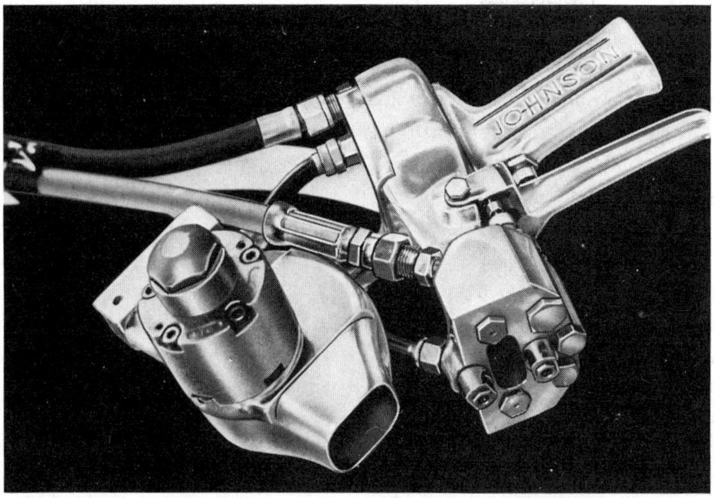

Fig. 20-14. The Johnson J-V Aggregate Feeder spray gun assembly has been modified to permit the application of mica in powder form. Thus, polyester resins may be applied to a surface with various ratios of chopped glass fibers and mica. Such techniques help to retain the aspect ratio of the mica flakes.

Fig. 20-15. The Glas-Mate (Ransburgh) spray gun in operation. Two separate ports feed dry powders such as mica into the spray zone. The glass fiber chopper may be operated independently.

or microballoons to the mica helps to fluidize the mixture and increase the packing efficiency.[62] Spray techniques are used widely as a means of reinforcing and increasing the stiffness of thermoformed acrylic sheets. Other types of equipment that are useful for processing mica filled composites, are described in Chapter 32 of this Handbook.

In thin sections, spray-up, injection, or compression molding usually favor parallel alignment of the mica flakes, an important requirement for optimum reinforcement. In thicker sections, parallel alignment may be more difficult to achieve unless the part is molded from preformed sheets which are plied up to the desired thickness.

Sheet molding compounds (SMC) can also be prepared using mica instead of chopped roving.[3,25] Chemical thickening agents such as magnesium oxide, zinc oxide, or calcium oxide may be employed to convert the prepreg into a non-sticky, leathery composition which is suitable for conventional, matched die molding processes. The prethickening stage is essential for preparing void free, high strength composites. Bulk molding compounds (BMC) may also be prepared in the same manner. Diisocyanates have also been successfully used as thickeners, particularly for liquid resins containing hydroxyl or carboxyl substituents. Liquid epoxy resins should also be "B" staged to a higher molecular weight and viscosity so as to ensure an adequate molding pressure during cure and prevent the resin from being squeezed out. Failure to control the viscosity of the resin at the molding temperature invariably results in weak specimens. Table 20-13 presents a comparison of the mechanical properties of HAR mica reinforced polyester SMC and epoxy resin SMC with a conventional glass fiber reinforced polyester SMC.

Mica composites are usually characterized by poor impact strength, usually below 53 J/m of notch (1 ft lb/in.). As expected, the Izod impact strength performance is dependent upon the direction of the impact with respect to the flake alignment, and highest results are obtained perpendicular to the mica plane. As a

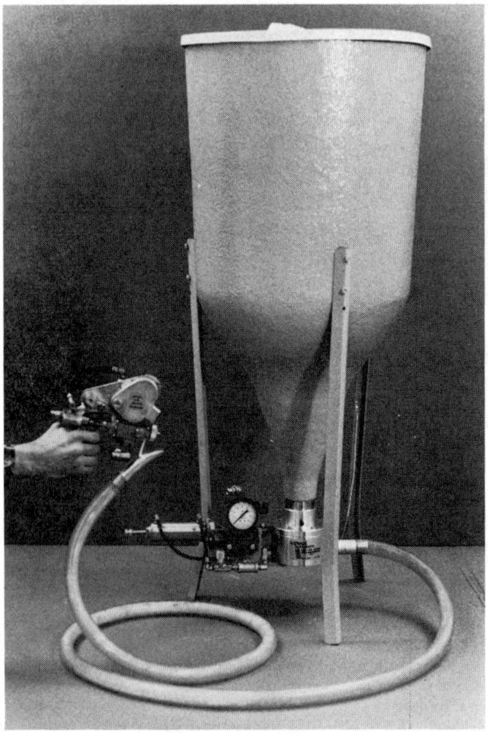

Fig. 20-16. The Venus Driadder accessory permits the conversion of a conventional spray gun assembly into a system that can accommodate powders such as mica.

result of the low impact strength, it is frequently advantageous to include a proportion of chopped glass fibers. Such combinations of fibers and flakes frequently impart a more favorable balance of properties to many reinforced composites. Figure 20-17 demonstrates that the combination of HAR mica with glass fibers tends to increase the impact strength considerably without significant loss of strength or modulus. Using such combinations, the Izod impact strength can be increased to nearly 530 J/m of notch (10 ft lb/in.).

The interlaminar shear strength of mica reinforced composites shows a linear relationship with the matrix shear strength, as shown in Fig. 20-18. The interlaminar shear strength was measured by the short beam method (ASTM D-2349-72). The values are comparable to those of short fiber reinforced composites.

Mica responds to treatment with silane coupling agents in the same manner as glass fibers. Wet strength and the long term retention of physical properties in the presence of moisture are significantly improved. Reactive silane coupling agents such as Dow Corning Z-6032 and Z-6031 have been found particularly effective in the case of polyester resins.

Table 20-13. Physical properties of SMC composites prepared from HAR mica and glass fibers.[3,25]

	HAR Mica SMC [60% by weight HAR mica]	Commercial Glass Fiber SMC [35% by weight glass]
Specific gravity	2.0	1.9
Flexural modulus,	48	11.8
GN/m^2 [10^6 psi]	[7.0]	[1.7]
Flexural strength,	197	200
MN/m^2 [10^3 psi]	[29]	[35]
Tensile modulus,	48	10
GN/m^2 [10^6 psi]	[7.0]	[2.0]
Tensile strength,	76	111
MN/m^2 [10^3 psi]	[11]	[16]
Interlaminar shear strength	25	–
MN/m^2 [10^3 psi]	[3.6]	
Izod impact strength		
[notched] J/m	34	720
[ft lb/in.]	[0.64]	[14]
Water absorption, %	0.25	0.5
Barcol hardness	60	60
Volume resistivity, ohm-cm	1.4×10^{15}	0.5×10^{15}
Surface resistivity, ohm	8.8×10^{13}	–

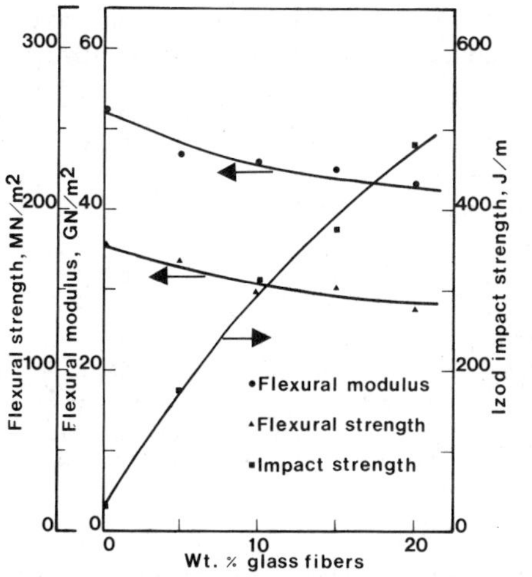

Fig. 20-17. Chopped strand greatly increases the impact strength and toughness of thermoset materials. In these experiments, the total weight percent of mica plus glass fibers was held constant at 50%. Thus at 20% glass fiber content, the Izod impact strength has reached 500 Joules/meter with only a slight sacrifice in strength and modulus.

Figure 20-19 illustrates the effect of several silane coupling agents upon the retention of flexural properties after several hours immersion in hot water.[3] The silane coupling agents do not significantly affect the flexural properties in the dry state.

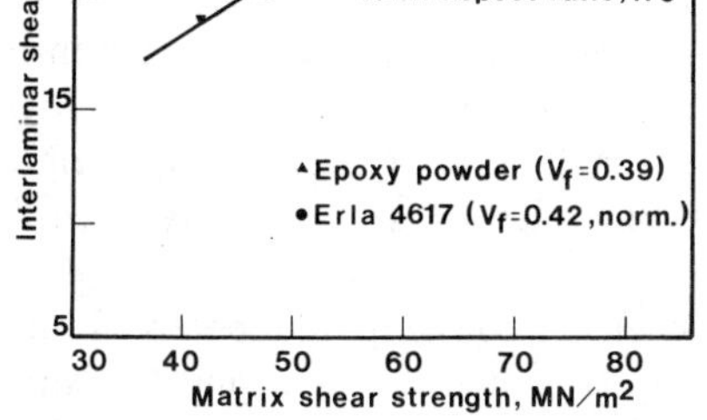

Fig. 20-18. An interesting observation is the linear relationship between the shear strength of HAR mica filled composites and the matrix shear strength for four different resins.

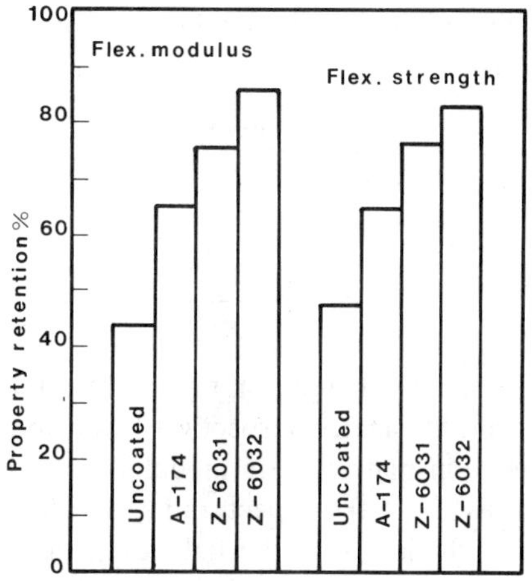

Fig. 20-19. Silane treatment of mica helps to retain the mechanical and electrical properties of mica composites during long term exposure to moisture. In these accelerated tests, samples were exposed to boiling water for a 2 hour period in order to determine the relative loss of strength and modulus. Dow Corning Z-6032 treatment offered the best property retention.

6.4 Theoretical

Several relationships have been proposed in order to relate the modulus and strength of a flake composite to the properties of the individual components. When the flakes consist of simple, rectangular ribbons of equal length, which are disposed in a perfect overlapping parallel array, then the situation is similar to that for a unidirectional, discontinuous fiber composite, and the modulus may be estimated from the rule of mixtures relationship

$$E_y = k E_1 V_1 + E_2 V_2 \qquad (5)$$

where E_y is the Young's modulus parallel to the direction of orientation of the ribbons and k approaches 1 for very long ribbons with large overlapping areas. The modulus of the flake and the resin matrix is given by E_1 and E_2 respectively, and the volume fractions of each component by V_1 and V_2 so that $V_1 + V_2 = 1$. When the ribbons are relatively short then the value of k may be less than 1. Using the approach originally developed by Cox[63] for short fibers. Padawer and Beecher[64] have developed a modified rule of mixtures equation which attempts to predict the composite modulus for short ribbons in a uniform array wherein

$$k = 1 - \frac{\tanh u}{u} \qquad (6)$$

and

$$u = \alpha \left[\frac{G_2 V_1}{E_1 V_2} \right]^{1/2} \qquad (7)$$

The subscripts 1 and 2 refer to the reinforcement and the matrix respectively, as before. G_2 is the resin matrix shear modulus and α is defined as the flake aspect ratio, which is the ratio of the length of the flake to its thickness. The composite stiffness is now seen to be a function of the aspect ratio of the flake and the shear modulus of the resin, as well as the volume fraction of the flakes.

For highly filled systems, the flakes usually overlap in a random fashion and the theory of Padawer and Beecher tends to overestimate the actual experimental results. Riley[78,25] has developed an expression for the random case which includes the effect of secondary flake interactions, wherein k is given by

$$k = 1 - \frac{\ln (u + 1)}{u} \qquad (8)$$

and u represents the same function as previously shown for the case of Padawer and Beecher. For well oriented samples, this expression is reasonably accurate.[25] A third relationship has been proposed by Halpin and Tsai,[66] which also overestimates the experimental results for mica composites. Flake alignment would also be expected to have a pronounced influence on the directional properties since similar effects have been observed by Charrier and Sudlow[67] in the case of fibers.

The tensile strength of a flake composite may also be related to flake aspect ratio. Piggott[79] modified the rule of mixtures equation to obtain

$$\sigma_x = k' \sigma_1 \left(1 - \frac{\sigma_1}{6 \alpha \tau_2} \right) V_1 + \sigma_2' \, V \qquad (9)$$

where σ_x is the ultimate tensile strength in the direction of alignment, σ_1 and σ_2' are the tensile strengths of the reinforcement and the matrix stress at fracture respectively, τ_2 is the shear strength of the matrix, and α is the flake average aspect ratio. The value of k' depends on the packing arrangement and varies between 1/2 (symmetrical array) to 2/3 (staggered array). For a random placement of parallel flakes, the value of k' is found experimentally to be 0.56 in the case of stainless steel disks.[68] Using this value, the calculated value of the strength of mica flakes was found to be 690 MN/m^2.[30]

Table 20-14. Value of ground mica sold by producers in the United States in 1971.[42]

Use	Short tons [2000 lb]	Value $U.S. [thousands]	Price $/lb
Joint cement	45,200	2980	3.30
Paint	26,800	2710	5.06
Roofing	17,800	669	1.88
Rubber	5,300	876	8.26
Plastics	480	93	9.70
Other[a]	24,000	956	1.99

[a]"Other" includes electronic insulation, thermal insulation, Christmas tree snow, annealing, well drilling, textile coating, and texture paint.

These theoretical relationships for strength and stiffness indicate that relatively large degrees of reinforcement are potentially available with mica if care is taken to maintain high aspect ratios and good alignment in the molded part. Techniques to accomplish this goal are still in the early development stages.

For a more detailed discussion of the theory of flake and ribbon reinforcements, see Chapter 21 in this Handbook.

7. OTHER FLAKE MATERIALS

7.1 Glass Flakes

Considerable attention has been focussed on the use of glass flakes as reinforcing elements in resin systems. Adams[69] compiled a literature survey on the use of glass flakes as reinforcing fillers in plastics in 1965. Narmco[70] was awarded a U.S. government contract in 1960 aimed at establishing the potential of flake reinforced composites as engineering structural materials. This was followed by a major development program at the Olin Mathieson Chemical Corp. on manufacturing methods for glass flake reinforced plastics.[71] Flake glass has been produced by stretching molten glass into a film, rolling glass beads or glass tape, blowing extruded ribbons of glass, drawing tape from a slit, or flame spraying frit. One of the most successful methods for producing high strength flakes consisted of expanding molten tubing into bubbles and then crushing the thin membrane into flakes. The radius of curvature of the flakes is sufficiently large that the flakes are almost planar and show good parallel alignment when fabricated into composites. Although it has been shown that 2 μ thick ribbons of E-glass which have been drawn from the melt can have tensile strengths between 1.1–1.4 GN/m^2, measurement of composite properties indicates that the effective strength of a glass flake is nearer 800 MN/m^2. This lower strength is not surprising in view of the fractured edges of the flakes which contain many flaws and weakened areas. So far, attempts to manufacture regular shaped discs with high strengths have not been successful. Table 20-15 represents the best results achievable with 2 μ thick E-glass flake laminates compared to glass cloth laminates at the same volume fractions. While there is an advantage in terms of increased modulus, the low tensile strengths of flake laminates compared to glass cloth laminates has discouraged further development. In addition, the impact toughness of flake laminates is relatively poor and the fabrication of void free laminates is difficult. For these reasons, it appears that the further development of high tensile glass flakes for reinforcement purposes has been discontinued.

Glass flakes are sold commercially for decorative purposes and for conferring special properties onto thermosetting resins, such as impermeability, scratch resistance, and chemical resistance. Such flakes are produced by calendering glass into a thin film, followed by fracturing. The aspect ratios of the sized fractions are usually below 40 and the flake thicknesses are from 76–152 μ. Although the tensile strengths of these flakes are unknown, such flakes would be much weaker than the 2-μ thick blown films just described. Hence, while these flakes are not particularly suitable for improving the mechanical properties of resin systems, they are used widely in protective coatings and for special decorative effects. Tables 20-16a and 20-16b list several stock sizes of glass flake available from Potter Industries and Owens/Corning Fiberglass. The flakes are obtainable in a variety of colors and may be partly or completely metallized in gold or silver.

The flakes are usually manufactured from chemically resistant glass (Type C) or an electrical grade glass (Type E). The Type C flakes are designed for use in corrosion resistant coatings, whereas the Type E is designated for use in electrical moulding compounds where they impart vapor impermeability, toughness, and improved adhesion to the substrate. Owens-Corning Fiberglas Corp. manufactures a range of screen sizes from 1/64 – 1/8 in. in thicknesses of approximately 0.00016 in. (4 μ).

7.2 Aluminum Diboride Crystals

One of the most detailed and comprehensive studies of flake reinforcement was conducted at the Carborundum Co. and included several

Table 20-15. Comparison of glass flake and glass fiber epoxy resin laminates.

	Glass Flake[69]		Glass Fabric (181 Cloth)	
Flexural strength, MN/m^2 (10^3 psi)	304	(44.1)	518	(75.1)
Flexural modulus, GN/m^2 (10^6 psi)	41	(5.95)	28	(4.06)
Tensile strength, MN/m^2 (10^3 psi)	172	(24.9)	500	(72.5)
Tensile modulus, GN/m^2 (10^6 psi)	41	(5.95)	29	(4.21)
Compression strength, MN/m^2 (10^3 psi)	359	(52.1)	408	(59.2)
Compression modulus, GN/m^2 (10^6 psi)	41	(5.95)	27	(3.92)
Glass volume, %	63		65	
Density, g/cm^3	2.01		2.05	

Table 20-16a. Glass Flake Data (Potter Industries)

Stock size[a]	U.S. Sieve No. [ASTM E-11-61]	Diameter[b] [mm]	Price range[c] 0–99 lb	[U.S. $/lb] 40,000 lbs plus
F 0407	6–10	3.36–2.00	0.46	0.23
F 0711	10–18	2.00–1.00	0.46	0.23
F 0913	14–25	1.41–0.707	0.46	0.23
F 1215	20–35	0.841–0.500	0.46	0.23
F 1418	30–50	0.595–0.297	0.46	0.23
F 1622	40–100	0.420–0.149	0.46	0.23
F 1920	60–140	0.250–0.105	0.46	0.23

[a]Stock sizes also apply to color coated or silver coated flakes.
[b]Flake thickness is between 0.762 and 0.1524 mm.
[c]Prices quoted on Jan. 31, 1975.

contributors. Economy, Wohrer, and Matkovich published a series of papers describing the synthesis of aluminum boride flakes,[72-77] and their use as reinforcing agents for both polymeric resins and metals. The interest in aluminum boride was sparked by the intense interest in 'whiskers' beginning about 1960. It was believed then that single crystal flakes might have several advantages with respect to single crystal fibers in planar reinforcement. Aluminum diboride was chosen because of its excellent mechanical properties, low density, and ease of crystallization into thin hexagonal flakes. The physical properties of aluminum diboride (AlB_2) are summarized in Table 20-17. The aluminum diboride flakes were recrystal-

Table 20-16b. *Prices*—Fiberglas Hammermilled "C" Glass Flakes No. TZ3.1.

Fiberglas* Hammermilled "C" Glass Flakes are thin glass flakes that have been hammermilled through a designated screen size. The flakes, manufactured from chemical resistant glass and having excellent moisture-resistant qualities, are designed for use as a moisture barrier in corrosion-resistant coatings.

	PRICE PER POUND			
Screen Size	2 to 7 Pallets	8 to 15 Pallets	16 pallets to 1 Less than TL/CL	Full Visible TL/CL
1/8"	$.83	$.82	$.80	$.78
1/32"	.85	.84	.82	.80
1/64"	.85	.84	.82	.80

Minimum release quantity: Two (2) Pallets of any one item. A $100 service charge applies to releases of lesser quantities.

Packaging: 50 pounds per carton

	Approx. lb Per Pallet
1/64 and 1/32 in.	1800
1/8 in.	1350

Truckload Quantity: 1/64 and 1/32 in. = 36,000 lb
1/8 in. = 27,000 lb

Carload Quantity: 1/64 and 1/32 in. = 43,200 lb
1/8 in. = 32,400 lb

Truck shipment not available in Zones 3 and 4.

Table 20-17. Physical and chemical properties of aluminum diboride flakes.[72]

Young's modulus,[a] GN/m^2 (10^6 psi)	480 (70)
Flexural strength, GN/m^2 (10^6 psi)	3.5–6.9 (0.5–1.0)
Density g/cm^3	2.95
Coefficient of thermal expansion, 10^{-7}/°C	1.6 [planar] 3.1 [transverse]
Thermal stability limits	500°C [in air] 1000°C [in nitrogen]
Chemical resistance	Unstable in acids and alkali, stable in water at 25°C, slowly attacked by water at 100°C

[a]Indirectly estimated from composite properties.

lized from molten aluminum at temperatures between 660–975°C, to which had been added finely divided aluminum boride powder. Boron or other boron containing compounds, e.g., boric oxide, boron trichloride, or boron trifluoride, could also be employed. The thin hexagonal flakes were generally recovered by filtration and then cleaned with refluxing isopropanol. Under ideal conditions, the dimensions of the flakes were between 1000–10,000 μ in diameter and 1–10 μ thick. Sieving, followed by elutriation with water, produced separated fractions with well defined aspect ratios between 37 and 430.

For aqueous elutriation of the aluminum boride flakes, Economy and coworkers[77] employed two different designs. In the first, water was introduced into the bottom of a 4-ft long glass tube with an inside diameter of 55 mm and containing a 22 mm outside diameter inner tube to serve as annulus. The water enters at the bottom, tangent to the diameter, thereby providing a spiralling upward flow.

In the second method, which was designed for the fractionation of larger quantities, three 8-ft tubes with inside diameters of 32, 46, and 85 mm were connected in series. Three different fractions could be recovered by using a constant flow of water. Each sample was carefully wet sieved prior to elutriation.

The classified AlB_2 flakes were subsequently used to study their reinforcement behavior in several epoxy resin systems and a phenolic resin. The best results were obtained with Union Carbide's ERLA 4617 cycloaliphatic epoxy resin with meta-phenylenediamine (m-PDA). These results have been replotted in Figs 20-20 and 20-21 to show the relative degrees of reinforcement in terms of the flexural strength and flexural modulus for different flake materials. It is apparent that aluminum diboride flakes impart an unusually high degree of strength and stiffness at the highest flake aspect ratios. For purposes of comparison, the volume loadings were maintained constant at 60% in each system. At very low aspect ratios, viz., below 30, the flexural properties become much less dependent upon the filler composition and conform to spherical particle reinforcement behavior. It is also apparent that as the modulus of the flake increases, relative to the modulus of the epoxy resin matrix, the aspect ratio needed to maximize reinforcement also increases proportionally. Thus, in order to achieve the full reinforcing potential of aluminum diboride flakes in an epoxy resin matrix, aspect ratios greater than 500 would be necessary. From the known strength and modulus of aluminum diboride flakes, the maximum

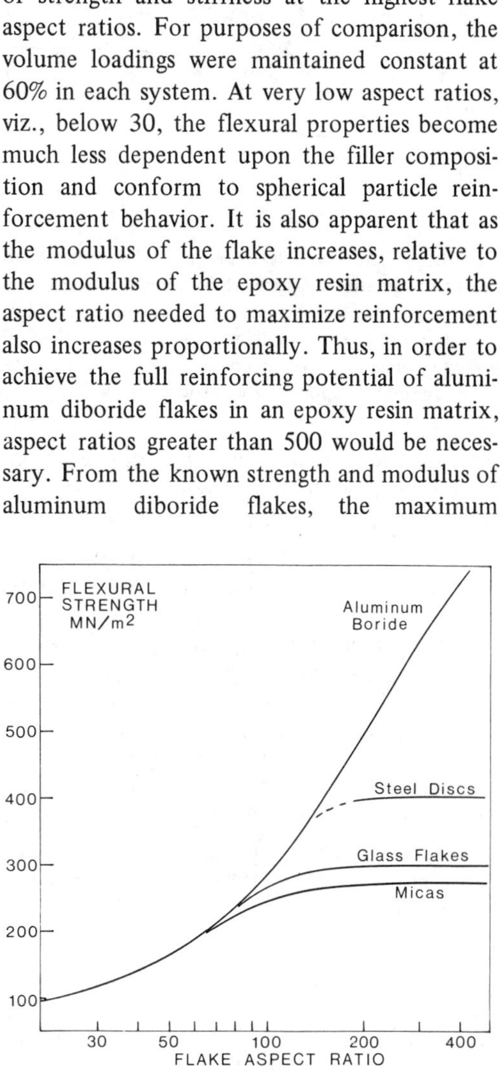

Fig. 20-20. The available experimental data has been used to predict the strength dependence of four flake composites as a function of the flake aspect ratio in an epoxy resin matrix. Each set of data has been normalized to a flake content of 60% by volume. At low aspect ratios, the reinforcement is governed by the relationships for spherical fillers. At high aspect ratios, the composite strength is limited by the strength of the flake.

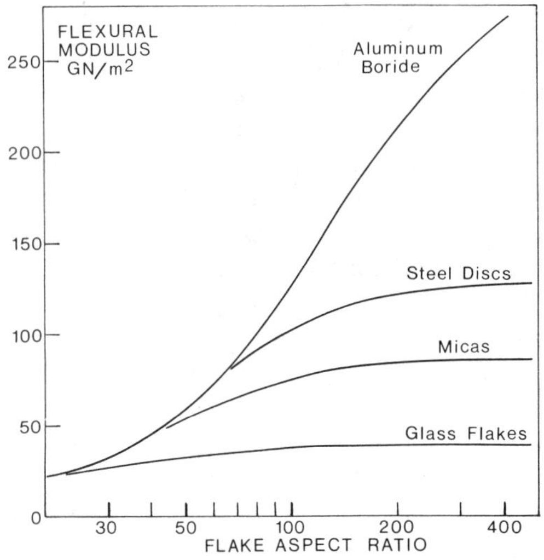

Fig. 20-21. The available experimental data has been used to estimate the aspect ratio dependence for four flake composites with respect to flexural modulus. The data has been normalized to a flake content of 60 volume per cent. The attractive properties of aluminum diboride composites are particularly evident.

flexural strength of a 60% by volume epoxy resin composite should approach 1035 MN/m^2 and the maximum flexural modulus should approach 290 GN/m^2 at aspect ratios above 500, according to the theoretical relationships developed by Riley[78] and Piggott.[79] Table 20-18 summarizes the results of Economy et al. for one particular aspect ratio (325) at 60% volume loading in an ERLA 4617/m-PDA resin system. High void contents (above 4%) tended to have a detrimental effect on composite strength. Unnotched Izod impact values showed very wide variations, the highest values occurring when the samples shattered on impact. At intermediate aspect ratios near 100, the Izod values tended to be higher and more uniform, perhaps due to a pullout mode of fracture.

Table 20-18. Mechanical properties of aluminum diboride flake composites.

Average flake diameter = 2.5 mm
Average flake thickness = 8 μ
Average aspect ratio = 325
Volume fraction of AlB_2 = 0.60

Tensile strength, MN/m^2 (10^3 psi)	452 (65.5)
Tensile modulus, GN/m^2 (10^6 psi)	293 (42.5)
Flexural strength, MN/m^2 (10^3 psi) (3-point)	644 (93.4)
Flexural modulus, GN/m^2 (10^6 psi) (3-point)	234 (33.9)
Compressive strength, MN/m^2 (10^3 psi)	451 (65.4)
Compressive modulus, GN/m^3 (10^6 psi)	276 (40.0)
Interlaminar shear strength, MN/m^2 (10^3 psi)	55.2 (8.00)
Sonic shear modulus, GN/m^2 (10^6 psi)	166 (24.1)
Izod impact strength, unnotched, J/m (ft-lb/in.)	90–4000 (1.7–75)
Density, g/cm^3	2.27

7.3 Steel Disks and Platelets

In an effort to understand the relationship between the properties of a flake and that of the corresponding flake composite, Suffrendini and Stevens,[70] and later Piggott and Glavinchevski[79] used well characterized steel disks or rectangular strips as the flake component and proceeded to measure the mechanical properties of the resulting composite. The circular or rectangular 'flakes' were die cut from stainless steel ribbon stocks, which had tensile strengths between 1.45 and 1.80 GN/m^2, depending upon the particular thickness and type of steel selected. These values were then accepted as the ultimate tensile strengths of the corresponding disks or rectangular strips. Various overlapping configurations were hand laid, producing a laminate which was several layers thick. A thermosetting resin (Metlbond 406) was used in the former case and a thermoplastic polycarbonate resin (Lexan) in the latter. The flake aspect ratios varied from 100–800. Circular flakes and rectangular flakes produced nearly identical composite strengths, so that the flake geometry did not appear to be particularly critical. Depending upon the particular flake arrangement, the value of k' in the rule of mixtures relationship (Equation 9) for a planar array falls between $2/3$ and $1/2$. In the experiments of Piggott, the value of k' was 0.56 ± 0.01 for a regular overlapping configuration. Consequently, at a volume fraction of 0.60, a well oriented flake composite will have a tensile strength approximately equal to $1/3$ of the flake strength (assuming the flake aspect ratio is above the critical value). This is a useful rela-

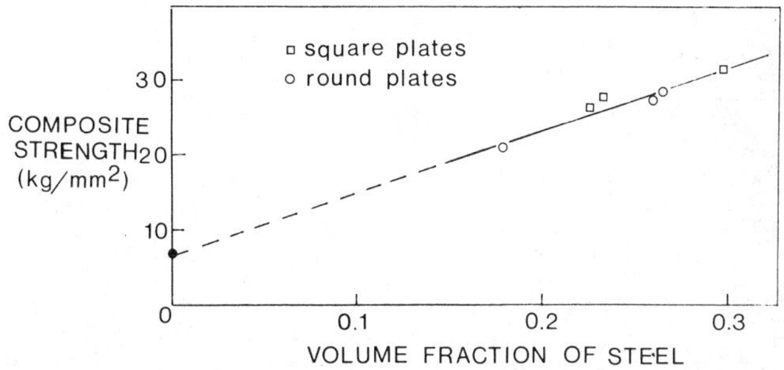

Fig. 20-22. The tensile strength of a polycarbonate resin containing round (○) and rectangular (□) steel disks is plotted versus the volume fraction of steel. The tensile strength of the polycarbonate resin is also shown (●).

tionship for estimating the practical limit of strength in a flake composite. The results of these investigations are presented in Figs. 20-22 and 20-23.

7.4 Alumina Flakes

White and Wood[80] have described the preparation of very thin, perfect crystals of α-alumina with aspect ratios approaching 50,000:1, using a vapor transport process. Transport occurs when a solution of Al_2O_3 in PbF_2 is allowed to evaporate freely in the temperature range between 1260–1380°C, i.e., near the boiling point of PbF_2 at 1293°C. A platinum gauze placed over the top of a platinum crucible provides a convenient support for the spontaneously nucleated crystals which can be easily harvested after each run. The resulting crystals grow as thin hexagonal platelets up to 6 cm^2 in area and 0.5–10 μ thick. The surfaces of the crystals are free from macroscopic growth steps and structural perfection was high, as evidenced by transmission electron microscopy and X-ray diffraction topography. Uniformity of thickness was also demonstrated by the constant interference color observed in reflection for the thinnest specimens. The platelets were highly flexible and could be bent almost double without breaking. Such crystals would be expected to have a high tensile and flexural strength comparable to sapphire whiskers. Although alumina platelets have not yet been employed for the purpose of reinforcement, the high strength and modulus of α-alumina should permit the fabrication of high strength composites comparable in mechanical properties to aluminum diboride flake composites with a higher density. α-Alumina (corundum) has a density of 4 g/cm^3. The crystals appear to be perfect hexagonal platelets without any observable dislocations.

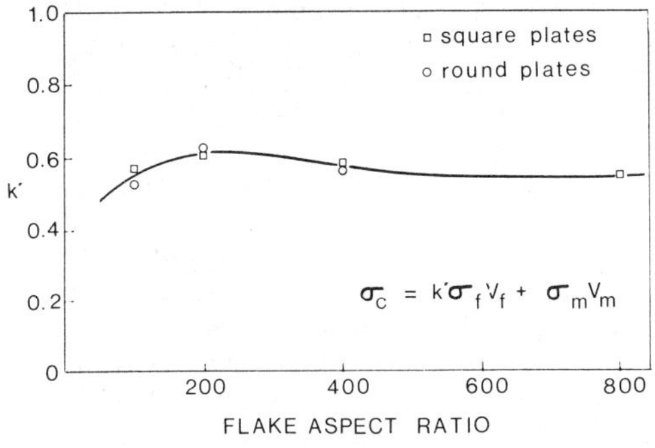

Fig. 20-23. The empirical parameter k′ is plotted versus the aspect ratio of steel disks or platelets in a polycarbonate resin. Above an aspect ratio of 100 the value of k′ is almost constant at 0.56 for a parallel array.

7.5 Silicon Carbide Flakes

In some cases, crystal modifications can be brought about by the addition of trace elements to a molten solution. Thus, hexagonal and triangular shaped crystal plates of β-silicon carbide may be grown from a solution of silicon saturated with carbon, contained in a graphite crucible which has a temperature gradient between 1530° and 1670°C.[80,81] The addition of 0.1 weight % tantalum produced polyhedrons and laths of β-silicon carbide with improved crystal surface quality. Silicon carbide platelets

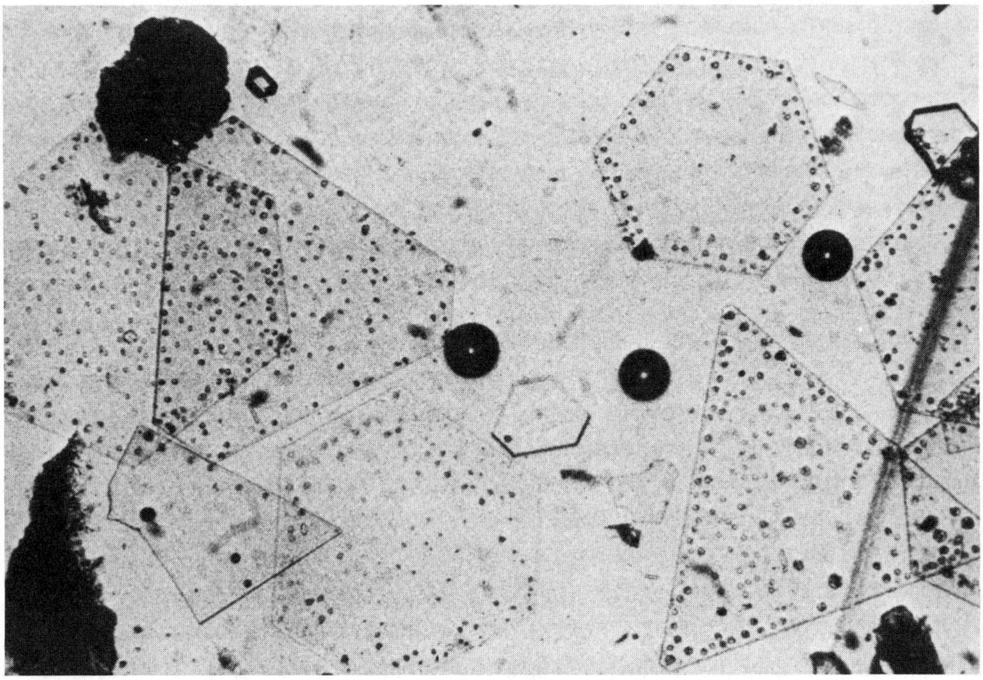

Fig. 20-24. Silicon carbide single crystals in platelet form reproduced from the studies of Zapf, Shaffer, and Corbett (83).

have also been grown by a proprietary sublimation process.[82] Typical β-silicon carbide crystals are shown in Fig. 20-24. In view of the known outstanding mechanical properties of β-silicon carbide whiskers, viz., tensile strengths of 28 GN/m^2 and modulus of elasticity values of 770 GN/m^2,[84] it was expected that composites made from silicon carbide flakes would possess equally attractive properties. In the published work of Zapf, Shaffer, and Corbett,[83] however, only low levels of reinforcement were obtained in epoxy resin systems. In retrospect, the lack of success was primarily due to the low average aspect ratios used in this study, coupled with broad aspect ratio distributions. Classification of these flakes by elutriation and sieving, in order to isolate the high aspect ratio component (above 100), most probably would have yielded composite mechanical properties approaching that of the aluminum diboride composites. As in the case of aluminum diboride, flake aspect ratios of at least 1000 would be required in order to reach their full reinforcement potential in epoxy resin systems. Somewhat lower aspect ratios would be adequate with more rigid polymers such as the polyimides.

References

1. Lusis, J., Woodhams, R. T., and Xanthos, M., "The effect of flake aspect ratio on the flexural properties of mica reinforced plastics," *Polymer Eng. Sci.* **13**(2): 139–145 (1973).
2. Maine, F. W., Shepherd, P. D., Golemba, F. J., Moskal, E. A., and Osborne, A. D., "A new family of reinforced thermoplastics," 28th Annual Technical Conference, Reinforced Plastics/Composites Institute, The Society of the Plastics Industry Inc. Section 5-A (1973).
3. Woodhams, R. T. and Xanthos, M., "Polyester resins reinforced with HAR mica," 29th Annual Technical Conference, Reinforced Plastics/Composites Institute, The Society of the Plastics Industry Inc., Section 4-E (1974).
4. R. T. Woodhams, Canadian Patent 893,163 (1972); U.S. Patent 3,799,799 (1974); British Patent 1,313,749 (1973).
5. Maine, F. W. and Shepherd, P. D., "Mica reinforced plastics: a review," *Composites* 193 (1974).
6. Locke, S. R., Lindsay, R., and Fenton, M., "The properties of Suzorite mica flake as a reinforcement in plastic composites," 30th Annual Technical Conference, Reinforced Plastics/Composites

Institute, The Society of the Plastics Industry, Section 4-D (1975).
7. Kauffman, S. H., Leidner, J., Woodhams, R. T., and Xanthos, M., "The preparation and classification of high aspect ratio mica flakes for use in polymer reinforcement," *Powder Technology* 9:125–133 (1974).
8. *Suzorite Mica Booklet*, Marietta Resources International, 277 Park Ave., New York, N.Y. 10017; Technical Services Dept., 1884 Mullins St., Montreal 104, Quebec, Canada.
9. W. K. Haller, U.S. Patent 3,087,482 (1958).
10. M. D. Heyman, Canadian Patent 451,223 (1948).
11. M. I. Cohn and R. D. Perdue, Reissued U.S. Patent 25,965 (1966).
12. M. J. de Bose, U.S. Patent 3,327,951 (1967).
13. H. H. Morris, K. L. Turner, U.S. Patent 3,206,127 (1965).
14. P. M. Brown, U.S. Patent 2,482,740 (1949).
15. G. de Senarclens, U.S. Patent 2,614,055 (1952).
16. M. Keuth, H. W. Rotter, and P. Guillery, U.S. Patent 2,906,649 (1959).
17. J. Ruzicka, U.S. Patent 3,719,329 (1973).
18. E. C. Kunz, N. C. Fletcher, and D. Ensminger, U.S. Patent 2,798,673 (1957).
19. C. H. Dye, U.S. Patent 3,240,203 (1966).
20. Orowan, V. E., *Zeitschrift für Physik* **82**:235 (1933).
21. D. C. Stavely, U.S. Patent 3,570, 755 (1971).
22. J. I. Miller, and K. H. Ivey, U.S. Patent 3,390,045 (1968).
23. F. C. Atwood, U.S. Patent 2,306,292 (1942).
24. J. W. Kraus, and F. R. Hurley, U.S. Patent 3,434,917 (1969).
25. Xanthos, M., "High Aspect Ratio Mica Reinforced Thermosets." Department of Chemical Engineering and Applied Chemistry, Univ. of Toronto, Ph.D. Thesis, (1974).
26. Heywood, H., *J.I.C. Chem. Eng. Soc.* **4**:17 (1948).
27. Coulsen, J. M. and Richardson, J. F., "Motion of Particles in a Fluid," "Chemical Engineering Unit Operations" Vol. **2**, Chap. 14, p. 481. Pergamon Press, London, 1955.
28. Sweco Canada Ltd., 22 Jutland St., Toronto, Canada.
29. Haultain, H. E. T., *The Infrasizer*, Univ. of Toronto Press, Second Edition, Jan., 1961.
30. Lusis, J., The Importance of Flake Aspect Ratio in Mica Reinforced Plastics. Dept. of Chemical Engineering and Applied Chemistry, Univ. of Toronto, M.A.Sc. Thesis (1970).
31. Kaye, B. H. and Jackson, M. R., "Felvation speeds powder fractionation," *Chem. Eng. News* (C&EN), p. 50 (Mar. 6, 1967).
32. Berard, *J. Can. Min. and Metal*, Bulletin (CIM), p. 106, July 1973.
33. Locke, S. R., Martin Marietta Laboratories, 1450 South Rolling Rd., Baltimore, Md. 21227.
34. Capes, C. E. and Coleman, R. D., "Particle size measurement by a powder film method," *Ind. Eng. Chem. Fundamentals* **12**:124 (1973).
35. Locke, S. R., Fenton, M. and Hawley, G. C., Suzorite mica flake: A new low cost mineral reinforcement for structural components. The first Industrial Minerals International Congress, Royal Lancaster Hotel, London, England, 1974.
36. Hawley, G. C. and Locke, S. R., High aspect ratio mica composites. 32nd Annual Conference, The Society of the Plastics Industry of Canada, Sept. 29–Oct. 1, 1974.
37. Hawley, G. C. and Bédard, L., Suzorite mica–the new polymer reinforcement, Eastern Reinforced Plastics Division, The Society of the Plastics Industry of Canada, Oct. 12, 1973.
38. Sharps, T. I., Mineral Industries Bulletin, Colorado School of Mines **5**(1):1–15 (1962); see also Horton, F. W. U.S. Bureau of Mines, Bulletin 6822 (1935).
39. Sax, N. I., *Dangerous Properties of Industrial Materials.* 3rd Edition, Van Nostrand Reinhold, New York, 1968.
40. Shell, H. R. and Ivey, K. H., Fluorine micas. Bulletin 647, Bureau of Mines, U.S. Dept. of the Interior, pp. 187–188 (1969).
41. Sharps, T. I., Colorado mica, Mineral Industries Bulletin. Colorado School of Mines, **5**(1):1–15 (1962).
42. Petkof, B., Mica, mineral facts and problems, Bureau of Mines Bulletin 540, United States Department of the Interior (1970).
43. Petkof, B., Mica, metals, minerals and fuels, Minerals Yearbook, United States Department of the Interior, Vol. 1 (1971).
44. Woodhams, R. T., Materials Engineering–The Next 100 Years, Centennial Lecture, Faculty of Applied Science and Engineering, Univ. of Toronto (1974).
45. Okuno, K. and Woodhams, R. T., "Mica reinforced polypropylene," *J. Polymer Eng. and Sci.* **15**(4):308 (1975).
46. Maine, F. W., A new family of reinforced thermoplastics–mica reinforced thermoplastics, 30th Annual Conference, Society of the Plastics Industry of Canada (1972).
47. Collins, W. T. and Kludt, J. L., New silane coupling agents for sphere and mineral reinforced engineering plastics, Reinforced Plastics/Composites Institute, 30th Annual Conference, Society of the Plastics Industry, Section 7-D (1975).
48. Steinberg, A. H. and Soliman, F. Y., Deck lids made from stampable nylon composite sheets., 30th Annual Technical Conf., Reinforced Plastics/Composites Institute, The Society of the Plastics Industry, Section 11-C (1975).
49. Mason, D., Solid phase forming of plastics sheet., Society of Plastics Engineers, Feb. 6, 1975.
50. Imperial Chemical Industries (ICI) Ltd., British Patent 1,220, 182 (1971).
51. W. R. Grace, British Patent 1,136,350 (1968).
52. Shapiro, E., Adhesion and Thermal Cycle Performance of Metals Electroplated on Mica Reinforced Plastics, Dept. of Chemical Engineering

and Applied Chemistry, Univ. of Toronto, M.A.Sc. Thesis (1974).
53. Mittler, R., Packing Volumes of Flakes and Spherical Fillers, Dept. of Chemical Engineering and Applied Chemistry, Univ. of Toronto, B.A.Sc. Thesis (1975).
54. Milewski, J. V., 28th Annual Technical Conference, SPE (1973); 29th Annual Technical Conference, SPE (1974); Ph.D. Thesis, Rutgers University (1973).
55. Toelcke G. A. and Larsen, J. V., Structural applications of reinforced polypropylene in the chemical process industries, 30th Annual Conference, Reinforced Plastics/Composites Inst., The Society of the Plastics Industry, Section 12-D (1975).
56. Golemba, F. and Woodhams, R. T., U.S. Patent 3,799,799 (1974).
57. Lee, H. and Neville, K. *Handbook of Epoxy Resins*, Chap. 21, p. 29, McGraw-Hill, New York, 1967.
58. Lauroesch, H. C., "Better ways to insulate stators with mica," *Power* pp. 41–42 (Apr. 1973).
59. Venus Products Inc., 1862 Ives Ave., Kent, Wash. 98031.
60. Jonson & Sons, 3225 N. Verdugo Rd., Glendale, Calif. 91208.
61. Ransburgh Corp., P.O. Box 88200, 3939 West 56th St., Indianapolis, Ind. 46208.
62. Stayner, V., Micro-mix platelets, spheres, rods, and particles for a new RP/C family. SPI 29th Annual Conf., Reinforced Plastics/Composites Institute, Section 10-C (1974).
63. Cox, H. L., *Brit. J. Appl. Phys.*, **3**:72 (1952).
64. Padawer, G. E. and Beecher, N., *Polymer Eng. Sci.* **10**(3):185 (1970).
65. Dingman, E. G., *SPE J.* **17**:981 (1961).
66. Halpin, J. C. and Tsai, S. W., Environmental factors in composite materials design. AFML TR 67-423; Ashton, J. E., Halpin, J. C., Petit, P. H., Primer on composite materials analysis. Technomic Pub. (1970).
67. Charrier, J. M. and Sudlow, M. J. *Fibre Science and Techn.* **6**:249 (1973).
68. Osborne, A. D., Shepherd, P. D., Maine, F. W., 32nd Annual Technical Conference, SPE Preprints, p. 401 (1974).
69. Adams, E. R., Bibliography on Glass-Reinforced Plastic Laminates and Reinforcing Glass Fibers and Flakes, Technical Information and Library Services, Ministry of Aviation (U.K.) TIL/BIB/97 (1965).
70. Suffredini, L., Narmco Industries Inc., Establishment of the potential of flake reinforced composites as engineering structural materials. Quarterly Report No. 1, Ad 265,192 (1961); Chia, C. Y., Duft, B. L., and Stevens, D. ibid. Quarterly Report No. 2, AD 265,263 (1961); Chia, C. Y., Duft, B. L., Stevens, and Horsley, J. ibid. Quarterly Report No. 3, AD 266,379 (1961); Suffredini, L., Stevens, D. W. Summary report, AD 265,927 (1959).
71. Aclin, J. J., Klahs, L. J., Manemeit, F. E., Snyder, A. D., et al., Olin Mathieson Chem. Corp., Development of manufacturing methods for glass flake reinforced plastics, AD 299,351 (1962).
72. Economy, J., Wohrer, L. C., Wosilait, A., and Yu, T., Development of high strength, high modulus AlB_2 flake reinforced composites. AFML-TR-72-74 (1972).
73. Economy, J., Matkovich, V. I., and Wohrer, L. "Single crystal flake reinforced composites," *SAMPE J.* **5**(1):31-36 (1969).
74. Economy, J., Wohrer, L. C., and Matkovich, V. I., "Single crystal flake reinforced composites," *SAMPE J.* **13**:83, (1968).
75. Economy, J., Matkovich, V. I., and Giese, R. F., *J. Am. Chem. Soc.* **86**:2337 (1964).
76. J. Economy, V. I. Matkovich, L. C. Worher. Canadian Patent 871,710 (1971).
77. Economy, J., Worher, L. C., and Matkovich, V. I., "AlB_2 flake reinforced composites," *SAMPE J.* **18**:340–68 (1973).
78. Riley, V. R. Interaction effects in fibre composites. Polymer Conf. Series, Univ. of Utah, June 1970.
79. Piggott, M. R. and Glavinchevski, B., "Steel disc reinforced polycarbonate," *J. Materials Sci.* **8**: 1373-1382 (1973).
80. White, E. A. D. and Wood, J. D. C., "The growth of highly perfect alumina platelets and other oxides by solvent vapour transport, *J. Materials Sci.* **9**:1999–2006 (1974).
81. Nelson, W. E., Rosengreen, A., Bartlett, R. W., Halden, F. A., and Meuller, R. A., Stanford Research Institute. AFCRL-66-579 (1966).
82. Carborundum Co., proprietary process.
83. Zapf, C. F., Shaffer, P. T. B., Corbett, W. J., et. al., Silicon carbide platelet reinforced epoxy composite investigation, AFML-TR-65-282 (1965).
84. General Technologies Corp. Brochure, 1821 Michael Faraday Drive, Reston, Virginia 22070, U.S.A.
85. Xanthos, M., Hawley, G. C., and Antonnaci, J., Suzorite mica/glass polyester laminates . . . 31st Annual SPI Conference, Section 22-B (1976).
86. Xanthos, M., Hawley, G. C., and Antonacci, J., Hot runner injection molded mica reinforced polypropylene. 34th Annual Technical Conf. SPE Preprint p. 582 (1976).
87. Narukawa, N., Kubotsu, A., and Akuyama, S. The structure and properties of mica reinforced plastics. Internal Report, Kuraray Co., Japan, June 20, 1973.

Section IX
Ribbon Reinforcements

21
Ribbon Reinforcements

S. T. Gulati
Corning Glass Works
Corning, New York

CONTENTS

1. ABSTRACT

The unique properties afforded by glass-ribbon as a reinforcing filament are described. These include mechanical properties (strength and stiffness), thermal properties (expansion coefficient), chemical properties (durability and corrosion resistance), and transport properties (permeation coefficient). Experimental results obtained from glass-ribbon plastic composites and their respective advantages and disadvantages are pointed out.* As an example of a specific application, the performance of a glass-ribbon composite pipe is compared with that of a conventional steel pipe. A brief review of ribbon and film composites, including materials other than glass, is given.

*The high planar isotropic strength and modulus of ribbon composites result in advantages in some applications over continuous fiber reinforced composites.

2. INTRODUCTION AND MANUFACTURING METHODS

The versatility of glass as a structural material is largely a result of the many shapes into which it can be formed. Among the many forming processes of interest is the drawing of continuous microsheet. Corning Glass Works has, for many years, formed microtape or glass ribbon with widths from 0.12-1.5 in. and thicknesses from 0.001-0.003 in. from Code 8871 (alkali lead silicate) glass for capacitor application. Glass ribbon is highly flexible and can be rolled up on spools much like textile ribbon. Most of the commercially produced ribbon can be wound safely to a diameter between 0.25–0.50 in.

The potential of glass-ribbon as a reinforcement for plastics was first reported by Humphrey,[1] and shortly thereafter by Halpin and Thomas[2] who established a firm theoretical foundation for ribbon reinforcement. The most appealing facet of ribbon as a reinforcing material, due solely to its geometry, is the isotropic stiffening and strengthening effect in the plane of the ribbons. This effective planar isotropic reinforcement is similar to the properties of randomly oriented fiber mat composites. Many applications requiring two-dimensional strengths and stiffness are found in aerospace, marine, automotive, material handling, and building materials industry. Specifically, ribbon composites may have a performance advantage in aircraft structures such as wings, ship hulls, tank cars, pipelines, automobile bodies, belted tires, building cladding, and other structural members in the construction industry.

The above advantage of ribbons in providing isotropic mechanical properties is realized at significant savings in fabrication cost and reinforcement volume. Ribbons do not need to be oriented like fibers to provide isotropic properties, thus reducing fabrication cost. Furthermore, due to their rectangular cross-section, they can be packed to much higher volume fraction than the fibers. For a given volume fraction, a quasi-isotropic fiber composite[3] is only 37% as stiff and 50% as strong as the ribbon composite. For similar mechanical properties, therefore, ribbon reinforcement would require 50-60% lower volume loading than fiber composites. Alternatively, ribbons only half as strong as fibers can result in composite strength comparable to that of quasi-isotropic fiber composite.

Usually, in order to make a quasi-isotropic continuous fiber composite, a relatively large number of unidirectional plies must be stacked together in a multi-directional pattern, such as 0°, ±45°, 90°. However, since some applications, especially in the aerospace industry, require a thinner isotropic structure than could be attainable in this manner, ribbon reinforcement becomes extremely attractive. Each unidirectional filament ply ranges from 0.002–0.006 in. in thickness and three to four orientations are needed to obtain quasi-isotropic properties. In contrast, planar isotropic structures can be attained in very thin ribbon composites since each layer of ribbon can be as thin as 0.0005 in.

The isotropic reinforcement provided by the ribbons is primarily due to their large aspect ratio (width/thickness), similar to the unidirectional properties of fiber composites. Transverse and shear properties as high as 90% of the longitudinal properties[1] were measured for aspect ratios > 100. Transverse strength of up to 60% of the rule of mixture value was measured by Lewis[4] with stainless steel ribbon of aspect ratio = 187 in SURLYN A matrix. Gray[5] worked with steel ribbon with an aspect ratio of 30 in silver matrix and obtained a transverse strength which was 50% of the longitudinal strength. Similar values were reported by Anderson et al.[6] and Anderson,[7] who used steel ribbon with an aspect ratio of 50 in aluminum matrix. More recently, Pollock and Arthur[8] have measured transverse strength of 50% of the longitudinal strength for 1% carbon steel ribbon having an aspect ratio of 40 in Sn–70% Pb matrix. Theoretical estimates of mechanical properties of ribbon composites have been reported by Padawar and Beecher,[9] Chen and Nielson,[10] and Chen and Lewis.[11]

One of the principal disadvantages of glass-ribbon composites is their notch sensitivity. In the presence of holes and slots, the strength degrades substantially. Transverse strength of 2 in. diameter tubes with two diametrically opposite holes of 1/8 in. diameter and of bars with freshly cut long edges was reduced by a factor of 2-5 due to stress concentration and the presence of secondary microcracking in the vicinity of holes and cut edges. However, etching in HF/HNO_3 solution reduces stress concentration by rounding off the flaw tip radius and restores the strength to a value corresponding to unnotched specimens. Another method of minimizing strength loss is to use smaller aspect ratio ribbon (~90). Notch sensitivity is not a major problem in the case of boron film composites. In fact, they have been used effectively to strengthen[12] bolt holes in filament reinforced structures.

Aside from mechanical properties, ribbon reinforcement offers superior expansion and transport properties in comparison with fiber reinforcement. As a direct consequence of isotropic elastic moduli and low thermal expansion coefficient of glass ribbon, the composite expansion is found to be isotropic. Moreover, in view of the large stiffness of glass in comparison with that of the plastic matrix, a small volume fraction of ribbon ($\sim$0.3) is sufficient to reduce the expansion coefficient of the matrix by one order of magnitude.[13] This is a highly desirable feature in those applications where the dimensional stability of composites is of importance.

The permeation coefficient of glass-ribbon reinforced composites is also considerably lower[14] than that of fiber composites, due to the tortuous permeation path afforded by large aspect ratio ribbon filaments. Brydges et al.[15] have recently developed a permeation model which indicates that glass ribbons of readily attainable aspect ratios and volume fraction offer 100–1000 fold improvement over the fibers. The permeation model was verified through mass spectrometer measurements of helium permeation through ribbon composites of various constructions. High permeation resistance is a key factor in applications such as storage containers, pressure vessels, and pipelines which store and transport precious fluids.

Yet another non-mechanical advantage of glass-ribbon composites is their chemical durability and corrosion resistance. In experiments conducted at Corning Glass Works, ribbon composites exposed to 30% sulfuric acid environment for several days gained weight at a lower rate than either the fiber composites or plain matrix (epoxy) specimens.

In view of the unique features of glass-ribbon composites discussed above, an immediate and practical application of such composites is a glass-ribbon reinforced plastic pipe which can be fabricated economically by winding matrix impregnated ribbon over a mandrel in a coil-winding machine. The properties of such a pipe are compared with conventional steel pipe in Section 7. It is shown that, except for stiffness, the glass-ribbon composite pipe surpasses the steel pipe in properties, although in terms of specific stiffness, the two are equivalent.

The chapter concludes with a brief literature review of work done on boron and graphite film composites.

2.1 Manufacturing Methods for Glass Ribbon

Glass ribbon is manufactured by two different methods, the choice of each being determined by the desired precision of cross-sectional geometry. In the first method, known as hot-draw, ribbon is drawn from the hot glass through an orifice of appropriate geometry made of platinum or other highly refractory material. The orifice slot has maximum width midway between its ends and is made progressively narrower as the ends are approached. When molten glass flows through the orifice, it is apparently stretched sideways by the drawing action of the glass passing through the end sections of the slot. This action, accompanied by application of a slight amount of cool air directed against the central region of the slot, results in a sheet of glass of nearly uniform thickness throughout its width.

The interior of the drawing die comprises a V-shaped well above the intermediate portion of the slot, terminating at either end in a somewhat deeper and triangular shaped well. In some cases, a more rigid form of drawing die is used with the wedge portion of the die projecting upward into molten glass. With this arrangement, there is less tendency for distortion of the die, since it is considerably more rigid against deformation due to the weight of molten glass. Also, the die slot being in the form of an inverted V enables the glass to flow freely from different levels in a vertical stream of ribbon.

The hot-draw apparatus is shown in Fig. 21-1 and consists of a furnace, orifice assembly, air jet carriage, ribbon guide block, and pulling rollers. Ribbon is drawn by a tractor with pulling rollers whose speed can be varied. The table carrying the rollers can be tilted to correct for misalignments which may cause the ribbon

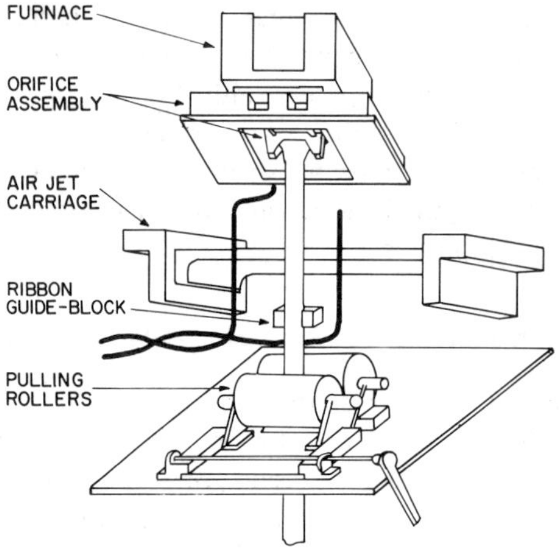

Fig. 21-1. Hot-draw apparatus for glass ribbon. (*Courtesy Corning Glass Works*).

to drift away from the center of the rollers, although this is minimized by the presence of ribbon guides. Usually, the ribbon is pulled through a catenary, similar to Vello tube making process, and ends up in the horizontal position. Pulling speeds are governed by the glass temperature and desired ribbon geometry. Higher glass temperature requires faster pulling speed and higher air jet pressure to maintain uniformity in section geometry. Figure 21-2 shows section geometry of ribbons of three different aspect ratios.

The second method of making ribbon is the redraw process, where an oversize blank with desired section geometry is heated to its softening point and then pulled. This process generally gives precise section geometry. Ribbons of several different aspect ratios are easily manufactured by this process. However, it involves two steps, one for preparation of blank or preform and the second for redraw operation. In general, higher pulling speeds are possible with the hot-draw process making it more economical. Both these processes are used by Corning Glass Works.

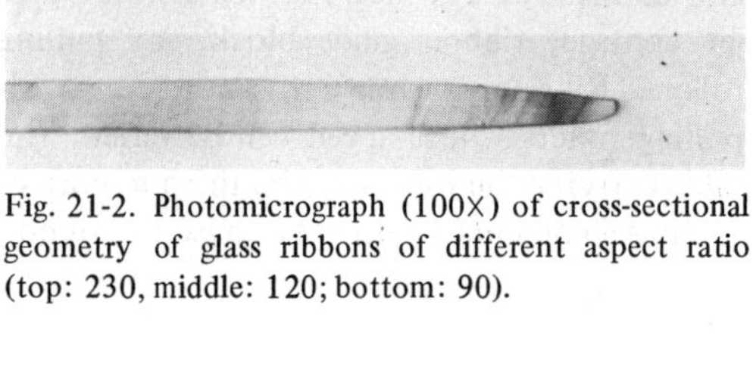

Fig. 21-2. Photomicrograph (100×) of cross-sectional geometry of glass ribbons of different aspect ratio (top: 230, middle: 120; bottom: 90).

2.2 Fabrication of Glass Ribbon Composites

Flat specimens in the form of bars and plates can be fabricated by conventional hand lay-up technique. Ribbons are laid side by side with minimal gap between the edges and are temporarily held together by scotch tape at the two ends of the ribbon sheet. Several sheets are prepared in this manner. A convenient matrix for construction of flat plates is SCOTCHWELD® or SURLYN®, available from 3M Company or E. I. Dupont Co. The former is a modified thermoset type matrix while the latter is a thermoplastic. Both are available as films with varying widths and thicknesses. The ribbon sheet mentioned above is placed in a mold with scotch tape ends sticking out. Then a film of matrix is placed and this process is repeated with alternate layers of ribbon and matrix until desired thickness of the composite is achieved. The mold is closed and placed in a press with heated plattens for curing.

One can also use prepregged ribbons which have been dipped in the liquid matrix, such as epoxy. The matrix, in such a case, is semicured to B stage to permit handling during fabrication. Ribbon sheets are made as discussed above and placed in the mold. The positioning of successive sheets of ribbon is critical in that it determines the stacking sequence and optimum "gap pattern" between ribbon edges. This is referred to as the stacking parameter, γ, discussed later in the chapter. It has a pronounced effect on the transverse strength of ribbon composites. Figure 21-3 shows a section through a ribbon composite plate fabricated in the above manner using Epon 828 Resin. The random sequence of gaps should be noted.

Fabrication of glass ribbon-epoxy tubes can be carried out in a coil winder shown in Fig. 21-4. The mandrel may be made of steel, or

Fig. 21-3. Photomicrograph (18X) of polished section through ribbon composite plate (Epon 828 resin).

aluminum with a teflon coating, to facilitate removal of the tube from the mandrel. Prior to winding, the mandrel is preheated to 50°C which speeds up subsequent curing. Also, during curing the mandrel expands less than the tube in the oven since it is preheated to 50°C. If, on the other hand, the mandrel is not preheated it would expand more than the tube during curing causing splitting of the tube. After the tube is cured, it is removed from the mandrel by quenching the assembly in cold water, which shrinks the mandrel. The tubes made in this manner exhibit good transparency, as shown in Fig. 21-5, and allow large volume fractions of glass. The ribbon used for tube fabrication was Corning Code 8871 with an aspect ratio of 120. Approximately 500 ft of this ribbon (0.219 in. wide X 0.0018 in. thick) are needed for a 10 in. long, 20-layer tube having a diameter of 2.5 in.

The wet-winding technique described above is not limited to tubular geometry. Winding can also be carried out on a square mandrel to obtain flat plates. To minimize air voids during fabrication, it is necessary to use a roller to make the ribbon, as it is laid down, contact the mandrel.

Fig. 21-4. Coil winder assembly for fabrication of ribbon composite tubes.

Fig. 21-5. Twenty-layer glass ribbon/epoxy wet-wound tube.

Table 21-1. Approximate glass compositions.

Composition (Wt %)	Corning Code 0080	Corning Code 8871	Corning Code 0211	Owens-Corning E Glass
SiO_2	73	42	65	54
Al_2O_3	1	–	2	14
B_2O_3	–	–	9	10
Li_2O	–	1	–	–
Na_2O	17	2	6.5	–
K_2O	–	6	7	–
MgO	4	–	–	4.5
CaO	5	–	–	17.5
PbO	–	49	–	–
ZnO	–	–	7	–
TiO_2	–	–	3	–

Courtesy: Corning Glass Works.

3 PHYSICAL PROPERTIES OF GLASS RIBBON

3.1 Glass Composition

The properties of glass ribbon are related to the glass composition. To date, ribbons have been made from three different glasses: soda lime (Corning Code 0080), alkali lead silicate (Corning Code 8871), and potash-soda-zinc borosilicate (Corning Code 0211). For comparison purposes we will also include E-glass, which is a lime aluminoborosilicate glass used by Owens-Corning for manufacturing fibers. The compositions of these four glasses are given in Table 21-1. It should be pointed out that ribbon glass compositions shown in Table 21-1 do not necessarily define the optimum glass for ribbon application as a reinforcement. Ribbon has been made from these glasses either due to ease of manufacturing or for applications other than reinforcement. It should be borne in mind that ribbon can also be produced from typical fiber glasses, such as E or S glass. The discussion of properties in sections 3.2 through 3.7 follows closely that given by Hutchins and Harrington,[16] where additional references may be found.

3.2 Mechanical Properties

The elastic modulus, Poisson's ratio, and specific gravity of the glasses mentioned in section 3-1 are given in Table 21-2. The strength is not included for two reasons. Since it is an extrinsic property, strength is a strong function of surface condition of the ribbon, which depends on the manufacturing process and the type of coating used to preserve the virgin surface. Secondly, any flaws introduced during manufacturing tend to orient themselves parallel to the draw or redraw direction, thereby making strength property anisotropic, i.e., the strength along the ribbon exceeds that in the width direction. The degree of anisotropy depends on the severity of oriented flaws. Of course, the ultimate goal is a flaw-free surface, which is coated immediately after the draw to minimize further degradation. If flaws do occur, they can be etched prior to coating. This step makes the process uneconomical but increases the strength significantly and makes it isotropic. Strengths of up to 300,000 psi (tensile) have been measured in properly protected ribbons of Code 0211 glass made by Corning.

The commercially available ribbon, made from Code 8871 glass, is primarily produced for

Table 21-2. Mechanical properties.

Glass Code	Elastic Modulus (10^6 psi)	Poisson's Ratio	Specific Gravity
0080	10.2	0.22	2.47
0211	11.2	0.21	2.53
8871	8.4	0.26	3.84
E Glass	10.5	0.22	2.58

Courtesy: Corning Glass Works.

Table 21-3. Rheological properties of glasses.

Glass Code	REFERENCE VISCOSITY TEMP, °C, FOR VARIOUS VISCOSITIES IN POISE			
	Strain point $10^{14.5}$	Annealing point 10^{13}	Softening point $10^{7.65}$	Working point 10^4
0080	473	514	696	1005
0211	510	550	720	1010
8871	350	385	525	785
E Glass	625	665	845	1070

Courtesy: Corning Glass Works.

capacitor application in the electronic industry where strength is not a critical requirement. This ribbon, which is not coated on the draw, has an average strength of 35,000 psi. Similarly, ribbon made from soda-lime glass by the redraw process is expected to have low strength unless its surface is protected by a suitable coating. Fortunately, ribbon reinforcement being three times as efficient as fiber, strength requirements for ribbon are not as stringent as for fibers. Since the surface area of a typical ribbon is several thousand times that of the fiber, ribbon strengths are going to be lower than those of fiber. Table 21-1 also shows that in view of high lead content, Code 8871 glass has high density and low stiffness, both of which make this glass unsuitable for reinforcement.

3.3 Rheological Properties

Viscosity plays an important role in all of the glass manufacturing processes. Ribbon or sheet drawing, in particular, is carried out at viscosities between 20,000–100,000 poises. The four important reference viscosity temperatures are listed in Table 21-3. Until recently, the temperatures corresponding to the four viscosities (from left to right) were defined as the strain point, annealing point, softening point, and working point, respectively. Of the 4 glasses listed, Code 8871 glass is the softest and E glass is the hardest.

3.4 Thermal Properties

The coefficient of thermal expansion of most glasses is significantly smaller than that of plastics. Also, glass is an order of magnitude stiffer than plastics. Thus, glass-ribbon reinforcement is very effective in lowering the expansion coefficient of plastics. The average expansion coefficients from 0–300°C and from room temperature to setting point (strain point + 5°C) are given in Table 21-4.

The room temperature specific heat for most glasses is between 0.075–0.230, a typical value being 0.19 cal/g°C. It is higher at elevated temperatures and approaches zero at 0°K. The

Table 21-4. Thermal properties.

Glass Code	THERMAL EXPANSION[a] (10^{-7} in./in./°C)		Specific Heat (cal/gm °C) at 25°C	Thermal Conductivity (cal/cm sec °C) at 25°C
	0–300°C	25°-Setting pt		
0080	93.5	104	0.19	0.0025
0211	74	84	0.18	0.0024
8871	102	110	0.11	0.0018
E Glass	52	70	0.18	0.0031

[a]See Table 21-12 for other thermal expansion data.
Courtesy: Corning Glass Works.

Table 21-5. Electrical properties of glasses.

Glass Code	Log_{10} VOLUME RESISTIVITY (Ω cm) 25°C	250°C	350°C	Power Factor (at 10^6 Hertz and 25°C)	Dielectric Constant	Loss Factor
0080	12.4	6.4	5.1	0.009	7.20	0.065
0211	>17	7.95	6.45	0.0043	6.23	0.0268
8871	>17	11.10	8.80	0.0005	8.40	0.0042
E Glass	>17	13.6	11.8	0.0012	6.4	0.0075

Courtesy: Corning Glass Works.

room temperature thermal conductivity of common glasses ranges from 0.0012–0.0030 cal/sec cm °C. Room temperature values for specific heat and thermal conductivity are given in Table 21-4.

3.5 Electrical Properties

The volume resistivity of glass depends primarily on the amount of alkali present and, to a lesser degree, on the glass network structure and previous thermal history of the glass. The volume resistivity is inversely proportioned to the alkali content. Similarly, the more slowly glass is cooled, the higher is its volume resistivity. Volume resistivity at three different temperatures is given in Table 21-5. Values greater than 10^{17} can not be measured accurately at room temperature. Also shown in Table 21.5 are the dielectric properties at 10^6 Hertz and 25°C.

3.6 Corrosion Resistance

Glass is generally considered to have excellent corrosion resistance in comparison with plastics. In many applications, this property of glass is of primary importance. Table 21-6 compares the corrosion resistance of four ribbon glass candidates. Glasses with high weathering resistance are rated 1; they show little, if any, weathering effects. Those rated 2 may occasionally be troublesome and a rating of 3 indicates that careful consideration of its use is necessary. Weathering is evidenced by a clouding of the glass surface or by electrical surface leakage. A glass rated 1 will lose less than 10^{-6} in./day when exposed to 5% hydrochloric acid at 95°C. Those rated 2 will lose between 10^{-6} and 10^{-5}; those rated 3 will lose between 10^{-5} and 10^{-4}; and those rated 4 will lose more than 10^{-4} in./day. The superiority of E glass in terms of corrosion resistance should be noted.

3.7 Optical Properties

The refractive indexes of the four glasses are given in Table 21-7. For most silicate glasses, its value is about 1.5. Code 8871 glass, however, due to its high lead content, and therefore high density, has a higher index of refraction. The reflection loss per surface is related to the refractive index and is typically 4% for most silicate glasses. Consequently, a perfectly clear glass transmits about 92% of the incident light. The absorption losses, in view of the thinness of ribbons, are also negligible. The values of surface reflectance and overall transmittance for the four glasses are given in Table 21-7.

4. MANUFACTURERS AND SUPPLIERS

Corning Glass Works is the primary manufacturer of glass ribbon. The composition that is readily available in ribbon form is Code 8871,

Table 21-6. Corrosion resistance.

Glass Code	Weathering	Water	Acid
0080	3	2	2
0211	2	1	2
8871	2	1	4
E Glass	1	1	2

Courtesy: Corning Glass Works.

Table 21-7. Optical properties.

Glass Code	Refractive Index (5893A°)	Surface Reflectance	Transmittance
0080	1.512	0.041	0.919
0211	1.526	0.043	0.915
8871	1.656	0.061	0.882
E Glass	1.548	0.046	0.909

Courtesy: Corning Glass Works.

alkali lead silicate, known for its superior dielectric properties. This ribbon is commercially produced by the hot-draw process for making capacitors. It is stacked alternately with aluminum foil, heated, and passed through rollers to fuse the assembly into a monolithic, hermetically sealed capacitor. Table 21-8 shows the cross-sectional dimensions of various commercially drawn ribbons of Code 8871 glass. Aspect ratios range from 80–800. Ribbon length available in a spool is also listed in Table 21-8. Additional information may be obtained by contacting Electronic Materials Department, Corning Glass Works, Corning, N.Y. 14830.

Corning also manufactures microsheet from Code 0211 glass for microscope cover slides. It is available in thicknesses from 0.002–0.020 in. and in widths up to 16 in. It has good durability and is easily drawn into sheets. Corning has produced, in limited quantity, microsheet as thin as 0.0003 X 20 in. wide for use in moisture barrier in plastic and paper containers. Glass ribbon, in limited quantity, has also been redrawn from Code 0211 blanks for experimental studies. It was 0.0014 in. thick and 0.125 in. wide. OEM Sales, Science Products, Corning Glass Works, Corning, N.Y. 14830, is a supplier of microsheet glass products.

Table 21-8. Code 8871 glass ribbon manufactured by Corning.

Article Code	Width (in.)	Thickness* (in.)	Length per Spool (ft)
679173R	0.121	0.0015	2000
679174R	0.121	0.0018	1600
679153R	0.121	0.0028	1100
679175	0.219	0.0014	2000
679176R	0.219	0.0018	1600
679187R	0.219	0.0028	1200
679232R	0.330	0.0013	1600
679188	0.462	0.0019	1600
679189	0.462	0.0024	1500
679013	1.000	0.0013	–
679075	1.500	0.0027	–

*These are nominal values. The thickness varies slightly from midwidth to the edge. For details contact Electronic Materials Department Corning Glass Works, Corning, N.Y. 14830.
Courtesy: Corning Glass Works.

Eakins and Humphrey[17] obtained a patent in 1969 which was assigned to DeBell and Richardson, Inc., Hazardville, Conn. This patent dealt with manufacture of glass ribbon by the redraw process and utilizing these ribbons for fabricating glass-resin structures by the filament winding technique. Use of ribbon of an aspect ratio of 30 ~ 50 creates a tortuous path for permeation of gases and fluids contained in the filament-wound structure. They used preforms of Code 0080 glass, approximately 3 in. wide and 0.050 in. thick, and produced ribbon with an aspect ratio of 37. They also produced glass ribbons with corrugated section, ribbons with grooved or ridged surfaces, and hollow ribbons, using a preform of appropriate geometry. The hollow ribbons have the advantage of providing a minimum density, glass-reinforced structure. It is not known whether DeBell and Richardson Inc. is still active in manufacturing ribbon.

5. FUTURE COST AND AVAILABILITY

Capacitor ribbon, made from Code 8871 glass, is commercially available in most sizes shown in Table 21-8. However, its present cost is considerably higher than that of E-glass fiber. It should be recalled that these ribbons are produced in limited quantity for a highly specialized application. Moreover, as mentioned previously, Code 8871 glass is dense and has a low modulus; it is not intended for reinforcement of plastics. Technologically, however, there is no reason why E glass ribbon can not be manufactured at high speeds and lower prices. Assuming large scale production, glass ribbon can be produced at prices comparable to fiber glass. It should be noted that the typical ribbon with a width of 0.25 in. and a thickness of 0.002 in. is equivalent to 4000 glass fibers in terms of cross-sectional area. Also, ribbon reinforcement offers some composite properties which are two to three times better than those afforded by fiber reinforcement. In view of these considerations, draw speeds for ribbon can be considerably lower than those for the fibers and yet the price of ribbon composites can be cost effective.

Glass ribbons are in a development stage. The current costs can be obtained by contacting Corning Glass Works and DeBell and Richardson Inc. at the addresses given in Section 4.

6. PROPERTIES OF GLASS RIBBON COMPOSITES

The inherent advantage of ribbon reinforcement is most easily understood by comparing their theoretical properties with those of the random fiber mat–which has isotropic properties in its plane. Table 21-9 shows the expected planar properties–elastic modulus (E), shear modulus (G), and strength (σ)–of random fiber mat and the ribbon composite.[3,18,19,20] It is clear that for a given volume fraction (V_f) of reinforcement, the moduli of random fiber mat are only 1/3 of those for the ribbon composite and the strength is only ½ of that of the ribbon composite. Thus, in those structural applications where planar isotropy is desirable, the use of ribbon reinforcement can result in substantial savings of volume loading. The additional advantage of ribbon reinforcement lies in the cost saving resulting from elimination of orientation of fibers in fabricating quasi-isotropic composites.

6.1 Elastic Properties

Glass ribbon composites are transversely isotropic and are characterized by five independent elastic constants (Fig. 21-6):

- E_L: Elastic modulus for all directions in the plane of isotropy (x, y)
- E_{TT}: Elastic modulus in the transverse direction (z).
- G_{LT}: Shear modulus in the longitudinal-transverse plane (x, y)
- G_{TT}: Shear modulus in the transverse plane (z)
- ν': Poisson's ratio relating elongation in the plane of isotropy to contraction in the thickness direction or vice-versa

The above elastic constants are related to constituent properties as follows:[2,10,21]

$$E_L = V_f E_f + V_m E_m \tag{1}$$

$$E_T = E_m \left[\frac{1 + J_e \eta_e V_f}{1 - \eta_e V_f} \right], \quad \text{finite aspect ratio} \tag{2}$$

Table 21-9.[a] Ribbon vs fiber composites.

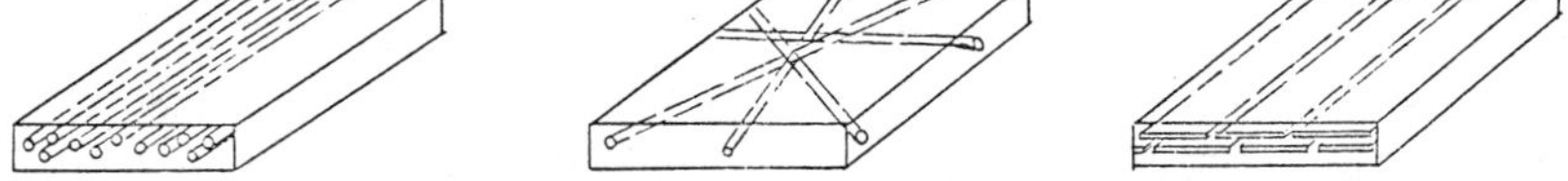

Property	Random fiber mat	Ribbon Composite
E (Elastic modulus)	$\frac{3}{8} V_f E_f$	$V_f E_f$
G (Shear modulus)	$\frac{1}{3} V_f G_f$	$V_f G_f$
σ (Tensile strength)	$\frac{1}{2} V_f \sigma_f$	$V_f \sigma_f$

[a]In these expressions, the contribution of the matrix has been neglected since it is small.

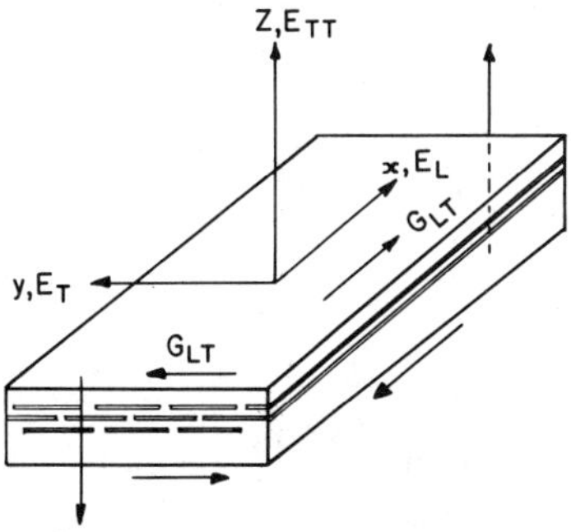

Fig. 21-6. Basic elastic and shear moduli of ribbon composites.

$$E_{TT} = \frac{E_m}{V_m + \left(\frac{E_m}{E_f}\right) V_f} \tag{3}$$

$$G_{LT} = V_f G_f + V_m G_m, \text{ large aspect ratio} \tag{4}$$

$$G_{LT} = G_m \left[\frac{1 + J_g \eta_g V_f}{1 - \eta_g V_f}\right], \text{ finite aspect ratio} \tag{5}$$

$$G_{TT} = \frac{G_m}{V_m + (G_m/G_f)\, V_f} \tag{6}$$

$$\nu' = \frac{\left\{\dfrac{E_f V_f \nu_f}{(1+\nu_f)(1-2\nu_f)} + \dfrac{E_m V_m \nu_m}{(1+\nu_m)(1-2\nu_m)}\right\}}{\left\{\dfrac{E_f V_f}{(1+\nu_f)(1-2\nu_f)} + \dfrac{E_m V_m}{(1+\nu_m)(1-2\nu_m)}\right\}} \tag{7}$$

where the subscripts f and m refer to reinforcement and matrix respectively and a = thickness b = width $\frac{a}{b} = \frac{1}{\text{aspect ratio}}$.

$$J_e = 2\left(\frac{a}{b}\right) \tag{8}$$

$$\eta_e = \left(\frac{E_f}{E_m} - 1\right) \bigg/ \left(\frac{E_f}{E_m} + J_e\right) \tag{9}$$

$$J_g = \left(\frac{a}{b}\right)\sqrt{3} \tag{10}$$

$$\eta_g = \left(\frac{G_f}{G_m} - 1\right) \bigg/ \left(\frac{G_f}{G_m} + J_g\right) \tag{11}$$

The principal Poisson's ratio ν which relates E_L and G_{LT} is also of importance. It may be shown from eqs. (1) and (4) that

$$\nu = \frac{\left(\dfrac{E_f V_f \nu_f}{1-\nu_f^2} + \dfrac{E_m V_m \nu_m}{1-\nu_m^2}\right)}{\left(\dfrac{E_f V_f}{1-\nu_f^2} + \dfrac{E_m V_m}{1-\nu_m^2}\right)} \tag{12}$$

Figure 21-7 shows the variation of E_L and E_T (modulus in the ribbon width direction) with V_f. The data was obtained for Corning Code 8871 glass ribbon, 2 mils thick, having an aspect ratio of 230 embedded in SCOTCHWELD® matrix. It is clear that both the inplane moduli, E_L and E_T, approach the rule of mixture value (solid line) providing the composite isotropic properties. The dependence of E_T on aspect ratio is shown in Fig. 21-8 for $V_f = 0.35$ and two different matrices. The solid lines are the predictions of Halpin-Tsai equations[21]. The rapid increase of E_T with aspect ratio should be noted. It follows that for $(E_f/E_m) \sim 100$, an aspect ratio of ~ 100 is sufficient to achieve isotropic elastic moduli in the plane of the ribbon composite.

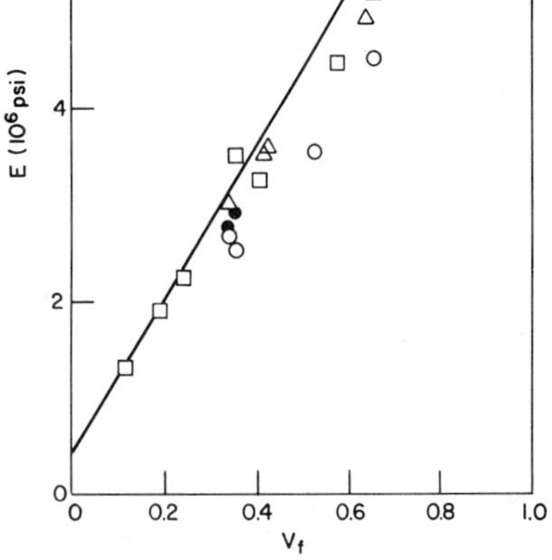

Fig. 21-7. Longitudinal (E_L) and transverse (E_T) elastic moduli as function of volume.

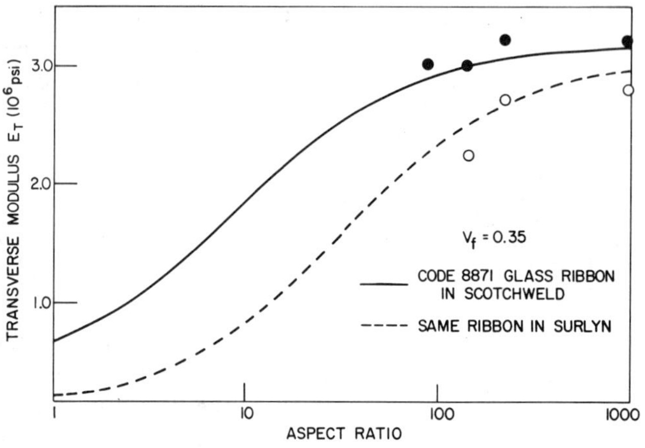

Fig. 21-8. Variation of transverse elastic modulus (E_T) with ribbon aspect ratio.

The values for E_{TT} have not been measured due to the difficulty in fabricating suitable specimens. Moreover, this stiffness is of little practical value when using ribbon composites with planar isotropy. The above equations, however, provide an estimate of 0.8×10^6 psi for E_{TT} for a glass-ribbon (AR = 230) - epoxy composite with $V_f = 0.40$.

The variation of shear modulus G_{LT} with ribbon aspect ratio is shown in Fig. 21-9 for $V_f = 0.35$. The data show that G_{LT} approaches the asymptotic value given by Eq. (4) for $(b/a) \approx 200$ and its magnitude is strongly dependent on the shear modulus of the matrix. The measurements for G_{LT} were carried out by twisting of square plates similar to the technique adapted by Whitney.[22] The values of G_{TT} were not determined due to difficulty in specimen fabrication but Eq (6) gives a value of 0.3×10^6 psi for G_{TT} for a glass-ribbon (AR = 230)-epoxy composite with $V_f = 0.40$.

The variation of ν and ν' with volume fraction, as predicted by Eq. (7) and (12), is shown in Fig. 21-10, where properties of Corning Code 8871 glass ribbon and epoxy matrix have been assumed. It should be noted that for low values of V_f, ν' differs substantially from ν, indicating the ineffectiveness of small quantity of ribbon reinforcement, but for $V_f \geqslant 0.3$ the two Poisson's ratios become almost identical, making the composite isotropic in so far as the Poisson's ratios are concerned.

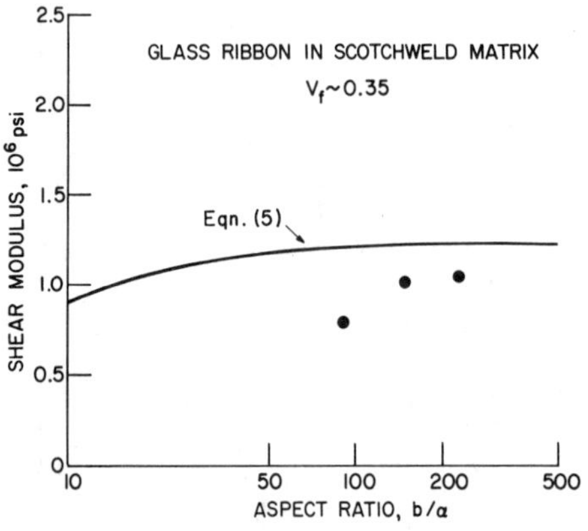

Fig. 21-9. Variation of shear modulus (G_{LT}) with ribbon aspect ratio.

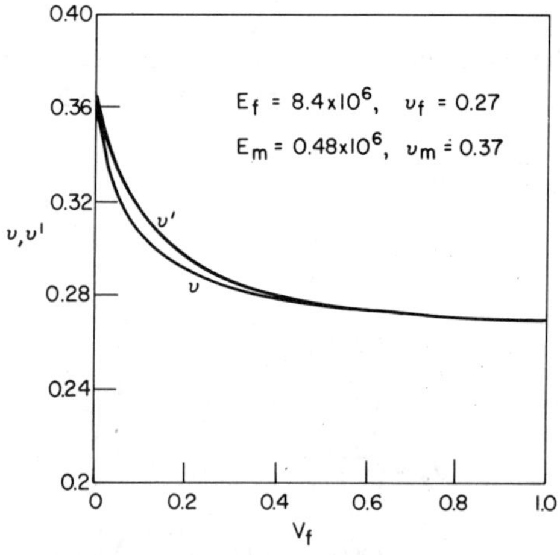

Fig. 21-10. Variation of Poisson's ratios (ν and ν') with volume fraction of ribbon.

6.2 Strength Properties

The longitudinal strength of ribbon composites is, to a reasonable accuracy, given by the rule of mixtures analogous to the strength of unidirectional fiber composites.

$$\sigma_L = V_f \sigma_f + V_m \sigma_m' \tag{13}$$

Figure 21-11 shows the measured strength values obtained from tensile tests on single ribbon wide test bars. The longitudinal strength is independent of the aspect ratio of ribbon.

The transverse strength, however, depends on many factors including:

a. transverse strength of ribbon
b. aspect ratio of ribbon
c. matrix properties (shear in particular)
d. stacking parameter
e. fabrication flaws

In theory, the ribbon strength should be isotropic; however, there is a finite probability of preferred orientation of surface flaws which are generally introduced during ribbon manufacturing processes. These flaws tend to be oriented in the longitudinal direction and consequently affect the transverse strength of ribbon. The net effect of such flaws shows up in the reduced transverse strength of ribbon composites. One

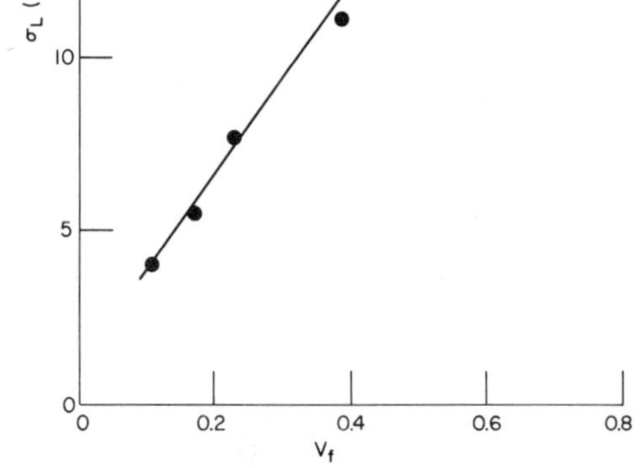

Fig. 21-11. Longitudinal composite strength as function of volume fraction.

of the possible ways to eliminate these flaws is to either coat the ribbon surface with a protective layer during drawing operation or to etch the ribbon in HF prior to fabrication of composites.

The transverse strength of ribbon composites is analogous to strength of unidirectional fiber composites with discontinuous short fibers. The stress transfer occurs at the edge of the ribbon via the matrix and the normal stress builds up to its maximum value at the mid-width of the ribbon. Thus, ribbons with large aspect ratio are more effective in providing isotropic strength assuming the ribbon itself is isotropically strong. During stress transfer, large shear stresses develop in the matrix, near the ribbon edge, and require large shear strength and interfacial bond strength of the matrix. In this connection, ribbon cross-section also plays an important role; e.g., sudden thickness changes should be avoided.

The stacking sequence is another important parameter which determines the effective reinforcement cross-section to carry the transverse load. Random stacking, in general, provides the best transverse strength. Fabrication flaws during composite manufacture can have marked effect on the transverse strength, as they are the source of stress raisers. A ductile matrix, with large strains, can forgive fabrication errors by yielding without failure.

Burst tests and torsion tests on thin walled tubes fabricated by winding glass ribbon through an epoxy coating bath and over a mandrel, provided the longitudinal and 45° strengths, whereas axial tension on similar tubes gave the transverse strength. The results are shown in Fig. 21-12 where strength is plotted vs. orientation. It is seen that the transverse strength is typically 50% of the longitudinal strength. The ribbon used in these tubes was narrow (AR = 120), indicating the need for larger aspect ratios. Moreover, crude strength tests on the ribbon showed that its transverse strength is lower than the longitudinal strength. An improvement in the transverse composite strength of the tubes was observed as the aspect ratio was increased to 230, as shown in Fig. 21-13. An asymptotic value of 70% of longitudinal composite strength was obtained for an aspect ratio of 230. This

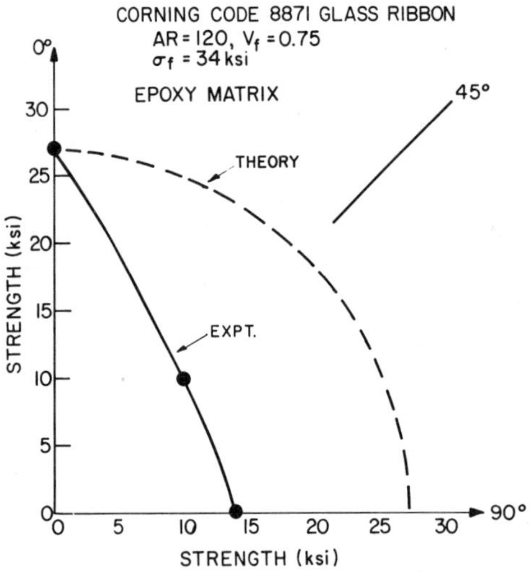

Fig. 21-12. Composite strength in different directions as determined by tube tests: theory vs. experiment.

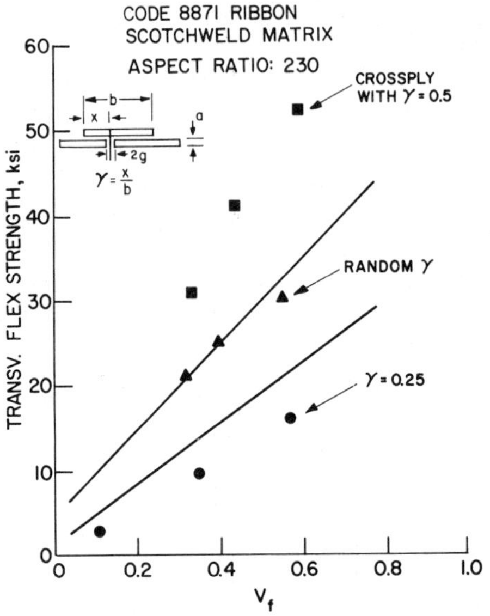

Fig. 21-14. Transverse flexural strength versus volume fraction as determined by concentric ring tests.

compares well with the predictions of Lovelace and Tsai[23] for discontinuous fiber composites. Other investigators[4,5,8,24] working with ductile steel ribbons in ductile matrix found the transverse strength to be 50% of the longitudinal strength. A simple expression relating transverse composite strength to the constituent properties is given by

$$\sigma_T = k\sigma_f V_f + \sigma_m' V_m \tag{14}$$

where k is the efficiency factor, also called strength reduction factor. Its value ranges from 0.5 to 0.7 and depends on the aspect ratio and matrix properties.

Flexural strength of ribbon reinforced discs was also measured in concentric ring test and the results are shown in Fig. 21-14. The failure was in the transverse direction for both unidirectional and crossplied ribbon composites. The effect of stacking parameter γ on the transverse strength is also shown in this figure according to which random stacking yields higher strength values. It is also worth noting that strength anisotropy of ribbon which is present in unidirectional ribbon composites is masked by cross-plying. Accordingly, the transverse strength is greater in the case of crossplied discs. The solid lines are the rule of mixture

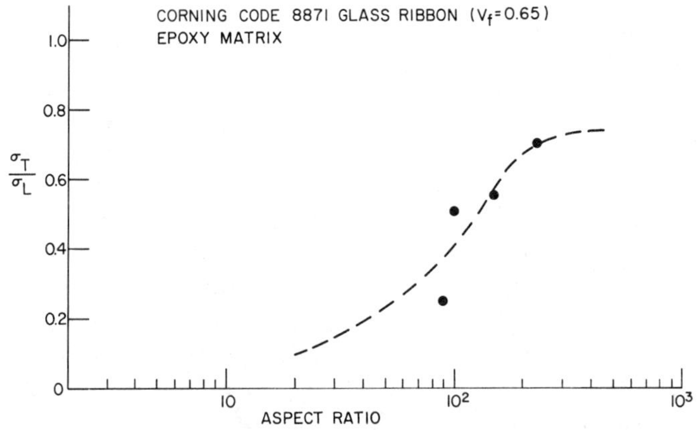

Fig. 21-13. Ratio of transverse and longitudinal composite strength as function of ribbon aspect ratio.

values using longitudinal tensile strength of ribbon. Properly speaking, flexural strength of ribbon which is ~40% larger than the tensile strength, should be used in the rule of mixture equation. Cross-plying combined with random stacking should yield the highest transverse strength.

6.3 Notch Sensitivity

The presence of notches and holes in composite materials degrades their strength due to stress concentration. This degradation is largely a function of the aspect ratio of ribbon and the matrix properties. Secondary microcracking is generally present at the periphery of the drilled hole and may propagate through the entire width of the ribbon containing the hole. To minimize strength loss, the aspect ratio should be kept low, the case of fiber reinforcement being the limiting one. However, with low aspect ratio, the transverse strength and stiffness are very low. Thus, a compromise is generally necessary with regard to the selection of aspect ratio, to obtain reasonably high strength and modulus values and to keep notch sensitivity down.

Table 21-10 lists the transverse strength values obtained in an axial pull test of ribbon composite tubes with and without holes. Two diametrically opposite holes of ⅛ in. diameter were drilled to create symmetric stress field (Fig. 21-15). The failure occurred at the section through the holes. In addition, a circumferential crack originated 3 ribbon widths below the holes in the case of AR = 90 but did not propagate completely around the tube. The results

Table 21-10. Transverse strength degradation due to presence of ⅛ in. diameter holes.
(Epoxy Matrix)

Aspect Ratio AR	Fiber Strength σ_f (Ksi)	Fiber Volume Fraction V_f	Transverse Strength σ_T (no hole) (Ksi)	Transverse Strength σ_T (with hole) (Ksi)
90	35	0.77	6.7	2.9
150	35	0.75	15.4	2.8

Test specimens were filament-wound glass-ribbon composite tubes—2.5 in. diameter and 10 in. long.

Fig. 21-15. Failure of tubes with two diametrically opposite holes in an axial tensile test.

show that the strength degradation factor is approximately 2 for the narrow ribbon (AR = 90) and 5 for the wider ribbon (AR = 150), pointing out the desirability of low aspect ratio. It should be noted that a stress concentration factor of 2 is also associated with holes in isotropic materials in the state of plane stress. Thus, an AR of 90 appears to be no worse than, say a fiber composite. Moreover, as shown in Sections 6.1 and 6.2, the transverse moduli for AR = 90 are near their asymptotic value and the transverse strength is 50% of the longitudinal strength.

Notch sensitivity was also measured for transverse bars cut from a composite plate, Fig. 21-16. The cut edges where glass ribbon was exposed had microcracks which served as stress concentrators during tensile tests on the bars. One way to minimize the loss of strength due to cut edges is to etch them, thereby eliminating or rounding off the microcrack tips. This was done for various times in HF/HNO_3 solution. Indeed, as Table 21-11 shows, the transverse strength improved with etching time, reaching an asymptotic value which was ⅔ of

Fig. 21-16. Failure of transverse bars in a tension test for determining notch sensitivity.

the longitudinal strength. It is also worth noting that, in the case of no etching, the transverse strength is only 24% of the longitudinal strength, indicating a stress concentration factor (SCF) of 4, similar to that obtained for tubes with holes.

Figure 21-17 demonstrates the ability of a ribbon-epoxy composite to store a large amount of fracture energy. A bar containing 100 layers of Code 8871 ribbon (AR = 180) in epoxy matrix was fabricated by hand lay-up in a steel mold. The volume fraction of glass was 68%. The bar was first tested in 4-point bending with ribbons in vertical position. Failure occurred at a stress of 33,000 psi (so that the maximum stress at the bottom of the beam was shared by all of the 100 ribbon edges) and was catastrophic. The second bar, fabricated in an identical manner, was tested in 4-point bending with ribbons in horizontal position. It too failed at 32,000 psi but the failure was not catastrophic. As can be seen in Fig. 21-17, the tensile crack did not propagate through the section but was dissipated in a series of interlaminar cracks.

Table 21-11. Tensile strength of transverse bars with chemically etched edges.

AR = 290; SCOTCHWELD Matrix

$V_f = 0.55$, $\sigma_f = 60$ ksi

Etch time (sec)	Transverse Strength σ_T (ksi)	Strength Ratio σ_T/σ_L
0	8	0.24
2.5	15	0.45
4	21	0.64
6	21	0.64
8	21.5	0.65

The test specimens, shown in Fig. 21-16, were 0.5 in. wide, 0.06 in. thick, and 3 in. long.

6.4 Thermal Expansion

The ribbon geometry, combined with the low expansion coefficient and high elastic modulus of glass, has a strong influence on the expansion coefficient of glass-ribbon composite. Glass ribbon is isotropic both in its expansion and the elastic modulus. Consequently, plane stress analysis[13] of glass ribbon composite subjected to thermal loading provides a simple expression for the expansion coefficient of the composite in terms of constituent properties.

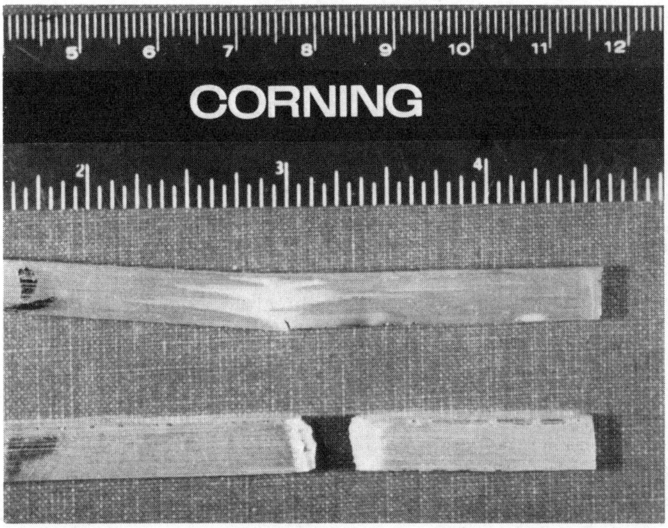

Fig. 21-17. 100-layer ribbon/epoxy bars showing effect of ribbon orientation on fracture characteristics (upper bar tested in flexure with ribbon in horizontal plane; lower bar with ribbons in vertical plane).

$$\alpha = \left[\frac{\left(\frac{E_f}{1 - \nu_f}\right) V_f \alpha_f + \left(\frac{E_m}{1 - \nu_m}\right) V_m \alpha_m}{\left(\frac{E_f}{1 - \nu_f}\right) V_f + \left(\frac{E_m}{1 - \nu_m}\right) V_m} \right] \quad (15)$$

In the above expression, α_f and α_m are the expansion coefficients of ribbon and matrix respectively. By virtue of $(E_f/1 - \nu_f) >> (E_m/1 - \nu_m)$ the composite expansion coefficient decreases rapidly as V_f is increased and approaches α_f for relatively small values of V_f. Thus, the large expansion coefficient of most thermoplastic and thermoset matrices may be reduced dramatically by reinforcing them with a small loading of glass ribbon. In this regard, it must be mentioned that both flakes[25] and fibers[26] are much less effective.

Expansion measurements were carried out for glass-ribbon composites using 2 different matrices: SCOTCHWELD and SURLYN. The necessary constituent properties for the calculation of composite expansion coefficient are shown in Table 21-12. The glass-transition behavior of the two matrices selected was observed between 50–60°C. Consequently, the expansion measurements, made on vitreous silica rod-type dilatometer in accordance with ASTM Designation E228, were restricted to the temperature range -150–50°C. To establish the isotropy of expansion coefficient, both longitudinal and transverse bars, with SCOTCHWELD matrix, were used for expansion measurements. The results, together with calculated values (solid line) using Eq. (15), are shown in Fig. 21-18. It is clear that transverse and longitudinal expansion

Table 21-12. Constituent properties for expansion calculations.

Material	Elastic Modulus E (10^6 psi)	Poisson's Ratio ν	Expansion Coefficient α (-150C to 50C) (10^{-7} in./in./°C)
Code 8871			
Glass Ribbon	8.4	0.26	87
SCOTCHWELD	0.29	0.45	720
SURLYN	0.08	0.495	1330

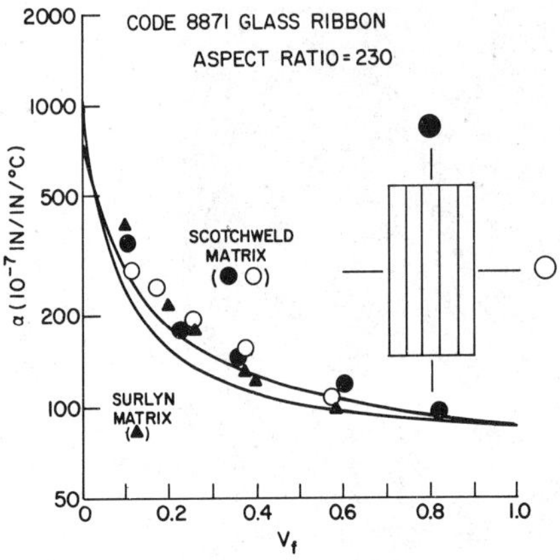

Fig. 21-18. Thermal expansion coefficient (−150–50°C) along longitudinal and transverse directions as function of volume fraction.

coefficients are reasonably close to the calculated value, thus establishing the isotropy of this property. In view of this, only the longitudinal specimens with SURLYN matrix were used and again the measured values agree with the theory. It should also be noted that the large expansion of thermoplastics such as SURLYN can be reduced by an order of magnitude by reinforcing them with glass ribbon having a volume fraction as low as 30%.

For comparison purposes, a quasi-isotropic bar fabricated from SCOTCHPLY, Type 1002, containing 44% volume of E-glass fiber gave an expansion coefficient in the temperature range −100–100°C of 136×10^{-7} in./in./°C. It may be verified, using Eq. (15) that a ribbon composite containing the same amount of E-glass ribbon in epoxy matrix would have an expansion coefficient of 86×10^{-7} in./in./°C which is 37% lower. Moreover, since the geometry of the ribbon permits the composite to be made with a much larger volume fraction of glass, a threefold reduction in the expansion coefficient of quasi-isotropic fiber composites is possible by going to ribbons. This feature is highly desirable from a dimensional stability point of view.

6.5 Chemical Durability and Corrosion Resistance

In applications such as pipes and storage tanks, chemical resistance is a highly desirable property. This property is controlled by the corrosion resistance of both the matrix and glass. Moreover, the tortuosity inherent in a ribbon composite construction offers a definite advantage over fiber composites with regard to corrosion resistance.

Although the effect of corrosion, in general, is to degrade the strength, strength measurements were not used to assess the severity of corrosion because strength is also a defect-

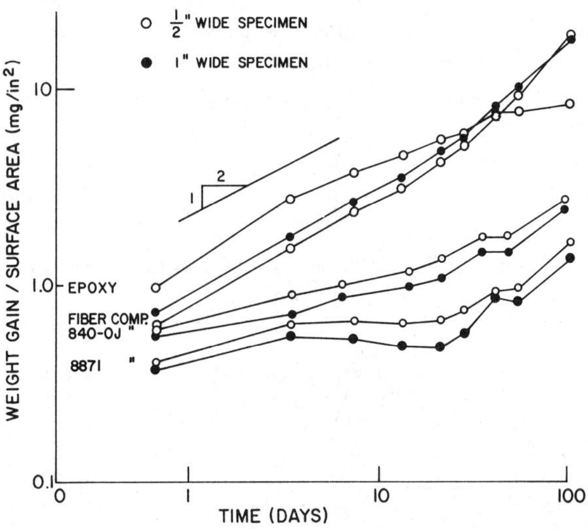

Fig. 21-19. Weight gain per unit surface area for epoxy resin, fiber and ribbon composites as function of exposure time in 30% sulfuric acid at room temperature.

sensitive property. It would be difficult to distinguish between defects generally present in the composite versus those brought about by corrosion. Consequently, chemical durability and corrosion resistance were measured by weight change experiments. Tests were conducted on matrix specimens, fiber composites, and ribbon composites in 3 corrosive environments: 30% sulfuric acid, ethylene dichloride (both at room temperature), and boiling water. The matrix used for all specimens was Shell Epon 828 epoxy with curing agent Z. Two ribbon glass compositions were studied: Corning Code 8871 and Code 0211. The former ribbon had an aspect ratio of 120 with $V_f = 0.74$ and the latter 90 with $V_f = 0.65$. Fiber composite specimen ($V_f = 0.33$) was fabricated from fiber glass cloth impregnated with the matrix and had the same dimensions as the ribbon composite specimen. Matrix specimens were cut from bars cast in open steel molds and cured at 80°C with post-curing at 150°C. Two specimen sizes were used: 0.5 in. wide × 0.045 in. thick × 3 in. long and 1.0 in. wide × 0.045 in. thick × 3 in. long.

After immersion for various times, the specimens were removed, rinsed in tap water, dried with a paper towel and weighed after 1 hour exposure to room conditions. Weight change per unit surface area, which is a measure of chemical durability, is shown in Fig. 21-19 for sulfuric acid environment. The general observation is that ribbon composites gained weight more slowly than either the fiber composite or plain resin. The fiber and resin specimens gained weight at a rate nearly proportional to the square root of time, indicating a diffusion controlled mechanism.

6.6 Permeation Resistance

Ribbon composites are less permeable than the fiber composites due to the more tortuous permeation path.[14] This advantage is a key factor in the application of ribbon composite for storage containers, pressure vessels, and pipelines. High permeability to fluids, in the case of fiber composites, has led to leakage of chemical storage tanks and loss of precious fuels. Consequently, an impervious liner is necessary when using fiber composite pipe.

The permeation coefficient of ribbon reinforced composites has recently been deduced by Brydges et al[15] using a simple analytical model which accounts for the ribbon geometry and the stacking parameter γ, introduced earlier. The composite permeation coefficient K_c is related to the matrix permeation coefficient K_m through

$$\frac{K_c}{K_m} = 1 \Big/ \left[1 + \frac{b}{2g} + \left(\frac{b}{a}\right)^2 \left(\frac{V_f^2}{1 - V_f}\right) \gamma (1 - \gamma)\right] \tag{16}$$

in which b is the ribbon width, a the ribbon thickness and $2g$ the matrix gap width between ribbons in the same layer. The above expression clearly shows the beneficial effect of the aspect ratio (b/a) of the ribbon. For aspect ratios >100 and $2g \sim a$, Eq. (16) can be approximated by

$$\frac{K_c}{K_m} = \frac{1}{(b/a)^2} \left(\frac{1 - V_f}{V_f^2}\right) [\gamma (1 - \gamma)]^{-1} \tag{17}$$

In the above equations, glass was assumed to be impermeable. The dependence of K_c/K_m on the aspect ratio for various volume fractions and γ values is shown in Fig. 21-20 using the approximate relation Eq. (17). The curved line is the exact solution given by Eq. (16). Of the 3 factors which affect the permeation coefficient (V_f, b/a, γ), the stacking parameter has the least effect. The aspect ratio, by far, has the largest effect making ribbon composites 100–1000 times less permeable than fiber composites ($b/a = 1$).

The above theory was verified by mass spectrometer measurements of helium permeation through composites of various constructions. The data are summarized in Table 21-13 and confirm the permeation advantages of ribbon composites. In all cases, a lower permeability was measured than was predicted by the theory. In the case of fiber composite, the agreement between theory and experiment was excellent. Also, the predictions of Eq. (16) for the case of fiber composite ($b/a = 1$) agree very well with the Halpin-Tsai equations,[21] as shown in Table 21-14. It is clear that fibers, even at the highest attainable volume fraction, can lower the per-

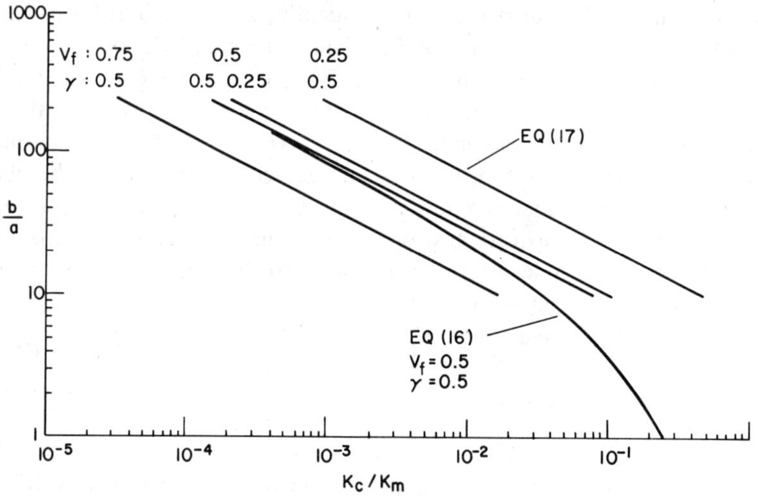

Fig. 21-20. Permeation coefficient of composites as function of ribbon aspect ratio (b/a), volume fraction (V_f) and stacking parameter (γ).

meation coefficient of the composite to at most 1/10 the matrix value. In contrast to this, Table 21-13 shows that ribbons of readily attainable aspect ratios and volume fraction offer 100–1000 fold improvement over the fiber case. The assumption of neglecting permeation through the glass ribbon was also examined and found to be justifiable in those cases where $K_m/K_g > 10^4$ and ribbon aspect ratio > 100. Code 8871 glass ribbon which was used for these experiments met these criteria.

7. APPLICATIONS—RIBBON COMPOSITE PIPE

One of the potential applications of ribbon composites is a pipe which might be fabricated

Table 21-13. Helium permeation results—theory vs experiment.

Specimen	Helium Permeation (measured) Cc (STP), mm/sec, cm², cmHg	K_c/K_m Measured	Theory (Eq. 17)	K_c (measured)/ K_c (theory)
Epon 828/Z	1.38×10^{-9}	–	–	–
SCOTCHWELD	6.45×10^{-10}	–	–	–
Code 8871 Glass	5.60×10^{-17}	–	–	–
Code 7740 Glass	1.00×10^{-11}	–	–	–
Ribbon Composites				
Epon 828/Z, $V_f = 0.62$				
$\gamma = 0.5, b/a = 264$	1.09×10^{-14}	7.9×10^{-6}	5.7×10^{-5}	0.14
SCOTCHWELD, $V_f = 0.22$				
$\gamma = 0.5, b/a = 121$	5.96×10^{-14}	9.2×10^{-5}	4.6×10^{-3}	0.02
SCOTCHWELD, $V_f = 0.54$				
$\gamma = 0.5, b/a = 121$	3.70×10^{-14}	5.7×10^{-5}	4.4×10^{-4}	0.13
Fiber Composite				
Epon 828/Z, $V_f = 0.45$				
(E-Glass Fiber)	3.03×10^{-10}	2.2×10^{-1}	2.9×10^{-1} (Eq. 16)	0.76

Table 21-14. K_c/K_m values for fiber composites.

V_f	Eq. (16)	Halpin-Tsai Solution
0.11	0.63	0.79
0.25	0.46	0.59
0.44	0.30	0.38
0.50	0.26	0.33
0.66	0.17	0.20
0.83	0.08	0.09

by winding a resin-coated glass-ribbon on the mandrel of a coil winding machine. This section will compare the performance of such a pipe with that of a conventional steel pipe. Pipe characteristics such as stiffness, strength, burst pressure, expansion, density, sagging, residual stresses, and permeation resistance are theoretically calculated as a function of the glass content and subsequently compared with those of a steel pipe. Three pipe sizes—2 in., 6 in., and 12 in. diameter—are considered.

7.1 Elastic Properties

For computational purposes, the assumed constituent properties are shown in Table 21-15. The elastic modulus of 15×10^6 psi and the isotropic strength of 100 Ksi for glass ribbon are considered practical within the realm of modern glass technology. The matrix properties listed in Table 21-15 correspond to those of a typical Nylon "Type 6" cast. The isotropic stiffness provided by ribbon reinforcement, assuming aspect ratio ≥ 100, follows the rule of mixtures:

$$E = V_f E_f + V_m E_m \qquad (18)$$

$$G = V_f G_f + V_m G_m \qquad (19)$$

Since $E_f < E_{steel}$, it is clear that a glass/resin composite will not outperform steel insofar as stiffnesses are concerned, when equal geometries and volumes are involved.

7.2 Longitudinal and Transverse Strength

Assuming continuous ribbon, the longitudinal strength of the composite would have similar dependence on constituent properties as the strength of a unidirectional fiber composite. It is therefore reasonable to take

$$\sigma_l = V_f \sigma_f + V_m \sigma_m \qquad (20)$$

The transverse strength is much more sensitive to fabrication, ribbon cross-section, and the matrix properties than the longitudinal strength. Lewis[4] reports that the transverse strength σ_t is bounded as follows:

$$0.5\,\sigma_l < \sigma_t < \sigma_l$$

Gray[5] gives an excellent discussion of the various factors which influence the transverse strength and concludes that the matrix shear strength is the governing factor. Due to the lack of a rigorous theoretical model for predicting transverse strength, we assume that it is also given by the rule of mixtures modified by a coefficient to reflect the efficiency of strength transfer:

$$\sigma_t = k(V_f \sigma_f + V_m \sigma_m') \qquad (21)$$

The value of k will, of course, depend on many parameters. Lovelace and Tsai[23] suggest 0.6–0.8 as a value for k. At low volume fractions of glass, the ribbon interaction may be negligible and the matrix shear stress small so that k would approach unity. However, as the volume fraction increases, the effect of various factors

Table 21-15. Assumed constituent properties.

Property	Glass Ribbon	Matrix (Nylon, Type 6 Cast)
Elastic modulus, E	15×10^6 psi	0.4×10^6 psi
Poisson's ratio, ν	0.27	0.35
Shear modulus, G	6×10^6 psi	0.15×10^6 psi
Strength, σ	1×10^5 psi	12,000 psi
Density, ρ	0.097 lb/in.^{3}	0.042 lb/in.^{3}
Thermal expansion coefficient, α	43×10^{-7} in./in./°C	900×10^{-7} in./in./°C
Maximum operating temperature, T	–	250°C
Aspect ratio, b/a	100	–

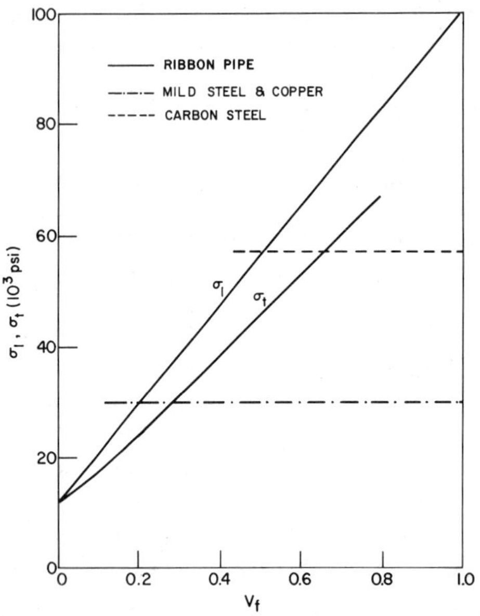

Fig. 21-21. Variation of longitudinal (σ_l) and transverse (σ_t) composite strengths with volume fraction.

becomes more pronounced and $k < 1$. It is almost impossible to predict the transverse strength for volume fractions of the order of 0.9–1.0. For our purposes, we have taken $k = 0.8$ and plotted the variation of longitudinal and transverse strengths with the volume fraction in Fig. 21-21. The yield strength of mild steel, copper, and carbon steel are also shown. To be competitive with carbon steel, a volume fraction of 70% would be necessary.

It should also be pointed out that for ribbon-wound pipes carrying pressure loading, the longitudinal strength (hoop strength) is of prime importance since the stress in the transverse direction (axial) is half of that in the longitudinal direction. If, on the other hand, the pipe is subjected to axial moment, in addition to pressure loading, the strength requirement in the axial direction becomes more critical and at times greater than that in the circumferential direction. In such cases, ribbon will have to be oriented so as to be parallel to the pipe axis as in the pultrusion process.

7.3 Density

The density of ribbon composites is given by the rule of mixtures.

$$\rho = V_f \rho_f + V_m \rho_m \tag{22}$$

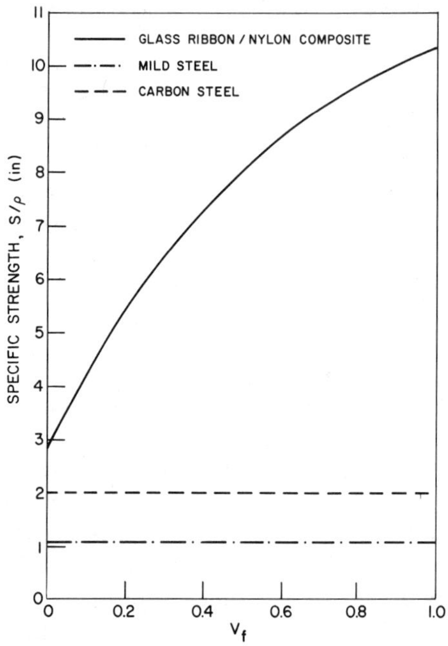

Fig. 21-22. Comparison of specific strength versus volume fraction for ribbon composite, mild steel and carbon steel tubes.

Assuming $V_f = V_m = 0.5$ and using the densities listed in Table 21-15, the composite density is found to be 0.07 lb/in.^{3}, which is 25% of the density of steel (0.283 lb/in.^{3}). Thus, ribbon composites have tremendous advantage in terms of light weight, which is desirable from the handling and transportation point of view. In view of their low density, the specific strength of ribbon composites as shown in Fig. 21-22 can be several times higher than that of mild or carbon steel.

7.4 Coefficient of Thermal Expansion

The composite expansion coefficient in the axial and hoop directions is given by

$$\alpha = \left[\frac{\left(\frac{E_f}{1 - \nu_f}\right) V_f \alpha_f + \left(\frac{E_m}{1 - \nu_m}\right) V_m \alpha_m}{\left(\frac{E_f}{1 - \nu_f}\right) V_f + \left(\frac{E_m}{1 - \nu_m}\right) V_m} \right] \tag{23}$$

Assuming the constituent properties of Table 21-15, a ribbon composite pipe containing 30% glass ribbon has an expansion coefficient of 110×10^{-7} in./in./°C compared with 120×10^{-7} in./in./°C for steel. Thus, the ribbon composite pipe with $V_f > 0.3$ would have lower expansion than either steel or copper ($\alpha = 160 \times 10^{-7}$).

7.5 Burst Pressure

For calculating the burst pressure, thin walled tube analysis is used. Accordingly, the hoop and axial stresses in an internally pressurized tube of mean diameter d and wall thickness t are given by

$$\left.\begin{aligned} \text{Hoop stress } \sigma_\theta &= \frac{pd}{2t} \\ \text{Axial stress } \sigma_z &= \frac{pd}{4t} \end{aligned}\right\} \quad (24)$$

where p is the internal pressure. Pipe strength in the hoop direction corresponds to composite longitudinal strength, σ_l, whereas the axial strength is equal to the transverse strength, σ_t. Since $\sigma_z = \frac{1}{2}\,\sigma_\theta$ and $\sigma_t > \sigma_l/2$, hoop stress is more critical. Steel pipes are generally rated at schedule 40 which permits a factor of safety of 5 for strength. Thus, the allowable burst pressure P_b for ribbon composite pipe is obtained by equating σ_θ with $\sigma_l/5$ which gives

$$P_b = 0.4 \frac{t}{d}\,\sigma_l \quad (25)$$

Since σ_l varies with volume fraction of glass, its value has been taken from Fig. 21-21. The variation of burst pressure with V_f is shown in Fig. 21-23 for 3 pipe sizes: 2 in., 6 in., and 12 in. diameter. Values of burst pressure are shown in terms of wall thickness t. The pressure carrying capacity of mild and carbon steel pipes is also shown in this figure and is based on the tensile yield strength of 57,000 psi for carbon steel and 30,000 psi for mild steel. It follows from Fig. 21-23 that a ribbon composite pipe with volume fraction greater than 0.55 can compete effectively with the carbon steel pipe. Such a pipe with wall thickness equal to that of a steel pipe will weigh only 25% of the weight of steel pipe.

The above analysis does not account for additional axial loading which might arise from the weight of soil above the buried pipe and may require greater axial strength. In such cases, it

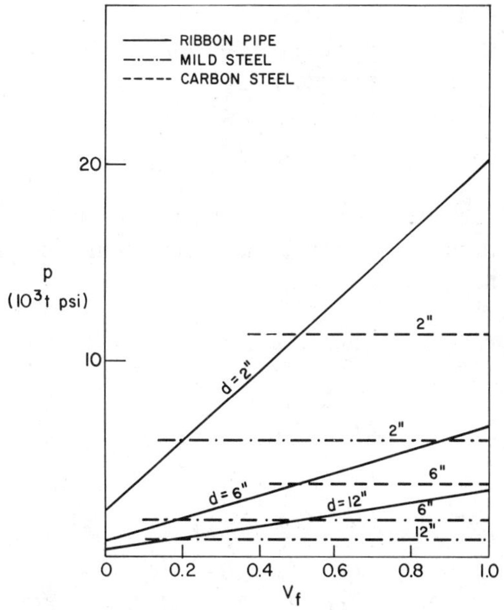

Fig. 21-23. Burst pressure for ribbon composite tube as function of volume fraction.

may be necessary to reorient ribbon so as to impart adequate strength in the axial direction.

7.6 Central Deflection of Supported Pipe

Very often, a continuous pipeline carrying food, chemicals, oil, etc., is supported by hangers at certain intervals and the pipe begins to sag midway between two hangers. At the very outset, we may conclude that by virtue of its lower stiffness, the ribbon composite pipe will deflect more than the steel pipe. However, calculations will show that the sag experienced by the composite pipe, although larger than that of steel pipe, is hardly detectable to the human eye and hence tolerable.

The fluid contained in the pipe is assumed to have a specific gravity of 1.5, which is a conservative figure for most applications. The span l between two hangers is assumed to be 10 ft. Treating the pipe as a beam carrying uniformly distributed load, the maximum deflection is given by

$$\delta = \frac{5}{384} \frac{wl^4}{EI} \quad (26)$$

where w is the weight distribution per linear inch of pipe and is composed of two parts—w_p, weight of the pipe, and w_f, weight of the contained fluid. In terms of ρ, the composite density given by Eq. (22), and ρ_f, the density of fluid, we may write

$$w = \pi d t \rho \left(1 + \frac{d}{4t}\frac{\rho_f}{\rho}\right) \tag{27}$$

The area moment of inertia I is primarily due to the glass content, matrix contribution being negligible. Thus

$$I = \frac{\pi}{64}(d_o^4 - d_i^4)V_f$$

$$= \frac{\pi}{64}(d_o^2 + d_i^2)(d_o + d_i)(d_o - d_i)V_f$$

$$= \frac{\pi}{64}(2d^2)(2d)(2t)V_f$$

$$= \frac{\pi}{8}d^3 t V_f \tag{28}$$

The assumptions leading to simplified expression for I are justified for thin-walled tubes. Substitution of Eqs. (27) and (28) into (26) together with $l = 120$ in. leads to

$$\delta = \frac{21.6 \times 10^6}{d^2 V_f}\left(1 + \frac{d}{4t}\frac{\rho_f}{\rho}\right)\frac{\rho}{E} \tag{29}$$

In the above equation ρ and E are the composite properties given by Eqs. (22) and (18) and $\rho_f = 0.054$ lb/in.^{3}. For the three pipe sizes considered, the wall thickness was assumed to be that of a schedule 40 pipe. These are listed in Table 21-16. The variation of maximum pipe deflection with glass content is shown in Fig. 21-24 for the three pipe sizes listed in Table 21-16. The values for steel pipe are also shown and are considerably lower. However, for 50% glass content, a 2 in. diameter pipe will deflect less than 3/8 in. over a 10 ft span. This corresponds to a slope of less than half of a degree and is hard to detect by human eye. Larger diameter pipes will deflect considerably less. It should be pointed out that for shorter spans there will be a further reduction of substantial amount in the maximum deflection, and if magnitudes mentioned above are acceptable, pipes with thinner walls can be designed resulting in further weight and cost saving.

Table 21-16. Schedule 40 pipe dimensions.

Mean Diameter (in.)	Wall Thickness (in.)
2	0.154
6	0.280
12	0.406

7.7 Residual Stresses

During the fabrication process, the matrix must cure at a temperature T_c. It experiences restraints against contraction during cooling from

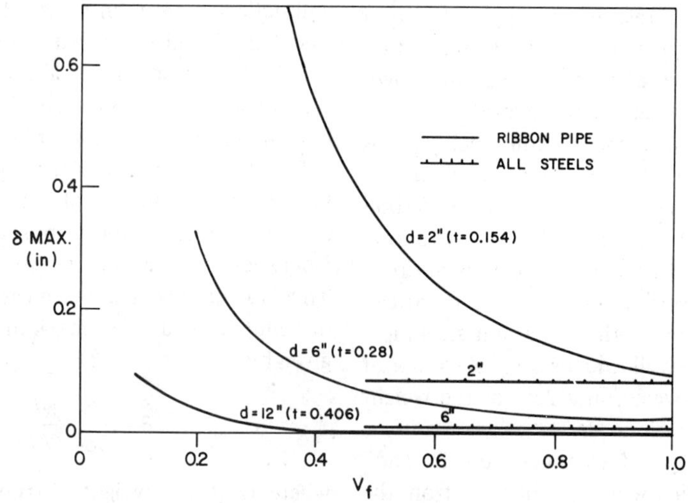

Fig. 21-24. Maximum deflection of ribbon composite pipes as function of volume fraction.

T_c to room temperature due to expansion and stiffness mismatch between the glass and matrix. Since the matrix is usually higher in expansion, it is put into tension. This residual tension weakens the matrix and reduces its effectiveness in transferring the load. As a result of tension in the matrix, the glass ribbon experiences a small amount of compressive stress.

The residual stresses arising from expansion mismatch are given by

$$\sigma_f = \left[\frac{K_f K_m(\alpha_m - \alpha_f)(T_R - T_C)}{K_m + K_f\left(\frac{V_f}{1 - V_f}\right)}\right] \tag{30}$$

$$\sigma_m = \left[\frac{K_f K_m(\alpha_f - \alpha_m)(T_R - T_C)}{K_f + K_m\left(\frac{1 - V_f}{V_f}\right)}\right] \tag{31}$$

where T_R is the room temperature,

$$K_f = \frac{E_f}{1 - \nu_f}$$

and

$$K_m = \frac{E_m}{1 - \nu_m}.$$

The residual stress in the matrix reaches its maximum value for low volume fractions of glass and thereafter remains substantially constant up to a volume fraction of 0.95. The flow temperature of Nylon "Type-6" ranges from 205-315°C. Assuming a cure temperature of 250°C, a tensile stress of 12,000-16,000 psi can occur in the matrix. Depending on the matrix, some of this stress may be relieved through relaxation, particularly in the absence of molding pressure during cooling. Such high stresses may result in matrix failure. When using thermosets, much of the stress is locked in and there is very little relaxation during cooling.

7.8 Resistance to Fluid Permeation Through Pipe Wall

Although this aspect of the performance of ribbon composite pipe cannot be compared quantitatively with the steel pipe, it does provide a qualitative measure and points out the longer life such a pipe will experience. Also, as will be clear, comparison with plastic pipes is made possible.

The measure of enhancement of permeation resistance is provided by examining the ratio of permeation coefficient of matrix, K_m, and composite, K_c, since the permeation coefficient is linearly related to the amount of fluid permeating through the pipe wall. This ratio is given by[15]

$$\frac{K_m}{K_c} = \gamma(1 - \gamma)\left(\frac{b}{a}\right)^2\left(\frac{V_f^2}{1 - V_f}\right) + \frac{b}{2g} \tag{32}$$

where the symbols have been defined in Section 6. For a pipe wound back and forth the

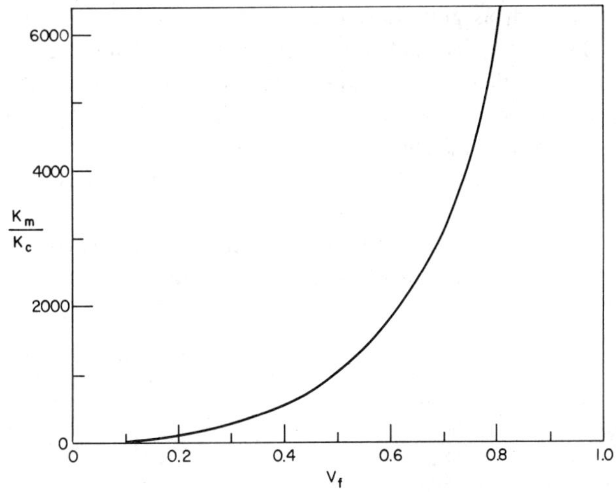

Fig. 21-25. Ratio of matrix and composite permeation as function of volume fraction.

average value of γ is 0.25. Assuming an aspect ratio of 100 and volume fraction ≥ 0.5, the first term of Eq. (32) is roughly 1000, whereas the second term may be between 10 and 50. Thus, the second term may be neglected for most practical cases. The dependence of K_m/K_c on glass content is shown in Fig. 21-25 where b/a has been assumed to be 100. It follows from this figure that a ribbon composite pipe is thousand fold less permeable than a pipe made from matrix alone. Consequently, there is less chance of moisture penetration and subsequent degradation of pipe strength, leading to longer pipe life. It should also be pointed out that this important advantage is not to be found in fiber reinforced pipes since they do not enhance the permeation resistance of plastics by more than a factor of 10.

The above analysis does not permit comparison on the basis of dynamic properties which must be handled experimentally. There is also the question of ductility offered by most metallic pipes. Undoubtedly, this property will be governed by the choice of matrix since glass must be considered brittle. Finally, the operating temperatures to which the ribbon composite pipe may be subjected will be lower than those for a steel pipe.

8. GRAPHITE AND BORON FILM COMPOSITES

Considerable attention has been given to more exotic reinforcing films, such as graphite and boron. In the early 70's, Pfizer, Inc. did some preliminary studies for the evaluation of graphite ribbon composites. This work, which was funded by The Army Materials and Mechanics Research Center, was recently reported by Froberg.[27] The following summary is based on Froberg's report.

Prepared by chemical vapor deposition technique, the pyrolytic graphite film ranges between 0.0001-0.0004 in. in thickness, with the nominal value of 0.0003 in. The graphite film has a tensile strength between 30 Ksi and 110 Ksi and a tensile modulus between 3×10^6 and 13×10^6 psi. It has an average strength of 60 Ksi and an average modulus of 7×10^6 psi. Its density is 1.6 g/cc. Thus, it compares well with glass ribbon in terms of specific strength and specific modulus, and has the potential of exceeding these preliminary values.

Graphite film composites enjoy the same advantages as glass ribbon composites. Thus, planar isotropy, more efficient use of reinforcement, superior specific strength and modulus, large volume loadings, and very thin laminates are distinct advantages which film composites have over corresponding filamentary composites. Since they can be made as thin as 0.003 in., graphite film laminates with 4-6 plies are very effective as skin reinforcement. For example, the addition of 5% graphite film composite increased the flexural modulus of an acrylic beam by 100%. Another widespread use of thin film composites is in providing local reinforcement for fiber composites in those locations where stresses are concentrated or deflections are large.

Graphite film properties have been utilized effectively in a multiple ply laminate. The composite tensile strength ranges between 30-60 Ksi and composite moduli between $4 \times 10^6 - 6 \times 10^6$ psi. For purposes of comparison, the typical room temperature properties of graphite filament composite with 52% volume loading are given in Table 21-17. Thus, graphite film composites compete favorably with quasi-isotropic filamentary composites. It should be borne in

Table 21-17. Properties of HERCULES graphite filament prepreg type 2002M and graphite film composites.

	Longitudinal	Cross-Plied	Angle-Plied (±45°)	Quasi-Isotropic	Graphite Film Composites
Tensile strength (Ksi)	104	53	21	40	30–60
Elastic modulus (10^6 psi)	24	14	2.5	9	4–6

Courtesy: Pfizer, Inc.

mind that in some applications where a very thin (0.003-0.005 in.) planar isotropic composite is required, graphite filaments which are 0.005 in. thick cannot be used since a minimum of 5-8 plies are necessary for a quasi-isotropic structure. This would mean a quasi-isotropic filamentary composite with a thickness of 0.040 in. or 20 times the thickness of a planar film composite.

It is worthwhile to compare graphite film composite properties with those of boron/polyimide composites which have been developed more extensively for a longer period of time. This comparison is made in Table 21-18, which also shows the properties of a well known aluminum alloy, 2024T-62. The Table clearly shows that graphite film composites compare favorably insofar as strength is concerned. The specific modulus, however, is lower and requires additional development such as hybrid composites, in combination with fiber reinforcement.

The boron film is formed[28,29] by evaporation and deposition of pure, elemental boron in a high vacuum on polyimide film substrate. The film is passed continuously over the crucible containing molten boron in the vacuum chamber. The deposited boron forms a smooth, black, glassy film in thicknesses up to 0.0003 in. The thickness of polyimide film substrate is 0.00025 in. The sheets thus formed by Norton Research Corporation* in 1970 were about 6 in. wide and up to 60 in. long. The microcrystalline structure of boron film resembles that of boron filaments. The interest and motivation in developing boron film stem from potentially high composite efficiencies, planar isotropy, and low density—all of which are necessary in buckling—and flutter—critical aircraft structures. The initial premise for developing these films was the inherent assumption that the films will have properties similar to boron fibers, i.e., a tensile strength of 400 Ksi and an elastic modulus of 55 $\times$ 10^6 psi. Assuming further that these properties can be utilized efficiently in a boron film/epoxy laminate, a composite strength of 200 Ksi and a modulus of 28 $\times$ 10^6 psi would be realized at a volume loading of 50%. Such properties would indeed be useful in aircraft structures.

However, such strength values have not, as yet, been attained. Working backward from the measured composite properties it was found that the film strength was 90 Ksi, instead of 400 Ksi, but the film modulus was 55 $\times$ 10^6 psi as expected. The drastic reduction in film strength is believed to be caused by the presence of periodic cracks which occur during the manufacturing process. The degree of cracking of the boron film is a function of the thickness of deposit. At 40% volume loading of boron, the distance between cracks varies from 0.25-1 in. and the film behaves like thin flakes of very large aspect ratio.

Boron film composites are fabricated by bonding multiple layers of film with a suitable adhesive. A number of resin systems including epoxy, phenoxy, and polyimide resins have been used to date. Epoxy resins have the advantage that they do not produce gaseous or liquid residues during curing. Polyimide systems provide good adhesion and high temperature properties. Mechanical properties of composites containing 39-44 volume % of boron, deposited on polyimide film and laminated with epoxy resin are given in Table 21-19.

One of the applications where boron film composites have shown good potential is the reinforcement of bolt holes in filamentary composites.[12,30] Unidirectional, bidirectional (±45°) and tridirectional (0°, ±45°) graphite fiber/

*Now National Research Division of Cabot Corporation, Cambridge, Mass.

Table 21-18. Comparison of properties of graphite film/epoxy composites with other materials.

Material	Volume Fraction	Tensile Strength (Ksi)	Elastic Modulus (10^6 psi)	Density (lb/in.^{3})	Specific Strength (10^3 in.)	Specific Modulus (10^6 in.)
Boron/epoxy	–	40	20	0.06	667	333
Al2024-T 62	–	62	10	0.10	620	102
Graphite film/epoxy	0.83	59	6	0.055	1060	108

Courtesy: Pfizer, Inc.

Table 21-19. Mechanical properties of boron film reinforced-epoxy composite.

Tensile strength	30–55 Ksi
Compressive strength	55–90 Ksi
Elastic modulus	20–22 × 10^6 psi
Shear modulus	8–9 × 10^6 psi
Interlaminar shear	12–14 Ksi
Density	0.062–0.065 lb/in.^{3}

epoxy resin composites were fabricated to form a series of double-lap bolted joint models. These were tested to measure the joint strength effectiveness of boron film colaminated locally at 6 and 10 volume % levels. Results showed that the addition of film plies to the highly directional fiber laminae increases the joint strength and stiffness up to 200%, the chief contribution of boron film being high bearing strength and resistance to shear distortion. The boron film was also effective in reducing the notch sensitivity of graphite fiber laminates, while at the same time substantially raising both the tensile strength and the work-to-fracture of the material.

Applications where boron and graphite films can provide optimum reinforcement include the reinforcement of bolt holes in filamentary composites, thin high performance laminates, skins for honeycomb structures, structures requiring high planar isotropic properties, such as in aircraft where buckling and flutter are important, aircraft wings, and tubular components with high specific strength and modulus.

References

1. Humphrey, R. A., Shaped Glass Fibers in *Modern Composite Materials,* ed. by L. J. Broutman and R. H. Krock, Addison-Wesley, Mass., 1967.
2. Halpin, J. C. and Thomas, R. L., "Ribbon Reinforcement of Composites," *J. Comp. Mat.,* 2 (1968).
3. Tsai, S. W. and Pagano, N. J., Invariant Properties of Composite Materials in *Composite Materials Workshop,* ed. by Tsai, Halpin, and Pagano, Technomic Pub. Co., Stamford, Conn., 1968.
4. Lewis, T. B., Ribbon Reinforcement in Composite Materials, Proc. 25th Ann. Conf. SPI., 1970.
5. Gray, R. M., Transverse Strength of Ribbon Reinforced Metal Matrix Composites, Proc. Symp. Metal-Matrix Composites, Met. Soc. AIME; 1969 (Govt. Document No. AD695-046).
6. Anderson, E., Lux, B., and Curssard, C., *Revue de Metallurgie,* 2 (1972).
7. Anderson, E., "Fatigue Behavior of Ribbon-Reinforced Composites," *J. Mater. Sci.,* 8 (1973).
8. Pollock, J. T. A. and Arthur, J., "Tensile Strength of Ribbon Reinforced Composites," *Mat. Science and Eng.,* **18** (1975).
9. Padawer, G. E. and Beecher, N., "On the Strength and Stiffness of Planar Reinforced Plastic Resins," *Polymer Eng. and Science,* **10** (1970).
10. Chen, P. E. and Nielson, L. E., "Mechanical Properties of Tape Composites," *Kolloid-Zeit. und Zeit. fur Polymere,* **235** (1969)
11. Chen, P. E. and Lewis, T. B., "Stress Analysis of Ribbon Reinforced Composites," *Polymer Eng. and Science,* **10** (1970).
12. Padawer, G. E., "The Strength of Bolted Connections in Graphite-Epoxy Composites Reinforced by Colaminated Boron Film," in Composite Materials: Testing and Design (Second Conf.), ASTM STP497, pp. 396–414. (1972).
13. Gulati, S. T. and Plummer, W. A., Thermal Expansion of Ribbon Reinforced Composites, Third Thermal Expansion Symposium; ed. by M. G. Graham and H. E. Hagy, Am. Inst. Physics, 1971.
14. Humphrey, R. A., "Glass Microtape" in *Materials Science Research, Vol. 2,* ed. by H. M. Otte and S. R. Locke, Plenum Press, N. Y., 1965.
15. Brydges, W. T., Gulati, S. T., and Baum, G., "Permeability of Glass Ribbon-Reinforced Composites," *J. Mat. Science,* **10** 2044-2049 (1975).
16. Hutchins, J. R., III and Harrington, R. V., "Glass" in *Encyclopedia of Chemical Technology,* ed. Kirk-Othmer, John Wiley & Sons, New York, 1966.
17. Eakins, W. J. and Humphrey, R. A., Glass-Resin Composite Structure, U. S. Patent 3,425,454 (Feb. 4, 1969).
18. Cox, H. L., "The Elasticity and Strength of Paper and Other Fibrous Materials," *Brit. J. App. Phys.,* **3,** 72 (1952).
19. Loewenstein, K. L., "Glass Systems," *Composite Materials,* ed. by L. Holliday Elsevier, 1966.
20. Bishop, P. H. H., An Improved Method for Predicting Mechanical Properties of Fibre Composite Materials, RAE Tech. Report No. 66245 (1966).
21. Ashton, J. E., Halpin J. C., and Petit P. H., *Primer on Composite Materials: Analysis,* Technomic Publ. Co., Stamford, Conn., 1969.
22. Whitney, J. M., "Experimental Determination of Shear Modulus of Laminated Fiber-Reinforced Composites," *Expt. Mech.,* 7 (1967).
23. Lovelace, A. M. and Tsai, S. W., "Composites Enter the Mainstream of Aerospace Vehicle Design," *Astonautics and Aeronautics,* pp. 56–61 (July, 1970).
24. Lewis, T. B., Nielson, L. E., and Hemmerly, D.M., Ribbon Reinforcement in Composites, 5th. St. Louis Symp. Adv. Composites, 1971.
25. Schapeny, R. A., "Thermal Expansion Coefficients of Composite Materials Based on Energy Principles," *J. Comp. Mat.,* **2** (1968).

26. Halpin, J. C., "Stiffness and Expansion Estimates for Oriented Short Fiber Composites," *J. Comp. Mat.*, **3** (1969).
27. Froberg, R. W., Research Study in Evaluation of Graphite Ribbon Composites, prepared for Army Materials and Mechanics Research Center, Report No. AMMRC CTR-72-28; Dec. 1972.
28. Beecher, N., Feakes, F., and Allen, L. R., Laminar Film Reinforcements for Structural Applications, Paper NR-3, 12th National Sampe Symposium, Western Periodicals Co., Hollywood, Calif., 1967.
29. Crawford, R. F., An Evaluation of Boron-Polymer Film Layer Composites for High-Performance Structures, Astro Research Corporation Report No. ARC R-276, also NASA CR-1114, Sept. 1968.
30. Padawer, G. E., Film Reinforced Multi-Fastened Mechanical Joints in Fibrous Composites, 13th. Structures, Structural Dynamics, and Materials Conference, San Antonio, Texas, April 10-12, 1972; AIAA Paper No. 72-382; American Institute of Aeronautics and Astronautics, 750 Third Ave., New York, N. Y.

GLOSSARY

Quasi-isotropic—A fiber composite with 0, ±45°, 90° oriented fibers to provide isotropic properties in the plane of the composite.

ACKNOWLEDGMENT

The author is grateful to Dr. W. T. Brydges for his assistance in the preparation of this manuscript, and to Mr. John F. MacDowell for constant encouragement in pursuing the important field of glass ribbon reinforcement.

Section X
Short Fiber Reinforcements

22
Wollastonite

L. W. Choate
Interpace Corp.
Willsboro, New York

CONTENTS

1. INTRODUCTION

Wollastonite is the only pure white mineral commercially available that is wholly acicular. It has a L/D ratio of 15:1. This crystalline acicularity allows Wollastonite to function as a resin reinforcing filler that contributes improved properties to plastic compounds, as do other short fiber fillers. Unlike fibrous talc and asbestos, Wollastonite is classified as a non-hazardous mineral filler. The allowable airborne dust level in handling Wollastonite is 15 mg/m^3 compared to a 2 fiber per cubic centimeter per eight hours count for asbestos and fibrous talc. Tensile and flex strengths of Wollastonite-filled systems compare favorably with most other filled systems. Under exposure to U/V and water immersion, the properties of Wollastonite filled systems are outstanding.

Wollastonite is a naturally occurring calcium silicate ($CaSiO_3$) that is rarely found in high purity deposits. What is believed to be the largest such deposit is located near Willsboro, New York and practically all commercial production of Wollastonite in the world is produced from this source. Plastic filler grades are available in carload lots for \$.03-.04/lb f.o.b. Willsboro, New York.

2. MANUFACTURING PROCESS

Wollastonite ore is extracted from an underground mine with methods adapted to preserving the purity of the final product. A high intensity magnetic beneficiation process removes metallic silicates associated with the ore. The beneficiated Wollastonite is reduced to most product sizes by pebble milling and air separation. The grades that are predominantly used as plastic fillers are produced in a proprietary design circuit that preserves the aspect ratio of the original crystal and breaks up the crystal bundles into individual fibers.

3. PROPERTIES OF WOLLASTONITE

3.1 Chemical Analysis

The chemical analysis of Interpace Wollastonite is:

SiO_2	50.9%	Al_2O_3	.25	TiO_2	.05
CaO	46.9	MnO	.10	LOI[a]	.9
FeO	.55	MgO	.10		

[a]Loss on ignition.

3.2 Physical Properties

Physical properties of typical Interpace Wollastonite are:

Chemical formula	CaS_1O_3
Molecular weight	116
Crystalline structure	acicular
Specific gravity	2.9
Coefficient of thermal expansion /°C	6.5×10^{-6}
Refractive index	1.63
Transition point (to pseudo-Wollastonite, °C)	1200
Melting point (°C)	1540
Solubility in H_2O (g/100 cc)	0.0095
Hardness (Mohs' scale)	4.5
Color	Brilliant White
Weight (lb/solid gallon)	24.2
Bulking value (gal./lb)	0.0413
pH (10% slurry)	9.9
Moisture content (%)	0.5

3.3 Particle Size Distribution

Figure 22-1 illustrates the particle size distribution curves of the available grades of Wollastonite. Grades C-6 and C-1 are normally used in ceramic applications. Grades P-1 and F-1 are designated as plastic filler grades and P-4 is used as a paint filler and pigment extender.

3.4 Photomicrographs of Filler Grades

Figures 22-2 and 22-3 are electron micrographs of grades P-1 and F-1. Note that the acicularity of the particle is preserved in even the finest crystal.

4. SUPPLIERS

Interpace Corporation, Willsboro, New York is the major supplier of Wollastonite in the world. Pfizer markets a calcium silicate, designated Wolcron, from a California source and small quantities are marketed in Europe by a Finnish producer.

Table 22-1 presents the latest price schedule for Wollastonite in truckload and carload quan-

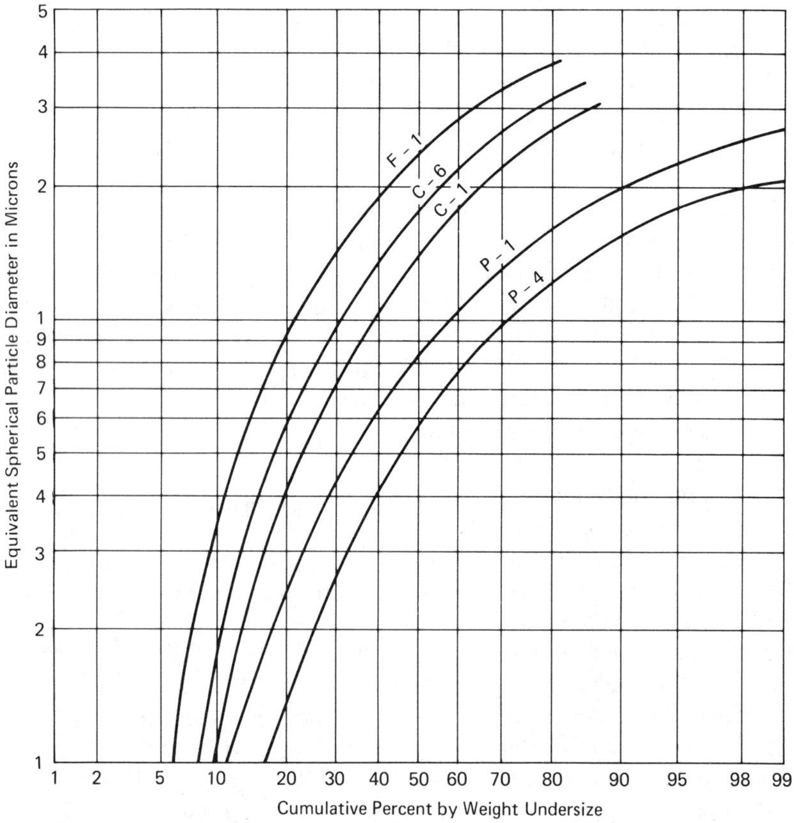

Fig. 22-1. Typical particle size distribution of Interpace Wollastonite grades by gravity-hydrometer method.

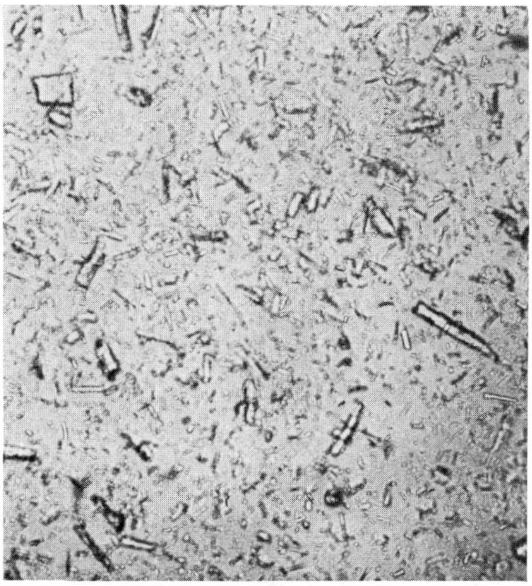

Fig. 22-2. Grade P-1. 100X.

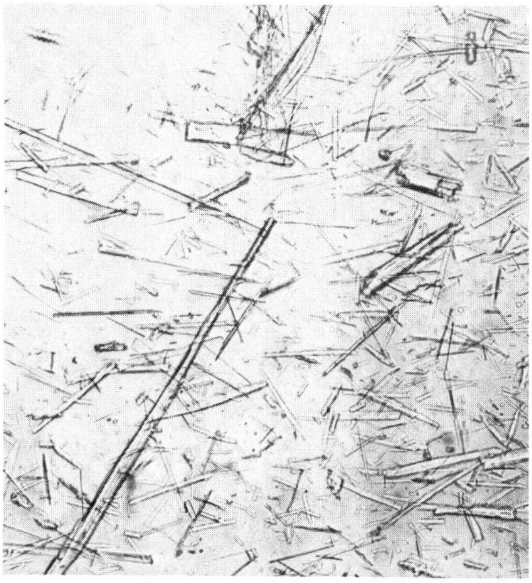

Fig. 22-3. Grade F-1. 100X.

Table 22-1. Price list for Wollastonite.

By Volume	P-1 Med. Paint Fine Ceramic	P-4 Fine Paint	F-1 Medium Fiber	C-1 Medium Ceramic	C-6 Coarse Ceramic	C-101 General All-Purpose
	GRADES OF WOLLASTONITE Cost per ton, f.o.b., Willsboro Plant					
Bulk carload	$ 54.00	$ 74.00	$ 74.00	$48.00	$46.00	$44.00
Carload or truckload, bagged	$ 60.00	$ 80.00	$ 80.00	$54.00	$52.00	$50.00
10,000 lb–min. T/L	$ 70.00	$ 90.00	$ 90.00	$64.00	$62.00	$62.00
2,000–9,950 lb	$100.00	$120.00	$120.00	$94.00	$92.00	$92.00

Terms: net 30 days

Packing: 50 lb bags can be shipped directly from the Willsboro, New York plant in carload or truckload lots. Lesser quantities are usually less expensive if ordered from an agency as listed below:

Pallet prices:	
Pok Pack (cardboard) disposable	$2.25
Disposable wood (single face)	$5.00
Reusable wood (double face)	$6.25

D. H. Litter Company, Inc.
30 Lowell Junction Road
Ballardvale, Massachusetts 01810
(617) 475-5315

D. H. Litter Company, Inc.
116 E. Sixteenth Street
New York, New York 10003
(212) 777-4410

Harwick Chemical Company
60 South Seiberling Street
Akron, Ohio 44305
(216) 789-9300

Harwick Chemical Company
800 Estes Avenue
Elk Grove Village, Illinois 60007
(312) 437-6560

George A. Rowley Company, Inc.
937 N. Front Street
Philadelphia, Pennsylvania 19123
(215) 627-3077

Harwick Chemical Company
7225 Paramount Blvd.
Pico Rivera, California 90660
(213) 723-9911

Harwick Chemical Company
1485 Bayshore Blvd.
San Francisco, California 94124
(415) 467-0505

Van Waters and Rogers, Inc.
P.O. Box 34749
Dallas, Texas 75234
(214) 239-9111

Van Waters and Rogers, Inc.
P.O. Box 14500
Houston, Texas 77021
(713) 644-1601

Table 22-2. Comparison of Wollastonite composites with other fillers.

		FORMULATIONS												
		WOLLASTONITE					WHITING							
		Corona		P-1		F-1	Treated		Untreated		Talc		Asb.	
	Phr ⟶	50	100	50	100	50	50	100	50	100	50	75	33	0
Wollastonite–Corona		900	1350											
” P-1				900	1350									
” F-1						900								
Whiting–Treated							900	1350						
” –Untreated									900	1350				
Talc											900	1150		
Asbestos													670	
(1) Polyresin P283LV		900	1350	900	1350	900	900	1350	900	1350	900	1550	2030	
Disperse at high speed. Add														
(1) Polyresin P283LV		900		900		900	900		900					2700
(2) Cobalt octoate (12%)		9.0	6.8	9.0	6.8	9.0	9.0	6.8	9.0	6.8	9.0	7.8	10.2	13.5
Total weight	lbs	2709	2707	2709	2707	2709	2709	2707	2709	2707	2709	2708	2710	2714
[a]Cost/lb	¢	15.4	14.0	12.8	10.2	12.9	13.5	11.3	13.5	11.3	13.5	12.4	14.5	18.0
Cost/cu ft.	$	13.60	14.25	13.35	10.30	11.50	11.80	11.15	11.80	11.20	12.05	11.80	12.15	12.80
Catalyzed with 1% MEK Peroxide, based on resin														
Physical property data														
Viscosity	poises													
Initial		39	1680	17	57	23	19	22	10	34	172	650	Over	4
1 week		49	1900	19	57	23	24	22	13	34	190	650	20,000	4
2 ” s		67	3200	19	65	23	24	23	13	34	190	650	”	4
4 ” s		134	9300	19	65	25	28	23	14	34	295	650	”	5
Thixotropy index														
Initial		X	X	1.1	1.8	1.2	X	1.1	X	1.1	X	1.6	X	1.0
4 weeks		X	X	1.2	1.8	1.2	X	1.1	X	1.1	X	1.6	X	1.0
Settling–2 wks	Score	9	10	5	5	4	4	7	8	10	10	9	9	X
Stability at 50°C time to gel	days	4	7	5	5	4	2	4	4	4	4	2	2	14

Pot life (+ catalyst)														
Gel time vs clear	%	100	130	210	180	210	60	180	85	210	155	180	165	100
Peak exotherm	°C	109	59	116	78	110	114	81	117	87	92	41	69	146
Time to peak	mins	25	40	46	51	50	21	51	21	51	48	58	76	21
Ease of casting	Score													
Pour		6	2	9	9	7	8	8	9	9	4	0	0	10
Air entrapment		10	9	8	8	8	10	9	10	9	8	2	2	10
Removal		10	8	8	8	8	10	8	10	8	8	2	2	10
Color	Score	8	8	8	8	6	8	7	8	7	6	6	2	X
Linear shrinkage	%	2.3	2.2	2.2	2.0	2.2	2.5	2.2	2.5	2.2	2.3	2.0	1.8	3.2
Hardness–Shore D		90	91	90	90	90	90	90	89	90	87	87	83	86
X–Not tested														
Mechanical property data														
Tensile strength														
Ultimate	psi × 10^3	5.7	4.7	5.9	5.0	5.4	3.9	2.7	6.2	4.1	4.3	3.5	4.4	8.0
Elongation	%	0.9	0.7	1.4	1.0	1.1	0.8	0.7	1.3	1.0	0.8	0.8	1.9	3.9
Modulus	psi × 10^5	6.7	7.2	4.4	4.4	5.7	5.1	5.1	4.8	4.8	5.6	5.6	2.9	2.1
Flexural strength	psi × 10^3													
Initial		7.4	6.4	8.5	5.7	8.1	6.1	3.6	6.0	5.4	4.7	4.7	5.4	10.4
24 hr–H_2O @ 100°C		3.6	3.3	5.8	4.0	4.2	4.0	3.6	4.1	4.1	3.1	3.1	3.7	3.0
24 hr–200°C		8.2	10.6	12.0	5.7	8.2	6.0	6.1	5.8	4.8	5.6	6.0	6.0	7.5
7 days–U/V		7.9	9.0	4.2	10.4	9.6	6.2	4.6	6.3	8.6	5.8	6.5	7.0	7.9
Flexural modulus	psi × 10^5													
Initial		8.3	10.1	5.1	8.3	6.7	5.9	7.9	6.3	5.9	5.0	4.9	2.4	4.6
24 hr–H_2O @ 200°C		2.4	4.5	7.0	7.0	6.2	2.7	5.7	2.9	5.7	2.1	3.7	3.8	2.9
24 hr–200°C		9.1	13.8	5.0	7.5	5.8	6.8	5.0	6.7	5.4	8.9	4.5	2.0	3.3
7 days–U/V		8.7	12.9	6.6	10.1	8.9	5.7	8.2	5.4	7.9	8.3	4.9	4.7	3.4
Compressive strength	psi × 10^3	18	17	21	21	18	16	21	16	22	13	12	21	15
Izod impact	psi	3.0	3.0	2.5	2.8	2.5	2.3	2.7	2.3	2.5	2.5	2.9	2.6	2.1
Deflection temperature	°C	48	50	51	54	54	47	49	48	52	49	49	58	47

Score

10 = Excellent or no change
9 = Very good or very slight
8 = Good or slight
6 = Fair/good or moderate
4 = Fair or considerable
2 = Poor or severe
0 = Very poor or extreme

Note: Castings aged 1 week before test.
1. Resinous Chemicals Corp.
2. Tenneco Chemicals, Nuodex Div.
3. U. S. Peroxygen Div., Argus Chemical Corp.
[a]Less Cobalt and Catalyst.

tities f.o.b. Willsboro, New York. Lesser quantities are available from the listed agents at their published prices.

5. PRICE HISTORY AND AVAILABILITY

Price history of the filler grades has followed the general inflation in the 1970's. The price per pound and a projection to 1980 are:

Grade	1970	1972	1974	1975	1976	1980
P-1	1.5¢	1.7¢	2.0¢	2.2¢	2.4¢	3.0¢
F-1	1.9	2.1	2.6	2.9	3.2	4.0

Ore reserves and expansion plans can only be discussed for Interpace's facility at Willsboro, New York. Proven ore at that location is 5.8 million tons with an additional 5 million tons under option. This is considered to be a 100 year reserve. Process facilities have a capacity of 80,000 tpy* of product mix. It is planned to expand to 100,000 tpy in 1977. A new mill capable of producing an additional 100,000 tpy is being designed for the future.

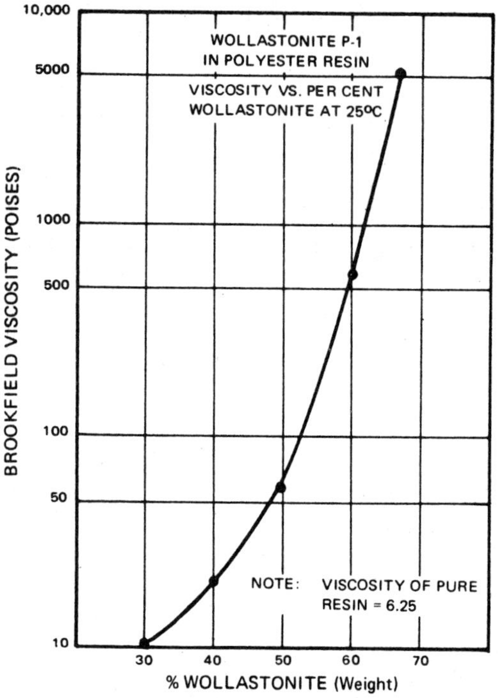

Fig. 22-4. Wollastonite P-1 in polyester resins.

6. GENERAL DISCUSSION OF CURRENT APPLICATIONS IN PLASTICS

All mineral filled systems using Wollastonite are proprietary and can only be discussed in general terms. Thermosetting polyesters, vinyl plastisols, nylon, and polypropylene utilize Wollastonite in compounding. Wollastonite offers improvements in tensile and flex strengths over other mineral fillers and is particularly moisture re-

*Tons per year.

Table 22-3. Wollastonite P-1 in vinyl plastisols for lower viscosity at high loadings, better color (at savings) in color pigment and improved resistance to heat aging.

A comparison of WOLLASTONITE with two kinds of fine particle size calcium carbonate showed that WOLLASTONITE causes less increase in viscosity at high loadings than does the calcium carbonate. The comparison was made at two inert levels, using Formula A. The pigments were dispersed in the plasticizer by mechanical agitation and resin added afterwards.

	Fineness N.S. Paint Scale 33 parts filler/ 100 parts resin	24-Hr. VISCOSITIES (in cps) 2 rpm at 75° F.*	
		33 parts filler/ 100 parts resin	66 parts filler/ 100 parts resin
WOLLASTONITE	3¾–4	8,000	9,800
Calcium carbonate (fine coated)	4½–5	16,000	60,000
Calcium carbonate (fine water-ground)	4½–5	21,100	92,000

*The viscosity readings of all compounds remained practically unchanged after 7 days of aging.

FORMULA A

Geon 121	100 wt. parts
DDP	22
DDA	44
Ferro 1212	3
Filler	33(66)

Table 22-4. Nylon 6: Effect of silane treatment on properties of a 70% Wollastonite-filled composite

Property	70% WOLLASTONITE Untreated[a]	A-1100-Treated[b]	Base Resin[c]
Flexural strength (psi)			
dry[d]	12,200	21,000	12,500
wet[e]	6,100	16,300	6,700
Flexural modulus (psi × 10^5)			
dry	9.6	9.9	2.7
wet	3.6	6.4	1.1
Tensile strength (psi)			
dry	7,600	11,900	9,000
wet	4,100	9,400	6,700
Tensile modulus (psi × 10^5)			
dry	14.3	15.1	3.8
wet	8.0	9.6	2.3
Deflection temperature (°F)[f]	349	356	133
Water absorption (wt. %)[e]	0.97	0.68	3.14
Dart impact (in-lb)	<4.5	11	>600

[a]Wollastonite, Grade P-1, Interpace Corporation.
[b]Union Carbide A-1100 silane at 1 wt. % based on filler content.
[c]Allied Plaskon 8201 ground to 50 mesh (300 μ).
[d]Tested after conditioning at room temperature and 35% R. H.
[e]After immersing 16 hr in water at 50°C.
[f]At 264 psi.

sistant. Silane coated Wollastonite offers unique properties that are expected to gain recognition in the next few years. Tables on the following pages compare Wollastonite-filled systems with those of other mineral fillers.

Table 22-2 compares the properties of Wollastonite-filled polyester castings with those of limestone, talc, and asbestos-filled polyester castings. The Corona grade of Wollastonite is now designated P-15 and is 100% minus 15 μs

Table 22-5. Nylon 6: Effect of silane treatment on physical properties of composites.

Property	50% CLAY[g] Untreated[g]	50% CLAY[g] Silane-Treated[h]	50% WOLLASTONITE[a] Untreated	50% WOLLASTONITE[a] A-1100-Treated[b]	Base Resin[c]
Flexural strength (psi)					
dry[d]	14,900	19,300	15,400	18,600	12,500
wet[e]	10,200	14,100	8,900	12,000	6,700
Flexural modulus (psi × 10^5)					
dry	7.9	7.2	8.3	7.7	2.7
wet	4.8	5.0	5.9	6.1	1.1
Tensile strength (psi)					
dry	8,300	10,600	8,900	10,500	9,000
wet	5,900	7,200	5,500	7,500	6.700
Tensile modulus (psi × 10^5)					
dry	9.7	9.2	11.2	10.3	3.8
wet	7.9	6.5	5.9	6.1	2.3
Deflection temperature (°F)[f]	307	304	302	284	133
Dart impact (in-lb.)	<4.5	<4.5	5.7	118	>600
Water absorption (wt-%)[e]	1.14	0.98	1.36	1.23	3.14

See footnotes a through f, Table 22-4.
[g]"Hydrogloss," Kaolin Clay, J. M. Huber Corporation.
[h]"Nulok 321," aminosilane-treated Kaolin Clay, J. M. Huber Corporation.

Table 22-6. Nylon 6,6: Effect of silane treatment and filler level on physical properties of composites.

	50% WOLLASTONITE		70% WOLLASTONITE		
Property	Untreated[a]	A-1100-Treated[b]	Untreated[a]	A-1100-Treated[b]	Base Resin[c]
Flexural strength (psi)					
dry[d]	13,500	17,100	14,500	17,700	14,300
wet[e]	9,600	12,400	8,200	15,400	8,900
Flexural modulus (psi × 10^5)					
dry	10.8	10.7	20.7	22.6	3.2
wet	6.4	7.0	12.2	16.2	1.7
Tensile strength (psi)					
dry	7,900	8,000	7,600	9,600	9,400
wet	6,300	7,600	5,300	9,400	8,000
Tensile modulus (psi × 10^5)					
dry	13.8	13.3	28.0	34.9	3.7
wet	9.6	9.9	18.0	22.8	2.6
Deflection temperature (°F)[f]	396	419	383	455	167

See footnotes, a, b, d, e, f, Table 22-4.
[c]duPont Zytel 101 ground to 50 mesh (300 μ).

in size. The current price for P-15 is \$.08/lb for all quantities.

Figure 22-4 illustrates the viscosity of Wollastonite loaded polyester resins at various concentrations. This graph indicates that Wollastonite P-1 can be used in polyester molding compounds at high concentrations without exceeding a practical working viscosity.

Table 22-3 compares loading and viscosity for Wollastonite with two types of calcium carbonate in vinyl plastisols. This indicates that Wollastonite provides a lower viscosity at high loadings.

Tables 22-4–22-8 compare the property advantages of silane-treated Wollastonite in various resin systems to untreated filled systems of

Table 22-7. Polyterephthalate: Effect of silane treatment and filler level on physical properties of composites.

	50% WOLLASTONITE		70% WOLLASTONITE		
Property	Untreated[a]	A-187-Treated[b]	Untreated[a]	A-187-Treated[b]	Base Resin[c]
Flexural strength (psi)					
dry[d]	11,800	16,200	7,400	11,600	11,800
wet[e]	11,200	16,100	7,300	10,900	11,700
Flexural modulus (psi × 10^5)					
dry	7.7	7.1	10.5	10.8	3.0
wet	6.9	7.2	8.9	9.5	2.8
Tensile strength (psi)					
dry	7,300	8,500	4,700	7,500	6,400
wet	6,200	8,500	4,500	7,400	6,400
Tensile modulus (psi × 10^5)					
dry	9.8	9.2	16.1	18.5	4.2
wet	8.0	8.8	12.2	17.3	3.2
Deflection temperature (°F)[f]	289	271	367	367	127

See footnotes a, d, e, f, Table 22-4.
[b]Union Carbide A-187 epoxy functional silane at 1 wt-% based on filler content.
[c]Eastman Plastics Tenite 6PRO, ground to 50 mesh (300 μ).

Table 22-8. Polycarbonate: Effect of silane treatment and filler level on physical properties.

	50% WOLLASTONITE		70% WOLLASTONITE		
Property	Untreated[a]	A-1100-Treated[b]	Untreated[a]	A-1100-Treated[b]	Base Resin[c]
Flexural strength (psi)					
dry[d]	11,900	14,400	7,200	9,400	12,600
wet[e]	9,900	12,400	7,200	9,400	12,500[g]
Flexural modulus (psi × 10^5)					
dry	9.4	9.2	18.8	18.9	3.0[g]
wet	8.7	9.2	18.7	18.8	3.0[g]
Tensile strength (psi)					
dry	6,400	8,100	5,700	5,600	7,800
wet	6,100	7,900	5,200	5,100	7,700
Tensile modulus (psi × 10^5)					
dry	11.5	11.9	27.6	24.5	2.8
wet	10.4	12.3	23.8	24.4	2.6
Deflection temperature (°F)[f]	266	271	252	252	268

See footnotes a, b, d, e, f, Table 22-4.
[c]Bayer Makrolon Type 3100L ground to 50 mesh (300 μ).
[g]Data from Ref. 4.

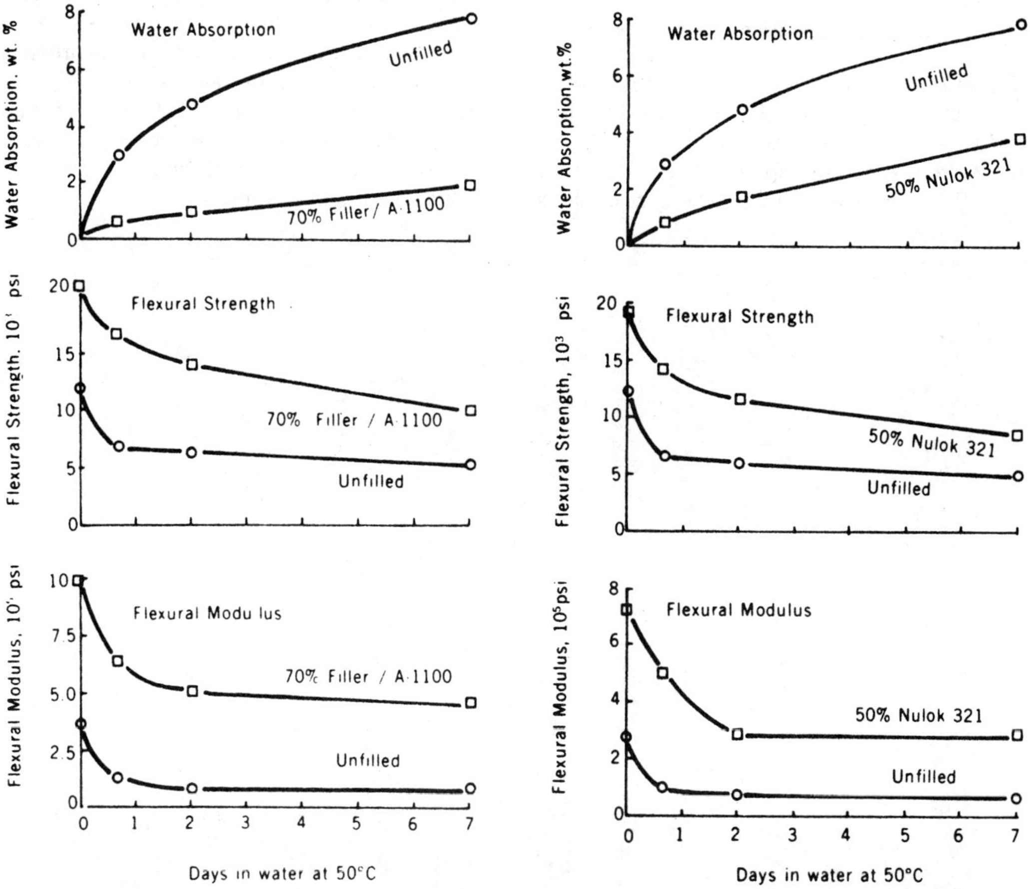

Figs. 22-5 and 22-6. Comparison of properties of Wollastonite and clay in nylon 6 after water immersion.

Wollastonite. Some comparisons are made with a popular treated and untreated clay filler. Figures 22-5 and 22-6 compare the properties of Wollastonite versus clay filled Nylon 6 after water immersion. The Wollastonite-filled system exhibits less water absorption and less loss of flexural strength and modulus.

Wollastonite competes with eastern talc, asbestos shorts, and mica as a mineral filler in plastics. It is estimated to have a 10-15% share of the mineral filler market. The growth rate for mineral filled plastics is projected at 15% per year through 1980.

References

1. Orenski, P. J., Berger, S. E., and Ranney, M. W., Silane Coupling Agents–Performance in Engineering Plastics, 28th SPL Annual Technical Conference, 1973.

Glossary: Significance of Tests

Viscosity–A measure of the consistency of the material. High viscosity generally results in difficult handling or casting.

Thixotropy Index–A liquid polyester resin tends to flow freely when poured. Fillers cause the resin to thicken and reduce the ability of the liquid to flow. The thixotropy index is a measure of this property, comparing the system at rest against itself when agitated.

Settling–All fillers tend to drop to the bottom of the mass during long periods of storage. If this happens rapidly or excessively, the result may be a non-uniform product when mixed.

Stability–60°C–An accelerated test to determine how long the filled resin can be stored before it must be discarded. Obviously, all the filled products have a relatively limited storage life–perhaps 2–6 months under ambient conditions.

Reactivity

Time to gel–A measure of relative working time after the catalyst is added. All require rapid handling to prevent gelling during use.

Exotherm–Chemically reactive resins will create heat during the reaction causing the temperature to rise, especially in bulk, as in the mixing container. Excessive temperature rise causes darkening of the product and can even ruin the product entirely. One major use for fillers is to reduce exotherm, thus making the resin safer to use.

Time to peak–A measure of the relative time required to reach the maximum exotherm temperature. The longer the time, the easier the compound is to use.

23
Asbestos

J. W. Axelson
Johns Manville
Denver, Colorado

CONTENTS

1. INTRODUCTION

Asbestos is the generic name for a group of hydrated metallic silicates. Chrysotile is a highly hydrated magnesium silicate with the formula $Mg_6[(OH)_4Si_2O_5]_2$ and comprises more than 95% of the world asbestos usage. It is available in bulk fibrous form of different lengths and also as a paper, yarn, cloth, and felt. As an inorganic mineral, asbestos is inert to weathering and to most chemicals except acids and strong bases. It is used in plastics as a reinforcing filler. It imparts flexural strength and modulus, improves creep resistance and heat deflection temperature, lowers thermal coefficient of expansion, controls resin flow properties, and has a low cost. Current cost ranges between 4.5¢/lb to over $1.00/lb for specially processed grades. Disadvantages are lowering of impact strength which can often be overcome, some processing difficulties, darker color, and need for more stabilization with some polymers. Special precautions are necessary when handling asbestos because it is known to cause asbestosis, bronchial cancer, and mesothelioma. The current threshold limit for asbestos is two fibers per ml longer than five μm on an 8-hour time-weighted average. Proper equipment design and dust control techniques allow asbestos to be used safely.

2. TYPES OF ASBESTOS AND METHODS OF PRODUCTION

There are two major classes of asbestos and, paradoxically, one contains only one type of asbestos but it is by far the most important and comprises over 95% of the total world production and use. This is chrysotile, the fiberized version of serpentine, which is the mineral class to which it belongs.

The second class of asbestos is called the amphiboles and contains five types of asbestos. However, only three are mined and used commercially and each has certain attributes which make it particularly suitable for certain uses.

Crocidolite comprises the amphibole used in the largest tonnage. It is the blue asbestos of commerce and is especially noted for its acid resistance. However, its major use is in the production of asbestos-cement products where it aids in the formation of the product. The second largest quantity of amphibole fiber produced is the amosite type, which has its largest use in insulating products due to its high heat

Table 23-1. Asbestos classes and types.

Class	Type	Chemical Formula
Serpentine		
	Chrysotile	$Mg_6[(OH)_4Si_2O_5]_2$
Amphibole		
	Crocidolite	$Na_2MgFe_5[(OH)Si_4O_{11}]_2$
	Amosite	$MgFe_6[(OH)Si_4O_{11}]_2$
	Anthophyllite	$(Mg, Fe)_7[(OH)Si_4O_{11}]_2$
	Tremolite	$Ca_2Mg_5[(OH)Si_4O_{11}]_2$
	Actinolite	$Ca_2(Mg, Fe)_5[(OH)Si_4O_{11}]_2$

resistance and ability to produce low density products. The last commercial variety of the amphiboles is anthophyllite. The greatest production of this was in Finland where it was used in asbestos cement products and also exported. However, the Finnish operation was closed down in 1975 for economic reasons. Anthophyllite is still produced in South Africa, South Carolina, and Arizona. At one time, it was the major ingredient in filled polypropylene but that use is now minimal and its usage in other plastics is also at a low level.

A summary of the classes and types of asbestos including idealized formulations are given in Table 23-1. Figures 23-1 through 23-6 are electron photomicrographs of the 6 kinds of asbestos. Since the majority of asbestos used in plastics is chrysotile, most of the following will be based on chrysotile unless specifically referring to other types.

Asbestos is classified by type and by fiber length. The most predominant grading system is the Quebec Screen Classification and most grading is based on some variation of this even

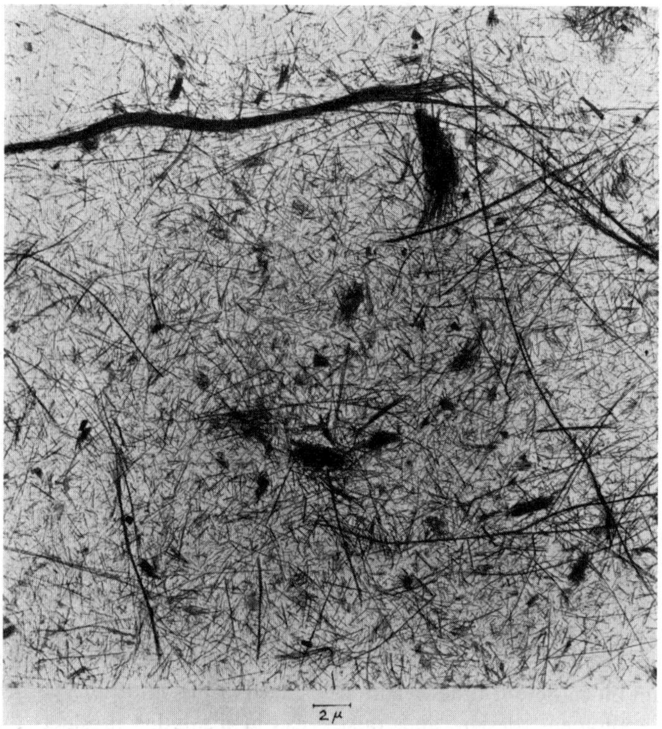

Fig. 23-1. Chrysotile-2500X.

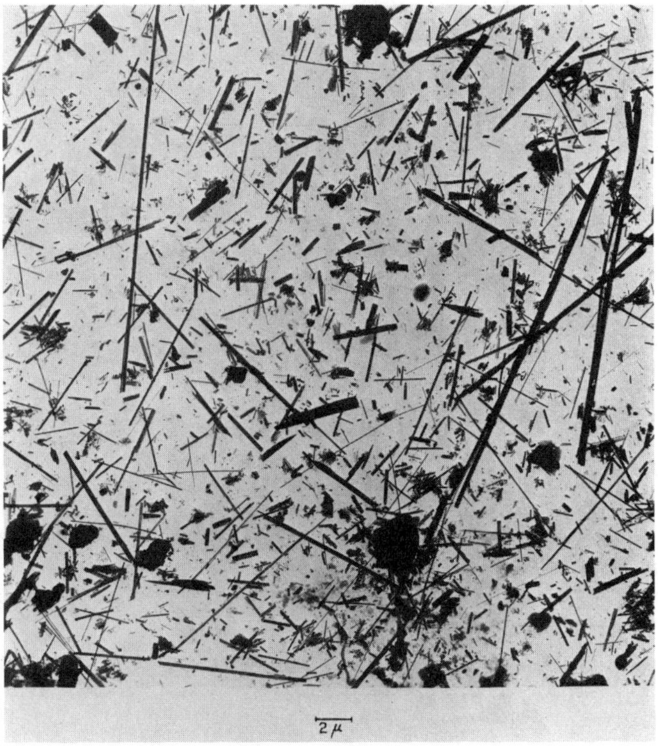

Fig. 23-2. Crocidolite-2500X.

Fig. 23-3. Amosite-2500X.

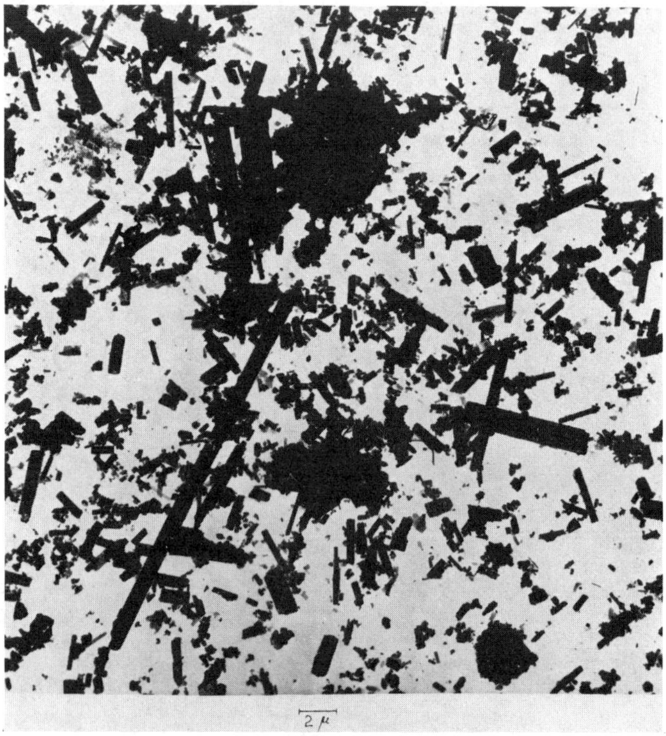

Fig. 23-4. Anthophylite-2500X.

Fig. 23-5. Tremolite-2500X.

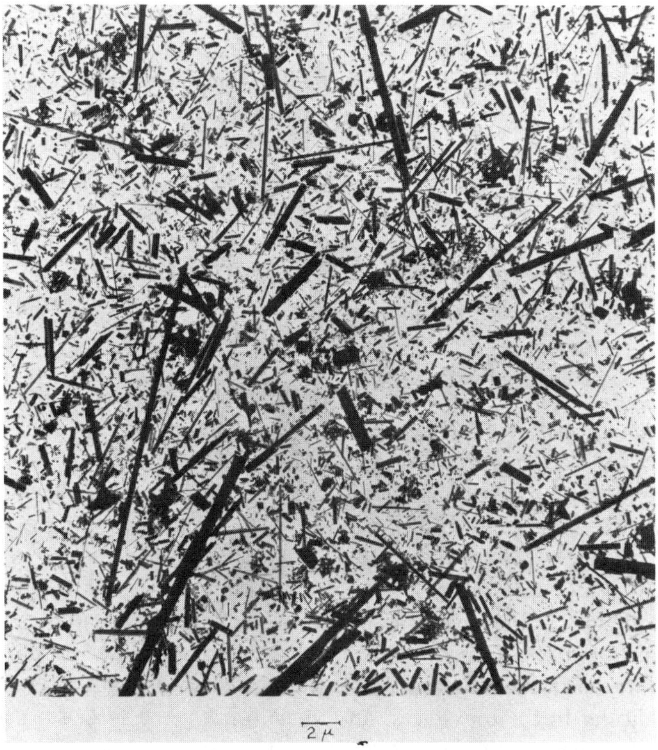

Fig. 23-6. Actinolite-2500X.

though the nomenclature may be different. Basically, the grading is done by dry screening a sample of asbestos. The relative amounts retained on each screen determines the grade of that particular fiber. Standard classifications have been set up by the Quebec Asbestos Mining Association. These are based on a number and letter nomenclature running from 1 to 8 and from A to Z in each number series although only a few of the letters are used in each. As the number increases from 1 to 8, the average fiber length becomes shorter, and as the letter decreases in the alphabet from A to Z the fiber grade becomes shorter in each number classification. Typical grades and their classification by Quebec Screen Analysis are given in Table 23-2.

In addition to the number-letter classification, most producers use another suffix number which denotes the amount of processing the fiber has had and, in general, the degree of fiber openness. Each producer uses a different system so each has to be identified separately.

Asbestos fiber is not sold in discreet lengths or mesh sizes as some other fibers and fillers but rather, each grade covers quite a spectrum of fiber lengths. The longer grades will always contain some very short fibers but the short grades will have none or very few long fibers. In order to put the grades in some kind of length perspective with regard to handling and use, Table 23-3 gives an approximate comparative length for the different major grade classifications.

Table 23-2. Grading of asbestos.

	QUEBEC SCREEN ANALYSIS, oz.			
Grade	½-In.	4 Mesh	10 Mesh	Pan
3K	7.0	7.0	1.5	0.5
3T	2.0	8.0	4.0	2.0
4D	0.0	7.0	6.0	3.0
4T	0.0	2.0	10.0	4.0
5D	0.0	0.5	10.5	5.0
5R	0.0	0.0	10.0	6.0
6D	0.0	0.0	7.0	9.0
7D	0.0	0.0	5.0	11.0
7R	0.0	0.0	0.0	16.0

Table 23-3. Approximate comparative lengths of asbestos grades.

Grade	Approximate Comparative Length, in.
2	5/8
3	1/2
4	3/16
5	1/8
6	1/16
7	1/32

Asbestos is a naturally occurring mineral and is mined by the usual techniques. The major mining is done by open pit methods but a considerable amount of asbestos is mined by underground techniques.

There are three modes of occurrence of asbestos. The most common is the cross-vein, in which a vein of fiber courses through the host serpentine rock with the fibers in parallel array essentially perpendicular to the rock faces. Another type is similar except the fibers are generally parallel to the enclosing rock structure and they are called slip fibers. Both types of occurrence can be present in the same deposit. Fiber lengths will generally vary from a few as long as 1½ in. to a large quantity in the short group 7 grade, with some deposits having very little of the longer grades. An economical deposit will generally have to have a minimum of 3–4% extractable fiber, and it is rare to find a deposit with over 6% usable fiber.

The third type of occurrence is quite rare and is called massive fiber. There are no veins of fiber but rather the whole deposit is fiber which is randomly oriented and gangue materials are interspersed in the ore body. Recoveries of 40–60% fiber are not uncommon from this type of deposit but the fiber is generally very short. Some of these deposits are large but the annual tonnage produced is relatively small.

For the conventional Canadian type ore, dry milling techniques are used almost exclusively for the separation of the fiber from the host rock. After blasting to obtain fragmented material, which can contain pieces up to 6 ft in diameter, the ore is fed through a primary crusher to reduce the particles to some smaller size. Screening and additional size reduction takes place down to about a 1 in. maximum size, which is fed into a dryer to remove surface moisture. At some point prior to the dryer, barren rock can usually be separated by size or magnetic methods to give a concentrated ore with higher fiber content. From the dryer, the ore goes to storage, which serves as a reservoir to feed the mill in case of a breakdown prior to that point, but also allows a distribution of the ore so that a more uniform quality of feed enters the mill.

The basic function of the mill is to separate the loose fiber from the rock, to process the fiber so it will meet specifications, and to package the final product. The loose fiber separation is accomplished by feeding the fiber and rock to a slightly inclined gyratory screen. As the material traverses the length of the screen, the fine rock and fiber go through the screen openings, the large rock migrates to the screen surface, and the loose, fluffy fiber floats on top where it can be sucked off by a vacuum hood at the end of the screen. The fiber goes to cyclone collectors and is fed to other parts of the mill for further processing.

The large ore particles are passed through hammer mills which are called fiberizers where more fiber is released. This material again goes over screens and the fiber is suctioned off to a cyclone collector. Many repetitions of this operation can take place.

Undersize material that goes through the screens may be processed further but the fiber content is generally low and of poor quality, and the material usually goes to a tailings pile.

Fortunately, the longer grades of fiber, which are in the widest veins, separate from the rock more readily so there is somewhat of an automatic classification in the mill. The longer grades come off first and the fiber recovered continues to get shorter and shorter.

The fiber from the collectors may go through additional screening devices to remove some of the fine material associated with it, and/or it may go to equipment which will separate or open the fiber more. Eventually it will go to mixers, where it is blended with other fiber to give a product with the desired characteristics.

Asbestos fiber is usually packaged in 100-

pound pressure packed bags. The density in the bag will range between 50-60 lb/cu ft. Paper, polyethylene film, and woven polypropylene are the most common bag materials. Special packaging is available when absolutely necessary.

Wet processing is also used to produce asbestos but only where there is a high fiber content, as with massive fiber, or after preliminary dry separation. The wet techniques used are screens, centrifugal cleaners, filtration units, dryers, and fiber reopeners.

A number of special treatments or types of processing are used for asbestos. Some consist of additives to improve the drainage characteristics in asbestos-cement production. Others will make the asbestos more thixotropic in systems such as polyester. Some special treatments have been applied to improve the physicals of asbestos filled plastics and to minimize the generation of airborne fines when handling asbestos. Special processing can consist of providing the fiber in pellet or flake form to contain the fines during handling, and/or to allow the asbestos to be handled more easily.

3. CHEMICAL AND PHYSICAL PROPERTIES AND FORMS OF USAGE IN PLASTICS

Since all types of asbestos are inorganic silicates, they have the usual good chemical, heat, and weathering resistance associated with these types of minerals. Chrysotile is a highly hydrated magnesium silicate and its structure actually consists of alternate layers of silica tetraheda and magnesium hydroxide. Therefore, it reacts readily with strong acids, leaving behind a network of silica. Chrysotile is little affected by bases but a strong caustic, like 70% potassium hydroxide at its boiling point, will attack this form of asbestos. Table 23-4 gives data on the resistance of the various types of asbestos to different acids and sodium hydroxide. There is no known data on the resistance of asbestos to solvents and other chemicals, but in its thousands of diversified uses there is no known evidence of any significant attack by these substances.

The resistance of asbestos to weathering is well known and it is one of the reasons for many of its uses. It is not affected by sun, oxidation, or ozone. It is essentially insoluble in water. Asbestos is often used because of its temperature resistance. One to two percent of absorbed water will be given off up to about 130°C. Above 200°C, the hydroxyl portion of asbestos will begin to be slowly evolved as water and this becomes very rapid for chrysotile above 600°C. This reaction is irreversible. The magnesium hydroxide is converted to magnesium oxide and the silica structure remains but the fiber loses strength and becomes brittle. At about 810°C chrysotile goes through a phase change to forsterite and fusion begins at 1520°C. Table 23-5 gives the weight loss for asbestos after heating to various temperatures and Table 23-6 shows the effect of temperature on the tensile strength of chrysotile.

Asbestos will adsorb water vapor under normal atmospheric conditions. In a humidity range of 50-70%, this will amount to about 1.5-2.0% and may approach 3% when the humidity is above 90%. Strangely enough, this

Table 23-4. Solubility of asbestos. Percent weight loss, refluxing 2 hours in 25% acid or caustic.

Fiber	ACID OR CAUSTIC				
Type	HCl	CH_3COOH	H_3PO_4	H_2SO_4	NaOH
Chrysotile	55.7	23.4	55.2	55.7	1.0
Crocidolite	4.4	0.9	4.4	3.7	1.3
Amosite	12.8	2.6	11.7	11.3	7.0
Anthophyllite	2.7	0.6	3.2	2.7	1.2
Tremolite	4.8	2.0	5.0	4.6	1.8
Actinolite	20.3	12.3	20.2	20.4	9.2

Table 23-5. % weight loss of asbestos after 2 hours at each temperature.

Temp. °C	Chrysotile	Crocidolite	Amosite	Anthophyllite	Tremolite
205	0.3	0.1	0.2	0.1	0.0
315	0.9	0.3	0.6	0.2	0.1
370	1.8	0.5	0.8	0.3	0.1
425	2.2	0.7	1.0	0.4	0.2
480	2.8	0.8	1.1	0.4	0.3
540	4.0	0.9	1.2	0.4	0.3
595	10.4	1.0	1.4	0.5	0.4
650	12.8	1.0	1.4	0.5	0.4
760	13.5	1.0	1.4	0.5	0.5

moisture adsorption does not appear to have any effect on the use of asbestos in plastics. It has been suspected many times but tests with pre-dried fiber have never had any effect on processing or product properties. Readsorption of water is instantaneous, so the experiment has to be made in a completely dry atmosphere or at an elevated temperature.

When chrysotile is wet by water, some of the magnesium hydroxide dissolves and it becomes quite caustic. Rather bad corrosion can take place on aluminum or other materials that cannot withstand caustic attack. A number of physical properties for the commercial types of asbestos are shown in Table 23-7.

The strength properties of asbestos, such as tensile strength and modulus of elasticity, are very good. Because of the small diameter of the ultimate fibrils of asbestos, it is difficult to get good measurements of the strength properties, so there is quite a range reported in the literature.

Contrary to some thinking, asbestos, and especially chrysotile, is not a very hard material and usually does not cause much wear to processing equipment. In some cases, a hard material such as quartz and some magnetite may be present and cause wear, but often these materials are quite fine and act more like polishing media.

One distinct advantage of chrysotile and crocidolite over other inorganic fibers is their degree of flexibility. They can be processed quite extensively without much deterioration of fiber length. However, the fiber bundles are subdivided more and more until they become microscopic in size and it appears that the fiber has been destroyed. A look at a fractured surface with a scanning electron microscope will show long fibers but of extremely small diameter. Figure 23-7 is a scanning electron microphotograph of the fracture surface of an asbestos filled PVC sample. Bundles and individual fibrils of asbestos can be seen.

Table 23-6. Tensile strength of chrysotile after 3 minutes at each temperature.

Temperature	Tensile Strength, psi	% Loss in Strength
Room	131,000	–
315 °C	120,000	8.4
425 °C	96,000	26.7
540 °C	78,000	40.5
650 °C	42,000	68.0

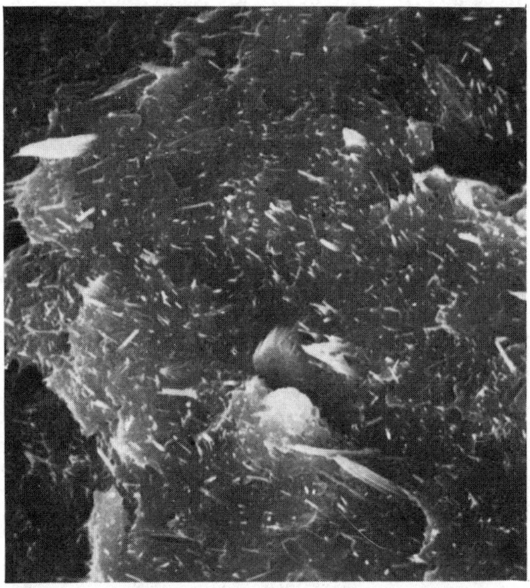

Fig. 23-7. Fracture surface of asbestos filled PVC 2000X, bundles and fibrils of asbestos visible.

Table 23-7. Physical properties of asbestos.

Property	Chrysotile	Crocidolite	Amosite	Anthophyllite
Color	White to gray	Blue	Brown	Brown to gray
Tensile strength, psi	300,000	500,000	160,000	350,000
Modulus of elasticity, psi	23.2×10^6	27.1×10^6	23.6×10^6	22.5×10^6
Hardness, Mohs	2.5–4.0	4.0	5.5–6.0	5.5–6.0
Flexibility	Good	Fair	Poor	Poor
Specific gravity	2.4–2.6	3.2–3.3	3.1–3.2	2.9–3.2
Specific heat, Btu/lb/°F	0.266	0.210	0.193	0.210
pH	10.3	9.1	9.1	9.4
Refractive index	1.50–1.55	1.70	1.64	1.61
Fibril, diameter, Å	160–300	600–900	600–900	600–900
Surface area BET m^2/g	17–60	9–10.5	8–9	6–7
Coefficient of cubical expansion, °F	5×10^{-5}	—	—	—
Charge in water	Positive	Negative	Negative	Negative
Isoelectric point	11.3–11.8	—	—	—

Asbestos is available in many forms and all of these are used to reinforce plastics. The strongest reinforced products are obtained by impregnating an asbestos paper or cloth with a resin incorporated as the binder during the manufacture, and resin contents up to 70% can be obtained this way. Impregnation of paper is somewhat difficult because the standard commercial grades of asbestos paper are not very porous. Special papers with higher permeability are usually made for this purpose. Chrysotile and crocidolite are the only asbestos fibers with any significant use in paper and cloth. Table 23-8 gives data on the mechanical properties of a few laminate systems.

Another way to improve the porosity of an asbestos sheet is to make it as a dry felt. This will impregnate very readily and can be made into a laminate. Although yarns are available, they are only used with plastics after being woven into a cloth or tape.

The largest quantity of asbestos is used in the plastics industry in bulk form. The fiber is added to the resin along with the other parts of the formulation to make a premix which can then be molded or extruded into the final part.

Much of the literature giving data on asbestos in plastics does not identify the form in which the asbestos is used so the true meaning of the data is obscure.

4. SUPPLIERS

Asbestos is available to the plastics industry from a number of major producers who supply direct to their major customers; smaller users are usually supplied through distributors. The major producers of fiber for the plastics industry are given in Table 23-9.

5. PRICE RANGE AND PRICE HISTORY

Asbestos has historically been sold by the producers on a F.O.B. basis with the customer paying the freight. The cost of shipping will vary with location, type of shipment, size of ship-

Table 23-8. Mechanical properties of pressure laminates reinforced with asbestos.

Laminate	Tensile Strength psi $\times 10^3$	Flexural Strength psi $\times 10^3$	Tensile Modulus psi $\times 10^6$
Phenolic resin/Crocidolite felt	28–31	57–58	3.25
Epoxy resin/Crocidolite felt	20–22	42–43	2.80
Polyester resin/Crocidolite felt	24–25	44–45	3.06
Melamine-Formaldehyde/Chrysotile fabric or paper	6–12	12–24	1.6–2.2

From *Materials in Design Engineering* (Ref. 1).

Table 23-9.

Company	Address	Type of Asbestos
Asbestos Corp., Ltd.	Suite 1940 Sun Life Building Montreal, Quebec	All grades of chrysotile
Bell Asbestos Mines, Ltd.	P.O. Box 99 Thetford Mines Quebec	All grades of chrysotile
Johns-Manville Corp.	Asbestos Fiber Sales Dept. P.O. Box 5108 Denver, Colo. 80217	All grades of chrysotile
Carey Canadian Mines, Ltd.	East Broughton, Quebec	All grades of chrysotile
Lake Asbestos of Quebec	Black Lake Quebec	All grades of chrysotile
Cassiar Asbestos Corp. Ltd.	85 Richmond St. W. Toronto, Ont.	Longer grades of chrysotile
North American Asbestos Corp.	200 S. Michigan Av. Chicago, Ill. 60604	All grades of amosite
Union Carbide Corp.	Mining and Metals Division 270 Park Av. New York, N.Y. 10017	Short "Calidra" chrysotile
Hedman Mines, Ltd.	P.O. Box 590 Timmins, Ont.	Short chrysotile
Special Asbestos Co.	3628 W. Pierce St. Milwaukee, Wis. 53215	All grades of crocidolite and amosite
Huxley Development Corp.	Time Life Building Rockefeller Center New York, N.Y. 10020	Short anthophyllite
Atlas Corp.	Coalings, Calif.	Short chrysotile
Vermont Asbestos Group	Hyde Park, Vt.	All grades of chrysotile

ment, and whether or not a commodity rate has been established. Average shipping costs for truckloads with established commodity rates will range from 1.5–2.5¢/lb.

The selling price of asbestos fiber is related to the length. All of the major producers sell at approximately the same price. A listing of the range within the various number classifications of the Quebec Screen Test is given in Table 23-10.

Asbestos fiber prices were very stable during 1960–1961 but showed about a 10% increase from 1965–1970. Since then, a world-wide shortage of asbestos has occurred and prices have increased significantly, with about a 50% increase in the price of a typical short grade used in plastics. Some of the longer grades, which are in shorter supply, have had an overall price increase of about 75%. These increases were brought about by the very small increases prior to 1970 and extremely large increases in the cost of mining and milling asbestos. A plot depicting price increases is shown in Fig. 23-8.

The world-wide shortage of asbestos is expected to continue. It will be particularly acute for the longer grades, such as Groups 4, 5, and 6, used extensively in the asbestos cement industry. The supply of fibers in Group 7 is ex-

Table 23-10. Asbestos fiber prices FOB Mine Location, May, 1975.

Grade	Price Range, ¢ per lb[a]
3	36.7–60
4	20.5–34.2
5	13.5–15.55
6	9.8
7	4.45–8.1

[a]The range of prices generally reflects different fiber lengths within a grade, but may reflect special processing or treatments.

pected to be better. There are two reasons for the long-range shortage. The world-wide demand for asbestos continues to grow. Conversely, the supply stays quite constant because new production essentially equals that which is lost when mines are closed because of depletion of the ore body. Price increases are expected to continue at the rate of 10-20% annually for the next few years because of increasing costs and the supply-demand picture. Any predictions beyond that time period would be very uncertain.

6. PROCESSING ASBESTOS WITH PLASTICS AND ADVANTAGES AND DISADVANTAGES OF USE

When asbestos is dry blended with a resin powder, a mix is obtained with poor flow properties and a low bulk density. This makes it very difficult to feed directly into an extruder, injection molder, compression mold, or any other type of forming device to obtain a good finished product. It is, therefore, common practice to precompound the asbestos with the resin.

One method used with thermoplastics is to dry blend, extrude, and pelletize or flake the product. This becomes the molding compound to be used in extruders or molding equipment. It is imperative that the pelletizing extruder be well vented so the pellets or flakes are essentially non-porous. Another technique is to blend the resin and asbestos in a Banbury and flake the product, which then is used for final production.

With powder thermoset resins, a relatively dry mix is made with the resin, asbestos, and a minimum of solvent. This is made into a preform which is dried and then compression molded in a hot press. It is also possible to make pellets of this "dry" mix and use them as a molding material directly in the mold after drying.

Liquid thermoset resins are mixed with asbestos in any common type of high viscosity mixer. It does not have to be high shear like a Banbury. The premix can be either extruded into plugs or used in chunks to fill a compression mold.

As with most materials, asbestos has some distinct advantages and disadvantages when used in plastics. Some of these are insignificant

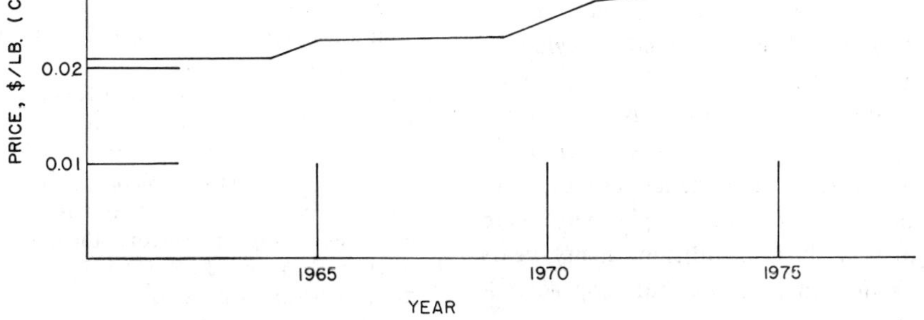

Fig. 23-8. Pricing pattern for typical filler grade asbestos F.O.B. mine location.

Table 23-11. Asbestos as a reinforcing fiber or filler.

Advantages	Disadvantages
High modulus	Lower impact strength
Good flexural strength	Increase in resin viscosity
Improved tensile strength	High density
Low thermal coefficient	Dark color
Improved heat deflection	Processing problems
Dimensional stability	PVC and PP heat stability
Heat resistance	
Low water absorption	
Easy flash removal	
Low cost	
Flow control in mold	
Minimum fiber degradation	
Good surface finish	
Chemical resistance	

in some systems and uses, but become very important in others. Table 23-11 lists these.

With thermoplastics the primary advantages gained with the use of asbestos are very significant increases in modulus and heat deflection temperature and a marked decrease in thermal coefficient. Modulus and flexural strength show marked increases when asbestos is used in thermosets. One of the major disadvantages with the use of asbestos is a loss of impact strength, especially with thermoplastics. However, by special treatments of the fiber or formula modifications, this effect can be eliminated or minimized (Ref. 2). Table 23-12 gives some values depicting the effect of asbestos additions on the physicals of typical thermoplastic and thermoset resins.

7. SAFETY, HEALTH AND THE FUTURE UTILIZATION OF ASBESTOS IN PLASTICS

Asbestos can be a hazardous material. It is known to cause asbestosis, bronchial cancer, and mesothelioma. However, medical evidence shows that generally, the health problems associated with exposure to airborne asbestos occur after long time exposure to excessive amounts of asbestos dust. This evidence comes from epidemiological studies of people who were exposed to very dusty conditions many years ago, and from animal studies utilizing accredited tests and exposures.

A Report of the Advisory Committee on Asbestos Cancers to the Directory of the International Agency for Research on Cancer at Lyon, France, October 6, 1972, states "The evidence of an exposure-response relationship based in part on past dust measurements and in part on the type of job within the industry suggests that an excess lung carcinoma risk is not detectable when the occupational exposure has been low. These low occupational exposures have almost certainly been much greater than that to the public from general air pollution."

A standard for *Exposure to Asbestos Dust* was published in the Federal Register, Volume 37 No. 110 on June 7, 1972. It set the present Occupational Safety and Health Administration Threshold Limit for airborne concentrations of asbestos in the work place as 5 fiber/ml longer than 5 μm time-weighted average or 10 fiber/ml longer than 5 μm ceiling concentration, both measured by the membrane filter/personal air sampler technique. These levels are being attained by the asbestos-using industry with good housekeeping and proper dust control practices, and in many cases much lower levels are reached. A new Threshold Limit of two fibers per ml became effective on July 1, 1976.

Asbestos will continue to be used in plastics. A number of potential new users are always investigating the utilization of asbestos in their new or old products. The asbestos industry will assist in providing information on proper handling systems so asbestos can be used safely.

Pertinent Literature and Patents

1. Cryor, R. E. "Asbestos Reinforced Plastics Resist Heat and Chemicals," *Materials in Design Engineering* (April 1966).
2. Axelson, J. W., "Asbestos in Plastics," *Polymer-Plast. Technol. Eng.*, 4:/(1), 93–109 (1975). A review of asbestos and its use in plastics with data on significant research including fiber treatments to improve impact strength.
3. Axelson, J. W., "Asbestos as a Reinforcement and Filler in Plastics," *Fillers and Reinforcements for Plastics, Advances in Chemistry Series 134,* American Chemical Society, Washington, D.C. pp. 16–28, 1974. Asbestos use in plastics is reviewed with emphasis on ways to overcome the difficulties.
4. Ancker, F. H., Azrak, R. G., and Berlolucci, M. D., "Coupled Chrysotile Asbestos Reinforced Thermoplastics," Union Carbide and Plastics Div., Bound

Table 23-12. Physical properties of asbestos filled plastics.

ASBESTOS WITH THERMOPLASTICS

% Asbestos	Flex Strength psi	Flex Modulus psi X 10^5	Tensile Strength psi X 10^3	Notched Izod ft lb/in.	Heat Distortion °F (264 psi)
		Polyethylene			
0	4170	1.54	3260	0.9	120
20	4560	2.82	2440	0.4	131
30	5840	4.21	3380	0.6	157
40	5440	6.44	3900	0.9	192
		ABS			
40–Long fiber J-M PLASTIBEST No. 20	10650	0.54	5855	1.42	190
40–Short fiber JM7D04	8765	1.11	4715	0.52	202
		Polystyrene			
0	5195	2.91	2585	1.42	179
40–Long fiber J-M PLASTIBEST No. 20	5180	6.24	3570	1.09	201
40–Short fiber J-M7D04	7410	5.97	2500	0.56	194
		Polypropylene			
27-Short asbestos UC RG144	10400	4.60	5900		187
27-Short asbestos UC RG600[a]	12900	9.92	7300		253
		Nylon			
0	12000	2.90	8400	1.31	165
15-RG144	22000	7.54	12400	0.94	304
15-RG600[a]	24800	8.34	14000	1.01	327
		Asbestos with Thermosets Phenolics			
0	8900	7.8	—	0.28	—
40	13500	14.4	—	0.20	—
60	12900	20.3	—	0.39	—

[a]RG600 has a special coupling treatment.

Brook, N. J. (Presented at SPE Antex, San Francisco, 1974). Describes a new treated asbestos to enhance the physical properties of thermoplastics.

5. Noga, E. A., and Woodhams, R. T., "Asbestos Reinforced Thermoplastics," *SPE Journal* **26**:/(9) 23–28 (Sept. 1970). Reviews the theoretical reinforcement to be expected with asbestos and how experimental results compare.

6. Xanthos, M., and Woodhams, R. T., "Polymer Encapsulation of Colloidal Asbestos Fibrils," *J. Appl. Poly. Sci.* **16**:/381–394 (1972). Free radical polymerization of highly opened and dispersed asbestos gives encapsulated fibrils to be used in composites.

ASTM Test Methods

Standard ASTM test methods for plastics are used for asbestos filled plastics. ASTM Subcommittee D30.03 is responsible for activity concerned with asbestos fiber and has written a number of test methods for fiber testing.

24
Micro and Short Fibers

John V. Milewski
Los Alamos Scientific Lab.
Los Alamos, New Mexico

CONTENTS

† Certainteed Products Corp.
Lionville, Pennsylvania

†† Alcoa Laboratories
St. Louis, Missouri

This chapter will review four microfibers: Franklin fibers, manufactured by Certain-teed Products Corp.; Processed Mineral Fiber manufactured by Jim Walters Resources, Inc.; Dawsonite, manufactured by Alcoa; and Fybex, manufactured by DuPont. The Franklin Fiber and Processed Mineral Fiber are both just emerging from the pilot plant stage and are now commercially available. Dawsonite is being held off the market pending health tests. Fybex was marketed for a few years and has been withdrawn from the market because of economic and marketing problems. Data on these latter two fibers is included to give the reader a better understanding of the microfiber field, because even if these fibers never become available, new microfibers of similar nature may be developed and, therefore, the data and processing information presented in this chapter can be expected to be applicable.

This chapter is compromised of four minichapters. They will be discussed in this order:

Processed Mineral Fiber	Dawsonite
Franklin Fiber	Fybex

PROCESSED MINERAL FIBER*

1. INTRODUCTION

Processed Mineral Fiber (PMF) is obtained from mineral wool (fiberized blast furnace slag) by a

*This section written by J. V. Milewski.

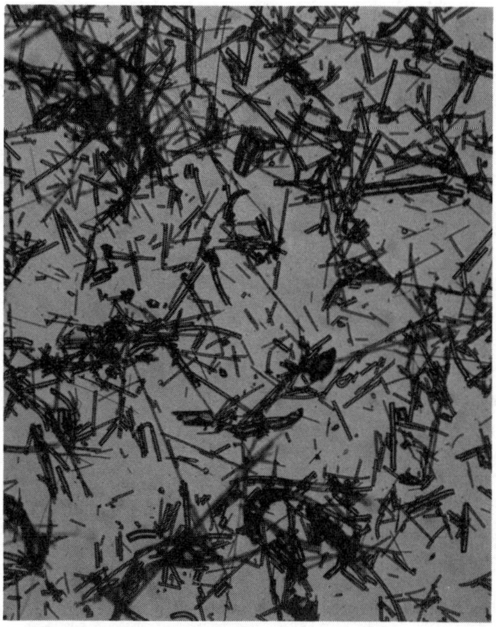

Fig. 24-1. PMF fiber illustrating typical fiber size and aspect ratio distribution–63X.

unique processing and classification technique. Its composition is basically 75% calcium silicate with the balance consisting of oxides of aluminum, magnesium, and other light weight metals.

Processed Mineral Fiber is a white to light gray, free-flowing, short fiber. Its diameter ranges from 1-10 μ (average 5 μ) and length averages 275 μ. The manufacturer gives an average range for the average aspect ratio of 40-60, as illustrated in Fig. 24-1. It is chemically stable and has a safety classification as "Inert Mineral Dust", in the same class as calcium carbonate and most clays. For this reason, it has been recommended for many applications as a replacement for asbestos and is useful in both thermoplastics and thermosetting plastics as a fibrous filler.

2. CHEMICAL COMPOSITION

Typical Analysis	% by wt.
SiO_2	42.1
Al_2O_3	8.1
CaO	35.4
MgO	7.8
Other Inorganics	6.6
	100.0

3. PHYSICAL PROPERTIES

Color	white to light gray
Fiber size	
diameter	range 1-10 μ; average 2-5 μ
length	average 275 μ
aspect ratio	average 40-60
Nonfibrous material	4%
Bulk density	25 lbs/ft^3
Specific gravity	2.7
Flexibility	brittle
Flammability	does not burn
Tensile strength (single fiber)	3×10^3 to 2×10^5 psi average 7×10^4 psi
Modulus of elasticity	15×10^6 psi
Heat resistance	good to 1400°F
Glass transition temperature, T_g	1400°F
Devitrification temperature, T_{dev}	1560°F
Melting point, T_{fus}	2300-2400°F
Material safety category	"Inert Mineral Dust"

4. SUPPLIER, GRADES, AND PRICES

PMF is available in pilot-scale quantities for experimentation and evaluation, from:

Jim Walter Resources, Inc.
3300 First Avenue, North
Birmingham, Ala. 35202
Telephone (205) 254-7000

Tentative prices are listed as:

Quantities, 1-50 lb @ \$0.21 lb
Car load or truck load @ \$0.15 lb

5. PROPOSED APPLICATIONS AND DATA

Recent tests at the Jim Walter Resources, Inc. Laboratory and by others indicate that PMF is useful as a fibrous filler in both thermoplastics and thermosetting plastic compositions. Also, extensive evaluation of PMF is underway in rubber. Other uses are being considered for applications in adhesives, sealants, caulks, and reinforcements in thermoplastic elastomers.

PMF has a low binder demand in liquid

Table 24-1. Physical and thermal properties of polypropylene filled with PMF

Property	LOADING (wt. %) 0	33	50
Tensile strength (psi)	4,810	4,520	4,070
1% Secant modulus (psi)	192,000	270,000	295,000
Izod impact strength (ft-lb/in. notch)	0.45	0.63	0.69
Flexural modulus (psi)	160,000	550,000	811,000
Heat distortion temperature (°F) (@66 psi)	236	266	268

Resin: Shell P-520 (general-purpose grade).
PMF: L/D = 58
Coal, Iron and Chemicals Division–A Jim Walter Company.

Table 24-2. Effect of aspect ratio of PMF in polypropylene.

Property	ASPECT RATIO 16	26	58
Tensile strength (psi)	4,150	4,480	4,520
1% Secant modulus (psi)	238,000	261,000	270,000
Izod impact strength (ft-lb/in. notch)	0.62	0.64	0.63
Flexural modulus (psi)	425,000	582,000	550,000
Heat distortion temperature (°F) (@66 psi)	253	261	266

Resin: Shell P-520 (general-purpose grade), 33 wt % loading.

Table 24-3. Comparison of milled glass fibers and PMF in polypropylene.

Property	PMF	Milled Glass Starch Coating	Milled Glass Heat Cleaned
Tensile strength (psi)	4,520	4,640	4,330
1% Secant modulus (psi)	270,000	295,000	258,000
Izod impact strength (ft-lb/in. notch)	0.63	0.74	0.72
Flexural modulus (psi)	550,000	678,000	557,000
Heat distortion temperature (°F) (@66 psi)	268	271	261

Resin: Shell P-520 (general purpose grade), 33 wt % loading.
Milled glass: OC-630, 1/32 in.
PMF: L/D = 58
Coal, Iron and Chemicals Division–A Jim Walter Company.

Table 24-4. Physical properties of a high impact polystyrene filled with PMF

	FIBER LOADINGS (phr)			
	0	50	100	ASTM Method
Izod impact strength (ft-lb/in. notch)	1.29	0.58	0.38	D-256
Tensile strength @0.2 in./min. (psi)	3,900	4,400	4,100	D-638
Heat deflection temperature @66 psi (°F)	181	196	200	D-648
Flexural modulus (psi)	297,000	750,000	1,200,000	D-790

Polystyrene: Shell-324
Processed mineral fiber: Aspect ratio = 59

plastics formulations and can be incorporated in plastics by methods now employed for other short fiber materials.

Data on PMF/plastic composites is compiled in the following tables. The first three tables contain data on PMF in polypropylene. Table 24-1 illustrates the improvement in secant and flexural modulus, Izod impact, and heat distortion temperature with increasing addition of PMF. Table 24-2 relates the effect of fiber aspect ratio on properties, while Table 24-3 compares some PMF composite properties to that of milled glass, starch coated and heat cleaned.

Table 24-4 gives data on PMF in Shells' polystyrene, illustrating significant improvement in flexural modulus with 50 and 100 phr of PMF. Heat deflection temperature and tensile strength are also somewhat improved, but impact strength drops substantially.

Because of the new and experimental nature of this fiber, little data is available. No doubt, with increased experimentation and evaluation of coupling agents, greater strength translation will be possible in the future.

FRANKLIN FIBER*

6. INTRODUCTION

Franklin Fiber is a trademark of Certain-teed Products Corporation for fibrous crystalline calcium sulfate in the hemihydrate, gamma anhydrite or beta anhydrite chemical form. Since the beta anhydrite form, which is commonly known as the *dead burnt* form, has a much greater resistance to hydration, it is the preferred form. Hereinafter, in describing the character and usage of Franklin Fiber, reference will be specially directed to the beta anhydrite, calcium sulfate anhydriate, form of Franklin Fiber.

Franklin Fiber has a white, lustrous appearance with a silky, soft feel. Microscopic examination shows fibrous crystals with an average length of 80 μ and a very uniform diameter of about 2 μ. (See Fig. 24-2.)

Franklin Fiber is chemically stable. It has low water solubility and exhibits no abnormal toxic effects in handling or production. Franklin Fiber has high temperature resistance and possesses good insulating qualities.

Fig. 24-2. Typical Franklin Fiber–500X.

*This section written by J.G. Winslow.

When used as a reinforcing filler in plastics, Franklin Fiber significantly improves the tensile strength and, in most cases, dramatically increases the modulus of elasticity. With improved tensile and flexural behavior, there is often a decrease in impact resistance. These strength characteristics are improved with suitable coupling and dispersing agents.

Besides its proven use in thermosetting and thermoplastic resins, Franklin Fiber has potential in such diversified matrixes as asphalt, mineral cements, paper, and paints.

The cost of producing Franklin Fiber, which is now available in development quantities, is between \$0.10–\$0.12/lb.

7. METHODS FOR PRODUCING THE MATERIAL

The formation of Franklin Fiber involves a crystal restructuring process in an aqueous environment at elevated pressure and temperature. Franklin Fiber is presently produced using a single batch reactor system. A water slurry of gypsum is first preheated and then pumped to a steam-jacketed reactor. The slurry is constantly agitated in the reactor to keep the gypsum in suspension as it is brought up to the reaction temperature and pressure. The reaction is completed within a few minutes. The reaction product is gravity-fed to a holding tank. From the holding tank, the now rather thick slurry is fed to a filter press that removes most of the water. The resulting fibrous cake from the filter press is broken up and heated in an oven to remove the remaining surface moisture. Thereafter, the fibrous cake is calcined to form the dead burnt, preferred form of Franklin Fiber. A continuous reaction method which is now under study should reduce production costs considerably.

8. SUMMARY OF PROPERTIES OF FRANKLIN FIBER

A. *Chemical*. Franklin Fiber has the same relatively inert chemical properties as normal orthorhombic beta calcium sulfate anhydrite.

Table 24-5. Weight loss data for Franklin Fiber and asbestos.

Temperature °F	PERCENTAGE OF WEIGHT LOSS (AFTER 2 HOURS) Franklin Fiber	Asbestos
800	0.40	0.30
1000	0.52	3.99
1400	0.65	13.43
1800	0.75	13.77

Its solubility in water is less than 1200 part/million at 22°C.

B. *Physical.* Franklin Fiber is an acicular unicrystal with tapered ends. The lengths of the individual fibers can be as long as 200 μ but average out to around 80 μ. The diameter or width of the fiber is a fairly constant 2 μ. With certain additives, the width of the fibers can be increased slightly which is helpful at times for better dispersion.

The true density of Franklin Fiber is 3 g/cm^3 and its bulk density is around 10 lb/cu ft.

The Franklin Fiber has a smooth, silky feel and when pressed has a tendency to mat.

C. *Optical.* Franklin Fiber is a lustrous white. If waste system gypsums are used in its manufacture, the impurity may impart color to the final product. The index of refraction of Franklin Fiber is 1.585.

D. *Thermal.* Temperature resistance of the Franklin Fiber is excellent, even superior to that of asbestos, as indicated by Table 24-5. Thermal conductivity is comparable, in equal density loading, to calcium silicate.

E. *Miscellaneous.* When tested on animals, under controlled laboratory conditions, Franklin Fiber showed no toxic effects.

9. MATERIAL SUPPLIER, COST AND AVAILABILITY

Franklin Fiber is available in pilot plant quantities from Certain-teed Corporation, Corporate R&D, P.O. Box 860, Valley Forge, Pa., 19482. Phone: (215) 678-5000. Its estimated produc-

Table 24-6. Dry physical properties of 30 wt. % Franklin Fiber reinforced nylon-6.

Properties	Test Method ASTM	Unit	Filled	Unfilled
Tensile strength	D638	kg/cm^2	835	635
Tensile modulus	D638	kg/cm^2	4.5×10^4	2.3×10^4
Tensile elongation at break	D638	%	2.3	182
Flexural strength	D790	kg/cm^2	1208	879
Flexural modulus	D790	kg/cm^2	4.0×10^4	2.2×10^4
Izod impact strength notched	D256	kg.cm/cm	3.4	4.72
Heat distortion temperature (18.6 kg/cm^2)	D648	°C	90.5	65.0

tion cost in pilot plant operation is $.10–.12/lb. Larger production will result in lower prices.

10. CURRENT APPLICATIONS

Franklin Fiber shows great promise as a reinforcing filler in many diverse mediums. It is, however, in plastics that a definite niche has been established for Franklin Fiber. This is due in part to a projected ever-increasing scarcity of fossil-based resins, abundant supplies, now and in the future, of Franklin Fiber's raw material, namely, gypsum, the strength Franklin Fiber imparts to the resins, the excellent thermal properties Franklin Fiber endows to plastic products, and the non-abrasive nature of Franklin Fibers on plastic molding equipment and its low toxic effects as a fibrous reinforcement as compared to other micro reinforcements.

Table 24-6 gives an overall performance of a thermoplastic resin with and without the inclusion of Franklin Fiber. An improvement in the impact properties should be possible with a suitable fiber coating.

A comparison study between Franklin Fiber and another mineral fiber, in both thermoplastic and thermosetting resins, was made and the results as shown in Table 24-7 are clearly favorable to Franklin Fiber.

To demonstrate that the beneficial effect of Franklin Fiber is due to its unique fibrous structure, a series of tests were undertaken using a calcium sulfate anhydrite in an orthorhombic powder form and Franklin Fiber as the reinforcement in acrylic films. The results, as shown in Table 24-8, dramatically indicate that, while the former adds no improvement over the unfilled control, an 18% loading with Franklin Fiber nearly doubles the tensile strength and increases the modulus fivefold.

Studies now underway with Franklin Fiber as a filler in asphalt have shown that the fiber has a decided effect on raising the asphalt's softening point (see Table 24-9). This effect of Franklin Fiber on asphalt has led to further exploratory work in its use in roof shingles, specifically those utilizing a glass fiber in place of organic based felt.

Because Franklin Fibers are highly resistant to heat degradation, they can be used as a reinforcement in metals, especially aluminum. The resulting aluminum composites are stronger

Table 24-7. Franklin Fiber versus Wollastonite: nylon 6,6-PBT (50 Wt. % Fiber)

	NYLON 6, 6		PBT	
	FF	Woll.	FF	Woll.
Flexural strength, psi	19,600	13,500	14,300	11,800
Tensile strength, psi	11,000	7,900	8,400	7,300
Heat deflection temperature, °F	469	396	385	289

Table 24-8. Properties of Franklin Fiber in acrylic films. (Rohm & Haas–F-10ES)

Material	Weight (%)	Tensile (psi)	Modulus (psi)
Control	0	330	1400
$CaSO_4$ Drying powder (Orthorhombic)	18	330	1300
Franklin Fiber	18	560	6900

Table 24-9. Properties of various calcium sulfate forms in asphalt.

Fillers	Percent by Weight Added	Softening Point (°F)
Control	0	104.5
Franklin Fibers	3.0	120.0
Franklin Fibers	5.6	128.5
Franklin Fibers	7.0	128.0
Franklin Fibers	18.8	163.5
Anhydrite powder	18.8	114.0
Gypsum	(Will not mix in hot asphalt)	

and stiffer, allowing them to be used more readily as a weight-bearing structural material.

11. FUTURE ECONOMICS

An undesirable by-product of many chemical processes is waste stream gypsum. In many cases, waste stream gypsum can be used for the production of Franklin Fiber rather than being discarded. This gives Franklin Fiber the potential of being a very low cost commodity reinforcement.

References

1. U.S. Patent 3,822,340, J. J. Eberl, et al. pp. 423–555.
2. "Inorganic Fibers Want to Fill Gaps in Insulation," *Chemical Week* (31) Apr. 17, 1974.
3. Hazleton Laboratories, Inc., *Toxicity Study-Calcium Sulfate Anhydrite Whiskers Fibers*–Final Report, 7/27/74.

DAWSONITE*

12. INTRODUCTION

Dawsonite, sodium aluminum hydroxycarbonate, in its synthetically manufactured acicular form, shows promise as a micro-fiber reinforcement for a variety of thermoplastics. Significant improvements in tensile and flexural strengths and moduli, higher heat deflection temperatures, and reduction in coefficient of thermal expansion are realized for thermoplastic composites containing dawsonite.

Over and above its reinforcing properties, dawsonite, by virtue of its endothermic decomposition at elevated temperatures into sodium aluminate, carbon dioxide, and water, provides flame retardant properties to composites.

13. METHODS OF PRODUCING DAWSONITE

The detailed process for producing dawsonite in crystalline fibers of controlled quality and size is proprietary to Alcoa. Bureau of Mines report RI-7664 describes a process for synthesis and characterization of dawsonite. This report states that dawsonite is made by a hydrothermal process in which the raw materials are digested, reacted, and precipitated. The raw materials are: sodium aluminate, sodium carbonate, and sodium hydroxide. These materials are reacted in solution at 175-200°C and 15 psig for 4 to 5 hours. After that period, the reaction vessel is cooled to 75°C and the water soluble carbonates are removed by three hot water, 70°C, rinses.

It appears reasonable to assume that the Alcoa process differs mainly in process controls and additives that in effect favor crystallization of acicular morphologies.

14. PHYSICAL AND CHEMICAL PROPERTIES OF DAWSONITE

Dawsonite, as recently manufactured by Aluminum Company of America (Alcoa) in pilot

*This section by P. V. Bonsignore (Aluminum Co. of America).

Table 24-10. Physical properties of Dawsonite microfiber crystals.

Chemical formula	$NaAl(OH)_2CO_3$
	Na_2O 21.53%
	Al_2O_3 35.40%
	CO_2 30.56%
	H_2O 12.51%
Diameter (D)	0.4–0.6 μm
Length (L)	15–20 μm
Aspect ratio (L/D)	25–40
Density	2.44 g/cm^3
Refractive index	1.53
BET N_2 surface area	15–17 m^2/g

plant quantities has the typical properties listed in Table 24-10.

An SEM photo of dawsonite microfiber crystals is shown in Fig. 24-3.

Of the physical properties listed in Table 24-10, those of special interest are the low density and the low refractive index compared to other microfiber reinforcements such as potassium titanate fibers. Low density (2.44 g/cm^3) means higher volume loadings at lower weight percent. Low refractive index (1.53) means low tinting strength in pigmented colored composites.

Dawsonite is a relatively stable compound toward most temperatures encountered in plastic processing operation. A thermogravimetric analysis (TGA) (Fig. 24-4) reveals the onset of a sharp weight loss occurring at about 300°C. This weight loss reflects the primary decomposition reaction:[1]

$$NaAl(OH)_2CO_3 \longrightarrow NaAlO_2 + CO_2 + H_2O$$

Accompanying this decomposition is a considerable absorption of heat as shown in the differential scanning calorimetry (DSC) curve in Figure 24-5. This endothermic decomposition at temperatures above about 300°C (575°F) accounts for dawsonite's ability to impart some degree of flame retardancy to plastics in which it is incorporated (see Section 17). Of additional interest is dawsonite's ability to offer some degree of smoke suppression and to act as an HCl scavenger for flame exposed PVC compounds.

15. MATERIAL SUPPLIER, COST, AND AVAILABILITY

Synthetic acicular dawsonite has been produced in pilot plant quantities in East St. Louis, Illinois, by the:

Aluminum Company of America (Alcoa)
1501 Alcoa Building
Pittsburgh, Pa. 15219
Phone: (412) 553-4545

Fig. 24-3. Dawsonite.

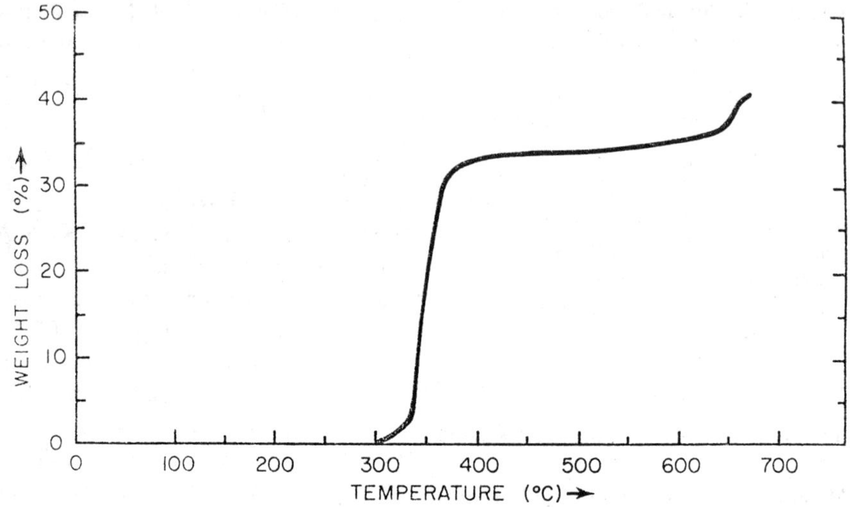

Fig. 24-4. Thermogravimetric analysis (TGA) of Dawsonite $NaAl(OH)_2Co_3$.

The status on availability is still in question because of continuing medical tests and pending regulations on allowable airborne fiber concentrations. The reader is advised to contact Alcoa directly for a report on the current availability and cost status of dawsonite.

WARNING: Breathing acicular dawsonite dust may cause serious bodily harm. Limited animal test data (mice) indicates potential tumorogenicity for surgically implanted dawsonite fibers. Acicular dawsonite must therefore be treated as carcinogenic.

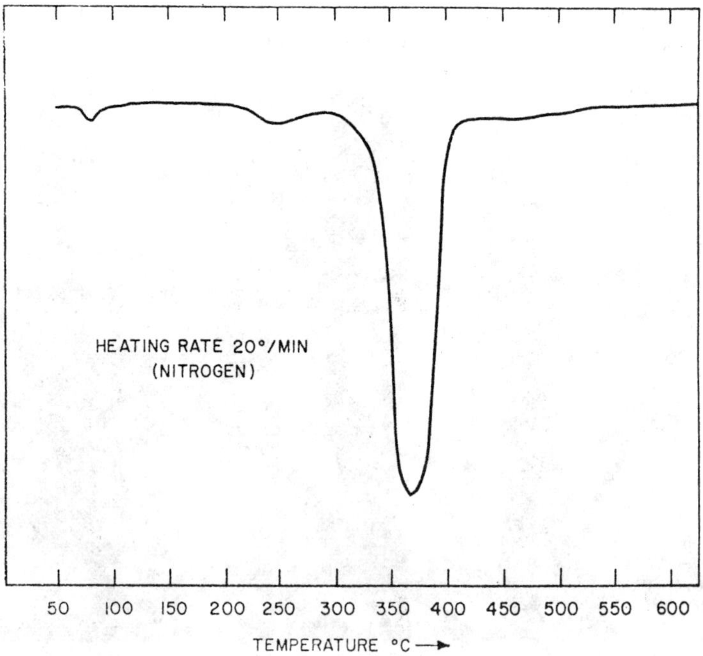

Fig. 24-5. DSC curve Dawsonite.

16. BEHAVIOR IN THERMOPLASTICS

16.1 Compounding

Compounding can have an important effect on the properties of a reinforced thermoplastic. The compounding operation should completely wet-out and uniformly disperse the fibers with a minimum reduction in fiber aspect ratio. The higher the compounded aspect ratio, the more efficient the reinforcement in improving resin properties. Fiber damage can be minimized by quickly melting the resin and wetting the fibers prior to applying compressive forces.[2,3]

Compounding of dawsonite in the respective thermoplastics was done by Banbury flux mixing. The thermoplastic was completely melted before the required amount of dawsonite was added. Levels of 0, 7.5, 15, and 30 wt % of dawsonite were studied in 3 different thermoplastics: polypropylene, rigid PVC, and ABS resin.

16.2 Physical Properties of Dawsonite/Thermoplastic Composites

16.2.1. Polypropylene. A typical injection moldable grade of polypropylene (PP), Hercules Profax 6523, was compounded in a Banbury mixer with dawsonite at levels of 0, 7.5, 15, and 30% by wt. The resultant compounds were then injection molded. Physical properties of these test samples are shown in Table 24-11.

16.2.2. Rigid PVC. The effect of dawsonite on the physical properties of a rigid PVC compound was likewise examined at 0, 7.5, 15, and 30% by wt loadings. The base PVC compound, Geon 87213, was supplied by B. F. Goodrich, who noted it to be a lead stabilized injection moldable compound based on Geon 80X6 PVC resin (0.65 IV). Banbury melt flux compounding was used, followed by sheeting, granulating, and injection molding of test specimens. Physical properties of rigid PVC/dawsonite composites are given in Table 24-12.

Rigid PVC, from the data of Table 24-12, appears to respond very favorably to reinforcement by dawsonite. Flexural modulus (stiffness) is increased some 300% while coefficient of linear thermal expansion is decreased to about 1/3 of its original value by incorporation of 30% dawsonite. For some applications of rigid PVC, such as vinyl siding, such increase in stiffness would mean thinner extrusion cross sections to achieve the same degree of stiffness. Similarly, the decrease in thermal expansion, and higher HDT, could permit easier mounting procedures while allowing wider latitude in dark color formulation for rigid PVC siding compounds. In common with other particulate

Table 24-11. Physical properties of Dawsonite/PP composites.

Property	ASTM Method	Units	DAWSONITE LEVEL (% BY WT) 0	7.5	15	30
Melt flow rate	D1238L	g/10 min	4.8	4.1	3.1	1.5
Deflection temperature	D648 @ 264 psi	°C	56	59	69	78
Coefficient of linear thermal expansion	D696	10^{-5} in/in/°C	7.3	6.1	5.0	4.1
Izod impact (notched)	D256A	ft-lbs/in notch	1.41	1.33	1.20	1.12
Izod impact (unnotched)	D256E	ft-lbs/in	10.3	7.9	4.9	3.5
Tensile strength	D638	psi	4290	4410	4515	4620
Tensile modulus	D638	10^5 psi	2.43	3.73	4.37	6.49
Flexural modulus	D747	10^5 psi	1.70	1.98	2.31	2.78

Table 24-12. Physical properties of Dawsonite/rigid PVC composites.

Property	ASTM Method	Units	DAWSONITE-LEVEL (% BY WT) 0	7.5	15	30
Melt flow rate	D1238-73F	g/10 min	4.7	3.5	3.2	2.7
Deflection temperature	D648 @ 264 psi	°F	152.6	155.3	159.8	163.4
Coefficient of linear thermal expansion	D696	10^{-5} in/in / °C	6.4	3.8	3.1	2.1
Izod impact (notched)	D256A	ft lbs/in (notch)	3.39	3.00	2.62	2.36
Izod impact (unnotched)	D256E	ft lbs/in	21.0	18.1	16.5	10.4
Tensile strength	D638	psi	5,647	5,900	6,100	6,200
Tensile modulus	D638	10^6 psi	0.41	0.54	1.43	1.60
Flexural strength	D790	psi	11,760	12,420	12,900	13,160
Flexural modulus	D790	10^6 psi	0.75	1.13	1.54	2.33

mineral fillers, impact resistance does decrease. The decrease in this particular PVC compound, however, does not appear to be particularly severe, about 3.4 unfilled as compared with 2.4 ft lb/in. with 30% by wt dawsonite, for notched Izod impact.

6.2.3. ABS Resin. The efficiency of dawsonite as a reinforcing agent for ABS resin at 0, 7.5, 15 and 30 wt % is given in Table 24-13. An electroplatable grade, Cycolac EP-3510 of Borg Warner Corporation was used. Test samples were injection molded from composites which had been melt fluxed in a Banbury mill.

In common with its behavior in PP and rigid PVC, dawsonite in ABS improves tensile and flexural strengths, increases HDT, and decreases thermal expansion.

Dawsonite addition should improve ABS's utility in a wide range of applications including electroplating. For electroplating durability, a plastic substrate should have a low coefficient

Table 24-13. Physical properties of Dawsonite/ABS composites.

Property	ASTM Method	Units	DAWSONITE LEVEL (% BY WT) 0	7.5	15	30
Melt flow rate	D1238G	g/10 min	2.18	1.43	1.17	0.83
Deflection temperature	D648 @ 264 psi	°F	178	182	189	196
Coefficient of linear thermal expansion	D696	10^{-5} in/in / °C	7.2	4.8	3.5	2.4
Izod impact (notched)	D256A	ft lbs/in (notch)	3.62	2.00	1.61	0.66
Izod impact (unnotched)	D256E	ft lbs/in	10.9	5.7	3.6	2.5
Tensile strength	D638	psi	5,928	6,466	6,667	6,848
Tensile modulus	D638	10^5 psi	3.44	6.47	7.16	8.24
Flexural strength	D790	psi	9,168	9,767	10,341	10,888
Flexural modulus	D790	10^5 psi	3.14	4.91	6.59	9.92

of thermal expansion, close to that of the deposited metal surface, to minimize thermal cycling problems such as bond failure.

17. FLAME RETARDANCY

Dawsonite decomposes at temperatures above 300°C with significant weight loss and heat absorption as previously indicated in Figs. 24-4 and 24-5.[1] This heat absorption can reduce the amount of heat available for pyrolyzing virgin plastic into combustible gaseous fuel elements. Dawsonite, therefore, behaves similarly to alumina trihydrate, $Al(OH)_3$, which has found significant utility as a flame retardant and smoke suppressive filler in reinforced polyester plastics.[4] The higher decomposition temperature of the dawsonite compared to alumina trihydrate (300-320°C vs 230-260°C) and its fibrous nature with attendant melt viscosity increases which limit loadings, are disadvantages for use as a primary fire retardant. It can be used as an adjunct to marginally flame retarded plastic systems with, however, the very desirable feature of improving char reinforcement.

17.1 Polypropylene

The response of PP (Hercules Profax 6523) to dawsonite was examined by the UL94 HB and ASTM D2863 oxygen index tests. The results are given in Table 24-14. Even with as high as 45% dawsonite in this grade of PP, oxygen index did not go much above 21.5. As little as 7.5% dawsonite did, however, prevent the separation of flaming drops.

18. CONCLUSIONS

Dawsonite microfibers can offer significant improvements in physical properties to a wide range of thermoplastics including polypropylene, rigid PVC, and ABS. These improvements include increased tensile and flexural strengths and moduli, higher heat deflection temperature, and reduced coefficient of linear thermal expansion. In addition to physical property improvements, a degree of improvement in flame retardance is noted for thermoplastic composites containing dawsonite.

ACKNOWLEDGMENTS

Grateful acknowledgment is extended to Mr. S. Mehta of Lowell Technological Institute who performed the extensive compounding, injection molding, and physical property evaluation under the technical guidance of Professors A. M. Crugnola and S. A. Orroth, Jr.

References

1. Huggins, C. W. and Green, T. E., "Thermal Decomposition of Dawsonite," *American Minerologist*, **58:** 548–550 (1973).
2. Speri, W. M. and Jenkins, C. F., *Polymer Eng. Sci.* **13:**(5), 409–414 (Nov. 1973).
3. Gras, D., *Plastics Technology*, pp. 40–43 (Feb. 1972).
4. Bonsignore, P. V. and Manhart, J. H., Proceedings of 29th Annual Technical Conference of the Reinforced Plastics/Composites Institute of the SPI, paper 23C, Feb. 1974.
5. Mathis, T. C. and Morgan, R. W., U.S. Patent 3,869,420 (to Monsanto), Mar. 4, 1975.

Table 24-14. Burning behavior of Dawsonite/PP composites.

Dawsonite (% by wt)	Dripping[a]	Smoke	Flame size	Rate of Burn[b] UL94 HB in./min	Oxygen index ASTM D2863
None	Flaming drips	Heavy black	Large	1.40, 1.33	18.5
7.5	N.D.	–	–	1.20	
15	N.D.	Less than 10	Medium	1.00, 1.03	20.0
30	N.D.	Much less than 1	Low	0.73, 0.87	20.5
45	N.D.	Almost none	Very low	0.47	21.5

[a] N.D. = Nondripping.
[b] All classified as slow burning, i.e., <1.5 in./min.

FYBEX*

19. INTRODUCTION

Fybex consists of single crystal, inorganic titanate microfibers with relatively high modulus and reasonably high strength, which qualifies them as reinforcements for plastics. Evaluation of these microfibers in a variety of resins, particulary themoplastics, show significant enhancement of mechanical, thermal, and physical properties.

Fybex, as a result of its small diameter, similar to other microfibers, has a specific advantage in compounding and molding, resulting from less fiber breakage during processing than is experienced with larger diameter fibers such as fiberglass. Moreover, parts molded with plastic which is reinforced with these fibers show surface smoothness and gloss equivalent to parts made with unfilled plastics, suggesting aesthetic as well as engineering applications.

Data are presented for reinforced thermoplastics. Typical results for a 30 wt. % loading of these fibers show twofold increase in tensile strength and fourfold increase in flexural modulus. Other advantages of Fybex microfibers reinforced composites are resistance to warp and moldability in thin sections.

20. MANUFACTURING PROCESS, PRODUCTS AND SUPPLIER

Fybex crystalline fibers are formed by a recrystallization process from a molten salt. They are formed by a process which blends TiO_2, KCl, and K_2CO_3 to yield $K_2Ti_4O_9$ after kilning to 900°C. Next, the $K_2Ti_4O_9$ is reprocessed by potassium leaching to form an octa-titanate, $K_2Ti_8O_{17}$. The product formed has the properties given in Table 24-15 and is illustrated in Fig. 24-6.

Fybex is a product of the DuPont Company and was available from E. I. DuPont De Nemours & Company (Inc.) Pigment Department Inorganic Fiber Division, Experimental Station, Wilmington, Delaware 19898. Marketing was suspended in 1975 pending further marketing studies and testing.

Table 24-15. Description of Fybex inorganic reinforcing titanate.

Diameter	0.10–0.16μ
Length: diameter	40:1
Density	3.2 g/cc
BET N_2 surface area	7–10 m^2g
Refractive index	2.35
Young's modulus	Estimated 40,000,000 psi
Tensile strength	Estimated >1,000,000 psi

21. COMPOUNDING TECHNIQUES

Most of the following information on compounding, properties, and application of Fybex has been obtained from a paper by Linsen and Regester.[1]

In compounding with Fybex and other microfibers, it is particulary important to use processing techniques that lessen the probability of breaking the fibers. Breakage usually occurs if the fibers are subjected to high compressive or mixing forces before the polymer is in the molten state. It is therefore necessary to allow

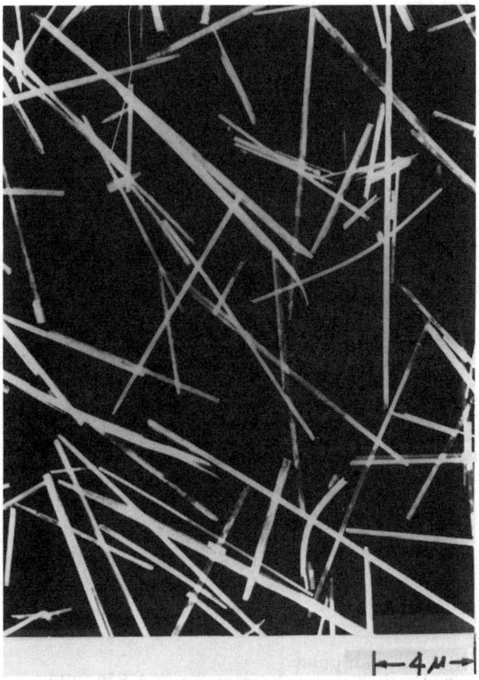

Fig. 24-6. Fybex. (*Courtesy E. I. Dupont*).

*This section by J. V. Milewski.

the polymer to soften before dispersion work is applied. This is best accomplished by premelting the polymer and then adding fiber, or heating the polymer/fiber blend while gradually increasing shear.

Prefluxing the polymer and then adding fiber is recommended when processing in a Banbury or a rubber (two-roll) mill. In extruders, hot, long feed sections should be used to heat the Fybex/polymer blend before high shear is applied. Another acceptable extrusion technique is to add only polymer in the feed section with fiber addition through downstream ports. This is particularly advantageous in twin-screw extruders. For additional information on processing resins with fibers and fillers (see Chapter 32).

In general, molding conditions for plastics reinforced with Fybex are similar to those for unreinforced counterparts.[2,3] Slightly higher temperatures are required because of the increased viscosity of the reinforced plastic. Melt toughness is increased and setup time decreased, resulting in faster cycles than with unreinforced resins.

22. PHYSICAL PROPERTIES OF "FYBEX" REINFORCED PLASTICS

In general, property improvement with Fybex in thermoplastics are comparable to those with glass fibers at equal weight loadings. The stiffness of plastics reinforced with Fybex is generally higher than glass fiber composites because the elastic moduli of the titanates are more than four times that of glass. Fybex has been particularly effective in reducing the coefficient of thermal expansion.

The effect of "Fybex" reinforcement on substrate expansion coefficient of ABS and polypropylene is shown in Figs. 24-7 and 24-8. The expansion coefficient changes most rapidly at low fiber loadings. High stiffness is achieved for composites with the preferred 10-15 wt. % fiber loading. This is particularly important for polypropylene, where the low modulus of the unreinforced resin can cause handling problems and distortion in hot applications. Fybex loadings up to 30 wt. % give parts which combine superior engineering properties with high quality appearance and performance. Thin parts with ultra-high stiffness are thus possible.

The aspect ratio of Fybex is less than the starting aspect ratio of short glass fibers used to reinforce thermoplastics; therefore, the strength of plastics reinforced with Fybex is generally slightly less than glass-reinforced plastics. Plastics reinforced with Fybex can be further strengthened by using coupling agents to improve the titanate to plastic bond. Impact strengths of the plastics reinforced with Fybex are generally at the level of the unfilled base plastic. Notch sensitivity appears to depend on the characteristics of the base resin, and Fybex

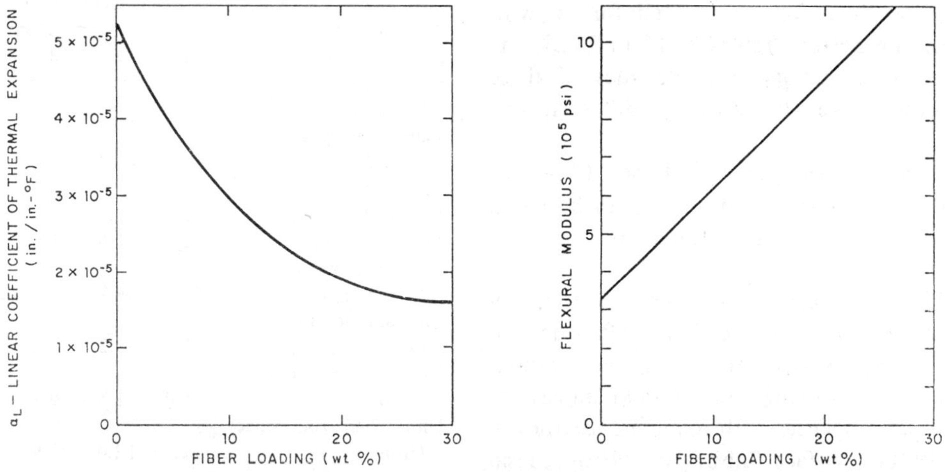

Fig. 24-7. Typical physical properties–"Fybex" reinforced ABS. (Plating grade; injection-molded bars 5″ × ½″ × ⅛″ Source: Ref. 3.)

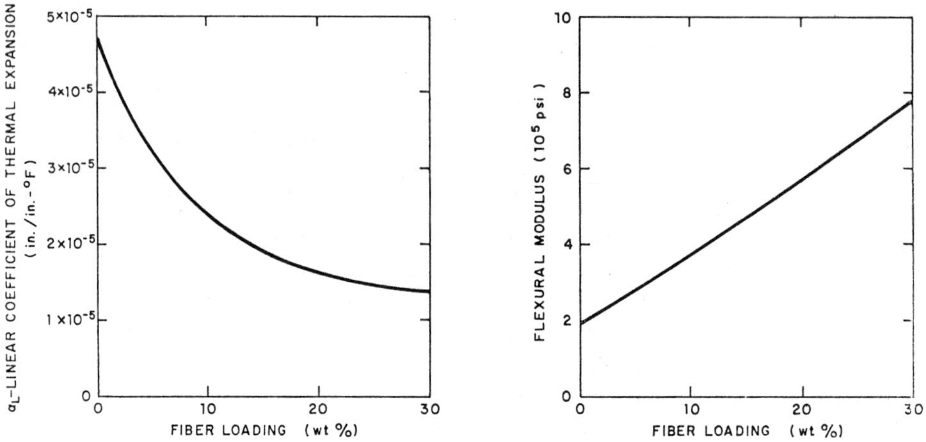

Fig. 24-8. Typical physical properties—"Fybex" reinforced polypropylene. (General purpose grade, MI = 15; injection-molded bars 5′ × ½′ × ⅛″ Source: Ref. 3.)

Table 24-16. Surface roughness of reinforced and unreinforced resins.

	AVERAGE SURFACE ROUGHNESS (microinches)	
	ABS[a] Composites	Polypropylene[b] Composites
Unreinforced resin	0.7	0.8
Titanate fiber reinforcement (20 wt. %)	1.0	1.5
"E" glass fiber reinforcement (20 wt. %)	>50	>10

[a]Plating grade.
[b]General purpose grade.
Source: Ref. 2.

neither improves nor degrades this effect. As stated previously, good dispersion is essential for good tensile and impact strengths. Typical property data for plastics reinforced with Fybex are given in Tables 24-16 to 24-23, and some of the unique characteristics of these composites are discussed in the following section.

Dupont's experience with Fybex (2 and 3) has shown Fybex to provide a high degree of reinforcement in addition to the other advantages.

While reinforcement of thermoplastics improves engineering properties, it often does so at the expense of other desirable features. Ideally, a reinforcing fiber should impart engineering properties with minimum sacrifice in moldability, surface finish, part isotropy, resin regrindability, machine wear, cycle time, and compatibility with pigments, flame retardants,

Table 24-17. Physical properties versus number of molding cycles[a]

	MOLD CYCLE			
	1	2	3	4
30/70 FYBEX/Nylon 6-6				
Tensile strength ($\times 10^3$ psi)	19.6	17.3	17.3	18.4
Tensile modulus ($\times 10^6$ psi)	1.4	1.4	1.4	1.3
Elongation (%)	2.2	1.6	1.6	2.1
Flexural strength ($\times 10^3$ psi)	31.8	29.7	30.0	29.0
Flexural modulus ($\times 10^6$ psi)	1.3	1.2	1.2	1.2
Notched Izod impact (ft-lb/in.)	1.4	1.5	0.9	0.8
Unnotched Izod impact (ft-lb/in.)	11.8	12.0	9.9	11.2

[a]100% regrind ratio.
Source: Ref. 1.

Table 24-18. Reinforcement uniformity versus fiber size.

Material[a]	Tensile Strength (X 10^3 psi)	Flexural Strength (X 10^3 psi)	Flexural Modulus (X 10^6 psi)	Notched Izod (ft-lb/in.)
30/70 FYBEX/Nylon 6-6				
Machine	18.5	27.8	1.83	1.71
Transverse	15.5	22.3	0.92	1.50
33/67 Glass Fiber/Nylon 6-6				
Machine	18.5	30.4	1.28	1.31
Transverse	11.9	17.3	0.67	0.80

[a]Molded into 3 in. X 5 in. X $^1/_{16}$ in. plaques.

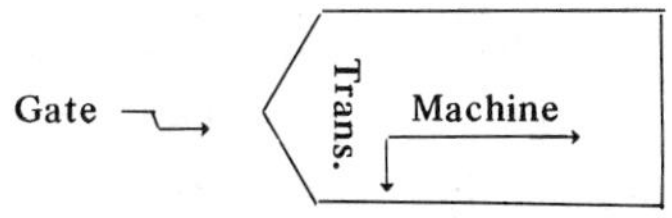

Source: Ref. 1.

Table 24-19. Fybex in engineering polymers.

Resin Fybex Loading	Polysulfone 25%	Modified Polyphenylene Oxide 20%	Nylon 40%
Tensile strength (X 10^3 psi)	14.0	9.0	26.0
Flexural modulus (X 10^5 psi)	10.0	8.8	19.0
Heat distortion temperature (°F at 264 psi)	–	270	480
α_L (in./in./°F X 10^{-5})	1.3	1.6	1.0
Notched Izod impact (ft-lb/in.)	1.8	1.3	1.8

Source: Ref. 1.

Table 24-20. Fybex in thermoformable ABS.

Sample	Tensile Strength (X 10^3 psi)	Flexural Strength (X 10^3 psi)	Flexural Modulus (X 10^5 psi)	Notched Izod (ft-lb/in.)	αL (in./in./°F X 10^{-5})	Room Temperature % Elongation
High impact ABS	4.4	7.0	2.4	7.5	6.1	>20
High impact ABS + 10% Fybex	5.0	7.8	4.2	7.0	3.1	>20
High impact ABS + 15% Fybex	5.8	9.0	6.0	3.0	2.3	>20

Source: Ref. 1.

Table 24-21. Fybex[a] in ABS extrusion grade system.

Fybex Loading (wt.%)	Tensile Strength (10^3 psi)	Flexural Modulus (10^5 psi)	Notched Izod (ft-lb/in.)	Expansion Coefficient (10^{-5} in./in./°C)
0	4.8	2.4	8.0	10.8
15	5.0	5.8	5.5	5.4
25	5.5	8.6	3.5	3.1

[a]DuPont trademark for inorganic reinforcing titanate fibers.
Source: Ref. 2.

and fillers. Failure to meet one or all of these criteria can exclude a fiber from an application, regardless of its reinforcing capability. Experiments have shown that Fybex provides a high degree of reinforcement in addition to the following advantages:

- Breakage resistance of Fybex during compounding and subsequent processing allow high regrind ratios or more regrind cycles without loss of physical properties. See Table 24-17.
- Surface smoothness of molded or extruded plastics. See Table 24-16.
- In addition to aesthetics, the small uniform fiber size of Fybex provides an important engineering feature to molded thermoplatics-isotropy, as indicated in Table 24-18.

Precisions Reinforced Moldings—Small gears, camera body components, electronic assemblies, and parts exposed to high thermal gradients frequently require dimensional stability beyond that attainable with unreinforce resins. Larger fiber reinforcement may not satisfy attendant requirements such as smoothness, isotropy, fast cycles, and others already mentioned. Fybex in engineering polymers provides durable, precise moldings that are often preferable to metals, as shown in Table 24-20.

The potential for Fybex type microfiber is good. Foam reinforcements have been investigated with Fybex, since theoretical calculations show it to be ideally sized for excellent foam reinforcement efficiency.

Vacuum formed ABS or PVC sheet reinforced with Fybex is another promising end use. Hot strength, high elongation, and toughness of extrusion grades of ABS and rigid PVC reinforced with Fybex suggest them as promising substitutes for sheet metal (Table 24-19). Struc-

Table 24-22. Fybex in 6,6 nylon, mechanical and thermal properties.[a]

Reinforcement Loading[b] (wt. %)	Tensile Strength (× 10^3 psi)	Flexural Modulus (× 10^5 psi)	IZOD IMPACT (ft-lb/in.)		Expansion Coefficient (10^5 in./in./°C)	HDT 264 psi (°C)
			Notched	Unnotched		
0	10.7	4.15	0.94	>60	8.4	66
30% Fybex[c]	20.0	13.25	1.10	16	3.0	225
30% Clay	12.4	6.75	1.24	48	6.1	89
20% Clay 10% Fybex[c]	14.0	7.90	1.15	28	5.0	168
10% Clay 20% Fybex[c]	16.5	9.45	1.10	22	3.7	175

[a]Dry-as-molded properties.
[b]1% Silane added based on reinforcement loading.
[c]DuPont trademark for inorganic reinforcing titanate fibers.
Source: Ref. 2.

Table 24-23. Relative wear rates of processing equipment.

Filler	Duration of Test (hr)	Part Weight Loss (%)	Relative Rate of Weight Loss
Glass fibers	200	0.066	1.0
Silica	100	0.069	2.1
Fybex[a]	200	0.006	0.1

[a]DuPont trademark for inorganic reinforcing titanate fibers.
Source: Ref. 2.

tural adhesives having high strength, low expansion, and toughness are other possible applications.

In summary, Fybex broadens the utility of thermoplastics. Its unique combination of properties qualified it for certain end-uses not now captured by reinforced thermoplastics.

References

1. Linsen, P. G. and Regester, R. F., "New Inorganic Fibers for Plastic Reinforcement," 27th Annual S.P.I. RP/CI, 1972.
2. Product Information Bulletin No. 1, "Reinforcement of Thermoplastics with Fybex Inorganic Titanate Fibers," Inorganic Fibers Division, Pigments Department, E.I. DuPont ou de Nemours and Co., Wilmington, Del. 19898, Nov. 1971.
3. Production Information Bulletin No. 2, "Fabrication and Electroplating of Fybex Fiber Reinforced Thermoplastics," Inorganic Fibers Division, Pigments Department, E.I. DuPont de Nemours and Co., Wilmington, Del. 19898, Nov. 1971.

25
Whiskers

John V. Milewski and Harry S. Katz

Los Alamos Scientific Lab.
Los Alamos, New Mexico

Utility Research Co.
Montclair, New Jersey

CONTENTS

1. ABSTRACT

Whiskers are fibers which have been grown under controlled conditions that lead to the formation of a single crystal. Their extremely small diameters allow little room for the defects that occur in larger crystals. The resultant highly ordered atomic structure approaches the material's theoretical inter-atomic bond strength. For example, tests on beta-silicon carbide whiskers have indicated tensile strengths ranging between 1-5-million psi and tensile moduli from 75-100-million psi. These values are far above those of the current high volume reinforcements, and represent the potential for producing the highest attainable composite strengths. Because of this potential, there had been great activity in research, development, and marketing of whisker materials during the 1960's. In recent years however, there have been very few active programs in this field. There are two main reasons for this situation. Most investigators were disappointed by the subtle technical difficulties in handling the available whisker materials, with the resultant poor translation of whisker properties in the final composite. Also, past cutbacks in development funds restricted work with these materials, which were very costly at that time. However, it is highly probable that processes for the production and handling of low-cost, high-quality whisker materials will be developed. This would lead to a resurgence of interest and progress toward the production of ultra-high-strength, whisker-reinforced-composites.

The capability for a sharp reduction in the cost of whisker materials is already evident, as shown by the recent test marketing by Silag of a high quality beta silica carbide at $20/lb.

2. INTRODUCTION

2.1 History

The first commercial source of whiskers was Thermokinetic Fibers, Inc. (TKF), a company that was started in 1962 with the intention of producing whiskers as their only product. From 1962 to 1968, TKF produced and offered for sale a wide variety of whisker products that included high-quality needle whiskers, wool and paper products of sapphire, silicon carbide, and silicon nitride. In 1966, TKF was merged with General Technologies, Inc., which was later acquired by Citco. About 1970, Citco sold the GTC assets to the major GTC officers and Versar Inc.; GTC is now a division of Versar Inc. They still market old TKF stock and special grades of some whisker products. Other major whisker suppliers that were active in the field are Horizons Inc., Carborundum Corporation, National Research Co., and Explosive Research and Development Establishment (ERDE) England. Lonza Co. of Switzerland was still supplying silicon carbide whiskers in 1975. A new source of whiskers is Silag, an affiliate of Exxon Enterprises N.Y., N.Y. (formerly Hulco of Salt Lake City, Utah). They currently have available a silicon carbide whisker product made from rice hulls. More details on these companies and their products are given in the section on suppliers and their products.

One of the major drawbacks in the development of whisker composite technology has been the price and availability of quality whisker products. During TKF's 5 years of major activity, 1963-1968, they produced about 50 lb of whiskers.

World production of whiskers during the late 1960's was less than 100 lb. Over 90% of those produced were *silicon carbide whiskers.* Philips Laboratories, Findhoven, The Netherlands, developed a process for the production of alpha-silicon carbide whiskers and licensed their process to Kempsten in Munich, Germany, and to National Research Co., Cambridge, Massachusetts, a former division of Norton Co., now owned by Cabot Corporation. During 1970, Kempsten produced from 20-50 lb of whiskers, mostly for internal use. Also during 1970, Norton's National Research Company produced about 10 lb of whiskers, and their product was priced at $1600/lb. The Cabot Corp., which has taken over this division, had hope of going commercial with a whisker product at about $300/lb, but this plan never materialized.

The Explosives Research and Development Establishment, Waltham Abbey, Essex, England developed a process for the production of beta-silicon carbide. During 1970, this group made about 25 lb of these whiskers. Their process has been licensed to Lonza Co., Switzerland, where about 20 lb of whiskers were produced during the 1970s, primarily for internal research and development work.

Also, in the late 60's, Carborundum Corporation introduced an alpha grade of silicon carbide whisker which they were able to produce in 100 lb quantities and sell at $250/lb. Unfortunately, the quality of alpha silicon carbide never approached that of beta silicon carbide in strength and uniformity of product. Therefore, this alpha SiC could not compete.

During 1975, Silag, formerly Hulco Corporation, in cooperation with Exxon Enterprises, produced a high beta silicon whisker carbide product from rice hulls. In less than one year they produced about 200 lb and were marketing it at an experimental price of $20.00/lb. As of this writing, production and marketing studies are being conducted to determine the future course of this high quality whisker product.

2.2 Methods of Manufacture

More than 100 materials, including metals, oxides, carbides, halides, nitrides, graphite, and organic compounds have been prepared as whiskers. Whiskers may be grown from supersaturated gas phases, from melts, from solutions by chemical decomposition or electrolysis, or by growth from a solid.

Whiskers grow by two different mechanisms—basal growth and tip growth. In basal growth, the atoms of growth migrate to the base of the whisker and extrude the whisker from the substrate. This type of growth has been observed on tin-plated structures subject to stress. On the other hand, sapphire, SiC, Si_3N_4, and AlN whiskers are formed by tip growth at temperatures high enough that the vapor pressure of the whisker or whisker forming material becomes significant. In tip growth, the atoms attach themselves to the tip of a growing whisker by various mechanisms.

Under ideal conditions, whisker growth rates of over 1 in. in 5 min. have been observed. The problem in quantity production is to optimize growth rate conditions in all areas of the reaction chamber by accurate control of temperature, gas flow rates, and other variables. When this is achieved, and large end uses develop, low-price, mass-production of whiskers will be economically feasible, because the basic raw materials are relatively inexpensive. Ultimately, a production process can be envisaged in which growth of whiskers is accomplished on a continuous belt passing through the furnace, being vacuumed off, air classified, plastic impregnated or metal coated, and then directly processed into the final composite. Presently, whiskers are grown by various batch-process techniques, and the whiskers are harvested and sorted by a hand operation.

2.3 Processes for Whisker Growth

Some metallic whiskers have been produced by sublimation or evaporation of the metal, which is transported in vapor phase to a lower temperature growth site, where under low supersaturation conditions, condensation produces whiskers. Zinc, cadmium, and other metallic whiskers have been grown in a simple chamber with a thermal gradient between the source and the growth site. An inert gas carrier may be used to help control the rate of whisker growth.

The hydrogen reduction of metal salts has been a frequent method for the production of metal whiskers such as nickel, iron, copper, silicon, and gold. The best reduction temperature is usually near or slightly above the melting point of the source material.[1,2,3]

Among oxides prepared as whiskers are Al_2O_3 (sapphire), MgO, MgO-Al_2O_3, Fe_2O_3, BeO, MoO_3, NiO, Cr_2O_3, and ZnO. A simple vapor transport method that can be used consists of heating the metal in a suitable atmosphere, such as wet hydrogen, moist inert gases, or air. For example, when a stream of moist hydrogen is passed over aluminum powder, aluminum oxide, or an intermetallic compound of aluminum which has been heated to 1300–1500°C, a mass of acicular sapphire is deposited in a cooler part of the furnace.

A large part of the whisker technology efforts has been directed toward alpha-Al_2O_3 (*sapphire*) whisker, because this material has excellent potential as a reinforcement in metal and polymer matrix composites. Webb and Forgeng investigated the growth of a-Al_2O_3 whiskers by passing wet hydrogen over a alumina boat (containing some silica), the boat filled with molten aluminum or $TiAl_3$. At operating temperatures between 1300–1450°C, large deposits of whiskers were formed around the boat after 2–20 hr, and the $TiAl_3$ gave a greater yield than the pure Al. Whiskers with lengths from 1–30 mm and hexagonal diameters of 3–50 μ were obtained.[4] Minor variations of this method have been used to produce research and development quantities of whiskers for composite fabrication. A problem that was encountered, and still must be considered in current whisker technology, was the lack of uniformity in each batch. Whisker type and size varied within each batch, and from batch to batch, and other growth forms such as ribbons and powders were observed. Often, a tight cocoon of whiskers forms in a manner that is difficult to disentangle and classify into discrete lengths for use in a composite. These problems spurred the development of the fiber classifier and other refining techniques which will be discussed in a later section of this chapter.

Since about 1969, the major activity in whiskers appeared to shift from Al_2O_3 to SiC. A main reason for this shift was the initial indication that SiC whiskers are easier to wet and bond in low-temperature-metal matrices and in polymer matrices. Also, SiC whisker growth is less susceptible to branching growth and thermal degradation. Therefore, there was a greater potential for producing a uniform product and for scale-up of production.

Philips Laboratories, Hindhoven, The Netherlands, developed a process for producing alpha-SiC. Bulk SiC was vaporized by heating under reduced pressure, and the whiskers formed upon nucleation sites containing lanthanum or another catalyst. Their process was licensed by National Research Co., Cambridge, Massachusetts, and Kempsten in Germany.

Beta-SiC whiskers have been grown by hydrogen reduction of methyltrichloro-silane at 1500°C onto carbon substrates.[5] Dr. Parratt and coworkers at ERDE, England, have explored the controlled use of combinations of chlorosilanes, CO, and CH_4 as source gases for the production of beta-SiC. Their process has produced relatively large quantities of good quality whiskers, and has been licensed by Lonza, Switzerland.

The many different crystallographic modifications of SiC differ only in the stacking arrangement of the atom planes. The sequence formed by stacking the Si—C double layer in the sequence ABCABC results in the structure of cubic beta-SiC. All other sequences are referred to as alpha-SiC. There have been laboratory tests that indicated no intrinsic difference in strength between the two forms. However, in commercial products, the alpha-SiC whiskers were poorer in strength and uniformity than beta-SiC. The difference may be attributed to the fact that the beta structure is unique and the production conditions must be more closely controlled than the conditions for producing the varying stacking arrangement of atom layers that can occur within the same batch of alpha whiskers.

A mechanism of crystal growth involving vapor, liquid, and solid (VLS) phases explains many observations of the effect of impurities in crystal growth from the vapor. The role of impurities is to form a liquid solution interface with the crystalline material to be grown and fed from the vapor throughout the liquid interface. The solution is a preferred site for deposition of feed from the vapor which causes the liquid to become supersaturated. Crystal growth occurs by precipitation from the supersaturated liquid at the solid-liquid interface.

A number of different whiskers have been grown by the VLS technique, including α-Al_2O_3, B, GaAs, GaP, Ge, MgO, $NiBr_2$, NiO, Se, Si, and SiC. Figure 25-1 and 25-2 illustrate the VLS growth processes and Figure 25-3 is a photograph of very uniform beta silicon carbide whiskers grown by Dr. J. V. Milewski at TKF.[6] The photo clearly shows the VLS catalyst ball on the top of each whisker.

Figure 25-1 illustrates the sequence of VLS growth as follows: Solid catalyst particle on left melts and forms the liquid catalyst ball. Carbon and silicon atoms in the vapor feed are extracted by the liquid catalyst, which soon becomes a supersaturated solution of carbon and silicon that precipitates silicon carbide on the substrates. As precipitation continues, the whisker grows, lifting the catalyst ball.

The advantages of VLS growth is that growth can be controlled and limited by the location and type of catalyst. A larger catalyst produces larger diameter whiskers. Figure 25-2 illustrates that the chemistry of the catalyst ball also effects the diameter of the whiskers formed, with some catalyst wetting the initial substrate over

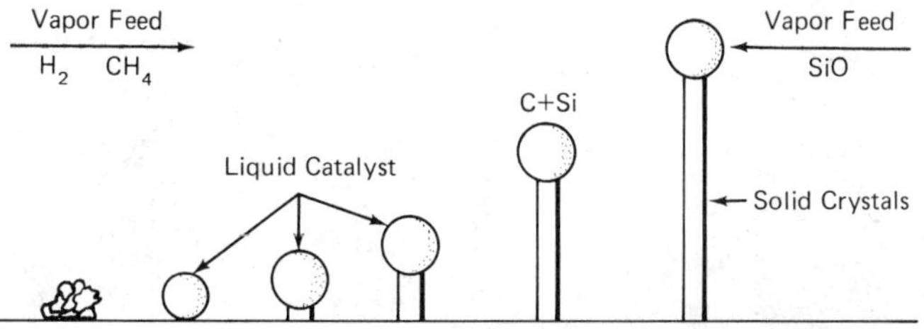

Fig. 25-1. V. L. S. (vapor feed, liquid catalyst, solid crystal) SiC whisker growth.

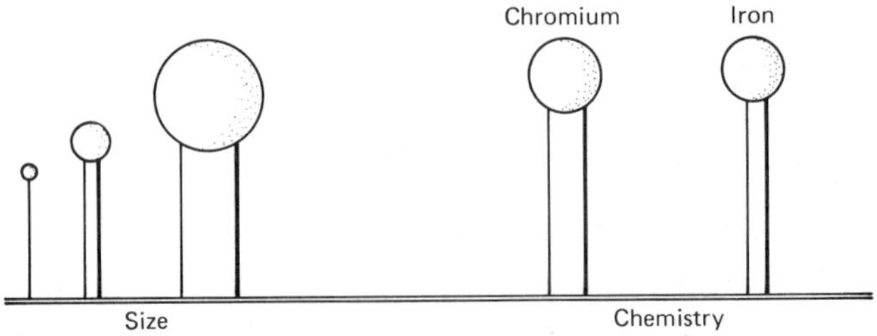

Fig. 25-2. Size and chemistry effects of the V. L. S. catalyst.

a wider range and producing a larger diameter whisker for the same size of catalyst.[6] Figure 25-3 illustrates the high uniformity of product produced by the VLS process.

Silag (formerly Hulco) is marketing a beta silicon carbide whisker product produced from rice hulls. The rice hulls contain a high silica content and upon coking, an intimate mixture of carbon and silica are formed. Upon further heating, the carbon and silica react to form a silicon carbide product, part of which is a fine diameter beta silicon carbide whisker. This submicron whisker is shown in Fig. 25-4.

3. BASIC PROPERTIES OF WHISKERS

3.1 What are Whiskers?

Whiskers are a generic class of materials having mechanical strengths equivalent to the binding forces of adjacent atoms. Whiskers are strong because they are essentially perfect crystals and their extremely small diameters allow little room for the defects which weaken larger crystals. Test data on silicon carbide whiskers record strengths greater than 4 million psi and a Young's modulus greater than 100 million psi. In comparing whiskers to conventional materials, the latter contain a multiplicity of grain boundaries, voids, dislocations, and imperfections, while the single crystal whisker which approaches structural perfection, has eliminated all such defects. Because of this internal perfection, the strength of whiskers is not strictly limited by its surface perfection. This is very significant in that it gives whiskers an unusual toughness and nonfriability in handling as compared to polycrystalline fibers and fiber glass.

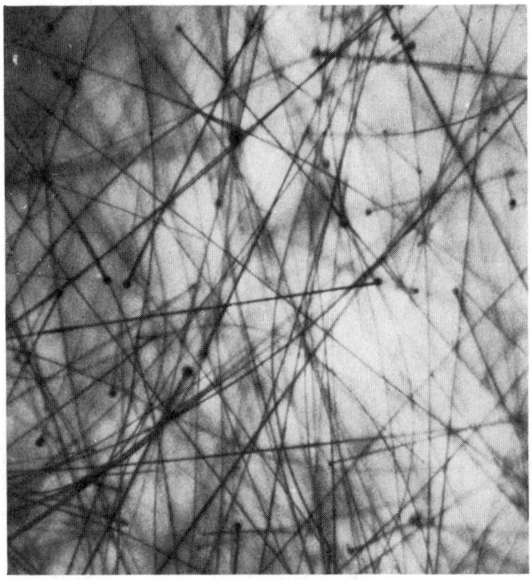

Fig. 25-3. Illustrating V. L. S. catalyst ball on ends of uniformly grown beta silicon carbide whiskers 100X.

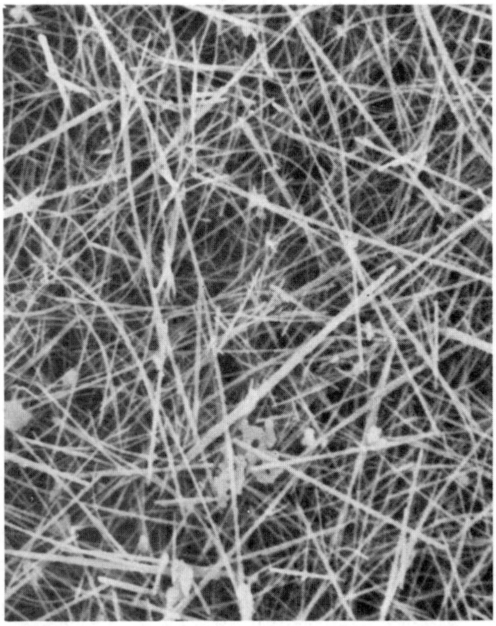

Fig. 25-4. ⊢⊣ = One micron. Magnification = 3,000X. *Sub-micron silicon carbide whiskers formed from rice hulls.* Courtesy: Silag

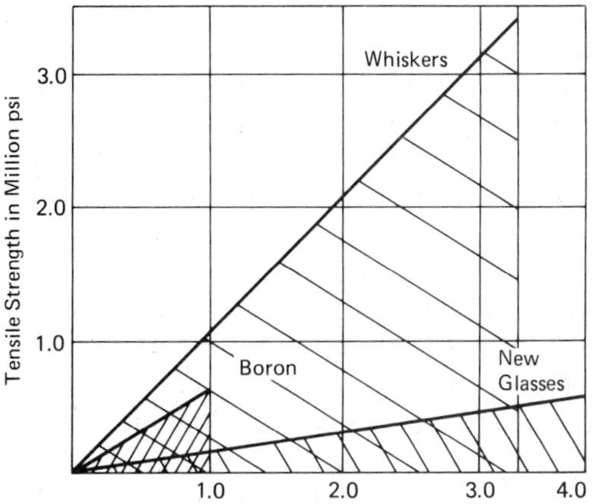

Fig. 25-5. Comparison of the strength of reinforcing fibers.

If the whisker's strength vs. diameter is plotted, it is found that as the whisker becomes smaller (and therefore even more perfect) its strength increases rapidly. In comparing the tensile strength of whiskers with those of other reinforcing fibers, such as glass and boron, a 5-10 fold increase in basic fiber strength is evident. Boron fibers have essentially the same tensile strength as E glass fibers, but the disadvantages of boron fibers are their larger diameter and reduced elongation, which limits processing due to a greater tendency for breakage during handling.

Whiskers combine the two best properties of glass and boron. They have the elongation of glass fibers (3-4%) and the modulus of boron (60-100 million or more psi), as shown in Fig. 25-5. The disadvantage of whiskers is that an improved technology for preparing discontinuous fiber composites must be developed in order to realize all their reinforcing potentials.

A whisker may be defined as a single crystal, in a fiber or elongated form, such that it has a minimum length to diameter ratio of 10:1, a maximum cross-sectional area of 7.9×10^{-5} in.^{2}, (corresponds to a circular cross-section of 0.010 in. diameter).

3.2 Physical Characteristics

Some of the most common whiskers and their properties are given in Table 25-1.

Whiskers are unique in that they can be strained elastically as much as 3% without permanent deformation, compared to less than 0.1% for bulk crystals. In addition, whiskers consistently exhibit much less strength deterioration with increasing temperature than the best conventional high strength alloys. This would be expected because of the absence of imperfections which are prone to permit slip. Sapphire whiskers have been found to retain tensile strengths as high as 100,000 psi at 3750°F.

Although whiskers appear to the eye as short fibers, they have very high length/diameter ratios (aspect ratios). The aspect ratio is important in reinforced plastics because of its obvious relationship to the amount of surface available for bonding to the resin and to the amount of load that can be carried by the composite.

No appreciable fatigue effects have been observed in whiskers. They can be handled roughly and can be milled and chopped or

Table 25-1. Physical properties of whiskers.

Material	Density g/cm^3	Melting point °F	Tensile strength psi $\times 10^{-6}$	Young's modulus psi $\times 10^{-6}$	Strength density in. $\times 10^{-6}$
Aluminum oxide	3.9	3780	2-4	70-150	14-28
Aluminum nitride	3.3	3990	2-3	50	13-21
Beryllium oxide	1.8	4620	2.0-2.8	100	31-43
Boron carbide	2.5	4440	1	65	11
Graphite	2.25	6500	3	142	37
Magnesium oxide	3.6	5070	3.5	45	27
Silicon carbide (alpha)	3.15	4200	1-5	70	9-44
Silicon carbide (beta)	3.15	4200	1-5	80-120	9-44
Silicon nitride	3.2	3450	.5-1.5	55	4.2-13

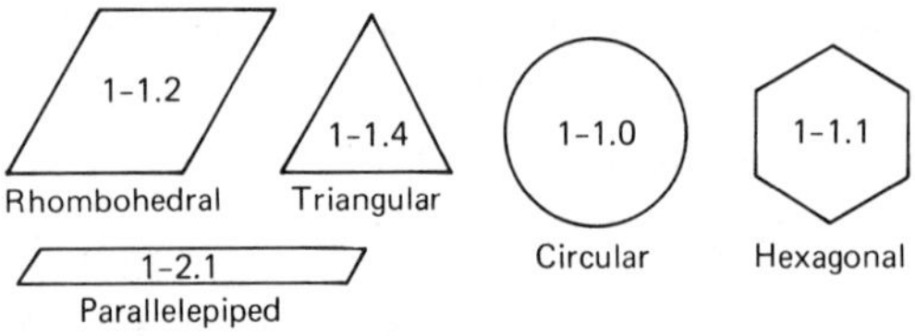

Fig. 25-6. Geometry of fiber cross-section.

otherwise worked in without having their strength impaired.

It is interesting to compare the whisker diameter with the diameters of various fibers such as boron and fiberglass. A bundle of about 100 whiskers 1 μ diameter would fit within 1 diameter of fiber glass–and about 10,000 whiskers within 1 diameter of a boron fiber. For a visual comparison of these fibers, see Fig. 25-6. In addition to the strength increase associated with small fibers such as whiskers, there is the possible added advantage of much greater number of fibers with a more uniform distribution in the matrix because of their small size.

The geometry of fibers has an affect on the aspect ratio. Specifically, a thin flat ribbon-shaped fiber has a larger girth or perimeter than the circumference of a round fiber with an equivalent cross sectional area. Round fibers have the minimum attainable surface area to volume ratio, hence any change in cross section geometry will increase the fiber's aspect ratio. Fiber glass and boron have round cross-sections, while whiskers have hexagonal, triangular, and thin ribbon cross sections. Some typical fiber cross-sections and the approximate aspect ratio increase with differing geometries are given in Fig. 25-7.

These new fiber cross sectional geometries are found in whiskers as a result of the unit cell crystal structure of the whisker. The basic crystal structure for sapphire is hexagonal. A C axis growth will produce a hexagonal whisker, while an A_I or A_{II} axis yields a parallelpiped whisker with a ribbonlike structure. Similarly the SiC unit cell structure is cubic and the (111) axis growth will produce almost triangular cross sectional whiskers.

3.3 Various Types and Forms of Whiskers and Whisker Products

Whiskers have been produced in a range of fiber sizes and in three forms. As grown

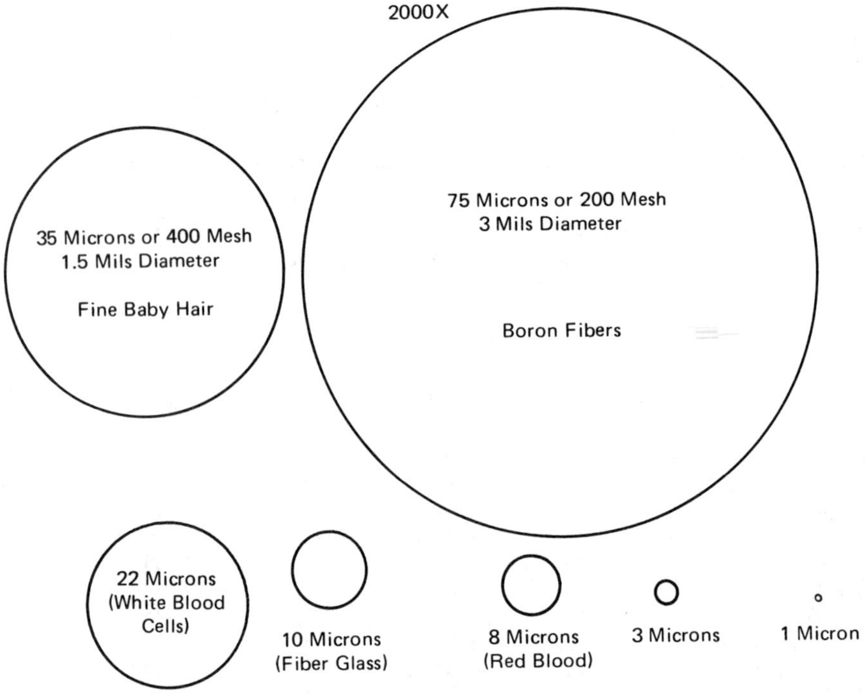

Fig. 25-7. Size comparison diameter of various fibers.

wool has a fiber cross-section ranging from 1-30 μ and aspect ratios of 500-5000:1. The bulk density is approximately 1 lb/cu ft, which is less than 1% solids. This is a very open structure suitable for vapor deposition of coating on the whiskers or direct use as an insulation sheet.

Loose fibers are produced by processing the as-grown wool mats in a blender to yield lightly interlocked clusters of fibers having aspect ratios of 10-200:1.

In felt or paper, the whiskers are randomly oriented and have fiber aspect ratios ranging from 250-2500:1. The paper, as felted, has approximately 97% void volume and a density of 4-8 lb/cu ft, and can be compressed to about 3% of its original thickness. The paper or felt contains no binder and consequently must be handled with care, but it does have sufficient integrity for impregnation and lamination.

Some special whisker products are metal coated whiskers, extrusion-oriented whisker-filled plastic filaments, and "Whiskerized" graphite fibers. These special intermediate products are important because they bring the compositor one step closer to his final products and save him developmental time on coating, alignment, or wetting techniques.

4. SUPPLIERS AND THEIR PRODUCTS

Table 25-2. Prior[a] suppliers types and prices of whisker materials.

Supplier	Type	Price $/lb[b]	Comments
Carborundum	α-SiC	250	No longer available. 500 lb produced in 1964-69. Contained impurities and low aspect ratio whiskers.
ERDE, England	β-SiC	–	Licensed Lonza.
General Technologies Corp. (Thermokinetic Fibers)	β-SiC α-Al_2O_3 Si_3N_4	9000 2700-13500 300-1500	Stock items, from TKF manufacturing on special order only.
Compagnie Francaise Thomson-Houston-Hotchkiss-Brandt	α-Al_2O_3	1300	Whisker program had 12 man staff in early '70's.
Electroschmeltzwerke Kempten, Munchen, Germany	α-SiC	250-300	Philips Process Licensee.
Horizons, Inc.	α-Al_2O_3	9000	Manufacture on contract.
Lonza, Switzerland	β-SiC	1000	ERDE Licensee.
LaPorte Industries, England	β-SiC	–	
Industrial Technical Institute, Osaka Lab., Japan	TiC	500	
	Current supplier		
Silag, Affiliate of Exxon Enterprises (formerly Hulco) 1251 Avenue of the Americas New York, New York 10020 (212) 974-2554	β-SiC	20-50	Currently producing and selling multipound quantities.

[a]These suppliers are listed because they are the most likely candidates for reinstating whisker production when activity in this area is renewed.
[b]Most of these are prices that existed at the time when these materials were available.

5. PRESENT STATUS AND FUTURE OUTLOOK

One cannot make a very precise projection for the sales of whiskers in the future, since there is no current established end use and the recent trend in this field has been a reduction of effort. However, there can be no doubt about the great sales potential of the ultrahigh strength composites that could conceivable evolve from whisker technology. A unidirectional oriented whisker composite with high modulus and a tensile strength approaching 1 million psi appears feasible. An even greater sales volume would be afforded by a whisker-reinforced compound that could be used in injection- or transfer-molding processes. These processes are used for about one third of all plastics currently molded, so the inclusion of even a low volume percent of whiskers would result in a large sales volume. The use of 50 v/o whiskers in a molding compound could eventually result in a material that is conducive to automated molding procedures and possesses a flexural strength two to three times stronger than the current highest strength injection- or transfer-moldable materials, which have less than 40,000 psi flexural strength. On the basis of these potentials, a renewed intense effort in whisker technology could lead to the development of sources for high quality whiskers at prices that are competitive with other high performance reinforcements, such as graphite and boron fibers.

6. FIBER PROCESSING, THEORY AND GENERAL APPLICATIONS

6.1 Methods of Fiber Preparation, Cleaning, and Classification

6.1.1 Cleaning. The material produced generally contains a wide range of aspect ratio fibers and a varying amount of non-fibrous debris. A general cleaning process whish is applicable to most raw whisker products is wet cleaning in a hydrocyclone. The hydrocyclone separates the large particles and very low aspect ratio fiber from the medium to long fibers. For more details on this subject see the book by Parrott.[7]

6.1.2 Fiber Classification. Classification of fibers is desirable, since various applications require specific aspect ratio materials. Screening of a fiber slurry does not yield a narrow range of fiber length because after the first layer of fibers is deposited on the screen, they effectively make the screen size smaller and subsequently trap shorter fiber, which again reduces the size of the screen hole. A rotary screening device, which eliminates this build up problem, was developed by Glass Development, Ltd. of England. It works on the principle of depositing a mono layer of fiber on a rotating cylindrical screen and back washes the fiber off the screen before the next layer of fiber is applied. This is done in a continuous manner. Two or more screens can be used in series or parallel arrangement as illustrated in Fig. 25-8. The short fiber goes through the first fine screen leaving the long and medium size fibers on the screen. These are backwashed off later in the cycle and recycled on a larger screen, which retains only the largest fiber and permits the medium size fiber to pass through. This combination gives three cuts of fiber lengths, a short, medium, and long. The actual fiber length and aspect ratio depends on the fiber diameter and screen mesh used in the experiment. A good reference to this process is given in Composites, 1969.[8]

6.2 Theory of Reinforcement

A rigorous theoretical analysis of the mechanism of discontinuous fiber reinforcements would fill a complete text in itself, so it is the authors' intent at this time to give a very generalized analysis of a discontinuous fiber reinforcement that will be helpful in understanding some of the further discussion in process development. For more detailed theory and references, see Chapter 3.

The modulus ratio of fiber to matrix affects the sphere of influence or reinforcing volume around each fiber and determines the optimum fiber length which, as a consequence, affects both the minimum fiber concentration and the L/D required. This is a complex interaction, but the following simplified points may help to visualize these factors.

The theory of reinforcement of discontinuous fibers assumes that the load or stress can be transferred through the matrix from fiber to fiber. The fiber is stronger than the matrix and has a higher modulus. When stress is applied to the composite, the fiber resists strain locally,

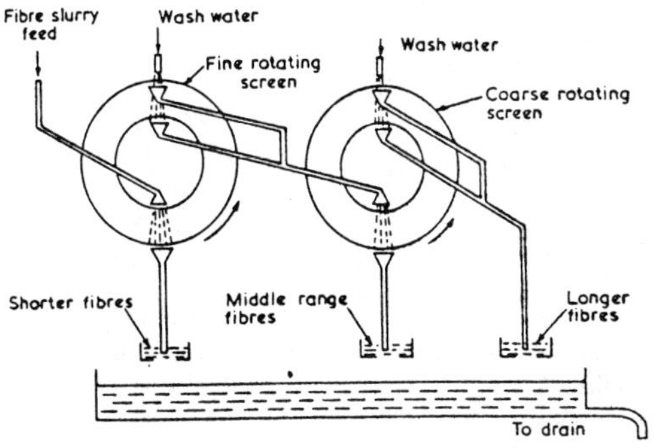

Fig. 25-8. Two wheels in series can be used to separate the fibres into three lengths in which case the second wheel has a coarser gauze. (*Courtesy IPC Science and Technology Press Ltd.*)

which induces a much higher stress in the fiber than the matrix around it.

This effect has been vividly illustrated in color photos, using birefringent plastic in some excellent work at Cornell by Schuster and Scala.[9] In Fig. 25-9A, the unshaded area surrounding the whisker represents a lower stress and strain. Figure 25-9B presents the authors' conception of what composite stress/strain lobes would look like with a number of fibers unidirectionally oriented. Note that the high stress areas, always found at the ends of the fibers, are reduced when they fall within the bounds of the low stress lobes near the center of adjacent fibers. Note also that the overall composite's strain lobe will develop an aggregate shape which will represent additive strain reinforcement in a manner illustrated. A final point is that these illustrations represent the stress/strain lobe as drawn at some discrete isostrain, e.g., at 50% of maximum, and that in actuality it is a continuation of a gradually dissipating strain.

The ratio of the modulus of the fiber to that of the matrix has a pronounced effect on the shape of the stress/strain lobe as is illustrated in Fig. 25-10[10] These figures were drawn assuming equal fiber strain and the optimum fiber length required by the shear bond and fiber strength. In these figures, one can observe the major differences in strain distribution effects that come about in different matrices.

In ceramics, for example, under a given stress in the whisker, the stress/strain lobe will be broad and short because the ceramic modulus is high and the strain is not easily dissipated. Another point to consider in ceramics is that the maximum stress in the whisker can be developed in relatively short fiber lengths if a high bond strength is obtained. Fortunately, high bond strengths are possible for whiskers in ceramic matrices.

Now, considering the other extreme of whiskers in a low-modulus matrix such as plastic, it can be seen that the stress/strain lobe is thin and long. This suggests that fibers with greater L/D's and also higher concentrations would be used in order to insure sufficient overlapping of the SS lobe. This idea, illustrated in Fig. 25-11, shows that fiber population must be over a certain minimum before significant reinforcement can be obtained. No attempt has

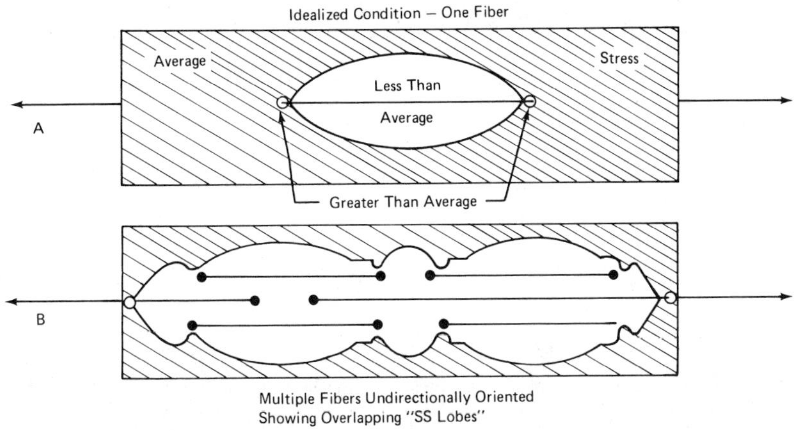

Fig. 25-9. Stress distribution around a reinforcing fiber.

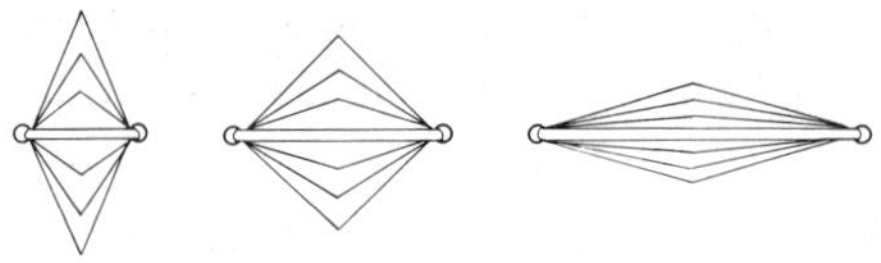

CERAMICS	*METALS*	*PLASTICS*
Ratio of the Modulus of the 2-1	Fibers to Matrix 5-1	100-1
L/D (Aspect Ratio) Required 10-20 to 1	for Optimal Reinforcement 20-50 to 1	30-100 to 1
Matrix Elongation Poor	Good	Excellent
Whisker to Matrix Bond Excellent	Poor Needs Coating	Good

Fig. 25-10. The effect of the fiber to matrix modulus ratio on the "SS" lobe.

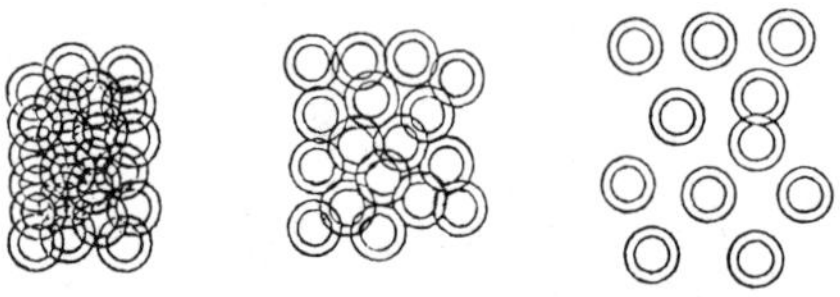

Fig. 25-11. The effect of fiber population on distribution of stresses. End view of "SS" lobe around unidirectional oriented fibers at different fiber population densities.

been made at this time to calculate these minimum concentrations, but an estimated value for ceramic is less than 1%, for metals a few percent, and it may be as high as 10% in plastics. This minimum population is affected by a number of other items not considered in detail, such as fiber diameter, fiber geometry, orientation, and the addition of other non-fiber materials, all of which have a pronounced effect on the "volume strain" produced.

6.3 Type and Percent of Whiskers to Use

A usual opinion of materials engineers regarding the use of whiskers is that they must be unidirectionally aligned in order to make them useful. Actually, this is not the case. Most applications of materials require bi- and tridirectional strengths as illustrated in Fig. 25-12.

Selection of the type and percent of whiskers for use in reinforced plastics will depend on the end product and the processing techniques to be used. For example a tridirectional reinforced molding compound, a bidirectional reinforced laminating sheet, and unidirectionally reinforced rods or wires all require different types of whiskers for optimum performance. (See Fig. 25-12.) It is obvious, then, that engineers must become familiar with the various types and with their advantages and disadvantages, in order to realize the maximum properties available in whiskers.

The end product and processing method will dictate the preferred diameter, length, and type of whisker to be used. For example, a very fine diameter high-aspect-ratio whisker should be used for making yarn to fabricate unidirectional composites. Short, low-aspect-ratio whiskers

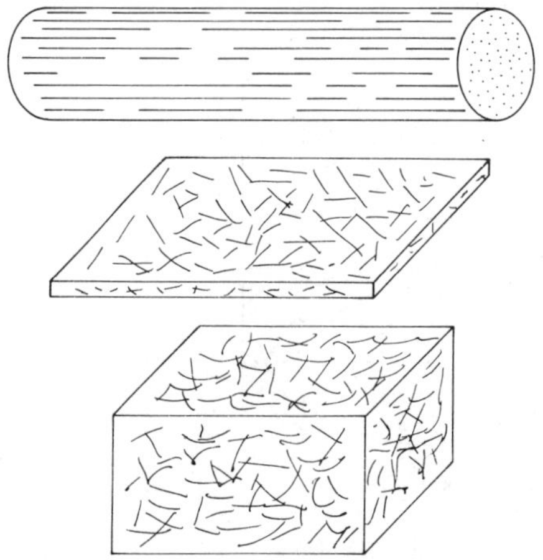

Fig. 25-12. Control orientation of whiskers in matrix.

will be ideal for interstitial interlaminer shear reinforcement. For producing sheet laminates having maximum planar tensile strength, whisker paper or felt should be used.

In making additions of whiskers, the following suggestions will serve as a guide:

1. For grain refinement and grain stitching in metals and ceramics, seed polymerization in plastics, and supplementary interstitial reinforcement of carbon, boron, and fiberglass composites, 1-5% addition of whiskers is recommended. This amount usually gives sizable percent increases in mechanical properties.

2. Generally, 5-50% addition of whiskers is recommended for random reinforcement of molding and casting compounds requiring three-dimensional strength. This amount usually gives a multiple increase in strength.

3. Generally 50-70% is recommended for partially oriented laminates for bidirectional strength. This amount usually gives a large multiple increase in strength.

4. Generally 70-90% is recommended for fully oriented composites such as impregnated yarn and oriented tape. This amount usually gives an order of magnitude increase in strength.

5. For a super strength, low density, bonded trusswork of whiskers, where only a sufficient amount of matrix is used to bond the whiskers together at their mutual points of contacts, 90-95% addition of whiskers is recommended.

7. METHODS FOR MAKING WHISKER COMPOSITES

7.1 Problems

Good composite fabrication techniques require the uniform distribution of whiskers either in a random manner or in the orientation required for the specific application. The most important consideration is that the whiskers be "individualized" and bonded by the surrounding matrix material. The physical properties of the fiber and the matrix can often present a problem in obtaining a uniformly dispersed, bonded fibrous structure. Factors such as surface tension, wettability, density differences, excess fiber length, and matrix viscosity impede the formation of uniform dispersions. Moreover, electrostatic surface attraction and excess fiber lengths cause a clustering of the fiber into balls and parallel groups which will result in island-like clusters in the final composite. With ordinary mixing techniques, the whiskers nearly always remain in the clustered state and produce poor composites (see Fig. 25-13).

Balling
caused by fiber length too long or fiber mod too low, allowing fibers to ball up in non-interconnecting groups of reinforcements.

Paralleling
caused by poor processing technique or raw material forms such as chopped fiber glass.

Grouping
caused by inefficient mixing, settling out in liquid matrix.

Fig. 25-13. Problems causing nonuniform distribution of reinforcing fibers.

7.2 Wetting and Impregnating Techniques

One mixing technique which allows for removal of adsorbed gases and water on the whisker surfaces and avoids the clustering problems will be described. It is obvious that with different resin systems this technique will not be possible, but it is given as an example of how to approach these problems. This process involves cleaning, wetting, and coating the whisker surfaces in a dilute resin system by first forming a uniform dispersion of individualized fiber in the dilute state, and then subsequently controlling the dispersions of the coated individualized fibers.

The following steps are recommended as a lab method for preparing test specimens.

1. Dry whiskers at 250°F or higher. (vacuum oven preferred)

2. Prepare the resin formulation. A typical formulation is shown below:

100 g	Shell 828 or equivalent
20 g	Shell activator Z
120 g	Resin-activator mix

3. Dilute part of above mixture in the following ratio:

10 g	Resin-activator mix
90 g	Acetone
100 g	Thinned resin-system (mix conventionally)

Coupling agents have been found helpful in improving composite properties, and may either be added to the solvent or to the resin formulation.

Impregnate whiskers by suspending them in a container of thinned resin solution. (Use approximately 1 cup of thinned mix to each gram of whiskers). Place container with whiskers in a vacuum chamber and evacuate until the vapor pressure of solvent is reached. Then hold at just above this pressure level to prevent excessive boiling of solvent. Vacuum bump a few times during this period by releasing the vacuum and then pumping down again. Then pour thinned resin mix and whiskers into a food blender (Waring or Osterizer) and mix for 15 sec at low speed and 15 sec at high speed for average 1–10 μ diameter whiskers. For smaller diameter whiskers, mix longer at high speeds and, conversely, shorter times for larger whiskers.

If twice the volume fraction of whiskers is used, double the mixing time. When the mixing is over, pour the slurry of fibers out into a fine metal screen, basket, or filter and let the excess resin and solvent drain through for about 5 min. Evaporate more of the solvent by suspending in an oven at 150°F for 15 min. Remove final remaining solvent in a vacuum desiccator.

At this point, the product is a relatively dry, 50-50 by volume whisker-resin, with all the whiskers individualized and their surfaces well wetted with resin. This mix now can be used directly for molding or adhesive applications or can be subsequently diluted with addition of resin for casting and potting application.

For more oriented structural application, the use of whisker paper is recommended. The process then must be modified after the impregnations by eliminating the mixing and including a vacuum paper mold in place of the steel basket to form the bidirectional felt. Impregnating of whisker yarn can be done on a spool or in a more or less continuous manner.

7.3 Casting Techniques

In a casting compound, the fiber resin bond is not as efficient as that obtained by high pressure molding. Consequently, longer whiskers are required then would be necessary with higher shear bonding strengths as indicated previously. For this application, the higher L/D whiskers found in whisker paper are recommended. The higher the percentages of whiskers, and the smaller the fiber diameter, the more difficult it will be to achieve good wetting. Therefore, in developing a casting formulation, it is recommended that initial efforts involve a low fiber concentration and a large whisker and then proceed to finer, stronger whiskers and higher fiber concentrations. For a given weight of fiber, there will be 100 times more 1-μ whiskers than 10-μ fiberglass fibers. Moreover, these small whiskers have about 20 times greater surface area to wet than does an equal amount of fiberglass.

Also, the normally flat geometry of sapphire whiskers, in effect, essentially doubles the aspect ratio. The problem resolves itself into one of the mechanics of wetting the extremely large surface area associated with fine diameter fiber whiskers. To obtain good quality casting and molding, use the generalized wetting and impregnating techniques as outlined in the section 7.2.

Other casting methods are based on the addition of low concentrations of loose whiskers to maintain casting fluidity to resin systems. The fibers are added by stirring in and then vacuum degassing. Such a system will not produce as high a degree of reinforcement as the long fiber wool, but substantial reinforcement has been produced by the co-dispersion technique previously described.

7.4 Transfer Molding

Transfer molding compounds are basically a tridirectionally stressed material and therefore must receive their reinforcement in all three axes. The best way to accomplish this is to have random distribution of the reinforcing fibers. This type of reinforcement is only possible with discontinuous fibers.

A problem common to nearly every molding compound available today is that it is impossible to obtain random and uniform distribution. "Balling," "paralleling," and "grouping," of the fibers occur as illustrated in Fig. 25-13. The requirements for sufficient shear bonding demand a long fiber, which is usually too long to allow random packing. Generally good shear bonding requires L/D ratios of about 100 to 1, while good random packing occurs with fibers that are less than 30 to 1. Also, the moduli of most such fibers are low, causing them to bend around bundles of other packed fiber, rather than allowing them to penetrate those bundles and achieve better intermeshing.

All of these factors make whiskers an ideal candidate material for development of a super-reinforced molding compound. With these factors in mind, and with what is known about fiberglass, a typical whisker reinforced plastic molding compound can be readily formulated.

A suggested whisker molding compound is:

	Volume Loading (%)
Resin	50%
Flow controlling and reinforcing sub-micron fiber mix, .2–1.0 μ in diameter.	20%
Reinforcing whisker, 1–10 μ loose needles.	30%
	100%

A small percentage of mold release agent, and other minor additives may be included in the formulation.

The two size ranges of whiskers used in this formulation serve to reinforce each other, and allow for both a good flow control and a more theoretical packing and distribution because the smaller fibers will fit between the larger ones. This also permits the use of a higher percentage of reinforcing fiber, up to 50 volume percent of the molded composite. Besides the unusually high strength expected, a composite of this type would have other very desirable properties, such as the capability of molding structures having a very small cross-section (less than 2 mils in thickness). In addition, good dimensional stability associated with the anticipated high modulus of >5 million psi, and excellent ability to mold around and to retain micrometallic inserts for electronic applications can be expected.

The advantages of using whiskers in tridirectional reinforced material are:

A randomly distributed fiber composite will have only 1/5th the strength of the equivalent volume loaded unidirectional fiber composite. Small diameter whiskers, with their extremely high strength of over 2 million psi, are ideal for bulk molding compounds since they will yield a high isotropic strength.

Whiskers are wet and bond well to plastics, especially at the high pressure used for most molding. Also, whiskers having a favorable geometry and high aspect ratio can be used in surprisingly shorter L/D than is generally considered necessary.

The higher modulus of whiskers gives sufficient rigidity for their size to enhance random distribution without the bundling and grouping common in fiberglass molding compounds. The smaller size of whiskers and their greater numbers allow a more homogenous distribution of the reinforcing strength.

8. UNIDIRECTIONALLY ORIENTED COMPOSITE

8.1 Various Methods for Unidirectional Orientation

Many different techniques have been studied for the production of unidirectionally reinforced plastics with whiskers. Some of these are listed below.

Spinning of whisker wool into yarn and then processing the yarn in a manner similar to continuous filament technology.

Making highly directional whisker paper (i.e., 90% x, to 10% y orientation), impregnating, "B" staging and cutting into strips, and then tape winding as is now done with asbestos paper.

Electrostatic, magnetic, or vibrational alignment of loose whisker needles, followed by a spraying with a thin film of plastic, alternating the alignment and spraying techniques to produce an oriented reinforced laminate.

Centrifugal casting of resin with loose whiskers, to force the whisker to line up in the outer hoop of the casting to produce a plastic hoop with peripheral orientation of whiskers.

Continuous extrusion of a whisker alginate slurry into a mild acid solution which gels the alginate to produce a monofilament or reinforced plastic rod with oriented whiskers. The alginate is then burned off and the unidirectionally oriented whickers are then reimpregnated with a matrix material.[10]

A batch process in which the whiskers are suspended in a slurry with glycerine and co-extruded through a convergent flow alignment slot nozzle, and are laid down simultaneously on a filtering screen. To produce a uniform layer of fiber, the nozzle travels rapidly back and forth over the screen. After sufficient fiber build-up is accomplished, the fibers are washed and can then be impregnated for compositing. See Fig. 25-14.[11]

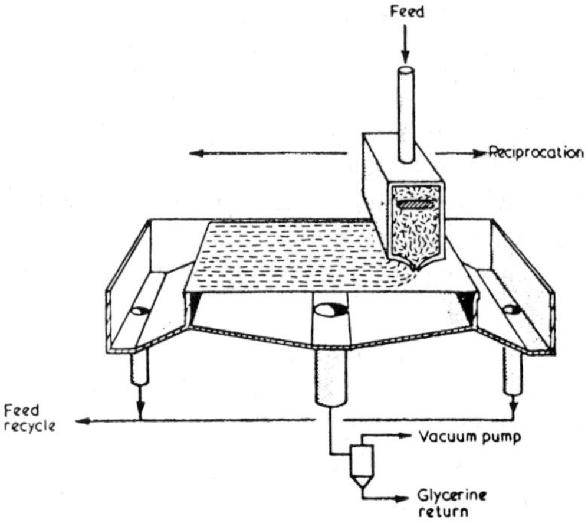

Fig. 25-14. Schematic diagram of the batch alignment plant showing deposition of aligned fibres in the glycerine onto a flat gauze filter. (Ref. 11 *British Crown copyright courtesy Controller, Her Britannic Majesty's Stationery Office.*)

Examples of the high strength that can be obtained in unidirectional whisker composites, are shown in Table 25-3 (12). Note that the ERDE whisker composite has a flexural strength of 200,000 psi and 20-million modulus. These are unusually high values, especially when considering the fact that this is a short fiber composite and demonstrates partial fulfillment of part of whisker's real potential.

Table 25-3. Unidirectional whisker/epoxy composites.

Whisker Source	% Whiskers by Volume	Flexural Modulus GN/m^2 (10^6 lb/in^2)	Flexural Strength GN/m^2 (10^6 lb/in^2)	Effective Stress GN/m^2 (10^6 lb/in^2)
Carborundum Co.	48	180–200 (26–29)	0.82 (0.12)	1.7 (0.25)
ERDE–Explosives Research and Development Establishment, England	35	140 (20)	1.4 (0.2)	3.9 (0.57)

Source: COMPOSITES (March 1970)[12].

8.2 Supplementary Reinforcement of Boron, Fiberglass and Carbon Filaments with Whiskers

Many of the failings of conventional continuous reinforced plastics can be attributed to the fact that the structures are too anisotropic and lack a homogenous distribution of strength. For example, a filament wound structure is designated with filaments wrapped in the direction of the principal stresses. The stresses, however, do not always fall perfectly in line with the axis of the fiber. This is usually caused by such things as imperfect or unstable winding patterns and non-axial stresses developing from torsion, bending, vibration, thermal changes, side loading, etc. All of these stray stresses cause premature failure in filament wound structures by overstressing the unreinforced resin-rich pockets around the fibers. This results in interlaminar shear failure. Therefore, a new application for the use of small diameter whiskers is in supplemental interstitial reinforcement of filament wound structures. These small fibers complete the reinforcing requirements by filling in the unsupported areas and by achieving a higher percentage theoretical packing density. This idea of using whiskers to reinforce the interstitial plastic phase and at the same time to increase modulus and strength of the matrix is applicable to fiberglass, boron, and carbon reinforced structures.

Four suggested ways of using whiskers for supplementary reinforcement for boron, fiberglass, and carbon are outlined next, (a more specialized case called "whiskerizing" with carbon fiber will be discussed later in the text):

Interlaminar shear in laminates where whiskers are added to the resin between layers of boron, fiberglass, or carbon.

Interlaminar shear in wet winding structures where whiskers are added to the resin before winding with boron, fiberglass, or carbon.

Filament winding tape where whiskers are incorporated in especially high stressed areas during winding.

Multiple coating of fibers (first, with sizing; second, with resin; third, with sub-micron whiskers to produce a barbed-wire type fiber.)

The examples in casting, transfer molding, and supplementary reinforcement were chosen because these general molding methods are presently being used extensively in fiberglass and boron reinforced plastics and, therefore, provide a good base to work from in the incorporation of whiskers. Under each category there are many other methods of producing similar effects, providing the proper adjustment in processing techniques is made to account for the small size and large surface area of whiskers.

8.3 Bifiber Composites ("Whiskerized")

The use of whiskers for supplementary reinforcement may play a big part in the initial commercial development of whisker composites, because a small percentage addition has a major effect on the composite properties. The large reinforcing effect from a small amount of whiskers easily justifies their price. A special form of supplementary reinforcement particular to carbon fibers, called "Whiskerizing," will be discussed.

Carbon-graphite filaments have excellent properties. However, one area for improvement is low shear strength in resin matrix composites, due to poor bonding. The problem is even more evident in the higher modulus materials. A solution to this problem was the addition of whiskers to the composites, not in the loose form as discussed earlier in the case with fiberglass and boron, but by growing the whiskers directly on the surfaces of the carbon fibers.

This process is called "whiskerizing" and it has the following effects on the carbon fibers:

1. It changes the surface of the carbon fibers.
2. It changes the surface chemistry of the carbon fibers.
3. It adds an integral bond between whiskers and the carbon fibers.
4. The projecting whiskers reinforce the interstitial plastic by improving its effective modulus and strength.

As a result of these physical and chemical changes to the carbon fibers, the bond strength (interlaminar shear) in general has been increased three to five times (Fig. 25-15).[11, 12]

More specifically, the interlaminar shear strength increases as a function of the percent of "Whiskerizing", as illustrated in Fig. 25-16.[14] Note how important it is to make the composites with a low void content. This is accom-

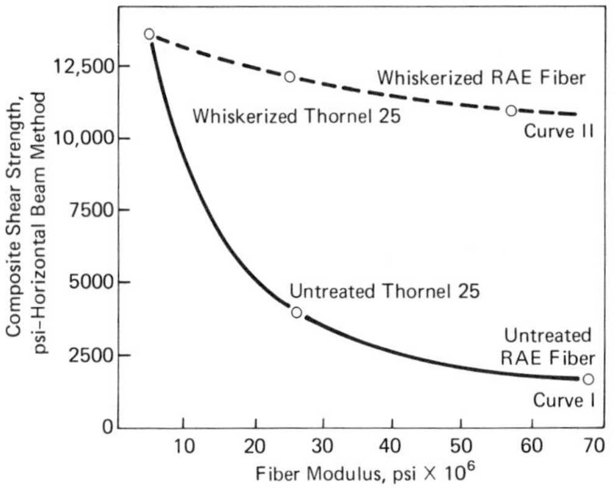

Fig. 25-15. Carbon fibers composite shear strength vs. fiber modulus (epoxy resin, amine cure). *(Courtesy Naval Ordance Laboratory.)*

plished by using solvent impregnation of the "Whiskerized" carbon lay-ups, solvent evaporation, "B" staging, and then compression molding to a calculated volume using whatever pressure is required to obtain full compaction.

The process of growing whiskers on carbon fibers is applicable to both batch and continuous methods. Fig. 25-17 shows the needle growth on a continuous "Whiskerized" graphite fila-

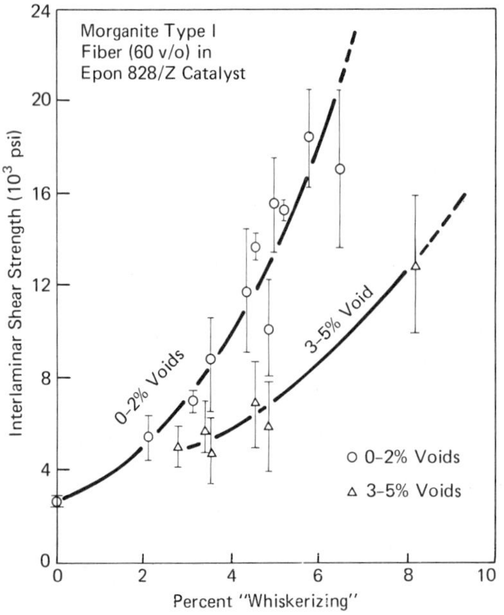

Fig. 25-16. Interlaminar shear strength vs. percent "whiskering."

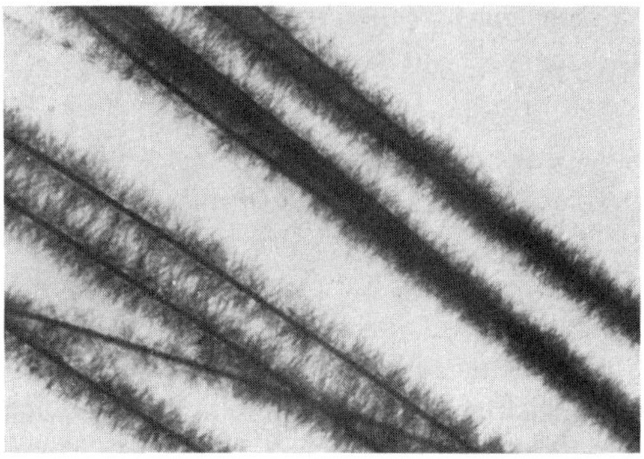

Fig. 25-17. "Whiskerized" RAE graphite fibers 200X.

ment.[18] The process of "Whiskerizing" can be applied to many other fiber systems. Whiskers have been grown on quartz, boron, SiC and tungsten filaments. This type of bifiber compositing will no doubt have many new and special applications as its full potential is realized.

9. FUTURE TECHNIQUES

9.1 General

Theoretically, it is possible to grow a whisker out of any material that will crystallize—ceramics, metals, and even plastics. As a result of this, many different types of experimental whiskers have already been grown for specialized compatible applications. In addition to reinforcing, the specialized whiskers will bring unusual electrical, optical, magnetic, ferromagnetic, dielectric, conductive, and super-conductive properties to the matrix to which they are added. As the demands of materials become more unique, so will the kinds and types of available whisker materials.

9.2 Cobweb Ultrasmall and Ultrastrong

The future of fine-fiber technology is foreshadowed by new experimental fibers called cobweb whiskers. These fibers are so small that they cannot be resolved with the best optical equipment; they are seen only as a blue cloud formed by light refraction similar to that which causes the sky to appear blue. Electron micrographs at 200,000X show the fibers to be 180

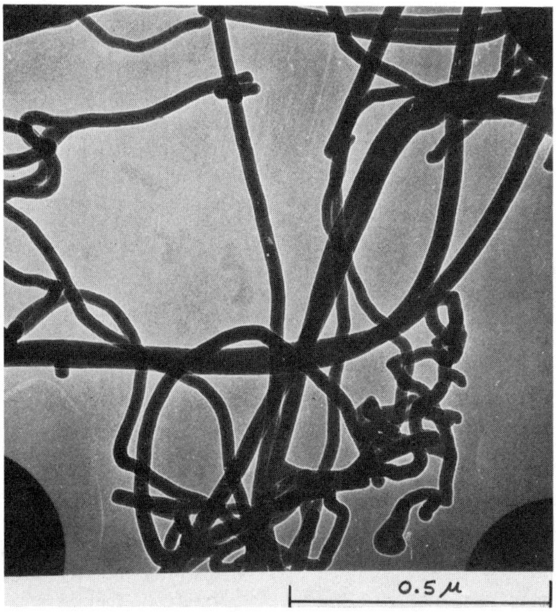

Fig. 25-18. Sapphire "cobweb" fiber 200,000×. Experimental whisker from T.K.F.

Angstroms in diameter, (Fig. 25-18). Their strength can only be estimated from extrapolated data. Theoretically, by growing whiskers smaller and smaller and, hence, making more perfect crystals, atomic strengths in the range of 50 million psi can be produced, bringing a new order of reinforcing strength to materials and design.

These magnitudes of strength in materials were first predicted over 45 years ago by Dr. A. Griffith a physical metallurgist[16] in England, and only recently been given serious consideration. When we begin to achieve these potentials, we will be at the beginning of a major technological revolution in super-strength materials.

9.3 Short Range Applications

The present high price of quality whiskers, and the projected prices in the next few years, will limit their applications to more or less specialized material problems.

The first commercial applications will probably be those that require a relatively low percentage of whiskers in relation to the total composite or those used to reinforce specific areas of weakness such as adhesive joints, weldings, hard surfacing, or interstitial reinforcements (as in the case for carbon fibers). Next to follow will be high priced small objects in specialized molding compounds used in electronics or medical application, such as dental and synthetic body components (bones, teeth, etc.), and other medical applications. These will then be followed by large aerospace and scientific applications, such as helicopter blades and high strength centrifuges. Finally, applications of whisker composites in wings, tail assemblies, and space capsules will be realized.

9.4 Long Range (Commercial)

With the advent of commercial production, whiskers may be used as a universal reinforcement in all types of materials and may find their way into many commonplace products. For example, whisker-reinforced plastics will be used extensively for body and basic structures in the automotive industry where they will save weight, and have twice the strength of steel, the modulus of titanium, and the density of aluminum.

The construction industry will require a whole new approach to design when using whisker reinforced plastics. These new reinforcements will allow new freedoms in which the structural member becomes part of the architectural designs, assuming phenomenal tensile strength and crack resistance in structures having extremely thin sections.

A big step in whisker reinforced material would be whisker yarn reinforced rubber. With the yarn woven into cloth and impregnated with rubber, lightweight super-pressure tires will be possible. Also, this rubber impregnated whisker cloth would greatly accelerate the growth of the inflatable structures industry for both air and undersea applications.

Last, but not least, the use of whiskers in the human body will bring a revolution to the area of dental bridge work, reinforced plastic bones, and reinforced flexible artery tubing.

10. FUTURE ACTIVITY IN WHISKER COMPOSITES

The main criteria for determining the future utility and sales volume of whiskers will be the level of properties that are attained in the whisker composites. There were many handicaps in the early efforts in whisker technology. There was a shortage in the availability of consistently high quality whiskers. Also, there was a poor understanding of the subtle difficulties in the handling of whiskers and subsequent composite fabrication. It is anticipated that this presentation may act as an incentive and guide to overcome these problems, and lead to the production of high quality whiskers and ultra-high-strength composites.

References

1. Brenner, S. S. and Sears, G. W., *Acta Met.*, **4:** 268 (1956).
2. Brenner, S. S. "Growth of Whiskers by the Reduction of Metal Salts," *Acta Met.*, **4:** (Jan. 1956).
3. Brenner, S. S., "Growth and Properties of Whiskers," *Science*, **128:** (3324) (Sept. 1958).
4. Webb, W. W. and Forgang, W. B., *J. Applied Physics*, **28:** 12 1957.
5. Ryan, C. E., *et al.*, Physical Sciences Research Papers No. 266, AFCRL-66-641, Sept. 1966.
6. Shyne, J. J., Milewski, J. V., Shaver, R. G., and Cunningham, "Development of Processes for the Production of High Quality, Long Length SiC Whiskers," AFML-TR-67-401, Jan. 1968.
7. Parrott, N. J., *Fibre-Reinforced Materials Technology*, Van Nostrand Reinhold Co., London, 1972.
8. Parrott, N. J., "The Rotary Fibre Classifier," *Composites*, **1:** (2), (Dec. 1969).
9. Schuster, D. M. and Scalla, E., "The Mechanical Interaction of Sapphire Whiskers with a Bi-Refringent Matrix," Report No. 166, p. 4 and 5, Cornell University, Dec. 31, 1963.
10. Parrott, N. J., "Whisker Alignment by the Alginate Process," *Composites*, **1:** (1) (1969).
11. Bagg, G. E. G., Evans, M. E. N, and Pryde, A. W. H., "The Glycerine Process for Alignment of Fibers and Whiskers," *Composites*, **1:** (2) (Dec., 1969).
12. Parrott, N. J., "Silicon Carbide as a Reinforcement". *Composites*, **1:** (3), 141–144, (March, 1970).
13. Prosen, S. P. and Simon, R. A. "Carbon Fiber Composites for Hydro and Aerospace," U.S. Naval Ordnance Lab, Oct. 1967.
14. Shaver, R. G., "Silicon Carbide 'Whiskerized' Carbon Fiber," AIChE Materials Symposium, Philadelphia Apr. 1–4, 1968.
15. Milewski, J. V., Shyne, J. J., and McGowan, H. C., "Interstitial Growth of Silicon Carbide Whiskers in Carbon and Graphite Filaments," Under Contract N-60921-67-C-0293, Sept. 1968.

Section XI
Continuous Filament Reinforcements

26
Fiber Glass

J. Gilbert Mohr
Johns Manville Fiber Glass Division
Waterville, Ohio

CONTENTS

1. INTRODUCTION

The product, fiber glass, is the result of conversion by fire of earth and oxide materials into a hot-melt, vitreous state and subsequent drawing into a filamentary aspect by high-speed attenuating mechanisms. The basic forms of fiber glass for reinforcement are two: continuous filament and staple fiber. A third form, not 100% applicable here, is comprised of blown fiber, which is the basic element of fiber glass wool. Wool products are utilized in filters, insulating products, and thin mats, one type of which forms the surfacing veils used in reinforced plastics/composites.

The fiber glass reinforcement types, after formation, are further processed into usable forms which are still only intermediates for final conversion into RP/C end products. These usable forms are designated as twisted and plied textile yarns, rovings, woven tapes or fabrics and woven rovings, chopped strand, chopped strand mats, combinations, and milled fibers. Specialty or related fiber glass products include such miscellaneous items as flake-glass, scrim fabrics, coated yarns and bonded cordage, bulk and wound fibers for filter media, fiber-optic elements, and many others.

Continuous filament fibers may be produced in a wide range of filament diameters from several different fiberizable compositions. When combined with the appropriate laminating resin, this fortuitous combination provides required versatility for a myriad of end-use applications. RP/C parts are sized from a 1 in. cube of extremely fine, precise dimensions through a 42-ft diameter × 40-ft high tank, capacity 414,600 gal., all made possible essentially because of the technology of fiber glass.

The structural and utility RP/C products so fabricated find outlets in a steadily growing number of applications in markets such as: automotive, appliance, aircraft, corrosion resistance, construction, electrical, marine, recreational, tooling, and transportation.

Hence, to summarize, as a textile and reinforcing material glass fiber provides the advantages of: a hard and impervious substance which is mildew and rot resistant, and which exhibits durability on exposure to corrosive chemicals; it does not relax its rigidly-held ionic bonds until a temperature of 1200–1500°F is reached; it has exceptionally high tensile strength, plus perfect elasticity, with an elongation of only 3% (no yield point); with the increased surface area per unit volume created by fiberizing, and the immediate chemical affinity of glass for (OH) groups available in the atmosphere, sites are provided for cross-linking and hence chemical bondings to specific polymers requiring reinforcement; glass fibers do not generate internal heat due to molecular friction during mechanical working as do metals; in the fibrous form, its stiffness factor, related to the minute filament diameter, permits processing on standard and applicable textile equipment without fiber breakage; and finally, numerous fiberizable glass compositions have been developed for specific applications, such as greater temperature resistance, higher electrical properties, resistance to acid or alkali, greater tensile strength or mechanical stiffness, and absorption of harmful radiation such as x-rays.

On the debit side, fiber glass applications are limited to a minor degree by: friability and brittleness which permits weakening from surface defects, thereby necessitating protective lubrication and/or surface treatments in processing; the fact that glass per se is a smooth-surfaced material without undulation or nodules that would provide mechanical locking sites for laminating; glass per se (and certainly no less glass fibers) is subject to static fatigue, causing the fibers to exhibit weakness when exposed to a sustained load near its innate elastic limit; and finally, glass fibers exhibit a lower-than-desired tensile modulus of elasticity.

2. HISTORY

A capsule summary of the history of fiber glass may be recorded by considering three major phases.

2.1 Invention and Development

In all probability, the original fiber glass was a creation of nature known as "Pele's Hair." This material resulted when high winds exerted a force on the leading edges of waves of vitreous volcanic dross during a lava flow. The lava was attenuated into a fine-fibered mass which was carried by the wind and deposited in trees. Strangely indicative of things to come, this fibrous mass was used by birds as reinforcement in construction of their nests. The phenomenon has been observed many places in the world, but notably in Hawaii from the Kilauea volcano, and from Mount Vesuvius in 1906.[1]

The first glass artisans in ancient Syria, Greece, and Egypt learned to draw fibers from a heated glass rod and to impress them while fluid onto the surfaces or stems of fabricated articles. This technique was used even prior to invention of the blowpipe in approximately 250 B.C. These fibers, as laid down, were either left in situ or further manipulated by reheating and tooling or scoring to form ingenious, artful decorations on the glassware of the day.

Similar and other related ways of incorporating fine-drawn glass fibers were used by Venetian glass makers in the 16th and 17th centuries (fine decorated, thin-walled "lace" glass stemware) and by the Romans (millifiori glass). English, German, and French glass producers (1650–1720) devised methods for drawing fine fibers from small melts or heated glass rods, and later (1832), glass threads from a hot glass gob onto a revolving wheel. The material so produced remained only a curiosity. Some was woven into fabric structures with other textile materials, and some was used for such decorative applications as the tails of birds in Christmas-tree ornaments.

Glass fibers drawn from heated rods were produced to make a dress and neckties and exhibited at a Chicago fair in 1893 by entrepreneur Edward Drummond Libbey and his technical genius, Michael J. Owens. A near-success by Libbey-Owens-Ford at a commercial fiber glass product occurred in the 1930's when colored glass fibers were laminated between sheets of flat window glass to form a glare-reducing structural panel. Unfortunately, these ventures were still classed in the novelty category, and the real expansion of the glass fiber industry still waited in the wings.

2.2 Commercialization

The Owens-Corning Fiberglas Corporation was founded in 1937, following 10 years of joint research between Owens-Illinois Glass Company and Corning Glass Works. This research program was especially geared to development of viable mass production methods for glass fibers, and resulted in workable technology for all three methods: continuous filament, staple fiber, and wool, or blown glass fibers.

The research was conducted at Purdue University. The original Owens-Corning producing facility was located at Newark, Ohio.

Original products were "white wool" insulation, coarse fibered air-filters, and continuous filament material which was used to weave a cloth subsequently impregnated with varnish for electrical laminates. The development of polyester resins in 1935 gave a natural impetus to the combination, and the reinforced plastics industry was at last down the ways for a first-class launching.

2.3 Expansion

The unfortunate incidence of World War II was fortunate indeed for the glass fiber reinforcing business. Successful applications of molded and laminated products were rapidly made. Development of molding methods particularly suitable to the two new materials, low-pressure curing resins, and the easily handled fiber glass reinforcement, proceeded with both speed and great ingenuity. The first reinforced plastic end-products were puncture-proof, jettisonable gas tanks, radomes, flack-suit body armor, and other non-metallic weight-reducing components for military aircraft, personnel, and land transportation.

Following the war, a concerted effort was made to adapt lessons learned on military applications to domestic production, although price-volume adjustments were necessary. Several domestic and foreign firms received licenses to manufacture from Owens-Corning Fiberglas Corporation. Research and development spurred on by lively competition among producers of reinforcing fiber glass products laid the groundwork and paved the way for a great upsurge in productivity beginning in 1965, and currently yielding well over 1 billion lb of reinforced plastic products annually. The improved technology brought about glasses with greater chemical durability, higher modulus of elasticity, and many other desirable properties. Improvements in resins and molding technology afforded fabricators the capability to produce commerical marine vessels over 150 ft in length, sophisticated beam and girder stock for structures in critical corrosive atmospheres, many elegant products for buildings and construction, automotive components produced at extremely high speed and with molded surfaces and paintability equalling their metal counterparts, and many others to be pointed out later. Licensing of the complex technology and production methods for fiber glass is presently continuing on a world-wide basis. To turn a phrase, the golden age of fiber glass and reinforced plastics is with us now, but in no way near retiree status.

3. METHODS OF PRODUCING GLASS FIBERS FOR REINFORCING

There are several basic methods employed for attenuating glass fibers from hot melts to yield products which have some role, large or small, in reinforcing plastics. These include continuous filament fibers, staple fibers, blown fibers for mats, and mechanically drawn fibers for mats. Each method is described individually here, together with a listing and brief description of the characteristic products resulting from each.

3.1 Continuous-filament Glass Fibers

The aforementioned historical process of drawing glass fibers from heated rods logically resolved to drawing fibers from hot melts. The high surface tension and slow rate-of-change of viscosity with temperature of melted glass permitted formation of stable meniscuses and enabled fibers to be formed and drawn from a suitable crucible after the natural flow of molten glass downward by gravity. The unique properties of the refractory metal platinum, including its ability to perform as an electrical resistor and to become incandescent at glass melting temperatures, give rise to its use as the most favorable melting and forming vessel, or "bushing", for drawing continuous filament fiber glass.

Marble Melt vs Direct Melt. In the "Neanderthal" era of fiber glass production, glass was melted in a separate furnace and formed into marbles, which were partially annealed and graded. Then they were fed into heated platinum bushings, where they were remelted and transformed into filaments, being drawn away rapidly at speeds approaching 12,000 ft/min. There is nothing magical about marbles except that they were easily formable mechanically, and rolled by gravity through grading and feeding contrivances.

As technology improved, the marble "remelting" step was superseded by installing in-line fiber producing bushings in the appropriately designed forehearths of a glass tank furnace, and the process known as "direct melt" glass fiber production evolved. Whereas marble bushings were originally capable of 7-9 lb/hr, production from larger direct melt bushings may now approach 100 lb/hr.

Continuous glass filaments produced commercially range from 2-3 μ to 25 μ in individual fiber diameters. Marble bushings are still utilized, being more satisfactory for fabrication of finer filaments. They also permit use of glass marbles, the quality and production efficiency of which have been predetermined, and thus, poor or undesirable glass can be eliminated.

Not so in direct-melt! In a production operation, the glass quality in a direct-melt furnace does not depart from desirable norms very often, but when and if it does, the fiber production obviously suffers. Also, by comparison, marble bushings obviously favor easy change to glasses of different composition. In summary, direct-melt is the preferred process because it eliminates the remelting step necessary in marble melt. Figures 26-1 and 26-2, respectively, illustrate schematically the marble and direct melt processes.

Function of the Bushing. The stages of reduction of batch into usable glass in a melting tank or furnace are three: melting, refining (or homogenization), and working (or processing). A

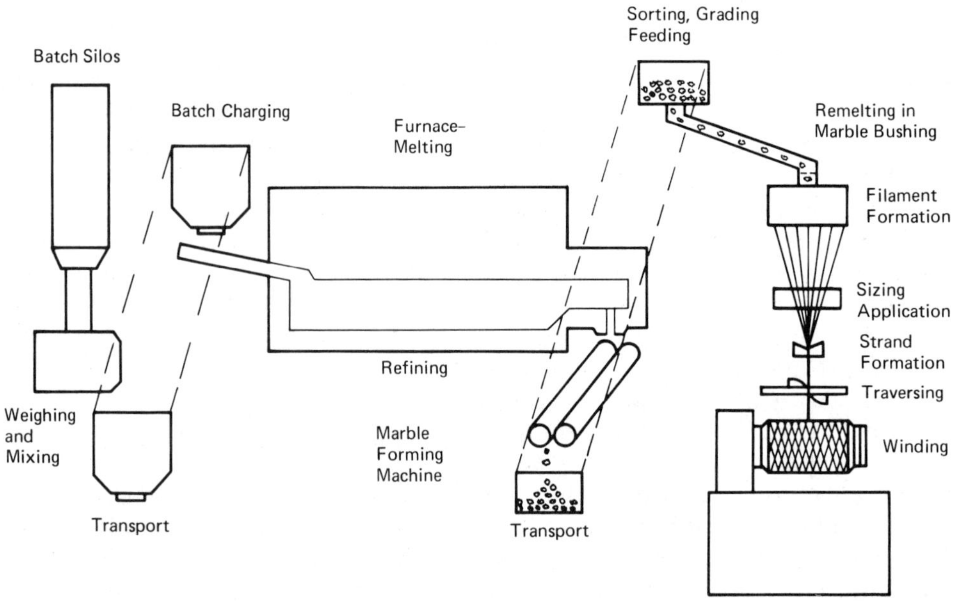

Fig. 26-1. Schematic diagram of marble process.

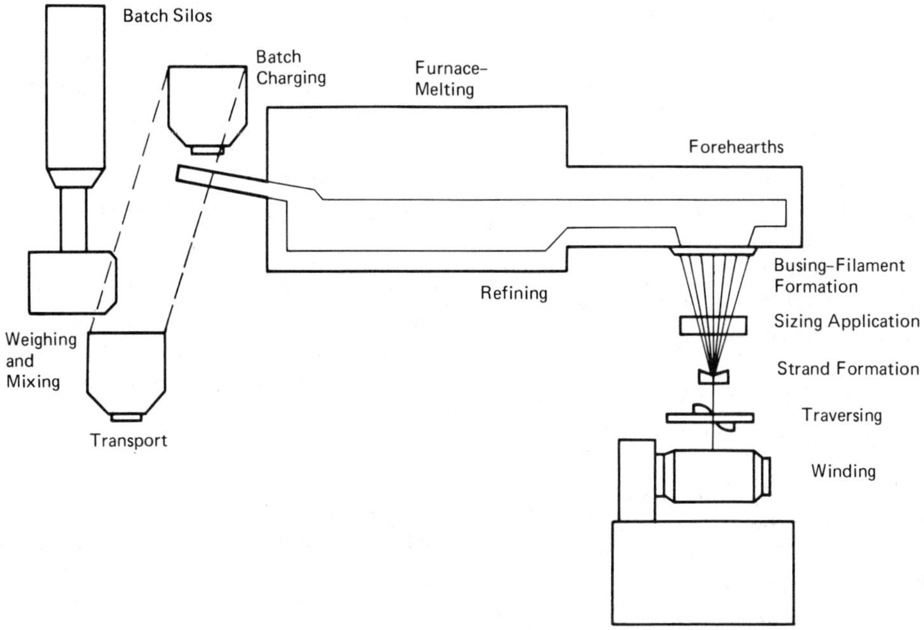

Fig. 26-2. Schematic diagram of direct melt process.

fiber glass bushing may be considered as a miniature melting unit since the glass is subjected to the same three phases (eliminating actual melting from batch) when progressing from the point of entry to the juncture at which filaments are formed. Some differences exist in design requirements for marble versus direct-melt bushings, the latter requiring less height because the glass delivered to them is already melted and at the approximate working temperature of the furnace (approximately 2400°F). Upon entering the bushing, the glass gravitates through a perforated baffle plate, inducing homogeneity, and thence is cooled further by entering air-exposed "tips" which are surrounded by cooling media. These act as a further heat-sink, lowering the glass temperature to approximately 2250°F (for "E" fiber glass composition, as an example). A representative marble bushing construction is shown schematically in Fig. 26-3.[2]

Strand Formation. The next step essential in glass fiber production is the formation of a "strand". A bushing may have a variable number of tips, depending on end-requirement, and each tip is capable of forming a filament. Through evolution, almost all bushings have been constructed with multiples of 204, the number designed into the original commercial production bushings.

Steps in strand formation comprise the following, below the bushing: (1) formation of a meniscus at each tip from which the filament is

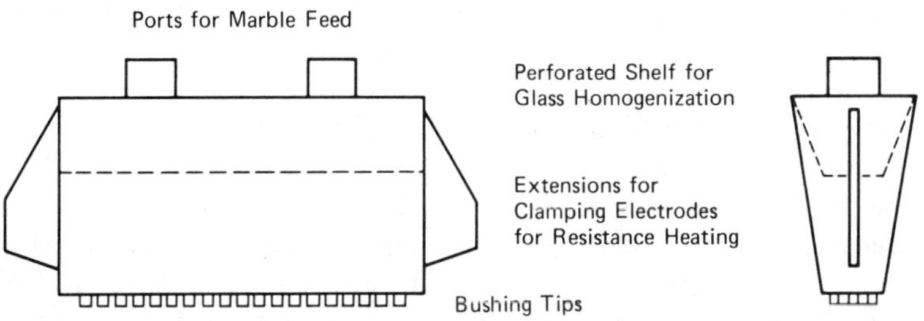

Fig. 26-3. Schematic diagram of construction of fiber glass bushing.

attenuated, (2) application of a water spray in a downward draft of humidity- and temperature-conditioned air, (3) application of an organic sizing composition, (4) collection into a strand using a suitable mechanical gathering device, and (5) winding onto a high-speed drum after passing over a traversing mechanism which creates a crossed, non-parallel orientation in the forming package. The elements required below the bushing for strand formation and winding are detailed schematically in Figs. 26-1 and 26-2.

Fiber production is a "batch" process, with each forming package requiring removal after it has reached optimum size. Semi-automatic winding mechanisms are in use which lessen the changeover time, however. The strand is formed at the gathering point, but does not actually preserve its identity until the full function of the binder or sizing ingredients is brought to bear. These proprietary formulations of organic chemicals contribute three main elements to the strand: (1) lubrication, both during and after application to the glass fibers, (2) formation of a film to bind the individual filaments together when the forming package of fiber is removed and dried, and (3) compatibility with the resin or types of resins contemplated for the end use of the fiber as plastics reinforcement.

Filament Diameters. The "denier," or strand length-to-weight relationship, hence filament diameter (classified as either yards per pound or Tex in the fiber glass industry) is determined or influenced by the following: glass composition, viscosity, tip diameter, number of tips or filaments, drawing temperature, efficacy of cooling, and speed of attenuation (or winding).

The complete gamut of filament diameters produced commercially as textile or reinforcing fibers is shown in Table 26-1. It is necessary to understand that the basic strand designation number is determined by dividing the nominal yards per pound for the output of a 204-hole bushing producing the particular filament size, B through U fiber by 100. Each filament size, designated by letter, has a small minimum to maximum tolerance range, not shown in the tabulation.

Coincident with the development of larger bushings, strands have been made available for end-use RP/C processing with an increasingly larger number of filaments. As the number of filaments increases, the strand weight also naturally increases. Hence, for strands containing multiples of 204 filaments, yards per pound will decrease proportionately, and the Tex value (grams per 1,000 meters) will increase proportionately.

Preparation for Processing. Glass fibers are treated with two distinct and separate classes of sizing materials: (1) a starch-oil emulsion type which has humectant properties, absorbing water to provide lubrication plus resistance to abrasion in such down-stream processes as twisting, plying, weaving, cording, etc., and (2) a plastic emulsion-cross-link type of sizing, generally for direct use in RP/C. The starch-oil size must be processed in areas above 55% humidity for best performance. It is later removed or coated for specific applications. The plastic emulsion type sizing must be oven-cured to drive off water, set the film, and bond the filaments into a strand.

Considerable technology has been rounded-out in making glass fibers compatible for matrix materials (resins) with which the fibers are ultimately to be combined. These treatments include chrome or silane finishes for fabrics (applied after weaving and heat-cleaning); resorcinol-formaldehyde treatments of glass yarns for tire cord; and chrome, silane, or organic titanate ingredients of sizings applied to continuous filaments specific to the particular matrix for which the fiber glass product is intended as reinforcement. The full import of these treatments will become more evident following the ensuing discussion of various fiber glass reinforcing products.

3.1.1. Continuous-Filament Glass Fiber Products. Almost all glass fiber products may be classed as intermediates, since they must undergo further processing or combination to reach their full potential value as a reinforcement.

Referring to Table 26-1, the strands formed using the finer filaments in the B to F range find application in decorative household fabrics, hence have little or no value as reinforcement, and are not of further interest here. The

Table 26-1. Description of basic continuous filament fiber designations with dimensions and corresponding length : weight relationships.

Letter Designation	Industry-assigned Strand Count–204 Filaments Only[a] approx. yd/lb 100	Nominal Filament Diameter[b] in.	μ	Nominal yd/lb for a 204 Filament Strand[c]	Corresponding TEX Value for a 204 Filament Strand[d]
B	900	.000125	3.19	136,375	3.4
C	600	.000175	4.45	65,525	7.6
D	450	.000225	5.72	38,700	12.8
E	225	.000275	6.99	25,590	19.3
F	170	.000325	8.26	18,200	27.2
G	150, 135	.000375	9.53	13,615	36.3
H	110	.000425	10.76	10,755	46.0
J		.000475	12.07	8,625	57.3
K	75	.000525	13.34	6,890	79.2
L		.000575	14.61	5,740	86.1
M		.000625	15.88	4,855	101.8
N		.000675	17.15	4,163	118.8
P	37	.000725	18.42	3,608	137.1
Q		.000775	19.69	3,155	156.7
R		.000825	20.96	2,783	177.7
S		.000875	22.23	2,475	199.8
T		.000925	23.50	2,212	223.5
U	18	.000975	24.77	1,990	248.5

[a]These strand counts represent the filament sizes most commonly produced on a commercial basis. The strand count is doubled for half the number of filaments and decreased by any factor used for a larger multiple number of filaments per strand.

[b]Filament diameter ranges: in. = ±0.000025, μ = ±0.64. One micron equals 0.00003937 in. (39.37 millionths of an in.). These filament diameters are also used to classify glass fiber types other than continuous filament, i.e., mats, sliver, etc.

[c]The values given for yd/lb of a 200 filament strand represent the middle of the range. The limits correspond to the range limits for filament diameter. The notation given for variation in strand count also applies here, i.e., a 400 filament strand of K filaments would average 3608 yd/lb (37's) and an 800 filament strand would average 1990 yd/lb (18's). Conversely, 200 filament D fiber (450's) would yield approximately 90,000 yd/lb (900's) with 100 filaments, and 180,000 yd/lb (1800's) for a 50-filament strand.

[d]TEX value is given because of its importance in attempts to establish a universal, international yarn measuring system. TEX is expressed as grams per 1000 meters (km), or equals $\frac{494420}{\text{yd/per/lb}}$.

bulk of reinforcing fibers utilized, then, fall in the G to P ranges, with some small amount of fiber below the G range finding specialty applications.

Generally speaking, finer fibers promote faster wet-out in resins, and also exhibit lower stiffness and hence lower resistance to mold conformity. However, the finer grade fibers are more difficult to produce.

Table 26-2 presents a brief description of the major types of glass fiber products for reinforcement of plastics. Included are statements of qualities or grades available and whether the reinforcement is generally suitable for room temperature cure, with open layup (HSB-high solubility sizing), high-temperature cure as in press molding (LSB-low-solubility sizing), or use in reinforcing thermoplastic resins (RTP). These designations will receive further clarification in a later Section discussing Reinforced Plastics (see Table 26-10).

3.1.2 Miscellaneous. Two products different in form but related to continuous filament fiber glass in function deserve mention here, and are included in Table 26-2 as reinforcing products.

Glass Beads. Hollow or solid glass spheres are manufactured by dropping minute glass fragments of controlled weight or screen size downward through a heated shot-chamber or equivalent. See Chapters 18 and 19 for

Table 26-2. Types of continuous-filament fiber glass reinforcing products (with related forms).

Type of Fiber Glass Reinforcement	Definition and Description	Range of Grades Available	General Types of Sizing Applied	General Usage in RP/C (and Secondary Uses if Any)
Twisted yarns	Single-end fiber glass strands twisted on standard textile tube-drive machinery	"B" to "K" fiber, 0.25–10.0 TPI, "S" or "Z," many fiber and yardage variations	Starch	Into plied yarns for weaving; many other industrial uses
Plied yarns	Twisted yarns plied with reverse twist on standard textile ply frames	"B" to "K" fiber, up to 4/18 ply–many fiber and yardage variations	Starch	Weaving industrial fabrics and tapes in many different cloth styles; also lighter decorative fabrics and heavy cordage
Fabrics	Plied yarns woven into a multiplicity of cloth styles with various thicknesses and strength orientations	"D" to "K" fibers, 2.5–40 oz/sq yd in wt	Starch size removed and compatible finish applied after weaving	Wet lay-up for open molding and also some press molding
Chopped strands	Filament bundles bonded by sizing ingredients, subsequently cured and cut or chopped into short lengths for further processing	"K" to "M" fiber, 1/8–1/2 in. or longer lengths, various yardages	LSB, RTP	Compression, transfer and injection press molding
Roving	Strands from forming cakes wound in parallel and untwisted manner into large packages using standard take-up machines; with more recent technology, unitized roving packages are wound directly in a single step	"G" to "P" fiber used. Strands 900 to 120 yd/lb and packages 5 to 450 lb wt. (up to 24 × 24 in. in size)	HSB, LSB, and RTP	Used in all phases of RP/C
Woven roving	Coarse fabric, bi-directional reinforcement, mostly plain weave, but some undirectional woven roving is produced	"G" to "P" fiber, fabric weights 15–48 oz/sq yd	HSB	Mostly wet lay-up, but some press molding
Chopped strand mats	Fibers from forming cakes chopped and collected in a random pattern with additional binder applied and cured; some "needled" mat produced with no extra binder required	"G" to "K" fiber, weights 0.75–6.0 oz/sq ft	HSB and LSB	Both lay-up and press molding
Mats, continuous strand (swirl)	Fibers converted directly into mat form without cutting with additional binder applied and cured	Nominally "K" fiber, weights 0.75–4.5 oz/sq ft	LSB	Press molding
Mat-woven roving combinations	Chopped strand mat and woven roving combined into a usable reinforcement with additional binder applied	38–62 oz/sq yd	HSB	Wet-lay-up to save time in handling
Milled fibers	Fibers reduced by mechanical attrition to short lengths in powder or nodule form	Lengths range 0.001–1/8 in. Several grades	None, HSB and RTP	Casting, potting, injection molding
Glass beads	Small solid or hollow spheres of glass	Range 1–53 μ diameter, bulk densities = hollow–0.15–0.38 g/cc; solid = 1.55 g/cc	Usually treated with cross-linking additives	Used as filler or weight reduction medium in casting, laminating and press molding
Glass flake	Thin glass platelets of controlled thickness and size	Nominally 0.004 in. thick	None	Used as a barrier or protective material by incorporating into resins used for corrosion resistant coatings.

detailed information on solid and hollow glass spheres.

Glass Flake. Glass flake is produced by a proprietary process. It is used as a resin filler for specialty applications. See Chapter 20 for information on flake composites.

3.2 Staple Fibers

Fibers which are noncontinuous in derivation, but which may be reduced to "strand" form are designated as staple fibers, and the individual composite strand is referred to as "sliver".

Manufacture. Staple fibers are manufactured by several different methods, almost all of which have as their primary fiberizing source a blown fiber bushing or crucible. This bushing is usually fed by marbles, and the glass compositions employed may be varied to cover a wide range of properties.

The staple fiber forming process requires collection of the blown fibers on a moving belt or other suitable surface, slightly drafting to induce alignment of the filaments, compacting or integrating the strand through a mechanical false-twister, and ultimately placing into a suitable package on a winder or take-up machine. An oil or sizing may be applied at some suitable location to abet manufacture and enhance strand properties (see Fig. 26-4).[3] Individual filament lengths in staple fibers may vary from 5-15 in. The filament diameter may be varied between E and K fibers.

Usage. Staple fibers are widely used in electrical insulation applications, for battery mat separators, and as filter media. Use in reinforced plastics has, in the past, been made in roving and woven roving products. This usage in the United States is extremely limited or nonexistent because of the low laminated strength values obtained.[4]

3.3 Blown Fiber (Veil) Mats

Light-weight fiber glass mats have found great utility in reinforced plastics for a variety of applications. Although they do not provide highest laminated strengths, the function supplied can not be duplicated by continuous-filament reinforcing fibers alone.

Manufacture. Figure 26-5 provides a schematic representation of a typical production method for blown fiber mats. The finished products are made in thicknesses varying from 10-30 mils (0.6-1.8 lb/100 sq ft). A chemically resistant glass composition is usually used. Filament diameters usually average 17 μ. A wide variety of binders are applied to suit end uses. These include none, acrylic, polyester, and others.

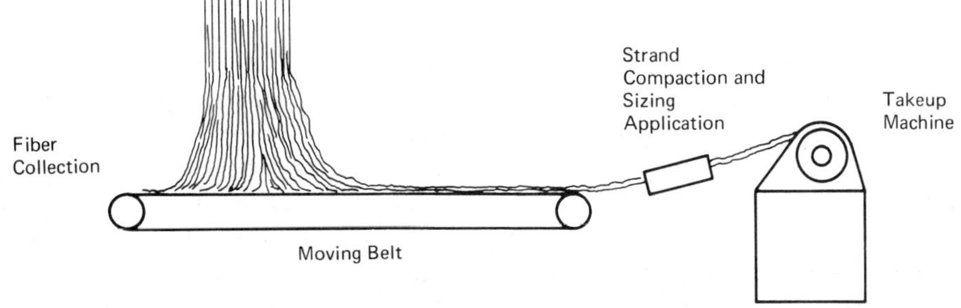

Fig. 26-4. Production of staple fiber.

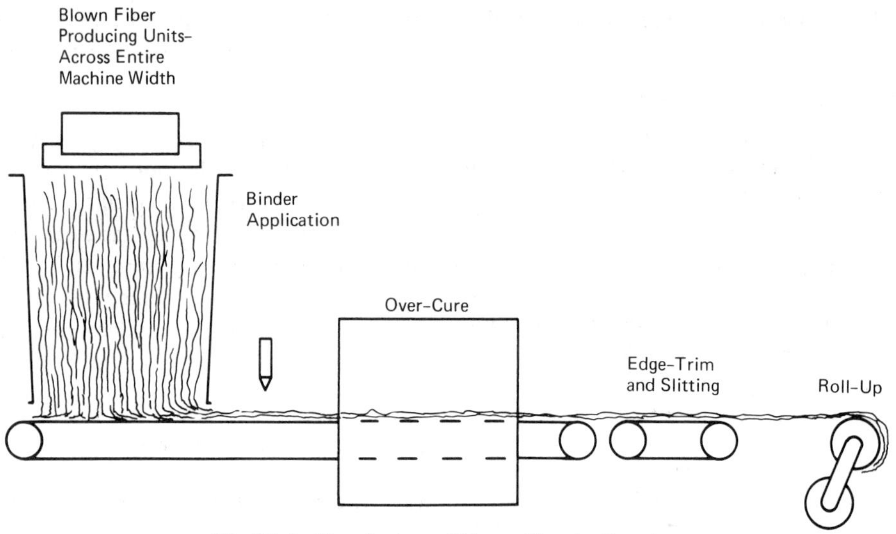

Fig. 26-5. Manufacture of blown fiber (veil) mats.

Usage. Surfacing or veil mats are applied to the inside of wet-layup composites (tanks, etc.) to provide a resin-rich, chemical resistant layer and to protect the underlying reinforcement layers, usually of "E" glass. They are also employed as a surface covering in press-molded items, preventing the reinforcing fiber pattern from appearing in the prime molded exterior.

3.4 Mechanically Attenuated Mats

The Modigliani process for mechanical attenuation of glass fibers has made possible adaptation of resultant mats to many useful applications in reinforcing plastics. The process is ingenious, and the resultant products are slightly more versatile with better uniformity of fiber distribution than blown fiber mats. This process has also been used, although unsuccessfully, for production of continuous filament reinforcing fibers.

Manufacture. A three-step process has evolved for conversion of the mechanically drawn fibers into mat form. These steps are illustrated schematically in Fig. 26-6. In step A, a gas-fired furnace (clay type with perforated Inconel bushing plate) is mounted on a structural frame to reciprocate parallel to and well above the main longitudinal axis of a large rotating drum 4-5 ft in diameter and 10-12 ft in length.

As the drum revolves, the fiber is drawn down by hand from the bushing and fed so as to be forced against the drum due to the low-pressure condition caused at the drum surface by the high speed of revolution of the drum. The Modigliani process is not capable of producing filaments as small in diameter as those in the continuous filament process. However, there is an advantage in that single filaments which break out from the bushing may be restarted around the winding drum.

As the drum revolves and the furnace reciprocates, a layer of glass fiber is built up on the drum surface. The winding angle is normally 8-9 degrees from centerline, but this may be changed even within a given wind to yield products of varied characteristics.

In stage B, after winding has been completed, the fibrous mass is cut at a right angle to the fiber direction and laid flat. Binder may be added to the glass fiber during build-up on the drum, but must not be permitted to dry at this phase.

In stage C, as shown, the fibers are attached to a bait and drawn transversely. Additional binder is applied, compression means are applied if a very thin flat mat is required, and the material is passed through a curing oven to set the binder.

Lower-melting "A" glass is usually used since the furnaces are charged with cullet (soda-lime

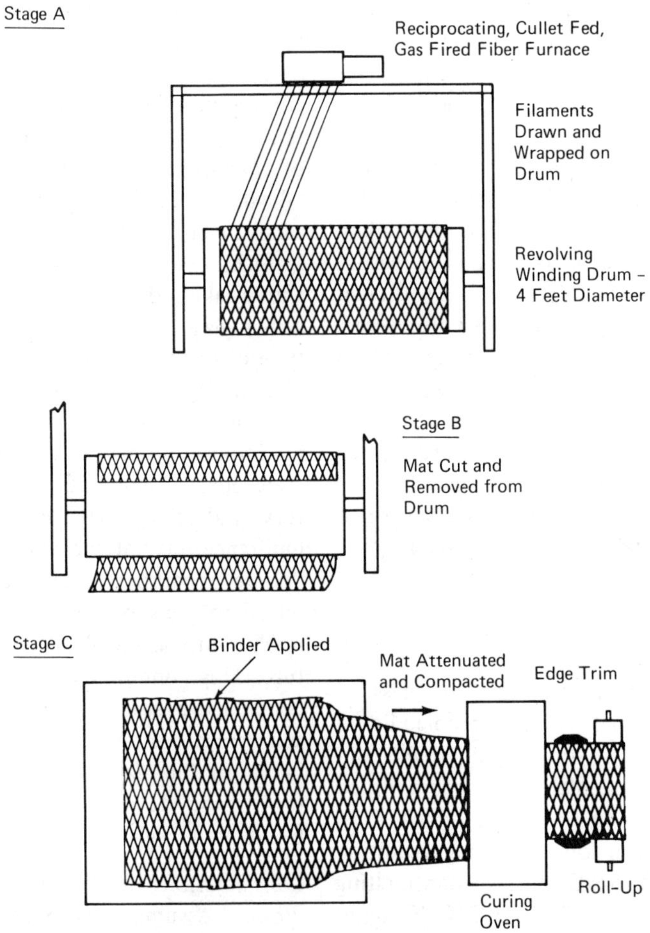

Fig. 26-6. Mechanical attenuation of glass fibers for mat production.

glass). Some glasses with high modulus or other exotic compositions have been used in the past, however, and laminates, not mats, were produced directly from the drum-wound fibers. Commercial production of mats by this process has persisted.

Mats from 10-60 mils thick may be fabricated by this process, and they are characterized by the typical "diamond" pattern formed between the glass filaments or residual fiber bundles. Filament diameters fall mostly in the L to P range.

Usage. The mechanical mats are competitive with the blown fiber mats for veil or surfacing mat applications. The mechanical types are slightly better for uniformity, and provide the persistence of the diamond-shape pattern in the laminate, enhancing the decorative aspects of the material. The thicker types, due to their superior uniformity, may actually be used as reinforcing mats in low-draft, low strength applications.

4. PROPERTIES OF GLASS FIBERS FOR REINFORCEMENT

4.1 Basic Considerations

The purpose of this brief discussion is to identify some of the basic chemical and physical changes that are brought about in glass-forming oxide materials when combined and subjected to melting temperatures, their behavior as a glass in the melted state, and the conditions which occur and prevail in processing during which the glassy melt becomes a usable rigid solid after cooling to room temperature. The discussion is intended as a background for a better understanding of the properties of glass fibers of the several different fiberizable compositions that will be considered.

Acknowledging that any one of the several facets of glass science touched upon here has been the subject of more than one life study,[5] no attempt will be made to explain the reasons for these fascinating occurrences. It is sufficient to say that the principles have been discovered, slowly developed by man, and brought to an advanced stage of beneficial utilization, of which glass fibers is by no means of minimum importance. To set the stage, the word "glass" leads somewhat of a double life, since it refers both to the material in the molten stage and also to the clear rigid solid that exists at room temperature.

Melting and Latent Compound Formation. The many definitions, long and short, large and small, that have been proposed for glass all agree that the material is an undercooled liquid with great propensity for crystal growth and compound formation. Melting is accomplished by a mutual solution of oxides which form eutectics or points of lowest melting. Addition of only 25% Na_2O to SiO_2 forms a new melting point, changing from 1730 to 793°C. Ternary, quaternary, and further multiple eutectics occur in any given oxide combination.

The compounds that may form and grow out of the melt are well-known chemical combinations of the oxides in varying proportions, and many occur elsewhere as natural minerals. Glass melted and brought directly to room temperature will usually turn out to be a clear rigid solid as intended. Crystal formation depends upon composition and is controlled by two factors: temperature-i.e., the glass composition in question must be held at a point below its liquidus temperature for mutual existence of crystal and glass phases, and below its solidus temperature for complete crystallization; as temperature decreases, glass viscosity increases, and hence crystal formation and compound growth are inhibited. In studying compound formation out of glassy melts, and determining liquidus and solidus curves, compositions must be kept at a given temperature for sometimes as long as one year or more to overcome the effects of viscosity and establish equilibrium. The physical chemistry of glasses (and ceramics) is well documented.[6] The result of crystals growing out of a glassy melt, termed "devitrification", is usually deleterious. These small nuclei or "stones" usually occur on the surface, interfere with production efficiency, and weaken the finished product. Most fiberizable glass compositions are subject to some type of devitrification, and must be cooled rapidly through the forming temperature range for crystallization to be avoided. Crystals induced by impurities may also grow out of a glassy melt due to unmelted batch, dissolved refractory, and other causes. As an adjunct and not a hindrance, crystals or compounds, by selective "phase separation", may be deliberately induced to grow out of a glassy melt and intermesh tightly as the glass cools to room temperature. This condition is induced by protracted, controlled post-forming heat treatment. The results are a desirable increase in strength, resistance to thermal shock, and other improved properties.[7] It remains for this technology to be adapted to glass fiber production.

Consideration of Differential Thermal Conditions. Assuming that bulk glass is successfully cooled through the critical zones without crystallization, the fact that it is a poor conductor of heat requires that care must be exercized to avoid excessive strains which would cause breakage. The method of preventing strain is "annealing," or gradually reducing temperature so that strains do not develop. In the solid form, glass has a viscosity of 10^{15} poise. At the softening point, strain will be released immediately: at the annealing point (10^{13} poise), strain will be released in 15 min. In controlled "chilling," "tempering," or "disannealing," the surface of a glass item, which is formed but held well above its annealing point in temperature, is rapidly cooled. By a unique process of combined contraction, flow, and contraction, the surfaces of the glass end up under a strong compressive force, held or maintained by the central portion of the cross-sectioned area under a strong tensile force.

This counteraction of forces strengthens the glass by factors of from 2.5-10, as determined by thermal coefficient of expansion and thickness. In fibers, the identical rearrangement occurs naturally and without the need for controlled cooling, because the filaments are not massive enough to set up breaking stresses. This thermal-shock treatment contributes substantially to the 40 or 50 to 1 tensile strength increase of glass in fibrous form over glass in the bulk form.

Properties vs Composition. The next most important consideration is the effect of the various glass forming ingredients on total properties, i.e., when melted as the glassy solution, and when rigid, as the usable glass product. Silica is the most important glass forming oxide. However, the sustained high temperatures required to melt large quantities of SiO_2 and free it from bubbles (due to its high melt viscosity) are commercially impractical. Hence, other oxides or fluxing agents are used, and this technology originally prompted the extensive investigation and use of phase diagrams. Silica (75%) and sodium oxide (25%) combine to form a true glass which is highly susceptible to water attack, even by the moisture in the atmosphere. Addition of CaO or CaO-MgO stabilizes this weakness for corrosion, but makes the glass susceptible to devitrification. Incorporation of a small quantity of alumina (Al_2O_3) directs the melt away from its tendency to devitrify, but adds again to its working viscosity.

Stabilization of glasses by oxide modifiers which will dissolve in the melt has been the basis for modern glass technology, and has resulted in many useful glass products. Much is known about the function of each usable oxide, and its effect in combination with other ingredients. The stages of oxidation and reduction and degree of acidity of the glass are well known. Edisonian research is no longer the rule in glass formulating. Also, many sophisticated testing systems and devices are currently available for meaningful analysis and evaluation.

Prior to a full-blown discussion of glass fiber compositions and their properties, Table 26-3

Table 26-3. Functions of the most common glass-forming oxides.

Glass Forming Oxide	Behavior in Glassy Melted State	Contribution to Finished Glass Properties
SiO_2	Most desirable glass-forming oxide, but requires extremely high melting temperatures, has high viscosity, and slow rate of bubble release	Very low thermal expansion
Na_2O, K_2O, Li_2O	Low viscosity and good fluidity	High thermal expansion, easily attacked by moisture
CaO, MgO	Intermediate viscosity, easily devitrified	Improved chemical durability and resistance to attack by water, acids, alkali
B_2O_3	Intermediate viscosity, acts as fluxing agent	Low thermal expansion
Al_2O_3	Increases melt viscosity	Assists in improving chemical durability
Fe_2O_3	Present as impurity unless used as colorant, lends some fluxing action, but absorbs infra-red and interferes with transfer of heat through melt	Discolors green in small quantities
ZnO	Slightly higher viscosity	Adds to chemical durability
PbO	Substantially reduces melt viscosity and acts as fluxing agent	Increases glass density and brilliance (light transmission), induces high thermal expansion
BaO	Intermediate glass viscosity	High density, some benefit in increasing chemical durability
TiO_2	Slightly higher viscosity	Improves chemical durability, especially alkali resistance

has been prepared to show both the melt qualities and finished glass properties imparted by each of the more commonly known major glass constituents. Please keep in mind that combinations are important, and will further modify individual oxide performances.

4.2 Properties of Fiberizable Compositions

To illustrate the importance of properties of glass melts, especially viscosity, a comparison is made with different types of bulk glass and fiber glass compositions. Glass intended for press molding or hand working is formulated to possess a long, slow rate-of-change of viscosity on cooling so that ample time is allowed for forcing glass to all extremities of the mold, or completing final shape manipulations before complete rigidity is reached. Conversely, glass intended for tube or rod drawing or mechanical blow-molding, is required to set fairly rapidly when cooling below the minimum working temperature, so that no distortion occurs after the final shaping has taken place, and also so that high production rates may be maintained.

An important requirement of fiber glass compositions is that they possess "short" viscosity performance akin to that of the second example given. This means setting up rapidly on cooling when drawn from the miniscus, but not so rapidly as to interfere with establishment of the required filament diameters.

Other properties of fiberizable glass compositions, particularly those at room temperature conditions, will be considered here. Table 26-4 shows a listing of 6 formulations representing a wide range of uses of fibers for commercial and specialty applications. The various types, designated by an industry-assigned letter nomenclature, are defined and properties discussed as follows: chemical properties, including composition and durability, followed by physical, thermal, optical, and electrical properties. Each composition may be studied by comparison and by reference to the chemical (oxide) composition, so that the reader may establish personal, scientific "bench marks" for the reasons the compositions perform as they do. Each of the various property classifications is discussed separately, with emphasis on the importance of the fiber glass composition in RP/C.

Classifications. Of course, many different fiber glass compositions have been developed and are used other than those listed. The 6 selected, however, were taken to represent a wide range of properties and performance. The list includes:

Designation	Description
A	a soda-lime glass similar to window or bottle glass with low thermal constants and poor chemical durability
C	a soda-lime-borosilicate glass with excellent chemical durability
D	a low-density glass with improved electrical properties
E	a lime-borosilicate glass, derived from a "Pyrex^R" composition, the real workhorse of general purpose reinforced plastics applications
S and YM-31A	high-strength and high modulus glasses developed to improve stiffness of RP/C laminates and moldings over that possible using "E" glass

Please note that, while the glass composition is extremely important, the functions of both resins and filler in an RP/C structure are equally important, and are also developed for specific applications. In a critical chemical process tank, "C" (chemical) glass is always laminated into the suitable resin on the surface to protect the main body of the structure, usually fabricated using "E" glass. On the other hand, in a DAP polyester molding compound, formulated for maximum electrical properties, the type filler used ($Al(OH)_3$) contributes much more to the ultimate molded part properties than does the fiber glass. Summarizing, the fiber glass industry has done its homework very well by making available many compositions for evaluation with the other RP/C elements to satisfy requirements and meet the demands which the reinforced plastics industry is continually challenged to meet.

Table 26-4. Fiber glass compositions and their comparative properties.*

Fiber Glass Composition	"A"	"C"	"D"	"E"	"S"	YM-31A
Description, definition, and characterization	Typical soda-lime-silica glass, limited for reinforcement due to poor resistance to water	Chemical glass–possesses improved durability, making it preferred composition for applications requiring corrosion resistance	Glass with improved dielectric strength and low density, developed for improved electrical performance	Borosilicate type, used for major share of all reinforcement applications	Glass with high tensile strength and modulus, developed for aerospace applications	Special composition developed for high modulus to impart greater rigidity to reinforced structures
Chemical properties: Chemical composition %						
SiO_2	72.0	65.0	74.5	54.0	65.0	53.7
Al_2O_3	0.6	4.0	0.3	14.0	25.0	
Fe_2O_3	–	0.2	tr	0.2	–	0.5
CaO	10.0	14.0	0.5	17.5	–	12.9
MgO	2.5	3.0	–	4.5	10.0	9.0
B_2O_3	–	6.0	22.0	8.0	–	(Li_2O–3.0)
Na_2O	14.2	8.0	1.0	{ 0.6	–	(BeO–8.0)
K_2O	–	–	1.5		–	(TiO_2–8.0)
ZrO_2	–	–	(Li_2O–0.5)	–	–	2.0
SO_3	0.7	0.1	–	–	–	(CeO–3.0)
F_2	–	–	–	0.1	–	–
Chemical resistance– 14 μ fiber % weight-loss– 1 hr boil-in						
H_2O	11.1	0.13	–	1.7	–	–
1.0*N* H_2SO_4	6.2	0.10	–	48.2	–	–
0.1*N* NaOH	15.0	2.28	–	9.7	–	–
Physical properties						
Specific gravity	2.50	2.49	2.16	2.54	2.49	2.89
Virgin tensile strength, psi	350,000	400,000	350,000	500,000	665,000	500,000
Tensile modulus elasticity, psi	9,800,000	10,000,000	7,500,000	10,500,000	12,500,000	15,900,000
Thermal properties						
Softening point, °F	1,300	1,380	1,420	1,555	1,580	N.A.–devitrifies
Coefficient of thermal expansion–in./in./°F × 10^{-7}	90	40	17	28	16	–
Optical properties						
Index of refraction	1.512	1.541	1.47	1.547	1.523	1.635
Electrical properties						
Dielectric constant, 72°F, 10^6 Hz	6.90	6.24	3.56	5.80	4.53	–
Loss tangent, 72°F, 10^6 Hz	0.0085	0.0052	0.0005	0.001	0.002	–

*Portion of data courtesy Owens-Corning Fiberglas Corp.

Chemical Properties. Bearing in mind the earlier emphasis placed on contribution of each oxide to glass properties, the chemical compositions (determined by wet quantitative analysis) may be examined to relate the importance of oxide percent to the different performance parameters. Also, please note that some new oxides not previously discussed have been incorporated into the "D" and "YM-31A" formulations. The actual chemical durability of "A," "C," and "E" glass as weight loss after a boiling test on 14 μ fibers shows that "C" glass is best by far, and the "E" glass is particularly poor in resistance to attack by acid. Given the same glass, a decrease in filament diameter will result in an increase of the chemical attack due to the extra surface area. It should be pointed out that chemical durability of glasses may be improved by surface treatment involving ion exchange and oxidation.[8] It is doubtful if this process has yet been adapted to glass in the fiber form.

Physical Properties. Physical properties listed in Table 26-4 include specific gravity (density), virgin fiber tensile strength, and tensile modulus of elasticity. Specific gravity is an important property for controlling glass quality right at the melting operation. Its sensitivity to even minor composition variations is readily detectable. In fact, to illustrate the advanced development of modern glass technology, density factors or constants have been assigned to each glass-forming oxide. The product of the percentage of each oxide in the glass formulation and its characteristic density factor are determined, and the accumulated total of these determinations for all oxide ingredients duplicates almost exactly the actual value for the glass density determined by test after the glass has been formulated and melted.[9] Also, to illustrate further the relation of glass density to molecular orientation, each comparative and/or sequential glass sample for density determination must be annealed according to a precisely programmed decreasing temperature cycle, so that consistent results are obtained.

Because weight/volume relationships are important in applications of reinforced plastic components for aircraft, transportation, and the like, specific gravity is important. The only fiberizable composition showing any significantly lower density is "D" glass. Hollow reinforcing glass fibers were developed several years ago, but apparently have not yet proven to be a major factor in weight reduction of laminates.

Virgin tensile strength and tensile modulus of elasticity are important physical properties of glass fibers, considering resistance to applied stresses. The more-than-adequate tensile strength of "E" glass is indicated in Table 26-4. Other glasses that are intended for similar applications, e.g., A, C, and D glass, are notable lower by comparison. In practice, from 70-80% of the "E" glass virgin tensile strength may be attained in a parallel-strand type laminate with 85% glass loading. Only "S" glass, a special high-performance composition difficult to fiberize, has a higher virgin tensile strength. Conversely, the tensile modulus of elasticity for "E" glass is low. This condition permits undesirable bending in laminated structures. Attempts to develop glasses with higher modulus and provide greater resistance to stress deflection in laminates have resulted in a class of compositions well represented by YM-31A glass (modulus of 15.9 million psi). These have proven valuable, but are superseded by reinforcements such as carbon and graphite (modulus 20-70 million psi) and /or beryllium or boron fibers (modulus 38 and 60 million psi respectively). Carbon and graphite fibers also have the advantage of extremely low density (1.4-1.9).

Another interesting physical property of glass not indicated in Table 26-4, applicable to both bulk and fibrous forms, is elongation. Glass is perfectly elastic, and recovers fully to original dimensions provided the extension has not exceeded 2.5-3.0%. If glass is extended beyond 3%, there is no yield point, and immediate fracture occurs. This condition is desirable for maximum strength output of the reinforcement. However, all RP/C components reinforced with fiber glass must be over-designed by a safe margin to avoid catastrophic failure.

Thermal Properties. Despite all the comparisons made of cooling characteristics of glasses, it is well to explain that approximately the same top melting temperature (2800°F+) is reached in commercial glass tank furnaces to reduce charged batch to molten glass. Specialty glasses or exotic melting methods may require different conditions and temperatures. A viscosity-vs-temperature curve may be determined for glasses by any of several tests. This curve will illustrate the rate-of-change of viscosity for the glass below the state of pure fluidity, and indicate whether the glass has a long uniform setting rate, or whether it is "short", setting more abruptly below a specific temperature. All of the various viscosity indicators, or thermal constants, (softening point, annealing point, strain point) have precise viscosity values included in their definitions, and relate to the glass viscosity curve. The softening point of the glass is one of the main indicators of the thermal hardness of the glass. Like glass density, softening point has become an important control test to be run frequently and at close periodic intervals. As related to fiber production, maintainance of uniformity in the softening point determinations assists in maintaining filament diameters within specification.

The effect of composition changes on softening point is readily noted in the tabulation, with the high alumina glasses ("E" and "S") showing the highest values.

The thermal expansion coefficient of a fibrous glass composition is important only when a parallel strand laminate is under consideration. Even then, the characteristics of the other ingredients, resin and filler, override the contribution of the glass to thermal expansion of the resultant laminate. The influence of the combined effect of boron and silicon oxides on thermal expansion is readily noted in the tabulation.

Optical Properties. When required, the ultimate clarity of reinforced plastic components is directly determined by matching the index of refraction of the glass with that of the resin, and, of course, eliminating opaque fillers. A glass with index of refraction of 1.47, such as the "D" composition, will never show near perfect clarity and high laminate translucency in a polyester resin system having an index of 1.55, as will "E" glass with an index of 1.547. Of course, a compatible or highly soluble sizing system must be used on the glass.

Additional purity of color in RP/C for high translucency applications may be obtained by maintaining the iron oxide, (Fe_2O_3) content as low as possible. This usually requires the chemical treatment with HCl of silica and alumina batch ingredients to remove Fe_2O_3. This treatment results in production of a color-clear fiber and frees the laminate from a characteristic transmission-lowering greenish cast.

It is well known in the glass industry that lead oxide, PbO, and potassium oxide, K_2O, combine in a melt to provide brilliance and clarity characteristic of art glass. An illustration related to, but not necessarily involving reinforcing fibers, is found in fiber optics. Using a glass of highest optical clarity, not only can the innermost cavities of the human anatomy be probed, and photographed, but also light signals converted from electrical impulses may be transmitted for miles without need for transforming or boosting.[10]

Electrical Properties. Glass is a good electrical insulating material by nature. The electrical properties of fiber glass contribute markedly to the high performance of molded RP/C parts in electrical applications. These applications include standoff insulators, arc-chutes, distributor caps, switchgear, etc. "E" glass was originally developed for enhancement of electrical resistance properties of molded parts. The improvement in electrical performance of the "D" composition over that for "E" glass may be readily noted in the tabulation. As stated previously, other RP/C ingredients, particularly fillers, contribute substantially to desirable electrical properties in finished molded parts.

One difficulty encountered in designing fiber glass reinforced plastic parts for electrical applications is in the discontinuous or interrupted nature of the glass-resin bond.[11] This condition results in substantial loss of electrical insulation resistance in high temperature and humidity atmospheres, while the wet laminate retains a high proportion (70-75%) of its original dry strength. It is postulated that this condition is improved somewhat by press-molding, where greater compaction is induced, and a higher

Table 26-5a. Basic manufacturers of fiber glass reinforcements and associated materials.

Name	Address	Phone Number
Certainteed Products Corp. Textile Glass Fibers	3000 Chrysler Road Kansas City, Kans. 66115	(913) 342-6625
Ferro Corp. Fiber Glass Div.	Fiberglass Road Nashville, Tenn. 37211	(615) 255-7471
Johns-Manville Sales Corp.	P.O. Box 5108 Denver, Co. 80215	(303) 292-3380
Nicofibers, Inc. Div. Nicolette Ind.	P.O. Box 26 Shawnee, Ohio 43782	(614) 394-2491
Owens-Corning Fiberglas Corp.	Fiberglas Tower Toledo, Ohio 43659	(419) 259-3000
PPG Industries, Inc.	One Gateway Center Pittsburgh, Pa. 15222	(412) 434-3131
Reichhold Chemicals, Inc.		
1. Glass Fibers Div.	P.O. Box 86 Bremen, Ohio 43107	(614) 569-4175
2. Chemglass Div.	P.O. Box 2148 Irwindale, Calif. 91706	(213) 334-7851
United Merchants & Mfrs., Inc. Uniglass Ind. Div.	1440 Broadway New York, N.Y. 10018 Plant–Statesville, N.C.	(212) 564-6000
3-M Co. Special Enterprises Dept.	3-M Center Saint Paul, Minn. 55101	(612) 733-1110
Cataphote Div. Ferro Corp.	P.O. Box 2369 Jackson, Miss. 39205	(601) 939-4612
Potters Ind., Inc.	377 Route 17 Hasbrouck Heights, N.J. 07604	(201) 288-4700
Emerson & Cuming, Inc.	Canton, Mass. 02021	(671) 828-3300

Table 26-5b. Weavers of industrial fiber glass cloth and fabrics.

Company	Address	Phone Number
Burlington Ind. Burlington Glass Fabrics Co. Div.	1345 Avenue of the Americas New York, N.Y. 10019	(212) 333-7322
Clark Schwebel Fiber Glass Corp.	245 Park Avenue New York, N.Y. 10017	(212) 986-3500
Hexcel Corp. Trevarno Div.	11711 Dublin Blvd. Dublin, Calif. 94566	(415) 828-4200
J. P. Stevens and Co. Glass Fabrics Div.	1185 Avenue of the Americas New York, N.Y. 10036	(212) 575-2000
United Merchants and Mfrs., Inc. Uniglass Industries Div.	1440 Broadway New York, N.Y. 10018	(212) 564-6000
Fiber Glass Ind., Inc.	Homestead Pl. Amsterdam, N.Y. 12010	(518) 842-4000

percentage of complete glass-resin bonding occurs.

5. MATERIAL SUPPLIERS AND PRODUCT TYPES*

Having considered methods of manufacturing glass fiber products, and the basic properties of the material, it is now advisable to provide information regarding sources and availability. In this section, details are provided to register both the manufacturers of fiber glass and related reinforcing materials plus actual product classifications and designations.

5.1 Manufacturing of Reinforcing Products

Table 26-5a lists the basic manufacturers of glass fiber reinforcements and associated miscellaneous materials. Table 26-5b lists the weavers of industrial fiber glass cloth and fabrics. The corporate headquarters of each firm are listed together with its telephone number as of the year 1976.

5.2 Product Designations

Table 26-6a presents basic reinforcing products made by all manufacturers according to application area. Drawn and blown glass fiber mat types by manufacturer type and application area are given in Table 26-6b. Lastly, construction and application data concerning related types of fiber glass fabrics are given in Table 26-6c.

6. PRICES AND AVAILABILITY†

Prices of fiber glass reinforcing products were exceedingly high (\$1.75-\$5.00/lb, average) during the periods of development and preliminary usage in classified and strategic applications during World War II. Immediately following the war, the prices broke to the lowest possible which the conditions and times would permit, and then continued slowly downward until the period 1960-1965. Since then, prices have eased gradually upward because of inflationary pressures and some material procurement difficulties, and the current 1976 prices are approximately double the low 1960-1965 values.

Any price trends that might be noted or prognosticated for the future would have to relate to probable slow continued inflationary states of the national and world economies.

The only serious shortage that has occurred in glass making raw materials has been with the boron oxide contributors. Domestic sources of borax and boric acid have become critical, and it has been necessary to use the less desirable colemanite, razorite, and related materials, some of which are also plentiful in countries outside the United States.

Table 26-7 presents extant (1976) prices of fiber glass reinforcing products and related materials. All prices are tabulated as a range of values from low to high. The lower prices represent the best price break in truckload or equivalent quanities. The higher values in the price ranges represent both the lower quanities, and in the case of the mats and fabrics, increases in grade (thickness, weight, or quality).

7. FIBER GLASS REINFORCED PLASTICS/COMPOSITES

Having considered in detail the contribution glass fibers make to reinforced plastic/composites, based on their fundamental properties, it is timely to proceed with actual applications of glass fibers and utilization in combination with the various plastic and filler ingredients available.

Glass is reputed to have been discovered or first noted by a Phoenecian sailor.[12] Natural plastics such as rosin have been known and used for centuries. Techniques for reinforcing have been used since the days of early Egyptian culture (straw in bricks).

The science of reinforced plastics/composites, however, is strictly a modern, up-to-date 20th Century affair. The amalgamation of a strong but friable reinforcing material with a weak but

*Acknowledgment with thanks is made for the assistance of Edward A. Turner, of Johns-Manville Sales Corp., in preparation of this section.

†This section prepared by the editors separately from the main writeup on fiber glass reinforcements.

Table 26-6a. Basic fiber glass manufacturer's reinforcing products according to type and application area.

RP/C Process or Application	Type Sizing or Resin	Certainteed St. Gobain Gobain	Chemglass Division RCI.	Ferro Corporation[a] Fiber Glass Division	Johns-Manville[b] Sales Corporation	Owens-Corning Fiberglas Corp.	PPG Industries Inc.
Polyester resin systems							
Spray	Chrome	R-82				825	
Up	Chrome/Silane	P-207		206, 208, 209		447, 477	526, 527, 539
Roving	Silane		709			881, 496, 496BB	535
Hand Layup Mat	Silane	M-113-247 M-123-247		803, 821 824	610, 630C 640B, CM104	M-700, M710 M-711	Type ABM AKM, ACM ABM-H
Press	Chrome	R-CR/6V, R-32		157	7301E	834	508, 510
Molding Roving (Pre-Form)	Chrome/Silane			226, 285 206, 207		877 Spun 447	539F
	Silane	R-82, P-207		385, 206X, 350	308J, 416, 613, 752E	858, 886, 881, 442 852 Spun	535, 541
Continuous Mat	Silane	Unifilo L/9 Hi Sol T/8 Lo Sol				M8600 M8605, M8610 M8624 (E1) (A)	
Woven Roving	Silane				1564, 1843, 2254, 2454	OC18-54P-475T to OC24-54P-475T	
Chopped Mat	Silane			GP		M-800	AMM-1, AMM-2
Chopped Strand	Chrome						CS40
(Molding compound, BMC)	Silane			606 (EL)	CS308, CS308A CS308C 417X, A20	833, 432, 400, 434 (EL), 847 (EL)	CS450 (EL) 1156, 3197, 3198
Mat Forming	Chrome						
Roving Plenum	Silane	K-931		203, 266, 307	7201E, 183E	855, 858	56WFF, 537 538, 56W 56WF
Long Cutter Roving	Silane	K-247		MS-267, 902		858, 866,	512, 512F, 512FF, 56W, 56WF, 56WFF

Roving	Chrome/Silane			206			
(For SMC)	Silane			350C	407, 407FX, 417F, 417X, 417X, 75CF	840, 433, 951, 952	516, 518 519
Pultrusion Roving		K-937	Series 300		7201J	431, 475, 474 (EL)	712, 713, 764 762, 1064, 764WT
Mat						M-8624	ACM
Pressure Bag Molding Mat	Silane	M-212				M850	
Vacuum Bag Molding Mat	Silane			MSB, MSHSB, MRHSB, GP, MSGP	630C	M710	
Centrifugal Casting Mat	Silane	M-123		HSB, MSHSB, MRSHSB, MSGP	630C	M-856, M-710	ACM
Roving	Chrome					834	508, 510
	Chrome/Silane					447, 477	
	Silane	K-247, P223	Series 100			886, 881	
Encapsulation Milled Fiber	Cationic Water			Type CATX Type HC	Vitro Strand G	731 739	
Filament Winding Roving	Silane	K-247 K-937	Series 200	307	7201J	431 475	713, 1064, 2016, 2029
Continuous Laminating Roving	Silane		Panel Roving		182	886, 881	59W 2124
Epoxy Resin Systems							
Filament Winding Roving	Silane		Series 200	777	7200J	901, 904, 836, 431 410	712, 1062 712S, 1080 1045 2016, 2029
Putty Compounding Chopped Strand	Silane				CS308A, CS308	411, 432	1156

Table 26-6a (Continued)

RP/C Process or Application	Type Sizing or Resin	Certainteed St. Gobain Gobain	Chemglass Division RCI.	Ferro Corporation Fiber Glass Division	Johns-Manville Sales Corporation	Owens-Corning Fiberglas Corp.	PPG Industries Inc.
Continuous Impregnating Roving	Silane					431	712S, 762
Thermoplastic Resin Systems							
Extrusion Compounding Chopped Strands	ABS				308A, 308C	885, 414	3130, 3526
	Acetal					409	3531
	Noryl (PPO)					497	3250
	Nylon	K-43			CS303, CS719	419	3536, 3535, 3531G
	Polycarbonate					409, 419, 418	3260
	Polyethylene				308A, 308C	415, 452	
	Polypropylene				308A, CS156 CS713	885, 832, 452	3130, 3148
	Polystyrene	P-343			308A, 308C	885, 414	3130
	Polysulfone					497	3535
	PVC				308C	885	3129
	SAN	P-303			308A, 308C	885, 414	3130, 3705
	Thermoplastic Polyester				CS-303	419	3531, 3303
	EPDM Rubber				CS-711		

[a] Purchased by Reichhold Chemicals.
[b] No commercial sales; all products being used internally.

Table 26-6b. Drawn and blown glass fiber mat types by manufacturer and application area.

	"Modiglass"[a] Glass Fibers Div. Reichhold Chemical, Inc.	Nicofibers[a] Div. Nicolette	Owens-Corning[b] Fiberglass Corporation
Matched-die molding surfacing mat, flat, low drape	Standard surfacing mat regular binder content CFRM	Surmat	M-514 trt. #233
Matched-die molding surfacing mat, contoured, high drape, low binder content	1/8 oz veil 1/4 oz veil	1. Veil-1/8 and 1/4 oz 2. Consurmat	M-514, trt. #234 M-514, trt. #236
Hand layup surface veil, corrosion applications	–	Surmat 100	Hand layup polyester M-514, trt. #236 Machine-made pipe M-514, trt. #234 and trt. #248 Epoxy corrosion resistant M-514, trt. #246
Electrical grade easy punchout printed circuit boards	CFSM CFRM	N-700 Conformat	Polyester molding M-514, trt. #236, trt. #248 Epoxy molding M-514, trt. #246
Decorative and lightweight reinforcing mat	CFRM	N-700 Conformat N-751 Conformat	Polyester molding M-514, trt. #235 (surfacing only)
Surfacing in pultrusion operations	–	N-720 Conformat	–

[a] All Products "A" Glass, Silane Treatments.
[b] All Products "C" Glass, Various Compatible Treatments.

Table 26-6c. Construction and applications data for fiber glass fabrics.

Pertinent data of fiber glass fabrics*

Weave Styles Plain (taffeta), basket, crawfoot satin, leno, mock leno, harness satin, twill, unidirectional or high-modulus weave. Many different finishes are applied for various resin compatibilities.

Fabric Weights 1.4–36 oz/sq yd, with special fabric weight down to 0.5 and as high as 48 oz/sq yd.

Fabric Thicknesses 0.0020–0.0405 in., with special fabrics thicknesses down to 0.0010 in. and as high as 0.0450 in.

Standard Fabric Widths 38 in., 44 in., 50 in., 60 in., with special fabrics woven to 72 in. Woven roving–38 in., 44 in., 50 in., 60 in., 72 in., and 144 in. Fabric tapes–1½–12 in. wide.

Uses for Fabrics in RP/C Aircraft and missile structures, boats and marine applications, tooling, fishing rods, automotive applications, pipes and tanks, custom molding, structural members, grinding wheels, and electrical laminates.

*Portion of data courtesy Uniglass Industries Div., United Merchants and Manufacturers, Inc.

adhesive resinous matrix was conceived in antiquity, but brought abruptly into today's focus by modern materials and processing technology, and by the demand for improved performance of modern manufactured products.

This brief section succinctly describes the markets for RP/C products and growth patterns, materials and equipment necessary for processing, types of fabricated RP/C parts, classified generally according to their various molding methods, and an analysis of where the industry is headed. An inside look at potential future products and molding methods is also included. Brevity is certainly required when at-

Table 26-7. Prices of fiber glass reinforcing and related materials.

Product	Approximate Range: Price per lb (1976)
Reinforcing	
Roving	$0.43-$0.52
Chopped strand	$0.47-$0.52
Chopped strand mat and continuous strand mat	$0.59-$0.74
Woven roving	$0.61-$0.72
Milled fibers	$0.50-$0.90
Veil and surfacing mats	
Veil mats	$0.95-$1.10
Surfacing	$1.10-$2.40
Electrical grade mats	$0.60-$0.93
Decorative and Reinforcing mats	$0.60-$2.75
Miscellaneous	
Hollow glass spheres	$0.88-$3.35
Solid glass spheres	$0.13-$0.24
Bulk glass flake	$0.80-$0.96
Glass fabrics	$0.35-$4.86 (/ lineal yd)

tempting to include a large amount of information in a limited space. Hence, liberal use of tabulations will be resorted to, and ample documentation will be provided.

7.1 Markets

Reinforced plastics/composites have realized annual growth rates summarized for 10-year periods in the accompanying Table 26-8. In Table 26-9 is presented a breakdown of the total poundage sold in each separate market for RP/C products (1975).

7.2 General View of Materials and Equipment

Reinforcements. The descriptions and definitions of fiber glass reinforcements from Section 3 are applicable here, and the uses for the different forms of fiber glass will become evident in the next discussion.

Table 26-8. RP/C growth rate.

Year	Annual Volume, lb
1945	7,500,000
1955	110,000,000
1965	340,000,000
1975	1,146,000,000

Table 26-9. Markets for RP/C and distribution (SPI classification).*

Market	% of Total (1975)
Aircraft and aerospace	2.2
Appliances and equipment	5.6
Construction	14.3
Consumer goods	5.1
Corrosion equipment	14.0
Electrical industry	6.5
Marine industry	24.8
Transportation	23.0
Other (tooling, etc.)	4.5
	100.0

*Some classifications include the following breakdowns: automotive, agricultural, business machines, house and home, materials handling, and recreational.

Resins. The resins used include polyesters, epoxies, phenolics, and the complete family of thermoplastics. Included with the thermosetting resins are all types of monomers, catalysts, promoters, modifiers, and other agents for performance improvement.

Fillers. Because fillers are one of the main subjects of this book, and thoroughly discussed in other chapters, they will not be considered here.

Equipment. Equipment for the preparation and processing of RP/C includes refrigerated storage vessels or compartments for resins and catalysts, dry storage for glass reinforcements, weighing and mixing equipment for batch preparation, mold-making facilities (whether commercial or in-house), and mold handling or support means, whether wet layup molds, throughput process, compression, transfer, injection or other types of press equipment for converting the ingredients into finished usable parts. Many of the methods were newly devised to accommodate the new materials. An appreciable battery of test equipment for evaluating materials, processing parameters, and finished products and their performance properties have also been developed, or adapted from similar industries. These are too numerous for detailed description, but will be alluded to as this discussion proceeds.

7.3 Molding Methods

Because the types of reinforced plastic parts produced vary so widely in size, complexity, function, and other parameters, it is more feasible to categorize the entire industry by molding methods.[13] These resolve neatly into five or six major groups and include all the finished RP/C products, although there is some minor duplication, i.e., one product type (piping) being fabricated by 2 or more different methods.

Prior to individual method descriptions, all molding methods and concomitant products produced are summarized in Table 26-10, and a correlation of fiber glass reinforcement types with applicable molding method is presented in Table 26-11.

Hand Lay-Up. This process was one of the first developed following the perfection and utilization of polyester resins. An open-type mold with no unrelievable undercuts is taken off a master mold, polished, release agent applied, and otherwise made ready. A catalyzed polyester gel coat is applied and cured.

Lay-up involves precutting and tailoring cloth, mat or woven roving reinforcement, placing it in the mold and applying the catalyzed resin, effecting thorough wet-out, eliminating gaps, bubbles, and voids, and building up a thickness of at least 0.10 in. The ratio is nominally 25% glass to 75% resin, and fillers are not a major ingredient except in the gel coats. Thicknesses may be built up to 1 in. or more with successive applications and extra supports may be molded in.

As noted in Table 26-11, the main large-volume products are boats and corrosion equipment. Large chemical processing tanks and other components up to 3,000 sq ft in area have been fabricated.

Spray-up. Spray-up evolved from efforts to create greater production speeds and more con-

Table 26-10. RP/C molding methods & typical products.

Processes	Typical products
A. Room temperature cure	
Hand lay-up	Corrosion equipment, boats, and machine guards
Spray-up, including rigidizing	Tub and shower units
Vacuum bag molding	Radomes
Tooling	Master molds, checking fixtures
Cold-press molding	Motor mounts, hollow ware, finished two sides
Casting, potting, encapsulation	Sinks, synthetic marble, electrical components
B. Intermediate cure	
Architectural panelling	Translucent and opaque, flat and corrugated panels, sandwich panels, foam structures
Centrifugal casting	Pipe and tubing
Pressure bag molding	Pressure cylinders and tanks
C. High temperature cure (compression molding)	
Premix and BMC molding	Distributor caps, electrical gear, shower bases
Preform and SMC molding	Automotive components, tubs, housings, radomes, aircraft components
D. Throughput processes	
Pultrusion	Rods, tubes, structural beams
Filament winding (including pre-impregnation)	Pipe and tubing
E. Reinforced thermoplastics	
Injection molding	Gears
Rotational molding	Hollow structures, tanks
Cold forming	Housing, appliance parts
F. Advanced composites	
Exotic methods and many combinations	Aircraft and aerospace parts

Table 26-11. Areas of usage of fiber glass reinforcing products.

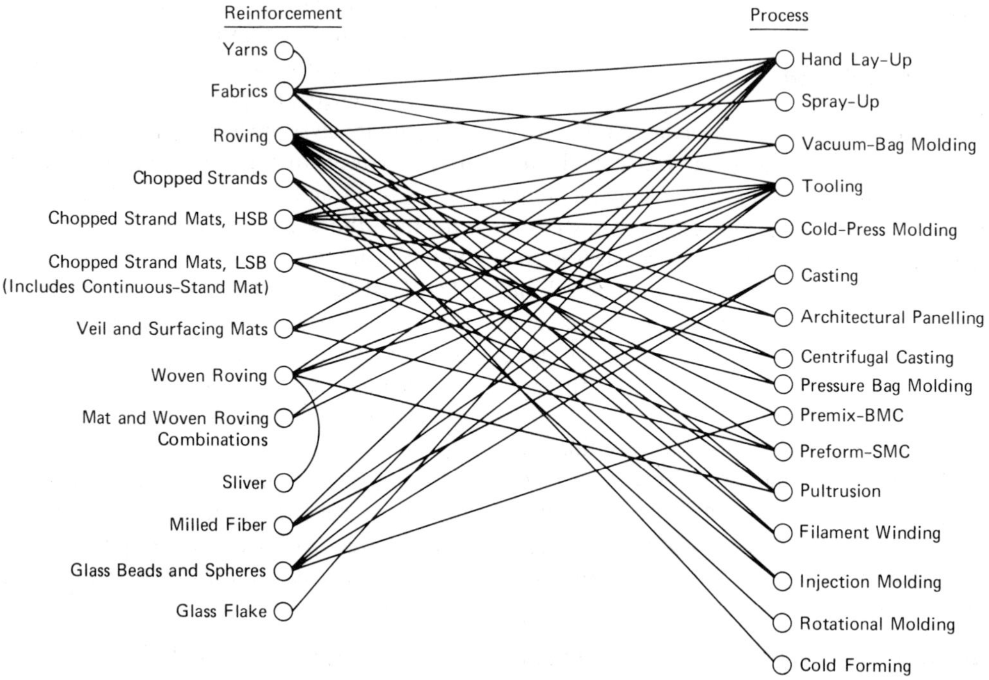

sistent product uniformity over the hand lay-up process. An especially-built mechanism pumps resin and catalyst, and chops fiber glass roving through a hand-held gun suspended at the end of a boom. The boom is rotatable and extensible so that the operator holding the gun may be efficiently moved around to perform the work successively on a battery of molds.

Gel coating is likewise employed as surfacing for spray-up. However, an innovation, "rigidizing," resulted in improved surfacing when vacuum-formed acrylic sheeting was used as a surface, with glass and resin sprayed into it. Mold costs are reduced, and improved product properties and performance result. Special decorative effects may be incorporated into the vacuum-formable plastic sheet. The spray-up process enjoys the same type applications, and is frequently used in conjunction with hand lay-up.

Vacuum Bag Molding. Usually cloth, but also chopped mat or woven roving may be laid up in a shell or paraboloidal type mold, the resin applied and uniformly squeezed out. A flexible plastic film then covers the entire lay-up, is clamped onto the mold flange and a vacuum applied. This introduces the equivalent of 13–14 psi pressure being applied to the layup while cure takes place. Autoclave and hydroclave cures may also be employed. Greater dimensional control and higher laminate density and quality may be obtained in the finished parts than for contact layup.

Sandwich structures with foam or honeycomb cores between two RP/C skins are commonly molded. The only limit to sizes of parts is the ability to construct and use molds. Radomes 12 or 15 ft in diameter and 4 ft deep are commonplace.

Tooling. Many different techniques and combinations of laminating, bondings, casting, etc., are employed in tooling procedures. The main objectives are forming master molds, fabricating molds to be used for production of parts, and making checking fixtures and many other RP/C elements for dimensional control.

Cold Press Molding. Parts are produced in this process by catalyzing the resin for room-temperature curing and laying up in a mold which has a matching half. Both halves may be gel coated. The mold halves are mounted in a press, and the molds closed to stops while cure takes place. Advantages are that the process provides two finished surfaces and is more rapid and consistent than open-mold, hand lay-up fabrication.

Casting, Potting and Encapsulation. In casting, resin and filler and sometimes reinforcement are mixed together and poured into open or closed molds to form lavatory sinks and related items. Mottled color effects may be incorporated into the resin to simulate marble and other finishes. Casting is also used in producing electrical components. Potting and encapsulation are similar processes.

Architectural Panelling. Production using a continuous layup, cure, and trim process yields one of the mainstays in the RP/C product household–near transparent, flat or corrugated architectural panelling. Either chopped roving or chopped strand mat is deposited uniformly into a controlled thickness of resin or film on a moving belt. Wetout devices complete the impregnation, a second film is added on top, and the sandwich passes through shapers or corrugators prior to gelling and curing in a continuous, throughput oven. The trimmed sheeting is either 0.040 or 0.065 in. thick, and special resin modifiers are incorporated to resist weathering.

Centrifugal Casting. Either open-end pipe and tubing, or closed-end containers are fabricated by placing combined reinforcement and resin in a rotating cylinder in a heated area. Advantages are many, and include good densification and elimination of bubbles, easy release by shrinkage away from the mold surfaces, and producibility of a wide range of diameters and lengths.

Pressure Bag Molding. A mold is fabricated to define the outside shape of the part, usually a cylindrical form. The layups (preformed end caps and L.S.B. mat plus resin) are placed into the mold, ends bolted, and an elastomeric bladder shaped to fit inside is inserted. Pressure is applied, forcing the bladder against the layup, inducing wetout, densification, and bubble release.

The mold, with pressure maintained, is passed through a hot water bath and the part cures. Water softening tanks, RP/C fire extinguisher containers, and filtration housings are made by this method. Higher pressure requirements in the end products are fulfilled by changing to filament winding methods.

Premix and BMC Molding. A mixture of chopped glass fiber (lengths ¼–½ in. or more), filler, and catalyzed resin is blended in a suitable compounding machine. The result is a paste-like material suitably termed "dough molding compound" by the British, but known better as "premix."

The premix is extruded, weighed, and molded in heated matched metal dies in a compression, transfer, or injection press. Typical parts range in size from 1 sq in. to 16 sq ft, and applications range in part size from small electrical components with molded-in inserts to large items such as industrial wash basins.

Part shrinkage and finished surface out of the mold have been important and originally influenced acceptance. Very recent developments involve modifications to the resin system affecting these parameters. Additives have been developed which prevent resin shrinkage out of the mold, and hence also the exposure of fiber and filler surface patterns. These additives also induce a synthetic gel or thickening of the resin prior to molding. This improvement results in a more uniform distribution of reinforcement and filler in the molded parts, and also better physical strengths due to freedom from voids, and higher densification.

Preform and SMC Molding. A page of technology was borrowed from felt hat manufacturing in the early days of RP/C fabrication. This yielded a method for collecting glass fibers (chopped from roving) onto a screen prepared in the shape of a part to be molded. Molding was carried out in large metal dies in a compression press. The chopped glass preform, bonded using a spray binder, was combined with a polyester resin and filler mix at the press and molded at 235-265°F, 250-500 psi pressure.

This process was used to fabricate all the large-sized, higher production rate RP/C parts such as washtubs, hard-hats, car bodies, and even boats up to 17 ft long.

The improved low shrink and thickened type resin technology indicated in the Premix-BMC discussion has enjoyed wide application in the preform molding field. In fact, the changes in technology made it possible to prefabricate the fiber glass, resin, and filler into a processable form designated "sheet molding compound–SMC."

For the first time, the RP/C industry had a material that could be taken off a roll, like sheet steel, fed directly to a press, and molded according to known methods. No longer was it necessary to pour fluid resin-filler mix laboriously over a shaped fiber glass preform at the press and hope for 100% successful molding results.

The difficult surface-finishing tasks originally associated with preforming were also eliminated. Parts 60–70 sq ft in area are about the largest possible using SMC,–not quite as large as those for the process of preform molding.

Pultrusion. The process for producing RP/C rod stock was re-dubbed "pultrusion" after technology improved to the point that solid, hollow, round, square, and rectangular cross-sectioned and even girder-type structural members could be continuously fabricated by impregnating roving and drawing it through shaping and curing dies. Improvement in glass fiber sizing technology enabled glass and polyester or epoxy resin to become almost as one.

Mat and fabrics are also used in pultrusion to vary the direction of fiber orientation. While slightly more expensive than their metal counterparts, pultruded structural shapes are indispensible in areas where corrosive chemicals are produced and handled. Ells, tees and I-beams up to 12 × 6 in. and more than ½ in. in thickness may be manufactured by this process. The use of radio-frequency curing has increased production rates, improved product quality, and reduced process rejects.

Filament Winding. Ancient methods of fortifying many types of structures using wire or cloth overwraps suggested the application of strong glass fibers circumferentially or axially wrapped. However, glass fibers can go one stage further. After impregnation with resin and curing, the filament-wound composite structure supports itself and needs no permanent substrate.

The ancient, wire-wrapped gun barrels or cloth-wrapped Egyptian mummies gave impetus to the development of methods for fabricating RP/C tanks and piping or large vessels completely and automatically wound as surfaces of revolution, with closed ends, port holes, etc. Filament-wound tanks up to 1500 sq ft in area, or cylinders 40 ft in diameter × 12 ft high are wrapped on a single mold. Pre-impregnated glass-resin stock may be used to gain higher and more uniform glass content.

The main advantage of filament winding is that glass contents of from 60–85% may be attained, making possible exceptionally high part strengths. Winding patterns may be varied to avoid unidirectional orientation and provide complete coverage.

The first piece of RP/C filament winding equipment was obviously some type of machine tool or lathe. However, exceptionally sophisticated winding equipment, all programmed automatically, has evolved for fabrication of the many types of parts now produced by the method.

Injection Molding of Thermoplastics. Thermoplastic resins for injection molding were first reinforced by forming an emulsion of the resin and coating fiber glass roving. The roving, with resin fused and set, was chopped to form moldable pellets which could be fed to an injection press in the conventional manner.

In later developments, melted thermoplastics were used to impregnate the roving, which was die-wiped, cooled, and cut into pellet size for molding. Both reinforced pellets and clear, unreinforced pellets could be combined at the injection machine to vary glass content as desired.

Dry, chopped fiber glass (chopped strand) and clear, unreinforced molding pellets are now combined immediately prior to molding, eliminating laborious mixing. However, the most recent technology for producing reinforced thermoplastic molding pellets is the Werner-

Pfleiderer method of feeding continuous-strand fiber glass roving into a twin-screw extension machine for combination with the required thermoplastic resin.

In injection molding, the required forms of resin and fiber glass are drawn from a hopper into a heated barrel, and thrust under high pressure into a cold, clamped cavity. The resin sets up rapidly and the part is ejected as the mold opens. Parts up to 300 sq ft in area (300 oz press) may be molded.

The incorporation of glass fibers into thermoplastic resins for injection molding substantially improves all properties, particularly tensile and flexural strengths, heat deflection temperature, and elongation. However, the process would become a more important factor in RP/C if it were less damaging to the glass fiber lengths. Dry glass strands ¼ in. long, starting out of the hopper and into an injection machine or pelletizer, are soon separated and end up in the molded part completely dispersed into individual filaments with a range in lengths somewhere between 0.001 and 0.030. It is postulated that, if some measure of the original strand configuration could be preserved, higher molded strengths would result and additional areas of use would open up for injection molded RP/C.

Rotational Molding. Chopped glass fiber lengths are directly charged together with a 35-mesh thermoplastic molding powder into the mold cavities of a rotational molding machine. In the standard heating cycle, the glass fibers are assimilated into the resin as it fuses, and they become firmly bonded into the part as the resin is cooled. Average glass contents up to 25% are possible, but loadings of from 1-10% are more frequently used. The presence of glass reinforcing improves all of the roto-molded physical strengths except impact resistance.

Parts molded using reinforcement include large containers for food storage and processing, housings, electrical components, and statuary.

Cold Forming. This interesting process for molding reinforced thermoplastics is a sort of SMC in reverse. Fiber glass roving is chopped and combined with the thermoplastic resin melt, at elevated temperatures, into a moldable sheet product. In order to convert the sheet into a molded product, precut, flat portions are reheated in a rapid temperature-rise oven on wires. They are then transferred directly to cold, matched molds in a stamping press, where they are immediately shaped and rapidly cooled. The molds are telescoping, and neither shearoff nor trimming to size after molding is required.

Small housings are produced very practically using this process, although molding of much larger parts up to 200 sq ft is also possible. Excellent molded physical strengths are obtainable using this method. However, the maximum service temperature is always limited by the type of thermoplastic resin, and is always lower than that for thermosets.

7.4 RP/C Parts Design and Molded Properties

Both parts design and molded properties are of great importance to potential purchasers and users of fiber glass reinforced plastic products. Unfortunately, each subject, if treated properly and to the fullest extent, would command the space of an individual volume in its own right. The reason is that each separate molding method yields a different type product with different glass content and with an individual area of application. This type of variability within the family of RP/C products requires completely separate treatment for full representation and understanding. Therefore, these two subjects will be treated briefly, and information presented so that the reader will be able to consult other more complete sources, if required.

Design Factors. Design parameters include such information as practical thickness and variations, recommended sidewall draft for proper release from molds, permissibility of undercuts, allowable diameter and placement of holes, extent or area of joining plates and location of screws or bolts, permissibility of ribs, bosses, molded-in labels or numbers, type of surface finish possible, and many other facets. These have been given a thorough review and are available for reference in a recent publication.[14]

Fiber glass reinforced plastics may be fabricated into almost any desired shape, with com-

Table 26-12. Representative properties of RP/C molded parts.

	REINFORCED POLYESTER/EPOXY			REINFORCED THERMOPLASTICS		
				INJECTION-MOLDED POLYPROPYLENE		COLD STAMPING POLYPROPYLENE
	Premix-BMC	Preform-SMC	Filament-winding, Pultrusion	Non-reinforced	Reinforced	Reinforced Moldable Sheet
Percent reinforcement	15	30	75	0	30	40
Reinforcement fiber length	¼ in.	1½ in.	Continuous	–	¼ in. start, 0.020 in. after molding	2 in.
Specific gravity	1.8	1.7	2.0	0.90	1.1	1.2
Flexural strength, psi	10,000	20,000	250,000	6,000	10,000	21,000
Tensile strength, psi	5,000	20,000	200,000	5,000	9,000	11,000
Tensile modulus elasticity, millions of lbs/sq in.	1.6	1.8	6.0	0.7	1.0	0.7
Compressive strength, psi	20,000	28,000	65,000	7,000	7,000	23,000
Izod impact strength-ft-lb/in. notch	2.0	20.0	50.0	1.5	2.5	10.0
Heat deflection temp. °F, 264 psi	350	400	385	140	275	310
Resistance to heat, continuous, °F	300	350	350	230	250	–
Arc resistance, seconds	200	185	–	–	–	–
Dielectric constant, 60 Hz	5.7	–	–	2.3	2.5	–
Dissipation factor, 60 Hz	0.015	–	–	0.0006	0.002	–

paratively great savings in tooling costs. They also have the versatility of cutting down drastically on the number of components required for a given assembly.

Molded Properties. In Table 26-12 are listed representative properties of a crosssection of fiber glass reinforced, plastic molded types. The lightweight nature of RP/C is indicated by the specific gravity, and it may be noted that the strength increases proportionately with increase in both fiber length and glass content in the laminate. The heat distortion and continuous resistance to heat limits are functions of the resin type used, but are also influenced slightly by the glass fiber length and content.

Also, because of the growth potential prognosticated for RTP "Engineered Plastics," Table 26-13 is presented to correlate and compare all properties in the complete range of reinforced thermoplastic resins. A qualitative scan is made of the base properties of all polymers considered. This is shown as a supplement to Table 26-13. Quick and easy reference has been made possible by abbreviated letter designations.

7.5 New Developments & New Directions

In this concluding section, the most notable of a series of unique new developments in processing methods and products is presented. Finally, a few "unofficial" observations are made on which the industry may move in to assure continuation of its growth and expansion.

General Press-Molding Technology. Presently, the largest presses for high speeed compression-molding are capable of producing flat sheets approximately 1.3 × 4 m. Other large presses for producing automotive components have a bed size of approximately 1.6 × 2.5 m with a capacity of 4,000 tons. At the 1975 Kunstoffe exhibit, Dusseldorf, Germany, Rudolf Meyer Co. showed an outsized clam-shell type press (no tie-bars) for molding RP/C. Platen sizes up to 3 × 10 m with clamping pressures up to 6600 tons are the largest design.

As discussed previously, development of SMC constitutes a turning point in RP/C progress, and has re-directed the industry toward higher production rates of more uniform parts, eliminating difficult compounding, and permitting use of a lesser-skilled labor force. Research personnel at PPG Industries recently developed two improvements in SMC: a. XMC—a process where fiber glass roving is impregnated with thickened, low-profile resin mix and filament-wound on a mandrel. Sections are cut, matured, and molded to yield parts with glass contents 75%, tensile and flexural strengths of 70,000 and 130,000 psi respectively, b. HMC—an easily-moldable blend of 65% chopped fiber glass and 35% resin (no filler) which, when processed, yields parts with tensile and flexural strengths of 30,000 and 50,000 respectively.

Press molding of high quality RP/C parts at lower pressures was made possible by ERM—elastic reservoir molding, a development of the John Z. DeLorean Corp. A "sandwich" type moldable composite is formed in which dry reinforcements are layered top and bottom around a flexible foam, impregnated with liquid catalyzed thermoset resin. Responding to comparatively low molding pressures, the resin is forced into the reinforcing fiber glass, the foam interliner reduces to 0.005 in. thickness, and heat from the matched dies induces ultimate cure. The process is available for licensing, and a large volume usage of the material is contemplated for manufacture of automotive components.

Liberal use is being made of the application of robot-arm, automatic devices for loading and unloading SMC charges and cured parts at the molding operation. The automatic "arm" is capable of reaching out as far as 88 in.

Aviation and Flying. The highly desirable light-weight plus high strength properties of fiber glass reinforced plastics are paying off in this field. Also, the many innovations made by the military are reaching commercial status. Several companies are producing and marketing all RP/C gliders, and at least 4, all-RP/C family-style commercial aircraft (2-4 places) have been introduced in the ten years from 1965-1975. Liberal use is made of pultruded material for landing gear springs. Also, RP/C, properly designed, may be fabricated into components which create, in flight, a minimum of air drag, making them highly efficient and desirable.

Table 26-13. Properties of fiber glass reinforced thermoplastic resins.

	Nylon-6	Nylon-6/6	Acetal	High Density Polyethylene	Polypropylene	Polybutylene Terephthalate	Polycarbonate
Properties of the Basic Polymers	Same as Nylon 6/6	TUF-E ABR-G MST-H CHR-G MBR-G WA-H PRO-E CO-M	MST-H CR-G ST-L DST-G FE-H CHR-G WR-G TS-P	CO-L CHR-G WR-G MST-L ST-L EPR-G SPG-L PRO-G	PRO-G CR-G TUF-G ST-M WR-G CHR-G	PRO-G HDT-L MST-M FE-G TUF-G WA-L CHR-G IR-L SP-L FL-M to H	CL-H TUF-G IR-E SC-H SP-H TS-H SP-P ABR-P
Molding parameters							
Injection pressure, psi	10–20000	12–20000	15–20000	10–20000	10–20000	10–15000	15–20000
Injection cylinder temp., °F.	450–550	525–550	375–410	350–550	450–525	450–520	550–650
Mold temp., °F.	150–200	150–200	200–250	75–125	90–150	100–250	150–250
Physical data							
Glass fiber, %	30	30	30	30	30	30	30
Specific gravity	1.36	1.36	1.63	1.18	1.13	1.53	1.43
Molding shrinkage, in./in.,							
1/8 in. section	0.003	0.004	0.004	0.003	0.004	0.003	0.001
1/4 in. section	0.005	0.006	0.005	0.0035	0.005	0.004	0.002
Water absorption, %, 24 hr @ 23°C	1.2	0.7	0.9	0.017	0.04	0.070	0.08
Mechanical properties							
Impact strength, Izod,							
notched 1/4 in.	2.0	2.1	0.9	1.3	1.9	1.4	1.7
unnotched 1/4 in.	12	13	5.0	4.5	12	11	11
Tensile strength, psi	23000	26000	12500	9000	14000	18000	20000
Tensile elongation, %	3.0	2.0	1.6	1.5	2.5	2.5	2.0
Tensile modulus, psi × 10^6	1.3	1.5	1.4	0.9	1.0	1.4	1.4
Flexural strength, psi	28000	36000	16800	11000	18000	27000	23000
Flexural modulus, psi × 10^6	1.1	1.3	1.1	0.8	0.7	1.2	1.1
Compressive strength, psi	22000	22500	11500	7000	12000	18000	20000
Hardness, Rockwell R	119	120	112	75	98	120	119
Electrical properties							
Dielectric strength, VPM, S/T	500	500	500+	500	500	500+	470
Dielectric constant, 1 MC, dry	3.8	3.8	4.0	2.7	2.7	3.8	3.4
Dissipation factor, 1 MC, dry	0.017	0.016	0.0055	0.008	0.001	0.015	0.0075
Arc resistance, sec.	120	120	182	140	125	130	120
Volume resistivity, ohm CM	10^{14}	10^{14}	10^{14}	10^{16}	10^{15}	10^{16}	10^{16}
Thermal properties							
Deflection temp., °F., @ 264 psi	400	480	325	250	295	415	300
@ 66 psi	420	500	330	265	315	420	300
Flammability	B	B	B	B	B	B	SE
Flammability, UL Sub. 94, 1/8 in.	SB	HB	HB	HB	HB	HB	VE1
Coefficient of linear thermal expansion, in./in./°F × 10^{-5}	1.75	2.0	2.3	2.7	2.0	1.4	1.3
Thermal conductivity, Btu/hr/ ft^2/°F/in.	3.5	3.5	2.2	2.5	2.3	1.4	2.2
Cost information							
Cost per lb cents	96	96	65	53	54	86	180
Cost per cu in., cents	4.6	4.6	3.8	2.2	2.0	4.6	9.0
Cost per 1000 psi tensile cents	3.8	3.6	5.2	5.9	3.9	4.8	10.6

Polysulfone	Polyphenylene Oxide	Polystyrene	Styrene Acrylonitrile	Acrylonitrile Butadiene Styrene	Modified ETFE Fluoro Plastic	Polyvinyl Chloride (rigid)	Polyurethane	Polyphenylene Sulfide	ASTM Test Method
CL-H SP-H TS-E SR-P ABR-L PRO-M	PRO-G to E WR-E SP-H FL-L SR-L ABR-P	CO-L PRO-E CL-E SP-L SR-L MST-M to L IR-L (brittle)	CO-L PRO-E CL-G SR-M to L SP-L MST-L	CO-L PRO-E SR-M TUF-E MST-M	PRO-G HAR-M ST-M ABR-M CHR-G TS-M EPR-G	HDT-L CHR-E WR-G EPR-G FL-L CR-G SP-L TS-L PRO-M to P	TUF-G IR-G AR-G ST-L MST-L	MP-H CHR-G TS-G FL-L SR-E ST-H MST-G (at hi-temp) HAR-G PRO-G SP-H MP-H TS-G (no oxidation)	
15-20000	15-20000	10-20000	10-20000	10-20000	3-20000	10-20000	10-15000	15-20000	
600-660	500-650	400-550	450-550	450-550	570-650	300-450	360-410	575-650	
200-300	450-600	100-160	100-160	150-200	–	300-415	60-150	100-350	
30	30	30	30	30	25	20	30	30	
1.46	1.36	1.28	1.31	1.28	1.80	1.54	1.45	1.53	D-792
0.001	0.004	0.001	0.001	0.001	.002-.030	0.001	0.003	0.002	D-955
0.002	–	0.001	0.001	0.0015	–	–	0.004	0.004	
0.23	0.06	0.06	0.15	0.16	0.022	0.008	0.15	0.02	D-570
1.3	1.0	1.1	1.1	1.3	9.0	1.6	2.0	1.4	D-256
10	–	2.1	4.0	5.0	–	–	15	6.0	
17000	17000	12000	19000	16000	12000	14000	17000	17000	D-638
2.1	6-0	1.0	1.6	1.8	8.0	4.0	3.5	1.3	D-638
1.2	1.2	1.3	1.6	1.2	1.2	1.8	1.05	1.6	D-638
23000	20000	16800	22000	19000	–	21000	23000	28000	D-790
1.0	1.1	1.2	1.4	1.0	0.95	1.0	0.80	1.4	D-790
22000	17900	17000	21000	15000	10000	12000	12000	24000	D-695
123	–	121	123	110	–	–	117 86 (Shore D)	122	D-785
430	600	450	500	400	425	–	450	350	D-149
3.65	2.93	3.2	3.6	3.5	3.4	–	4.5	3.5	D-150
0.005	0.0009	0.002	0.008	0.009	0.015	–	0.014	0.001	D-150
100	120	50	60	75	–	–	90	120	D-495
10^{15}	10^{17}	10^{16}	10^{16}	10^{15}	10^{15}	–	10^{11}	10^{16}	D-257
362	310	210	212	230	410	180	185	500	D-648
370	310	225	225	240	510	180	270	500+	
SE	SE	B	B	B	–	SE	B	SE	D-635
VEO	–	HB	HB	SB	–	–	HB	VEO	
1.4	2.2	2.0	1.9	1.7	3.2	1.7	2.4	1.4	D-696
2.2	1.1	2.1	2.0	1.5	–	–	2.8	2.1	C-177
180	110	50	61	68	750	75	200	200	
9.0	4.8	2.2	2.7	3.0	48.7	4.1	9.6	17.8	
10.6	6.5	4.2	3.2	4.3	62.5	5.4	11.8	11.8	

Table 26-13 Supplement - Code for description of properties of base thermoplastic resins.

Code or Symbol	Property	Code or Symbol	Property
ABR	Abrasion resistance	SR	Solvent resistance
CHR	Chemical resistance	SP	Softening point
CFR	Coefficient of friction	SPG	Specific gravity
CL	Clarity	ST	Stiffness
CO	Cost	SC	Stress cracking
CR	Creep resistance	TUF	Toughness
CRT	Crystallization tendency	TS	Thermal stability
CST	Continuous service temperature	WA	Water absorption
		WR	Water resistance
DIS	Discoloration		
DST	Dimensional stability		
EPR	Electrical properties		Qualitative grading nomenclature
FE	Fatigue endurance	E	Excellent
FL	Flammability	H	High
HAR	Hardness	G	Good
HDT	Heat deflection temperature	M	Medium
IR	Impact resistance	L	Low
MBR	Microbiological resistance	P	Poor
MP	Melting point		
MST	Mechanical strengths		Example CO – L = low cost.
MV	Melt viscosity		
PRO	Processability		

RP/C in Construction. Although performing this duty well and most aesthetically, the larger RP/C parts used in building construction have, to date, filled the function of fascia only. Unfortunately, increased product costs and ability to support maximum loads without cold-flow are in a direct relationship. However, by ingenious use of self-supporting designs using domes, arches, and paraboloids with inside and outside finished sandwich construction, many interesting small home designs have been successfully marketed.

RP/C continues to do its job in construction very ably, and Kalwall Corporation has perfected a system for use as an insulated energy collector cover in solar heating units.

Corrosion Resistant Applications. Expanded usage of RP/C components continues in the chemical process fields. Parts are made by almost all processing methods. The largest parts are, of course, made by either hand lay-up or filament winding. Pultruded structural members, spray-up, and compression-molded parts are widely applicable. The chief criterion is selection of the correct resin and proper laminate construction to provide the required corrosion resistant service. A new volume, published by the non-metallic section of NACE[15], ably presents in graph form data concerning the resistance of RP/C components to some 1425 different chemicals. Comparison with many other materials is included.

Notable recent applications include several types of corrosion resistant pumps with both rotor and stator parts of RP/C, and also, assemblies which float in a waste-water treatment pond, supporting both delivery and out-take piping so as not to damage a special film lining the pond.

Reinforced Foams. The record to date indicates that almost any plastic material (thermosets or thermoplastics, but usually thermoplastics) can be foamed either by induced gas or by chemical blowing agents. Foaming agents also may be combined with ingredients in the new reaction-injection molding process.

Fortuitously, it has been determined that incorporation of glass fibers into the foamable resins from 2 to 30% by weight provides reaction sites for more rapid and thorough generation of the foaming agents. This results in faster cures, smaller and more uniform foam cells, and better cell distribution for improved finished part properties.

The great advantage of one-cycle foaming operations is that they produce directly, in one stroke and without a two or three stage operation, a structure with a greater cross section and hence higher modulus of elasticity. Tennis racket frames and car body parts are examples, and also included are 1 shot panels for containerized truck trailers. The panels may measure up to 45 ft in length, and weigh up to 2,000 lb each.

For large, contoured, polyester-foam-polyester structures, such as boat decks, an interesting woven fiber glass reinforcement has been developed. Mölnycke Co. of Göteborg, Sweden, produces a woven roving fabric with loops 1 in. in diameter, on 1 in. centers, extending from 1 surface. Cut shapes of this fabric are laminated to the top and bottom RP/C panels of a boat deck or other structure, separated by a distance approximately the diameter of the loops. Foamable polyurethane resin is injected into the cavity and allowed to foam and cure. The gap is filled, and the foam becomes substantially reinforced by surrounding and encapsulating the loops.

Cure Systems and Related Mechanisms. Many chemical systems have been researched which were intended to trigger more effectively the reactions between catalysts and thermosetting resins, basis for 85-90% of all RP/C production. However, two significant recent advances, using any of the most effective chemical curing systems available, are concerned with the mechanics of applications of a heating means more effective than the present standard near-infra red energy sources.

Radio-frequency curing systems are being used on pultrusion processing lines. This type of excitation generates cures which occur simultaneously throughout the entire laminate cross-section, instead of merely moving from the surface inward. Faster pulling speeds and higher product quality are thus made possible.

Very recently, announcement was made of the use of a pulsating ultraviolet energy source, reputed to cure laminates close to 1 in. thick in approximately 4 min. Xenon Corporation is the innovator. Possibilities include production rates for pultrusion of from 10-20 ft/min. instead of the current 5 ft/min. or less, and curing wet layup or SMC parts in a press mold after U.V. exposure and without any additional heating required in the press equipment. The overall ultimate effect would be a welcome speedup of RP/C processing, lowering finished product costs, and enhancing the competitive position.

Use of organic titanates has been proposed to induce wetout in fiber glass reinforcements intended for hand layup, or any use in fact where it is necessary to combine them with liquid resins. These compounds have already been used to advantage on fillers, and show great promise as agents to improve speed and thoroughness of wetout of reinforcements.

New Directions

Combinations. The first commercially important sheathing or cladding of an RP/C structure with a chemically resistant or more protective surfacing layer was developed in 1960. Using a small quantity of resin additive or modifier with a solvent action, a polyester-fiber glass layup was bonded to PVC to form chemical hoods, stacks, and ducting. The PVC skin protected the stronger fiber glass structural portion from corrosion.

More recently, the process was refined by vacuum-forming a decorative acrylic or other thermoplastic sheeting into a desired shape (hand lavatory, shower and bath unit, or edged wall panel), and spraying up fiber glass-polyester on the unfinished side. Mold costs and cleanup problems were reduced, and the characteristic strength and lightness of weight of RP/C were used to advantage in creating unique decorative and utilitarian objects.

Wood, plywood, masonite, rigid foam, and other porous or semi-porous materials have been used as substrates or cores for RP/C skins. Glass piping has been over-wrapped with a layer of filament-wound RP/C for protection against shattering. Metal tanks for service below ground level have been coated with fiber glass to protect them from deterioration due to galvanic action in heavily populated, urban areas.

Hence, it should be stressed that this type of combination of RP/C with other materials should be diligently pursued where cost and performance show a true advantage.

Injection Molding. Reinforcing of injection-molded thermoplastic resins as a process became an entity well after the time that injection molding procedures, equipment, and technology had become well established. By easing in the back way, so to speak, reinforcements were required to live by the governing molding and press conditions. The few innovations made were in the methods for combining glass and molding resins.

The fact that glass fiber reinforcement starts out at $^3/_{16}$ or $^1/_4$ in. in length and ends up in a shorter aspect ratio, with the filament bundles completely separated and the filament lengths averaging somewhere between 0.001 and 0.030 in., simply does not speak well for the process.

Admittedly, the geometry of most RTP molded parts–sharp corners and angles– dictates uniformity of the reinforcement and resin mix. Also, the small diameters or cross-sections of nozzles, sprues, runners, gates, and the near zero screw-to-barrel I.D. tolerance induce breakup of the reinforcement. It is also acknowledged that the screw or plunger configuration is necessary for uniform heating of the resin prior to molding, avoiding a bulky, thick cross-section which would allow nonuniform heating. However, it should be pointed out that a great amount of energy is lost in the mechanical work required to cause the fibers to break up catastrophically. In so doing, glass gets the reputation of being highly abrasive to machine equipment and difficult to handle.

The injection molding process has also been successfully adapted to the molding of fiberglass reinforced thermoset plastics. In the case of thermosets, fiber length reduction may be even less than observed with thermoplastics, especially in the barrel and screw zones. However, in the injection molding of all fiber reinforced composites, the preservation of the fiber length is a major concern. Although much progress has been made toward this goal, there will be a continued need for cooperation between molders, compounders, and glass manufacturers in order to investigate means for preserving the original fiber lengths throughout the compounding and molding process. For a discussion and information on equipment for processing fiber reinforced plastics with minimum fiber breakage, see Chapter 32.

Additional Resin-Glass Combinations. In order to keep up with inflationary trends, prices have been kept down by eliminating process steps and making labor and material-saving combinations of glass, resin, and filler. SMC is a prime example. It has already been proposed and attempts have been made to formulate glass-resin molding compounds by incorporating fibers at the glass forming operation directly into the polymeric matrix. It is suggested that this course of action be reviewed and re-undertaken.

Inorganic Polymers. The disparity in thermal end-points that exists between glass and organic polymers for continuous elevated-temperature service truly limits the usage of RP/C. The presently used organic polymers are the restricting factor. Even the present high demand for reinforced plastics would increase considerably if the restrictive temperature barriers were removed.

Even ablative structures accomplish their high-temperature resistance by decomposition of the organic fraction, leaving a char layer that remains intact only due to the high tensile strength of the reinforcement which was originally incorporated.

The answer is undoubtedly in inorganic polymers. Several of these have been proposed and unfortunately are still curiosities of the laboratory bench. Aluminum phosphate was used to encapsulate fiber glass reinforcing fibers and fabricate radomes. Other phosphate compounds, reportedly resilient enough to be substituted for polymeric resins, have been known for quite some time, but are not yet commercially viable.[16] Metal and carbon-filber reinforced ceramics have high temperature resistance but are friable. Use of magnesium-oxychloride cements, made reinforceable and resilient, as substitutes for organic polymers, was recently announced as available on a licensee basis by Arthur D. Little Co.

Expansion. While many users of fiber glass and other reinforced plastics solicit quotations from molders for parts they need, a continuing, energetic, and creative sales effort is required to sustain industry growth. "Bread and butter" sales for existing items are required to maintain steady incomes, but innovation and looking diligently for new applications must be regarded as a concomitant function. RP/C can be made faster, better, and cheaper than any other materials, and it will certainly perform better.

References

1. Oleesky, S. S. and Mohr, J. G. *SPI Handbook of Reinforced Plastics*, 1st ed., Van Nostrand Reinhold, 1964, p. 117.
2. Loewenstein, K. L., *The Manufacturing Technology of Continuous Filament Glass Fibres,* Elsevier Scientific Publishing, 1973, p. 104.
3. Carroll-Porczynski, C. Z., "Inorganic Fibres," National Trade Press, Ltd., 1958, p. 13.
4. Beach, N. E., *Plastic Laminate Materials,* Foster Publishing, 1967, p. 33.
5. Weyl, W. A. and Marboe, E. C., *The Constitution of Glasses*, Vol I, II, and III, Interscience Publishers, 1962.
6. Levin, E. M., Robbins, C. R., McMurdie, H. F., "Phase Diagrams for Ceramists," American Ceramic Society, 2 Vols, 1964 and 1969.
7. U.S. Patents No. 3,819,387, No. 3,785,833, and No. 3,785,834.
8. U.S. Patent No. 3,816,107.
9. Morey, G. W., *The Properties of Glass*, Van Nostrand Reinhold, 1938, p. 224.
10. Di Domenico, M., Jr., "Wires of Glass," *Industrial Research Magazine,* (August, 1974), p. 50.
11. Mohr. J. G., "Use of the Differential Expansion Technique for Study of Internal Strain in Reinforced Plastic Structures," 15th Annual SPI RP/C Institute Proceedings, 1960, Section 10-E.
12. Scholes, S. R., *Modern Glass Practice,* Industrial Publications, 1941, see Frontmatter.
13. Mohr, J. G., et al, *SPI Handbook of Technology and Engineering of Reinforced Plastics/Composites*, Van Nostrand Reinhold Co., 1973.
14. Bralla, J. G., *Handbook of Production Design*, McGraw-Hill, 1978, see section on Design of RP/C Products.
15. Hamner, N. E., "Non-Metals Section, Corrosion Data Survey," National Association of Corrosion Engineers, 1975.
16. Stone, F. G. A., and Graham, W. A. G., *Inorganic Polymers,* Academic Press, 1962, p. 1, 68, and 189. See also Holliday, L., *Ionic Polymers,* Halsted Press, 1975.
    Also Buckley, J. D., "Advanced Materials: Composites and Carbon, American Ceramic Society Symposium," Editor and Compiler, American Ceramic Society, 1973.

26A
Basalt Fibers

R. V. Subramanian and H. F. Austin
Washington State University
Pullman, Washington

1. INTRODUCTION

Basalt fibers possess the desirable properties of glass fiber reinforcements and have some additional advantages. These include better temperature resistance, better adhesion to epoxy resins, and superior resistance to alkali solutions. The chemical composition of the fibers is determined by the native basalt rock that is used as raw material, which can show small variations from the typical composition represented by: SiO_3 :50%, Al_2O_3 :15%, TiO_2 :3%, FeO:11%, Fe_2O_3 :2%, MnO:0.2%, CaO:9%, MgO:5%, K_2O:1%, Na_2O:3%, $P_2O_5 < 1\%$. The strength and moduli of virgin basalt fibers have been found to be comparable to those of E-glass fibers produced and tested under the same conditions in the laboratory. The fibers are brown in color because of their iron content, which may prevent their use in some applications. At the present time, they are not being produced commercially in the United States, but a prototype unit for their production has been set up at Washington State University. It is expected that, based on this study, a pilot plant for the production of larger quantities will soon be in operation. Present estimates indicate that the cost of basalt fibers will be somewhat less than that of glass fibers.

2. MANUFACTURE

Basalt fibers are readily made by the same techniques and equipment used for the production of glass fibers. Both continuous fibers and staple fibers can be produced from basalt rock. Staple fibers that average 30 mm in length and that are useful for mats and polymer reinforcements are produced by a "nozzle-blow" process. (Abstract 9.2.1). This consists of a melting tank furnace that feeds an electrically heated platinum-rhodium bushing containing a large number of nozzles. A steam jet (or compressed air jet) is located below the nozzles and creates an air stream that moves parallel to the direction of extrusion. This pulls the basalt fibers from the nozzles and tears them apart into different lengths. The fibers fall onto a porous drum which is under vacuum. Plants for the production of 5-10 tons/day of staple basalt fibers by this process, have been built in Germany. Superfine fibers, 0.2-4 μ diameter, which are excellent as thermal insulation have reportedly been manufactured in Russia.

Continuous basalt fibers are produced in standard glass fiber production equipment. The rock is melted and fed to electrically heated platinum-rhodium bushings containing 200-400 openings. The fibers are pulled from the apertures by winding on a drum. Size of the fiber is controlled by the temperature of the melt (1250-1400°C) and speed of the collecting drum. The usual size is 10-15 μ. Desired sizes and coatings of various kinds can be put on the fiber before it is wound on the drum. The type of coating is determined by the intended application of the fiber. Silanes and conventional sizes can be applied by the usual methods. (Abstracts 8.2.4, 8.3.4, and 8.3.5)

The use of basalt rock as a raw material has several advantages. The rock is available in large quantities in each of a number of locations in the United States. Even though basalts from different locations vary somewhat in composition, each location has uniform composition and almost all will produce fiber of comparable properties. No mixing of ingredients prior to melting is necessary and only crushing and washing is needed to prepare the raw material. The selection of the appropriate

Table 26A-1. Chemical composition and strengths of fibers from basalts obtained from various locations.

Sample number, rock flow, source	X-6, Lolo flow, Whitman County, Wash.	RC-11, Imnaha, Rocky Canyon, Idaho	K-9068, Elephant Mt., Saddle Mt., Wash.	K-9064, Middle Yakima, Saddle Mt., Wash.	0-2, Sweet Home, Ore.	K-9048, Pomona, Saddle Mt., Wash.	RC-3, Imnaha, Rocky Canyon Idaho	K-9017, Lower Yakima, Saddle Mt., Wash.	BCR-P, Yakima Flow	E-glass[b] Owens-Corning
Analysis, %										
SiO_2	49.10	49.41	50.02	50.49	50.50	50.86	51.41	53.61	54.50	52.20
Al_2O_3	13.80	17.92	13.29	13.62	16.00	15.18	15.14	15.14	13.60	14.80
TiO_2	3.16	2.41	3.48	3.06	2.17	1.69	2.22	1.84	2.20	0.00
Fe_2O_3	2.00	2.00	2.00	2.00	–	2.00	2.00	2.00	2.00	0.30
FeO	11.98	9.66	13.27	12.62	13.80	9.21	11.32	9.60	10.50	0.00
MnO	0.21	0.17	0.21	0.22	–	0.19	0.21	0.18	0.18	0.00
CaO	9.43	9.06	8.59	8.48	10.00	10.62	9.33	8.43	6.92	18.70
MgO	5.25	5.64	4.28	4.45	4.30	6.49	5.05	4.98	3.46	3.30
K_2O	1.26	0.81	1.35	1.54	0.35	0.80	0.68	1.14	1.70	0.00
Na_2O	3.09	2.57	2.95	2.93	3.20	2.62	2.28	2.73	3.27	0.30
P_2O_5	0.68	0.35	0.55	0.58	–	0.33	0.36	0.35	0.36	0.20
B_2O_3	0.00	0.00	0.00	0.00	0.00	0.00	0.00	0.00	0.00	10.20
Temperature of drawing, °C	1325	1300[a]	1250	1250	1250	1250	1250	1250	1360[a]	1250[a]
Speed of drum (rpm)	515	250	370	515	370	250	370	515	250	250
Number of samples	40	29	30	18	25	20	20	20	31	23
Average diameter of fiber (μ)	13.0	12.2	13.5	9.0	11.4	11.8	11.3	10.2	12.1	12.2
Average tensile strength										
(GPa)	1.97	1.99	2.13	2.23	2.08	2.08	2.25	2.45	1.97	2.52
psi	286,000	288,000	309,000	323,000	302,000	301,000	326,000	355,000	285,000	365,000
Young's modulus										
(GPa)	82.76	77.93	77.93	87.59	90.34	82.76	87.59	87.59	71.03	81.38
(millions psi)	12.0	11.3	11.3	12.7	13.1	12.0	12.7	12.7	10.3	11.8

[a] Air jet used.
[b] Produced in-house from Owens-Corning E glass marbles.

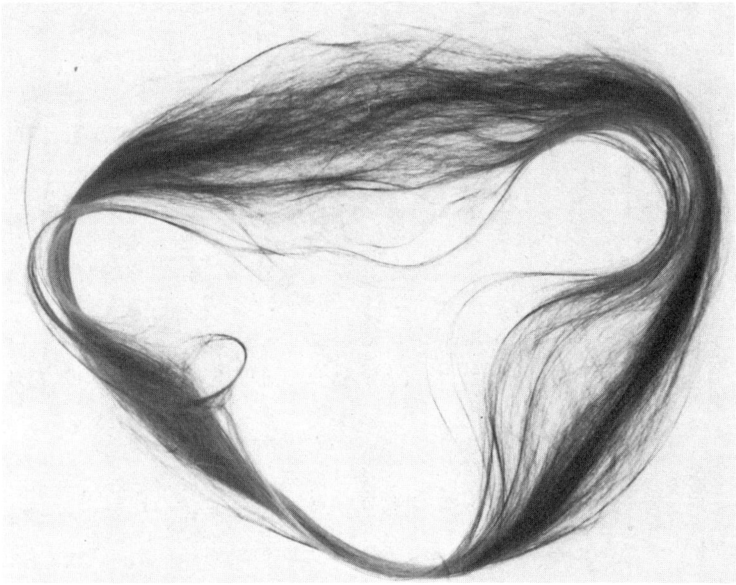

(a)

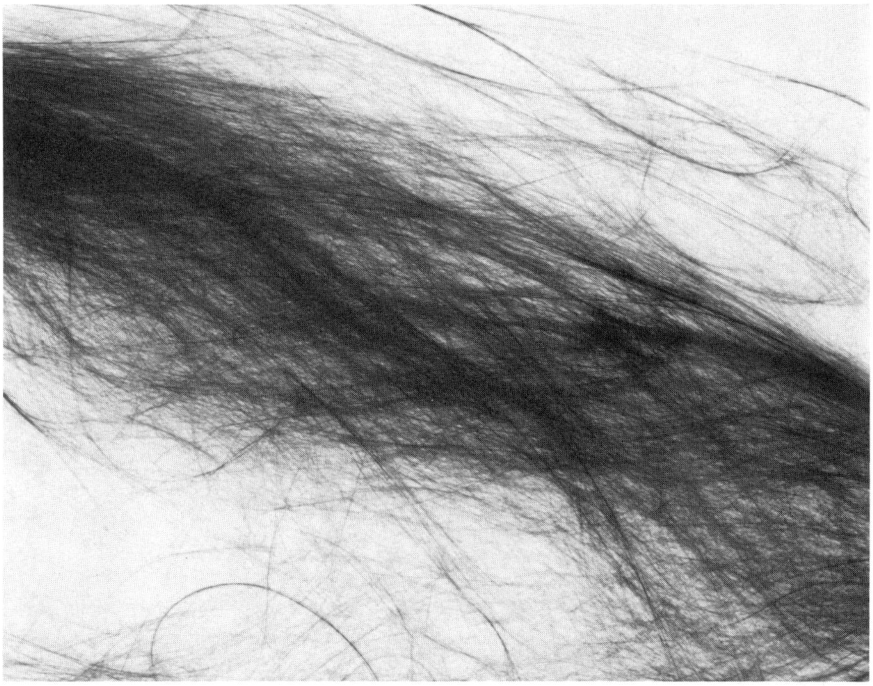

(b)

Fig. 26A-1. A skein of basalt fibers (a) and an enlarged section showing uniformity of the fibers (b).

basalt rock will enable the production of fibers of specific properties (e.g., alkali resistance).

3.1 PROPERTIES

Chemical compositions and strengths of fibers from basalts from various locations as shown in Table 26A-1. A skein of basalt fibers with a close-up view showing uniformity of fine fibers is shown in Fig. 26A-1. Figure 26A-2 shows the tensile strength of basalt fibers drawn at various temperatures.

4. SUPPLIERS

At the present time there are no commercial sources of basalt fibers in the United States. It is hoped that a pilot plant will soon be in operation for the production of basalt fibers. Further information may be obtained from the Department of Materials Science and Engineering, Washington State University, Pullman, Wash., 99164. Present indications are that the cost of basalt fibers should be less than that of comparable glass fibers. In Germany, basalt fibers are being produced by:

Basalan-Isolierwolle Gmbh & Co.,
KG. Industriestr. 4
4403 Hiltrup
Erst Deutsche Basalt-wolle KG.
A. Messler
8712 Volkach
and in Scotland by:
Cape Insulation Ltd.
Kerse Road,
Stirling, Scotland FK77RW

5. HISTORY AND FUTURE

Basalt fibers have been and are being produced in Russia and Germany. As long ago as 1959, the Germans reported on installations for the production of mineral fibers from basalt (Abstract 8.2.1). More recently, in 1973, the official Russian news agency, 'Novosti', reported that "use of basalt fibers, made from basalt rocks, is being developed extensively in the Soviet Union. Several large plants for fiber output are being built."

Experimental production of basalt fibers from basalt of the Pacific Northwest at Washington State University has shown the feasibility of commercial production of quality fibers from basalt resources occurring in the United States. The ready availability of abundant supplies of raw material and the resultant decrease in cost of production compared to glass, should provide the incentive to initiate a basalt fiber facility. This cost decrease should arise from not having to purchase, store, and mix many individual ingredients as is necessary to produce glass.

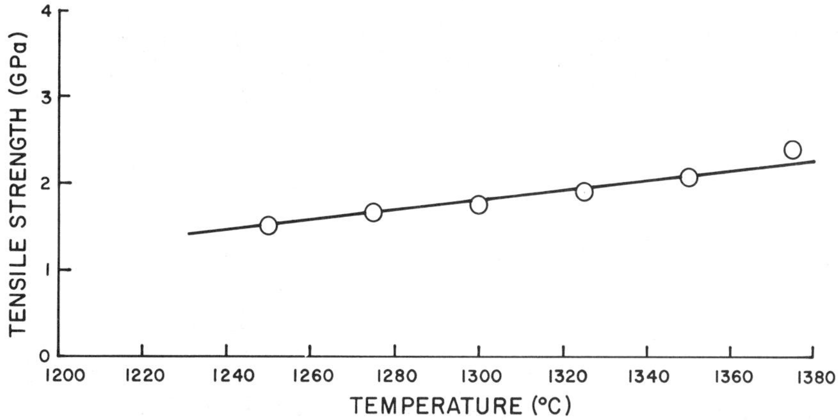

Fig. 26A-2. The variation of tensile strength of virgin basalt fibers as a function of drawing temperature. (Ref. 8.3.4.)

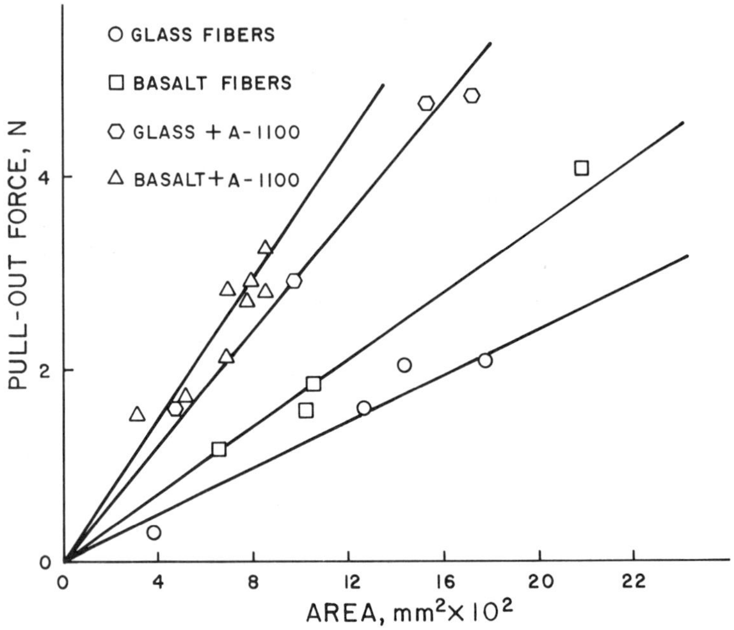

Fig. 26A-3. Pull-out force for fibers embedded in epoxy matrix plotted against area of contact. (Ref. 8.3.4.)

6. APPLICATIONS

The effectiveness of basalt fibers as reinforcement for polymers has been studied with epoxy resins (Abstracts 8.3.3 and 8.3.4). The results show that basalt fibers can be substituted for glass fibers in all composites where their brown color is not objectionable. This conclusion is based upon tests of adhesion of basalt and glass fibers to an epoxy matrix and upon determination of the relative improvements in adhesion and composite properties as a result of treatment with silane coupling agents.

The force required for pull-out of fibers embedded in an epoxy matrix in single-fiber adhesion tests is plotted in Fig. 26A-3. It is seen that basalt fiber adhesion to the epoxy resin is somewhat superior to the adhesion of E-glass fibers. This superior adhesion is true for both silane coated and uncoated fibers. The addition of silane adhesion promoting coatings increases the tensile strength and tensile modulus of basalt fiber-epoxy composites. The data in Figs. 26A-4 and 26A-5 illustrate these increases when Union Carbide Silane A-1100 (γ-aminopropyl trimethoxy silane) was used. The matrix resin was a DGEBA resin (EPON

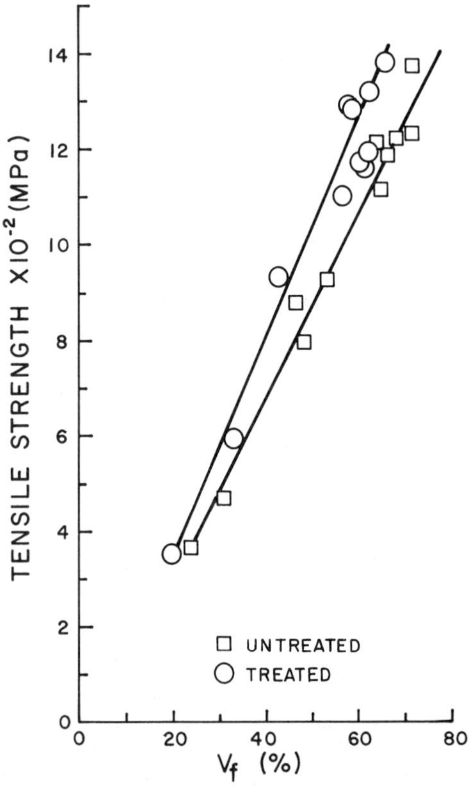

Fig. 26A-4. Tensile strength of epoxy resin composite (DGEBA cured by metaphenylenediamine) reinforced by untreated and treated (Silane A-1100) basalt fibers. (Ref. 8.3.4.)

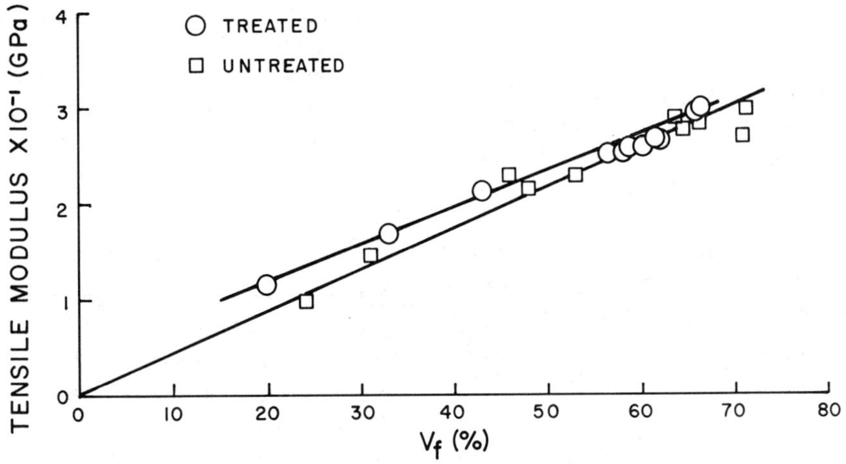

Fig. 26A-5. Tensile modulus of epoxy resin composite (DGEBA cured by methaphenylene-diamine) reinforced by untreated and treated (silane A-1100) basalt fibers.

828) cured with metaphenylene diamine (MPDA).

The similarities are thus much more striking than differences between the 2 fibers in reinforcement of thermoset polymers. Tensile strengths of the fibers are almost identical and the elastic modulus of the basalt fibers is a little higher, which could indicate its preferential use in some applications.

7. UTILIZATION

The possible uses of basalt fibers are numerous: air filters, heat insulation at high temperatures, insulation of vibrating surfaces, sound insulation, reinforcement in concrete, as mineral paper or cardboard, reinforcement of thermoset resins with better adhesion than glass fiber, reinforcement of polyester resins, as a building material with emphasis on the replacement of asbestos. The Russians have reported the production of microfine fiber (0.4-2 μ diameter), which is in the size range to compete successfully with asbestos fibers. The current search for substitutes for asbestos, to alleviate the health problems connected with its use, should lead to extensive testing of microfine basalt fibers. If basalt fibers are found suitable as a substitute for asbestos fibers, the potential uses are very large. The superior alkali resistance of basalt fibers indicated in laboratory tests so far, makes this highly probable.

The alkali resistance of basalt fibers promises an excellent potential for use in the reinforcement of concrete. These fibers are very much cheaper to produce than the special alkali-resistant glasses that have been formulated and tested.

8. ABSTRACTS OF LITERATURE

8.1 Review.

Raff, R. A. V., *Engineering/Mining Journal*, pp. 76-77 (Feb. 1974).

An excellent review article that lists 50 pertinent references to the literature on basalt fibers and presents a clear picture of the technology and uses throughout the world and the potential for its production in the United States.

8.2 Production of Fibers

9.2.1 Kaswant in Kastellaun, "Modern Installations for the Production of Mineral Fibers using the Pressure Nozzle Process," Sprechsaal für Keramic, Glass Email, **91**, pp. 577-86 (1958); *CA* **53**, 7464 (1959).

This is a most detailed description of a unit for the commerical production of 10 tons of basalt fiber per day. The unit is fed with basalt chips and produces staple fiber from a platinum-rhodium nozzle. Detailed cost figures including wages, materials, and power costs are given but will require revision in order to be useful today.

8.2.2 Dubroskie, V. A., Rychko, V. A., Bachilo, T. M., Lysyuk, A. G., "Basalt Melts for the Forming of Staple Fiber," *Steklo i Keramika*, No. 12, pp. 18-20 (Dec. 1968).

The most important properties of basalt melts for fiber production are viscosity and crystallization characteristics. Staple fibers can be produced from melts whose viscosity at 1400°C is at least 90 poise and at 1250°C below 350 poise. The upper limit of crystallization of the melts is 1250°C.

8.2.3 Darenskie, V. A., Yu, N. Dem'yanenko, Kozlovskii, P. P., Manhurnet, K. V., Kukarkin, A. I., Dzhugaryan, R. T., and Badalyan, K. S., "Producing Basalt Staple Fiber," *Steklo i Keramika*, No. 1, pp. 38-40 (Jan. 1968); *CA* **68**, 195933 (1968).

A description of a production plant in Armenia SSR gives some details of the equipment used. The temperature of the melt is 1490 ±10°C, the fusion level in the melting tank is maintained at 195 mm. The level of the melt over the nozzle is 155-160 mm. The platinum-rhodium feeder is electrically heated and has 10 nozzles, each having a diameter of 3.0 ±0.2 mm. The pressure of the compressed air fed to the blowing head is 4 atm and the optimum distance from the nozzle to the entry into the blowing head is 100 mm. The output is 2100 kg/day.

8.2.4 Subramanian, R. V., Austin, H. F., Raff, R. A. V., Sheldon, G., Dailey, R. T., and Wullenwaber, D., "Use of Basalt Rock for the Production of Mineral Fiber," Pacific Northwest Commission, Annual Report, June 1975, College of Engineering, Washington State University, Pullman, Wash. 99164.

A detailed description of an experimental unit for the production of continuous basalt fibers from basalt found in Washington is given. Equipment for testing physical properties of the fibers is also described. The latter half of the report gives a comprehensive economic analysis, market survery, and cost estimate for a plant to produce basalt fibers.

8.3 Properties of Basalt Fibers

9.3.1 Andreevskaya, G. D. and Plisko, T. A., "Some Physical Properties of Continuous Basalt Fibers," *Steklo i Keramika*, No. 8, pp. 15-18 (Aug. 1963); *CA* **60**, 1470 (1964).

Elastic modulus of 13-18 μ fibers was determined by a frequency method and was found to be 9230 kg/mm^2. Strength of fibers in relation to diameter were measured and given: 8 μ–260 kg/mm^2, 10 μ–200, 13 μ–125, 14 μ–105, 15 μ–100, 16 μ–98, 17 μ–92.

8.3.2 Dubovkaya, T. S. and Kosmina, N. E., "Chemical Stability of Staple Basaltic Fibers," *Strukt. Sostav. Svoistva Formovanie Steklyanykj Volokon*, 1968, pp. 1, 140; *CA* **73**, 112426 (1970).

Basalt fibers were found to be more stable in the presence of alkali than acids.

8.3.3 Subramanian, R. V., Austin, H. F., Wang, T. J. Y., "Use of Basalt Rock for the Production of Mineral Fiber," Pacific Northwest Regional Commission Annual Report (Sept. 1976), College of Engineering, Washington State University, Pullman, Wash. 99164.

A study of process parameters in the production of basalt fibers, properties and performance of basalt fiber reinforcements in composites, and modification of basalt fiber properties for reinforcement of polymer matrices.

8.3.4 Subramanian, R. V., Wang, T. J. Y., and Austin, H. F., "Reinforcement of Polymers by Mineral Fibers from Basalt Rocks" Sampe Quarterly, pp. 1-10 July 1977.

27
High Modulus Organic Fibers

D. L. G. Sturgeon and R. I. Lacy
E. I. duPont de Nemours & Co., Inc.
Wilmington, Delaware

CONTENTS

---

1. INTRODUCTION

High performance organic fibers can be used to advantage when lighter weight, more stiffness, and higher tensile strength composites than obtainable with E-glass or S-glass reinforcement are desired. Aramid composites will exhibit slightly lower inter-laminar shear and flex strength, higher flexural and compressive modulus, but about half the compressive strength of glass items made with the same volume fraction of glass reinforcement in the same resin. Weight savings over glass result from the lower specific gravity of aramid fibers (1.45 g/cc) versus E-glass (2.55 g/cc). Higher stiffnesses reflect a Young's modulus up to 19MM psi for some aramids like Kevlar® 49, compared to 10×10^6 psi for E-glass, and 12.5×10^6 psi for S-glass. Aramid fiber reinforced resin composites are more electrically and thermally insulating than their glass counterparts, more damped to mechanical and sonic vibration, and they are also radar and sonar transparent. These high modulus organic fibers retain the processibility normally associated with conventional textiles despite their outstanding mechanical properties. This leads to wide versatility in the form of the reinforcement that can be obtained (e.g., yarns, rovings, woven and knit goods, felts, papers, etc.). Composites of these fibers have low creep, high impact resistance, very low notch sensitivity, and excellent creep rupture characteristics.

Present cost of high modulus aramid fibers is higher than E-glass and equivalent to some grades of S-glass on a unit weight basis. Price differences versus glass are reduced by about half on a per-unit volume basis by virtue of the lower density of the aramids.

The term "aramid" was approved by the Federal Trade Commission of the United States to denote a special class of aromatic polyamide fibers. The category "aramid" presently includes "Kevlar", "Kevlar" 29 and "Kevlar" 49, and Nomex®, all products of the du Pont Company. Other high modulus organic fibers are reportedly under development. Whether any of these new products will qualify under the generic "aramid" remains to be established. Aramid fibers have high temperature capabilities by organic fiber standards, having no melting point and decomposing at temperatures in excess of 400°C. They are not ceramic as are graphite and boron fibers and therefore cannot be used to reinforce metals. Their high temperature capability, however, is sufficient for effective reinforcement of many widely used organic matrices, since resins generally have lower temperature tolerance than these fibers. Long hold-up times at the melt temperature required for processing some thermoplastics may result in some deterioration of fiber properties. A continuous service temperature limit in composites of about 180°C appears to be reasonable for present aramid fibers. This might possibly exclude reinforcement of very high performance polyamide resin compositions rated for continuous service up to 300°C. These fibers possess excellent chemical resistance. For example, "Kevlar" 49 is more resistant than E-glass to hydrofluoric acid and sodium hydroxide, but less resistant than glass to hydrochloric acid. Common industrial solvents, fuels, and lubricants at ambient temperature are innocuous to these fibers.

2. PRODUCT FORMS

As explained in Section 1, the designation "aramid" applies not only to fibers with primarily reinforcing capability, like "Kevlar", but also to others with certain textile and flame retardant attributes, like "Nomex". Since the interest of the reader is in reinforcing fibers, what is said in this chapter is meant to apply to those high modulus aramids like "Kevlar" 29 and "Kevlar" 49 whose primary function is the reinforcement of elastomers, thermoplastic resins, and thermoset resins. Presently, "Kevlar" 49 is the leading candidate for plastics reinforcement and is available in forms similar to those of glass fiber.

Table 27-1a. "Kevlar" 49 yarn.

Denier	Yield (yd/lb)	Number of Filaments
195	22,895	134
380	11,749	267
1420	3,144	1000

Table 27-1b. "Kevlar" 29 yarn.

Denier	Number of Filaments
200	134
400	267
1000	666
1500	1000

Table 27-1c. "Kevlar" 49 roving.

Denier	Yield (yd/lb)	Grade
4560	979	Aerospace
7100	629	Commercial

2.1 Fiber and Roving

Commercially available fiber and roving sizes of "Kevlar" 29 and "Kevlar" 49 are shown in Table 27-1. Denier (den.) is the textile unit defined as the weight in grams of 9000 m of yarn. The equivalent "yield" (yds/lb) more common in the glass fiber literature is given for comparison. Sizes range from 195 den. (22,895 yd/lb) to 7100 den. (629 yd/lb), with heavier deniers reportedly under development for pultrusion applications.

2.2 Woven Fabrics

The style number, construction, type of weave, basis weight, and thickness of typical fabrics of "Kevlar" 49 are shown in Table 27-2. Basis weights ranging from 1.8-15.0 oz/yd^2 are currently available. Many of the fabrics of "Kevlar" 49 are designed to permit a ply-for-ply substitution of a standard glass fabric. "Kevlar" 49 Style 181, weighing 5.0 oz/yd^2, for example, can replace an equal volume, equal thickness per ply, glass fabric Style 181 or Style 1581

Table 27-2. Typical "Kevlar" 49 woven fabrics.

Style	Basis Weight (oz/yd^2)	CONSTRUCTION ENDS/INCH (YARN DENIER) Warp	Fill	Weave	Thickness (mils)
120	1.8	34(195d)	34(195d)	Plain	4.5
143	5.6	100(380d)	20(195d)	Crowfoot	10
181	5.0	50(380d)	50(380d)	8-HS	11
243	6.7	38(1140d)	18(380d)	Crowfoot	11
281	5.0	17(1140d)	17(1140d)	Plain	10
285	5.0	17(1140d)	17(1140d)	Crowfoot	10
328	6.8	17(1420d)	17(1420d)	Plain	13
335	6.8	17(1420d)	17(1420d)	Crowfoot	13
1033	15.0	40(1420d)	40(1420d)	8 X 8 Basket	24
1050	10.5	28(1420d)	28(1420d)	4 X 4 Basket	20

weighing 8.9 oz/yd^2. "Kevlar" 29 is also available in a variety of fabric constructions and widths from a number of commercial weavers.

2.3 Prepregs

As with most fiber reinforcements, B-staged epoxy impregnated unidirectional tapes, broadgoods, and fabrics are often used. These forms of aramid materials are available from most commercial impregnators.

2.4 Nonwoven Fabrics, Felts and Batts

The textile processibility of high modulus aramid fibers has allowed the development of nonwoven fabrics to satisfy certain performance requirements, such as the isotropic reinforcement of sheet structures (e.g., circuit boards), and the development of lightweight thermal insulation (e.g., for aircraft). Developmental quantities of nonwoven fabrics in basic basis weights ranging from 0.7-14.0 oz/yd^2 have been reported, as well as insulation batts of basis weights 0.40 lb/ft^3 and 0.80 lb/ft^3.[1] At the end of 1976, however, these items had not become commercial.

2.5 Hybrids

Unidirectional tapes, broadwovens, and fabrics that combine aramid fibers with other reinforcing fibers such as glass and graphite have been developed. A description of "Kevlar" 49/graphite fabrics is given in Table 27-3. Products combining aramids with other reinforcing fibers are available in the form of tapes from several commercial prepreggers and in the form of woven items from specialty weavers.

2.6 Chopped Forms

Relatively little work has been done to date on the development of high modulus aramid chopped fiber molding compounds. Most successful work to date has involved epoxy chopped lengths of $>1/2$ in. in resins. Cutting of "Kevlar" 29 and 49 is significantly more difficult than

Table 27-3. Graphite/aramid hydrid fabrics. (Fiberite Corp. Data)

Style	Basis Weight (oz/yd^2)	Thickness (mils)	Weave	Construction Ends/inch Warp X Fill	YARN TYPE IN Warp	Fill
105	10.2	12.3	8H Satin	24 X 24	75% "Thornel" 300 25% "Kevlar" 49	75% "Thornel" 300 25% "Kevlar" 49
107	9.7	12.0	8H Satin	24 X 24	50% "Thornel" 300 50% "Kevlar" 49	50% "Thornel" 300 50% "Kevlar" 49

cutting glass, and some of the conventional tools (e.g., chopper gun) designed to cut glass are ineffective with "Kevlar". No hand-held tool for the cutting of unimpregnated roving of "Kevlar" has been devloped by late 1976. Epoxy preimpregnated B-staged strands are more easily cut than dry fiber, and epoxy resin chopped fiber molding compounds for compression molding applications are available commercially. Polyester resin chopped fiber molding compounds using high modulus aramids similar to those popular with glass (BMC, SMC, etc.) are presently in a very early stage of development. In a similar early stage of development are chopped fiber forms of high modulus aramids useful for the reinforcement of thermoplastic resins. Some notable successes have been reported in laboratory preparation of compression molded thermoplastics of "Kevlar" 49.[2,3] Developmental work in the polyester and thermoplastic area is generally proceeding with prechopped "Kevlar" additions to the resin, since the toughness of these aramids precludes cutting by shearing in conventional extruders.

3. PRODUCT PROPERTIES

3.1 Fiber and Roving

3.1.1 Chemical. Table 27-4 shows the effect of certain chemicals on the yarn tensile strength of "Kevlar" 29 and "Kevlar" 49. Tensile modulus is generally significantly less degraded than tensile strength after exposure. It can be seen that, except for very strong acids and bases, these fibers are relatively immune to a variety of chemicals. Industrial solvents, lubricants, fuel, and water under the conditions tested have no effect on the fiber. Significantly more detailed information on the chemical resistance of "Kevlar" 29 and "Kevlar" 49 are available from the fiber manufacturer. Chemical resistance of aramid fiber reinforced composites is given in Section 6.1.8.

3.1.2 Physical. High modulus, continuous aramid yarns consist of multiple continuous single filaments of circular cross-section brought together to form bundles of different sizes according to the desired denier. Dishevelment and catenary are carefully avoided in the process. Yarns and roving from the fiber manufacturer currently have no twist, but twist can be added to improve processibility. The diameter of the individual filaments is 0.00046 in. (~12 μ). The density of the fiber is 1.45 g/cm^3.

The stress-strain curves of yarns of "Kevlar" 29 and 49 are compared to those of galvanized, improved, plow steel wire and Dacron® polyester in Fig. 27-1. The properties of these materials are compared in Table 27-5. Note that the high modulus aramids have an order of magnitude higher tensile modulus than industrial nylon, and up to two-thirds that of steel wire. In reporting the strength of reinforcing fibers, it is important to designate the gage length, and whether single filaments, yarns, or impregnated strands were measured. This is particularly true when reporting the strength of brittle, inorganic fibers whose strength can be strongly gage sensitive. The bare yarn properties shown in Table 27-5 reflect a gage length of 10 in. Yarn deniers of "Kevlar" 29 and 49 were 1500 and 1420 den., respectively, tested at two turns per in. The effect of twist on the tensile properties of high modulus aramid fibers is shown in Figs. 27-2 and 27-3. If resin impregnation as per ASTM D2343 strand test is used, the tensile strength of "Kevlar" 29 and 49 will

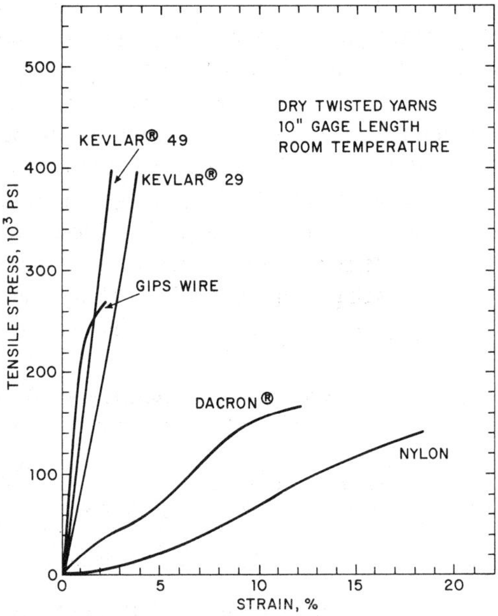

Fig. 27-1. Yarn stress-strain curves.

Table 27-4. Stability of "Kevlar"® 29 and "Kevlar"® 49 in chemicals.

Chemical	Conc. %	Temperature °F	Time hrs	STRENGTH LOSS, % "Kevlar" 29	"Kevlar" 49
Hydrochloric acid	37	70	24	N.A.*	0
" "	37	70	1000	83	N.A.
Hydrofluoric acid	10	70	100	12	8
Nitric acid	1	70	100	18	5
Sulfuric acid	10	70	100	14	N.A.
" "	10	70	1000	N.A.	31
Sodium hydroxide	50	70	24	N.A.	10
Ammonium hydroxide	28	70	1000	10	N.A.
Acetone	100	70	24	0	0
Dimethyl formamide	100	70	24	N.A.	0
Methyl ethyl ketone	100	70	24	N.A.	0
Trichloroethylene	100	70	24	N.A.	1.5
"	100	190	387	7	N.A.
Ethyl alcohol	100	70	24	0	0
Jet fuel (JP-4)	100	70	300	0	4.5
" "	100	390	100	4	N.A.
Brake fluid	100	70	312	2	N.A.
"	100	235	100	33	N.A.
Transformer oil (Texaco #55)	100	140	500	4.6	0
Kerosene	100	140	500	9.9	0
"Freon" 11	100	140	500	0	2.7
"Freon" 22	100	140	500	0	3.6
Tap water	100	212	100	0	2
Sea water (Ocean City, N.J.)	100	–	1 yr.	1.5	1.5
Water at 10,000 psi	100	70	720	0	N.A.
Water-superheated	100	280	40	9.3	N.A.
Steam-saturated	100	300	48	28	N.A.

*NA = Not Available

reach approximately 525×10^3 psi. These very high strengths, combined with the low density of the high modulus aramid fibers, result in very high specific tensile strengths as shown in Fig. 27-4. Notice that the specific tensile modulus of the organic fibers is intermediate between those of glass fibers and those of the stiffer graphite and boron reinforcements. The strength and modulus of yarns of "Kevlar" 29 and "Kevlar" 49 at several temperatures are shown in Figs. 27-5 and 27-6. The effects of prolonged exposure to high temperatures in air are shown in Figs. 27-7 and 27-8. The creep performance of twisted "Kevlar" 29 and "Kevlar" 49 aramid

Table 27-5. Properties[a] of bare reinforcing yarns.

	Nylon	"Kevlar" and "Kevlar" 29	E-Glass[b]	"Kevlar" 49	Steel
Specific gravity	1.14	1.44	2.55	1.45	7.86
Tensile strength (10^3 psi)	145	400	220	400	285
Tensile modulus (10^6 psi)	0.8	12	10	19	29
Elongation (%)	18	4	3	2.4	2

[a]10-in. gage length, bare yarn, 3 tpi added to organics.
[b]Owens-Corning Fiberglas Corp. publication No. 1-GT-1375-C.

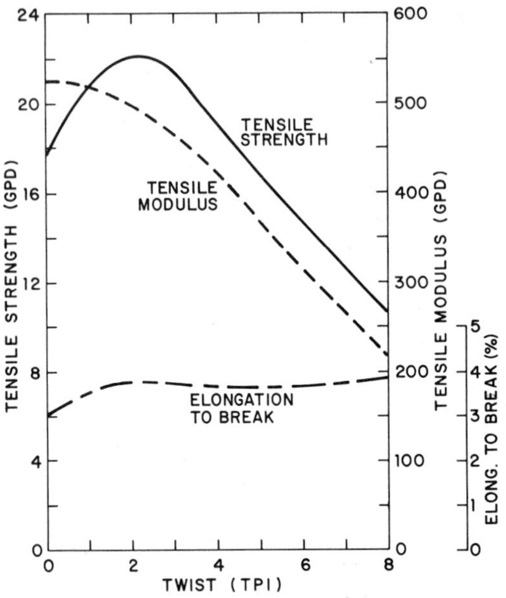

Fig. 27-2. Influence of twist on properties of Kevlar® 29 1500 denier yarn.

at 50% of ultimate yarn tensile strength at room temperature is shown in Fig. 27-9 after 25 weeks, while the stress rupture behavior of "Kevlar" 49 under varying percentages of ultimate tensile strength is compared to nylon in Fig. 27-10. Further insights into the perfor-

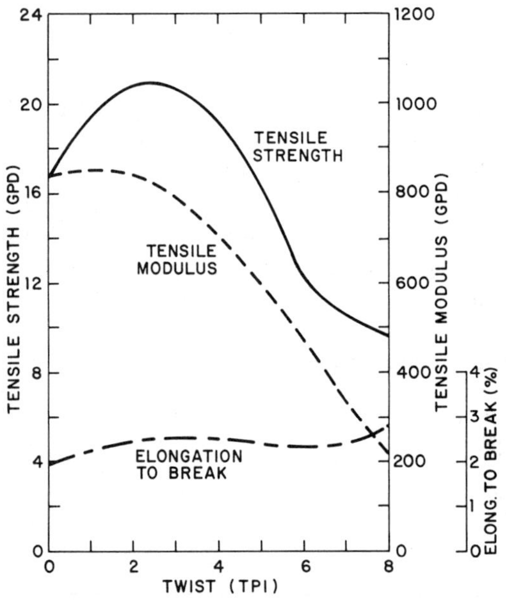

Fig. 27-3. Influence of twist on properties of Kevlar® 49 1420 denier yarn.

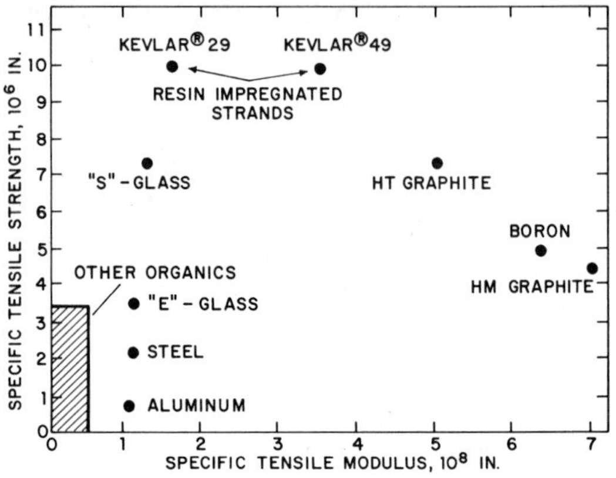

Fig. 27-4. Specific tensile strength and tensile modulus of fibers.

mance of high modulus aramid fibers under a variety of loading conditions can be gained from information on the behavior of composites made from these fibers that is presented in Section 6.3.1.

3.1.3 Thermal, Optical, and Others. Information on selected thermal, optical, and other properties of fibers are given in Table 27-6. Because of its practical significance, the ultraviolet (UV) resistance of aramids deserves separate mention. The data in Fig. 27-11 show how

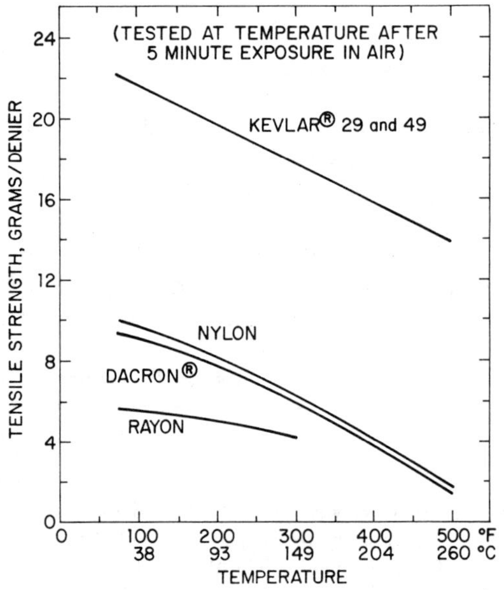

Fig. 27-5. Effect of temperature on yarn tensile strength. (Tested at temperature after 5-min. exposure in air.)

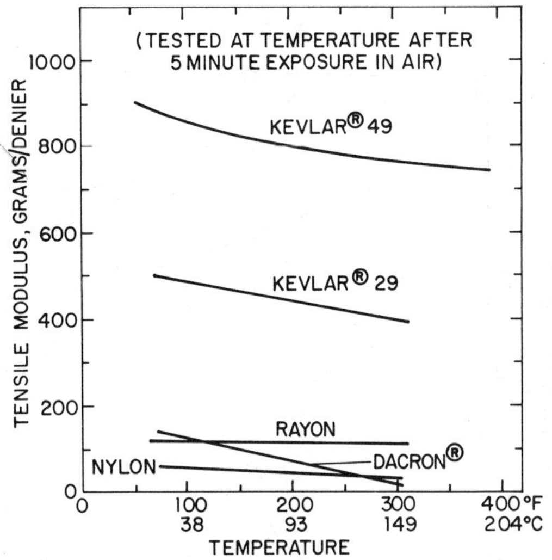

Fig. 27-6. Effect of temperature on yarn tensile modulus. (Tested at temperature after 5-min. exposure in air.)

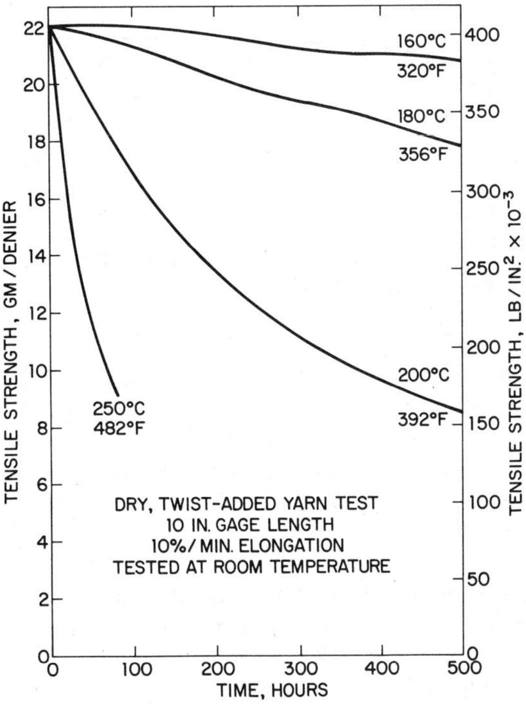

Fig. 27-7. The effect of temperature on the tensile strength of Kevlar® 29.

strength of unprotected "Kevlar" deteriorates with Xenon lamp UV exposure. Data in Table 27-7 indicate how this degradation can be minimized or totally prevented due to the self-shielding characteristic of the fiber that makes a substantial cross-section of material like a rope

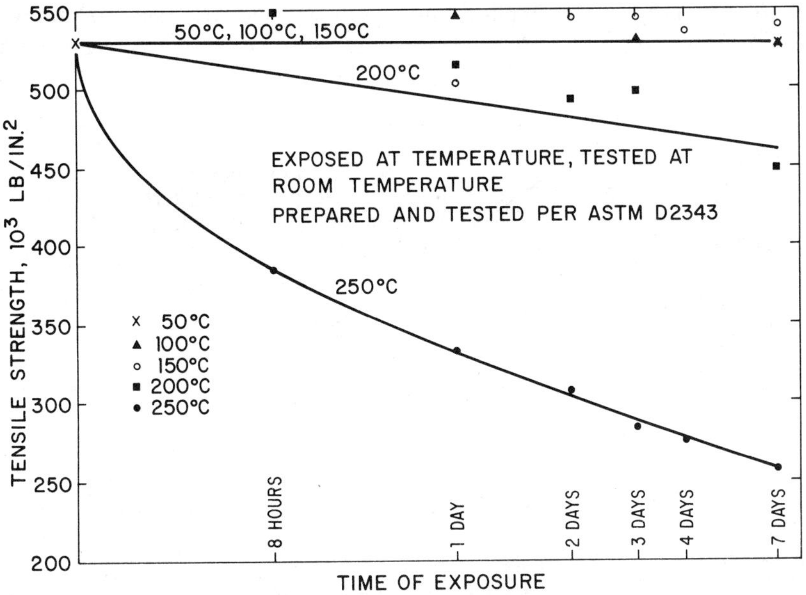

Fig. 27-8. The effect of elevated temperature exposure on the tensile strength of Kevlar® 49/epoxy roving strands. (Exposed at temperature, tested at room temperature prepared and tested per ASTM D2343.)

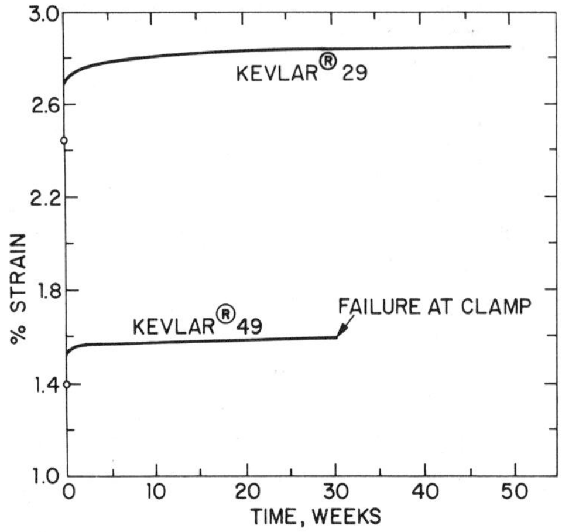

Fig. 27-9. Creep of twisted Kevlar® 29 and 49 aramid yarns at 50% of ultimate tensile stress, room temperature.

significantly more UV resistant than a single yarn. The benefit of jacketing with more UV resistant materials is also illustrated in the table. Multiple ply composites made with aramid reinforcement are relatively immune to UV degradation due to their thickness and the possibility of incorporating UV absorbers to the resin matrix. The user, therefore, should be aware of the need for UV protection of aramid fibers, particularly in very thin sections, but proven technology exists to accomplish this economically when required.

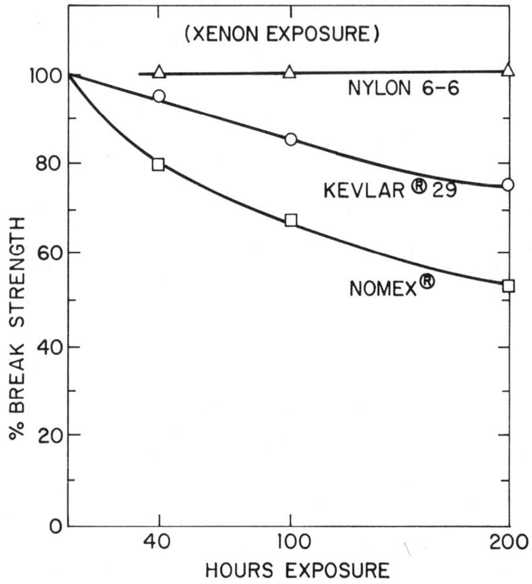

Fig. 27-11. Comparative light stability of yarns. (Xenon exposure effect on break strength of yarns.)

3.2 Fabrics

3.2.1 Wovens. The properties of selected woven fabrics of "Kevlar" 49 are compared to glass fabrics in Table 27-8. Similar properties of a variety of fabrics made with "Kevlar" 29 are available from commercial weavers. Generally, the high modulus aramid fabrics will exhibit a lower basis weight and a higher tensile strength

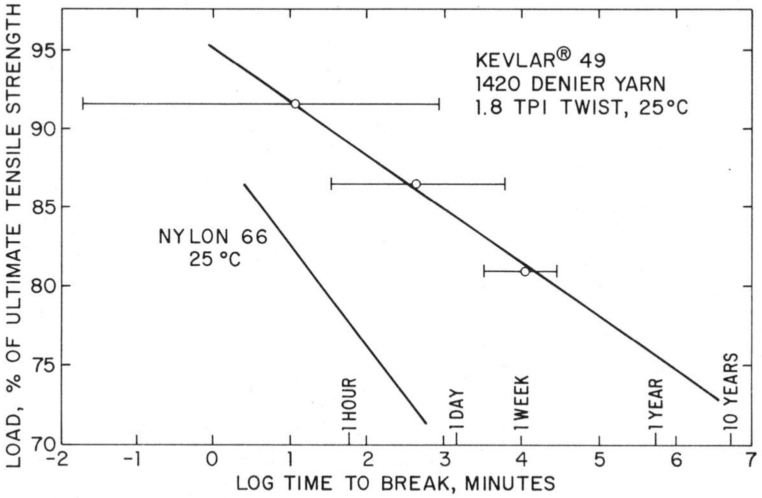

Fig. 27-10. Stress-rupture of Kevlar® 49 yarn. (Time to failure under continuous static load.)

Table 27-6. Miscellaneous thermal and optical properties of "Kevlar"® 49.

Long-term use temperature in air	320°F (160°C)
Decomposition temperature	930°F (500°C)
Shrinkage	2.2×10^{-4}%/°F (4×10^{-4}%/°C)
Thermal coefficient of expansion	
Longitudinal, 0–100°C	-1.1×10^{-6} in./in.°F (-2×10^{-6} in./in.°C)
Radial, 0–100°C	$+33 \times 10^{-6}$ in./in.°F ($+59 \times 10^{-6}$ in./in.°C)
Thermal conductivity, room temperature = 0.19 g/cm^3	
Heat flow perpendicular to fibers	$0.285 = \frac{\text{Btu-in.}}{\text{hr ft}^2\text{°F}}$
Heat flow parallel to fibers	$0.334 = \frac{\text{Btu-in.}}{\text{hr ft}^2\text{°F}}$
Specific heat, room temperature	$0.34 \frac{\text{Btu}}{\text{lb°F}}$
Heat of combustion	$15{,}000 \frac{\text{Btu}}{\text{lb}}$
Electron radiation	No effect
Vacuum outgassing	2.28% weight loss
Refractive index	
Parallel fiber axis	2.0
Perpendicular fiber axis	1.6
Moisture regain, Equilibrium at 55% R.H., 72°F	3.5%
Regain rate at 55% R.H.	0.3%/hr avg.

Table 27-7. Ultraviolet stability of "Kevlar"® 29 and "Kevlar"® 49.

Material	Size	Twist tpi	Weather-Ometer Xenon Lamp Exposure, hr	BREAK LOAD, LB Before	BREAK LOAD, LB After	Strength Loss %
"Kevlar" 29	1500 den. bare yarn	2.1	200 Dry	72.9	53.5	27
"Kevlar" 49	1420 den. bare yarn	1.8	200 Wet-dry	71.4	53.5	25
			500 "		45.0	37
"Kevlar" 49	7100 den. bare roving	0.9	200 Wet-dry	331	258	22
			500 "		220	34
"Kevlar" 49	7100 den. urethane impreg. strand	0	200 Wet-dry	344	228	34
			500 "		185	46
"Kevlar" 49	7100 den. urethane impreg. strand	0.9	200 Wet-dry	356	285	20
			500 "		235	34
"Kevlar" 29 core, nylon jacket	3/8" diameter "Uniline" rope	–	200 Wet-dry	9,220	9,270	0
			1000 "		9,330	0
"Kevlar" 29 core, dacron® jacket	5/16" diameter braided rope	–	1000 Wet-dry	6,880	6,260	6
"Kevlar" 29	1/2" diameter 3-strand rope	–	Florida exposure-Hialeah			
			6 months	14,400	13,000	10
			12 "	"	11,600	19
			18 "	"	9,950	31
			24 "	"	9,940	31

Table 27-8. Comparison of fabrics of glass and "Kevlar" 49.
Basis weight—ASTM D1910
Tensile strength per raveled strip test—ASTM D1682

	Style	Basis Weight (oz/yd^2)	Average Tensile Strength (lb/in.)
"Kevlar" 49	120	1.80	250
"E"-Glass	120	3.16	120
"Kevlar" 49	181	5.00	700
"Kevlar" 49	281	5.00	650
"E"-Glass	181	8.90	340

than their glass fiber counterparts at the same thickness.

3.2.2 Nonwovens and Felts and Insulation Batts. The properties of high modulus aramid nonwovens and felts are shown in Table 27-9.[1] The properties of composites made with some of these nonwoven aramids are given in Section 6.1.7. Experimental insulation batts made with these fibers have also been reported.[1] Performance of these batts is described in Section 6.1.9.

3.3 Hybrid Tapes and Fabrics

Hybrid tapes combining high modulus aramid fibers with other reinforcing fibers are available, as said in Section 2.5, from selected preimpregnators. The description of selected hybrid "Kevlar" 49/graphite fabrics are shown in Table 27-3. The mechanical properties of composites reinforced with hybrids tapes and fabrics are given in Section 6.3.1.

Table 27-9. Physical properties of nonwoven fabrics.

Style No	Basis Weight oz/yd^2	Nominal Thickness mils	Machine/Cross Direction Strip Tensile 1 in. X 3 in. Sample ASTM D1910, lb/in.
1200	0.7	<2	1.8/1.1
1210	1.0	3 ± 1	2.2/1.8
1215	1.5	15 ± 5	11/9
1230	3.0	<100	<20/<20
1270	7.0	60 ± 10	>80/>80
1290	14.0	120 ± 120	>100/>100

4. SUPPLIERS

At the end of 1976, high strength and modulus aramid fibers were made commercially only by E. I. du Pont de Nemours & Co., Wilmington, Delaware, under the registered trademark Kevlar® 29 and Kevlar® 49. Research on high strength and modulus organic fibers has been reported by other fiber manufacturers.[4] Fabrics and prepregs of "Kevlar" are available from a number of commercial weavers and impregnators.

5. PRODUCT PRICING

5.1 History

The first high performance aramid fiber designed for the reinforcement of plastics to be sold commercially was PRD-49, now called "Kevlar" 49, made by E. I. du Pont de Nemours & Co. That product was introduced in January of 1972 at $50.00/lb. Late in 1972 the price was reduced to under $20.00/lb for some of the fiber, roving, and fabric items that made up the "Kevlar" 49 product line. Today, some yarn and rovings are selling under $10.00/lb.

5.2 Outlook

A significant price decrease has taken place since the introduction of high performance organic fibers. No long range pricing forecasts have been made public by the fiber manufacturers.

6. APPLICATIONS, FABRICATION METHODS, AND COMPOSITE PROPERTIES

6.1 Applications

The unique combination of properties found in aramid fibers has led to their use in a wide range of commercial and industrial applications. Substantial penetration of the tire and the mechanical rubber goods markets (belts, hoses, etc.) is expected for the aramids. This section will outline a number of non-rubber reinforcing uses.

6.1.1 Aircraft. The potential for weight savings and increased stiffness are the performance advantages that have led to adoption of aramid fibers instead of glass fibers in the interiors of commercial aircraft.[5] Exterior components on wide body jets continue logging successful flight time on commercial routes in a NASA-sponsored flight service evaluation program.[6,7] Confidence built in these programs has led to the inclusion of aramid fibers in some designs of the UTTAS, AAH helicopter, and the YF-17 lightweight fighter prototype.[8] Equivalent or better mechanical performance than the incumbent material at substantial weight savings was the driving force in these applications. Additionally, low radar profile and good impact characteristics have been important in the development of drone components,[9] and hybrid composite jet engine fan blades,[10] respectively.

Non-abrasive wear characteristics are important in the compression molded, jet engine blade tip rub strips made with a "Kevlar" 49 chopped fiber molding compound. Impact resistance, combined with lightweight, are the reasons for qualification in wide bodied jet cargo compartment liners. The greater fracture toughness of composites reinforced with aramid fiber, or combinations of these with graphite versus those reinforced with only the more brittle inorganic fiber, has been established quantitatively.[11,12] Flammability characteristics that are of paramount importance in uses like aircraft interiors have been defined for high modulus aramid fibers and found to exceed present and proposed FAA specifications when precaution is taken to use appropriate flame retardant resins.[13]

6.1.2 Pressure Vessels. It has been demonstrated that an aramid can be a superior fiber for the filament winding of internally pressurized vessels. Performance factors (PV/W) that relate the pressure at burst (P) to the volume (V) and weight (W) of the vessel of 1.8×10^6 in. have been obtained routinely in vessels of up to 8 in. diameter[14] and this performance has translated into successful firing of full-size, solid propellent, rocket motor cases.[15]

The excellent retention of tensile strength of an aramid at cryogenic temperatures has motivated extensive development work funded by NASA in liquid H_2 and O_2 tankage for the Space Shuttle.[16] In addition, its lower creep and excellent creep rupture characteristics are important in the use of aramids as a filament wound, pressure vessel reinforcement (Table 27-10). Their non-Griffith flaw sensitive behavior under long-term loading at high percentages of their ultimate tensile strength allows for higher design allowables which, in turn, defray some of their price premium over glass. Excellent tension-tension fatigue performance further enhances the cost effectiveness of high performance aramids in these uses.

Table 27-10. Filament wound pressure vessels.

	"Kevlar" 49	S-Glass
Composite density (lb/in.^{3})	0.044	0.039
Fiber volume (%)	65	65
Composite tensile strength, 10^3 psi	223,300	196,900
Specific strength, 10^6 in.	5.08	2.85
Relative specific strength	1.78	1.0
Composite modulus, 10^6 psi	13.2	8.8
Specific modulus, 10^8 in.	3.00	1.28
Relative specific modulus	2.34	1.0
Relative cyclic performance		
90% of ultimate	10	1
60% of ultimate	>3	1
Stress rupture life (hr)		
80% of ultimate	209	1
90% of ultimate	1	Fails

6.1.3 Composite Armor. A function that illustrates the versatility of aramid fibers is the ballistic protection they offer in composite form as well as in unimpregnated fabrics and felts. Properly designed aramid fabrics can provide twice the ballistic protection of nylon at about equal weight, or equivalent performance at almost half the weight. Furthermore, composite armor reinforced with aramid fibers will provide equivalent ballistic protection to glass fiber composite armor at about two-thirds the weight. This weight savings has led to the adoption of aramid fiber composites in various aircraft and helicopter armor applications. Jet engine turbine blade containment rings in the form of unimpregnated or properly engineered composite armor are being developed.[16] Protective gloves, aprons, vests, helmets, jackets, boots, etc., are in various stages of development utilizing the cut and penetration resistance of these high performance fibers.

6.1.4 Marine. As in most aircraft and aerospace applications, requirements for improved stiffness with weight economy also apply to many boat hulls—especially those designed for racing. Aramid fibers are providing these advantages as well as improved damage tolerance in offshore power boats, bass boats, tunnel boats, ocean racing yachts, small sailboats, canoes and kayaks. While the overall weight is fixed in some boat classes, the use of aramid fibers instead of glass offers the potential to redistribute the weight for maximum racing efficiency. Combinations of "Kevlar" 49 with glass fiber in woven or chopped mat form have been popular in many marine applications.[17,18]

6.1.5 Sporting Goods. High performance fibers have made significant commercial penetration in segments of the sporting goods market. Graphite, boron, and aramid fibers are being used where they can provide an athlete with a competitive advantage versus other materials. Weight savings, combined with extra stiffness and impact strength, have been the main reasons for using aramid fibers in applications such as canoes, kayaks, golf shafts, archery bows, paddles, tennis rackets, fishing rods, snow skis, and track and field equipment.

Scarcity of good wood has motivated work with aramid fibers to upgrade the performance of lower grade woods. The successful combination of "Kevlar" 49 and wood has been demonstrated in hockey sticks, paddles, and tennis rackets.

6.1.6 Pultruded Products. Aramid fibers are being successfully used to prepare a number of pultruded products. In addition to the weight savings and increased stiffness versus glass outlined in Table 27-11, good tension-tension fatigue and electrical properties are observed. Rods, channels, box beams, and other shapes can be pultruded with aramid fibers by themselves or in combination with other fibers.

6.1.7 Electrical. Electrical properties of aramid fabrics in a typical flame retardant epoxy resin are seen in Table 27-12, compared to those made by the identical method using glass fabrics in the same resin. The aramid composites are characterized by one unit lower dielectric constant, equivalent loss factor, and similar electrical insulation values than glass. These properties, combined with adequate high temperature durability and excellent creep rupture characteristics, have motivated development of electrical rotor banding tapes. The wear characteristics of aramid fibers has been of interest in the development of insulation wedges for large electrical generating equipment.

Aramids exhibit lower dielectric constant than glass at frequencies of 1 KHz to 10 GeHz. Thus, they outperform the inorganic fiber up to and including the radar transmission range. Lightweight radomes and electromagnetic windows can be made to advantage with aramid fibers. The relative radar transparency of aramid rein-

Table 27-11. Properties of pultruded rods.

	REINFORCEMENT	
Property	"Kevlar"® 49	Glass
Rod diameter	3/8 in.	3/8 in.
Fiber volume	60.0	58
Flex modulus, 10^6 psi	9.6	5.7
Weight, g/ft	29.3	45.5

Table 27-12. Electrical properties of "Kevlar" 49 and glass fabric laminates in FR-4 resin systems.

	"Kevlar" 49	Glass
Dielectric constant (ASTM D-150, 10^6 Hz)		
Conditioned 72 hr/23° C/50% RH	4.12	5.15
" 24hr/23°C/water	4.15	5.18
Dissipation factor (ASTM D-150, 10^6 Hz)		
Conditioned 72 hr/23°C/50% RH	0.0239	0.0208
" 24hr/23°C/water	0.0247	0.0212
Dielectric strength, volts/mil (ASTM D-149)		
Conditioned 72 hr/23°C/50% RH		
+ 24hr/23°C/water		
Thickness, mils–29.4	959	–
70.6	663	–
75.7	–	654
35.7	–	785
Volume resistivity, ohm/cm (ASTM D-257)	5×10^{15}	2×10^{15}
Surface resistivity, ohm/square (ASTM D-257)	5×10^{15}	3×10^{15}
Arc resistance, sec.	125	123
Samples–Fabric style	120	128
Plies	22	11
Thickness, mils	68	75
v/o fiber	50	46

Conditions: Tests at R.T. after samples had been conditioned at 73°F for 24 hr at 50% RH, unless otherwise noted.

forced plastic articles has suggested its use where low radar profile is important.

Good dielectric properties, coupled with a negative thermal expansion coefficient, have been of interest to circuit board manufacturers who have performed extensive evaluations on aramid fibers in woven and nonwoven forms for rigid, flexible, and multilayer boards. The negative thermal expansion is useful in providing dimensional stability to the circuit board by restraining the possible thermal expansion of the resin matrix. Good tear and crease resistance, as well as tolerance of solder temperatures, have stimulated these developments. The weight saving over glass that invariably accrues from the use of aramids is of further value in airborne electronics.

6.1.8 Corrosion Resistant. A wide variety of fuels, lubricants and solvents have no deleterious effect on aramid laminates. The retention of strength under prolonged exposure to these environments has been reported in detail elsewhere.[19] The relative insensitivity to hydrolytic attack compared to glass fiber composites is a prime example of the corrosion resistance of aramid laminates in very strong acids and bases. "Kevlar" 49/polyester laminates have shown superior performance to glass analogs in caustic environments, and "Kevlar" 49/fluorocarbon matrix laminates have shown good resistance to HF environments up to concentrations of 48% at room temperature. No advantage over glass has been demonstrated for aramid laminates in HCl and HNO_3 environments.

6.1.9 Ropes and Cables. Organic fibers such as nylon or Dacron® polyester are being used successfully in the soft cordage market, but their properties limit their ability to perform in applications requiring high strength and low stretch, such as wire ropes and electromechanical cables. Aramid fibers in either the dry or resin impregnated form cannot only compete in these applications, but they also offer a significant weight advantage (Table 27-13). The impreg-

Table 27-13. Strength of various "Kevlar" ropes and cables.

Material	Diameter (in.)	Break Strength (lb)	Weight lb/100 ft	Elongation %	Strength Conversion[a] %
3-Strand rope					
"Kevlar" 29	0.5	14300	8.1	16	30
"Dacron" T-67	"	6400	8.0	28	29
Nylon T-707	"	7100	6.6	40	36
Parallel lay					
"Kevlar" 29 ("Parafil")	0.28	3650	2.5	2.6	83
"Kevlar" 49 ("Parafil")	"	3320	2.5	1.8	75
Polyester ("Parafil")	"	1700	2.4	7.0	84
"Kevlar" 29 ("Uniline")	0.63	38000	15.2	–	85
Polyester ("Uniline")	"	16000	15.1	–	78
Nylon ("Uniline")	"	16000	13.2	–	82
Braid					
"Kevlar" 29 (med. pick)	9/16	34000	10.8	8	53
"Dacron" (med. pick)	"	16000	10.0	–	59
"Kevlar" 29 (long pick)	1/4	7100	1.3	5	92
Nylon (short pick)	"	2000	1.6	–	41
Wire rope construction					
"Kevlar" 29 (1 X 7)	1/8	2500		4	65
Galvanized (1 X 7)	"	2100	3.1	2.5	80
"Kevlar" 29 (7 X 7)	3/8	17000	4.7	–	62
Galvanized (7 X 7)	"	13100	23.6	–	71

[a]Conversion based on 375×10^3 psi for "Kevlar," 163×10^3 psi for polyester, 145×10^3 psi for nylon, 275×10^3 psi for steel—helix angle not considered.

nated strands are handled like steel wire on conventional stranding, braiding, and cabling equipment. The free length of "Kevlar" yarns that will be self-supported in both air and water as compared to steel, are shown in Table 27-14. This means that substantially increased payloads are possible with high performance aramid fibers in long cables, such as those used in the oceanographic and aerospace markets.[20,21] Additionally, corrosion resistance, non-conductivity, and improved handleability have suggested applications in oil rig, buoy, and sonobuoy mooring,[22] air and sea towed antennas,[23] helicopter hoists and anchor pendant lines, balloon tethering, antenna guys,[24] running and standing rigging, etc.

6.1.10 Coated Fabrics. Table 27-15 shows Hypalon® synthetic rubber coated, nylon fabric (5.1 oz/yd^2) intended as air support shelter material, compared to an aramid analog that utilizes fabric of less than half the basis weight (2.1 oz/yd^2). The aramid item is 20% lighter, stronger, and more tear resistant. Based on laboratory experience, it is estimated that one acre of Teflon®-coated, unidirectional aramid fabric would weigh 4500 lb compared to over

Table 27-14. "Free" length comparison.

(Length at which strength member breaks of its own weight = tensile strength/density).

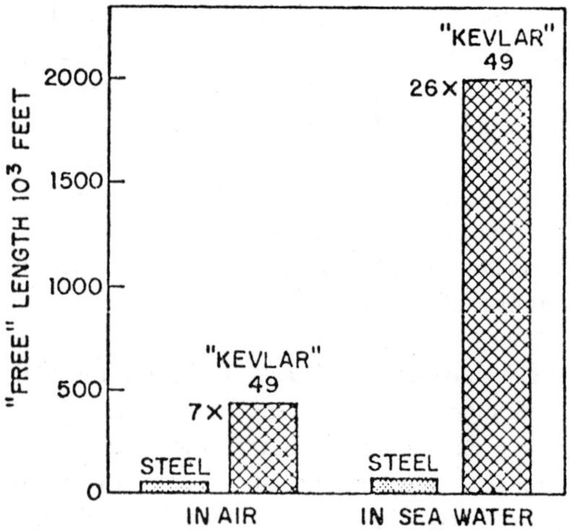

Table 27-15. Air supported shelter material.

	Nylon	"Kevlar" 29
Fabric properties		
Weight, oz/yd^2	5.1	2.1
Tensile strength grab method (W × F), lb	380 × 375	215 × 230
Burst–Mullen, psi	800	930
Coated[a] fabric properties		
Weight, oz/yd^2	15.0	11.3
Tensile strength grab method (W × F), lb	300 × 300	380 × 335
Tongue tear strength (W × F), lb	20 × 20	20 × 25
Burst–Mullen, psi	840	900

[a]Hypalon® synthetic rubber.

14,000 lb for a comparable glass fabric item of equal strength. Work in tethered balloons[25] has shown that multi-ply laminates of aramid fabrics offer significant strength-to-weight improvements, are less permeable, and have equal or better abrasion resistance than conventional "Dacron" polyester reinforced counterparts. Skirts for air cushioned vehicles, specialized conveyor belting, inflatable forms for concrete, arrester webbing for aircraft, and inflatable boats and storage tanks are examples of other uses under development.

6.1.11 Insulation. Equivalent or better insulation than obtained with glass combined with weight savings of up to 30% have been demonstrated with aramid insulation batts.[26] These are nonwoven fibrous assemblies lightly impregnated with appropriate flame retardant resins that give them adequate consistency. Flammability characteristics of these aramid fiber products substantially exceeds specifications. Thermal conductivities in solid laminates made with thermoset resins also are lower than their inorganic fiber counterparts (Table 27-16). Aircraft insulation, liquified natural gas containers, and refrigerator truck trailers have been proposed based on this kind of performance.

Table 27-16. Thermal conductivity.

	Density lb/ft^3	$K = \frac{BTU\text{-}ft}{hr\text{-}ft^2\,°F}$ at 125°F	BURN 12-sec (in.)	BURN Drip	BURN Glow (sec)	SMOKE Time to 16% O.D.[b] (min)
• **Aircraft insulation batts**						
Material						
"Kevlar" 49/phenolic 80/20	0.80	0.306	1	No	No	14
"Kevlar" 49/acrylic 80/20	0.80	0.290	1	No	No	10.5
Glass/unknown resin 80/20	1.20	0.390	←	Unavailable	→	
"Kevlar" 49/phenolic 80/20	0.40	0.350	1	No	No	11
"Kevlar" 49/acrylic 80/20	0.40	0.355	1	No	No	7.5
Glass/unknown resin 80/20	0.60	0.350	←	Unavailable	→	
FAA specification 25.853	–	–	6	No	1	2

• **Fabric reinforced composites[c]**

Material	Heat Flux Direction	THERMAL CONDUCTIVITY $\frac{Cal\text{-}cm}{sec\ cm^2\,°C}$	THERMAL CONDUCTIVITY $\frac{BTU\text{-}ft}{hr\ ft^2\,°F}$[a]
"Kevlar" 49 S-120/epoxy	Across fabric layers	4.95×10^{-4}	0.124
"Kevlar" 49 S-120/epoxy	Parallel to warp	21.0×10^{-4}	0.525
"Kevlar" 49 S-181/epoxy	Across fabric layers	4.87×10^{-4}	0.122
Glass fabric/epoxy[d]	Across fabric layers	10.0×10^{-4}	0.250

[a]Measured at 212°F
[b]O.D.–optical density.
[c]Nominal 50 v/o fiber.
[d]*Materials Engineering, Materials Selector Issue*, Oct. 1969.

Table 27-17. Typical properties of epoxy molding compounds.

	REINFORCEMENT	
Property	"Kevlar"® 49	Glass
Fiber volume, %	50	47
Fiber length, in.	1/2	1/2
Specific gravity	1.32	1.86
Tensile strength, psi	28,500	14,400
Tensile modulus, 10^6 psi	2.96	2.23

6.1.12. Others. *Audio.* The utility of lightweight sheet structures comprising low basis weight, thin woven and nonwoven "Kevlar" 49 products in speaker cones has been explored. The high specific modulus of these resin impregnated and cured composites has resulted in high acoustical output. Increasing concern in removing flammable components from household radio and TV appliances has further accelerated the evaluation of these inherently flame retardant aramid structures in these uses.

Molded Products. The inherent cut resistance of aramid fibers has slowed their penetration into molded products. However, the formation of a chopped fiber molding compound by epoxy coating of the aramid fibers and subsequent cutting is being done on a commercial basis. The data in Table 27-17 shows that the expected improvements in tensile and modulus as well as the savings in weight are observed when compared to a similar glass molding compound.

6.2 Fabrication Methods

We will describe in this section fabrication methods that are applicable to high modulus aramid fiber composites. In most cases, procedures will be similar to those employed in the manufacture of glass fiber reinforced products; however, the organic nature of aramid reinforcing fibers will require, in some cases, modifications in methods as described below.

6.2.1. Drying Procedures. Aramid products, as shipped by the fiber manufacturer, will contain a small amount of moisture on the surface of the yarn. The reaction of this adsorbed water with certain constituents of resin systems during cure can decrease composite physical properties, especially interlaminar shear strength. To date, only anhydride-cured epoxides have been identified as being moisture sensitive. Amine- and dicyandiamide-cured epoxides do not appear to be sensitive to absorbed moisture on aramid fibers.

To prevent composite property degradation when using anhydride-cured epoxides, it is recommended that aramid products be thoroughly dried prior to impregnation. This can be accomplished by exposing bobbins of yarn, balls of roving, and rolls of fabric to 120°C (250°F) for a minimum of 16 hrs. After drying, aramid products will regain moisture with exposure to ambient conditions and thus should be resin impregnated within 1 hr after drying. If this time limitation is not practical, material can be passed through opposed banks of infrared heat lamps (or other appropriate drying devices) just prior to resin treatment.

6.2.2. Prepregging. Aramid prepregs are made using virtually the same procedures used for glass yarns and fabrics. Experience indicates that coaters have not had to alter production equipment or conditions beyond the normal range of their operating variables to produce satisfactory products of "Kevlar" 49.

Fabrics of "Kevlar" 49 have been successfully coated by the popular procedure of dipping in solvent solutions of epoxy resin, followed by tower drying to remove the solvent, and in some cases to partially advance the resin. Ambient equilibrium moisture on fabric has not had an adverse effect on fiber-matrix bonding for the resin systems commonly used in solvent solutions. Therefore, drying of fabrics prior to production of fabric prepregs using solvent solutions has generally not been required. Another method for producing fabric prepreg is to combine the fabric with hot melt, epoxy films. Simple lamination of film to fabric is a practical procedure for the preparation of laboratory-scale prepreg samples and can be adopted for larger-scale production quantities.

As with glass prepregs, resin content, volatiles, flow characteristics, and tack requirements are dictated by the layup method, tailoring, and

Table 27-18. Mechanical properties of hand-layup "Kevlar" 49 and fiberglass fabric reinforced epoxy composites.

Material and Fabric Style	Specific Gravity	Actual Volume %	Normalized Volume %	Tensile Strength (psi)	Tensile Modulus (10^6 psi)	Flexural Strength (psi)	Flexural Modulus (10^6 psi)	Interlaminar[a] shear strength (psi)
"Kevlar" 49/epoxy								
Style 243	1.31	44.6	40	81,400	5.92	45,200	4.35	5250
281	1.29	36.0		72,400	3.76	45,100	3.93	5120
285	1.28	32.5		72,600	3.96	50,600	3.46	4470
328	1.34	54.4		53,700	2.95	20,300	2.32	2950
1050X	1.28	32.1		74,400	3.71	43,900	3.29	4610
1033X	1.30	38.6		54,300	3.54	34,600	2.77	4200
"E"-Glass/epoxy								
8.9 oz/yd² cloth. (style 181)	1.67	34.9		46,800	2.49	–	2.68	–
18 oz/yd² woven roving	1.82	45.9	↓	42,800	2.71	57,700	2.35	6030

[a]Shear strength results are not normalized.

curing procedure. In general, prepregs of "Kevlar" 49 are made using the same volume percent resins as with glass and thus the weight percentage of resin is higher due to the lower density of "Kevlar" 49 (1.45 g/cm³ vs 2.55 g/cm³). Information on the optimum curing procedures are provided by the prepreg manufacturer. Most commercial impregnators offer prepregs of aramid products.

6.2.3 Wet Lay-Up. A wide variety of standard wet lay-up polyester and epoxy resins has been successfully used with "Kevlar". However, as with glass fiber, some resin systems produce laminates of superior stiffness and interlaminar shear strength. Epoxy matrices generally translate the fiber properties better than do polyesters, as shown in Tables 27-18 and 27-19. Tests have shown that "Kevlar" wet-out in hand-layups is optimum when the resin system viscosity is 300–700 centipoise during application. Many wet lay-up resin systems designed for glass fiber contain relatively high amounts of flexibilizers to improve impact resistance of the cured laminate. Since "Kevlar" is generally superior to fiberglass in impact, and aramid composites generally do not exhibit brittle failure, users have typically cut in half the

Table 27-19. Mechanical properties of hand-layup "Kevlar" 49 and fiberglass fabric reinforced polyester composites

Material and Fabric Style	Specific Gravity	Actual Volume %	Normalized Volume %	Tensile Strength (psi)	Tensile Modulus (10^6 psi)	Flexural Strength (psi)	Flexural Modulus (10^6 psi)	Interlaminar[a] shear strength (psi)
"Kevlar" 49/polyester								
Style 281	1.26	37.0	40	64,500	3.49	34,700	2.74	4420
285	1.24	29.8		65,800	3.69	42,300	3.18	3680
328	1.29	46.4		54,800	2.67	20,200	2.23	2890
1050X	1.23	25.6		69,700	4.52	42,000	3.14	2850
1033X	1.27	39.9		50,800	3.14	22,600	2.22	3230
"E"-glass/polyester								
8.9 oz/yd² cloth (style 181)	1.63	37.0		44,100	2.93	52,500	2.42	5700
18 oz/yd² woven roving	1.71	39.9	↓	45,500	2.99	58,300	2.06	4470

[a]Shear strength results are not normalized.

amount of flexibilizer they would normally use with glass.

It is recommended that a prospective user of "Kevlar" make a small test laminate with the planned fabric layup schedule and the selected resin system. This will provide practice in visually determining wet-out—assessment of degree of impregnation is more difficult with the yellow, translucent fabrics of "Kevlar" than with transparent glass—and will allow simple testing to assure adequate interlaminar shear strength. Optimum interlaminar shear strength is obtained by laying up the entire laminate in one operation, and not allowing layers to completely cure before adding subsequent layers. Two wet layup resin systems, one an epoxy and the other a polyester, that reportedly have given good composite properties are:

Dow "XD-7818" epoxy	100	PBW[a]
Jefferson "Jeffamine" D-230 Catalyst	37.5	"
Jefferson "Accelerator" A-398	4.0	"
Dow "Derakane" 411-45 polyester	100	PBW
MEK peroxide catalyst	1.0	"
Cobalt naphthanate accelerator	0.3	"
Dimethyl aniline accelerator	0.05	"

[a]PBW = Parts by weight.

6.2.4. Filament Winding. Filament winding procedures used for glass fiber and roving are directly applicable to the high modulus aramids. Drying prior to impregnation, as explained in Section 6.2.1, may be required for maximum interlaminar shear strength with some resins. Aramid fibers can become electrostatically charged by running over guide surfaces. In order to improve handleability, static eliminators are used, such as those that are effective in preventing ballooning of filaments in a yarn or static repulsion of nearby threadlines. The drying procedures of Section 6.2.1 should be followed with anhydride-cured epoxies, while amine-cured systems, as said before, are relatively insensitive to the equilibrium moisture content on the yarn. Interlaminar shear strengths of filament-wound composites of "Kevlar" 49 at 60-65% by volume fiber will range from a high of 9-10,000 psi in some resin systems to 5-7,000 psi in others. High burst strengths have been obtained, however, both at the high and low level of shear strength. Three specific recipes of filament winding resins used with "Kevlar" 49 are given below.

Shell "Epon" 828	100	PBW[a]
"Nadic" methyl anhydride (NMA)	90	"
Dimethylbenzylamine (BDMA)	1	"

Stir ingredients. Heat to 50°C to reduce viscosity. Maintain 50°C during winding. Cure 2 hr/150°C + 16 hr/170°C. Shell "Epon" 815 may be substituted without a heated bath.

Applied Plastics "APCO" 434	100	PBW
Applied Plastics "APCO" 2347	7.5	"

Use at room temerapture. Cure 4 hr/54°C + 4 hr/93°C + 2 hr/150°C.

Fiberite "X-506" is a precatalyzed system. Resin requires refrigeration when not in use. Allow resin to warm to room temperature before use. Cure 1 hr/100°C + 1 hr/130°C + 4 hr/175°C.

[a]PBW = Parts by weight.

A broader spectrum of epoxides, curing agents, and reactive diluents identified as to chemical type, tradename, popular abbreviation and manufacturer is given in Table 27-20. The epoxides in Groups I and II can be cured with any of the curing agents 1-4. The epoxy in Group III is best cured with agents 5 or 6. The epoxies of Group I are sometimes diluted with diluents 1 or 2. Specific formulations will depend on processing requirements, e.g., viscosity, pot life, etc.

6.2.5. Pultrusion. Pultruded parts of "Kevlar" 49 can be prepared by passing continuous roving through a resin impregnating bath, then through a heated die to shape the part, and then curing, all in one continuous operation. The operation is entirely analogous to that used with glass roving. "Kevlar" 49 may be used interchangeably with glass, replacing all or a portion of the continuous glass roving in a given part.

Fiber Selection. A 22,720 denier roving of Kevlar-49 which has a yield of 197 yd/lb, is comparable to a 123 yd/lb E-glass roving (such as PPG Type 764, 30-end) on a volume basis (i.e., 1 yd of aramid 22,700 denier roving contains approximately the same volume of fiber as does 1 yd of the glass roving). Other pultrusion rovings of "Kevlar" varying only in the yield are being evaluated, these are 6000

Table 27-20. Filament winding.

EPOXIDE			CURING AGENT				REACTIVE DILUENT[a]		
Group	Trade Name	Manufacturer	Type	Trade Name	Manufacturer	Abbrev.	Type	Trade Name	Manufacturer
I	XD-7818	Dow							
	"Epon" 828	Shell					1. Butyl glycidyl ether	BGE	Dow
	"Epikote" 828	Shell	1. Nadic methylanhydride + benzyldimethylamine	"Araldite" 906 DY-062	Ciba-Geigy	NMA BDMA	2. 1, 4-Butanediol diglycidylether	RD-2	Ciba-Geigy
	DER-331	Dow							
	"Araldite" 6010	Ciba-Geigy	2. Methylene dianiline	"Araldite" 972	Ciba-Geigy	MDA or DDM			
			2. Methylene dianiline	–	Anchor	MDA or DDM			
II	"Epon" 826	Shell	3. 4, 4′-Diaminodiphenyl methane	DEH-50	Dow	–			
	"Epikote" 826	Shell							
	DER 330	Dow	4. 4, 4′-Diaminodiphenyl methane	–	Shell	–			
	"Araldite"	Ciba-Geigy							
III	XD-2793	Ciba-Geigy	5. Hexahydrophthalic anhydride	HT-907	Ciba-Geigy	HHTA			
			6. Methylene dianiline	"Araldite" 972	Ciba-Geigy	MDA			

[a]When required.

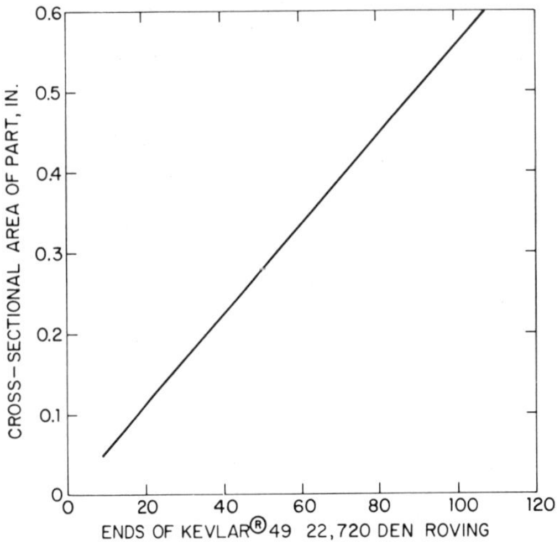

Fig. 27-12. Kevlar® 49 pultrusions. (Cross-sectional area of part vs. roving ends needed for 50 v/o fiber.

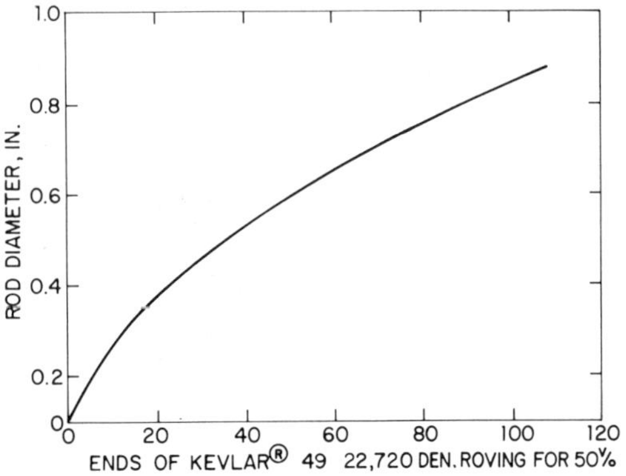

Fig. 27-14. Kevlar® 49 pultruded rod. (Roving ends needed for 50 v/o fiber in various diameter rods.)

denier roving and 11,400 denier roving. All of these rovings have the same mechanical properties and the same compatibility with polyester or epoxy resins.

For selection of the appropriate number of roving ends to be used in a given part, the plots of Figs. 27-12, 27-13 and 27-14 can be used.

Resin Selection. In most cases, the typical polyester resins used for glass pultrusion are satisfactory for use with "Kevlar" 49. For optimum shear properties, resins based on bisphenol A are recommended (commercially available examples include "Derakane" 411-45, "Atlac" 382-05, or "Atlac" 4010). Epoxy resins in a variety of formulations are satisfactory.

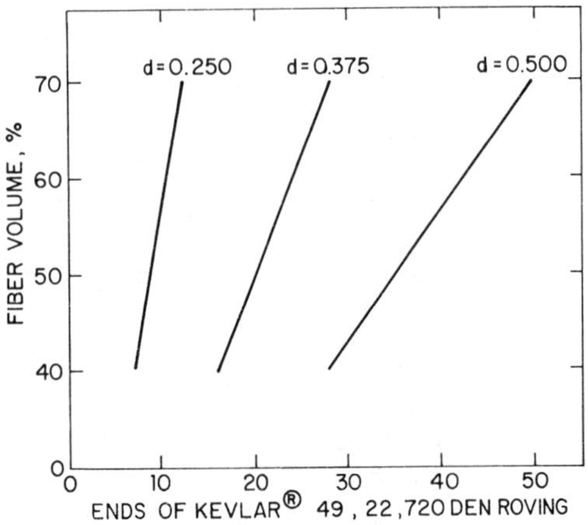

Fig. 27-13. Kevlar® 49 pultruded rod. (Fiber volume vs. roving ends for various diameters.)

Properties. Property improvements to be expected when "Kevlar" 49 is incorporated into a pultruded part as a replacement for all, or part of, E-glass fiber include increased stiffness and lighter weight. Typical values for pultruded rods are shown in Table 27-21. The measured values of flexural modulus versus span/depth ratio for both "Kevlar" 49 and E-glass rods of the same fiber volume are shown in Fig. 27-15. This plot illustrates the importance of making flexural modulus measurements for quality control purposes at a fairly high span/depth ratio (>50/1) if one hopes to measure true stiffness values.

Tensile strength values of 202,000 psi have been measured on pultruded strips of "Kevlar" 49 (45 mils) containing approximately 55 v/o fiber.

Table 27-21. Comparison of "Kevlar"® 49 and E-glass pultruded rods.

	"Kevlar"®	Glass
Diameter, in.	3/8	3/8
Fiber volume, %	60	58
Tensile modulus, 10^6 psi	10.1	6.3
Flexural modulus, 10^6 psi	9.6	6.3
Weight, g/ft	28	45
Weight savings, %	38	–
Matrix resin	Polyester	Polyester

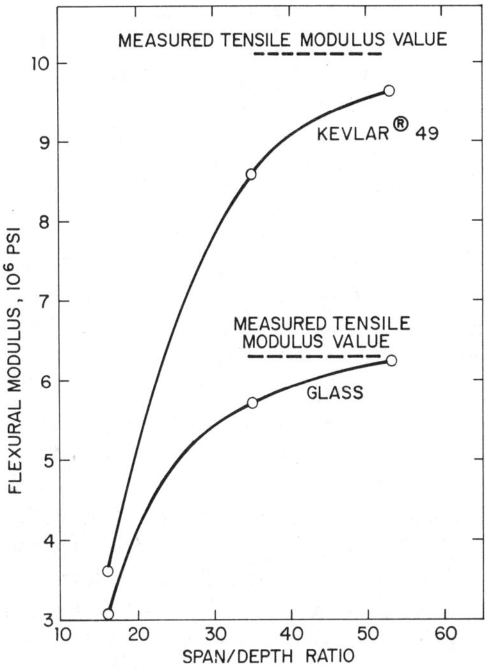

Fig. 27-15. Flex modulus vs. span/depth ratio for 3/8 in. pultruded rod. (58–60 v/o fiber.)

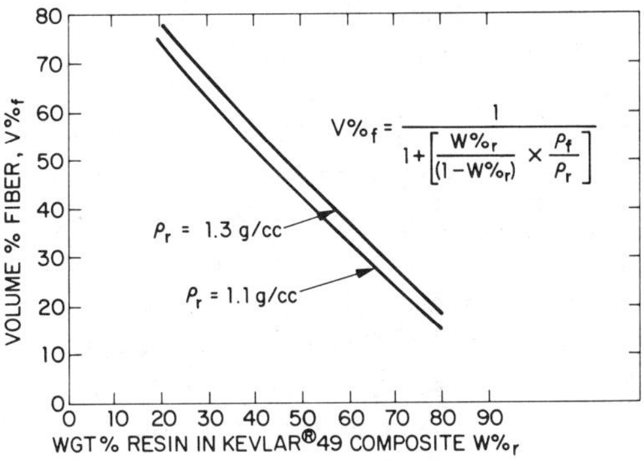

Fig. 27-17. Fiber volume fraction vs. resin weight fraction for Kevlar® 49 composite.

6.2.6. Prevention of Composite Warpage. When fabrics of glass or "Kevlar" 49 become skewed or bowed, usually due to misalignment of the filling fibers, they tend to produce warped laminates (Fig. 27-16). In order to minimize warping, it is important to handle fabrics carefully to keep skewness to a minimum, i.e., keep the warp and fill yarns as perpendicular (orthogonal) as possible.

6.2.7. Sandwich Construction. The conventional procedures used for glass/resin faced "Nomex" aramid, aluminum, foam, or wood core sandwich panels are applicable to aramid reinforcements.[27]

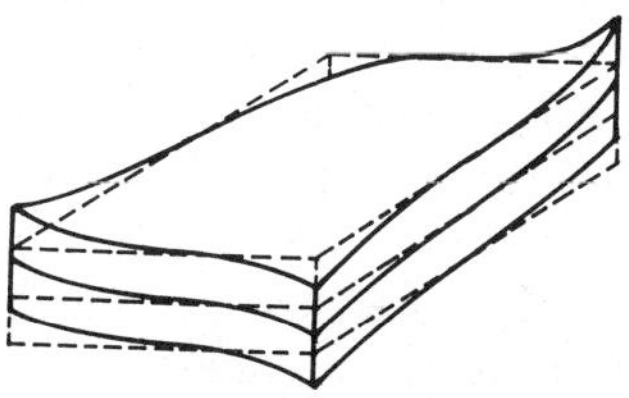

Fig. 27-16. Distortion of a cured laminate due to fabric skewness.

6.2.8. Curing of Composites. The standard vacuum bag, press, and autoclave cures used with glass fiber reinforcement are also used with high modulus aramids.[27]

6.2.9. Resin and Fiber Content Determination. For meaningful property comparisons between composites reinforced with fibers of different densities, it is usual to compare them at equal fiber *volume* contents. The plots of Fig. 27-17 facilitate the calculation of the different weight percents of glass fiber and "Kevlar" 49 fiber required for the same fiber volume percent. Figure 27-18 helps to determine the volume percent "Kevlar" 49 that will result for a given weight percent resin in the composite and vice versa. Note that this conversion takes into account the density of the resin (ρ_r). These conversions assume no voids or particulate fillers

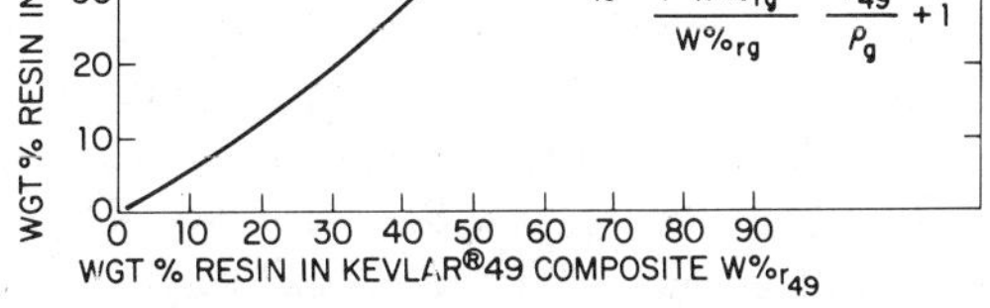

Fig. 27-18. Weight % resin glass composite vs. weight % resin for Kevlar® 49 composite of equal fiber volume fraction.

in the composites. That is, materials are assumed to be made entirely out of fiber and resin.

For epoxy composites, the digestion procedures of ASTM 3171, "Fiber Content of Reinforced Resin Composites," are applicable to high modulus aramids. For glass reinforced polyester composites, it is usual to burn off the resin to determine fiber volume. Since aramid fibers will not survive this treatment, alternate methods must be employed. If the number and basis weight of fabric plies in a given layup are known, the total weight of fabric divided by the weight of the composite can give an estimate of the weight percent fiber. This calculation assumes zero voids in the composite. If the fiber and the layup is in such form that it cannot be simply calculated, then the density of the cured composite (ρ_c) must be determined. With the density of the fiber (ρ_f) and that of the resin (ρ_r), the fiber volume content can be calculated as

$$\text{Volume percent (\%) fiber} = \frac{\rho_c - \rho_r}{\rho_f - \rho_r} \times 100$$

Again, the above formula will be exact only if there are no voids in the composite.

6.2.10. Cutting of Fabrics and Prepregs. Aramids are inherently "tough" materials due to their organic nature and combination of physical properties. Yarns, rovings, fabrics, woven rovings, and prepregs of "Kevlar" 49 can be cut and trimmed using sharp, clean tools. Scissors must have low tolerances between cutting surfaces. The serrated scissors described below work best, preventing the material from slipping out from between the cutting surfaces. It is recommended that when cutting prepregs, the tool be cleaned often with acetone to prevent resin buildup and loss of cutting action. For yarns, roving, fabrics less than 7 oz/yd^2, serrated scissors like the Model TAI-LDS made by Technology Associates, Inc., Wilmington, Del., or upholstery, carpet, and canvas shears like the Wiss No. 4 I.S. (12 in.) made by J. Wiss and Sons Co., Newark, N.J., are effective. A list of special tools and recommendations for their use can be obtained from the aramid fiber manufacturer.

6.2.11. Machining of Composites. All of the machining operations (sawing, milling, drilling, turning, countersinking, chamfering, grinding, and routing) required for construction of composite parts have been successfully accomplished with composites reinforced with "Kevlar" 49. Work continues to optimize tooling which will make production of these composites as simple and inexpensive as for similar glass parts. Tooling, speeds, and feeds recommended for specific operations are available from the fiber manufacturers.

The machining characteristics of "Kevlar" 49 composites should be thought of as similar to those of wood. To obtain optimum results, shearing action using sharp blades with low heat generation is desirable. Particular attention should be given to the cutting of the outermost layer of the composite. Fibers in this layer either must be restrained by external backing or the shearing action must be directed toward the interior of the composite. Adherence of resin to cutting surfaces appears to be the cause of apparent dulling of tools and methods are being developed to minimize this problem.

Drilling. A special, spade type drill has been developed by du Pont that will cut "Kevlar" 49 reinforced solid laminates and honeycomb sandwich panels with clean entrance and exit surfaces without sacrificial backing. Conventional drills can also be used, but firm sacrificial backing must be provided at the exit surface. Drills which are designed for wood (twist drills, flat ended) are best, but special pointed drills can be used if cutting edges are sharp. Firm backup of the composite eliminates fuzzing and delamination at the hole exit. Backup can be obtained by using "Micarta," brass, or by leaving the glass peel-ply on the composite until after drilling. Moderate to high drill speeds should be used.

Band sawing can be accomplished by cutting with a fine-tooth blade, using water as a coolant. Some fuzzing remains on the edges after such a cut, but it may be removed by light, wet sanding. A saber saw blade has been developed which cuts the outermost fibers on both sides of the laminate toward the interior. Circular sawing using a metal slitting saw has been highly successful. Cutting production parts with power shears may be another practical method for "Kevlar" 49 laminates. Saw cutting and trimming generally cannot be accomplished using diamond wheels developed for glass laminates.

Grinding of test specimens can be accomplished simply. Samples are generally stacked and clamped together. Diamond or alumina composition wheels have given the best results. Sanding "Kevlar" 49 laminates requires wet paper for cooling and to prevent buildup of waste between abrasive particles. Positioning of the composite so that the cutting action is always directed towards the interior of the material will assure clean trimmed edges.

More information on machining of "Kevlar" 49 composites is available from du Pont. Specialty tools are available from Technology Associates, Inc., Wilmington, Del.

6.2.12. Safety. The manufacture of reinforced plastics involves, in addition to the reinforcing fiber, the handling of dozens of resins, solvents, curing agents, accelerators, thixotropic agents, etc. All require the user to become familiar with their respective hazards. The following are some recommendations on the safe handling of "Kevlar" aramid fiber.

No particular health hazards from skin contact with "Kevlar" are known. A few people may develop mild to moderate erythema (redness of the skin) on prolonged contact. It is recommended that anyone who routinely handles "Kevlar" and develops such symptoms should either refrain from further continual contact or wear appropriate protective clothing such as gloves. No evidence of sensitization was found in patch tests on the skin of 185 human subjects.[27]

Laboratory vapor inhalation tests conducted by E. I. du Pont de Nemours & Co. indicate that one of the in-process solvents, hexamethylphosphoramide (HMPA) used in the manufacture of "Kevlar," will produce cancer in rats. There is no evidence that this solvent or its vapor has caused cancer in humans. Reportedly, "Kevlar" fiber retains amounts of this solvent so minute and securely held within the fiber structure that there is no hazard to individuals handling the fiber.[28] The fiber manufacturer suggests that good industrial ventilation and safety practices be followed in the processing of "Kevlar" fiber into desired end products and in the disposal of waste therefrom.

Analysis on the possible effects of inhaling fly which may be generated in some operations, such as high-speed winding, is incomplete. Inhalation of any fine fibrous material, including "Kevlar," should be avoided.

"Kevlar" has unusually high strength, modulus, and cutting resistance compared with other yarns. Caution should be exercised in cutting, splicing, or handling yarns of "Kevlar" during processing. Manual cutting and splicing should be attempted only with stationary yarns to avoid possible injury from entanglement in a moving yarn or fabric, which could pull an operator into machinery or cause cuts.

6.3 Composite Properties

6.3.1. Thermoset Resins. *Epoxy* (Unidirectional). The properties of unidirectional composites made with "Kevlar" 49 are shown in Table 27-22, both in the direction of fiber reinforcement [0°] and transverse to the fiber reinforcement [90°]. Notice that 0° compressive strength of unidirectional composites of "Kevlar" 49 is significantly lower than the tensile strength in that same direction. The contrasting stress-strain curves in tension and compression along the fiber direction are shown in Fig. 27-19. The yielding behavior of this high modulus aramid fiber in compression leads to stress-strain curves in bending that are nonlinear beyond the strain at which compressive yielding begins on the upper-most surface of the bar for the loading scheme shown (Fig. 27-20). This yielding behavior, which is similar to the behavior of metals like aluminum, accounts for some of the higher Charpy impact strength recorded for "Kevlar" 49 than for more brittle fiber reinforcements such as graphite and boron (Table 27-23).

The creep behavior of "Kevlar" 49/epoxy unidirectional composites is shown in Fig. 27-21, while the stress rupture behavior of these composites is compared to that of S-glass in Fig. 27-22. Tension-tension fatigue data of

Table 27-22. Unidirectional composite lamina properties.

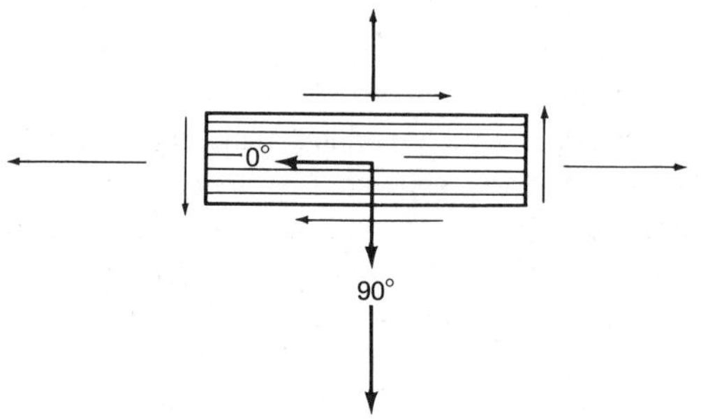

	Glass	"Kevlar" 49
Density, lb/in.^{3}	0.075	0.050
Tensile strength 0° (10^3 psi)	160	200
Compressive strength 0° (10^3 psi)	85	40
Tensile strength 90° (10^3 psi)	5.0	4.0
Compressive strength 90° (10^3 psi)	20	20
In-plane shear strength (10^3 psi)	9	6.4
Interlaminar shear strength (10^3 psi)	12	14
Poissons' ratio	0.30	0.34
Tens. and comp. modulus 0° (10^6 psi)	5.7	12
Tens. and comp. modulus 90° (10^6 psi)	1.3	0.8
In-plane shear modulus (10^6 psi)	0.5	0.3

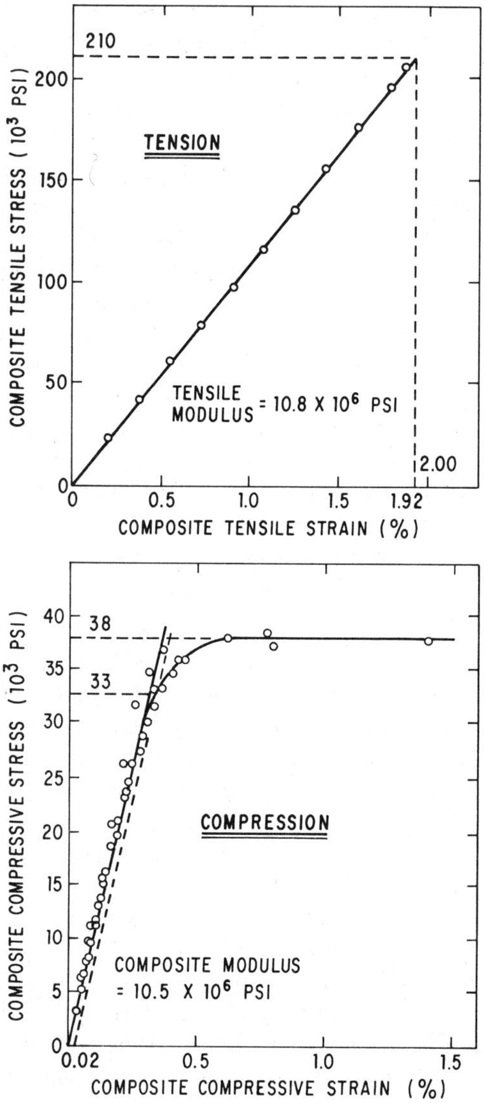

Fig. 27-19. Typical 0° unidirectional Kevlar® 49 composite stress-strain curves.

Table 27-23. Unidirectional composite charpy impact strength.

	Maximum Impact Strength ft-lb/in.^{2}
"Kevlar" 49/epoxy	150
Graphite/epoxy	5-20
Boron/epoxy	40
"E"-Glass/epoxy	275
Aluminum	220

Impact transverse to fiber alignment unnotched specimen.

Impact strength of composites reinforced with "Kevlar" 49 is dependent on test method. Composites reinforced with "Kevlar" 49 are highly resistant to ballistic and missile impact.

high modulus aramid composites compared to other fiber reinforcements in epoxy matrices are shown in Fig. 27-23, while the fracture toughness values of 0/90/±45 composite laminates made from unidirectional plies are shown in Table 27-24.[11] Coefficient of thermal expansion of unidirectional composites reinforced with "Kevlar" 49 are shown in Table 27-25, while the thermal conductivity of these composites along and across the fiber reinforcement are in Table 27-26.

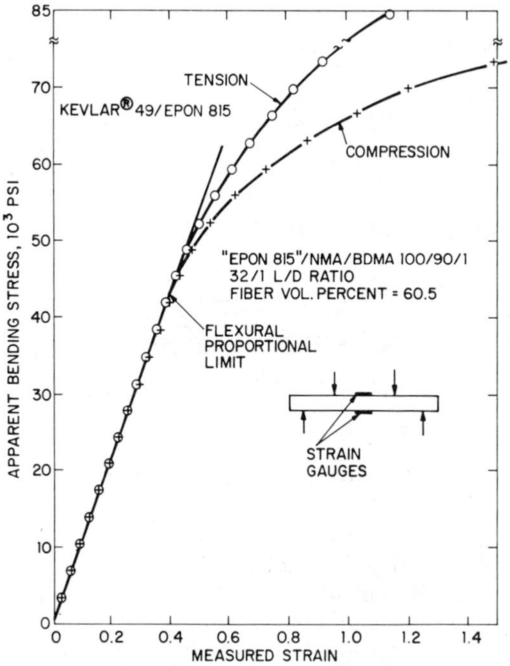

Fig. 27-20. Stress-strain curves of the outermost compressed and extended layers of a unidirectional composite in 4-point bending.

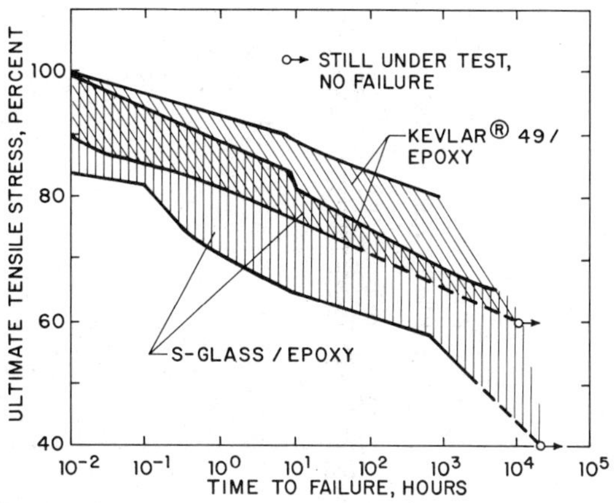

Fig. 27-22. Stress-rupture behavior of Kevlar® 49 and s-glass unidirectional composites.

The interlaminar shear strength of unidirectional composites, as measured by the short beam shear test described in ASTM D2344, is often taken as a measure of adhesion between fiber and resin. A comparison of the different levels of interlaminar shear strength obtained with different resin systems and "Kevlar" 49

Table 27-24. Fracture toughness values and strength retention factors for ceramics, metals and 0/90/+45 composite laminates.

Material	K_c ksi-in.$^{1/2}$	K_q ksi-in.$^{1/2}$
Al-2024-T6	73	–
Al-7075	27	–
18 Ni Maraged S Steel	100	–
Ti-6Al-4V	196	–
Alumina	4	–
Graphite/epoxy	22	–
E-Glass/epoxy (from unidirectional tape)	14	16
"Kevlar" 49/epoxy	23	30
"Kevlar" 49/epoxy (181-style fabric)	19	23

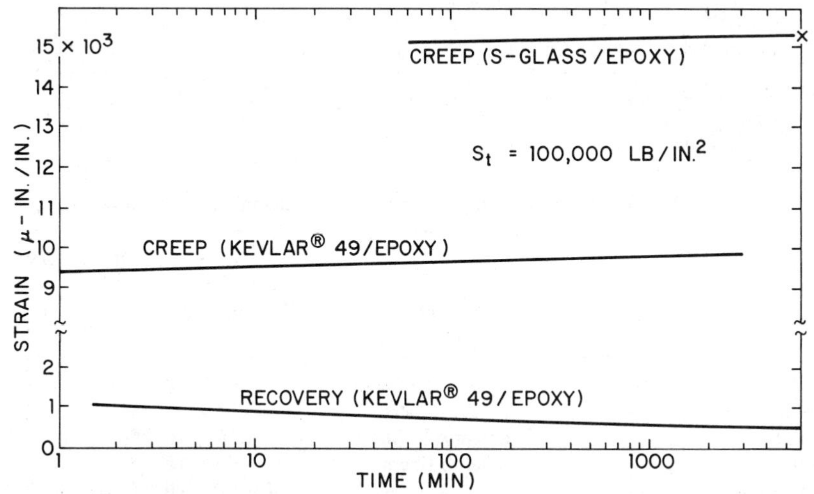

Fig. 27-21. Creep and recovery behavior of Kevlar® 49 and s-glass unidirectional composites.

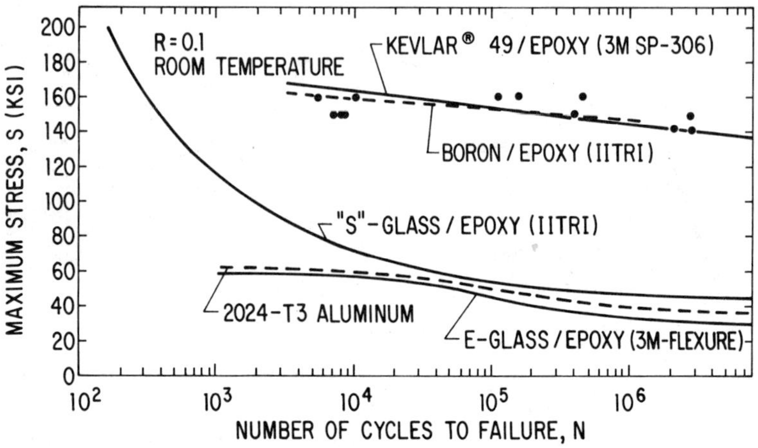

Fig. 27-23. Tension-tension fatigue behavior of unidirectional composites and aluminum.

Table 27-25. Coefficients of thermal expansion for unidirectional composites reinforced with "Kevlar" 49.

Ply Orientation	Resin	Expansion Coefficient (in./in./°C × 10^{-6})	Temperature Range (°C)
0°	Ferro CE-3305	−4 ± 1	−79 to +100
0°	Ferro E-293	−4.1	−195 to +120
0°	Du Pont 5134	−2.3	−79 to +100
0°	3M SP-306	−3.5	−18 to +66
50%, 0° 50%, ±45	Ferro CE-3305	−2.5 ± 0.5	−79 to +100
±45°	Ferro CE-3305	−1.25 ± 0.25	+20 to +100
±60°	Ferro CE-3305	+34	−79 to +100
90°	Ferro CE-3305	+57	+20 to +100
90°	Ferro E-293	+58	+90 to +175
90°	Ferro E-293	+35	−195 to +120
90°	3M SP-306	+69	−18 to +66
Resin only no fiber	Ferro CE-3305	+64.5	+20 to +100

CE-3305 data courtesy Teledyne-Ryan Corp.
ASTM D696-44 test method.
60 volume percent fiber.

Table 27-26. Thermal conductivity of unidirectional epoxy composites.

Fiber Form	v/o Fiber	Heat Flux Direction	THERMAL CONDUCTIVITY Cal-cm / sec cm²°C	THERMAL CONDUCTIVITY Btu-ft / hr ft²°F
"Kevlar" 49	54.0	Transverse to fibers	3.4 × 10^{-4}	8.58 × 10^{-2}
"Kevlar" 49	54.0	Parallel to fibers	40.3 × 10^{-4}	1.01
Graphite	–	Transverse to fibers	15.0 × 10^{-4}[a]	37.5 × 10^{-2}
Graphite	–	Parallel to fibers	400 × 10^{-4}[a]	10
Epoxy resins	–	–	4.0–32 × 10^{-4}[b]	10–80 × 10^{-2}

[a] *Fiber Science and Technology*, p. 129, Mar. 1970.
[b] *Materials Engineering, Materials Selector Issue,* Oct. 1969.

Table 27-27. Unidirectional composites reinforced with "Kevlar" 49—interlaminar shear strength.

Short beam method—ASTM D2344
55-65 volume percent fiber

Manufacturer	Epoxy Resin or Prepreg Code	Shear Strength (psi)
American Cyanamid	BP-907	7100–8700
Applied Plastics	APCO 434	9000
Ferro	E-293	9000
Ferro	CE-3305	11–12,000
Fiberite	X-505	8100
Fiberite	X-506	8300–10,100
Fiberite	X-934	7300
3M	SP-306	9800–10,500
3M	SP-307	8200
3M	SP-308	10,000
Shell	"Epon" 815	7500–9500
Shell	"Epon" 828	8500–9700
Shell	ERX-67	4600
Shell	ERX-68	3800
Union Carbide	ERL-2256	5000–6700
Union Carbide	ERLA-4617	4000–6800
U.S. Polymeric	E-702	9100
U.S. Polymeric	E-787	9000
U.S. Polymeric	E-7119	9000

are shown in Table 27-27. Note that almost two-fold improvement can be obtained depending on the resin used. Regardless of the magnitude of the original interlaminar shear strength, high modulus aramid composites made with epoxy resin retain a high percentage of their original strength after exposure to a variety of environments, as shown in Tables 27-28 and 27-29. Another useful property of high modulus aramid composites is their ability to damp both mechanical and sonic vibration. The damping ratios of "Kevlar" 49/epoxy composites versus other materials are shown in Table 27-30.

Table 27-28. Environmental stability of "Kevlar" 49/epoxy unidirectional composites.

3M SP-306 tape
60 volume percent fiber
Short beam method—ASTM D2344

Environment	Interlaminar Shear Strength, psi
Control	10,520
Water boil, 4 hr	9,290
Salt water, room temperature, 100 hr	9,770
Texaco "Abjet" K-40 jet fuel room temperature, 100 hr	10,480
"Skydrol" lubricating oil room temperature, 100 hr	10,460

Table 27-29. Environmental stability of "Kevlar" 49/epoxy unidirectional composites.

American Cyanamid BP-907 epoxy
60 volume precent fiber
Short beam method—ASTM D2344

	Interlaminar Shear Strength, psi
Control	7060
Water boil, 4 hr	7150
Ammonium hydroxide, room temperature, 100 hr	6560
Texaco "Abjet" K-40 jet fuel room temperature, 100 hr	6950
"Skydrol" lubricating oil room temperature, 100 hr	7040
Urine, room temperature, 100 hr	6780

As was said in Section 3.1.3, multi-ply composites reinforced with aramid fibers are relatively immune to UV degradation due to their thickness and the possibility of incorporating UV absorbers to the resin matrix.

Table 27-30. Vibration damping.

Material	Damping ratio
"Kevlar" 49/epoxy	
Unidirectional (0°)	
3M Co. SP-306, PR-286 resin	0.86×10^{-2}
181-Style fabric (0°)	
Hexcel F-161 resin	0.84×10^{-2}
181-Style fabric (0°)	
Hexcel F-155 resin	0.74×10^{-2}
Aluminum[a] (2024-T3)	$0.02–0.09 \times 10^{-2}$
Epoxy ("Epon" 815)[a]	$1.17–1.93 \times 10^{-2}$
E-glass/epoxy[a] (3M Co. 1002) (0°)	$0.16–0.42 \times 10^{-2}$

Band width at half power points under forced oscillation at 40–70 hz.

$$\text{Damping ratio} = \frac{\text{Damping coefficient}}{\text{Critical coefficient}} = \frac{1}{2}\ \text{loss factor}$$

[a]Schultz, A. B. and Tsai, S. W., "Dynamic Moduli and Damping Ratios in Fiber-Reinforced Composites," *J. Comp. Mat.*, 2:/3 (1968).

Table 27-31. "Kevlar" 49 style 181 fabric with typical 350°F cure epoxy mechanical properties—warp direction, 50 V/O

Property	Value
Density	0.048 lb/in.^{3}
	1.33 g/cm^3
Volume percent fabric	50%
Tensile strength	75,000 lb/in.^{2}
R.T., dry	517 MN/m^2
R.T., after 2-hr boil	73,000 lb/in.^{2}
	503 MN/m^2
R.T., after 1/2 hr at 350°F (177°C)	70,000 lb/in.^{2}
	483 MN/m^2
300°F (150°C)	50,000 lb/in.^{2}
	345 MN/m^2
Tensile modulus	4.5 X 10^6 lb/in.^{2}
R.T., dry	31,030 MN/m^2
R.T., after 2-hr boil	4.0 X 10^6 lb/in.^{2}
	27,580 MN/m^2
R.T., after 1/2 hr at 350°F (177°C)	4.5 X 10^6 lb/in.^{2}
	31,030 MN/m^2
300°F (150°C)	3.5 X 10^6 lb/in.^{2}
	24,130 MN/m^2
Tensile strain-to-failure	1.7%
Compressive strength	25,000 lb/in.^{2}
R.T., dry	172 MN/m^2
R.T., after 2-hr boil	23,300 lb/in.^{2}
	158 MN/m^2
350°F (177°C)	15,000 lb/in.^{2}
	103 MN/m^2
Compressive stress (0.02% offset)	12,000 lb/in.^{2}
	82.7 MN/m^2
Compressive modulus	4.5 X 10^6 lb/in.^{2}
R.T., dry	31,030 MN/m^2
350°F (177°C)	4.0 X 10^6 lb/in.^{2}
	27,580 MN/m^2
Flexural strength	50,000 lb/in.^{2}
R.T., dry	345 MN/m^2
R.T., after 72-hr water boil	35,000 lb/in.^{2}
	241 MN/m^2
350°F (177°C)	30,000 lb/in.^{2}
	207 MN/m^2
Flexural modulus	4.0 X 10^6 lb/in.^{2}
R.T., dry	27,580 MN/m^2
R.T., after 72-hr water boil	3.3 X 10^6 lb/in.^{2}
	22,750 MN/m^2
350°F (177°C)	3.0 X 10^6 lb/in.^{2}
	20,680 MN/m^2
Flexural stress (0.02% offset)	25,000 lb/in.^{2}
R.T., dry	172 MN/m^2
R.T., after 72-hr water boil	18,000 lb/in.^{2}
	124 MN/m^2
Interlaminar shear strength[a]	4000–10,000 lb/in.^{2}
R.T., dry	27.6–68.9 MN/m^2
R.T., after 72-hr water boil	3500–8700 lb/in.^{2}
	24.1–60.0 MN/m^2
In-plane (rail) shear strength	16,000 lb/in.^{2}
R.T., dry	110 MN/m^2

Table 27-31. *(Continued)*

Property	Value
In-plane (rail) shear modulus	0.30 X 10^6 lb/in.^{2}
R.T., dry	2070 MN/m^2
Tensile strength	30,000 lb/in.^{2}
	207 MN/m^2
Tensile modulus	1.1 X 10^6 lb/in.^{2}
	7580 MN/m^2
Compressive strength	18,000 lb/in.^{2}
	124 MN/m^2
Compressive modulus	1.0 X 10^6 lb/in.^{2}
	6890 MN/m^2
Compressive stress (0.02% offset)	6500 lb/in.^{2}
	44.8 MN/m^2
In-plane (rail) shear strength	32,000 lb/in.^{2}
	221 MN/m^2
In-plane (rail) shear modulus	3.0 X 10^6 lb/in.^{2}
	20,680 MN/m^2

[a]Highly dependent on resin.

(Fabrics). A summary of the mechanical properties of fabric composites made with "Kevlar" 49 and epoxy resin are shown in Tables 27-31 and 27-32. Tensile, compressive, flexural, and shear strengths and moduli are displayed for Style 181 fabric in typical 350°F cure and 250°F cure epoxides. A comparison of fabric laminates made with E-glass and "Kevlar" 49 is given in Table 27-33. It should be noted that as with unidirectional composites, the tensile stress-strain curve of "Kevlar" 49 fabric composites is linear to failure while the compressive and flexural stress-strain curves show

Table 27-32. "Kevlar" style 181 fabric with typical 250°F cure epoxy mechanical properties—warp direction, 50 V/O

Tensile strength	70,000 lb/in.^{2}
	483 MN/m^2
Tensile modulus	4.3 X 10^6 lb/in.^{2}
	29,650 MN/m^2
Compressive strength	25,000 lb/in.^{2}
	172 MN/m^2
Flexural strength	50,000 lb/in.^{2}
	345 MN/m^2
Flexural modulus	4.0 X 10^6 lb/in.^{2}
	27,580 MN/m^2
Interlaminar shear strength	6000–8000 lb/in.^{2}
	41.4–55.2 MN/m^2
Bearing strength	45,000 lb/in.^{2}
1/8 in. diameter pin	310 MN/m^2
1/4 in. diameter pin	207 MN/m^2

Table 27-33. Comparative fabric laminate performance.

1/8 in. thick panels, style 181 fabric, 350°F epoxy resin matrix, autoclave molded (75 psi)

	"E"-Glass	"Kevlar" 49
Laminate density (lb/in.^{3})	.065	.048
Tensile strength (10^3 psi)		
R.T., dry	63	75
R.T., wet	55	73
300°F (150°C) (1/2 hr)	50	50
Compressive strength (10^3 psi)		
R.T., dry	62	25
300°F (150°C) (1/2 hr)	61	15
Modulus (tension and compression) (10^6 psi)		
R.T., dry	3.6	4.5
R.T., wet	3.2	4.0
300°F (150°C) (1/2 hr)	3.4	4.5

Table 27-34. Fabric composite of "Kevlar" 49 interlaminar shear strength.

Short beam method–ASTM D2344; 45-50 fabric volume content.

Manufacturer	Epoxy Resin or Prepreg Code	Shear Strength (psi)
American Cyanamid	BP-907	8-9000
Dexter	Hysol 9703	8000
	Hysol 9704	5-8000
Du Pont	"Corlar" 5134	7-9000
Fiber-Resin	FR-8600	6500
Fiberite	934	4000
	505	8100
	506	8300
Hexcel	F-155	6-8000
	F-161	4-5000
U.S. Polymeric	E42-1	4-5500
	E-715	6-7000

a definite yield point. The interlaminar shear strength of high modulus aramid fabric composites depends on the epoxy resin used in their preparation. Table 27-34 shows the wide range of short beam shear strengths measured per ASTM D2344 that are obtained with several epoxides. The bearing strength of epoxy composites of "Kevlar" 49 compared to their glass counterparts are shown in Tables 27-35 and 27-36. These are important in designing attachments.

The thermal conductivity and expansion characteristics of "Kevlar" fabric reinforced composites are presented in Section 6.1.9, while the electrical properties of these materials are presented in Section 6.1.7. The weathering characteristics of aramid fabric composites are illustrated in Table 27-37. The tensile strength of the epoxy laminates are shown to remain essentially unchanged by either wet or dry UV exposure. The aramid reinforced polyester laminates are equal in strength retention to glass fiber controls made with the same resin.

(Hybrids). The hybrid constructions with high modulus aramids developed to date are those that combine "Kevlar" 49 with graphite. While there has been significant development work on these materials, only a few are available commercially.[29,30,31] The combination of "Kevlar" 49 and Union Carbide's "Thornel" 300 in both unidirectional tapes and fabrics has been well characterized and is available commercially.

Table 27-35. Bearing strength of fabric reinforced epoxy composites—1/8 in. diameter pin.

COMPOSITION				BEARING STRESS 10^3 psi				
V/O	Fiber	Type	Construction Relative to Test Direction	Yield	4% Offset	Max.	Failure Mode	Sample Thickness (in.)
54	"Kevlar" 49	181	Warp in test direction	18	29	45	Bearing	0.154
50	"Kevlar" 49	181	1.5 mil Ti foil between layers	36	47	–	Pin failure	0.155
50	"E"-Glass	181	Warp in test direction	26	40	–	Pin failure	0.163
78	"Kevlar" 49	143	0°-90° cross-ply	23	30	40	Bearing	0.121
70	"Kevlar" 49	143	0°, 90°, ±45°	25	33	45	Bearing	0.132

Table 27-36. Bearing strength of fabric reinforced epoxy composites—¼ in. diameter pin.

COMPOSITION				BEARING STRESS				
V/O	Fiber	Type	Construction Relative to Test Direction	Yield	4% Offset	Max.	Failure Mode	Sample Thickness (in.)
54	"Kevlar" 49	181	Warp in test direction	12	22	30	Bearing	0.154
50	"Kevlar" 49	181	1.5 mil Ti foil between layers	13	23	49	Bearing	0.155
50	"E"-Glass	181	Warp in test direction	21	–	(50)	Tabs pulled off	0.163
78	"Kevlar" 49	143	0°–90° cross-ply	11	18	31	Bearing	0.121
70	"Kevlar" 49	143	0°, 90°, ±45°	14	24	38	Bearing	0.132

Table 27-37. Effect of ultraviolet exposure on the tensile strength of "Kevlar"® 49 and fiberglass fabric reinforced composites.

Exposure	Kevlar® 49/ Epoxy[a]	Kevlar® 49/ Polyester[b]	E-Glass Polyester[c]
None	67,900[d]	43,700	37,000
No UV 120°F, 93% R.H. 1584 hr	70,000	41,700	31,600
Weather-ometer 500 hr	63,800	46,800	30,700
Weather-ometer 1000 hr	76,000	33,600	34,900
Fade-ometer 870 hr	68,500	38,300	31,100
Weather-ometer 100 hr + 72 hr Salt water soak + 600 hr fade-ometer	68,600	37,300	31,700

[a]2 plies style 181 + 1 ply style 120/Hexcel "F-155"; 55 vol. %.
[b]2 plies style 281 + 1 ply style 120/"Dion" 6908; 40 vol. %.
[c]2 plies style 181 + 1 ply style 120/"Dion" 6908; 40 vol. %.
[d]Each average of 3 axial tensile breaks.

Properties of unidirectional, cross-ply and fabric composites are shown in Tables 27-38 and 27-39.

(Chopped fiber). The properties of chopped fiber molding compounds made with "Kevlar" 49 are compared to similar glass compositions in Table 27-40. The effects of bundle size at constant length and of chopping length at constant bundle size are shown in Fig. 27-24. The most prevalent method of manufacture of high modulus aramid chopped fiber molding compounds to date has been compression molding. The mechanical properties of lab-scale compresssion moldings of "Kevlar" 49 versus graphite and glass analogs indicate significant performance pluses for high modulus aramids.[2,3] Injection molding the pre-cut "Kevlar" 49 has been successfully demonstrated, but mechanical properties so far have not supported the significant price premium over glass. As discussed in Section 2.6, cutting of high modulus aramids is significantly more difficult than glass and therefore conventional bulk molding compound and sheet molding compound preparation have used pre-cut yarn or roving.

Table 27-38. Properties of unidirectional "Thornel" 300, "Kevlar" 49 and hybrid composites—nominal fiber volume fraction = 0.60.

Percentages of "Thornel" 300/ "Kevlar" 49 Fibers	Specific[a] Gravity	TENSION Modulus (10^6 psi)	TENSION Ultimate Stress (ksi)	COMPRESSION Stress at 0.02% Offset (ksi)	COMPRESSION Ultimate Stress (ksi)	FLEXURE Stress at 0.02% Offset (ksi)	FLEXURE Ultimate Stress (ksi)	Short Beam Shear Stress (ksi)	Prepreg Cost $/lb
100/0	1.60	21.1	227	98.4	146	233	233	13.2	60
75/25	1.56	17.4	186	68.8	136	181	197	11.0	48
50/50	1.51	15.7	176	59.9	99.8	120	160	8.1	35
0/100	1.35	11.2	183	26.4	41.5	49.2	91.9	7.1	10

[a]Data supplied by Fiberite Corp.

Table 27-39. Properties of hydrids—"Thornel" 300/"Kevlar" 49.

(50 volume percent fabric)

Material and Percentage of "Thornel" 300/"Kevlar" 49	Modulus (10^6 psi)	Ultimate Stress (ksi)	Modulus (10^6 psi)	Prop. Limit (ksi)	Stress @ .02% Offset (ksi)	Ultimate Stress (ksi)	Modulus (10^6 psi)	Prop. Limit Stress (ksi)	Stress @ .02% Offset (ksi)	Ultimate (ksi)	Short Beam Shear Stress (ksi)
[0]–Tapes											
100/0	21.1	227	20.3	48.6	98.4	146	17.0	233	233	233	13.2
75/25	17.4	186	16.8	40.7	68.8	136	15.3	161	181	197	11.0
50/50	15.7	176	14.4	43.7	59.9	99.8	12.6	97.8	120	160	8.1
0/100	11.2	183	10.4	15.9	26.4	41.5	8.78	34.0	49.2	91.9	7.1
[0/90]–Tapes											
100/0	11.0	111	9.82	–	–	131.4	10.4	100	123	141.6	–
75/25	9.83	93.4	8.91	24.0	46.4	86.1	8.66	64.0	84.7	123	–
50/50	8.04	82.8	7.29	11.75	22.6	53.5	6.83	44.2	57.4	96.7	–
0/100	5.52	87.7	5.16	5.74	10.89	22.2	4.94	20.0	26.0	59.9	–
Balanced fabrics											
100/0	9.53	69.2	10.2	20.1	48.7	85.1	8.44	95.7	95.7	95.7	5770
75/25	8.92	68.1	10.1	16.5	34.5	40.2	7.06	57.0	70.6	84.5	4660
50/50	7.44	61.2	8.33	8.00	23.7	35.0	6.13	32.7	43.0	64.9	4150
0/100	5.41	83.1	5.59	5.60	11.8	23.9	4.18	23.0	26.6	60.1	3770

Polyester. Mechanical properties of high modulus aramid fibers in polyester resin can be as high as with epoxy resins manufactured according to the same methods, if the specific polyester is carefully selected (Table 27-41). The properties of several hand-layup fabric composites of "Kevlar" 49 are compared to their fiberglass counterparts in Table 27-19. It should be noted that the interlaminar shear strength of high modulus aramid fiber reinforced polyesters are generally lower than those of their glass counterparts. The level of interlaminar shear strength obtained with "Kevlar" 49 is strongly dependent on the type of polyester used.[32]

Phenolic, Polyimide and Others. Significantly less data exists on aramid fiber reinforced composites made with phenolic or polyimide resins (Table 27-42) than with epoxy or polyester. However, interest in phenolic composites as possibly lower smoke, more flame retardant materials for weight critical applications like

Table 27-40. Comparison of molding compounds "Kevlar" 49 Fiber and epoxy resin versus U.S. polymeric EM-7302 (glass/epoxy).

	"Kevlar" 49	EM-7302
Fiber denier	380	–
Fiber content, % of vol.	50	47
Fiber length, in.	0.5	0.5
Specific gravity of CFMC	1.32	1.86
Tensile strength, psi	28,500	14,400
Tensile modulus, 10^6 psi	2.96	2.23
Flexural strength, psi	35,500	29,100
Flexural modulus, 10^6 psi	2.67	2.33
Izod impact strength, ft-lb/in.	17.4	19

Table 27-41. "Kevlar" 49 style 181 fabric with heat and pressure cured polyester resin—mechanical properties—warp direction.

(Properties will vary widely depending on choice of resin.)

Density	0.047 lb/in.^{2} (1.30 g/cc)
Volume percent fabric	40%
Tensile strength	60,000 lb/in.^{2} (414 MN/m^2)
Tensile modulus	3.5 × 10^6 lb/in.^{2} (24,130 MN/m^2)
Flexural strength	30,000 lb/in.^{2} (207 MN/m^2)
Flexural modulus	3.0 × 10^6 lb/in.^{2} (20,680 MN/m^2)
Flexural stress	14,000 lb/in.^{2} (96.5 MN/m^2)
Interlaminar shear strength	3000–4000 lb/in.^{2} (20.7–27.6 MN/m^2)

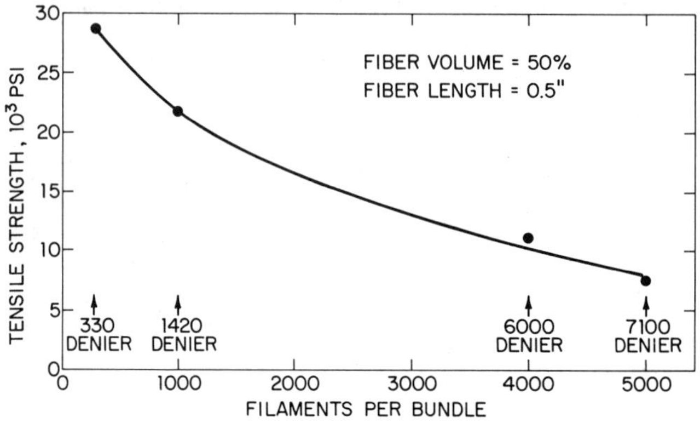

Fig. 27-24a. Tensile strength vs. fiber bundle size Kevlar® 49/epoxy chopped fiber molding compounds.

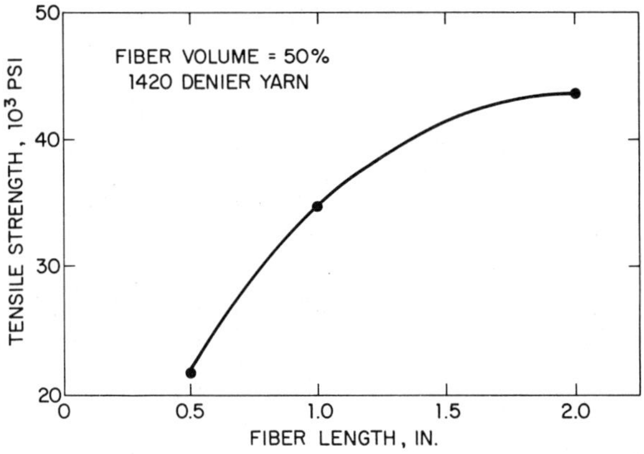

Fig. 27-24b. Tensile strength vs. fiber length Kevlar® 49/epoxy molding compounds.

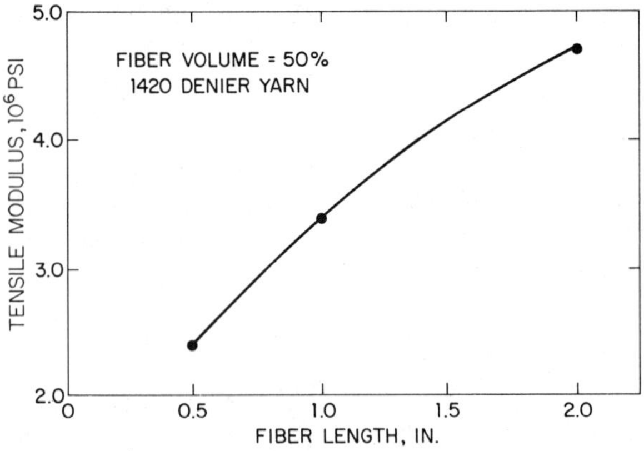

Fig. 27-24c. Tensile modulus vs. fiber length Kevlar® 49/epoxy molding compounds.

Table 27-42. "Kevlar" 49 style 181 polyimide resin—mechanical properties—warp direction 50 V/O.

Fabric volume percent	55%
Tensile strength	65,000 lb/in.^{2} 448 MN/m^2
Tensile modulus, R.T.	4.1 × 10^6 lb/in.^{2} 28,270 MN/m^2
450°F (230°C)	3.6 × 10^6 lb/in.^{2} 24,820 MN/m^2
Compressive strength	20,000 lb/in.^{2} 138 MN/m^2
Compressive modulus	4.0 × 10^6 lb/in.^{2} 27,580 MN/m^2
Flexural strength, R.T.	40,000 lb/in.^{2} 276 MN/m^2
450°F (230°C)	28,000 lb/in.^{2} 193 MN/m^2
550°F (290°C)	18,000 lb/in.^{2} 124 MN/m^2
Flexural modulus, R.T.	4.0 × 10^6 lb/in.^{2} 27,580 MN/m^2
485°F (250°C)	3.4 × 10^6 lb/in.^{2} 23,440 MN/m^2
Interlaminar shear strength	2200 lb/in.^{2} 15.2 MN/m^2

interiors of aircraft has motivated development into this area with high modulus aramids.

6.3.2. Thermoplastic and Elastomer Matrices. The reinforcement of thermoplastic resin matrices, primarily polyurethanes, with high modulus aramid is being pursued in developing high strength-to-weight ratio cables to compete with steel wire. Performance information in these specialized applications for composite materials can be found in Refs. 33 and 34.

The coating of high modulus aramid fabrics with elastomers like noeprene, PVC, Teflon®, and others, has been motivated by the promise of high specific strength and tear resistance in boating and architectural uses. Properties of these specialized composite structures can be found in Ref. 35.

7. PROJECTED UTILIZATION

As was described in detail in Section 6.1, the combination of properties offered by high strength and modulus aramid fibers has led to their use in a number of rubber (tire, V-belt, hose, etc.), plastic (epoxy, polyester, PVC, etc.), and elastomer (neoprene, urethane, etc.) reinforcement uses. Growth of aramid fibers in these markets is expected to occur at the expense of less cost effective incumbent fibers.

The very high specific tensile strength of aramid reinforced composites has suggested their use in more unconventional applications such as high performance flywheels. Advanced flywheels are in an early state of development, but if their promise as energy storage devices with quick charge-up capability is confirmed in service testing, they would increase significantly the fuel economy of transportation vehicles and represent a large volume market for aramid fibers.[36] First results have been encouraging, with some "Kevlar" 49/epoxy flywheels outperforming S-glass fiber analogs.

The higher resistance of aramid fibers to chemically basic pH has suggested their use instead of glass in chopped fiber reinforced concrete. Early results by a number of investigators have been positive, but cost effectiveness in this use remains to be established.

References

1. Sturgeon, D. L. G., Wagle, D. G., and Wolffe, R. A., "Kevlar® 49 Woven and Nonwoven Fabric Composites—Performance and Applications," 19th National SAMPE Symposium and Exhibition, Buena Park, Calif., Apr. 23–25, 1974.
2. Blumentritt, B. F., Vu, B. T., and Copper, S. L., "The Mechanical Properties of Oriented Discontinuous Fiber-Reinforced Thermoplastics, I. Unidirectional Fiber Orientation," *Polymer Engineering and Science*, **14**: (9), 633 (Sept. 1974).
3. Blumentritt, B. F., Vu, B. T., and Copper, S. L., "The Mechanical Properties of Oriented Discontinuous Fiber-Reinforced Thermoplastics, II. Random In-Plane Fiber Orientation," *Polymer Engineering and Science*, **15**: (6), 428–436 (June 1975).
4. Preston, J., "High Strength/High Modulus Organic Fibers," *Polymer Engineering and Science*, **15**: (3), 199 (Mar. 1975).
5. Stone, R. H., "Advanced Organic Fibers for L-1011 Interior Structures," National SAMPE Technical Conference, Los Angeles, Calif., Apr. 10–12, 1972.
6. Paschal, D. R., "The Successful Use of Composites in the L-1011 Tristar Commercial Transport," Symposium on Reinforced Plastics in Aerospace Applications, The Plastics Institute, London, Apr. 5–6, 1973.
7. Wooley, J. G., "Flight Service Evaluation of PRD-49/Epoxy Composite Panels in Wide-Boiled Commercial Transport Aircraft," First Annual Flight Service Report, LR-26580, NASA Contract NASI-11621, July 1974.

8. Click, H. F., "Applications of Graphite and Aramid Composites on the YF-17 Prototype Fighter," The Sixth National SAMPE Technical Conference, Vol. **6**, pp. 352–360, Oct. 1974.
9. Dickard, H. E., et al., "Design and Fabrication of An RPV Wing Panel with PRD-49/Epoxy Skins," AFFDL-TR-73-31, Mar. 1973.
10. Wise, C. E., "Turbofan of the Future," *Machine Design*, pp. 20–25 (Aug. 22, 1974).
11. Beaumont, P. W. R., et al., "Methods for Improving the Impact Resistance of Composite Materials," ASTM Symposium on Foreign Object Impact Behavior of Composites, Philadelphia, Pa., Sept. 20, 1973.
12. Zweben, C., "Fracture of Kevkar® 49, E-Glass and Graphite Composites," ASTM Symposium on Fracture Mechanics of Composites, Gaithersburg, Md., Sept. 25, 1974.
13. Moore, J. W., et al., "High Modulus Organic Fiber Composites in Aircraft Applications," *Composites*, **4**: (1), 34–38 (Jan. 1973).
14. Hoggatt, J. T., "Thin-Metal Lined PRD-49-III Composite Vessels," NASA CR-13455, Jan. 1974.
15. Wright, O. C., "Development of a High Performance PRD-49-III Filament-Wound Chamber for Trident I (C4) Third Stage," AIAA/SAE 9th Propulsion Conference, Las Vegas, Nev., Nov. 5–7, 1973.
16. Chiao, T. T., et al., "Organic Fiber/Epoxy Pressure Vessels," *SAMPE Quarterly*, **5**: (3), 17–27 (Apr. 1974.
17. Sturgeon, D. L. G., et al., "Kevlar® 49 Composites for Sporting Goods and Recreational Equipment," 19th National SAMPE Symposium and Exhibition, **19**, 304–319 (Apr. 1974).
18. Riewald, P. G., et al., "Kevlar® 49 Hybrid Composites for Commercial and Aerospace Applications," 30th Annual Conference of the SPI, Reinforced Plastics/Composites Institute, (Feb. 1974).
19. Moore, J. W., et al., "Recent Advances in PRD-49 Composites," 1972 National SAMPE Technical Conference, Palo Alto, Calif., Oct. 17–20, 1972.
20. Bowers, W. E., "High Strength-to-Weight Cable for Deep Ocean Projects," Proceedings of the International Wire and Cable Symposium 1973, pp. 279–285, Atlantic City, Dec. 1973.
21. Hightower, J. D., et al., "Development of PRD-49 Composite Tensile Strength Members," American Society of Mechanical Engineers, Ocean Technology Division, Winter Annual Meeting, Detroit, Nov. 11–15, 1973.
22. Swenson, R. C., et al., "Design and Construction of Cables for Sensor Systems," *Sea Technology* (Oct. 1973).
23. Ties, T. S., "Submarine Tow Cables," *Depthlines*, ITT-Cable Hydrospace Division, **2**: (1), 4, (Feb. 1974).
24. Scala, E., "High-Strength Filaments for Cables and Lines," American Society for Testing and Materials, Special Technical Publication 521, pp. 390–408, 1973.
25. Niccum, R. J., et al., "Investigation of Kevlar® Fabric Based Materials for Use with Inflatable Structures," NASA CR-132411, Apr. 1974.
26. Sturgeon, D. L. G., et al., "Kevlar® 49 Woven and Nonwoven Fabric Composites-Performance and Applications," 19th National SAMPE Symposium and Exhibition, **19**, 229–243 (Apr. 1974).
27. Kevlar® 49 Data Manual, E. I. du Pont de Nemours & Co., Inc., Textile Fibers Department, Kevlar® Special Products, Wilmington, Del. 19898.
28. E. I. du Pont de Nemours & Co., Textile Fibers Dept., Industrial Fibers Div., communication dated Sept. 24, 1975.
29. Riewald, P. G. and Zweben, C., "Kevlar® 49 Hybrid Composites for Commercial and Aerospace Applications," 30th Annual Conference of the SPI Reinforced Plastics/Composites Institute, Washington, D.C., Feb. 1975.
30. Henshaw, J. and McElman, J., "Selective Reinforcement with Boron for Cost Effective Composite Structures," AIAA/ASME/SAE, 13th Structures, Structural Dynamics and Materials Conference, San Antonio, Tex., Apr. 1972.
31. Fogg, L. D., "Hybrid Composite Design for Damage Tolerance," 2nd Conference on Fibrous Composites in Flight Vehicle Design, Dayton, Ohio, May 1974.
32. Miner, L. H., Wolffe, R. A., and Woodrick, J. V., "Kevlar® 49 Aramid, A New Material for Boat Hull Construction," The Second Chesapeake Sailing Yacht Symposium, pp. 8-1 to 8-18, Annapolis, Md., (Jan. 18, 1975).
33. Wilkins, G. A., "Performance Characteristics of Kevlar® 49 Tension Members," 1975 International Conference on Composite Materials, Geneva, Switzerland (Apr. 7–11) and Boston, Mass. (Apr. 14–18).
34. Riewald, P. G. and Venkatachalam, T. K., "Kevlar® Aramid Fiber for Rope and Cable Applications," Marine Kevlar® Cable Workshop, Offshore Technology Conference, Houston, Tex., May 6, 1975.
35. Sturgeon, D. L. G. and Wardle, M. W., "Advanced High Specific Strength Materials for Space Enclosures," IASS World Congress on Space Enclosures, Montreal, Canada, July 4–9, 1976.
36. Sturgeon, D. L. G., "Potential Advantages of Kevlar® 49 in Flywheel Applications," 1975 Flywheel Technology Symposium, University of California at Berkeley, Nov. 10–12, 1975.

ACKNOWLEDGMENT

Much of the information contained in this Chapter was generated by past and present members of the Kevlar® Special Products Marketing and End-Use Research groups, in the Textile Fibers Department, of E. I. du Pont de Nemours & Co. The authors wish to thank them for their anonymous contribution.

28
Boron Filaments

Val Krukonis
Fiboron, Inc.
Lexington, Massachusetts

CONTENTS

1. INTRODUCTION

Boron filament possesses a combination of strength, stiffness, and density unequaled by any other continuous filament form; typical physical and mechanical properties of the filament are given in Table 28-1.

These outstanding properties of boron were first determined at Texaco Experiments, Inc. in 1959. The potential of such a material for advanced aircraft applications was recognized by the Air Force Materials Laboratory, which funded a large-scale research and development effort to advance the filament from the laboratory stage to a production level.

During the early 1960's, boron was a laboratory curiosity priced at over $5,000/lb, but today boron filament production has expanded to a level of over 45,000 lb/yr, primarily for use on the two advanced military aircraft, the F-14 and F-15. The current cost of boron filament has dropped to about $200/lb. About 90% of the filament produced is converted into a continuous boron/epoxy tape, which is the raw material form employed in fabricating the high stiffness stabilizer sections of the F-14 and F-15. (Boron/epoxy tape is comprised of about 65 weight percent boron and costs about $160/lb.) In addition to its use in fabricating the empennage components on the F-14 and F-15, boron/epoxy is being used in the B-1 development program for fabricating the horizontal stabilizer and for reinforcing the steel longerons; if the B-1 reaches production status, it represents a market for boron filament of over 100,000 lb during the early 1980's. Boron filament is also converted into boron/aluminum composite forms; one of the most important applications of the material is in the tube members used to reinforce the mid-fuselage structure of the Space Shuttle Orbiter.

During the mid-1960's, much of the development emphasis centered on only the technical merits of boron. For example, effort was directed to the fabrication of components with

Table 28-1. Properties of boron filament.

Property	Value
Tensile strength	500 Ksi
Modulus of elasticity	58 million psi
Density	2.6 g/cc
Diameter	4-, 5.6-, and 8-mils

lower weight and superior performance than components of titanium or aluminum; however, continued work brought out the fact that boron/epoxy composite structures could not only be technically superior, but they could also be cost effective structures. As was stated above, the current cost of boron/epoxy tape is about $160/lb, and, for comparison, titanium costs about $10/lb. However, most of the titanium structures (and similarly, aluminum or steel structures) on an advanced aircraft are machined from large forgings, and frequently the final metal component comprises only 10% of the weight of the initial forging. The remainder consists of turnings and scrap generated by extensive and costly machining operations, operations which can result in a finished component whose final cost is many hundred dollars per pound. In contrast, boron composite components can be fabricated to a given shape, with very little scrap. For example, based upon a minimum part thickness requirement, it was found that the high initial materials cost of boron/epoxy could be a lesser consideration than the cost of the extensive machining of an all-metal structure. Hence, boron composites for a specific design and required performance could excel, not only because of their low weight and technical performance, but because of the relatively low fabrication cost as well. Current research effort is being directed to reducing still further, filament costs and composite component manufacturing costs in order to extend the use of boron to large scale use in commercial aircraft structures.

The strength and stiffness properties of boron composites are also being used in commercial products such as tennis rackets, golf shafts, and bicycle frames; to a smaller extent, the hardness of boron filament is being directed to industrial applications in drills, high temperature brushes, and honing tools. Still other studies, investigating the use of boron in energy production, saving, and storage devices have been undertaken. The reinforcement of flywheels, high speed centrifuges, and superconducting magnets and generators, which experience very high stresses, is receiving increasing attention. Based upon the current success in some of these applications, the boron filament market is projected to be well over 100,000 lbs/yr in the 1980's.

2. MANUFACTURE OF BORON FILAMENT AND BORON PRODUCT FORMS

Boron filament is produced by a process called Chemical Vapor Deposition. The chemical reaction consists of the reduction of boron trichloride by hydrogen on an electrically-heated, continuously moving, tungsten substrate. The vapor deposition of boron was first reported by Weintraub in 1911,[1] but it was work at Texaco Experiments, Incorporated, that demonstrated that a continuous high strength, high modulus, low density fiber reinforcement material could result from this vapor deposition process.[2] Because of the potential importance of such a material as a reinforcement for advanced composites in aircraft applications, the United States Air Force Materials Laboratory funded further research and process development effort at Texaco and also at a number of other laboratories. During the early and mid-1960's, United Aircraft Research Laboratory, Avco Systems Division, General Electric Research Laboratory, Monsanto, and Battelle Memorial Institute, among many others, carried out boron filament process development and fracture analysis studies. By the late 1960's, however, only two companies in the United States were still involved with boron filament research and production, viz., Avco Systems Division and Hamilton-Standard, a division of United Aircraft Corporation. (The Hamilton-Standard plant was subsequently purchased by Composite Technology, Inc. A short history of CTI is given in Section 4.)

2.1 Chemical Vapor Deposition Process

A schematic diagram of a basic deposition unit is shown in Fig. 28-1. The reactor module consists of a glass deposition tube which is fitted with gas inlet and outlet ports, two mercury-filled electrodes, a variable-DC power supply connected to the two electrodes, a tungsten substrate let-off system, and a boron filament take-up unit. Tungsten substrate, typically 0.0005 in. in diameter, is drawn through the reactor and heated to about 1300°C by the DC power supply. (Prior to entering the deposition reactor, shown schematically in Fig. 28-1, the tungsten substrate is first passed through a short "cleaning" stage, not shown in the figure, in which the substrate is heated to candescence in a hydrogen atmosphere in order to remove surface contaminants and residual lubricants which are used in the drawing of tungsten wire.) A stoichiometric mixture of boron trichloride and hydrogen is introduced at the top of the reactor. At the 1300°C temperature level, a mantle of boron is deposited on the tungsten via the reaction

$$BCl_3 + 3/2H_2 \longrightarrow B + 3HCl.$$

Exhaust gases consisting of HCl, unreacted H_2, BCl_3, and intermediate species are removed through the outlet port at the bottom of the reactor and are processed to remove the unreacted BCl_3. Typical diameter of the filament exiting from the reactor is 0.004, 0.0056, or 0.008 in., depending upon the drawing rate.

A number of chemical reactions occur simultaneously during the formation of the filament; for example, boron is deposited via the halide-hydrogen reaction indicated above, and by-products such as B_2H_6, $HBCl_2$, and B_2H_5Cl are formed in the gas phase in addition to the HCl indicated by the stoichiometric relation. In the core of the filament, a solid state reaction between the tungsten and the boron mantle occurs to form a number of tungsten boride forms; boride phases which have been identified in boron filament are δ-WB, W_2B_5, and WB_4.[3,4,5]

Figure 28-1 also shows the temperature profile of a filament which is heated with a DC power supply; the temperature rises rapidly and falls gradually as the filament passes through the reactor, and a combination of complex chemical and physical phenomena are responsible for the particular shape of the profile. The sharp rise in the temperature as the substrate enters the reactor results from the chemical reactions occurring in the core, viz., low resistivity tungsten is converted to higher resistivity tungsten boride phases. The subsequent increase in the diameter of the boron mantle causes the temperature to fall because of resistance considerations, i.e., the resistance decreases because the cross sectional area increases and because the high temperature resistivity of boron is about the same as that of tungsten boride. The highest temperature along the profile in Fig. 28-1 is called the deposition, or "hot spot" temperature. The temperature at the reactor outlet has been measured to be about 300°C lower than the hot spot. A photograph of one hundred boron filament reactors which produce about 9,000 lb/yr in Avco's plant is shown in Fig. 28-2.

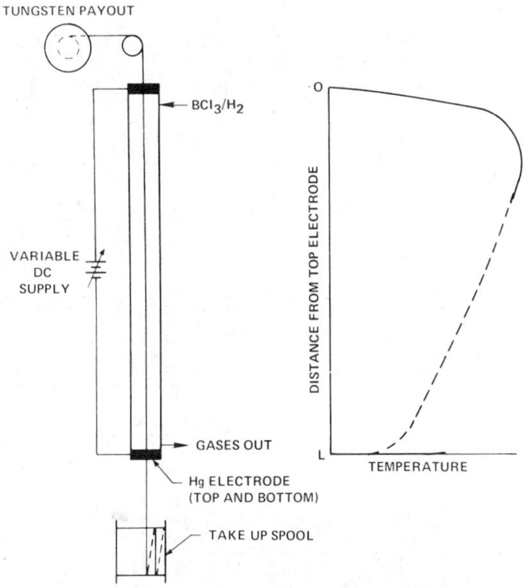

Fig. 28-1. Schematic diagram of boron filament reactor; temperature profile in single stage reactor.

2.2 Other Boron Filament Processes

Besides the current chemical vapor deposition process utilizing the reduction reaction between

Fig. 28-2. Photograph of boron filament plant.

boron trichloride and hydrogen, other boron filament formation processes were tested in the mid-1960's. A major effort was directed to a lower temperature chemical vapor deposition process utilizing diborane, B_2H_6, as the reactant gas.[6] The initial motivation for use of diborane as the boron source was the replacement of the (then) very expensive tungsten with a lower cost substrate. Diborane can be decomposed to form amorphous boron at a temperature below 600°C, and hence, at such a temperature level it was found possible to utilize lower cost silica as a substrate. However, in order to operate the process, it was necessary to render the silica substrate conductive by first coating it with carbon so that it could be heated resistively. Colloidal carbon solutions and pyrolytic carbon were tested as coatings on the surface of the silica.

A number of problems limited the scale-up of the diborane-silica substrate process. First, it was found that the tensile strength of diborane-based boron filament was not as high as that produced from boron trichloride;[7] further, the cost of tungsten continued to decrease as the boron filament process was expanded from laboratory scale to the 10,000 lb/yr level in the late 1960's, providing less impetus to develop the diborane process. Finally, the difficulty in operating a large scale diborane-silica process was demonstrated when an explosion occurred in one such plant.

Another boron filament formation process which received substantial development effort in the mid-1960's was based upon a liquid extrusion process.[8,9] Molten boron, either as a pure melt or in admixture with a melting point reducing additive, was forced through a die and rapidly cooled in an effort to form high strength filament. One of the technical problems that was encountered in producing a continuous molten stream was the absence of a viscosity index which precluded drawing or "spinning"; other practical problems, such as boron's extreme reactivity with most die and crucible materials, also limited the experimental development. Finally, it was never demonstrated that boron solidified from the liquid could be a high strength form. The molten boron work was not successful, and the halide reduction process practiced by Avco and CTI remains today the only successful manufacturing method for producing high strength boron filament.

2.3 Conversion of Boron Filament into Boron/Epoxy and Boron/Aluminum Products

Boron filament is converted into boron/epoxy tape (or prepreg) and into boron/aluminum, the two primary material forms used in fabricating composite structures. Boron/epoxy prepreg is produced by collimating and coating large arrays of filaments with epoxy resin using either hot melt or film techniques, the filament comprising about 65% by weight of the tape. Figure 28-3 is a photograph of Avco's tape manufacturing equipment, and a photograph of 3 in. wide boron/epoxy prepreg is shown in Fig. 28-4. The material is used to lay up composite components using the fabrication sequence developed in the 1950's for fiberglass structures and which has been described in other publications.[10,11] Prepreg of 3, 6, and 48 in. wide continuous boron/epoxy is available from both manufacturers.

Fig. 28-4. Three inch wide boron/epoxy tape.

The majority of boron/aluminum sheet and plate is consolidated by a process called diffusion bonding. Arrays of filament are collimated on aluminum alloy foils and are hot-pressed under conditions of 5000 psi pressure and a

Fig. 28-3. Equipment for manufacturing boron/epoxy tape.

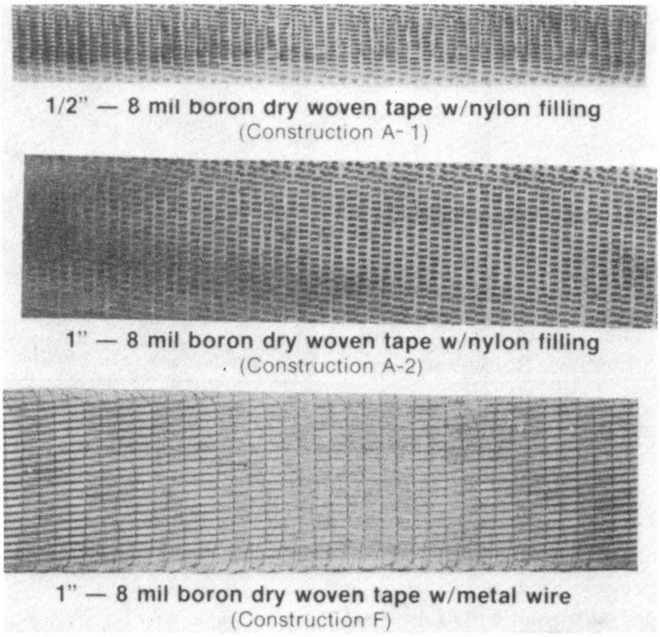

Fig. 28-5. "Dry-woven" boron tape.

temperature of 965°F for a period of about 1 hr. The high pressure, high temperature conditions cause the aluminum foils to deform, to encase and bond to the boron filaments, and to bond to each other.

The boron/aluminum material possesses several advantages over boron/epoxy. For example, the use temperature of boron/aluminum can be extended to about 700°F, and boron/aluminum composites can, in many operations, be fabricated using a number of common metal working processes.

One recently developed boron raw material form consists of "dry woven" tape, i.e., the product consists of collimated arrays of filament held in the transverse direction by glass fiber or aluminum wire. These dry tape forms offer some advantages in shipping and fabrication; for example, refrigerated storage, which is necessary to maintain the epoxy reactive in boron/epoxy prepreg, can be eliminated, and its use in boron/aluminum composite manufacturing can eliminate several costly fabrication steps. A photograph of one dry woven tape form produced by CTI is shown in Fig. 28-5. The dry woven form is receiving increasing attention, especially since widths of 48 in. wide material are becoming available.

3. PROPERTIES AND FORMS OF BORON FILAMENT AND BORON COMPOSITES

The properties of boron filament, viz., its high strength, high modulus, and low density, are not matched by any metal or by any other continuous reinforcement filament being used in organic- or metal-matrix composites. Boron is produced in several diameters for a variety of applications, and besides its applications in high strength, high stiffness structures, some of the nonstructural properties of the filament are receiving attention.

3.1 Mechanical Properties

In Table 28-2, the mechanical properties of 50 volume percent boron/epoxy composites are compared to those of fiberglass/epoxy, graphite/epoxy, titanium, and a high strength aluminum alloy. The data of the table show that the mechanical properties of unidirectional boron/epoxy composites surpass those of fiberglass/epoxy and graphite/epoxy composites, and, except for shear strength, the properties of boron/epoxy far surpass the properties of aluminum and titanium. Graphite filament is a near-competitor of boron; however, the

Table 28-2. Comparison of mechanical properties of selected aircraft materials.

	MECHANICAL PROPERTY			
Material	Tensile Strength (Ksi)	Tensile Modulus (Ksi)	Compressive Strength (Ksi)	Shear Strength (Ksi)
Boron/epoxy (50 v/o)	220	30	450	15
Glass/epoxy (60 v/o)	175	7	100	15
HS Graphite/epoxy (60 v/o)	200	18	200	15
HM Graphite/epoxy (60 v/o)	140	30	150	15
Aluminum (2024)	60	10	60	35
Titanium (Ti-6A1-4V)	150	16	150	80

graphite filament forms currently available exhibit either a high strength and low modulus (HS), or a low strength and high modulus (HM). In compressive strength, an important requirement of materials used in fabricating critical aircraft components, the table shows that boron/epoxy exhibits a factor of three over the metals and the high strength graphite.

A typical tensile strength histogram of boron filament is given in Fig. 28-6. Current production-grade boron filament usually averages about 500 Ksi, although as can be seen,

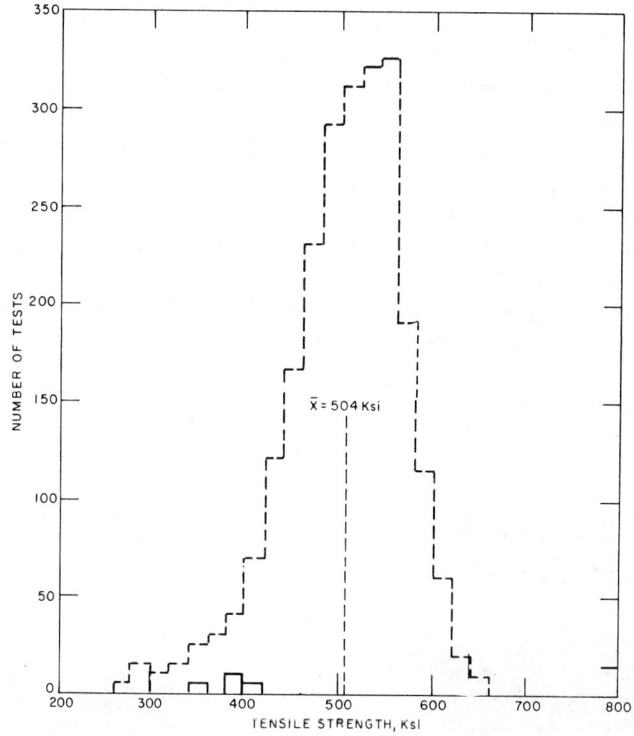

Fig. 28-6. Tensile strength histogram of boron filament.

some low strength and some very high strength material is also obtained in routine production. Boron is almost always utilized in combination with a resin or metal matrix in composite structures, and, therefore, it is most meaningful to evaluate the mechanical properties of the composite as a more realistic measure of the filament properties. A low strength "tail" of 2 or 3% does not affect the filament's high performance in a composite, nor, similarly, does the 3-5% very high strength contribution result in any significant increase in the composite strength. Based upon over 7 years of large scale production experience, it is to be pointed out that the process has been developed to the point where the properties of boron/epoxy shown in Table 28-2 are always attained in fabricated structural parts.

3.2 Internal and Surface Characteristics

The surface of 4-mil boron filament exhibits the well-known "ear of corn" structure and a photomicrograph of the surface is shown in Fig. 28-7. The kernels are a result of nucleation and growth phenomena. Boron deposits selectively at certain sites along the die grooves on the tungsten substrate. These first deposits of boron act as further deposition sites which grow outward from the substrate to form growth cones whose ends terminate in the surface kernels seen in Fig. 28-7. A photomicrograph of a polished cross section of 4-mil boron is also shown in Fig. 28-7. The core is comprised of an admixture of W_2B_5 and WB_4, although the individual phases cannot be distinguished optically. As was stated earlier, another boron phase, δ-WB, has been identified. This phase is formed early in the process, but the phase disappears when the tungsten core is reacted further.

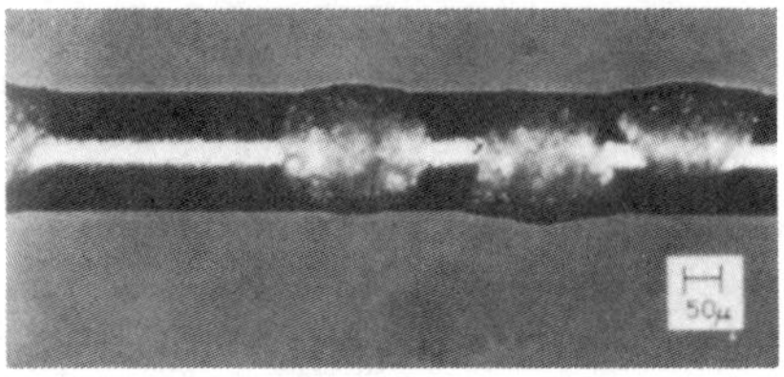

Fig. 28-8. Crystal flaw on boron filament produced at temperature above 1300°C.

Certain flaws contribute to, and are responsible for, the low strength "tail" in Fig. 28-6; these flaws, which are either anomalous growths in the filament or internal stresses which are created during the deposition process, have been covered in detail in other works and will not be described in detail herein.[5,12] As one example, Fig. 28-8 is a photomicrograph of a crystal flaw which results when too high a deposition temperature is reached during deposition. Other flaws caused by impurities in the mercury or in the gas phase can result in localized anomalies, but the low strength contributions of such contamination have, over the many years of process development, been almost completely eliminated.

Figure 28-9 is a photomicrograph of a flaw which can influence most markedly the tensile strength of boron filament being produced today. This flaw has been called a "crack tip" because it appears to be a crack in the cross section of the filament.[5] The strength of boron which possesses the crack tip flaw can be as low as 100 Ksi, but control of the parameters of production can minimize the appearance of the low strength flaw. The crack is not a pre-existing flaw in boron filament, although many authors have assumed it to be.[5,13,14] It is the understanding and elimination of these flaws which is the predominate factor which has taken boron out of the laboratory and has placed it in

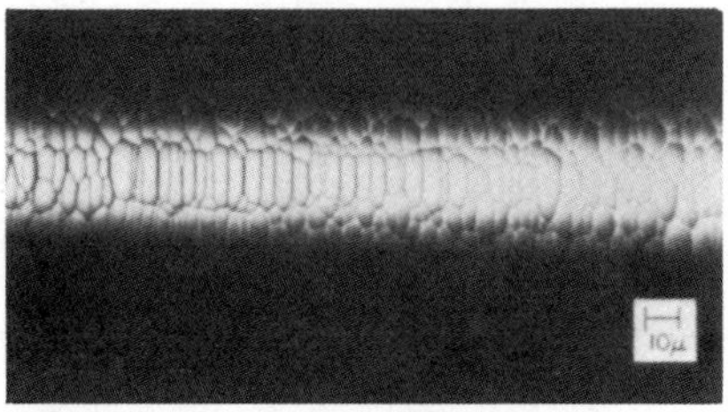

Fig. 28-7. Surface and cross-section of boron filament.

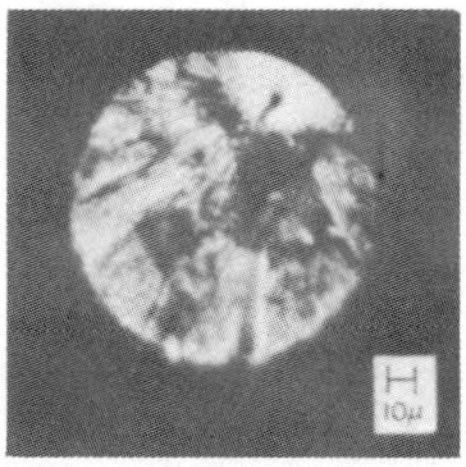

Fig. 28-9. Crack tip flaw in boron filament.

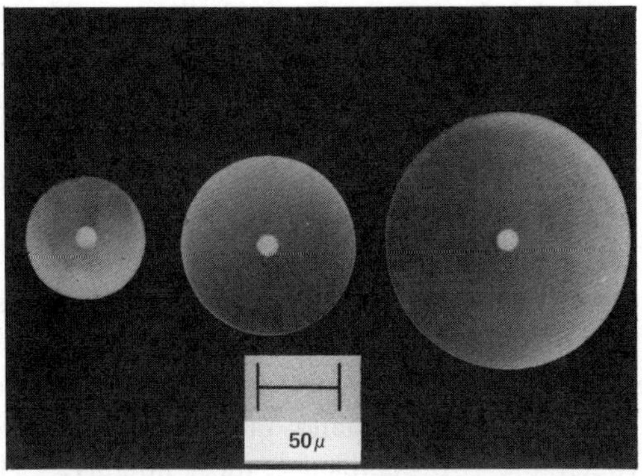

Fig. 28-10. Comparison of cross-sections of 4-, 5.6-, and 8-mil boron.

the category of a state-of-the-art production material.

3.3 Filament Sizes and Forms

Several sizes of boron filament are being produced at the present time, and the most common diameters are 4-, 5.6-, and 8-mils. The initial motivation for producing larger diameter boron involved costs, and this aspect is discussed later in Section 5. Figure 28-10 shows a cross section of the three filament sizes, with 0.5 mil tungsten used as the substrate for all sizes. The larger sizes are used in metal matrix composites also because of technical considerations. In early tests on boron/aluminum composites, it was found that the use of 4-mil boron resulted in low transverse strength because of splitting of the filament, but the use of 5.6-mil boron eliminated the splitting problem.[15] For reasons not yet understood, the internal stresses in 5.6-mil boron do not result in low transverse strengths in boron/aluminum composites (although, if the temperature profile of the filament is altered, 5.6-mil boron can also be made to split).[16]

Boron deposited on a carbon substrate is another boron form currently being researched (and henceforth, B/W will be used to designate the "standard" boron when there is any chance of ambiguity with boron-on-carbon, B/C). B/C filament is deposited on a 1.3 mil diameter carbon monofilament substrate, and the material is receiving concerted development effort at the current time primarily because of its projected lower cost aspects.[17,18,19] B/C is produced in a reactor essentially the same as that shown schematically in Fig. 28-1, except that prior to entering the deposition reactor, the carbon substrate is first coated with a thin layer of pyrolytic graphite.[18] The strength of B/C filament is only slightly lower than that of B/W, but the modulus is about 10% lower because of the large area of the carbon core. However, the specific modulus, i.e., the ratio of modulus to density of B/C is essentially the same as of B/W. Two sizes of B/C filament are currently being manufactured, viz., 4-mil and 5.6-mil, with both forms totaling about 200 lbs/year. A cross section of the two B/C filament sizes is shown in Fig. 28-11, and the carbon substrate, the PG layer, and the boron mantle are distinguishable. The mechanical properties of B/C in epoxy are about the same as those of B/W but, B/C, both of 4-mil and 5.6-mil diameter currently suffers from low transverse strength in an aluminum matrix caused by splitting.[20] Research directed to understand and circumvent the phenomenon is being carried out at Avco.

The other properties of boron filament, e.g., its hardness, thermal conductivity, thermal

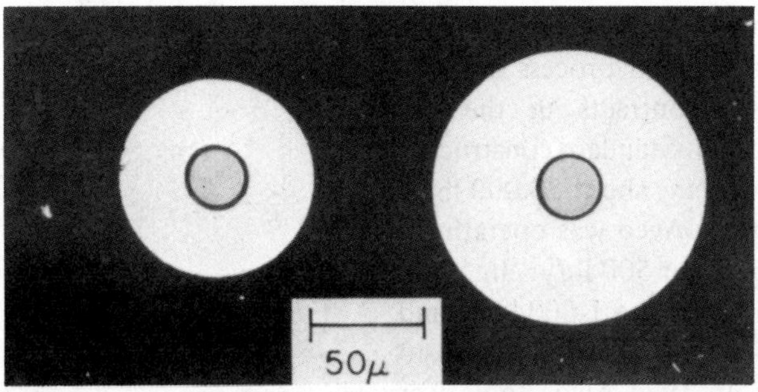

Fig. 28-11. Comparison of cross-sections of 4- and 5.6- mil boron-on-carbon.

expansion, etc., have in the past played not too important a part in applications and product development. The hardness of boron filament is about 2800 kg/mm^2, and the coefficient of thermal expansion is about 3×10^{-6}/°F. In the late 1960's, the hardness of the filament was utilized in the development of a honing tool by Barnes Drill, Inc, and currently, the hardness property of boron filament is receiving increasing investigation by a number of companies for applications in various types of drill bits and abrasives. Similarly, the combination of low thermal expansion, low thermal conductivity, and high stiffness is causing boron to receive increased development effort for future composite reinforcement structures for ultra-low temperature superconducting generator and magnet applications. Applications are explored more fully in Section 6.

4. CURRENT MANUFACTURERS OF BORON FILAMENT

There are two U. S. manufacturers of boron filament, namely, Avco and CTI, and their respective addresses follow:

Composite Technology, Inc.
6 Mill Street
Broad Brook, Conn. 06016
(203) 623-2611
Avco Systems Division
Lowell Industrial Park
Lowell, Mass. 01851
(617) 452-8961

The following brief history traces the chronology of both plants. Although during the mid 1960's many companies were initially funded by AFML to conduct boron research and development, Hamilton-Standard and Avco carried out the major boron filament process development and production contracts in the late 1960's. In 1968, Hamilton-Standard constructed a plant capable of producing about 20,000 lb/yr and at that same period, Avco was operating a pilot plant producing about 500 lb/yr. In 1969, Avco enlarged its pilot plant to 1,000 lb/yr and subsequently in 1970 constructed a facility to produce 9,000 lb/yr. In 1974 Alcoa took over the operation of the Hamilton-Standard facility with the formation of a wholly-owned subsidiary, Composite Materials Corporation (CMC). After two years of operation, Alcoa announced its intention to cease the manufacture of boron. Recently, Composite Technology, Inc. has taken over the plant.

The current total production of boron filament is about 45,000 lb/yr; Avco manufactures about 30,000 lb and CTI about 15,000 lb/yr. There are some differences in the reactor design and manufacturing philosophy which exist between the two companies; most notably, CTI uses reactors with two stages, whereas Avco uses the single stage reactor which was shown schematically in Fig. 28-1. Figure 28-12 is a schematic diagram of the two-stage reactor developed by researchers at United Aircraft Corporation and to which CTI has exclusive

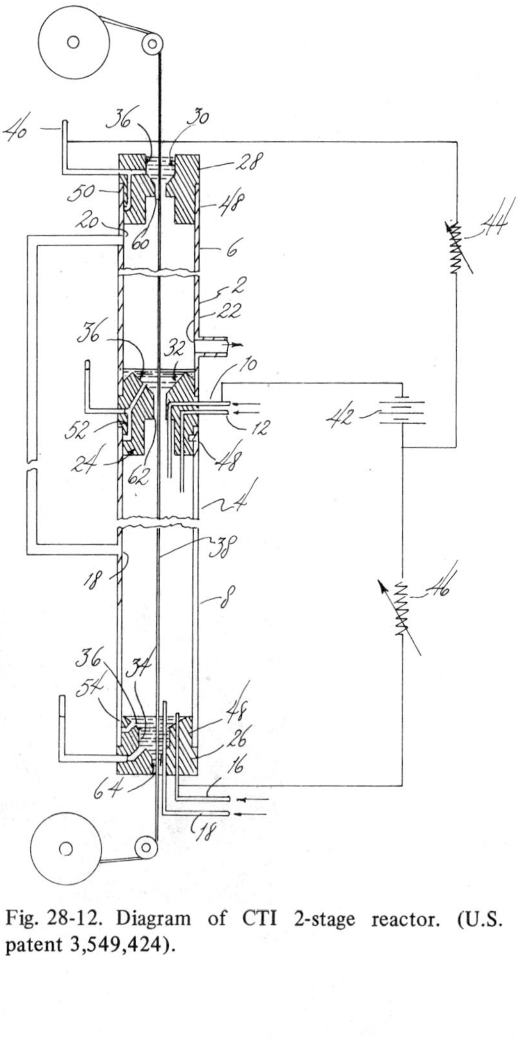

Fig. 28-12. Diagram of CTI 2-stage reactor. (U.S. patent 3,549,424).

license.[21,22] The two-stage reactor is comprised essentially of 2 reaction sections connected in series, each equipped with its own power supply. The second power supply heats the lower section of the filament independently, thus eliminating the drop in temperature shown in Fig. 28-1. The major advantage of the two-stage reactor is its higher throughput, e.g., the drawing rate of 4-mil boron in a two-stage reactor is about 17 ft/min, whereas a single stage reactor operates at only about 9 ft/min. Hence, the initial capital costs and the labor and depreciation cost contributions for a plant operating with two-stage reactors are lower than for a plant using single-stage reactors.

As a means of circumventing the drop in temperature and increasing deposition rate in their reactors, Avco developed augmented VHF power, i.e., VHF power selectively introduced at the larger diameter portion of the filament. Specifically, Avco produces its 8-mil filament in a VHF reactor because the production rate of 8-mil filament in a single stage DC reactor is economically prohibitive. The capital costs of VHF-augmented reactors, however, are much greater than those of a single- or two-stage reactor, but can be justified based on labor and throughput considerations.

For completeness here, only two other companies produce boron filament in small amounts, viz., Societe Nationale des Poudres et Explosifs (SNPE) in Paris, using two stage units[23] and another boron facility using two-stage reactors is in operation in the Soviet Union.[24]

5. COST AND AVAILABILITY

5.1 Historical Review

When boron filament was first produced at levels large enough to be called development quantities, the cost was as high as $5000/lb. By the mid-1960's, the price of boron filament had been lowered to about $300/lb because of improved process development, decreased raw materials costs, expanded volume, and learning curve benefits. In 1967 the tungsten substrate contributed the single largest fraction of the cost—almost $100/lb of 4-mil boron.

5.2 Current Boron Cost Situation

At the present time, boron filament sells for about $200/lb. The major product form is boron/epoxy tape and in large volume, viz., about 10,000 lb/yr, it sells for about $160/lb. (For a quick rule of thumb, here, boron/epoxy tape is, on a per pound basis, about ¾ the cost of boron filament.) For next-decade, large production volume, boron filament prices of below $100 are projected. Nearer term prices, in the 1978–1979 time frame, of about $150/lb are projected for filament based upon either larger diameter boron or the use of a carbon substrate. Concerted research and development effort being carried out by both Avco and CTI is directed to decreasing the manufacturing costs in order to remain reasonably competitive with other composite reinforcement materials such as graphite, mentioned earlier, and with the newer fiber forms, Kevlar[25] and FP.[26]

Although boron possesses outstanding mechanical properties not equalled by any other fiber, its prices must be weighed against the price and technical merits of other fibers in each application; at some (but not easily definable) price level, considerations of the fabrication cost advantages of boron coupled with its technical superiority may dictate the use of boron rather than other fiber forms even when the latter are available at a lower price. This trade off is a complex one and is different for each application. For example, in a given aircraft structural application, graphite might be preferable because of the particular cost-performance requirements. On the other hand, based upon all the data obtained to date, boron/aluminum has shown superior, repeatable state-of-the-art properties for applications in advanced turbojet engines, especially where there is a requirement for high temperature operation.

A simplified manufacturing cost breakdown for boron filament is given in Table 28-3.

The costs given above are based upon operating a new plant constructed in 1975 and producing 20,000 lb/yr. Even though tungsten has dropped almost a factor of two in price since 1968, the substrate still contributes over 25% of the current-day costs for 4-mil boron, and the rationale for producing larger diameter boron or for using a carbon substrate becomes

Table 28-3. Manufacturing Cost[a]
(of 4-mil boron filament)

Cost Element		Contribution $/lb	
Materials			
Tungsten substrate	$55		
Boron trichloride	30		
Hydrogen	12		
Other gases, chemicals	5		
		102	
Production labor		55	
Depreciation		50	
Total			$207

[a]Costs include all overhead and SG&A; the depreciation schedule is for a plant built in 1975.

quantitatively clear. For example, 1 lb of 4-mil boron is comprised of about 70,000 linear ft, while 1 lb of 5.6-mil boron is comprised of about 35,000 ft (ignoring second order density differences). Thus, the larger diameter filament contains only one-half the tungsten of 4-mil filament. Similar mathematics apply to 8-mil boron which contains only one-quarter the tungsten (and, of course, only one-quarter the tungsten cost). Based upon only the substrate savings, the manufacturing cost of 8-mil boron decreases to about $160/lb, but, in addition, other reductions in production, testing, and other handling labor accrue from the large diameter material. For example, in both the Avco process, which utilizes VHF power augmentation, and in the CTI 2-stage process, the production rate of 8-mil filament is higher than that of 4-mil boron, and, hence, depreciation and labor costs for larger diameter boron are further reduced relative to 4-mil filament. (Although, as stated earlier, capital costs for a VHF-augmented reactor are higher than for a single stage DC reactor, the throughput of 8-mil filament more than counteracts the additional amortization contribution.)

Use of carbon substrate for B/C filament represents the other major cost reduction potential. B/C filament development, which was first started in 1968, established that the manufacture was technically feasible.[17] The tensile strength of B/C was found to average 475 Ksi, similar to that of B/W; the modulus, however, averaged only about 52 million psi (compared to 58 million for B/W), but on a specific basis, i.e., on a modulus to density ratio basis, B/C was about equal to B/W. Currently, Avco is being funded on an AFML program to integrate the carbon substrate production and B/C manufacture in one facility.[27] Estimates based on the current pilot plant that is in operation at Avco indicate that carbon substrate will contribute about $25 to a pound of 4-mil boron compared to the $55/lb contribution of tungsten shown in Table 28-2.

Other technical and economic benefits derived from the use of carbon substrates have also been found in the work at Avco. For example, 4-mil B/C could be produced at a rate about 40% higher than B/W.[18] Thus, not only does the carbon substrate represent a $25/lb reduction, but the 40% higher production rate represents a potential commensurate decrease in labor and depreciation. On the other hand, at larger diameter boron filament sizes, the substrate cost differences between tungsten and carbon become less distinguishable, and CTI is directing its cost reduction emphasis to improve the process for producing large diameter filament, especially since the large diameter boron is being established as the leading filament for aluminum matrix applications. At the present state of development, boron is the only proven fiber for an aluminum matrix; furthermore, it is a common opinion of the boron industry that boron/aluminum applications will eventually be much larger than boron/epoxy.

5.3 Raw Materials Availability

The 3 basic materials for producing boron (B/W) are tungsten, boron trichloride, and hydrogen. No problem in future availability exists for any of the materials. However, for historical interest, in 1965 a problem had been envisaged in the future availability of tungsten because much of the tungsten ore originates from Asian countries. The supposition was later shown to be incorrect; for example, the total yearly U.S. production of tungsten products (from both U.S. and Asian ore) is about 16,000,000 lbs.[28] Much of the tungsten is converted to tungsten carbide forms, but as an emphatic example of availability, *if only one week's* production of tungsten were diverted to boron substrate, 4,000,000 lb (!) of (4-mil) boron could be manufactured therefrom.

Boron trichloride originates from borax and chlorine and, as far as boron filament manufacture is concerned, there is an almost-infinite supply of both. A total of about 500,000 lb of boron trichloride is used at both plants, whereas the yearly production of borax is about 2,000,000,000 lb[28] and chlorine about 18,000,000,000 lbs.[29] The total amount of hydrogen utilized in producing boron at both plants is miniscule compared to that which is flared at even 1 large U.S. oil refinery. About 1000 SCF of hydrogen is used to produce 1 lb of boron filament at the Avco plant;[18] the CTI plant is equipped with total hydrogen recycle, and hence, hydrogen requirements are almost an order of magnitude less.

6. CURRENT AND FUTURE APPLICATIONS OF BORON COMPOSITES

6.1 Military Aircraft and Aerospace

The primary use for boron composites is in the fabrication of weight and stiffness critical components of military aircraft, primarily for the F-14 and F-15. A schematic diagram of the F-14 Tomcat is given in Fig. 28-13, and the shaded portions of the aircraft indicate the locations of boron/epoxy on the horizontal stabilizers of the empennage structure. Further, as the data in Fig. 28-13 show, if the stabilizers had been constructed of titanium, they would have weighed about 950 lb, but with the use of the stronger, stiffer, and lighter boron composite material, the equivalent performance stabilizer weighs only 776 lb. In addition, the decrease of almost 200 lb on the tail structure provides enhanced performance and maneuverability for the aircraft. Figure 28-14 shows a schematic of the F-15 Eagle with the shaded portions pointing out the location of boron/epoxy on both the horizontal and vertical stabilizer, similar weight savings have been achieved on that aircraft. The total usage of boron filament for these 2 aircraft is 35,000 lb for 1976, and the future usage on these 2 aircraft will be about the same until the F-14 production schedule is finished.

The B-1 bomber, currently in the development stage, (as compared to *production* status of the F-14 and F-15) is employing boron composites in the fabrication of several components, e.g.,

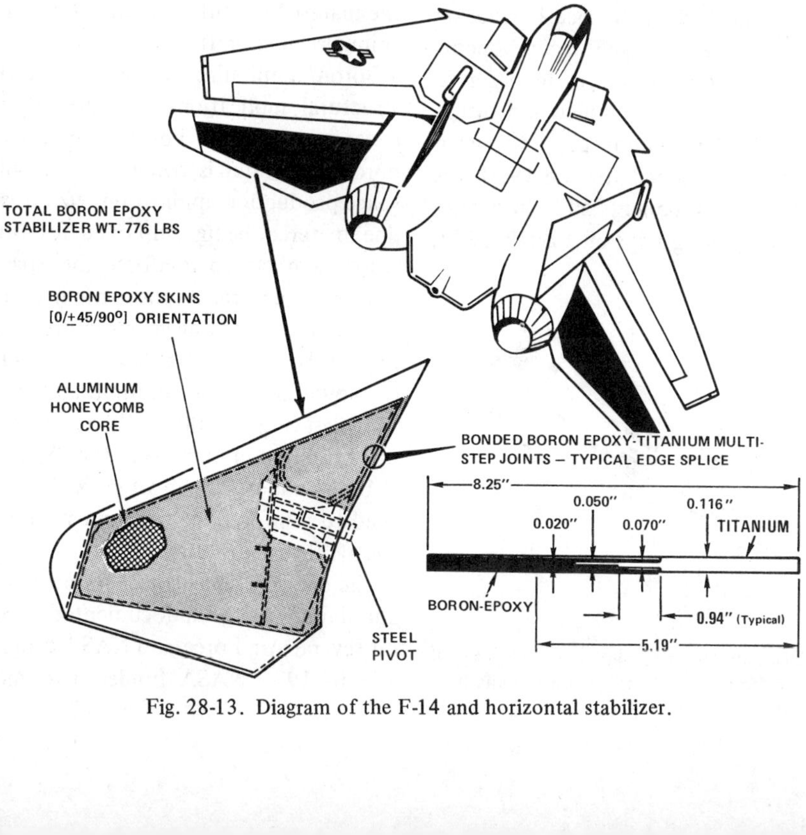

Fig. 28-13. Diagram of the F-14 and horizontal stabilizer.

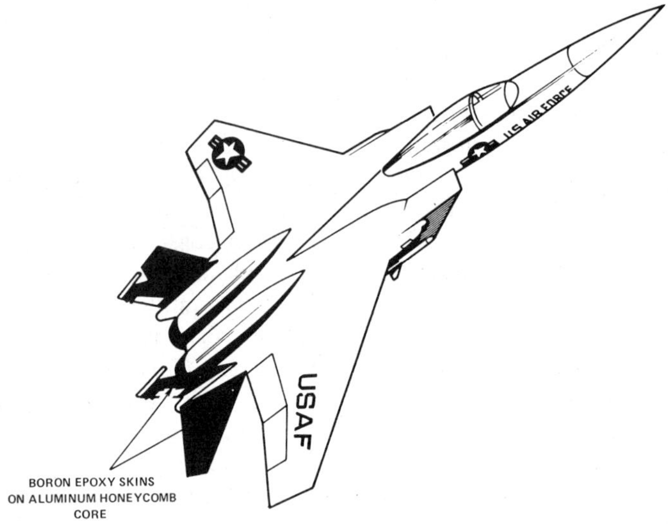

Fig. 28-14. Diagram of F-15, showing boron/epoxy horizontal and vertical stabilizers.

for the reinforcement of the main load-carrying longerons and on the horizontal stabilizers. A diagram of the aircraft and the location of the potential boron components is given in Fig. 28-15.

In addition to the large scale usage on the F-14 and F-15, other aircraft were retro-fitted with boron composite structures during the late 1960's and early 1970's, NASA/Langley Research Center funded Flight Service Evaluation programs to study the use of boron components on a variety of aircraft structures in order to evaluate the long-term performance and future applicability of boron composites to advanced commercial aircraft planned for the 1980's. A partial list of such components and aircraft include boron/epoxy wing flaps on a Boeing 707,

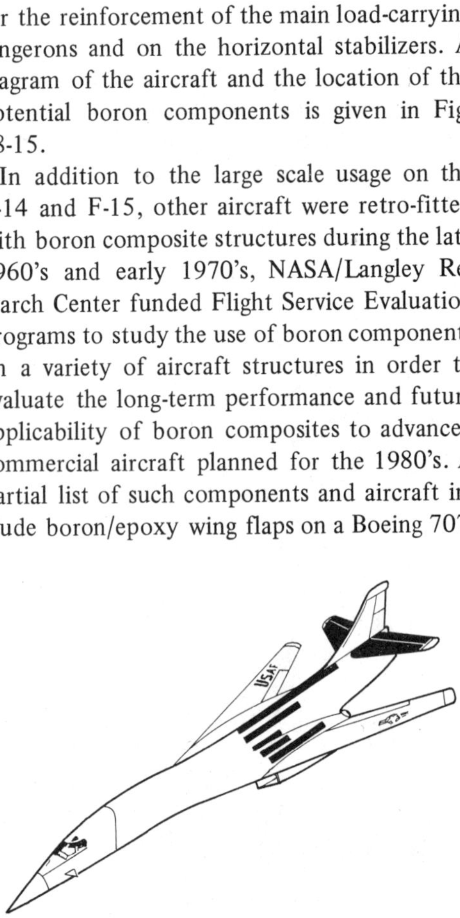

Fig. 28-15. Diagram of B-1, showing locations of boron/epoxy components currently in development.

wing slats on 11 C-5A aircraft, a boron reinforced wing box structure on C-130 transports, and a tail structure on the CH-54 helicopter. Boron was verified as a primary structural material by the Air Force, but NASA is taking a more conservative approach and is investigating the long-term service life of boron components in less critical applications on both military and current commercial aircraft in order to evaluate the material for future applications on advanced commercial aircraft.

Boron/aluminum is being investigated in structural applications, but currently to a much lesser extent than boron/epoxy. At present, boron/aluminum is being used in only 1 aerospace production application, viz., it is the baseline material being employed for tube-shaped truss members to reinforce the Space Shuttle Orbiter structure. A photograph of a partially completed mid-fuselage structure is shown in Fig. 28-16. Over 200 tubes from 2–6 ft long are employed in its construction, and the resultant weight savings is 300 lb. (In addition to the use of boron/aluminum on the Space Shuttle Orbiter, about 200 lb of boron/epoxy will be used to reinforce the titanium strut members in the aft-thrust structure.)

The superior mechanical properties of boron/aluminum have been documented in other work on several Air Force and NASA programs,[30,31] and in 1975 NASA funded two more Flight

Fig. 28-16. Photograph of partially completed Space Shuttle Orbiter, showing boron/aluminum tubes in mid-fuselage structure. (*Courtesy General Dynamics*)

Service Evaluation programs to investigate the long-term performance of boron/aluminum composites.

The use of boron composites in the production of commercial aircraft has not yet been defined, nor will it be defined for several years, but NASA/Langley has, in 1976, undertaken large programs at several of the prime aircraft manufacturers to investigate composite structures for their cost effective considerations. Both NASA and the Air Force are funding programs to evaluate "hybrid" structures, i.e., components fabricated from a combination of composite materials, such as fiberglass and graphite, or graphite and boron. One important hybrid structure currently in development is the horizontal stabilizer of the B-1 in which the more expensive boron/epoxy is used to selectively reinforce the stabilizer at critical points with graphite used for the bulk of the component. Such synergistic hybrid structures, wherein cost and performance factors of both materials can be optimized, represent a large potential market for boron.

6.2 Sporting Goods and Industrial Markets

In 1973, boron penetrated the sporting goods market, and the first large application was the reinforcement of graphite golf shafts. Over 3,000 lb of filament reached this segment in both 1974 and 1975. This hybrid structure, i.e., the combination of boron and graphite similar to its counterpart in aircraft components, utilizes boron for selective reinforcement of the highly stressed lower region of the graphite shaft.

In 1974 and 1975, tennis rackets became the

predominate market for boron/aluminum. In 1975 over 1,000 lb of the composite was produced for just one racket, the AMF/Head, Arthur Ashe Competition 2, and this segment of the boron/aluminum market was a factor of two greater than the rest of the military and aerospace market combined.

Other tennis rackets and cross-country skis are being developed to take advantage of boron's unique strength and stiffness. Again, the emphasis in expanding the markets for boron in both epoxy and aluminum matrices is directed to selective reinforcement applications in which only 1-2 oz of the expensive boron material can be used to provide requisite stiffness or strength. For example, the AMF/Head racket mentioned earlier utilizes a few tenths of an ounce of boron/aluminum composite in the throat section to enhance its performance.

Current and future industrial uses for boron filament will use boron's hardness properties for applications in tool bits and drills and its high modulus and low thermal expansion properties for applications in ultra-centrifuges, superconducting generators, and other high speed, high stress rotating machinery. The Cryogenics Division of the National Bureau of Standards is evaluating the properties of boron at liquid helium temperatures (4°K) for important potential applications in superconducting generators and magnets.[32] As an example of the market for advanced composites in such systems, one superconducting generator system which will be operational in 1985 with ERDA funding and which was discussed at a recent composites meeting in Washington, weighs *11 million lbs* as now designed with steel, but the weight and size could be reduced to less than half if it were selectively constructed of advanced composites.[33] In these applications, the use of boron can again not only be technically superior but cost effective because heat loss and cryogenic cooling requirements can be reduced.

The abrasive properties of boron are also receiving increasing attention. Barnes Drill Company has already developed a honing tool using 5.6-mil boron, and currently, General Dynamics Corporation, Borad, Inc., and Skyline Industries, are developing boron filament core drills and similar tools. In this application, boron is competing with more costly diamonds, which also have a higher hardness. The shortcoming of diamonds is the geometric limitation of approximately equiaxial particle shapes that have a tendency to "pull out" during use, as the matrix is worn down. However, the boron filament form overcomes this limitation and has greater resistance to "pull out." Again, raw materials cost, fabrication, and service life will dictate the choice and design of a component, although in this rare instance boron is *less* expensive than the material it can replace.

6.3 Closing Remarks

When boron filament's high strength and stiffness properties were first demonstrated, it was heralded as the "greatest single material advance in 3000 years."[34] Applications in high performance, and initially high cost, aircraft structures pointed out the technical superiority of boron, but learning curves also showed that fabrication cost benefits could be derived by using the newer concepts of composite component "build up" rather than metal component "machine down". Continued design and fabrication development in the 1970's showed that composite structures of boron could even be of lower cost than conventional metal components.

The early success of boron in many of the aircraft applications motivated some segments of the Air Force to predict that "all-boron" aircraft would be built utilizing millions of pounds of filament per year. These million pound applications have not developed, especially in light of the availability of some of the less expensive fiber forms which can be utilized in conjunction with boron to produce more cost-effective structures. The first production application, the F-14 Tomcat, utilized 6,000 lb/yr in 1969. Based upon current development effort, and based upon designs on the drawing board, a boron filament requirement of 100,000 lb/yr can easily be reached in the 1980's.

For future new designs or applications, the cost of boron in the total life-cycle, i.e., the raw material, component fabrication, and system operation steps, must be weighed carefully against the more common metal structures, because, again, the high cost of boron can frequently be outweighed by savings in fabrication and operating costs. As the energy and materials situation becomes more acute, consideration of total life cycle costs will play a more important part in component design and fabrication, and initial acquisition cost considerations will become less influential. These factors will increase boron's usage significantly.

References

1. Weintraub, E., *J. Ind. Eng. Chem.* **3**: 299–301 (1911).
2. Talley, C. P., *J. Appl. Phys.* **30**: 1114–1115 (1959).
3. Chretien, A. and Helgorsky, J., *Compt. Rendu* **252**: 742–744 (1961).
4. Galasso, F. S. and Paton, A., *Trans. AIME* **236**: 1751–1752 (1966).
5. Wawner, F. E., Jr., "Boron Filaments," in *Modern Composite Materials* (eds. L. Broutman and R. Krock), Addison-Wesley, Reading, 1967, pp. 244–269.
6. Reeves, R. B. and Gebhardt, J. J., Preparation of Boron Filament on Fused Silica Substrate by Decomposition of Diborane, 10th SAMPE Symposium, San Diego, Nov. 9–11, 1966.
7. Wawner, F. E., Jr. and Satterfield, D. B., "Strength, Uniformity and Fracture Characteristics in Boron Filaments," in *Boron Filament Process Development,* AFML-TR-67-120, Vol 1A, Appendix C, Texaco Experiment Inc. Richmond, Va. 1967.
8. Kimpel, R. F. and Moss, R. G., Research on Continuous, High Quality Filament by the Melt Process AFML-TR-66-185, Aerojet-General, July 1966.
9. Kloepper, D. and Witt, R., Boron Filament Manufacture in Space–A Literature Feasibility Study, Proceedings of Space Processing Manufacturing, Oct. 21–22, 1969, Marshall Space Flight Center, Huntsville, Ala.
10. Monroe, S. F. and Chitwood, B. E., Structural Laminate Bag Molding Process, in *Handbook of Fiberglass and Advanced Plastics Composites*, pp. 306–334 (G. Lubin, ed.), Van Nostrand Reinhold, New York, 1969.
11. Sampson, R. N., Laminates, *Reinforced Plastics and Composites* in *Handbook of Plastics and Elastomers*, pp. 5-1–5-111 (C. A. Harper, ed.), McGraw-Hill, New York, 1975.
12. Line, L. E., Jr. and Henderson, U. V., "Boron Filament and Other Reinforcements Produced by Chemical Vapor Plating," in *Handbook of Fiberglass and Advanced Plastics Composites* (ed. G. Lubin) New York-London, 1969, pp. 201–236.
13. Layden, G. K., *J. Mat. Sci.* 8: 1581–1589 (1973).
14. Adler, R. P. and Hammond, M. L., *Appl. Phys. Letters* **14**: 354–356 (1969).
15. Kreider, K. G., Dardi, L., and Prewo, K., *Metal Matrix Composite Technology,* AFML-TR-71-204, United Aircraft Corp., East Hartford, Conn., 1971.
16. DeBolt, H. and Krukonis, V., *Low Cost Boron Filament,* AFML-TR-75-29, Air Force Materials Laboratory, Wright-Patterson Air Force Base, Ohio, 1975.
17. DeBolt, H. E., Diefendorf, R. J., Gruber, P. E., Joo, L. A., Krukonis, V. J., McKee, J. A., and Mehalso, R. M., *Lower Cost High Strength Boron Filament,* AFML-TR-70-287, Avco Corp., Lowell, Mass., 1971.
18. DeBolt, H. E., Krukonis, V. J., McKee, J. A., Prescott, R., and Wawner, F. E., *Development and Demonstration of a Low Cost Boron Filament Formation Process,* AFML-TR-72-271, Avco Corp., Lowell, Mass., 1972.
19. McKee, J. A., DeBolt, H. E., Krukonis, V. J., Prescott, R., and Sharpe, G. E., *Production Process for Large Diameter Carbon Filament Substrate,* AFML-TR-74-141, Great Lakes Research Corp., Elizabethton, Tenn., 1974.
20. Melnick, P. and Toth, I., Development of Boron (on Carbon)–Aluminum Composites, AMMRC-CTR-73-48, TRW, Inc. Cleveland, Ohio, Dec. 1973.
21. U. S. Patent 3,549,424, Method for Producing Filamentary Boron, C. Rice, United Aircraft Corp.
22. U. S. Patent 3,574,649, Process for Producing Continuous Boron Filament, R. Fanti, et al, United Aircraft Corp.
23. Morin, D., Societe Nationale des Poudres et Explosifs (Paris), private communication.
24. Lutsau, V. G., Mechanical Engineering Research Institute (Moscow), private communication.
25. Kevlar, DuPont's Aramide Fiber.
26. FP, DuPont's polycrystalline alumina fiber.
27. DeBolt, H., Krukonis, V., Prescott, R., McKee, J. A., *Carbon Monofilament Production,* IR-424-5(I), Avco Systems Division, Lowell, Mass., 1975.
28. Commodity Data Summaries 1976, Bureau of Mines, United States Department of the Interior.
29. *Chemical & Eng. News*, p. 35 (May 3, 1976).
30. Dolowy, J. F., Jr., Diffusion Bonded Boron Aluminum Sheet & Plate, AFML-TR-73-30 DWA Composite Specialties, Inc., Chatsworth, Cal., Mar. 1973.
31. Dolowy, J. F., Jr., Evaluation of Forming and Manufacturing Characteristics of Boron Aluminum Materials, NAS8-29081, DWA Composite Specialties, Inc. Chatsworth, Cal., Mar. 1973.
32. Kasen, M. B., Mechanical and Thermal Properties of Filamentary-Reinforced Structural Composites at Cryogenic Temperatures, 2. Advanced Composites, Cryogenics **15**, p. 701 (1975).
33. Zuirys, E., 2nd Workshop on Metal Matrix Composites, DARPA, Task T-115. Washington, D.C., Oct. 1975.
34. Schriever, B. A., "A Look at the Future," Presented at 26th Conference of Aviation Writers' Association, Bal Harbour, May 1964.

29

Carbon-Graphite Filaments

Harry S. Katz
Utility Research Co.
Montclair, New Jersey

CONTENTS

1. INTRODUCTION

The high-modulus carbon-graphite filaments have the potential of dominating the composite field in the distant future. Their eventual low cost and superior modulus-to-weight and strength-to-weight ratios will lead to their usage as substitutes for fiberglass composites and metals. There are a large number of different types of products within this class of materials that range from fiber mats, with low strengths and moduli to continuous filaments with high tensile strengths and moduli. The current high modulus filaments range up to a modulus of 75 million psi, and have tensile strengths of about 300 Ksi. The high strength types have a tensile strength of about 400 Ksi and moduli that range between 30 to about 50 million psi. The fiber mats have a tensile strength of about 15 Ksi and modulus of 5 million psi. The density of these products usually ranges between 1.6-1.9 g/cm^3. Prices of high performance filaments dropped rapidly from the initial level of \$350/lb in the early 1960's to a current level of about \$20 for a general purpose grade of continuous filament. There are predictions that a quality filament derived from a pitch precursor will be available in the near future at less than \$10/lb. The lower priced mat materials have been selling at \$8/lb and have been used because of their unique properties, such as high electrical conductivity and low thermal expansion.

Among the outstanding properties of carbon-graphite filaments are high modulus and strength, low density, high electrical conductivity, low thermal coefficient of expansion, low coefficient of friction, and excellent resistance to most environmental exposure conditions and chemicals. The combination of desirable properties of these materials will lead to their widespread use in most industries.

Although high performance carbon-graphite filaments have been proven cost effective in many applications, their relatively high current cost must be considered for any potential application. Another possible disadvantage is the fact that this is a new material, and many materials engineers or designers may not have the experience to take full advantage of its potential. Due consideration must be given to the low interlaminar shear strength of many types of filaments. Also, the differential thermal coefficients of expansion of the resin and the filament may cause crazing problems in some end products, especially those molded at very high temperatures or exposed to both high temperature and high humidity conditions. The impact strength of the composite may be low. These and other problems that can be considered birth pains of a revolutionary new material, will undoubtedly be resolved by technical progress in carbon-graphite filament composites.

In many applications, carbon-graphite filaments may be combined with another filament, such as Kevlar 49, glass, or boron in order to form a structure that takes advantage of the best features of both types of filaments. These hybrid structures have been utilized in aerospace components and sporting goods.

2. PRODUCTION METHODS

Carbon filaments have been produced for more than 90 yrs. In 1879, Thomas Alva Edison produced the first practical, electric, incandescent lamp, which had a filament of carbonized sewing thread in an evacuated glass globe. The original lamp element consisted of a relatively fragile carbon filament, 1 mm in diameter. These lamps were marketed in 1880. U. S. Patents 223,898 and 230,309 describe the production of carbon filaments in lamp manufacturing. At about the same time, an English scientist, Sir Joseph Wilson Swan, produced lamp filaments by carbonizing cotton threads that had been parchmentized in sulfuric acid. In 1883, Swan was granted a British patent for a process to produce fine cellulose filaments, which he had developed as a means toward smaller diameter carbon filaments. After about 1910, the lamp industry turned to the use of tungsten filaments, and the production of carbon filaments for this application was terminated.

The renewed interest in carbon-graphite filaments started in the late 1950's. The advent of jet engines, proposed faster and larger aircraft, and other requirements of advanced technology led to an intensive search for improved structural materials. Laboratory tests of graphite whiskers showed values of 3 million psi tensile strength and 140 million psi tensile modulus.[1] This is a strength-to-weight ratio, specific strength, over 30 times greater and specific modulus over 17 times greater than the structural metals, as shown in Table 29-1.

Research aimed toward the production of continuous graphite filaments was initiated by the Air Force Materials Laboratory. In a parallel effort during 1957, Dr. A. Shindo of the Government Industrial Research Institute, Osaka, Japan, developed a process for manufacturing carbon-graphite filaments from polyacrylonitrile (PAN) filaments. Two patents, Showa 37-4405 (1962) and Showa 38-12375 (1963), evolved from this research and were licensed to Tokai Electrode Co., Nippon Carbon Co., and Nitto Spinning Co. Also, during the 1960's, patents were granted to other groups for methods to produce carbon-graphite filaments. British Patent 1,034,542, granted to Union Carbide in 1966, described a process for continuously graphitizing a textile thread, and discussed the method of strengthening the yarn by stretching at a temperature of at least 2200°C. Rayon was the preferred precursor, but PAN was also mentioned. British Patent 1,110,791, granted to the National Research Development Corporation (NRDC) in 1968, stressed the use of PAN precursor and cited values of 260,000 psi tensile strength and 60 million psi Young's

Table 29-1. Properties of graphite whiskers vs structural metals.

	Density lb/in.^{3}	Modulus psi $\times 10^6$	M/D in. $\times 10^6$	Tensile Strength psi $\times 10^3$	TS/D in. $\times 10^6$
Graphite whiskers	0.081	142	1750	3000	37
Titanium 6A1-4V	0.16	15	94	155	.97
Aluminum 7075	0.10	10	103	82	.81
Maraging steel	0.26	27	93	319	1.1

Table 29-2. Selected carbon-graphite filament patents.

United States	Japan
223,898	Showa 37-4405
230,309	38-12375
2,957,956	38-20609
3,011,951	39-16681
3,011,981	41-15728
3,053,775	43-893
3,107,152	43-4550
3,116,975	44-21175
3,235,323	
3,285,696	United Kingdom
	894,458
3,305,315	1,016,351
3,313,596	1,034,542
3,313,597	1,110,791
3,378,345	
3,392,216	Belgium
3,399,252	678,679
3,404,061	
3,412,062	France
3,443,899	1,430,803
3,449,077	
3,454,362	
3,461,082	
3,462,289	
3,462,340	
3,503,708	
4,005,183	

modulus. Claim 1 of this invention is: "Carbon fiber made by the conversion of organic polymer fiber, and which has been given a high ultimate tensile strength and a high value of Young's modulus by having been submitted at some stage of the conversion to the combined effect of heat and longitudinal tension." There have been a large number of licensing and cross-licensing arrangements. A list of pertinent patents is given in Table 29-2.

The basic elements for producing carbon-graphite filaments from rayon are shown in Fig. 29-1 (U. S. Patents 3,305,315 and 3,235,323).

The rayon may be subjected to a chemical treatment before the first stage exposure in the oxidizing furnace at about 500°F. The chemical bath can be an aqueous ammonium chloride solution or a dilute solution of phosphoric acid in denatured ethanol, and serves to reduce the time for the low-temperature step from hours to about 5 min. Note that the rayon-precursor process requires that the filament be stretched after the intermediate carbonizing step.

In the PAN-precursor process, which was the method stressed by the British manufacturers, it was found preferable to restrain or stretch the filaments during the low-temperature oxidation step in order to improve the modulus.

Table 29-3, from British Patent 1,110,791, shows the effect of the initial stretch and the graphitizing treatment on the final modulus of the PAN-precursor filament.

The oxidation of the PAN filaments in air is a diffusion-controlled process. During this process, the filaments add about 8% by weight of oxygen, which may crosslink the parallel oriented chains of PAN and prevent their relaxation or contraction during the later, high-temperature exposures.

The effect of flaws on the strength of carbon filaments produced from acrylic precursors has been reported by D. J. Thorne, Rolls-Royce Limited. The tensile fracture of a carbon filament occurs almost exclusively at a detectable flaw, which may be at the surface or internal. Surface flaws, which produce the greatest reduction in observed filament strength, may be removed with controlled surface etching by ion bombardment, wet chemical, or gaseous oxidation etching.

A study that supports the theory that the

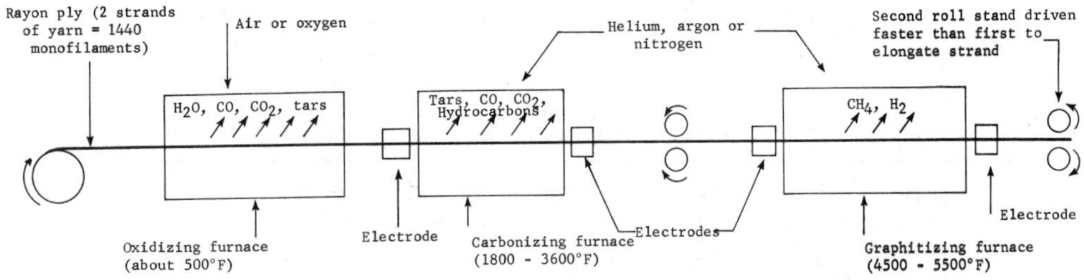

Fig. 29-1. Production line for carbon-graphite filaments from rayon.

Table 29-3. Factors affecting PAN-precursor graphite filament modulus.

Load Applied to a Yarn of 100 Filaments of 2½ Denier Polyacrylonitrile Fibers, for 24 hrs at 220°C for Initial Treatment by Complete Oxygen Permeation	Length Change During 220°C Treatment	PROPERTIES OF THE FIBER AFTER SUBSEQUENT CARBONIZING TO 1000°C IN AN INERT ATMOSPHERE		PROPERTIES OF THE CARBON FIBER AFTER HEAT TREATING TO 2500°C IN AN INERT ATMOSPHERE	
		Tensile Strength psi	Youngs Modulus Axially of Fiber psi	Tensile Strength psi	Youngs Modulus Axially of Fiber psi
Nil	−40%	100×10^3	13×10^8	80×10^3	30×10^8
10 g	−12%	100×10^3	16×10^8	100×10^3	38×10^8
20	+2%	120×10^3	20×10^8	120×10^3	47×10^8
30	+15%	200×10^3	21×10^8	200×10^3	53×10^8
40	+36%	200×10^3	21×10^8	200×10^3	60×10^8

practical tensile strength is determined by flaws, involved the treatment of commercial carbon fibers, which had been made from a PAN precursor, with liquid bromine or bromine solutions in nitromethane.[2] Upon comparison of the original treated fibers, definite changes were evident in the macrostructure of the treated fibers. Some of the treated fibers showed tensile strength improvements up to 50%.

Internal flaws often originate in the organic precursor, which can contain voids and contaminating inorganic and organic particles. Because of the high degree of stretching of the polymer during carbon filament production, the internal defects are all elongated. The presence of these flaws results in a great decrease in fiber breaking strength with increasing length. The closed micropore volume of the carbon filament can be as high as 15%, while that of the original PAN is less than 1%. If internally flaw-free carbon filaments could be prepared with a flaw-free surface, strengths of 600,000–800,000 psi and breaking strains of 2–3% at usual lengths could be anticipated. An ultimate "ideal tensile strength of 26×10^6 psi for carbon fiber" as an optimistic value has been discussed in theoretical terms.[2] This upper limit appears unattainable in light of the more usual assumption that a level of strength may be attained that is about one-tenth that of the modulus. Since graphite whiskers have a modulus of 140-million psi, it would seem justifiable to assume that the optimistic upper limit for the strength of graphite filaments is 14-million psi.

There are a large number of precursors that have been considered for the production of carbon-graphite filaments. The list would include cotton, PAN, rayon, polyvinylidene chloride (saran), aromatic polyamide, polybenzamidazole, polyoxadiazole, polyphenylene, lignin, crosslinked polyethylene, and various pitches from PVC, petroleum, or coal tar. At the present time, commercial high modulus graphite filament is produced primarily from PAN precursor, and rayon-derived material has a minor share of the market. Low modulus carbon fibers from a low cost pitch precursor have been produced in Japan for a number of years. Recently, Union Carbide Corporation has been marketing a high modulus mat and continuous filaments that have been made from a pitch precursor.[3] Figure 29-2 is a photograph of these filaments.

The cross-sectional shape of graphite filaments from rayon precursor is irregular or crenulated, as shown in Fig. 29-3, in contrast to the circular cross-section of the pitch or PAN precursor materials, as shown in Fig. 29-4. The latter filaments are about 250–1000 Angstroms in diameter, and the turbostratic graphite crystallites are about 50 Angstroms in size.

A serious problem that was noted with early high modulus graphite filament composites

Fig. 29-2. Pitch precursor carbon filaments–240X. (*Courtesy Union Carbide Corp.*)

was extremely poor interlaminar shear strength (ILS), in the order of 2000-4000 psi. Surface treatments for the filaments were developed in order to improve this important property to values in the order of 10,000 psi, and these usually involved surface oxidation or halogenation.

3. PROPERTIES OF CARBON-GRAPHITE FILAMENTS

3.1 Chemical Properties

Carbon-graphite filaments are relatively inert to most chemicals and resistant to corrosion under ambient conditions. The surface is subject to attack by strong oxidizing agents and halogens, especially at elevated temperatures. This has been a means for surface treatment of the high performance filaments in order to improve the interlaminar shear strength, which was a severe problem with early graphite reinforced composites. In general, carbon and graphite are resistant to common alkaline solutions at all concentrations and temperatures, and to the aqueous solutions of most inorganic salts up to their boiling point. Immersion of carbon filaments in the following reagents showed no significant loss in tensile strength or modulus at 50°C: 50% v/v hydrochloric acid, sulfuric acid and nitric acid; 50% w/v sodium hydroxide solution; 50% w/w hy-

Fig. 29-3. Cross section of rayon based graphite fiber–2000X.

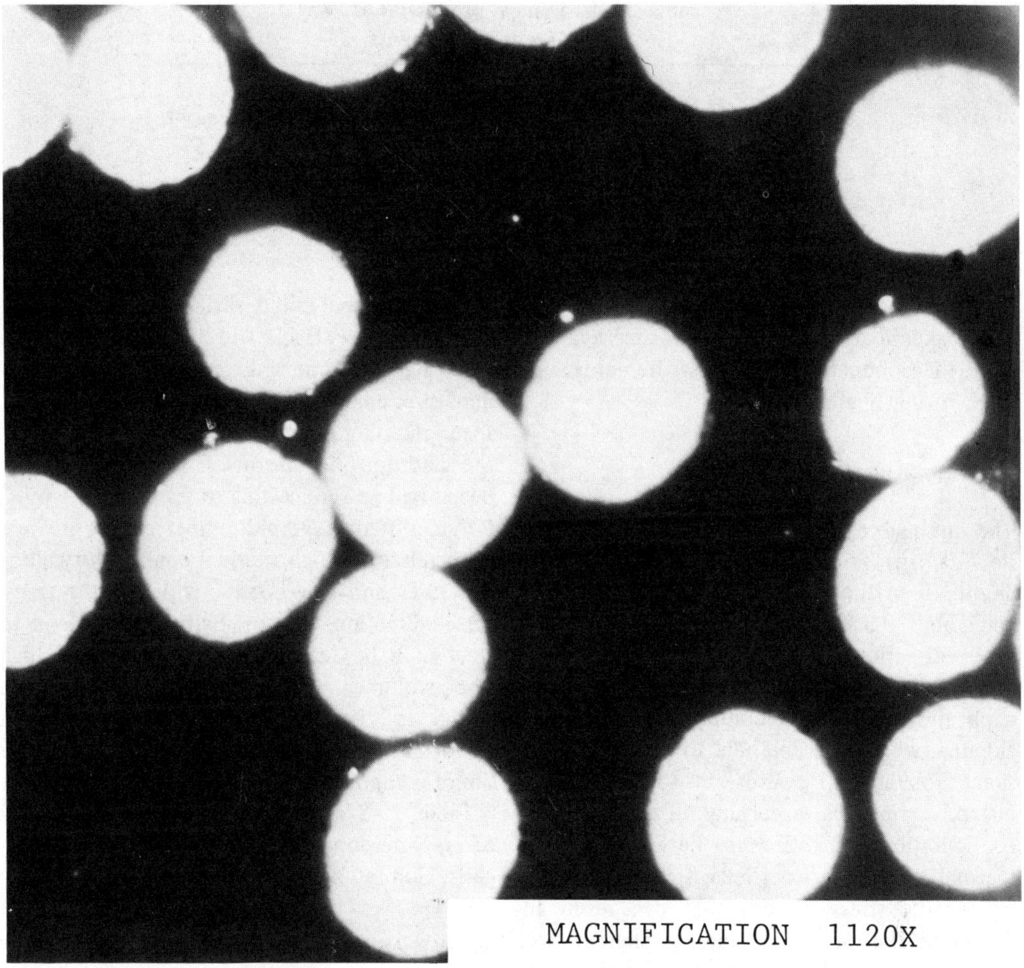

Fig. 29-4. Cross section of pan based graphite fiber—1120×.

pophosphorous acid; and orthophosphoric acid. At 20°C, the fibers are resistant to glacial acetic acid, 90% w/w formic acid, 32% w/v benzene sulfonic acid, and 0.880 ammonium hydroxide. The fibers react with most molten metals.

3.2 Physical Properties

There are many types of carbon-graphite filaments that are commercially available. The main types of high performance materials derived from PAN precursor are usually divided into three classes, designated as high modulus, high strength, and the Type A or Type III filaments. Typical properties of these materials are shown in Table 29-4.

The filament diameter is usually about 7.5-8 μ.

3.3 Electrical Properties

A unique property of carbon-graphite filaments is their relatively high electrical conductivity. Union Carbide's pitch precursor "Thornel" Mat Grade VMA filaments have a diameter of 9 μ, and electrical resistivity of 12×10^{-4} ohm-cm. The mat, which is 0.4 in. thick, has an electrical resistivity of 7000×10^{-4} ohm-cm and its thermal conductivity is 0.24 BTU in./hr/ft^2 °F. Because of its low electrical resistivity, the mat can be used as a surface veil in a glass fiber sheet molding compound where it provides a conductive surface that can be electrostatically

Table 29-4. Typical properties of PAN carbon-graphite filaments.

Fiber Type	Density lb/cu in.	Tensile Strength × 10^3 psi	Tensile Modulus × 10^6 psi
High modulus	0.0700	250–325	50–60
High tensile	0.0635	350–450	35–42
A or III	0.0628	275–325	28–35

painted. The filamentary form of the product also makes it applicable for resistance heating, since the conductivity is almost the equivalent of Nichrome heating wire.

3.4 Thermal Properties

The thermal expansion coefficient of graphite filament is approximately -0.55×10^{-6} in./in./°F in the direction of the filament axis and about 9.3×10^{-6} in./in./°F in the transverse direction. The low, slightly negative longitudinal expansion can be an advantage in applications such as a composite for a space antenna, which must be held to a close dimensional tolerance. However, in some polymer matrix composites, especially in the cross-ply configuration, the difference between the two thermal expansion coefficients can cause an undesirable stress pattern that can result in microcracking of the laminate. Thermal expansion characteristics must also be considered when designing the method for attachment of a high modulus graphite filament composite to a metal structure.

Because of the low thermal expansion of carbon-graphite filaments, they may be used to produce zero mold shrinkage. This permits tight tolerance thick and thin sections to be fabricated without sinks or warpage. The standard sink characteristic of an automotive grade, sheet molding compound is about 0.5 mil, but the addition of Thornel P carbon fibers has been used to reduce this to less than 0.1 mil.[4] Polymer matrix graphite fiber composites exhibit relatively high thermal conductivity along the fiber and low conductivities in the transverse direction. Conductivities equivalent to steel may be reached along the fiber if high fiber volumes are used in the matrix. The ability to dissipate heat is thought to be an important characteristic in increased fatigue life of components such as compressor blades.

Table 29-5 shows the thermal conductivity of graphite composites as compared with fiberglass, aluminum, and steel.

Table 29-5. Thermal conductivity.

Type Material	CONDUCTIVITY (Btu-ft/hr-ft^2-°F) 0°	0°, ±45, 90°
Graphite AS	6–10	2–3
Graphite HTS	12–20	3–6
Graphite HMS	28–35	6–12
Fiberglass	2	0.2
Aluminum	80–125	
Steel	9–27	

Courtesy: *Hercules, Inc.*

3.5 Miscellaneous Properties

Carbon-graphite filaments exhibit excellent resistance to creep. The creep rupture resistance of several materials is compared in Table 29-6. The HTS graphite filament-epoxy 0°, 90° laminates have longer creep rupture times at 250°F than those of 7075-T6 aluminum at room temperature and at the same stress level.

The impact resistance of carbon-graphite filament composites is generally considered low. This factor should be given consideration in the design of any component that requires high impact strength. A comparison of the impact characteristics of graphite composites with those of metals is shown in Table 29-7.

Carbon-graphite filament composites have vibration damping that is superior to aluminum and steel. They have low coefficients of friction, and excellent dry lubrication characteristics. This property, in combination with the

Table 29-6. Creep rupture resistance of composites.

Material	Orientation (Degrees)	Tensile Strength (ksi)	Creep[a] Load (%)	Creep Rupture (hr)
S-Glass	0	260	85	0.01
			60	60
			51	10^4
Kevlar-49	0	200	80	2
			70	150
			60	10^4
AS graphite	0	200	80	>1000[b]
HTS graphite @ 250°F	0/90	83	90	455
7075-T6 aluminum		70	96	100
		70	93	350

[a]Percent of static ultimate.
[b]No failure in 1000 hr.

Table 29-7. Damage tolerance levels, notched charpy impact.

Material	Impact Energy (ft-lb), Onset of Plastic Deformation	Total Impact Energy (ft-lb)
2024-T6 aluminum	0.05	65
7075-T6 aluminum	0.60	3.0
17-7 PH stainless steel	0.75	67.0
4130 steel (R_C20)	1.60	102.0
6AL-4V titanium	2.10	10.5
4340 steel (R_C55)	3.40	6.0
3501/AS Hercules composite	3.40	6.0

Courtesy: *Hercules, Inc.*

Table 29-8. Coefficient of friction of various materials.

Material (No Lubrication)	μ
Steel on HTS epoxy composite	0.15–0.18
Steel on UHM[a] epoxy composite	0.13–0.16
Steel on steel	0.50
Steel on nylon 6/6	0.25
Steel on nylon 6/6 (20% glass)	0.18
Steel on nylon 6/6 (20% graphite)	0.10
Steel on steel (lubricated)	0.20

[a]75×10^6 psi modulus graphite fiber.
Courtesy: *Hercules, Inc.*

Table 29-9. Wear resistance of various materials.

Material (No Lubrication)	K[a] (10^{-7})
Copper on copper	300,000
Zinc on zinc	200,000
Copper on low carbon steel	10,000
Low carbon steel on copper	5,000
Steel on steel	5,000
Bakelite on bakelite	200
Graphite composite on steel	10
Steel on graphite composite	1
Metal on metal (lubricated)	1–10

[a]Wear coefficient $K = \dfrac{\Delta V(\sigma)}{L(P)}$.
Courtesy: *Hercules, Inc.*

high thermal conductivity that permits dissipation of frictional heat, leads to excellent, non-lubricated bearings that have low wear rates.

Table 29-8 shows the coefficient of friction of various materials. Note that the value for steel on graphite epoxy composite is even lower than lubricated steel on steel.

A study[5] of wear coefficients has shown that non-lubricated graphite composites can be as resistant to wear as lubricated metals, as shown in Table 29-9. Therefore, carbon-graphite filament composites should be outstanding for use as self-lubricating bearings and seals, and liners for journal bearings. The use of a high temperature polymer matrix, such as a polyimide, makes the composite suitable for high temperature bearings.[6]

4. MATERIALS SUPPLIERS, TYPES, AND CURRENT COSTS

There are six U. S. suppliers of carbon-graphite filaments, and their addresses and phone numbers are listed below.

1. Celanese Advanced Engineering Composites
   26 Main St.
   Chatam, N. J. 07928
   (201) 635-2600

2. Great Lakes Carbon Corp.
   299 Park Avenue
   New York, N. Y. 10017
   (212) 935-7405
   Also distributor for Prepregs manufactured by:
   Fothergill & Harvey, Ltd.
   Composite Materials Division
   Littleborough, Lancashire OL15 9QP, England

3. Hercules Inc.
   Industrial Systems Dept.
   Wilmington, Del. 19899
   (302) 575-5000
   Bacchus, Utah plant (801) 250-5911

4. Morganite Modmor Inc.
   401 N. Ashe Avenue
   P. O. Box 347
   Dunn, N. C. 28334
   (919) 892-8081

5. Stackpole Fibers Co.
   Foundry Industrial Park
   Lowell, Mass. 01852
   (617) 454-0409

6. Union Carbide Corp.
   Carbon Products Division
   270 Park Avenue
   New York, N. Y. 10017
   (212) 551-3407

There are suppliers of carbon-graphite filaments in Japan. Toray Industries, Inc. has a sales agreement with Union Carbide Corp. Mitsubishi International Corp. markets their line of Kureha Carbon Fiber. Their sales office is at 277 Park Avenue, New York, N. Y. 10017.

Toho Beslon Co., Ltd produces carbon-graphite fiber in Japan and has a sales and technology agreement with Celanese Corp.

The trend toward cost reductions of continuous filaments appears to be accelerating, since there have been recent announcements by Hercules that their new Magnamite AS-3 is available at $18/lb and by Union Carbide that their Type P grade is $20/lb.

Table 29-10 compares some properties of commercial carbon-graphite filaments, and indicates the cost range of these products.

Table 29-11 shows typical unidirectional properties of composites fabricated from some of the commercial filaments.

The preferred method for obtaining the highest performance graphite filament composites has involved the use of unidirectional tape that has been preimpregnated with the matrix resin. This use of "prepregs" enables better quality control of the resin and resin content. There are a number of companies that have been suppliers of these prepregs, and their names and addresses are listed below.

1. Fiberite Corp.
   501–559 W. Third Street
   Winona, Minn. 55987
   (507) 454-3611

2. Fothergill & Harvey, Ltd.
Lancashire, England
U. S. Representation by Great Lakes Carbon Corp.

3. Hercules Inc.

4. Hexcel Structural Products
11711 Dublin Blvd., Dublin, Calif. 94566
(415) 828-4200

5. 3M Industrial Specialties Div.
3M Center, St. Paul Minn. 55101

6. Narmco Materials Inc.
600 Victoria St., Costa Mesa, Calif. 92627
(714) 548-1144

7. Reliable Manufacturing Inc.
10910 Talbert Avenue
Fountain Valley, Calif. 92708
(714) 968-3361

Table 29-10. Properties of commercial carbon-graphite filaments.

Company	Trade Name, Property or Type	Tensile Strength (psi × 10^3)	Modulus of Elasticity (psi × 10^6)	Elongation at Break (%)	Filament Diameter (μ)	Density g/cc (lb/in.^{3})	Specific Tensile Strength (psi × 10^6 in.)	Specific Modulus (psi × 10^6 in.)
Celanese Corp.	Celion GY50 $/lb–65							
Celanese Corp.	Celion GY70 $/lb–75–110	250	77		8	1.96 (.071)	3.5	1085
Celanese Corp.	Celion 3000 $/lb–32–40	440	36	1.2	7	1.76 (.064)	6.9	567
Great Lakes Carbon Corp.	Fortafil CG-3	250	25			1.80 (0.065)	3.8	384
Hercules Inc.	AS $/lb–32–50	410	30–34	1.3	7.8-8.1	1.80 (0.065)	6.2–6.4	495
Hercules Inc.	HTS $/lb–65	400	34-37	0.1	7.4-7.8	1.80 (0.065)	6.0–6.2	542
Hercules Inc.	HMS $/lb–75–90	340	50–55	0.5	7.2–7.5	1.88 (0.068)	5.0–5.1	775
Stackpole Fibers Co.	Panex 30 $/lb–30–35	400	30	1.3	8	1.74 (0.063)	6.3	476
Stackpole Fibers Co.	Panex 30Y/800d $/lb–38–43	225	38			1.75 (0.063)	3.6	603
Union Carbide Corp.	Thornel Mat VMA $/lb–8	200	35		9	1.99 (.072)	2.8	486
Union Carbide Corp.	Thornel 300 $/lb–32–40	360	34		6.9	1.76 (0.064)	5.6	530
Union Carbide Corp.	Thornel Type P $/lb–20	200	55			1.99 (0.072)	2.8	764

Table 29-11. Unidirectional composite properties of commercial carbon-graphite filament prepregs.

Company	Celanese Corp.			Great Lakes Carbon Corp.		Hercules Inc.			Stackpole Fibers Co.
Trade name, Property or types	GY-50 65 v/o	Celion-3000	GY-70 70.5 w/o	Fortafil CG-3 60 v/o	Fortafil CG-5 60 v/o	AS	HTS	HMS	Panex 30 60 v/o
Tensile strength (psi × 10)	125	270	90	150	140	225	170	105	225
Tensile modulus (psi × 10^6)	30	22	42			20	23	29	18
Ultimate Elongation (%)	0.4	1.2	0.2			1.2	–	.55	
ILS (psi × 10^3)	7–10	14–18	8.6	11	10	18	–	14	17
Flexural strength (psi × 10^3)	145	270	115	170	160	270	–	140	250
Flex modulus (psi × 10^6)	27	20	37	15	23	18	–	25	18
Compressive strength (psi × 10^3)			90	130	120	140[a]			
Compressive modulus (psi × 10^6)			42						
Density (g/in.^{3})									1.58
(lb/in.^{3})						.056	.057	.059	

[a]crossply

8. U. S. Polymer Division of Hitco
   1600 W. 135th St.
   Gardina, Calif. 90249
   (213) 321-8080

A rough estimate of the cost of a commercial epoxy prepreg is about 5%/lb increase in cost above the filament cost per pound, for unidirectional tape. For fabric prepregs, the estimate would be from 10–15% above the filament cost. In the case of high cost resins, such as some polyimides that cost currently about $100/lb, there would be a great increase in cost for the prepreg as compared with the filament cost.

5. HISTORY AND FUTURE OF COST AND AVAILABILITY

The great expenditures in development of the manufacturing processes, and the need for good quality control, were among the factors that led to the high prices for the initial high modulus carbon-graphite filaments. During 1970, the average price for continuous filaments was about $325/lb, and the price for meter length materials was about $175/lb. During 1975, there was a wide spread of prices because of the many different types of products, and while some types of filaments were still above the $300/lb level, the average selling price of high modulus filaments was about $50/lb. As an example of the rapid trend toward lower prices, Type III or Type AS fiber is currently being quoted at $32/lb and has been projected to $10. with large scale production. Hercules recently announced their new AS-3 continuous tow at $18/lb. Union Carbide's recently marketed pitch precursor, continuous filament has been projected at a price level of $10/lb or less as soon as a high production market develops, which they anticipate will be reached by 1981. Their new Type P continuous yarn is $20/lb. The mat materials, which are currently selling for $8/lb, are projected to reach about $4/lb in the near future.

As noted in the preceding tables, a number of companies make filaments of the three types designated as high strength, or high modulus, or the Type A or Type III. The prices of each of these types of materials from the different companies will be competitive, and can often be negotiated when the required volume is high.

At the high end of the available high modulus materials is Celanese Celion® GY-70 with a listed modulus of 77 million psi. This is a premium product, and the current price schedule for GY-70, 3 in. wide continuous dry tape, is \$75/lb for 100-2500 lb. Special quotations may be obtained for quantities above 2500 lb.

The basic factors in the anticipated low cost for graphite filaments may be understood by the following considerations:

The cost of commercial PAN filament is about 60¢/lb, and the cost of the special yarns or tows favored as precursors are higher. Since there is a high weight loss during conversion to carbon, the precursor cost per pound of product can range from \$1-\$2 or higher. The cost for rayon precursor is two to three times this figure, because of even lower conversion ratio for rayon. The major costs involved in the production of graphite filaments are those related to processing labor and equipment, and quality control.

The low-temperature stabilization step is costly. An important goal of ongoing studies is direct graphitization, without going through the intermediate carbonization step. Cost and time reductions for each step in the production cycle for PAN can be anticipated, and will be a factor in the continual reduction of the price of high modulus graphite filaments.

The cost of a suitable raw petroleum pitch precursor is about 10¢/lb. However, this product must be refined in order to make it suitable for fiber production. Conversion to carbon is usually over 65% and may be as high as 90%, in contrast to the much lower yields for the organic filament precursors.

When production volume for the continuous fiber from pitch precursor reaches several million lbs/yr, the price is projected at \$5/lb, and when the multimillion pound markets develop, a \$2-\$3/lb selling price can be attained.[3]

6. CURRENT APPLICATIONS

Even though carbon-graphite filaments have only been available for about a decade, they have already been utilized in every usual type of organic matrix.

Table 29-12. Properties of reinforced thermoplastics.
Properties of Nylon 6/6 composites.

Property	ASTM Method	Unreinforced Resin	Reinforcing Fiber (%) 20 Carbon	30 Carbon	40 Carbon	20 Carbon, 20 Glass	40 Glass
Specific gravity	D792	1.14	1.23	1.28	1.34	1.40	1.46
Water absorption (%)	D570						
24 hr		1.6	0.6	0.5	0.4	0.5	0.6
Equilibrium after continuous immersion		8.0	2.7	2.4	2.1	–	3.0
Mold shrinkage, 1/8 in. average section (10^{-3} in./in.)	D955	15.0	2.0–3.0	1.5–2.5	1.5–2.5	2.5–3.5	3.5–4.0
Tensile strength (psi)	D638	11,800	28,000	35,000	40,000	34,000	31,000
Tensile elongation (%)	D638	10	3–4	3–4	3–4	3–4	2–3
Flexural strength (psi)	D790	15,000	42,000	51,000	60,000	49,000	42,000
Flexural modulus (10^6 psi)	D790	0.4	2.4	2.9	3.4	2.8	1.6
Shear strength (psi)	D732	9,600	12,000	13,000	14,000	13,000	12,000
Impact strength, Izod (ft-lb/in)	D256						
notched, 1/4 in.		0.9	1.1	1.5	1.6	1.8	2.6
unnotched, 1/4 in.			8.0	12.0	13.0	16.0	19.0
Heat-distortion temperature at 264 psi (F)	D648	150	495	495	500	500	500
Coefficient of thermal expansion (10^{-5} in./in. °F)	D696	4.5	1.40	1.05	0.80	1.15	1.4
Thermal conductivity (Btu-in./hr-ft^2 °F)	C177	1.7	5.5	7.0	8.5	6.4	3.6
Surface resistivity (ohms/sq)	–	10^{15}	25–30	3–5	1–3	–	10^{15}

Table 29-12 (*Continued*)

Property	ASTM Method	Polysulfone			Polyester		
		Unreinforced	30% Glass	30% Carbon	Unreinforced	30% Glass	30% Carbon
Specific gravity	D792	1.24	1.45	1.37	1.32	1.52	1.47
Water absorption (%)	D570						
24 hr		0.20	0.20	0.15	0.08	0.06	0.04
Equilibrium after continuous immersion		0.60	0.58	0.38	–	0.45	0.23
Mold shrinkage, ⅛ in. average section (10^{-3} in./in.)	D955	7-8	2-3	2-3	17-23	3-4	1-2
Tensile strength (psi)	D638	10,200	18,000	19,000	8,000	19,500	20,000
Tensile elongation (%)	D638	50-100	3-4	2-3	10	3-4	2-3
Flexural strength (psi)	D790	15,400	24,000	25,500	13,000	28,000	29,000
Flexural modulus (10^6 psi)	D790	0.39	1.20	2.05	0.35	1.35	2.00
Shear strength (psi)	D732	9,000	9,500	7,000	8,000	8,000	
Impact strength, Izod (ft-lb/in.)	D256						
notched, ¼ in.		1.2	1.8	1.1	0.3	1.6	1.2
unnotched, ¼ in.		60.0	14.0	4-5	25	9-10	4-5
Heat-distortion temperature at 264 psi (F)	D648	345	365	365	155	430	430
Coefficient of thermal expansion (10^{-5} in./in.)	D696	3.1	1.4	0.7	5.3	1.2	0.5
Thermal conductivity (Btu-in./hr-ft^2 °F)	C177	1.8	2.2	5.5	1.1	3.2	6.5
Surface resistivity (ohms/sq)		10^{16}	10^{16}	1-3	10^{16}	5×10^{15}	2-4
Flammability	UL Subj 94	VE-O	VE-O	VE-O	–	–	–

6.1 Thermoplastic Composites

The following two companies have product lines that include carbon-graphite filament reinforced thermoplastic compounds:

Liquid Nitrogen Processing Corp.
412 King Street
Malvern, Pa. 19355
(215) 644-5200

Fiberite Corp.
Winona, Minn. 55987
(507) 454-3611

Table 29-12 compares carbon and glass fiber filled Nylon-6/6, polysulfone, polyester, polyphenylene sulfide, ethylene tetrafluoroethylene, and vinylidene fluoride-PTFE.[7]

Although most of the early industrial activity on carbon-graphite filament composites involved thermoset resins, such as the phenolic matrices for ablative components, and epoxy matrices for ultra-high modulus and high-strength aerospace components, the anticipated predominance of thermoplastic composites has begun to emerge. The high volume and low cost of thermoplastic injection molding and extrusion processes have been an incentive for progress with engineering thermoplastic resins.

Carbon-graphite fiber reinforcement is now available in molding compounds of nylon, polysulfone, thermoplastic polyester, polyphenylene sulfide, polycarbonate, polypropylene, ethylene tetrafluoroethylene copolymer, and vinylidene fluoride tetra-fluoroethylene copolymer. Content of the fiber can usually be chosen at the 20, 30, or 40% by weight levels. Nylon 6/6 is available as a 20%-glass, 20%-carbon fiber hybrid formulation. The cost of these compounds is currently \$10/lb and higher, depending on the matrix and quantity purchased. However, it is anticipated that there will be a rapid increase in the sales volume of these materials, so that the price level of many types will be under \$5/lb by 1979.

Compared with glass-fiber reinforced compounds, the carbon fiber composites generally have greatly improved tensile and flexural

Table 29-12 (*Continued*)

Polyphenylene Sulfide			Ethylene Tetrafluoroethylene			Vinylidene Fluoride-PTFE	
Unreinforced	30% Glass	30% Carbon	Unreinforced	30% Glass	30% Carbon	Unreinforced	20% Carbon
1.34	1.56	1.45	1.70	1.89	1.73	1.76	1.77
0.20	0.04	0.04	0.020	0.018	0.015	0.04	0.03
–	–	0.10	–	–	0.24	–	–
10	2	1	15–20	2–3	1.5–2.5	–	2.5–3.5
10,800	20,000	27,000	6,500	14,000	15,000	5,500	12,300
3–4	3–4	2–3	150	4–5	2–3	100	3–4
20,000	29,000	34,000	10,000	19,000	20,000	7,000	17,200
0.60	1.60	2.45	0.20	1.05	1.65	0.20	1.10
–	–	–	6,000	7,000	7,000	6,000	7,500
0.3	1.4	1.1	N.B.	7–8	4–5	2.3	2.6
3–4	8–9	5–6	N.B.	17–18	10.0	–	7–8
278	500	500	165	460	465	190	295
3.0	1.3	0.6	4.2	1.7	0.8	8.5	3.0
2.0	2.8	5.2	1.65	2.8	5.6	2.1	3.0
10^{16}	10^{16}	1–3	5×10^{14}	10^{15}	3–5	5×10^{14}	3–5
VE-O	VE-O	VE-O	VE-O	VE-O	VE-O	VE-O	VE-O

strengths and moduli, but the impact strengths are lower. Thermal expansion of the carbon formulations is lower, and mold shrinkage ranges from about one-half that of glass-reinforced compounds to one-fifth that of the non-reinforced resin. Thermal conductivity of the carbon composites is about twice that of the equivalent glass-reinforced formulation. The much higher electrical conductivity of the carbon composites can be advantageous for many applications, such as when electrostatic painting is required or for electromechanical components when it is desirable to dissipate static charges. The electrical conductivity makes these composites unsuited for applications that require a good insulator.

Carbon fiber reinforcement reduces the coefficient of friction of thermoplastic resins to an extent that is determined by the type of resin. However, the higher the degree of graphitization of the fiber, the lower will be the coefficient of friction, with the type of matrix material being less of a factor. The coefficient of friction of carbon fibers against steel is about 0.25 as compared with 0.8 for glass fibers. Also, polymers reinforced with glass fibers cause at least ten times more wear on a relatively soft counterface, such as mild steel, than the equivalent carbon fiber composite.

Table 29-13 presents data on the wear, frictional, and limiting pressure-velocity (PV) values for a number of reinforced thermoplastics.[7] Note the low wear factor of 13 for 30%-carbon-fiber-reinforced polyvinylidene PTFE copolymer and 17 for 30%-carbon-fiber-reinforced ETFE copolymer, which is attributed to "the excellent adhesion of these melt-processable fluorocarbon resins to carbon fibers."[7]

The carbon fibers that are currently being used in thermoplastics have a tensile strength of about 300,000 psi and 30-million psi modulus. The fibers are surface treated to improve wetting and adhesion of the resin matrices.

Tables 29-14 and 29-15 show the properties of Fiberite Corporation's carbon-graphite filament reinforced Nylon 6/6, and polyphenylene sulfide.

Table 29-13. Wear, frictional, and PV properties of reinforced thermoplastics.

Base Resin	Fiber Content (%) and Type	Wear Factor[a] At Equilibrium	Coef of Friction Static	Coef of Friction Dynamic	Limiting PV 10 fpm	Limiting PV 100 fpm	Limiting PV 1,000 fpm
Nylon 6/6	0	200	0.20	0.28	3,000	2,500	2,500
	30 Glass	75	0.21	0.20	12,500	10,000	7,500
	30 Carbon	20	0.16	0.20	21,000	27,000	8,000
Polysulfone	0	>2,500	–	–	5,000	5,000	3,000
	30 Glass	90	0.22	0.18	–	–	–
	30 Carbon	75	0.17	0.14	8,500	8,500	6,000
Polyester	30 Carbon	24	0.12	0.15	11,000	12,500	3,000
Polyphenylene sulfide	0	540	0.30	0.24	2,500	3,000	4,000
	30 Carbon	160	0.23	0.20	8,000	20,000	8,000
ETFE	0	6,000	0.32	0.40	–	–	–
	30 Glass	20	0.38	0.36	7,500	10,000	13,000
	30 Carbon	17	0.24	0.22	14,000	28,000	14,000
Vinylidene fluoride-PTFE	30 Carbon	13	0.10	0.13	–	–	–

[a]A wear factor of $K = 1$ indicates a test condition that produces a wear volume of 1 $in.^3$ of material in 1 hr at a load of 1 psi and a velocity of 1 fpm. Wear data presented here (to be multiplied by 10^{-10}) were determined on a 0.900-in. ID × 0.125-in. OD plastic thrust washer operating at 50 fpm under a 40-psi load, against a 1040 steel washer with a 12-rms finish.
Courtesy: *LNP Corp.*

6.2 Thermoset Composites

A list of companies that supply carbon-graphite filament prepregs has been given in a previous part of this chapter. These prepregs have usually involved the use of an epoxy resin, because epoxies have excellent properties, such as superior adhesion, low shrinkage during cure, versatile formulation capabilities that result in a wide choice of processing characteristics and final properties, and they are readily "B" staged for long-term storability of the prepreg. The epoxy matrix can be cured as low as 250°F. As the high performance composite technology advances, and prices become lower and more competitive, there will probably be an increased usage of the lower priced thermoset polyester matrices. Also, there may be an anticipated increased usage of high temperature capability thermoset matrices, such as the polyimide resins.

Table 29-16 shows the mechanical properties of a laminate made from Hercules graphite prepreg, type AS/3501-5. The reinforcements are Hercules continuous type AS, high strength, graphite filaments. The letter "S" in the designation AS indicates a surface treatment to increase the composite shear and transverse tensile strength. The epoxy resin matrix, 3501, was developed by Hercules to operate in temperature environments of 350°F. Type AS/3501-5 prepreg has been recommended by Hercules for general-purpose structural applications where high strength, high modulus, and low weight are required.

Table 29-17 presents some mechanical properties of Hercules Graphite Prepreg Type 3501 HMS, which utilizes the high modulus surface treated filament.

The epoxy composites have generally been limited to a maximum use temperature of 350°F. Polyimide resins have been proposed for long-term durability at about 500°F and short-term, about 1,000 hr or less, up to 600°F. Table 29-18 shows the properties of various graphite filaments in a polyimide resin at 250°C (482°F), as compared with the properties at room temperature.

One of the main early uses of carbon-graphite filaments was in ablative components for the rocket industry. In these applications, a phenolic matrix was usually employed, because of the hard char formed and low erosion rate during the extremely high temperatures encountered in rocket nozzles and heat shields.

The ultimate in high temperature capability is afforded by carbon-carbon composites, where carbon is formed as a matrix for the carbon-graphite filament reinforcement. The aerospace industry has made use of this type of composite in projects such as the Space Shuttle. Fabrication and properties of carbon-carbon composites are indicated in the following summary.[8] PAN precursor carbon fibers were used to mold a composite with a char-yielding resin, preferably a phenolic resin, which was then carbonized by heating to 1000°C in an inert atmosphere. High modulus fiber composites, densified by resin re-

Table 29-14. Properties of Fiberite carbon-graphite fiber reinforced Nylon 6/6.

Compound Properties	Average Values
Color	Black
Injection pressure, psi	10–20000
Injection cylinder temp., °F	525–550
Mold temp., °F	150–200

Properties of Injection Molded Specimens	Base Resin	RTP 283	RTP 285	RTP 287	ASTM Test Method
Permanence					
Carbon-graphite fiber, %	0	20	30	40	
Specific gravity	1.14	1.23	1.28	1.33	D-792
Molding shrinkage, in./in. 1/8 in. section	0.014	0.0025	0.002	0.002	D-955
1/4 in. section	0.018	0.003	0.002	0.002	
Water absorption, %, 24 hrs @ 23°C	0.7	0.5	0.5	0.4	D-570
Mechanical					
Impact strength, IZOD, notched 1/4 in.	1.0	1.1	1.5	1.6	D-256
unnotched 1/4 in.	15	9.0	11	12	
Tensile strength, psi	11000	26000	34000	36000	D-638
Tensile elongation, %	60	4.0	3.0	2.0	D-638
Tensile modulus, psi $\times 10^6$	0.40	2.4	3.2	3.7	D-638
Flexural strength, psi	16000	41000	45000	50000	D-790
Flexural modulus, psi $\times 10^6$	0.42	2.0	2.4	2.8	D-790
Compressive strength, psi	4900	20000	24000	25000	D-695
Hardness, Rockwell R	118	120	120	120	D-785
Electrical					
Dielectric strength, VPM, S/T	600	NA[a]	NA	NA	D-149
Dielectric constant, 1 MC, dry	3.6	NA	NA	NA	D-150
Dissipation factor, 1 MC, dry	0.01	NA	NA	NA	D-150
Arc resistance, sec	135	NA	NA	NA	D-495
Volume resistivity, ohm CM	10^{15}	100	75	35	D-257
Thermal					
Deflection temp., °F, @ 264 psi	160	490	500	500	D-648
@ 66 psi	370	500	500	500	
Flammability	SE	B	B	B	D-635
Flammability, UL Sub. 94, 1/8 in.	VE2	HB	HB	HB	
Coefficient of linear thermal expansion, in./in./°F $\times 10^{-5}$	4.4	1.1	0.9	0.8	D-696
Thermal conductivity, Btu/hr/ft²/°F/in.	1.7	5.0	6.5	8.0	C-177

[a]NA–Not Applicable.
Courtesy: *Fiberite Corp.*

Table 29-15. Properties of Fiberite carbon-graphite fiber reinforced polyphenylene sulfide.

Compound Properties	Average Values
Color	Black
Injection pressure, psi	15–20000
Injection cylinder temp., °F	575–650
Mold temp., °F	100–350

Properties of Injection Molded Specimens	Base Resin	RTP 1383	RTP 1385	RTP 1387	ASTM Test Method
Permanence					
Carbon graphite fiber, %	0	20	30	40	
Specific gravity	1.3	1.38	1.42	1.46	D-792
Molding shrinkage, in./in. 1/8 in. section	0.010	0.0015	0.001	0.0005	D-955
1/4 in. section	0.009	0.002	0.001	0.0008	
Water absorption, %, 24 hrs @ 23°C	0.02	0.02	0.02	0.02	D-570
Mechanical					
Impact strength, IZOD, notched 1/4 in.	0.4	0.8	1.2	1.2	D-256
unnotched 1/4 in.	1.8	3.0	4.0	4.0	
Tensile strength, psi	9500	22000	25000	26500	D-638
Tensile elongation, %	1.6	0.75	0.5	0.5	D-638
Tensile modulus, psi $\times 10^6$	0.63	2.5	3.7	4.5	D-638
Flexural strength, psi	14000	27000	31000	34000	D-790
Flexural modulus, psi $\times 10^6$	0.55	2.1	2.5	3.5	D-790
Compressive strength, psi	16000	24000	26000	27000	D-695
Hardness, Rockwell R	120	122	123	123	D-785
Electrical					
Dielectric strength, VPM, S/T	380	NA[a]	NA	NA	D-149
Dielectric constant, 1 MC, dry	3.1	NA	NA	NA	D-150
Dissipation factor, 1 MC, dry	0.0009	NA	NA	NA	D-150
Arc resistance, sec.	20	NA	NA	NA	D-495
Volume resistivity, ohm CM	10^{16}	75	40	30	D-257
Thermal					
Deflection temp., °F, @ 264 psi	275	500	500	500	D-648
@ 66 psi	300	500+	500+	500+	
Flammability	SE	SE	SE	SE	D-635
Flammability, UL Sub. 94, 1/8 in.	VEO	VEO	VEO	VEO	
Coefficient of linear thermal expansion in./in./°F $\times 10^{-5}$	2.7	1.1	0.89	0.78	D-696
Thermal conductivity, Btu/hr/ft²/°F/in.	2.0	2.1	2.5	3.3	C-177

[a]NA–Not Applicable
Courtesy: *Fiberite Corp.*

impregnation, produced composites with a modulus of 200 GN/m², 29-million psi, and strength of 0.8 GN/m², 116 ksi, after heat treatment at 2600°C. High strength fibers gave similar strengths after heat treatment at 1600-1800°F. Densification by chemical vapor deposition of carbon led to much higher strengths, greater than 217 ksi. The mechanical properties of the carbon-carbon composites were virtually unaffected by temperatures in the order of 1600°C, and the composites exhibited very little creep at stresses of 70% of the ultimate strength.

In the preparation of carbon-carbon composites, various char forming resins may be used,

Table 29-16. Typical composite properties of Hercules magnamite graphite prepreg type 3501/A5.

Properties or Characteristics	Unit of Measurement	VALUES AT °F	
		77°	350°
0° Tensile strength[a]	psi	235,000	b
0° Tensile modulus[a]	psi × 10^6	20.7	b
Elongation at failure	%	1.12	b
90° Tensile strength	psi	8,200	b
90° Tensile modulus	psi × 10^6	1.35	b
Elongation at failure	%	0.62	b
0° Flexural strength	psi	242,000	160,000
0° Flexural modulus	psi × 10^6	17.4	17.0
Interlaminar shear	psi	17,900	8300
Fiber volume	%	62	62
Density	lb/in.^{3}	0.057	0.057
Void content	%	<2	<2

[a]These composite properties may be adjusted to other fiber volume contents by direct ratio. The manner in which the other properties change may not be linear with fiber volume content.
[b]Values not presently available.
Courtesy *Hercules Inc.*

such as petroleum pitch or furfuryl resin[9] and any form of carbon-graphite filament, fabric, or felt.[10]

7. CURRENT AND PROJECTED UTILIZATION OF CARBON-GRAPHITE COMPOSITES

Carbon-graphite filament composites were introduced and initially used primarily in the aerospace industry, where its high price could be justified because each pound of weight reduction can result in great take-off and in-service cost savings. For example, a pound saved in the fan blade of an aircraft engine eventually saves 5–8 lb as lighter supporting structures become feasible and lead, in turn, to lighter pylons, wings, and fuselage.[11] Benefits that can result are increased range, increased payload, lower power requirements, and lower fuel consump-

Table 29-17. Mechanical properties of Hercules graphite prepreg type 3501 HMS. (4 lots)

Item	AVERAGE VALUES[a]			Fiber Volume F_v
	RT	250°F	350°F	(%)
0° Flexural strength (ksi)	189	173	163	59
0° Flexural modulus (msi)	27	26	26	59
Short beam shear (ksi)	8	8	6.5	59
0° Tensile strength (ksi) (3 lots)	165	–	–	62
0° Tensile modulus (msi)	29.8	–	–	62
Cured ply thickness (mils/ply)				
Autoclave	5.2	–	–	
Press cure	4.8			

[a]Flex and SBS data are press cured. 0° tensile data is autoclave cured.

Table 29-18. Typical properties of various graphite filaments in a polyimide resin matrix.

	HM-S	I-S	HT-S	2-S	A-S	T3000 B
Flexural strength @ R.T. GPa (psi)	0.95 (137,000)	0.85 (123,000)	1.60 (232,000)	1.50 (217,000)	1.50 (217,000)	1.60 (232,000)
Flexural strength @ 250C GPa (psi)	0.85 (123,000)	0.75 (108,000)	1.30 (188,000)	1.30 (188,000)	1.20 (174,000)	1.30 (188,000)
Flexural modulus @ R.T. GPa (psi $\times 10^6$)	190 (27.5)	190 (27.5)	130 (18.8)	115 (17.0)	110 (16.0)	120 (17.4)
Flexural modulus @ 250C GPa (psi $\times 10^6$)	185 (27.0)	185 (27.0)	118 (17.0)	105 (15.0)	100 (14.5)	110 (16.0)
ILSS @ R.T. MPa (psi)	60 (8700)	55 (8000)	110 (16000)	95 (13,700)	110 (16,000)	95 (13,700)
ILSS @ 250C MPa (psi)	42 (6100)	30 (4300)	60 (8700)	55 (8000)	60 (8700)	55 (8000)

Courtesy *Fothergil & Harvey Ltd.*

Table 29-19. Value of weight saved in aerospace structures.

Value/lb Saved ($)	Application
10,000–15,000	Space Shuttle
10,000	Synchronous orbit satellite
1,000	Near orbit satellite
200–500	SST
150–200	Fighter plane
150–200	Boeing 747
100–200	Aircraft engines
100	Commercial planes
50–75	Transport type aircraft

tion. Table 29-19 is an estimate of the value/lb saved in aerospace structures.

The most frequently used metals in the aerospace industry are aluminum and titanium. The initial cost per pound of these metals is much lower than the current cost of high performance graphite filament composites. However, when an aircraft part is fabricated from aluminum or titanium, high strength forgings are generally required, and a large amount of the metal, often from four to twelve times the final part weight, is machined away as scrap. In contrast, the scrap rate for the prepreg tape fabrication process may be in the order of 10–30%. Therefore, the actual cost per fabricated pound of aerospace structural metals and the graphite filament

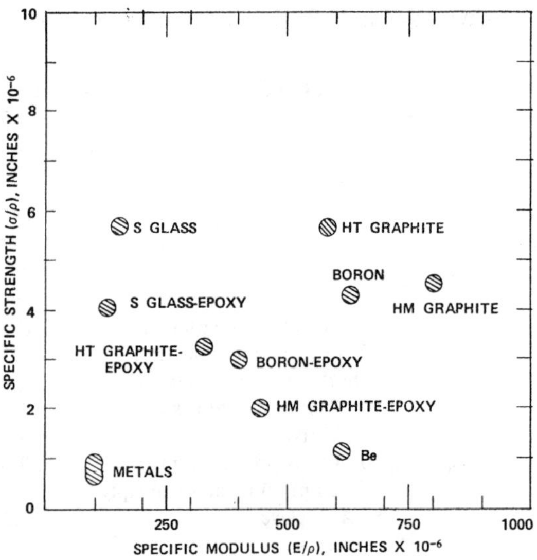

Fig. 29-5. Comparison of specific strength and specific modulus of aerospace materials. (*Courtesy Hercules, Inc.*)

composites will be much closer than the comparison of raw materials costs.

The major factor in the rapid growth of carbon-graphite filament composites in the aerospace industry is the great improvement that these composites offer in specific strength and specific modulus as compared with the structural metals. This is shown in Fig. 29-5.

McDonnell Aircraft Company, Division of McDonnell Douglas Corporation, has announced that the F-18 Fighter, which is being developed jointly by McDonnell and Northrop Corporation for the navy, will utilize composite materials for 10% of the total structural weight. This is substantially more than the 2.1% of boron filament or graphite filament-epoxy composites that are in the McDonnell F-15, which is now in production.[12]

The airframe of Northrop's YF-17 advanced technology demonstration aircraft, lightweight fighter, includes 900 lbs of composite assemblies. Figure 29-6 shows the structural applications on the YF-17. Carbon-graphite filament composites have also been utilized on the F-14 and B-1 aircrafts, the C-4 Trident missile, and many other aerospace structures.

The excellent properties of the carbon-graphite filament composites have attracted the attention of the sporting goods industry, and has resulted in a rapid acceptance and growth rate in this field. In 1973, about 140,000 lb of graphite-epoxy prepreg was used in the manufacture of golf club shafts.[13] A golf club with a composite shaft retails for about $75 as compared with $40 for one with a steel shaft. However, the reduced weight of the composite shaft, 2.5 oz vs. 4 oz, permits a lower overall weight, and a little more weight in the head, which leads to a possible increased drive of 5–10 yards.

Fishermen will also pay a premium for fly rods. As costly as carbon fibers are, only about $2 worth is used in a fishing rod for which anglers will pay $150.[14] About 35,000 graphite fishing rods were made and sold during 1975.

A 1976 issue of a tennis magazine had several ads advertising the superior performance of graphite filament, reinforced rackets. During 1975, about 50,000 tennis rackets were produced and represented about 12,000 lb of graphite fiber prepreg.

Other applications where carbon-graphite filament composites are being utilized are racing bicycles, skis, canoes, kayaks, racing yachts, and bows and arrows.

The widespread use of carbon-graphite filaments in the sporting goods field has been a

Fig. 29-6. VF-17 advanced composed components. (*Courtesy Hercules, Inc.*)

great help in increasing the sales volume and resulting in the decreased price level.

In many applications, there is a choice in the type of starting material to be used, regarding whether the unidirectional tape or a woven fabric should be chosen. For attaining the high-highest strength or modulus, the unidirectional tape had been recommended because of concern that the bends and fiber-to-fiber contact in a fabric would lead to greatly reduced properties. However, the fabric approach presents some economies in materials and fabrication costs. Recent studies have shown that it may be advantageous to use fabric prepreg in many applications.[15]

Much recent work has been spent on the development of combinations of carbon-graphite filaments with another filament, such as Kevlar 49, glass, or boron. The dual structure can be synergistic in exhibiting the better properties of each of the filaments. This type of hybrid structure has been used in many aerospace and industrial parts. A particularly beneficial combination consists of graphite filaments and Kevlar 49, which has thermal expansion propeties that are compatible with those of graphite. The Kevlar 49 contributes lower cost, lower density, and superior toughness and damage tolerance. This hybrid has light weight, stiffness, and low cost, and is particularly promising for aircraft applications[16,17] and has also been used for the manufacture of golf club shafts.

As the price of carbon-graphite filaments decreases, this type of composite will find applications in all industries, including textile equipment, packaging containers and equipment, and household furniture. It will be especially useful for any equipment involving oscillating, reciprocating, and rotating parts. Therefore, it will be essential for anyone involved in the fabrication or use of plastics components to be completely familiar with the latest progress in carbon-graphite filament reinforced composites. The expanding interest and rapid growth in this field was evident in the July 1977 issues of the technical magazines, *Plastics Engineering*, *Modern Plastics*, *Plastics World*, and *Plastics Technology* when each contained lengthy articles that reported "breakthroughs for carbon" and "carbon fibers come on strong."

References

1. Bacon R., *Applied Physics,* **31**, 383 (1960).
2. Deitz, R. D. and Vaughan, W. H., "The Fracture and Breaking Strengths of Single Carbon Fibers Before and After Bromine Treatment," SPI Reinforced Plastic/Composites Institute 21-A, 1974.
3. Forsyth, R. B., "Low Cost Continuous Fibers From A Pitch Precursor," 20th National SAMPE Symposium, 1975.
4. "Zero Mold Shrinkage Made Possible with Carbon Fibers," *Plastics Design & Processing* 6 (May 1976).
5. Berg, C. A., Barta, S., and Tirosh, J., "Friction and Wear of Graphite Fiber Composites," *Research of the N.B.S.–C. Engineering and Instrumentation,* Vol 76C (Jan.-June, 1972).
6. Sliney, H. E., and Jacobson, T. P., "Graphite Fiber-Polyimide Composite Rod End Bearings for High-Temperature High-Load Applications," NASA Tech Brief B75-10151, Oct. 1975.
7. Theberge, J. E., Arkles, B., and Robinson, R., "Carbon Fibers Add Muscle to Plastics," *Machine Design* (Feb. 7, 1974).
8. Hill, J., Thomas, C. R., and Walker, E. J., "Advanced Carbon-carbon Composites for Structural Applications," Plastics Polymer Conference Supplement, 6, 122–30 (1974).
9. Adams, D. F., "Transverse Tensile and Longitudinal Shear Behavior of Unidirectional Carbon-carbon Composites," *Material Science English,* **17** (1), 139–152 (1975).
10. Pierson, H. O. and Northrop, D. A., "Carbon-felt, Carbon-matrix Composites," *Composite Materials,* 9, 118–137 (1975).
11. Sattar, S. A., Stargardter, H., and Randall, D. G., "Development of JT8D Turbofan Engine Composite Fan Blades," *J. Aircraft,* **8**, 648 (Aug. 1971).
12. "Composite Material Use in Jet Fighters Grows at McDonnell Aircraft," *American Metal Market/Metal Working News* **27**, (Aug. 2, 1976).
13. "Now Fiber Composites Also Reduce Costs," *Iron Age* (June 16, 1975).
14. "Aerospace Fibers Score High in Sports," *Business Week* 28H, 28J (Aug. 9, 1976).
15. Shibata, N., Nishimura, A., and Norita, T., "Graphite Fiber's Fabric Design and Composite Properties," *SAMPE Quarterly* 25–33, (July 1976).
16. Zweben, C. and Norman, J. C., "Kevlar 49/Thornel 300 Hybrid Fabric Composites for Aerospace Applications." *SAMPE Quarterly* 1–10 (July 1976).
17. Hamersveld, J. V. and Fogg, L. D., "Producibility Aspects of Advanced Composites for an L-1011 Aileron." SAMPE Journal 6–13 (May/June 1976).

30

Aluminum Oxide and Other Ceramic Filaments

John V. Milewski

Los Alamos Scientific Lab.
Los Alamos, New Mexico

CONTENTS

---

1. INTRODUCTION

Ceramic fibers are continuous fibers of metal oxides that feature combinations of properties not previously available. The major advantages are resistance to extremely high temperatures (2500-3000°F), coupled with higher modulus and excellent compression strength. Their ceramic composition gives them exceptionally good chemical resistance and their small fiber diameter lends flexibility and workability to the fiber. Because of these features, they can be combined into a strand and woven or braided into fabrics. Many of these fibers were developed specifically for the insulation market, but because of unusual properties, they are being considered for unique reinforcing applications.

They are not noted for high tensile strength, which ranges from 50 kpsi for the weakest types to about 250 kpsi for the stronger grades.

Some typical applications are high temperature resistant continuous conveyor belts, jackets, heat shields, high temperature filtration, reinforcement of metals and ceramics, and in reinforced plastics in such products as radomes, sporting goods, chemical ware, aircraft components, and brake lining or friction materials.

2. SUMMARY OF PROPERTIES

2.1 DuPont's Fiber FP

DuPont's fiber is a round, continuous, multifilament, polycrystalline alumina yarn having essentially 100% alpha alumina composition. The small diameter, 20 μ, permits bending of the extremely stiff yet flexible fiber over small radii while its yarn form can be handled easily. The properties are given in Table 30-1.

Figures 30-1 and 30-2 show the variations of fiber strength and modulus with temperature.

Table 30-1. Properties of Fiber FP.

- Tensile strength 200–230 kpsi (¼ in. filament gage length)
- Tensile modulus 55 × 10^6 psi
- Strain to failure ~0.35%
- Melting point 2045°C
- Density 3.95 g/cc (0.143 lb/in.^{3})
- Fiber diameter 15–25 μ
- Form Continuous multifilament
- Strength retention after 1000°C, Air atmosphere, 92%

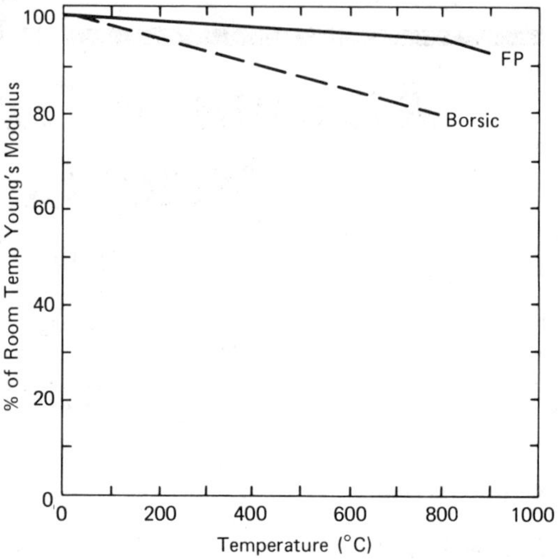

Fig. 30-2. Young's modulus at high temperature. (Data from Ref. 1.)

2.2 3 M's Ceramic Fiber Products

3M ceramic fibers are continuous fibers of metal oxides. Major advantages, in addition to the continuous nature of 3M ceramic fibers, include strength and resistance to extremely high temperatures. They are flexible, round, dense, transparent, and essentially chemically resistant. They have excellent compressive strength and can be internally colored as opposed to being dyed or color coated.

3M ceramic fibers can be tailor-made, within limits, to meet specific requirements. Because the product is continuous and strong, it can be easily combined into a strand and woven or braided into fabrics and sleeving on conventional fiberglass production equipment.

The properties of three ceramic fibers currently under investigation are given in Tables 30-2, 30-3, and 30-4.

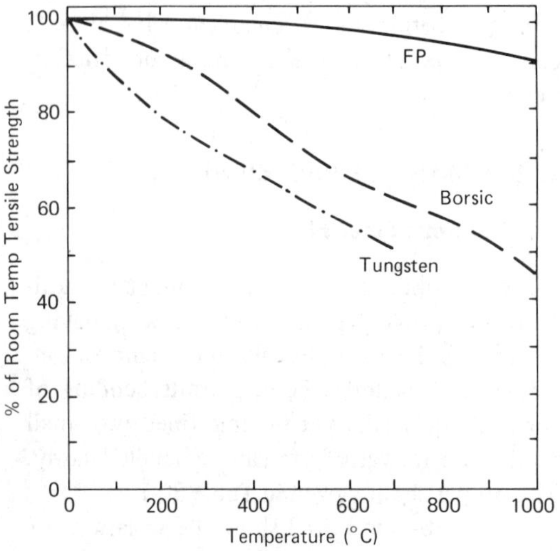

Fig. 30-1. Fiber strength at high temperature. (Data from Ref. 1.)

2.3 ICI'S Saffil Fibers

"Saffil" fibers are unique. No other fiber combines uniform fine diameters, virtual absence of non-fibrous impurities, excellent flexibility and resilience, with such high degrees of refractoriness and chemical inertness. They are microcrystalline and hence do not undergo thermal devitrification. Individual fibers are microporous and this enhances the adsorptive and related surface effects characteristic of alumina and zirconia. The bulk material has a non-irritant, silky handle. Initial screening tests on animals have shown no demonstrable toxic effects.

The "Saffil" fiber manufacturing process permits close control over fiber diameter, and a mean diameter has been selected to optimise key properties. "Saffil" fibers have mean diameters around 3 μ and a relatively narrow range compared with other inorganic fibers, as illustrated by the photographs (Fig. 30-3).

Fine fibers have important technological advantages such as exceptionally good thermal insulation properties (more interstices per unit weight), a large surface area for bonding if the fibers are to be used in a composite refractory material and, in general, better handle and resilience.

Table 30-2. 3M brand ceramic fiber data.
AB-312 fiber

Composition	Alumina-boria-silica
Forms	Continuous yarn, roving, chopped fibers, bulk fibers, fabric, braided products, prepreg
Appearance	Smooth, round, dense, transparent, continuous
Average denier (390 filament strand)	860
Color	White
Density	.0904 lb/in.^{3} (2.50 g/cc)
Average diameter	.44 mil (11 μ)
Surface area	$< 4.8 \times 10^3$ ft^2/lb ($<$1.0 m^2/g)
Tensile strength	250×10^3 psi (1720 mN/m^2)
Tensile modulus of elasticity	22×10^6 psi (152×10^3 mN/m^2)
Elongation at break	1.2%
Specific tensile strength	2.7×10^6 in. (6.85×10^6 cm)
Specific tensile modulus of elasticity	245×10^6 in. (615×10^6 cm)
Extended use temperature	2370°F (1300°C)
Short term use temperature	2910°F (1600°C)
First liquid phase	3090°F (1700°C)
Liquidus temperature	>3270°F (>1800°C)
Thermal conductivity	
@1000°F, 3 lb/ft^3 density	2.5 Btu in./hr ft^2°F
@1600°F, 3 lb/ft^3 density	6.2 Btu in./hr ft^2°F
Suggested uses	Thermal insulation, reinforced plastics, heat shielding, continuous high-temperature conveyor belts

Available Grades and Forms. Four basic grades of fiber are available to meet the wide range of possible needs.

"Saffil" zirconia fiber
  Standard grade (S)
  High Temperature grade (HT)
"Saffil" alumina fiber
  Standard grade (S)
  High Temperature grade (HT)

The standard and high temperature grades differ in the nature and amount of crystal phase stabilizer.

Summary of Typical Properties.

"Saffil" zirconia fiber		
Fiber density	g/cm^3	5.6
Melting point	°C	>2500 (4500°F)
Maximum use temperature		
HT grade	°C	1600 (2910°F)
S grade	°C	1400 (2550°F)
Specific heat	cal/g°C	0.14
Tensile strength	MN/m^2	0.7×10^3 (100×10^3 psi)
Specific tensile strength	m^2/s^2	13×10^4
Young's modulus	MN/m^2	1×10^5 (15×10^6 psi)
Specific modulus	m^2/s^2	2×10^7
Mean diameter	microns	3
Surface area	m^2/g	5
Hardness (Moh)		6
Shot content		Negligible
"Saffil" alumina fiber		
Fiber density	g/cm^3	2.8
Melting point	°C	2000 (3600°F)
Maximum use temperature		
HT grade	°C	1400 (2550°F)
S grade	°C	1000 (1830°F)
Specific heat	cal/g°C	0.25

Table 30-3. 3M brand ceramic fiber data.
AC-02 fiber

Composition	Alumina-chromia-silica
Forms	Continuous yarn, roving, bulk fibers, braided products
Appearance	Smooth, round, dense, transparent, continuous
Average denier (390 filament strand)	795
Color	Light green
Density	.101 lb/in.^{3}(2.80 g/cc)
Average diameter	.40 mil (10 μ)
Tensile strength	200×10^3 psi (1380 mN/m^2)
Tensile modulus of elasticity	23×10^6 psi (159×10^3 mN/m^2)
Elongation at break	1.2%
Extended use temperature	2550°F (1400°C)
Short term use temperature	2910°F (1600°C)
Tensile strength after 16 hr at 1500°C	80×10^3 psi (550 mN/m^2)
Weight loss after 16 hr at 1500°C	0.6%
Thermal conductivity	
@1000°F, 3 lb/ft^3 density	2.6 Btu in./hr ft^2°F
@1600°F, 3 lb/ft^3 density	5.8 Btu in./hr ft^2°F
Suggested uses	Wire and cable insulation, thermal insulation, reinforcement for plastics

Tensile strength	MN/m^2	1×10^3 (150×10^3 psi)
Specific tensile strength	m^2/s^2	40×10^4
Young's modulus	MN/m^2	1×10^5 (15×10^6 psi)
Specific modulus	m^2/s^2	4×10^7
Mean diameter	μ	3
Surface area	m^2/g	100
Hardness (Moh)		6
Shot content		Negligible

"Saffil" is a trade mark of Imperial Chemical Industries Limited for inorganic fibers.

2.4 High Silica and Quartz Fibers

(Abstracted from Reference (4) by Hugh Shulock)

High Silica. The term *high silica* can be used to describe any high purity glass. However, for use in reinforced plastics, high silica may be defined as a high purity glass of 95%-plus purity SiO_2, produced by a leaching process. High silica fibers and fabrics are flexible materials similar in appearance to, and produced from, fiber glass.

Table 30-4. 3M brand ceramic fiber data.
ZS-11 fiber

Composition	Zirconia-silica
Forms	Continuous yarn, roving, chopped fibers, bulk fibers, fabric
Appearance	Smooth, round, dense, transparent, continuous
Chemical resistance	Inorganic alkali and acid, organic solvents
Average denier (390 filament strand)	2060
Color	White
Density	.134 lb/in.^{3} (3.70 g/cm^3)
Average diameter	.56 mil (14 μ)
Surface area	$<4.8 \times 10^3 ft^2/lb$ ($<1.0 m^2/gm$)
Tensile strength	150×10^3 psi (1030 mN/m^2)
Tensile modulus of elasticity	14.0×10^6 psi (96.5×10^3 mN/m^2)
Elongation at break	1.1%
Extended use temperature	1830°F (1000°C)
First liquid phase	3090°F (1700°C)
Liquidus temperature	>3630°F (>2000°C)
Suggested uses	Thermal insulation, reinforced ceramics and concrete, high temperature filtration, ablative

Fiber glass, which has a silica content of 65%, is subjected to a hot acid treatment that removes virtually all of the impurities and leaves the silica intact. This is commonly called the leaching process.

Most textile forms (chopped fiber, mat, yarn, and fabric) are available in high silica. Fabrics are available in various weights and thicknesses; with purities from 95-99.4% SiO_2.

Quartz. The word quartz can also mean high purity glass, but we will define quartz fibers as those fibers produced from high purity (99.95% SiO_2) natural quartz crystals. The crystals are formed into rods from which filaments one-fifth the diameter of human hair are drawn. Up to 240 filaments are combined to form a flexible high strength fiber, which is made into yarn and woven into fabric. All textile forms, (including chopped fiber, mat rovings, cordage, sleevings, tapes, and fabrics) are available in quartz.

Quartz fiber with a purity of 99.95% SiO_2, exclusive of binder, retains virtually all of the characteristics and properties of solid quartz.

Physical and Mechanical Properties. High silica and quartz materials have higher strength to weight ratios than most other high temperature materials, with quartz having approximately 5 times the tensile strength of silica. High silica and quartz yarns are perfectly elastic, and their elongation at break is approximately 1%.

The physical properties of the fibers, yarns, and fabrics made from high silica and quartz are given in Table 30-5 and a typical chemical analysis is given in Table 30-6.

Thermal Properties. Since high silica and quartz are so similar in nature, they also have similar thermal characteristics. The major variable is a higher melt viscosity for quartz, because of its higher silica content (see Fig. 30-4). High silica and quartz do not melt or vaporize until the temperature exceeds 3000°F. At continuous temperatures in excess of 1800°F, both forms of silica will begin to devitrify into a crystallized form known as Christobalite. This conversion tends to stiffen the material, but causes no change in its physical form or insulating properties.

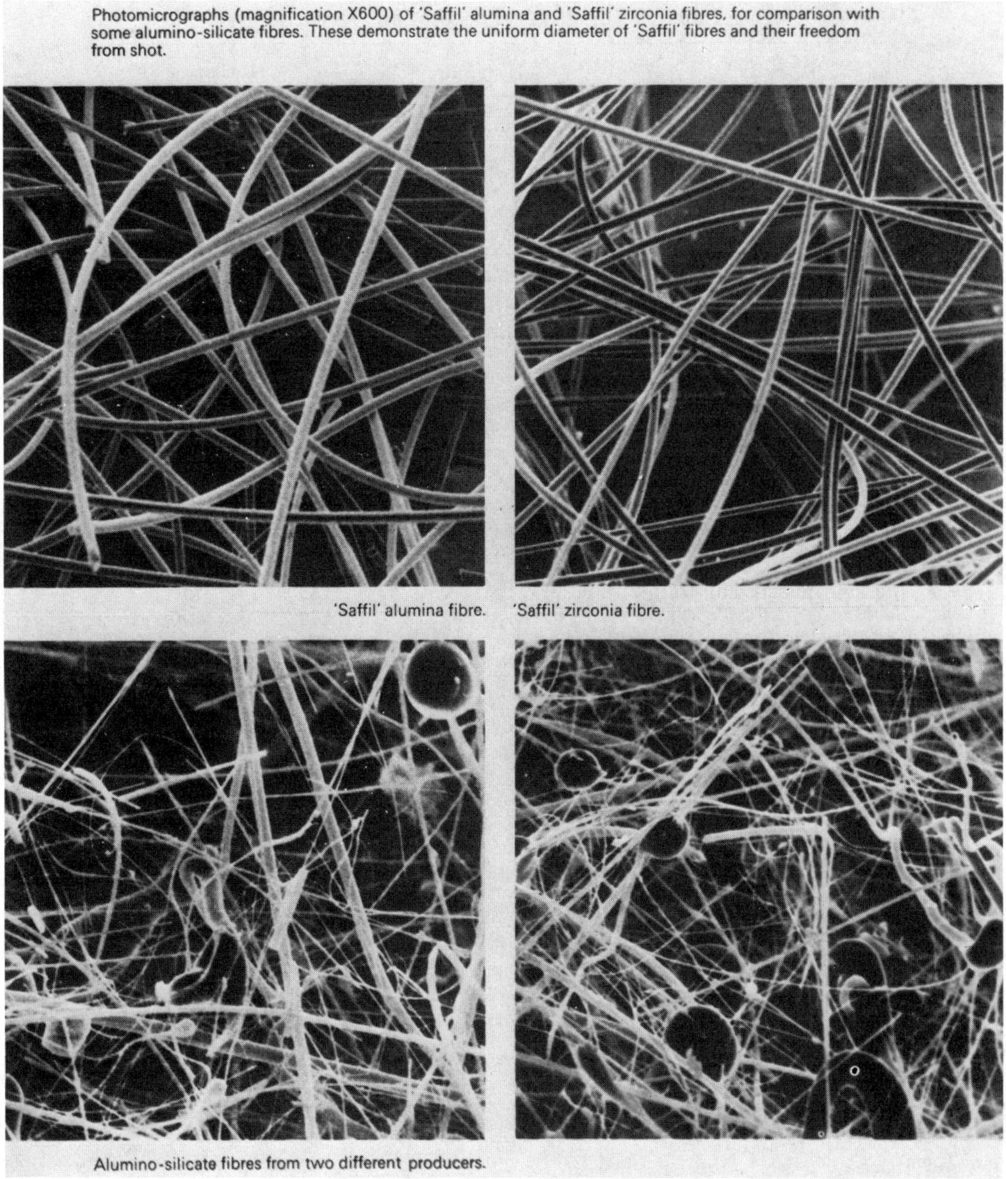

Fig. 30-3. (*Courtesy ICI–USA*)

Table 30-5a. Physical properties–Fibers.

Property	High Silica	Quartz
Filament diameter (in.)		
Yarn and fabric	.0004	.0004
Mat	.00005–.0004	.00005–.0006
Filament tensile strength (psi)		
@ Room temperature		130.000
@ 400°F		99.000
Specific gravity	1.74	2.2
Hardness–Moh scale		5–6
Young's modulus (psi)		10×10^6

Table 30-5b. Physical properties—Yarns.

Property	High Silica		Quartz	
Type number	#100	300–2/2	#300–2/4	#300–4/4
Nominal diameter (in.)	.020	.005	.010	.014
Approximate yd/lb	3050	7,500	3750	1875
Minimum breaking strength (lb)	2.5	3.0	5.0	11.0
Linear shrinking	12%	1%	1%	1%

Table 30-5c. Physical properties—Fabrics.

Property	High Silica		Quartz	
Type Number	#82	#84	#581	#570
Wt–oz/sq yd	10.3	18.5	8.4	19.5
Thickness (in.)	.013	.026	.011	.027
Thread count–warp		50	57	38
fill		40	54	24
Breaking strength (lb)–warp	30	70	185	480
fill	25	45	170	400
Weave	8H satin	8H satin	8H satin	8H satin
Areal shrinkage % (max)	5	5	1	1
Moisture content %	5	5	1	1
pH	4.0–7.0	4.0–7.0	6.0–8.0	6.0–8.0
Silica content %	98–99.2	98–99.2	99.9+	99.9+
Surface area sq m/g	20–150	20–150	1–5	1–5

Table 30-6. Typical chemical analysis.

Element	High Silica (high purity)	Quartz
SiO_2% (exclusive of binders)	99.23	99.95
Phosphorous % (ppm)	0	3
Sodium (ppm)	4	9
Potassium (ppm)	3	5
Lithium (ppm)	1	1
Boron (ppm)	205	10
Calcium (ppm)	5	23
Magnesium (ppm)	6	2
Strontium (ppm)	0	0
Iron (ppm)	16	3
Titanium (ppm)	2,800	12
Aluminum (ppm)	900	100
Manganese (ppm)	1	2
Copper (ppm)	2	1
Cadmium (ppm)		.5
Antimony (ppm)		.5
Chromium (ppm)	150.	0

From J. P. Stevens and Co., Inc., Central Research Laboratory, Garfield, N.J.

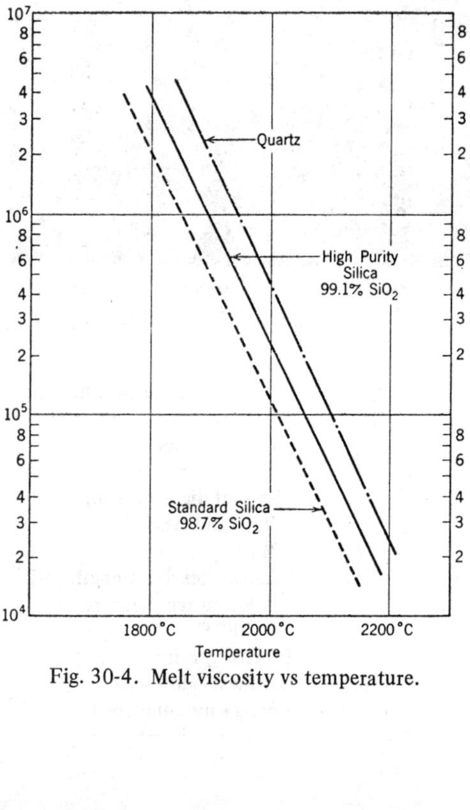

Fig. 30-4. Melt viscosity vs temperature.

Resistance to thermal shock is excellent. Silica products can be heated to 2000°F and rapidly quenched in water without any apparent change.

Chemical Properties. High silica and quartz have similar chemical properties. They are not affected by halogens or common acids in liquid or gaseous state, with the exception of hydrofluoric and hot phosphoric acids. Silica products are not recommended for use with alkalies, either hot or cold. Weak alkali solutions can be used for certain applications. Silica is not soluble in water or organic solvents.

3. SUPPLIERS AND COST

3.1 Fiber FP

At present, fiber FP is at an experimental stage and hence availability is limited. The initial price is expected to be about $25/lb. The ultimate price will depend on such factors as market volume and process improvements. Current status on availability and cost can be obtained by contacting E. I. duPont de Nemour and Co., Pioneering Research Laboratory, Textile Fiber Department, Experimental Station, Wilmington, Del. 19898. (302) 722-3663.

3.2 Ceramic Fibers

The 3M ceramic fibers are at a development stage and hence availability is somewhat limited. The initial price is in the $5–$10/lb range. Information on availability and current prices can be obtained by contacting the 3M Company—Ceramic Fiber Products, New Business Venture Division, 3M Center, St. Paul, Minn. 55101. (612) 733-9064.

3.3 Saffil Fibers

ICI's Saffil Fibers are being produced in semicommercial quantities in low density mat form. The 1976 price range was between $9.75–$10.75/lb, FOB Wilmington, Del. Detailed information on current prices can be obtained from: ICI America Inc., New Ventures Development, Wilmington, Del. 19899. Phone 302 575-3000.

3.4 High Silica and Quartz Suppliers, Materials and Cost

The suppliers of these silica products in the specific forms that are available are given in Table 30-7.

Addresses of these companies are listed below:

Names and Address of Suppliers:
Alpha Associates, Inc.
2 Amboy Ave.
Woodbridge, N. J. 07095
(201) 634-5700

General Electric Co.
Quartz and Chemical Products Section
24400 Highland Road
Richmond Heights, Ohio 44143
(216) 266-3561

Haveg Corp.
Subsidiary Ind. Hercules Inc.
910 Market Street
Wilmington, Del. 19899
(302) 995-3000

Hitco
1600 W. 135th St.
Gardena, Calif. 90249
(213) 321-8080

Johns Manville Sales Co.
Greenwood Plaza
Denver, Colo. 80217
(303) 770-1000

J. P. Stevens and Co., Inc.
1185 Avenue of the Americas
New York, N. Y. 10036
(212) 575-2000

Costs and Pricing. The principal raw material used to produce fiber glass fibers from which high silica is made is sand, available abundantly at approximately 15¢/lb. Quartz fibers are produced from natural pure fused quartz crystals that cost approximately $5/lb. The great disparity in raw material costs is not always reflected in the final product. There are great differences in manufacturing costs. High Silica is sold in a price range of $4–$50/lb depending on the form and the quantity purchased. Quartz products are priced from $25–$95/lb, again depending on quantity and

Table 30-7. Silica products and suppliers.

Form	High Silica	Quartz
Woven fabrics	Haveg Corp. Hitco Corp. J. P. Stevens and Co., Inc.	J. P. Stevens and Co., Inc.
Woven tapes	Hitco Corp.	Alpha Associates, Inc.
Yarn	Hitco Corp.	Alpha Associates, Inc. General Electric Co. J. P. Stevens and Co., Inc.
Chopped fibers	Hitco Corp.	J. P. Stevens and Co., Inc.
Roving	–	General Electric Co. J. P. Stevens and Co., Inc.
Cordage	Hitco Corp.	Alpha Associates, Inc.
Sewing thread	Hitco Corp.	Alpha Associates, Inc. Dodge Industries, Inc.
Mat	Haveg Corp. Hitco Corp. Johns Manville Co.	General Electric Co. J. P. Stevens and Co., Inc. Johns Manville Co.
Wool		General Electric Co. J. P. Stevens and Co., Inc.
Braided sleeving	Hitco Corp.	Alpha Associates, Inc. Thermo Electric Co.
Knit fabrics	–	Alpha Associates, Inc.
Knit tapes	–	Alpha Associates, Inc.

form. It is sometimes less expensive in the final product to use the more expensive but stronger, more flexible, and more easily processible quartz product than the less expensive high silica. Because of the great difference in strength, it is sometimes possible to use considerably less reinforcement with quartz than silica. If thermal protection is all that is required, then high silica is the obvious choice. Comparative prices on the most widely used products in the largest volume categories are shown in Table 30-8.

3.5 Refractory Fibers

Although the prime objective of this book is reinforcements and fillers, the author feels that the subject of ceramic fibers would not be com-

Table 30-8. Cost comparison.

HIGH SILICA		QUARTZ	
Product Description	Price	Product Description	Price
Silica fabric–heavy Wt–18.5 oz/sq yd Thickness–0.26 in. Strength–50 lb/in.	$6.15/sq yd	Quartz fabric–heavy Wt–19.5 oz/sq yd Thickness–0.27 in. Strength–480 lb/in.	$48.25/sq yd
Silica fabric–light Wt–10.3 oz/sq yd Thickness–.013 in. Strength–30 lb/in.	$4.00/sq yd	Quartz fabric–light Wt–8.4 oz/sq yd Thickness–0.11 in. Strength–185 lb/in.	$21.00/sq yd
Silica yarn Teflon binder Yield 3000 yd/lb Breaking strength average 2.5 lb	$18.00/lb or $.006/yd	Quartz yarn Teflon binder Yield 7500 yd/lb Breaking strength average 3.0 lb	$29.00/lb $.004/yd

plete without a mention of the kinds and sources of the principal refractory fibers available in the U.S. These fibers are primarily used for insulation, but could take on applications as a moderate reinforcement or secondary filler to complete a packing requirement as the art of packing becomes more of a science in the future.

Suppliers of Other Ceramic Fibers Use as Insulation

Product name	Kao Wool
Description	Aluminum silicate fiber wool and other insulation products.
Supplier	Babcock and Wilcox Co. Refractories Div. Old Savannah Rd. Box 923 Augusta, Ga. 30903 (404) 798-8000
Product name	Fiberfrax
Description	Aluminum silicate fiber wool
Supplier	Carborundum Co. Research and Development Div. Box 337, Niagara Falls, N.Y. 14302 (716) 278-2000
Product name	Cerafiber
Description	Aluminum silicate fiber wool
Supplier	Johns Manville Corp. Greenwood Plaza Denver, Colo. 80217
Product name	Zircar Product Inc.
Description	Zircona and alumina insulation
Supplier	Zircar Product 110 North Main Street Florida, N.Y. (914) 651-4481

4. PROCESSING DATA, CURRENT AND PROJECTED APPLICATIONS

4.1 Fiber FP Application Data

The following properties are important features of duPont's new FP fibers:

- Excellent retention of modulus and strength at elevated temperatures
- High compression strength
- Extremely high temperature stability
- Chemically resistant, hard refractory ceramic
- Fabricability–in continuous yarn form.

The continuous forms of FP Fiber make it amenable to low cost processing techniques for making forms such as rods, tubes, billets, beams, and engine blades containing high fiber volume loading in the order of 60%.

4.2 3M Brand Ceramic Fibers Application Data

The unique combination of properties of 3M Ceramic Fibers suggest many applications not satisfied by existing fibers. The AB-312 fiber, for example, can withstand continuous temperatures of up to 2600°F and short-term temperatures of up to 3000°F. It has a modulus elasticity of 22-25 $\times 10^6$ psi, a tensile strength of about 250 $\times 10^3$ psi, and can be colored.

Insulation, reinforcement, and decorative markets are among many market areas being explored for 3M Ceramic Fibers in fiber, yarn, and fabric form.

The product can be used for decorative materials where fireproofing and color permanence are necessary, in thermocouple wire insulation, in high temperature resistant continuous conveyor belts, gaskets, heat shields and high temperature filtration, reinforcement of metals and ceramics, and in reinforcement of plastics in such products as radomes, sporting goods, and aircraft structural components.

Composite property data for AB-312 Fiber is given in Tables 30-9, 30-10, 30-11, and 30-12. Excellent translation of fiber properties in the composite is evident in all cases.

4.3 ICI's Saffil Fiber Application Data

A wide range of applications is foreseen. Areas of particular importance are high temperature thermal and acoustic insulation and high efficiency filtration. Other applications include use as a catalyst support and in the reinforcement of composite products such as friction materials.

The fibers are produced in a staple form and their length is several centimeters. The primary physical form of "Saffil" fibers is a loose wool which is capable of further conversion to other end forms including:

Chopped fiber	a range of short staple fibers of aspect ratio 10–1000

Table 30-9. Composite data—AB-312 Fibers in PR 286 epoxy resin[a].

Volume fiber loading	50%
Composite density	.073 lb/in.^{3} (2.02 g/cm^3)
Young's modulus in the fiber direction	11.0 × 10^6 psi (75.8 × 10^3 mN/m^2
Specific modulus	150 × 10^6 in. (380 × 10^6 cm)
Young's modulus transverse to the fiber direction	1.7 × 10^6 psi (11.7 × 10^3 mN/m^2)
Tensile strength in the fiber direction	135 × 10^3 psi (862 mN/m^2)
Specific tensile strength	1.7 × 10^6 in. (4.30 × 10^6 cm)
Tensile strength transverse to the fiber direction	3 × 10^3 psi (20.7 mN/m^2)
Interlaminar shear strength	14 × 10^3 psi (96.5 mN/m^2)
Compressive strength	>135 × 10^3 psi (>862 mN/m^2)
Flexural strength in the fiber direction	180 × 10^3 psi (1240 mN/m^2)
Coefficient of thermal expansion in the fiber direction	3.5 × 10^{-6} in./in./°F (6.3 × 10^{-6} cm/cm/°C)
Coefficient of thermal expansion transverse to the fiber direction	17.0 × 10^{-6} in./in./°F (30.6 × 10^{-6} cm/cm/°C)

[a]3M PR 286 epoxy resin system.
A data source: 3M Ref. 2.

Table 30-10. Composite data—AB-312 fabric in PR 286 epoxy resin[a].

Weave	Unidirectional[b]	Bidirectional[c]
Flexural modulus warp direction	8.8 × 10^6 psi (60.6 × 10^3 mN/m^2)	7.0 × 10^6 psi (48.2 × 10^3 mN/m^2)
Flexural strength warp direction	110 × 10^3 psi (7580 mN/m^2)	80 × 10^3 psi (5510 mN/m^2)
Volume percent fibers in composite	48.5	54.0
Volume percent fibers in warp direction	40.0	27.5

[a]3M PR 286 epoxy resin system.
[b]Fabric contains 83% of the fibers in the warp direction.
[c]Fabric contains 51% of the fibers in the warp direction.
Data Source: 3M Ref. 2.

Table 30-11. Composite Data—AB-312 Fibers in PR 286 epoxy resin[a]—unidirectional fatigue properties[b].

$\frac{\sigma \text{ Max}}{\sigma \text{ Ult}}$	N Cycles
0.71	660,000
0.51	9,000,000[c]
0.40	6,000,000[d]

[a]3M PR 286 epoxy resin system.
[b]$R = \frac{\sigma \text{ Max}}{\sigma \text{ Min}}$ = 0.1, 20 hz, 40% fiber volume fraction.
[c]95% Static ultimate residual strength.
[d]100% Static ultimate residual strength.
Data Source: 3M Ref. 2.

Blankets	in a range of thicknesses and bulk densities
Needled blankets	blankets needled between organic scrims
Strip	narrow sections of blanket
Paper	conventionally manufactured papers of various weights containing organic and inorganic binders alone or in combination

Boards	rigid or semi-rigid, containing various proportions of "Saffil" fiber with inorganic and/or organic binders and possibly fillers
Vacuum formed shapes	rigid or semi-rigid shapes such as cones and tubes with inorganic and/or organic binders and possible fillers (see Fig. 30-5)
Temporarily compressed forms	novel forms of densified boards and shapes containing organic binders which burn out at a high temperature allowing the fiber to recover much of its original volume
Textile forms	yarns, rope, cloth, etc. are under development

Fig. 30-5. "Saffil" fibers formed shapes. (*Courtesy ICI–USA*)

Not all of these forms are commercially available. "Saffil" fibers offer users a material that can be formed by novel processes, and can provide solutions to unusual product requirements that could be accomplished by use of previously available materials.

The author sees increasing use of Saffil filler in plastic composites as the price becomes more competitive with other reinforcements. The combination of fine diameter, high modulus, and long lengths are unavailable in any current reinforcing fiber. This combination can be used in applications not possible with short microfibers or other large diameter, high modulus fibers.

One suggested application is the utilization of the extremely low cost, forming technique such as fiber vacuum forming. This forming method permits the rapid formation of large complex shapes on a very low cost screen mold. (See Fig. 30-4.) The formed parts are subsequently impregnated with a low viscosity resin solution,

Table 30-12. Composite data—unidirectional impact characteristics.

Materials[a]	Apparent Flex Strength	Total Energy per Unit Area	Ductility Index
	(KSI)	$\frac{\text{ft. lbs.}}{\text{in.}^2}$	
E-glass[c] epoxy	73	114	0.4
Kevlar[c] epoxy	142	124	1.6[b]
HMS graphite[c] epoxy	125	3.8	0
3M fiber[d] epoxy	126	39	0.47

[a]Specimen thickness ranged from 0.11–0.15 in.
[b]Based on maximum stress, not point of initial nonlinearity.
[c]Beaumont, P. W. R., et al, "Methods for Improving the Impact Resistance of Composite Materials", to be published in ASTM STP 568, 1975.
[d]AB-312 fibers in 3M PR 286 epoxy resin system.
Date Source: 3M Ref. 2.

Table 30-13. Typical laminate properties—high silica and quartz filaments in phenolic resin.

Property	HEAVY WEIGHT FABRIC High Silica	HEAVY WEIGHT FABRIC Quartz	LIGHT WEIGHT FABRIC High Silica	LIGHT WEIGHT FABRIC Quartz
Fabric description				
Oz/sq yd	18.5	19.5	10.3	8.4
Finish	None	A–1100	None	A–1100
Weave	8H satin	5H satin	8H satin	8H satin
Laminate preparation				
Resin–phenolic	V-204[a]	91-LD[b]	SC-1008[c]	V-204[a]
Cure and molding pressure	60 min at 325°F		60 min at 325°F	
Post cure	180 min at 325°F pressure 250 psi		180 min at 325°F pressure, 250 psi	
Laminate properties				
Flexural strength, psi × 10^3				
Average at 75 ± 5°F	33.3	71.7	36.6	95.4
Average at 500 ± 10°F ½ hr	24.8	35.5	21.1	60.0
Tensile strength psi × 10^3				
Average at 75 ± 5°F	23.1	56.8	22.7	72.1
Average at 500 ± 10°F ½ hr	19.1	50.0	15.3	47.0
Specific gravity	1.72	1.73	1.62	1.80
% resin	30.3	34.0	37.0	30.7
Thickness–in.	.124	.127	.152	.118
Number of plies	6	6	12	12

From J. P. Stevens and Co., Inc., Central Research Laboratory, Garfield, N.J.
[a]Barrett Division Allied Chemical Co.
[b]Cincinnati Testing Labs, Division Studebaker-Packard
[c]Monsanto Chemical Co.

Table 30-14. Laminate comparison of mechanical and electrical properties for various resin binders.

Property Fabric description	QUARTZ FABRIC Phenolic[a] V-204	Silicone[b]	Epoxy[c] Epon 828/CL	Teflon[d]	PBI AF-R-100
Oz/sq yd	8.4	8.4	8.4	8.4	8.4
Weave	8H satin	8H satin	8H satin	8H satin	8H satin
Finish	A1100	None	A1100	Teflon	A1100
Mechanical					
Flexural, psi × 10^3					
dry at 75° ± 5°F	95.4	34.0	98.6	–	–
Tensile, psi × 10^3					
dry at 75 ± 5°F	72.1	34.0	79.2	–	–
Electrical					
Dielectric constant					
X-band, dry-room temp. (9,375 megacycles)	3.86	2.93	3.470	2.47	3.360
Loss tangent					
X-band, dry-room temp.	0.040	0.00098	0.0092	0.0007	0.0034

[a]U.S. Polymeric Inc. data.
[b]Coast Manufacturing and Supply Co. data.
[c]Ferro Corp. and Grumman Aircraft Engineering Corp. data.
[d]Custom Materials, Inc. data.

and the excess solution is allowed to drain from the highly porous felt material. Curing the resin then forms a strong, low density product, which is a resin bonded, truss-like structure.

As Saffil prices become competitive with other reinforcement filaments, they will provide the potential for creative ideas in new processing and new product development.

4.4 High Silica and Quartz Fiber Application and Compositing Data

Properties of Composites. High silica and quartz can be used with most resin systems, and are easily impregnated on most conventional coating equipment, either vertical or horizontal. Preimpregnated fabrics are easily cut, either straight or on a bias, for use in tape winding or in lay-up techniques. The lighter weight fabrics are usually more flexible, and are normally used on highly contoured parts. Fabricated parts can also be made from chopped squares or with molding compounds made from shredded fabrics or chopped fibers. High silica materials are usually used without a finish or a binder. However, many types of resin compatible finishes and modifying additives can be applied to silica fabrics.

Quartz fabrics are generally supplied with chemical finishes compatible with the resin system to be used. Resin compatible finishes substantially improve laminate physical properties and reduce moisture absorption. Most finishes, binders, or additives that are available on fiber glass fabrics can be applied to quartz. Standard commercial binders are available for phenolics, epoxies, polyimides, silicones, polyesters, and fluorocarbon resin systems.

All property data presented in Tables 30-12, 13 and 14 were produced under laboratory conditions using flat panels prepared in accordance with MIL R-9300 with unfinished silica fabrics. These values are for comparative reference purposes only.

Uses. High silica and quartz are both used in a wide variety of similar products. The choice of type of raw material to use is generally dictated by a combination of performance requirements, manufacturing needs, and cost. (See section on Costs and Pricing.)

Table 30-15 lists the most widely known end uses and product recommendations, but should

Table 30-15. Comparison of mechanical and electrical properties, silicone resin laminates.

QUARTZ VS D GLASS VS E GLASS

Property	E Glass[a]	D Glass[b]	Quartz[c]
Style number	181	181	581
Oz/sq yd	9.0	9.0	8.4
Finish	None	None	None
Mechanical			
Tensile strength psi, dry–room temp.	39,500	25,000	34,000
Compressive strength psi, dry–room temp.	27,300	18,300	24,000
Flexural strength psi, dry–room temp.	–	41,800	36,000
Flexural modulus psi $\times 10^6$, dry–room temp.	–	2.90	2.80
Electrical			
Dielectric constant X-band, dry–room temp.	3.946	3.60	2.93
Loss tangent X-band, dry–room temp.	.0082	.002	.00098

[a]Summary Card, "Mechanical Properties of Aerospace Fiberglas Fabric Laminates," Owens Corning Fiberglas Corp. Office of Aerospace and Defense, April 1966.
[b]Trevarno, F., "130 Silicone Data Bulletin #54b," Coast Manufacturing and Supply Co., January 1965.
[c]Trevarno, F., "Structural Materials Data Bulletin #109," Coast Manufacturing and Supply Co.

Table 30-16. Product recommendations.

End Use	High Silica	Quartz
Ablative		
Re-entry heat shields		
Tape wrapped	Fabric–18.5 oz/sq yd	Fabric–19.5 oz/sq yd
Filament wound		Roving–20 end
Woven-three-dimensional		Yarn–to order
Nose cones		
Compression molded	Fabric–18.5 oz/sq yd –10.3 oz/sq yd	Fabric–19.5 oz/sq yd –8.4 oz/sq yd
Filament wound		Roving–20 end
Woven–three-dimensional		Yarn–to order
Nozzles and exit cones		
Tape wrapped	Fabric–18.5 oz/sq yd	Fabric–19.5 oz/sq yd
Filament wound	Cordage–1/8 in. diameter	Roving–20 end Cordage–1/8 in diameter
Compression molded	Fabric–18.5 oz/sq yd –10.3 oz/sq yd Chopped squares–1/2 in. squares Chopped fiber–1/4 in. length	Fabric–18.5 oz/sq yd –8.4 oz/sq yd Chopped squares–1/2 in. squares Chopped fiber–1/4 in. length
Structural		
Fins and struts		
Lay-up (vacuum or auto-clave molded)		Fabric–8.4 oz/sq yd
Filament wound		Roving–20 end
Compression molded		Fabric–8.4 oz/sq yd
Electrical		
High pressure laminates		Fabric–5.6 oz/sq yd
Antennas		
Lay-up (vacuum or autoclave molded)		Fabric–8.4 oz/sq yd
Filament wound		Roving–20 end
Radomes		
Lay-up		Fabric–8.4 oz/sq yd
Filament wound		Roving–20 end
Thermal		
Blast shields		
Compression molded	Fabric–18.5 oz/sq yd Mat Chopped fiber–1/4 in. length Chopped squares–1/2 in. squares	Fabric–19.5 oz/sq yd Mat Chopped fiber–1/4 in. length Chopped squares–1/2 in. squares
Separators		
Compression molded	Fabric–18.5 oz/sq yd	

not be treated as a complete listing. New products and end uses are being developed continually, and the product manufacturers should be contacted for the most current information.

References

1. Dhingra, A. K., Champion, A. R., Krueger, W. H., "Fiber Reinforced Aluminum and Magnesium Composites," E. I. duPont presented, 1st Institute for Defense Analysis Workshop on Metal Matrix Composites, Washington, D.C., July 1975.
2. 3M Company's Product Bulletin "New 3M Ceramic Fiber Products Continue Where Others Leave Off."
3. ICI's Product Bulletin #NV/84/IED/103/274 "Saffil Fibers As High Temperature Materials."
4. Shulock, H., "High Silica and Quartz," *Handbook of Fiberglass and Advanced Plastics Composites*, Chapter 9, ed. G. Lubin, Van Nostrand Reinhold, 1969.

31
Metal Filaments

John A. Roberts
Brunswick Corp.
Skokie, Illinois

CONTENTS

1. INTRODUCTION

Metal fibers as fillers or reinforcements for plastics possess two distinguishing features, namely, metal characteristics, and a controlled fiber geometry.

The fiber geometry lends itself to fabrication of a variety of product forms, such as those associated with paper making and textile processing techniques. The paper making process involves low aspect ratio (length to diameter) fibers and produces porous fibrous sheets. Standard textile processing, which requires high aspect ratio fibers, leads to products in the form of tow, sliver, yarn, carded or air layed web, woven and knit structures. These fibers are available in a variety of metals and alloys and in various conditions resulting from heat treatment or working. As a consequence there is a choice of the wide range of products, and a high degree of flexibility in the application of the product to metal/plastic composites. Advantages can be realized either by the ease of fabrication of the composite or designing the composite to have optimum properties related to the metal fiber characteristics.

The disadvantages of metal fibers for fillers and/or reinforcement can be generally related to their cost and density.

Metal fiber costs vary greatly depending on base material cost, fiber diameter and length, product form, and specific properties required. The range is from less than \$1/lb for steel wire wool to \$50/lb for stainless steel of uniform fine diameter and consistent mechanical properties.

A definite functional need, utilizing some unique metal characteristic and/or fiber geometry advantage, must be identified in order to justify the incorporation of the metal fiber into a plastic matrix.

Metal fibers have properties and characteristics which are dependent on both the material and the process by which they are manufactured. Shape, cross-sectional area uniformity, surface roughness and surface chemistry variations can be functions of the fabrication process while chemical, physical and mechanical properties are generally similar to those of the bulk material.

2. METHODS FOR PRODUCING METAL FIBERS

2.1 Mechanical Processes

Mechanical processes for producing metal fibers can be divided into two groups, those which involve attenuation, e.g., wire drawing, and those which rely on the sub-division of a prefabricated metal form, e.g., foil slitting, broaching, shaving, turning, etc.

2.1.1. Attenuation Processes. Conventional wire drawing has provided a source of metal fiber for many years; however, this cumbersome process requires multiple reduction steps leading to low machine production rates at the smaller sizes (less than 0.003-0.005 in. diameter) and results in an extremely expensive product. Applications of these smaller sizes are limited mainly to the electrical and electronics industry.

Modifications of the wire drawing process to allow many wires to be reduced in size at the same time, and commonly referred to as 'bundle drawing', have eased the economic situation considerably. A schematic illustration of the basic concept is shown in Fig. 31-1. Bundle drawing processes routinely provide a product in continuous filament form and equivalent diameter in the range of 0.0005 in. At such diameters, the metal fiber product is consistent with manipulation on conventional textile processing equipment which further adds to the economic attraction.

Bundle drawing was pioneered by Brunswick Corp. and the process has recently been licensed to Bekaert AB. The process is flexible in terms of the variety of metals and alloys which can be fiberized and the ultimate fiber

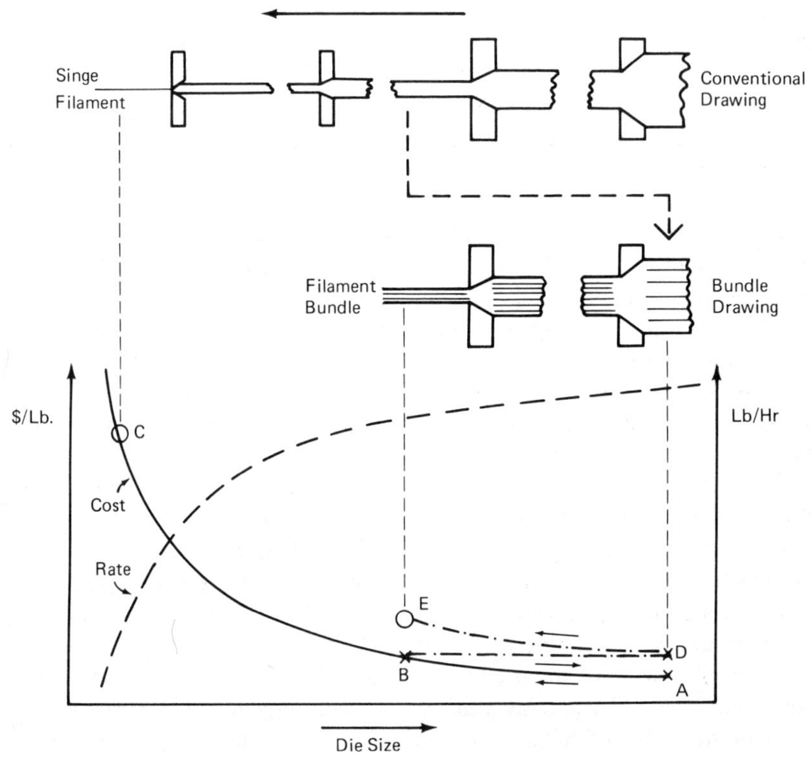

Fig. 31-1. Schematic illustration of the fundamental economic advantage of bundle drawing fine metal filament. To arrive at the same filament diameter, conventional wire drawing follows the cost curve ABC, while bundle drawing operates only in the cost insensitive region and follows the discontinuous curve ABDE.

diameter. Certain metals, e.g., niobium and tantalum, have been reduced to the sub-micron level (less than 0.00004 in.). The constraints on the process lie in the selection of suitable matrix materials, which serve to separate the individual fibers during the attenuation process. The matrix material must have several specific properties. First, it must be compatible with the reduction process, providing uniform plastic flow and consequently a uniform fiber product. Secondly, the matrix material must be chemically, mechanically, or thermally separable from the fiber without causing degradation to the properties of the fiber itself.

The bundle drawing process has been applied to a wide range of metals and alloys, a general listing of which follows:

Ferrous base alloys
 –stainless steels
Nickel and cobalt base super alloys
Titanium and titanium alloys
Refractory metals
 –tantalum, niobium
Precious metals
 –silver, gold, platinum

Fibers produced by the process are in a diameter range from sub-micron (less than 0.00004 in.) to 100 μ (approx. 0.004 in.), which is the upper limit corresponding to the economically competitive process of conventional wire drawing. The fibers are characterized by a degree of surface roughness and non-uniform shape. The product can be additionally characterized in terms of a coefficient of variation (standard deviation/mean) of area of cross section, which can be controlled to some degree by the producer. The coefficient of area variation depends on process parameters including:

- volume fraction of fiber and matrix materials
- mechanical property match of fiber and matrix materials
- degree of working between anneals
- fiber diameter required.

In addition, certain materials such as steels must be of exceptionally high purity to prevent serious degradation in fiber properties resulting from non-ductile inclusions. Such effects are dramatically illustrated in Fig. 31-2 for two steels of differing purity but processed identically. If similar data are presented in a different way, Fig. 31-3, it becomes obvious that the length of the fiber being tested is important and a critical length can be defined at which the fiber effectively exhibits bulk properties.

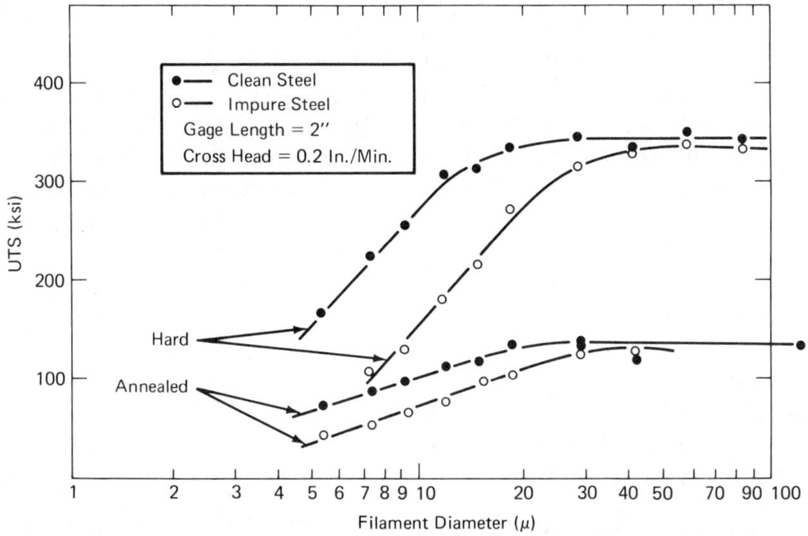

Fig. 31-2. Tensile strength vs filament diameter for 300 series stainless steel filament manufactured from a clean heat of steel and a heat containing a relatively large number of non-metallic inclusions.

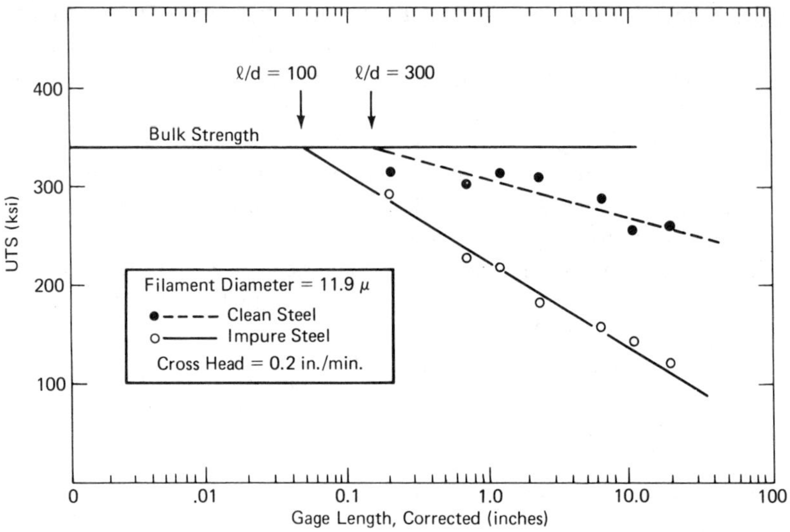

Fig. 31-3. Tensile strength vs gage length for 12μ stainless steel filament manufactured from the heats used to generate the data in Fig. 31-2.

2.1.2. Material Sub-division Processes. In this category of mechanical fiberizing processes are included slitting or shearing of foil, broaching, and shaving. All of these processes rely in principle on sub-dividing wrought or cast metal. Depending on the techniques used, products with a wide distribution of length, shape, and cross section area (equivalent diameter[a]) can be expected, together with considerable variation in mechanical or structural soundness.

The foil slitting approach has been widely applied, and a number of patented processes or equipment designs are described in the literature. The product reflects the inherent properties of the foil from which it is made; one dimension is governed by the foil thickness while the width is controlled by the degree of resolution of the slitting or shearing operation. Fiber has been produced in materials such as Al, Cu, and steel with cross section dimensions of 0.001-0.002 in. by 0.0005 in. (foil thickness) but more typically the product is in the size range 0.002-0.005 in. by 0.001-0.002 in. by 0.500 in. long if slit normal to the foil length. Much longer fibers are feasible if slitting is along the length of the foil. Economy in processing is achieved by slitting or shearing a stack of foils simultaneously.

[a]Equivalent diameter is defined as the diameter of a circle whose area is identical to the cross sectional area of the fiber.

The broaching and shaving processes differ from the foil slitting techniques in that they continuously remove material (fiber) from the surface by employing a suitably shaped cutting tool. The product in this case generally bears a shape (cross section) associated with the cutting tool profile, typically approximating a triangle or wedge, the magnitude of the cross section being a function of the bite employed.

Patented processes and machine designs utilizing shaving heads to accommodate hot rolled steel rod stock produce fiber on a tonnage basis and the product is familiar to many as steel 'wool.' The size range classification used by the industry gives an approximate indication of cross section in terms of the maximum dimension or width.

Classification	Mean fiber width (in.)
0000	0.0006-0.0010
000	0.0008-0.0013
00	0.0012-0.0022
0	0.0016-0.0024
1	0.0018-0.0026
2	0.0040-0.0060
3	0.0070-0.0095

Federal Specification FF-S-740a Oct. 1965.

A fiber, designated Type A, produced by a proprietary process by Brunswick Corp. and incorporated in the product Feltmetal® fibermetal (FM 1100 series), bears many similarities to the broached or shaved product described above but with a much shorter length and generally finer diameter. The Feltmetal® fibermetal product is a sintered porous structure of controlled density and pore size and is available in a wide selection of metals and alloys. The fiber structure has been well characterized and comprehensive literature is available on properties, applications, machining, forming, and joining. The FM 1200 and 1300 series of fibermetal product embody fibers made by different processes and in increasing fiber diameter. The B-type fiber of the 1200 series is characteristic of a product produced by conventional shaving or broaching while the C-type fiber incorporated in the 1300 series is derived from a wire product.

2.2 Melt Spinning and Related Processes

Melt spinning in the following context is taken to encompass all processes which form a fiber directly from the liquid phase. Particular melt spin processes, by analogy with standard manufacturing methods for synthetic organic fiber and glass fibers, could be expected to be the most economical and also consistent with large volume production. However, certain inherent properties of liquid metals, such as the relatively high surface tension together with the low viscosity, the first order transformation for liquid to solid, and the crystalline nature of the metallic state, preclude the direct application of polymer or glass melt spinning technology to metals. Many techniques to circumvent the technical difficulties of metal fiber forming from the melt have been developed and may, for further discussion, be conveniently divided into the following two groups.

1. Processes involving the solidification of a liquid jet emerging from an orifice into a fluid medium with:
   a. simple heat transfer from the emerging jet to the fluid surroundings (Pond, United Aircraft, Brunswick, Rolls Royce)
   b. a secondary surface reaction to effect jet stabilization prior to solidification of the core metal (Battelle, Brunswick, Monsanto, United Aircraft, Clevite)
   c. encapsulation of the molten stream with a second material, generally of a glassy nature (Taylor[a], Brunswick/ADL, Battelle).
2. Processes relying on solidification by direct contact with a moving solid heat extraction surface on which either a liquid metal jet impinges or where a solid surface is brought into momentary proximity with a reservoir of liquid metal. No orifice is employed in this process.

Consider first the group employing jet solidification in a fluid medium. The basic requirement in this type of process is to extract heat from the jet and thus promote solidification before jet instabilities develop. The destabilizing forces acting on the jet, such as surface tension, aerodynamic drag, and mechanical vibration, result in a tendency for the liquid stream to break up into discrete lengths or droplets. Therefore, the process parameters such as jet velocity, heat transfer environment, orifice design, etc. must be carefully adjusted for each type of metal fiber desired.

Processes employing several variations of the technique have been successful in producing fiber in a variety of metals and alloys, but over a rather limited size range. It would appear at this time that a diameter limitation is inherent in the process with a realistic minimum being in the region of 0.001 in. (25 μ). The limitation is, in major part, a consequence of an inability to develop significant attenuation in the fiber after it exits from the orifice. (The glass encapsulation process referred to above allows attenuation but suffers from the high glass to metal ratio required for process control). With this restriction, the process becomes economically unattractive at very small fiber sizes because of the low mass flow rate (pounds of fiber per orifice per hour) and the cost of maintaining small orifices. A further difficulty is experienced with compatability of both crucible and orifice materials with higher melting point and

[a]No orifice employed in original Taylor concept.

more corrosive or reactive metals. In spite of the shortcomings of the process, many metals have been made for commercial application or test marketing. Tin, zinc, aluminum, and various alloy filaments have been produced in experimental quantities down to 0.0015 in. diameter and in continuous lengths. Steel has been processed on a pre-pilot plant scale for tire cord application with individual filaments of 0.004 in. diameter.

The second group of liquid metal processes for fiber forming to be described is generally referred to as the melt extraction techniques. The processes were developed by Battelle Development Corp. and eliminate some of the serious limitations of the jet solidification method described above. The Crucible Melt Extraction process circumvents the need for an orifice by providing a revolving heat exchange disc, the periphery of which just dips under the surface of a molten pool of metal in a crucible. Solidification takes place on the disc periphery and a fiber is continuously extracted as the disc rotates. The fiber is released as a result of the centrifical force experienced following exit from the liquid pool. Staple fiber can be produced by introducing notches on the periphery of the mold or heat exchange disc which result in gaps in the solidification and consequently a series of controlled length fibers.

A second version of the melt extraction process involves the use of a pendant drop of liquid metal, supported by surface tension, and into which the spinning heat exchange disc is inserted. This process eliminates the need for both an orifice and a crucible, thus offering the potential of fiberizing highly reactive and high melting point metals in an economical fashion.

The effective diameter of fibers produced by the melt extraction process would appear to have a practical lower limit of approximately 0.001 in. The cross section shape can vary from near circular to cresent depending upon the velocity and contour of the mold periphery and the depth of insertion into the molten metal. The smaller the fiber, the greater the tendency for a circular cross section to result.

The structure of the materials produced by melt extraction is generally fine grained close to the cooling surface and dendritic beneath the surface, which is consistent with the rapid cooling rates of the casting process. Claims by the developers indicate that cooling rates can approach those associated with 'splat cooling.' Under such conditions, certain appropriately alloyed metals are observed to exhibit an amorphous structure and embody many unique properties.

2.3 Miscellaneous Metal Fiber Forming Processes

Attempts to obtain metals in fibrous form by an economical process or with unique properties have lead investigators to explore many avenues in addition to those already described. Their efforts have met with mixed success. Examples of four such investigative or developmental areas will be taken as respresentative of this miscellaneous group of processes.

2.3.1. Electro-plating. The electroplating approach to fiber forming is based on the reasoning that building up a fiber by deposition rather than attenuating a massive form would be a logical approach to economical filament manufacture. A number of novel designs for a suitable mandrel on which to plate the filament appear in the literature. Of particular interest is the concept described by Newton, which is a cylindrical drum constructed to define a helical, electrically conducting groove so that continuous deposition and wind up can be achieved by rotating the drum. The method was used to produce Ni filament of roughly rectangular cross section 0.001 in. × 0.0005 in. The mechanical properties reflect the extremely fine grain size, with filaments exhibiting a measured tensile strength of approximately 200,000 psi. Further development of the process is required but the recent introduction of a commercial process for electroformed iron strip gives reason to direct new attention to the concept.

2.3.2. Vapor Plating. The application of vapor plating techniques to fiber forming again uses the logic of the electroforming approach. There are two general process classifications, one where the composition of the deposited material is identical to the vapor phase (e.g., Al deposited from Al vapor), the second where the deposited material composition is the result of a vapor or gas phase reaction (e.g., Ni from Ni $(CO)_4$). In both cases, deposition takes place on a substrate but in the latter process the substrate can be heated to promote decomposition, as might be the case in the reduction of $TiCl_4$ to Ti.

After deposition has occurred, the option exists of carrying the substrate material into the final product application or removing it by thermal, chemical, or mechanical means. The direction taken depends very much on the final application and the compatibility of the substrate material with the end use environment. For example, if the substrate is an organic fiber and the plated product is intended for high temperature filtration (as was the case with Ni fiber webs manufactured by Fram Corp.), it is necessary to pyrolyze the product prior to use in order to avoid possible gaseous contamination by the filter.

Vapor plating is used to prepare a variety of fibrous or fiber-like products, many of which are commercially available. These products include the metallized thread used in decorative textile application. Aluminum deposited on glass fiber or plastic film, which is subsequently slit to form a fiber or ribbon, is used for such diverse applications as radar reflection and tinsel for decorative purposes. In the case of the high modulus boron filament (typically 0.004 in. diameter), the substrate material is generally 0.0005 in. diameter tungsten wire, but since the tungsten represents only a small fraction of the cross section ($<5\%$) and boron is non-metallic in properties, this product will not be considered in the metal fiber category. Boron filaments are discussed in Chapter 28.

A highly specialized technology involving vapor deposition and decomposition processes has been developed for the production of metal and ceramic whiskers. These materials are fibers but composed essentially of one crystal. The conditions for producing whiskers require that growth on the substrate or condensing surface be highly directional and confined to limited regions. Most of the recent developments of the technology have related to the high modulus ceramic materials. It does not appear that commercial processes for large volume metal whiskers at economical manufacturing cost are available although the potential of such fibers, in view of their exceptional mechanical properties, justifies continued process development. Experimental samples of iron, copper, and tin whiskers have been produced. Ceramic whiskers are discussed in detail in Chapter 25.

2.3.3. In Situ Decomposition of Inorganic Compounds. Many attempts have been made to apply chemical decomposition, such as the reduction of oxides and salts, to form fiber directly. Union Carbide Corp. developed processes based on the concept of infiltrating an organic filament precursor with a solution of the desired metal salt and then chemically reducing the compound to form a metal filament. Claims were made for an exhaustive list of potential fiber products with diameter, length, and form corresponding to the organic fiber precursor.

2.3.4. Slurry Forming. The concept for fiber forming using a slurry involves combining a metal powder with suitable carrying agent to allow the forming of a filament by extrusion or coating of a substrate. For high mechanical integrity in the filament, this process may be followed by removal of the binding agent and sintering of the powder particles.

Spindletop developed a hollow filament product in the 1–10 mil diameter range based on the above technology and claimed that their process was applicable but not restricted to stainless steels, Ni, and Co. More recently, a class of electrically conducting fibers, some containing metal powders rather than carbon black, have been developed to control static electricity in textiles. These fibers are made by one or other of the processes described above. Teijin has produced a silver coated nylon and although the electrical conductivity is low, since it involved only point contact between powder particles, it is sufficient to reduce the build up of static charge in textiles.

3. SUMMARY OF PROPERTIES

3.1 Fiber Properties

The chemical, physical, and mechanical behavior of metal fibers as a class of materials will generally approximate the bulk material properties. Exceptions may arise due to the influence of the fiberizing process or as a result of the diminished cross section (diameter) and increased length.

The outstanding characteristics of metals and alloys are summarized in Table 31-1. Detailed descriptive literature is available in the standard metal reference works, such as 'Metals Handbook' Volume 1, Edition 8 published by the American Society for Metals.

Deviation from bulk material properties which occur as a result of the fiberizing process are briefly summarized in Table 31-2. The property alteration may or may not be advantageous, and in many cases the magnitude of the effect can be controlled by modifications of the processing technique employed. It is also conceivable that post forming operations can be carried out which could radically change the property characteristic of interest. Heat treatment, for example, can be employed to cause recrystallization and hence eliminate the preferred crystallite orientation associated with wire drawing where such a fiberizing process is employed. In addition, heat treatment can promote diffusion and homogenization and thus change the reactivity of a fiber with minor surface chemistry alteration due to processing. Still further examples of the effect of heat treatment on fiber properties would be the transition of cold worked 302 or 304 type stainless steel from a ferromagnetic to a paramagnetic state, or the development of ductility in a cold worked fiber by inducing recrystallization.

The effect of diminished cross section and increased length on a specific fiber property can have far reaching impact. The particular or intended application must be fully analyzed in order to choose the optimum fiber product. A brief listing of some of the properties where physical size may have a major influence is presented in Table 31-3.

Table 31-1. Outstanding characteristics of metals by general category.

Metal or Alloy	Outstanding Characteristics
Carbon and alloy steels	High strength, high modulus, low cost, ferromagnetic properties
Stainless steels	Good corrosion resistance
Nickel and cobalt base superalloys	Strength at elevated temperatures, oxidation resistance
Refractory metals and alloys, e.g., Ti, Ta, Nb	High melting points, good corrosion resistance
Copper and copper alloys	High electrical and thermal conductivity
Aluminum and aluminum alloys	Low density, high electrical and thermal conductivity, low cost, can be anodized to produce extensive color range
Precious metals, e.g., Pt, Ag, Au	High electrical and thermal conductivity, the noble metals are particularly corrosion resistant

3.2 Fiber Geometry

In the foregoing discussion on fiber properties, the importance of the geometrical characteristics of the individual fibers is clearly brought out. The quantification of such characteristics is not simple, and no attempt will be made here to apply any detailed ranking. It is useful, however, when considering the application of fibers in composite products, to have a guide as to the cross section shape and uniformity as well as the surface topography. These characteristics can change dramatically from process to process and from material to material and in some cases from one supplier to the next. An attempt is made in Fig. 31-4 to schematically illustrate fiber cross sections that would typify the various fiberizing processes listed. No size (diameter) comparisons between fibers of the different processes should be drawn from the figure. Where two or more shapes are shown for the same process, this is either indicative of the flexibility of the process to control the cross section or an inherent characteristic associated with the absolute physical dimensions. A series

Table 31-2. Property alteration associated with the fiberizing process.

Fiberizing Process	Possible Deviation from Bulk Material Properties
Conventional wire drawing	Strongly developed wire texture (preferred orientation of crystallographic directions with respect to wire axis)—particularly in unannealed or hard drawn filament. Such preferred orientation effects can give rise to directionality of the basic properties of the metal in the filament product
Bundle drawing	Preferred orientation effects can be expected with consequent directionality of properties. Surface chemistry alteration can occur and hence changes in initial reactivity. Small cyclical variations in cross section area can be present, resulting in an apparent property change, for example, in mechanical strength, which is based on the 'weakest link' (smallest cross section area) in the length being measured
Foil shearing or slitting	Fiber properties will reflect the parent foil properties which may or may not show preferred orientation. If preferred orientation is present, the fiber properties will depend on the direction of shear, that is, either parallel or perpendicular to the rolling direction in the foil
Shaving and related processes	The occurrence of partial fractures in fibers during processing results in weakened areas along the fiber length which constitute an apparent change in mechanical strength
Melt spin	Surface chemistry change associated with the product of the designed reaction to promote stabilization during fiber forming
Melt extraction	Extreme cooling rates produce fine grain size (particularly on the side of fiber in contact with extraction wheel) and can produce amorphous like structures which will result in major property alteration. This latter phenomenon is exploited to the fullest in a product called 'Metglas' but which is not generally categorized as a fiber. The amorphous state is metastable and the unique properties disappear on heating

of scanning electron microscope photographs of representative samples of fibers produced by the various processes are presented in Figs. 31-5 through 31-19. Photomicrographs are shown for two levels of magnification in most cases in order to highlight the surface detail (high magnification) and provide a general field of view. The latter contains many fibers, allowing for a judgment in the range of size and shape and to some degree the "straightness." A brief description and the source of the sample is given in each case.

Table 31-3. Size related property alteration.

Property	Effect of Diminished Cross Section and Increased Length.
Corrosion	High surface to volume ratio provides for relatively large reaction surface and hence limited life in corrosive environment
Electrical conductivity	Small cross section provides efficient use of material for high frequency electrical conduction since the fiber radius can approach the skin depth
Mechanical properties	Sphere of influence of defects (e.g., inclusions in steel) associated with certain fibers approximates to the fiber diameter and results in reduced or degraded mechanical properties for the free fiber
	Whiskers (single crystal fibers) can be essentially dislocation free and approach theoretical strength

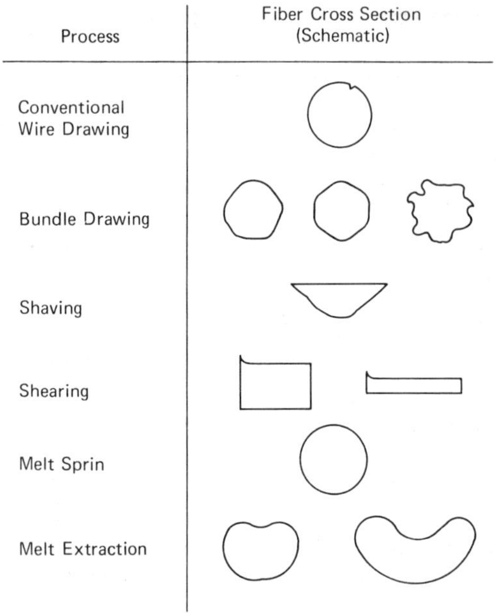

Fig. 31-4. Typical fiber cross sections.

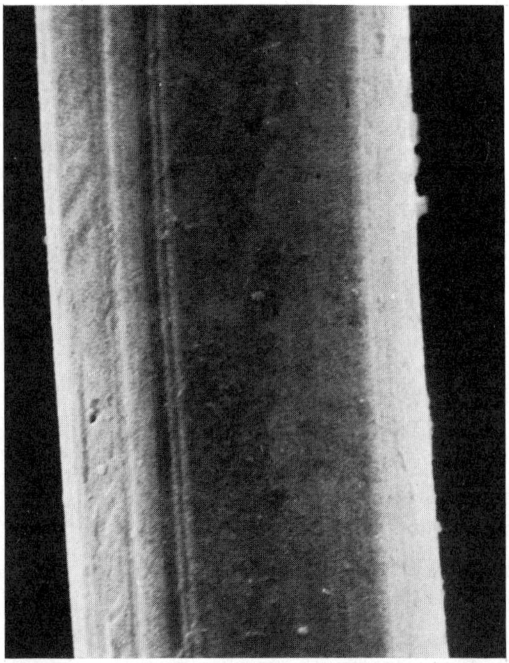

Fig. 31-5. Conventional wire drawn 304 stainless steel, 0.001 in. diameter (nominal). Burnished surface with minor draw die defects (grooves). Mag. 1700X.

Fig. 31-6. Bundle drawn 304 type stainless steel showing pronounced longitudinal markings and surface irregularities. Filament diameter 0.0003 in. (8μ). Mag. 2200X.

Fig. 31-7. Grouping of bundle drawn filaments illustrating good uniformity. Product manufactured by Brunswick Corp. and identified as BRUNSMET MF Al. Mag. 500X.

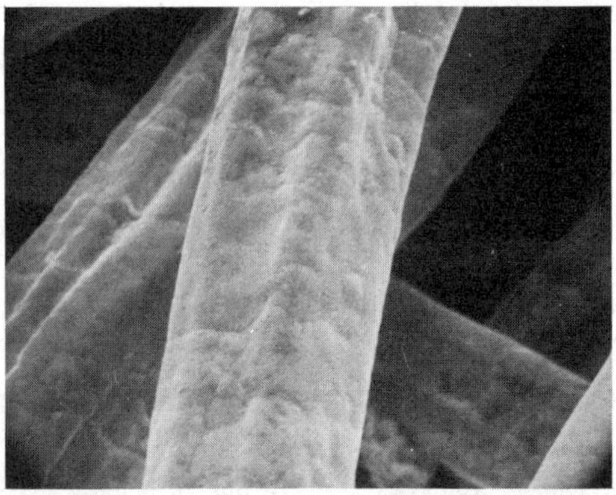

Fig. 31-8. Bundle drawn 304 type stainless steel, diameter 0.00078 in. (19μ). Annealed condition. Thermally etched grain boundaries are clearly visible. Mag. 1800X.

Fig. 31-9. Grouping annealed bundle drawn filament, product manufactured by Brunswick Corp. and identified as BRUNSMET® MF Al. Mag. 450X.

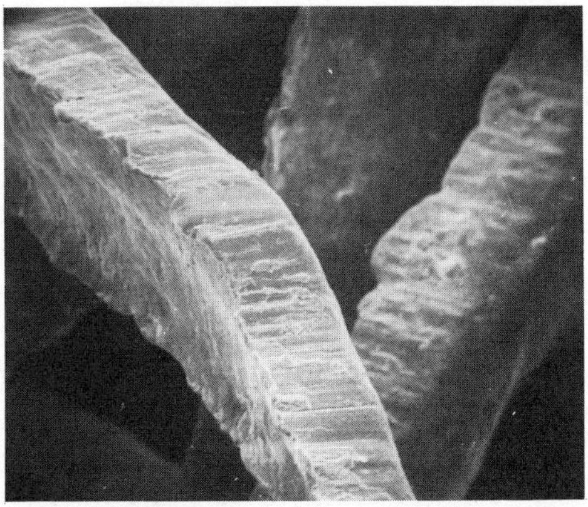

Fig. 31-10. Sheared low carbon steel fiber. Dimensions 0.002 X 0.002 X .062 in. Flowed metal caused by the shearing action is clearly visible. Product originally manufactured by Fiberfil Div., Rexall Chemical Co. and identified as Filamets™. Mag. 450X.

Fig. 31-11. Group of sheared fibers showing variation of metal flow from fiber to fiber. Mag. 180X.

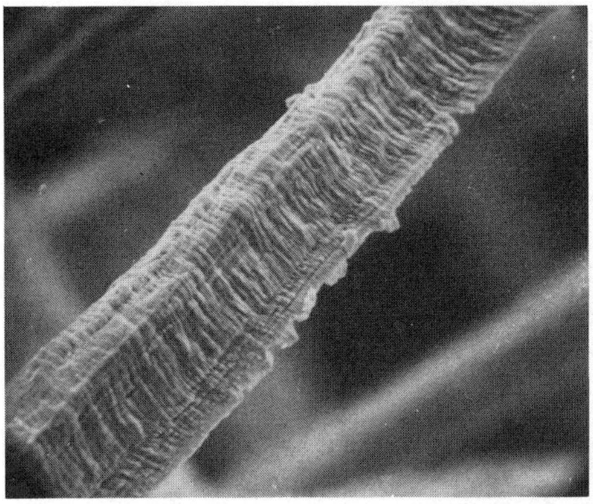

Fig. 31-12. Shaved low carbon steel fiber, 0000 grade. 50% of fibers falling within width limits 0.0006 to 0.0010 in. Deformation marks are visible transverse to the fiber length and also incipient fracture or weak points can be identified. Product manufactured by American Steel Wool Manufacturing Co. Mag. 1000X.

3.3 Product Properties Related to Form

The discussion to this point has dealt with the metal filament product as a raw fiber. This product form is not necessarily the most desirable for end use application. Other fiber based industries have evolved a vast array of first, second, and even third generation products by developing suitable fiber manipulating technology. In many cases, the same technology can be applied to the metal fiber field. This is particularly true when the metal fiber geometry and properties approximate those of

Fig. 31-13. Random array of shaved fibers illustrating significant width to thickness ratio and size variability. Mag. 435X.

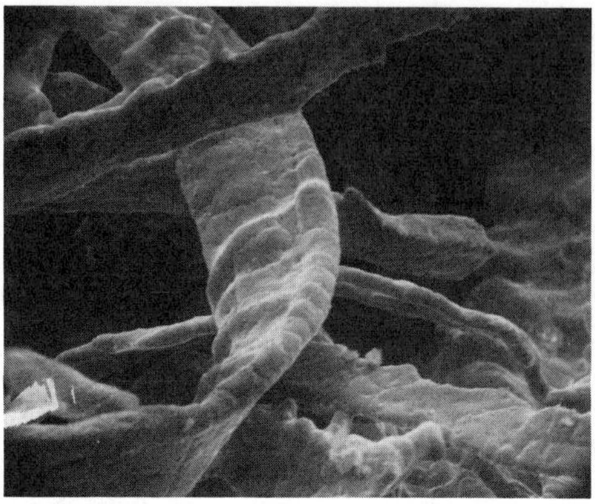

Fig. 31-14. Feltmetal fiber structure detailing surface characteristics of the Type A stainless steel fiber. Product manufactured by Brunswick Corp. and identified as Feltmetal FM 1100 series. Mag. 900X.

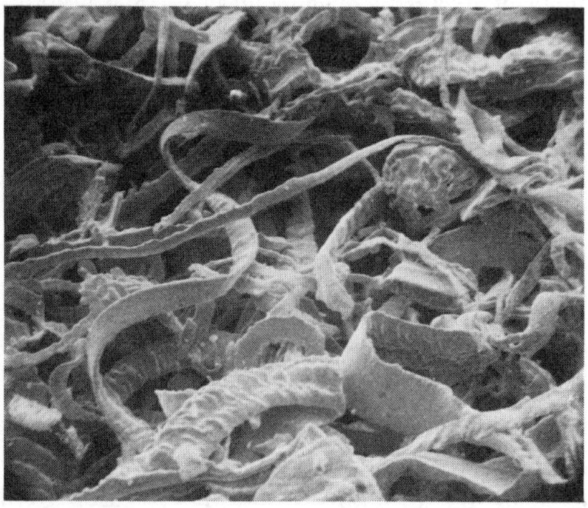

Fig. 31-15. Detail of the surface of Feltmetal FM 1100 series metal fiber structure showing wide range of fiber sections and dimensions. Mag. 180X.

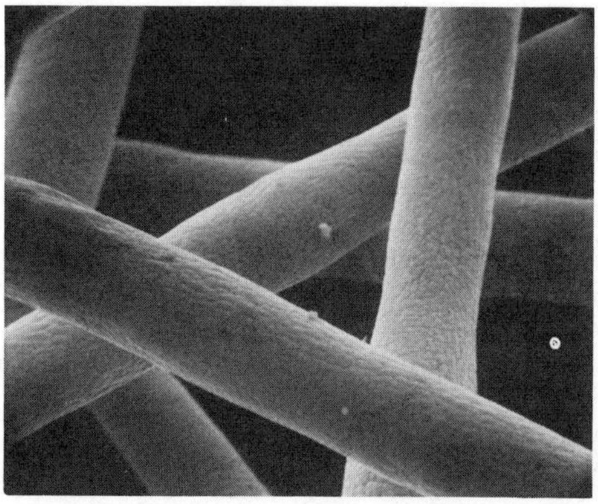

Fig. 31-16. Melt spun aluminum filament showing characteristic surface detail associated with solidification process. Nominal diameter 0.002 in. Produced by Brunswick Corp. Mag. 450X.

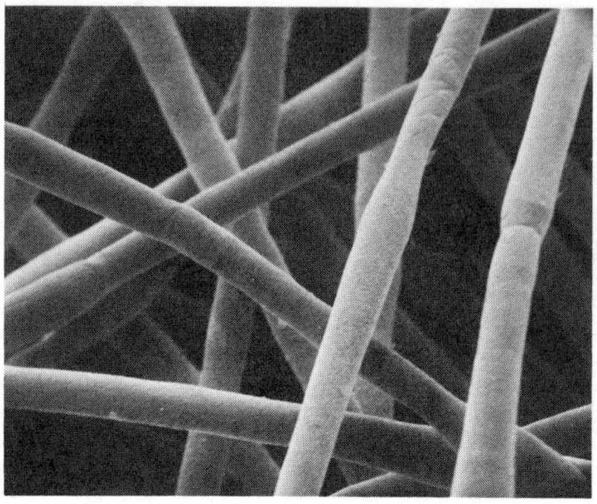

Fig. 31-17. Array of melt spun aluminum fibers showing regions of diminished diameter associated with liquid jet instability during the filament forming process. Mag. 190X.

the natural, synthetic, or glass fibers for which the manipulating techniques and machinery were originally developed. Fiber handling methods can be broadly divided into two categories, primarily associated with fiber length. Short fibers (approximately one half in. or less) are generally handled in a slurry (paper making techniques) while long fibers can be processed by a variety of machinery involving spinning, carding, and airlaying. Each machine produces a new product form or an intermediate form of a more complex final product. In some cases, minor modification of equipment may be

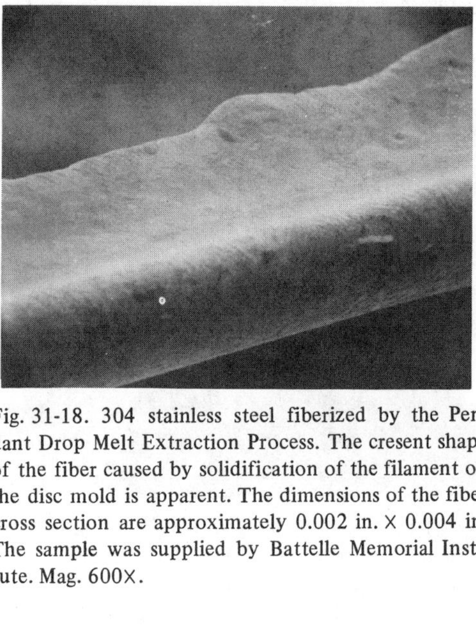

Fig. 31-18. 304 stainless steel fiberized by the Pendant Drop Melt Extraction Process. The cresent shape of the fiber caused by solidification of the filament on the disc mold is apparent. The dimensions of the fiber cross section are approximately 0.002 in. X 0.004 in. The sample was supplied by Battelle Memorial Institute. Mag. 600X.

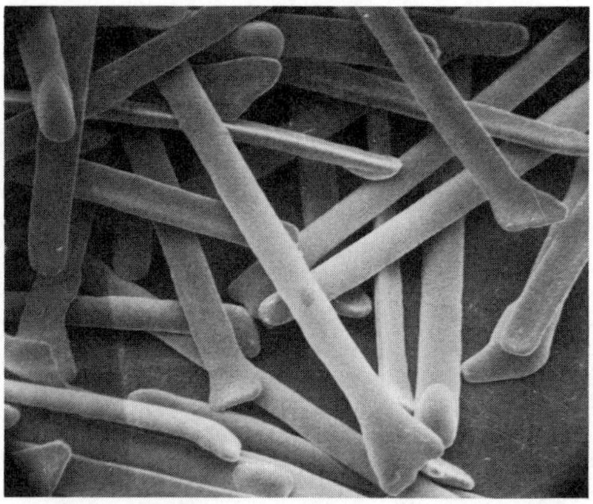

Fig. 31-19. Array of 0.125 in. long PDME process fibers showing dog bone leading end and rounded trailing end. (Nominal width 0.006 in.). Mag. 40X.

necessary to accommodate the metallic properties of the fiber, such as hardness or abrasive character. Also, it is imperative that, in certain cases, completely new process steps such as annealing or heat treatment be incorporated to either facilitate processing or achieve the desired end product properties.

A simplified flow diagram to illustrate the process steps for the more significant product forms is shown in Fig. 31-20. Metal fibers made by a variety of filament forming techniques have been successfully processed in the forms indicated but, of course, the raw fiber must be consistent with the processing techniques to achieve a satisfactory result.

Properties associated with, or specific to, the major product forms are outlined below.

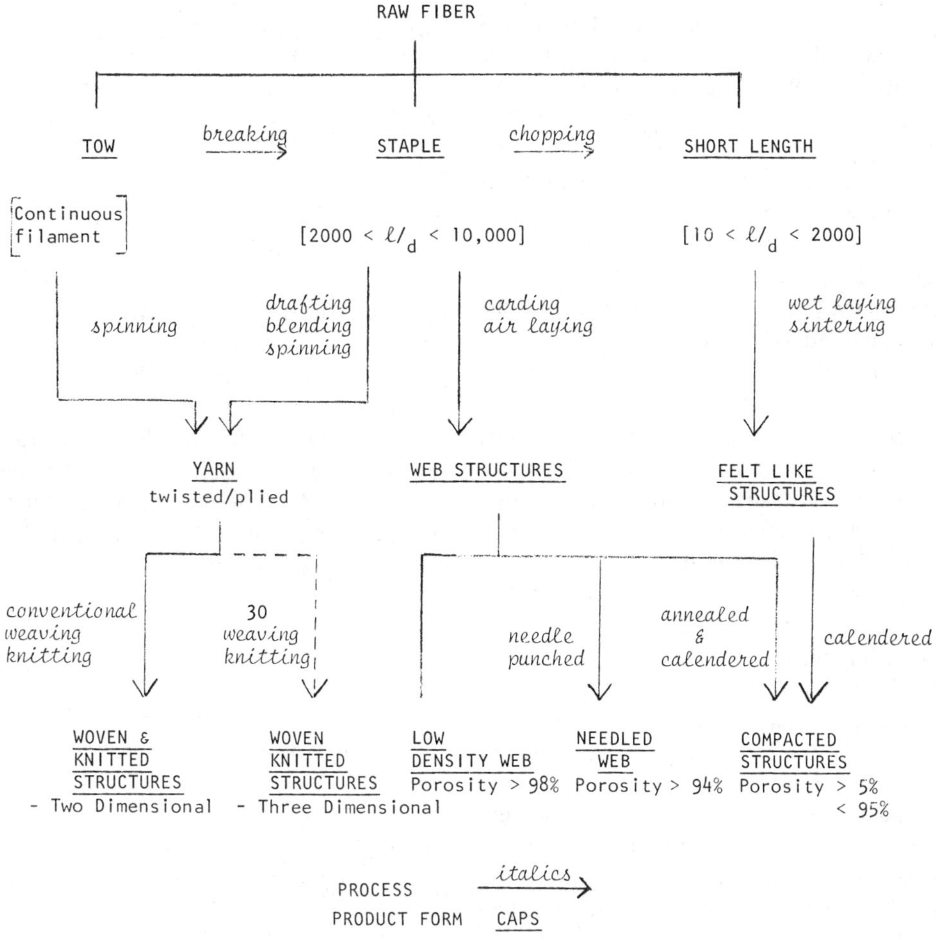

Fig. 31-20. Product form relationships.

3.3.1. Yarn. Continuous filament, staple, blends, and plies properties are a function of twist, staple length, where applicable, and inter-fiber friction. In a properly designed yarn, the breaking load can approach the cumulative breaking load of the individual filaments.

3.3.2. Woven and Knitted Structures. Properties can be expected to be highly anisotropic. Optimizing the principal property in the desired direction or directions for a particular end use application can be achieved by design of the structure.

3.3.3. 'Felted' or Web Structures (Non Wovens). Structures in the low density range (>98% porosity) are available with long fiber lengths only and represent the ultimate in low density porous metal structures. Mechanical properties are a function of density but handle-able products in the 99+% porosity range are practical. Typical web compressibility curves are shown in Fig. 31-21. It can be seen that annealing and sintering of the manufactured web improves the initial or low load characteristics of the web, due probably to fiber bonding (welding). At the high load levels this tendency is reversed and inter-fiber friction and the inherent high individual fiber strength, of the as-manufactured web, predominates.

In the intermediate and higher density range represented by the needled and compacted structures, properties can vary widely in accord with the extremes of density and the type and condition of the parent fiber. By compacting (calendering or pressing) lower density web with suitable annealing and sintering steps, controlled pore size, pore size distribution, and final density can be achieved reproducibly.

Certain properties of such structures as they relate to application in composite materials are worthy of particular note. However, it should be remembered that once the structure is incorporated into a composite, the specific component property can be radically altered. For example, data on the thermal conductivity of a web in air (which can be important in composite fabrication processes) will require extensive correction once substitution of the gas by a solid is made. Likewise, the apparent electrical conductivity of a compacted fibrous structure may be markedly altered, particularly if the second component of the composite has a significant electrical conductivity.

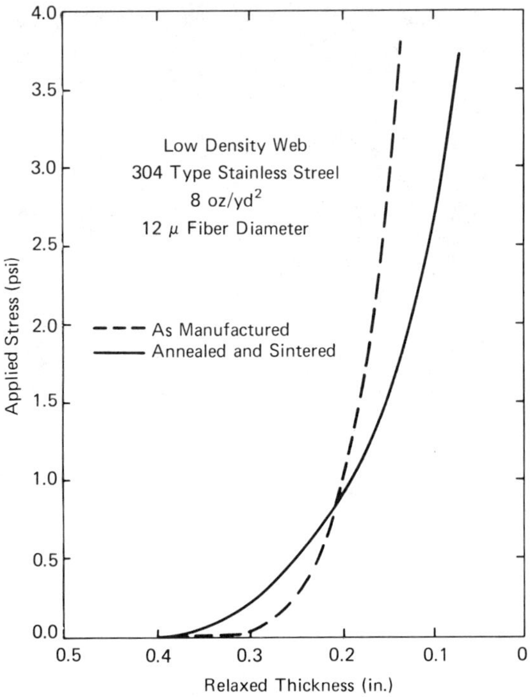

Fig. 31-21. Compressibility curves for low density web (98+% porosity). Load applied normal to face of web.

Electrical and thermal conductivity of fibrous web structures have been measured on Feltmetal® and compacted web of stainless steel and a nickel/chrome alloy fiber produced by the bundle drawing process. The thermal conductivity data, measured through the thickness of the materials, is consistent with the relationship derived by Koh and Fontini:

$$\frac{\lambda}{\lambda_0} = \frac{1 - \xi}{1 + 11\xi^2}$$

where

λ = porous structure thermal conductivity
λ_0 = bulk material thermal conductivity
ξ = porosity

A similar relationship holds true for the electrical resistivity of fibrous metal structures:

$$\frac{\rho_0}{\rho} = \frac{1 - \xi}{1 + 11\xi^2}$$

Table 31-4. Thermal conductivity.

Material	Density (% bulk)	Specimen Thickness (in.)	Thermal[a] Conductivity Btu/hr/fr/°F	Thermal Conductivity (% bulk)
Ni	17	0.50	0.36	1
Stainless steel	46	1.06	0.28	2.3

[a]Measurement taken through thickness.

where

ρ_0 = bulk material resistivity
ρ = porous structure resistivity
ξ = porosity

Electrical measurements made perpendicular and parallel to the porous structure surface of certain samples indicate a significant anisotropy. It is, therefore, important to be consistent in measurement techniques when transposing from electrical to thermal conductivity using the above equations and those referred to in the original texts.

A specific example of thermal conductivity data obtained for two samples of Feltmetal® composed of Type A nickle and stainless steel fiber with mean diameter approximately 15 μ is presented in Table 31-4.

The tensile strength of fibrous porous structures will be dependent on a large number of parameters including:

fiber diameter
fiber length
fiber mechanical properties
degree of sinter
degree of preferred orientation
density of structure
compacting process.

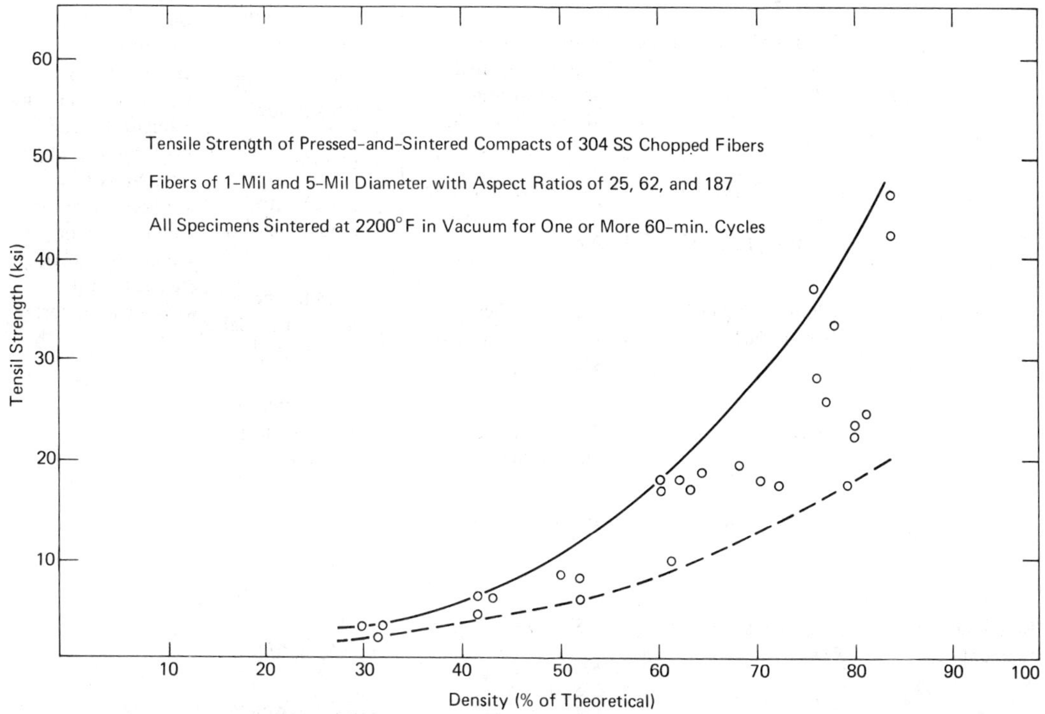

Fig. 31-22. Tensile strength vs density for stainless steel porous structures in the upper density range.

A series of measurements made on 304 stainless steel porous structures of varying density is shown in Fig. 31-22. The data points lie within a broad envelope, the upper bound representing the optimum conditions for high strength in the structure, e.g., high aspect ratio and hard sinter. The data serve to indicate the necessity of identifying and controlling material and process parameters in order to achieve reproducible properties.

4. MATERIAL SUPPLIERS

The following list of suppliers is not meant to be exhaustive, but rather representative of the various metal fiber manufacturing methods. It should be understood that fiber metallurgy is an emerging technology with strong incentive to improve products both from the performance and economic standpoint. As a result, new products and processes are continually being developed.

Process	Suppliers (Representative Listing)	Product Description	Price and Availability	Remarks
Conventional fine wire	National Standard Co. Worcester, Mass. or Niles, Mich.	High carbon and stainless steels	Except for the exotic metals and alloys, price is governed mainly by wire diameter below 0.010 in. As a rough guide one can estimate price as follows:	Special product for refractory concrete and conventional concrete reinforcement N-S Fiber Wire 0.004–0.017 in. diameter 0.250–1.500 in. long
	Fort Wayne Metal, Inc. Fort Wayne, Ind.	Superalloys Ni, Au		
	General Electric Co. Schenectady, N.Y.	Tungsten	$d^a > 10$ mil \$.50–10/lb $2 < d < 10$ mil \$10–100/lb	
	Hoskins Manufacturing Co. Detroit, Mich.	Ni-base alloys, superalloys, high temperature resistant alloys, Cu alloys	$0.7 < d < 2$ mil \$100–1000/lb $d < .7$ mil > \$100/lb	Examples of volume useage would be: steel tire cord wire, Al and Cu magnet wire, W incandescent filament, Precious metal thermocouple wire, Ni base alloy resistance wire for potentiometers stainless steel, Al, etc., as used in woven wire industry
	Kawecki-Berylco Ind. Inc. Hazelton, Penn.	Be (0.002–0.125 in.) Ta Nb	(Small quantity and assuming good ductility)	
	Wilbur B. Driver Co. Newark, N.J.	Cu alloys, heat and corrosion resistant steels	Availability depends on material; where high volume usage is established, immediate availability is common	
	Driver Harris Co.	Ni-base alloys, superalloys		
	Western Electric Co., Inc. New York, N.Y.	Cu, Al		
Bundle drawn filament	Bekaert Steel Wire Corp. New York, N.Y.	S.S. tow (8 μ)	Stock approx. \$50/lb	
		S.S./nylon blended yarn	Stock approx. \$10/lb	

Process	Suppliers (Representative Listing)	Product Description	Price and Availability	Remarks
Bundle drawn filament (*continued*)	Brunswick Corp. Skokie, Ill.	S.S. tow (8 and 12 μ) Ni base superalloy	Stock approx. $50/lb	Many other materials and product forms have been developed and can be discussed with the supplier.
		Feltmetal® sintered porous structure[b] FM 1100 FM 1200 FM 1300 FM 100	Price range approx. $20-100/sq ft depending on material, density, thickness, and fiber type	
Shaving, shearing, slitting	American Steel Wool Manufacturing Co., Inc. Long Island City, N.Y.	Low carbon steel Stainless steels (Ferric)	Typical: low carbon steel wool Grade $/lb 0000 1.14 0 0.73 3 0.72 Lots of 250 lb (pads)	A number of products made by these processes have made an impact in the metal fiber field but have since been withdrawn or their source is not known by the author. Notable in this group is the Rexall Fiber which was available in Al, Cu, Steel, etc. and found application in various composites.
	International Steel Wool Corp. Springfield, Ohio	Aluminum		
	Purex Corp. Carson, Calif.	Usually 7 grades by filament size (width) (see text) 0.0008–0.0080 in.		
Melt spin	Monsanto	Alloy steel–tire cord		Program discontinued-technology available for license
	National Lead	Lead filament 'Plumbers Wool'	< $1/lb	
Melt extraction	Battelle Development Corp. Columbus, Ohio			Developer and licensor of melt extraction processes
	Wilbur B. Driver Newark, N.J.	Ni, Co, and Ti alloys	Will be producing in near future	Pilot stage
	National Standard Niles, Mich.	Ferrous alloys, carbon steel, stainless steel, 0.005-0.030	Estimated cost 0.005 in. steel $0.50/lb,	Setting up for production, Currently supplying

Process	Suppliers (Representative Listing)	Product Description	Price and Availability	Remarks
Melt extraction (*continued*)	National Standard (*continued*)	in. diameter 0.250–3.00 in. long $MELTEX_{TM}$	0.030 in. steel $0.20/lb stainless steel $1.75/lb	.016 in. stainless steel .018 in. carbon steel both 1 in. long
	Ribbon Technology Corp. Columbus, Ohio	Carbon steel 0.016–0.020 in. diameter 1.3 in. or 0.75 in.	$.28/lb stock	
		Stainless steel various grades dimensions as above	$1.25–1.95/lb stock	
Other:				
Splat cooling	Allied Chemical Co.	Metglass® Fe-Ni alloys Be-Ti-Zr alloy	Currently approx. $100/lb Projected price $1–5/lb depending on alloy	Metal with amorphous structure. Unusually high strength
		Ribbon 0.003 in. thick (max) 0.050–0.100 in. width		

[a]d = diameter.
[b]FELTMETAL® is a porous metal structure which is composed of fibers made by a variety of processes and is included here purely for convenience. (See text for fiber identification).

5. COST AND AVAILABILITY

In view of the large number of suppliers and the extensive range of products available both in material composition and product form, the task of enumerating in detail cost and availability would not be practical at this stage. However, a few general guidelines might be pertinent. It is safe to say that where a product has found significant utility, if necessary, outside the plastics reinforcement field, it is likely to be readily available at low cost. Many examples of such materials exist, some of which are cited in the preceding section. Steel wool is a household as well as industrial item, manufactured in great volume with a low cost base material. Advantage of the efficient processing can be carried into new technology and new uses such as providing thermal conductivity in brake linings or rubber goods. A second point to bear in mind when evaluating cost of a fiber product is to consider its value in use. For example, if the objective is to control static electricity in an organic fiber textile product, then only a minute quantity of metal fiber is needed to provide a satisfactory level of conductivity, allowing what appears initially as a high-priced metal fiber to be used economically in a low cost product.

6. HISTORICAL AND FUTURE PRICE AND AVAILABILITY

The development of metal fiber technology is in very early stages. The breadth of potentially fiberizable materials and the many processes by which the product can be produced have yet to be fully explored. The direction for optimization, whether it be based on economy or unique properties, has not yet reached the stage at which basic business projections can be made with confidence. In summary, the metal fiber business is in an embryonic stage where only certain isolated applications have proved to be attractive in a business sense.

As with all new materials or material forms, the acceptance of the product will depend on two basic factors; namely, economy of scale and value in use. Either one or both factors may be important in a particular application. It is likely that, with broad exposure and vigorous product application programs, metal fibers will become significant across a wide range of applications, including their combination with polymeric materials. The current general trend seems to be associated with the conversion of generally low cost metals and alloys already enjoying wide acceptance in the conventional forms. Particular attention has been directed to the ferrous metals, e.g., stainless steels for their obvious advantages in terms of corrosion resistance, which becomes very important in fine fibers. More recently, significant emphasis is being applied to the plain steels where the low base material cost and the newly developed, potentially economic liquid metal fiber forming processes offer the possibility of multi-tonnage usage.

The combination of metal wires with polymeric materials has resulted in the development over the years of many important products. Perhaps the most striking example is the steel cord pneumatic tire currently accepted as the premium product in the industry. Most of these products use wire in a size range of 10-mils and below. The limitation to this upper size range is, in most cases, clearly based on economic considerations. In fact, there may be significant technical advantages in many products if smaller diameter fibers were employed. Such advantages may accrue from:

- improved flexibility
- longer life (fatigue)
- improved bonding (higher surface area/volume)
- more uniform dispersion for equivalent concentration (more homogeneous)

At the present time, applications of metal fibers have been generally restricted to those products where the value in use is compatible with current pricing. These restrictions have necessitated the recognition of the unique properties of the fiber, whether they be mechanical, electrical, magnetic, or thermal. In general, the metal fiber product has, to the present time, not found significant application associated with mechanical properties, either tensile strength or elastic modulus. The greatest application has been in the other property areas involving electrical or thermal conduction and magnetic interaction. Products developed to exploit the high thermal conductivity of metal fibers range from gaskets to plastic dies.

The development of metal fiber/plastic composites based on the interaction of the fiber with electromagnetic radiation has led to products with applications ranging from shielding to military camouflage. The electromagnetic shielding application is associated with cable terminations and components and relies not only on the intrinsic electrical and magnetic properties of the fiber but also on the fiber to fiber contact and distribution within the plastic matrix. In the case of the camouflage application, a suitably pigmented and cut PVC fabric material is provided with the radar signature associated with foliage by the incorporation of a carefully engineered steel fiber less than 0.001 in. diameter.

7. FUTURE UTILIZATION

The previously discussed potential for the fiberizing of metals and alloys with the associated vast spectrum of physical properties makes it difficult to speculate on future utilization. The main consideration will most probably be conversion cost. Low cost would further expand those applications involving the improved electrical and thermal conductivity of plastic materials, and the method of incorporation and processing could give desirable anisotropy to their properties. Improved or new fiber properties may be a second evolutionary development, which could result in a rapid increase in demand by either altering the value in use or providing a special property in the composite. An economical process for the manufacture of metal whiskers, where the strength approaches theoretical values, would constitute a major breakthrough and could provide the potential of ultra-high-strength composite structures. The development of a fiber form of the titanium-nickel alloys, which exhibit a pronounced thermally activated memory effect, and their incorporation in a resilient matrix material, could establish a base for a range of unique products. The recent discovery of 'glassy' metals and the future development of a fibrous form will provide a completely new class of materials. These fibers, with their amorphous structure, would offer properties quite foreign to conventional crystalline metals and their characterization would, of necessity, precede composite design and product development.

Section XII Procedures, Equipment and Utilization Technology

32

Procedures, Equipment and Utilization Technology

Benjamin M. Walker
Walker Engineering Associates
Madison, Connecticut

CONTENTS

1. INTRODUCTION

In order to take advantage of the benefits derived from the use of fillers and reinforcements, the molder must be aware of the special considerations that are involved as compared with the processing of unfilled plastics. Molders who have not been involved with filled plastics may be concerned with their higher melt viscosity and abrasiveness, which may lead to molding problems, lower production rates, and equipment erosion. The uniform dispersion of the filler can also cause difficulty. During recent years, there has been great progress in the effective utilization of fillers and reinforcements, and fast and economical production rates have been achieved in both the compounding and molding of composite materials. In some cases, this has been accomplished by minor modifications of existing equipment, such as the use of hard surface coatings on wear surfaces. In other cases, special equipment has been developed for more effective utilization of these materials. Examples are the new multicomponent spray guns that can spray polyester resin, catalyst, filler, and chopped fibers simultaneously, and the high production rate pultrusion lines.

Compounding and processing objectives are generally the same whether fillers, reinforcements, or combinations of these are added to a resin. Both types of fillers must be uniformly dispersed in the polymer, and good wetting of fillers by the polymers is also desired. However, the processing problems with fillers and reinforcements are different in several important respects, and the compounding and processing equipment and procedures may be different. These two classes of fillers are therefore treated separately as to equipment and utilization procedures.

Types of compounds included are plastisols, filled thermoplastics, filled thermosets, and epoxy, polyester, and other liquid resins. Topics discussed included criteria for equipment selection, types and sources of equipment, and special notes regarding equipment design and use.

2. CRITERIA FOR SELECTION OF EQUIPMENT

2.1 Mixing Equipment

2.1.1. General Discussion. In compounding with fillers or reinforcements, the objectives are generally the same. Both types of additives must be uniformly dispersed in the polymer, and good wetting of additives by the polymers is also desired.

The type of equipment best suited for mixing an additive with a resin will depend upon the nature and proportions of the ingredients used and the physical characteristics of these materials and of the final mix. This will include particle size, viscosity of resin, and viscosity of the mixture. Temperature is also important, as a room temperature blending of a solid powder resin with a filler may only require a dry tumbling operation. The same resin at a higher temperature may be a viscous liquid, and require a more intensive kneading or milling type of mixing.

The forming process to be used with the compound will also affect the thoroughness of mixing and type of mixer required. A preliminary dry blending of granules and filler may be sufficient if the compound is later to be screw extruded or injection molded in a reciprocating screw machine, because these processes will provide a great deal of additional mixing. Compression or transfer molding, or plunger injection molding will require a more intimate mixture of materials to insure a more uniform product.

2.1.2. Fillers. Resins may be dry-mixed at room temperature, and this may vary from a dry tumbling of powder and pellets, to fine grinding or milling the pellets for more thorough mixing. The same resin pellets or powder may be heated to an elevated temperature where they become viscous fluids, and a more intimate dispersion of filler and resin may then be obtained. The type of mixer required for this higher temperature mixing will depend upon the viscosity of the melt and the amount of power required to mix it.

With thermosetting resins, such as phenolics, alkyds, epoxies, and polyesters, it is important to control the temperature during mixing. Room temperature mixing is used, and it is important to avoid heat build-up in grinding and mixing, which would cause precure of the resins.

2.1.3. Reinforcement Fibers. When fiber reinforcements are mixed with resins, different procedures may be required than for powdered fillers. Fibers are most effective in providing reinforcement in compounds when they maintain a high length to diameter ratio. It is, therefore, desirable to minimize the breaking up of fiber length during mixing. Such intensive, high shear mixers as Banbury, roll mills, or screw extruders must be used judiciously with such compounds. Some breakdown in fiber length by such mixers may be acceptable, but excessive breakdown may reduce the effectiveness of the fiber reinforcement. (See Fig. 32-1, photo of fibers with mixing, and Fig. 32-2,

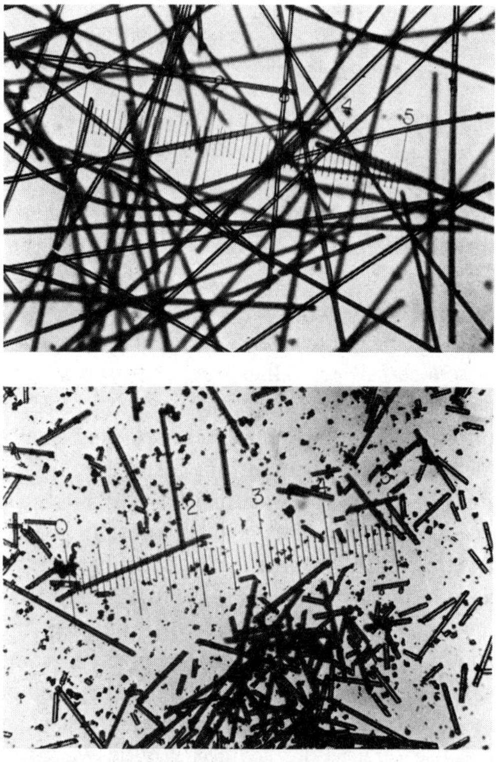

Fig. 32-1. Effect of mixing equipment on glass fiber length.
Upper–using twin screw extruder
Lower–using intensive mixer. (*Courtesy Baker Perkins, Inc.*)

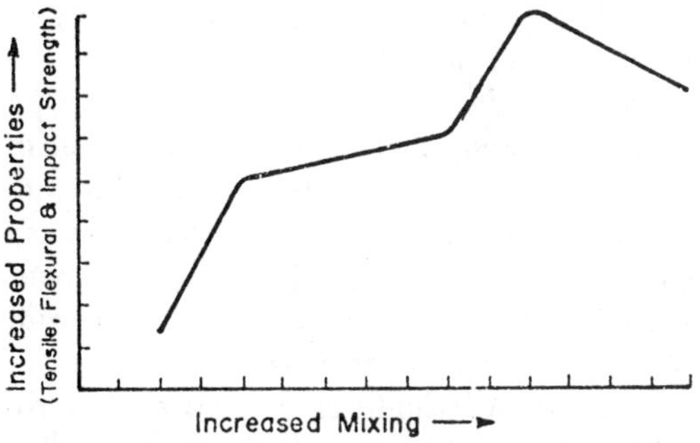

Fig. 32-2. Effect of degree of mixing on properties. (*Courtesy Liquid Nitrogen Processing Co.*)

curve of properties vs mixing). An optimum balance is required between good mixing and wetting of fibers with the resin and retention of fiber length.

For retention of maximum fiber length, special processes have been devised. For example, continuous strands of glass fibers may be fed through an extruder die, and thermoplastic melt extruded over the fiber. Other processes, especially adapted to liquid resins, involve combining fibers in thread, cord, mat, or fabric form with liquid resins applied with spray guns, rolls, or doctor blades.

Note: Patent litigation is presently in process on this subject of glass-filled, thermoplastic resins and their processing, and its status should be checked.[23]

2.2 Processing Equipment

2.2.1. General Notes. Plastics compounds containing fillers and/or reinforcements may be processed with the complete range of processing equipment used for unfilled plastics. This includes screw extruders and compression, transfer, and injection molding machines, and other forming methods. Processes are essentially the same, but the details of conditions may be somewhat different. These differences are due to such factors as increased viscosity of the compound melt at processing temperatures, with the corresponding decrease in flow rate and the need for using higher pressures.[1,2,19,20,22,24,25]

2.2.2. Wear. Wear of equipment or molds is essentially the same in processing filled or unfilled compounds, provided the compound contains a well dispersed filler and the molds are properly designed. If some abrasive filler, such as glass fibers, is not well dispersed, or if mold gates are too small, excessive wear may occur.

2.2.3 Feeding. Feeding of fillers, reinforcing and non-reinforcing, to the plasticating unit may present problems due to bridging of material in the feed hopper. More positive feed equipment and procedures may be necessary. Due to the abrasive nature of some of the fillers, as well as the need for retention of the aspect ratio of reinforcing fillers, special techniques of downstream feeding into the melts are frequently used.

2.2.4 Drying and Venting. Drying and venting to insure a void-free product is even more important with filled and/or reinforced materials than with unfilled compounds. The fillers and reinforcements used are of small particle size, and the resulting large surface area of the particle favors adsorbtion of moisture. This, combined with the natural tendency of many such materials to absorb moisture, makes it especially important to use well dried material and/or to provide for venting of the processing equipment.

3. MIXING EQUIPMENT

3.1 Description of Types of Mixers

3.1.1. Intensive Internal Mixers. *Banbury.* (Name registered by Farrel Co.) This is a high shear mixer consisting of two contra-rotating, spiral-shaped blades encased in segments of cylindrical housings, intersecting so as to have a ridge between the blades. Blades may be cored for circulating heating or cooling media. It is used for a wide range of plastics and rubber compounds on a batch basis.

Farrel Continuous Mixer. An adaptation of the Banbury mixer to provide continuous, rather than batch, operation.

Plasticator. (Plastimat by Day Mixing Co.) A high-shear machine designed to accomplish mixing similar to the Banbury.

Turbulent Mixer. A high speed cylindrical blender for very rapid mixing of solids/solids and solids/liquids. Useful for combining liquid polyesters and glass fibers.

Roll Mills. Different speed rolls placed close to each other provide high shearing action to the plastics materials and compounding ingredients being mixed.

3.1.2. Dry Mixers and Blenders. *Ribbon Blender.* Consists of helical, ribbon-shaped blades rotating close to the edge of a U-shaped vessel. Used for relatively high viscosity fluids and dry blends such as PVC calendering and extrusion compounds.

Cone Blender. Cone shaped vertical tank has screw flight agitator which rotates on its own axis while orbiting the periphery of the tank. Used for mixing plastisols and organisols, and as a mixer/dryer for powdered and pelletized plastics. (See Fig. 32-3.)

Kneader. Mixes with a pair of intermeshing blades, often in the shape of the letter Z, used for working plastics masses of semi-dry or rubbery consistency.

Double Arm Mixer. Two agitators rotate in a tank. Tangential or overlapping types of action may be used, depending upon the viscosity of the materials being mixed.

Vertical Intensive Mixers. Comprise a propeller-like impeller rotating at high speed in the bottom of a stationary container, continuously recirculating the materials between closely spaced stationary and rotating pins. Used for dry-blending resins such as PVC with plasticizers and other additives.

Ball Mill. A cylindrical or conical shell rotating about a horizontal axis, partially filled with a grinding medium, such as flint pebbles, ceramic pellets, or metal balls. Material to be ground is added, the shell rotated, and the cascading pellets reduce the particle size by impact on the material.

Double Planetary Mixer (Ross). Two rectangularly shaped blades revolve around the tank on a central axis. Simultaneously, each blade revolves on its own axis at twice the speed of central rotation. Effective for liquid/solid mixes.

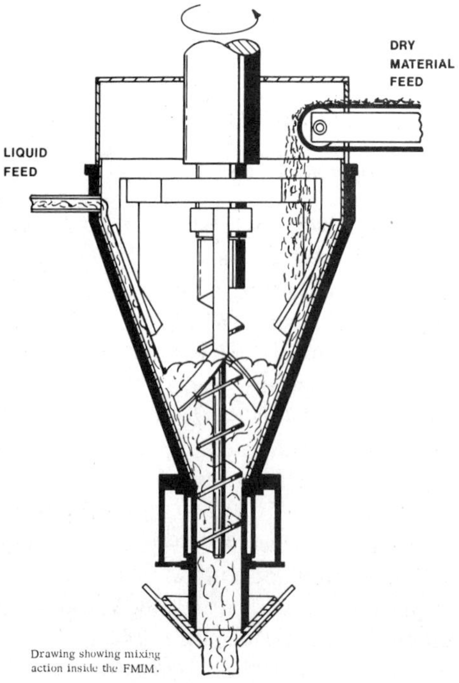

Fig. 32-3. Medium intensity mixer (Patented). (*Courtesy Farrel Company Div., USM Corp.*)

3.1.3. Rotary Tumblers. *Drum Tumbler.* Cylindrical drum rotates end over end or about an inclined axis to thoroughly blend materials. Commonly used to mix plastic pellets with color concentrates and/or regrind.

Conical Dry Blender. Consists of two hollow cones joined at their bases by a short cylindrical section, mounted on a horizontal shaft passing through the sides of the cylindrical section. Material is mixed by cascading, rolling, and tumbling actions as the cones rotate.

3.1.4. Liquid Mixers. *Vertical Cone Blender.* A vertical conical tank with screw flight agitator which rotates on its own axis while orbiting the periphery of the tank. Especially effective for liquid/solid mixes.

Paddle Mixer. Single or double propellers or paddles on a shaft rotating inside of a tank at low or high speeds. Used for suspending solids in a liquid medium.

Intensive Stirrer. A high speed impeller blade in a baffled mixing tank to produce high shear mixing in dispersion of solids in liquids.

3.1.5. Extruders. *Single Screw Extruder.* With screw designed for compound mixing this is

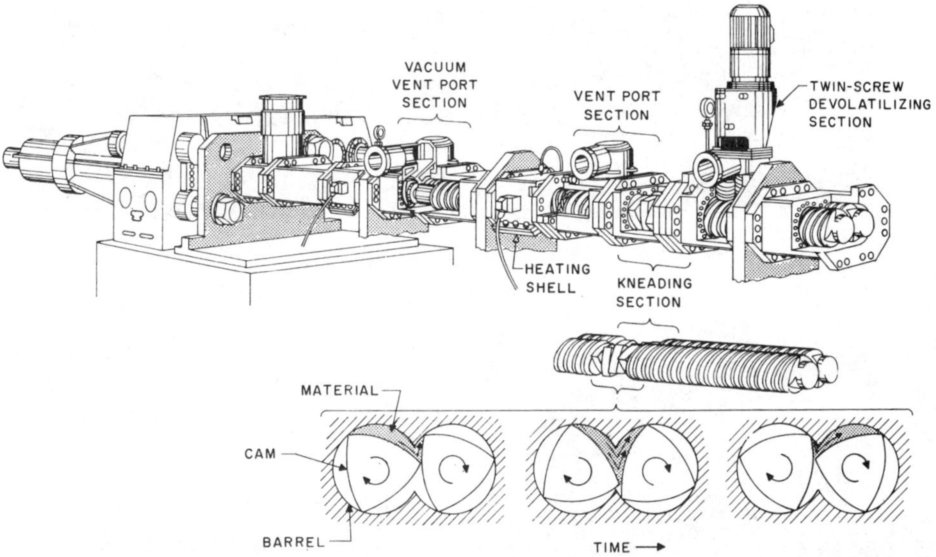

Fig. 32-4. Twin-screw compounding extruder with co-rotating intermeshing screws. (*Courtesy Werner & Pfleiderer Corp.*)

commonly used for compounding fillers and resins.

Twin Screw Extruder. Uses an extruder with twin, co-rotating screws designed with sections for feed, high shear, and high intensity mixing, and discharge on a continuous basis. May give more effective mixing than a single screw extruder. (See Fig. 32-4.)

Mixer Extruder. Combines mixing, such as kneader blades, shear cone, or special mixing screw, with an extruder for continuous delivery of the mixed compound. (See Fig. 32-5.)

3.1.6. Other Types. *Motionless Mixer.* Low shear devices with no moving parts, in which melted polymer is pushed by a ram or screw feed through a baffled section which divides and recombines the melt flow. Effectiveness is a function of material melt viscosity and design of the mixing section, often resulting in improved quality of mixing. (See Figs. 32-6 and 32-7.)

3.1.7. Special RP Equipment. *Pultrusion.* A process for combining long continuous strands of liquid, resin-impregnated, reinforcing material, such as polyester/glass fibers, are pulled through a steel die and heating chamber to yield continuous lengths of material with high unidirectional strength. (See Fig. 32-8.)

Filament Winding. A method for combining continuous, resin-impregnated reinforcing fibers by coating and winding on to the final product shape in a continuous process. Yields high reinforcement/resin binder ratios and very high structural strength. (See Figs. 32-9, 32-10, and 32-11.)

Spray-Up. A general term for several processes

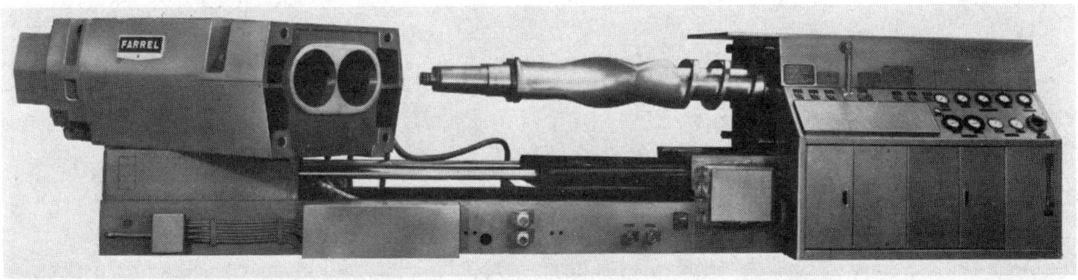

Fig. 32-5. Continuous mixer extruder (Patented). (*Courtesy Farrel Company Div., USM Corp.*)

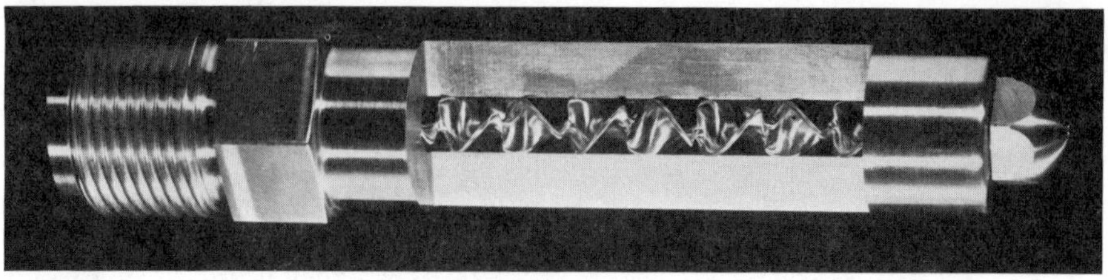

Fig. 32-6. Motionless mixer–injection molding "super nozzle." (*Courtesy Kenics Corp.*)

Fig. 32-7. Motionless mixer–injection molding "mixing head." (*Courtesy Koch Engineering Co.*)

using a spray gun, including simultaneous mixing and spraying of resin and chopped reinforcing fibers and/or fillers onto the mold or mandrel. (See Fig. 32-12.)

Matched Metal Die Molding. Process for forming reinforced plastics articles with high fiber/resin ratios, similar to metal stamping. One variation is the use of precut mats of reinforcing material placed in the mold, liquid resin then poured over the mat. Another variation is the use of preimpregnated fibrous mat in the mold. This may be referred to as "SMC" or "prepreg." Still another variation uses "premix," which has a high content of

Fig. 32-8. Pultrusion–standard glastruder. (*Courtesy Goldworthy Engineering, Inc.*)

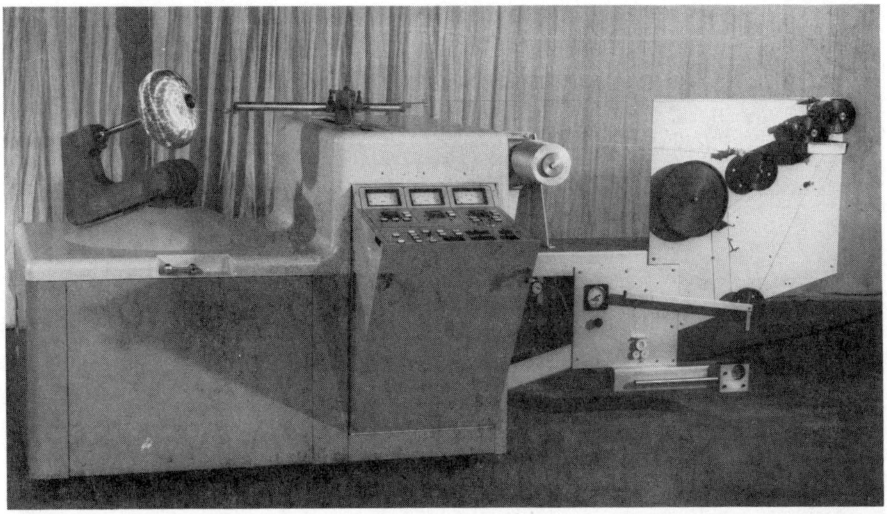

Fig. 32-9. Laboratory filament winder. (*Courtesy Goldsworthy Engineering, Inc.*)

chopped fibers compounded with polyester resin.

3.2. Summary of Mixer Types and Uses

Table 32-1 gives a summary of the various types of mixers and their uses.

3.3 Suppliers of Mixing Equipment

These listings are based upon information supplied by equipment manufacturers. More extensive listings of suppliers are given in the Refs. 5, 16, 17, and 18.

3.3.1. Intensive Internal Mixers.

J. H. Day Co.–Plastimat, high-speed turbulent mixers, roll mills

Farrel Co.–Banbury, FCM continuous mixers, roll mills

Charles Ross & Son Co.–Roll mills

Teledyne Readco–High intensity mixers

Fig. 32-10. Filament winder-polar orbital winder with gantry traverse. (*Courtesy Goldsworthy Engineering, Inc.*)

Fig. 32-11. Filament winder–tank erector. (*Courtesy Goldsworthy Engineering, Inc.*)

Fig. 32-12. Dual spray gun and chopper. (*Courtesy Binks Manufacturing Co.*)

Table 32-1. Summary of mixer types and uses.

	1	2	3	4	5	6	7
Intensive Internal Mixers							
Banbury		X	X	X	X	X	
Farrel continuous mixer		X	X	X	X		X
Plasticator		X	X	X	X	X	
Vertical intensive mixer		X				X	X
Turbulent mixer	X	X	X			X	
Roll mill			X		X	X	
Dry Mixers and Blenders							
Ribbon blender	X	X	X			X	
Cone blender		X				X	
V-shape blender		X				X	
Kneader, double arm mixer	X	X	X			X	
Vertical high intensity	X	X				X	
Ball mill				X		X	
Double planetary	X					X	
Tumblers							
Drum tumbler		X				X	
Conical dry blender		X				X	
Liquid Mixers							
Vertical cone blender	X					X	
Paddle mixer	X					X	
Intensive stirrer	X					X	
Extruders							
Single screw			X		X		X
Twin screw			X		X		X
Mixer extruder			X		X		X
Other Types							
Motionless mixer			X				X

KEY: 1. Liquid resins/solids (plastisols, organosols, thermosets)
2. Solids, cold (dry blends)
3. Solids, hot
4. High shear mixing, grinders, cold
5. High shear mixing, hot
6. Batch mixing
7. Continuous mixing

NOTES:

1. *Dry Blending* of fillers and/or pigments may give sufficient mixing if some additional mixing takes place in subsequent processing, as in extrusion or injection molding with a reciprocating screw type molding machine.

2. *Dry Blending* of reinforcements may be desirable to minimize shear in mixing and reduce breakdown of fiber length.

3. Heat sensitive or thermosetting materials usually require room temperature mixing conditions. The equipment used must be capable of being maintained at ambient temperature during mixing.

4. The many variations of these types of equipment by many manufacturers requires that the specific end use required be thoroughly discussed and explained with the manufacturers before purchase.

3.3.2. Dry Mixers and Blenders.

Paul O. Abbe Inc.–Ribbon blenders, Rota-Cone blenders, double shaft mixers, ball mills

J. H. Day Co.–Ribbon blenders, Nautamixer

Farrel Co.–Medium intensity mixers

Feeco International, Inc.–Ribbon mixer, rotary mixer

General Machine Co. of N.J.–Cone blenders, V-shape blenders

S. Howes Co., Inc.–Ribbon mixers, vertical and horizontal batch mixers

Patterson Industries, Inc.–Double cone Thoroblender, ribbon and paddle mixers

Pfaudler Sybron Corp.–Conical dryer-blender

Prodex, HPM Div., Koehring Co.–High intensity mixer

Charles Ross & Son Co.–Kneader extruder, double planetary mixer

Teledyne Readco–Double arm mixers

3.3.3. Tumblers.

Feeco International, Inc.–Rotary mixers

General Machine Co. of N.J.–Cone blenders, V-shape blenders

Patterson Industries, Inc.–Double cone Thoroblender

3.3.4. Liquid Mixers.

Conn and Co.–Intensive stirrers, vertical mixers

J. H. Day Co.–Nautamixer, Daymax

LFE Corp.–Portable paddle mixers

Farrel Co.–Medium intensity mixers

Feeco International–Paddle mixers

S. Howes and Co., Inc.–Vertical and horizontal batch mixers

Kinetic Dispersion Corp.–Kady mill–high-speed rotor dispersion

Oakes Machine Corp.–Continuous mixers for plastisols and latices

Patterson Industries, Inc.–Ribbon and paddle mixers

Pfaudler Sybron Corp.–Paddle mixers

Charles Ross & Son Co.–Double planetary mixers

3.3.5. Extruders and Mixer Extruders.

Baker-Perkins Inc.–Twin screw continuous mixer

Charles Ross & Son Co.–Kneader extruder
Teledyne Readco–Continuous processor
Werner & Pfleiderer Corp.–Twin screw compounding extruder

3.3.6. Motionless Mixers.

Armorflite-Northeast & Southeast Welding & Machine Co.–Intromix for extruders
Kenics Corp.–Thermogenizer
Koch Engineering
Charles Ross & Son Co.–Melt homogenizer for extrusion, IsoNozzle for injection molding

3.4 Names and Addresses of Mixer Manufacturers

Abbe, Paul O., Inc., 509 Center Ave., Little Falls, N.J. 07424
Armorflite-Northeast Welding & Machine Co., Inc., 662 Cross St., Malden, Mass. 02148; Southeast Welding & Machine Co., Inc., 150 North Wilson Ave., Cocoa, Fla. 32922; Bimetallix, Inc., 798 Broadway, Revere, Mass. 02151
Baker-Perkins Inc., 1000 Hess St., Saginaw, Mich., 48601
Conn and Co., James St., Warren, Pa. 16365
Day, J. H. Co., LeBlond Div., 4932 Beech St., Cincinnatti, Ohio 45212
Farrel Co., Div. USM Corp., Ansonia, Conn. 06401
Feeco International, Inc., Route No. 1, 3913 Algoma Road, Green Bay, Wis. 54301
General Machine Co. of N.J., Inc., 55 Evergreen Ave., Newark, N.J. 07114
Howes, S. Co., Inc., 1939 Miller St., Silver Creek, N.Y. 14136
Kenics Corp., One Southside Rd., Danvers, Mass. 01923
Kinetic Dispersion Corp., 45 Boxwood Lane, Buffalo, N.Y. 14225
Koch Engineering, 4111 E. 37th N., Wichita, Kans. 67220
LFE Fluids Control Div., 100 Skiff St., Hamden, Conn. 06514
Oakes Machinery Corp., 235 Grant Ave., Islip, N.Y. 11751
Ross, Charles & Son, Inc., 710 Old Willets Path, Hauppage, N.Y. 11787
Teledyne Readco, 901 S. Richland Ave., York, Pa. 17405
Werner & Pfleiderer Corp., 160 Hopper Ave., Waldwick, N.J. 07463

4. PROCESSING EQUIPMENT

4.1 General Notes on Utilization

4.1.1. Materials of Construction. The same equipment, such as screw extruders or injection molding machines, which is used for unfilled materials, may also be used for processing filled or reinforced compounds. However, these filled materials have higher viscosities and often higher processing temperatures and pressures. Also, processing under non-optimum conditions of mixing uniformity and tool design may lead to increased rates of wear of equipment. It is therefore deisrable to use tougher and more resistant materials of equipment construction than is otherwise sometimes the case. Some of these are:

Barrels. Nitrided medium alloy-medium carbon steels, heat treated to hardness of Rc50-63.

Nitriding steels, such as nitraloy 135 or 135M, heat treated to bore hardness of over R15N-90.

Bimetallic barrel designs, combining a high alloy, hardened steel sleeve and replaceable liner of more resistant compositions. Liners of Xalloy, Wisconsin Centrifugal, and other such metals are used, and may be shrunk or cast in the barrel. This construction is superior in corrosion and abrasion resistance, but its cost is also greater. The choice will be made in a particular application after balancing initial cost and overall processing economics.

Screws. Nitrided steels the same as for barrels.

Medium alloy-medium carbon steels with Union Carbide "Stellite" applied over the diameter of the screw flights where they are most subject to wear.

Stainless steels, hardened, are sometimes used for improved corrosion resistance, but are more expensive than nitrided steel screws.

Screw Tips, Barrel Heads, Extrusion Dies.

Medium carbon steels, polished or plated
Nitrided steels

4.1.2. Injection Molding Machine Requirements. *Machine Types.* Both plunger and screw type injection molding machines may be used for filled and/or reinforced plastics compounds. However, a screw type machine, either 2 stage screw or reciprocating screw, is recommended for best uniformity and control of material properties and product tolerances.

Injection Pressure. An injection pressure of at least 20,000 psi should be provided for the machine to accommodate the higher melt viscosities encountered with filled materials. Injection speed control is also desirable.

Screw. A general purpose screw is acceptable for all resins except PVC. For PVC, a standard low shear PVC screw with 2:1 compression ratio should be used.

Clamp. A clamping force of 5-10 tons/sq in. of projected area is used to accommodate the higher injection pressures used.

Machine Capacity. With the higher pressures and temperatures used, and the lower rates of flow encountered, machine capacity will be only 50-75% of the capacity as rated for unfilled materials, such as polystyrene.

Drying Materials. Facilities should be provided for drying hygroscopic filled materials before processing. A *dehumidifying oven* with dry hot air passing through it may be used prior to processing. Material is spread in a thin layer over trays and baked for 2-4 hrs at a temperature of at least 250°F. A *hopper drier* may also be used.

Mold Temperature. Mold temperatures for filled and/or reinforced compounds are usually higher than for unfilled resins. Higher mold temperatures result in better flow of the melt and a better surface finish. Standard mold temperature control equipment is satisfactory.

Nozzle. A large diameter, heated nozzle with reverse taper of 2-3° is recommended. A positive cut-off valve should be used with low viscosity resins, such as nylons or when molding in a vertical machine. Nozzle diameter may be 7/16-1/2 in. for most fiber reinforced thermoplastics. A hot nozzle extension will aid molding efficiency by maintaining melt temperature as close as possible to the part being molded.

4.2 Special Notes on Injection Mold Design

4.2.1. Sprues. These should be large, as short as possible, highly polished, and have a good draft angle of at least 3°. Radius of the sprue bushing should be 1/8-1/4 in. greater than the radius of the nozzle to provide a contact line rather than a broad surface.

4.2.2. Runners. Cold slug wells should be used where possible to trap the leading cold resin from the nozzle, and permit hot resin to enter the runner and cavity. They should be made with a back taper to permit pulling the part. Runners should be full round with larger diameters the longer they travel. As reinforced resins tend to solidify at the walls of the runner, this will maintain a maximum free flow path through the center of the runner. A diameter of 1/4-3/8 is recommended for most parts. Streamlined flow should be provided as nearly as possible, but when right angles are used, a cold slug well should be provided. The runner system should be as short as possible and highly polished. In multicavity molds, a balanced runner system is important to obtain part to part uniformity. Where possible, cavities should be by-passed with the main runner, and branch runners should be as short as possible.

4.2.3. Gates. Gates should be located to achieve a balance of flow in all directions, if possible. When flow is primarily in one direction, reinforcing fibers will tend to become oriented in that direction. The result will be differences in properties and shrinkage in different directions in the part, and a resulting tendency towards poor dimensional tolerance control and warping. Where uniform flow is not possible, the gate should be located so that flow direction is along the axis of the most critical dimension.

Gates should be as large as possible, but may be as small as 0.040 in. in diameter. Short land lengths from 0.020-0.040 in. are recommended. Gates should be located at thick sections of the part to minimize sinks or voids. The number of gates used should be kept to a minimum to avoid unnecessary weld lines. Gates are usually round or rectangular, but diaphragm or flash gates may be used. The size of round or rectangular gates should be from ⅔ to full width of the cavity wall. Small pinhole gates should be avoided.

4.2.4. Weld Lines. Weak or prominent weld lines in filled materials may be minimized by *proper mold design.* One technique is the use of overflow wells in the mold, allowing *venting* and cold surface material to be removed from the cavity, and promoting the intermixing of fibers and resins. Another procedure is the use of a cold well by-pass extension of the main runner, allowing a branch runner to deliver hot resin into the cavity. Hot runner molds will also promote good bonds.

4.2.5. Venting. Good venting of the cavity is desirable for a rapid and uniform fill. Vents should be located wherever air may be trapped in the cavity as the resin flows in.

4.3 Special Processing Equipment

4.3.1. Pultrusion. This process, developed by Goldsworthy Engineering, Inc., combines long continuous lengths of reinforcing fibers, such as glass, with liquid resins such as polyester or epoxy resins. It provides a high reinforcement/resin ratio, and the result is a very strong, resilient continuous shape, which has found many unique applications, such as fishing poles. Graphite fibers and fibers of duPont's Kelvar are also used with, or in place of, glass fibers.

4.3.2. Fiberglass Spray-Up Systems. Several systems are marketed for combining catalyzed polyester resins with glass fibers in a continuous spray operation. As a spray gun mixes and dispenses the mixed liquid resin, this resin is combined with chopped glass fibers in the desired proportions. This mixture is applied to a surface or into a mold to cure in the desired final shape. This process is well adapted and widely used for making structural parts, such as boat hulls, shower stalls, and other large, integral structures.

4.3.3. Matched Metal Die Molding. This is done with reinforcing fiber/liquid resin mixtures in which liquid resin, catalysts, glass fiber, and any other additives are combined into a one compound molding material. This may be done using sheet molding compounds (SMC) or a dough-like *premix* material. Applications are widespread, and are similar to those with spray-up systems. It may have advantages of closer control of structural dimensions, tolerances, and finish, but may be more expensive, depending upon the specific applications.

4.3.4. Filament Winding. This consists of winding a continuous, resin-coated strand of reinforced fiber or roving on a mandrel having the structural shape desired. By winding the fibers in a predetermined pattern, a very strong, thin, light structure may be obtained. Here again, the liquid resins may be polyester, vinyl ester, or epoxy. The very high structural strength and stiffness values resulting are due to the high glass fiber/resin ratio, and the precise orientation of these fibers within the structure to obtain maximum utilization of their strength.

4.4 Suppliers of Special Processing Equipment

Pultrusion Equipment.

Goldworthy Engineering, Inc., 2917 Lomita Blvd., Torrance, Calif. 90505

Gatto Machinery Dev. Corp., 45 Rabro Drive, Hauppage, N.Y. 11787

Spray-Up Equipment.

Binks Mfg. Co., 9201 W. Belmont Ave., Franklin Park, Ill. 60131

Cook Paint & Varnish Co., P.O. Box 389, Kansas City, Mo. 64141

Johnson & Sons, 3225 N. Verdugo Rd., Glendale, Cal. 91208

Lincoln St. Louis, Div. McNeil Corp., St. Louis, Mo. 63120

McLean-Anderson Inc., 120 W. Melvina St., Milwaukee, Wis. 53212

Miles, A. L., Fiberglas & Plastic Supply, 4060 Wyne St., Houston, Tex. 77017

Ransburg Corporation, Box 88220, Indianapopolis, Ind., 46208

Shyodu Instrument Co., 197 King St., Brooklyn, N.Y. 11231

Venus Products, Inc., 1862 Ives Ave., Kent, Wash., 98031

Matched Metal Die Equipment (SMC and Premix).

Blue, E. B. Co., 651 Connecticut Ave., S. Norwalk, Conn. 06854

Brenner, I. G. Co., 100 Manning St., Newark, Ohio 43055

Dake Corp., 930 Robbins Rd., Grand Haven, Mich. 49417

Engineering Technology Inc., 145 W. 2950 S., Salt Lake City, Utah 84115

Finn & Fram Inc., 8400 San Fernando Rd., Sun Valley, Calif. 91352

Grace, W. R. & Co., Marco Chemical Div., Linden, N. J. 07036

Littleford Inc., 414 Airport Rd., Cincinnatti, Ohio 45226

McLean-Anderson Inc., 120 W. Melvin St., Milwaukee, Wis. 53212

Filament Winders.

Goldsworthy Engineering, Inc., 2917 Lomita Blvd., Torrance, Calif. 90505

American Barmag Corp., P.O. Box 15262, Charlotte, N. C. 28202

Vermont Instrument Co., Inc., 62 Overlake Park, Burlington, Vt. 05401

5. PROCESSING FILLED COMPOUNDS

5.1 Injection Molding Conditions

The *general procedures and conditions* for molding are similar for filled and unfilled plastics compounds. However, filled materials have higher viscosities and poorer melt flow than unfilled materials, requiring higher processing temperatures. Also, fiber reinforced compounds may show orientation of fibers as the melt flows into the mold cavity, resulting in anisotropy, or properties which are different in different directions in the part. This orientation may result in problems, such as poor dimensional stability, warpage, and undesirable effects upon properties of the part. Special attention to molding conditions may be needed.

The effect of *melt temperature* depends upon the polymer, properties sometimes increasing with temperature, in other polymers decreasing with temperature, and in still other polymers, independent of temperature. High temperatures will also affect different polymers in regard to polymer degradation, melt viscosity, shear sensitivity, and crystallinity. The final properties obtained are a composite result of all of these factors. Generalities are therefore difficult, and the supplier of the particular polymer should be consulted for his processing recommendations.

Dimensional Instability due to excessive orientation may be influenced by molding conditions, and minimum orientation and maximum dimensional stability will be favored by the following:

Material Temperature	Hot
Mold Temperature	Hot
Part Cooling Time	Slow
Injection Pressure	Low
Ram Forward Time	Short
Part Thickness	Thick
Gate Size	Small
Fill Rate	Fast

References

1. Theberge, John E., "How to Process Glass Fiber Fortified Thermoplastics," *Plastics Design and Processing*, **13:** (1), 14-17, (Jan. 1973); and (2), 19-20 (Feb. 1973).
2. Jakopin, Stan, "Compounding of Fillers," *Advances in Chemistry Series*, Number 134, Am. Chem. Soc., 1974.
3. Mack, Wolfgang, "Continuous Compounding–Where Twin Screws Fit," *Plastics Technology*, Feb. 1975.
4. Quillen, S. C., "Mixing," *Chem. Eng.* **61**; p. 204 (1954).
5. Updegrove, L. B., "Selecting Mixers and Compounders," *Plastics World*, pp. 53-57 (Nov. 18, 1974).
6. Seamon, R. G. and Merrill, A. M. *"Machinery & Equipment for Rubber & Plastics"* (1952).
7. Bolen, W. R. & Colwell, R. E., "Intensive Mixing," SPE 14th ANTEC (1958), 1004.
8. Bergen, J. T., "Mixing & Dispersing Processes," *Processing of Thermoplastic Materials*, ed. Bernhardt, Van Nostrand Reinhold, New York, (1959), pp. 409 ff.
9. Schlick, W. R., Hagen, R. S., Thomas, D. P., and Musselman, K. A., "Critical Parameters for Direct Injection Molding of Glass Fiber Thermoplastic Powder Blends," SPE Tech. Paper, 1967, p. 929.
10. Goldsworthy, B., "Continuous Pultrusion," *Modern Plastics*, p. 28 (June 1975).

11. Skoblar, Sandra M., "What's All the Fuss About Motionless Mixing," *Plastics Technology*, **11**; p. 37–43 (1974).
12. Dietz, G. R., "The Art of Formulating Plastisols," *Plastics Eng.*, pp. 50–51 (Apr. 1974).
13. DuBois, J. Harry, "Equipment for Compounding Plastics Materials," *Plastics Machinery & Equipment*, Part I, **4**; (7), 17–20 (July 1975); Part II, **4**; (8), 19–20 (Aug. 1975).
14. Hall, Alan, "In-house Compounding," *Plastics World*, pp. 34–36 (Apr. 21, 1975).
15. DeLuca, Robert M, "Mixing and Compounding Equipment," *Modern Plastics Encyclopedia*, pp. 500–501 (1974–75).
16. "Compounding Equipment," *Modern Plastics Encyclopedia*, p. 845–846, (1974–75).
17. "Mixing and Blending Equipment," *Plastics Technology, Manufacturing Handbook and Buyers Guide*, pp. 237–238 (1974/75).
18. Listing of Mixing Equipment Mfrs., *Chemical Engineering Equipment Buyers Guide*, pp. 388–389 (1975–76).
19. "Fortified Polymers, Processing," Liquid Nitrogen Processing Co., Bulletin 203–769, p. 3.
20. "Molding and Engineering Data for Reinforced Thermoplastics Processors, Equipment Recommendations," Fiberfil Corp., Form 34-7154, p.11.
21. Pickens, S. H., "Pultrusion–The Accent is on the Long Pull," *Plastics Engineering*, pp. 16-21 (July 1975).
22. Olmsted, B. A., "How Glass-fiber Fillers Affect Injection Machines," *SPE Journal*, **26**; (2), 42–43 (Feb. 1970).
23. U.S. Patent 2,877,501, R. Brandt, "Glass-Reinforced Thermoplastic Injection Molding Compound and Injection Molding Process Employing It," Mar. 1959 (to Fiberfil).
24. "Molding Filled Material is a Lot Easier Now," *Modern Plastics*, No. 7, pp. 44–45. (July 1975).
25. Avery, Donald, "Wear - The Unseen Causes," *Plastics Technology*, pp. 31-36 (Oct. 1975).
26. Saltzman, Gilbert A., "Wear Data Aid Selection of Barrel-Screw Combination," *Plastics Design and Processing*. pp. 16–18 (Oct. 1975).

Section XIII
Appendix

Appendix
Metric Conversion Tables

Harry S. Katz and Harold E. Brandmaier
Utility Research Co. Montclair, New Jersey
Burns and Roe Paramus, New Jersey

The conversion of the United States from U.S. Customary units to metric units is inevitable. The United States is the only major industrial country in the world which has not fully adopted the metric system. The metric system is more logical than the present U.S. system of measurement and is, for this reason, employed in many laboratory procedures.

In 1954, the General Conference on Weights and Measures adopted the MKSA system of units based on the meter, kilogram, second, ampere, degree Kelvin, and the candela. In 1960, the system was named the International System of Units or SI. The major advantage of SI is that there is only 1 unit for each physical quantity. As an example, such diverse volume units as barrel, fluid ounce, gallon, liter and cubic foot are replaced by meter^3. Note that while meter and liter are the current preferred spellings in the United States, metre and litre are also used.

Table A1 lists the 9 fundamental SI quantities and the corresponding unit and SI symbol for each. From these, the mechanical, thermal, electrical and other quantities shown in Table A2 were derived. Tables A2, A4, and A5 contain representative quantities considered useful to those using this handbook. More extensive information can be found, for example, in ASTM standard E-380-74; "Using the SI system of Units," F. S. Conant, *Rubber Chemistry and Technology*, March/April 1975; and "Brief History of Measurement Systems with a Chart of the Modernized Metric System," National Bureau of Standards Special Publication 304A, October 1972.

That the units in Table A2 are derivable from the fundamental units in Table A1 can be readily demonstrated. Thus, the SI unit of force is the newton (N), equal to kg · m/s^2. From this, the unit of energy, heat, or work, the joule (J), is defined as N · m. Continuing, the SI unit of power, the watt (W), is given as J/s and finally the unit of electromotive force or voltage, the volt (V), is watts/ampere (W/A). The remaining mechanical, thermal, electrical, and magnetic quantities in Table A2 can be derived from these 4 quantities. Generally,

Table A1. SI units.

Quantity	Unit	SI Symbol
Basic units		
Length	meter	m
Mass	kilogram	kg
Time	second	s
Electric current	ampere	A
Thermodynamic temperature	Kelvin	K
Amount of substance	mole	mol
Luminous intensity	candela	cd
Supplementary units		
Plane angle	radian	rad
Solid angle	steradian	sr

Table A3. SI prefixes.

Mutliplication Factor	Prefix	SI Symbol
10^{-18}	atto	a
10^{-15}	femto	f
10^{-12}	pico	p
10^{-9}	nano	n
10^{-6}	micro	μ
10^{-3}	milli	m
10^{-2}	centi	c
10^{-1}	deci	d
10^{1}	deka	da
10^{2}	hecto	h
10^{3}	kilo	k
10^{6}	mega	M
10^{9}	giga	G
10^{12}	tera	T

Table A2. Derived units.

Quantity	Unit	SI Symbol	Formula
Force	newton	N	$kg \cdot m/s^2$
Energy, heat, work	joule	J	$N \cdot m$
Power	watt	W	J/s
Electromotive force, voltage	volt	V	W/A
Electric capacitance	farad	F	$A \cdot s/V$
Electric field strength	volt per meter	–	V/m
Electric inductance	henry	H	$V \cdot s/A$
Electric resistance	ohm	Ω	V/A
Frequency	hertz	Hz	1/s
Luminous flux	lumen	lm	$cd \cdot sr$
Illuminance	lux	lx	lm/m^2
Magnetic field strength	ampere per meter	–	A/m
Magnetic flux	weber	Wb	$V \cdot s$
Magnetic flux density	tesla	T	Wb/m^2
Pressure, stress	pascal	Pa	N/m^2
Quantity of electricity	coulomb	C	$A \cdot s$
Specific heat	joule per kilogram-kelvin	–	$J/kg \cdot K$
Tear strength, adhesion	newton/meter	–	N/m
Thermal conductivity	watt per meter-kelvin	–	$W/m \cdot K$
Dynamic viscosity	pascal-second	–	$Pa \cdot s$
Kinematic viscosity	square meter per second	–	m^2/s
Radioactive source activity	disintegration per second	–	(disintegration)/s

Table A4. Frequently used U.S. to SI conversion factors.

Quantity	U.S. System	Multiplied by =	SI
Length	foot	$3.048\ 000 \times 10^{-1}$	m
	inch	$2.540\ 000 \times 10^{-2}$	m
Velocity	inch per second	$2.540\ 000 \times 10^{-2}$	m/s
Acceleration	foot per second^2	$3.048\ 000 \times 10^{-1}$	m/s^2
Area	foot^2	$9.290\ 304 \times 10^{-2}$	m^2
	inch^2	$6.451\ 600 \times 10^{-4}$	m^2
Volume	foot^3	$2.831\ 685 \times 10^{-2}$	m^3
	inch^3	$1.638\ 706 \times 10^{-5}$	m^3
	U.S. fluid ounce	$2.957\ 353 \times 10^{-5}$	m^3
	U.S. gallon	$3.785\ 412 \times 10^{-3}$	m^3
Volume flow rate	U.S. gallon per minute	$6.309\ 020 \times 10^{-5}$	m^3/s
Section moment	inch^4	$4.162\ 314 \times 10^{-7}$	m^4
Mass	pound mass	$4.535\ 924 \times 10^{-1}$	kg
	short ton	$9.071\ 847 \times 10^{2}$	kg
Mass flow rate	pound-mass per minute	$7.559\ 873 \times 10^{-3}$	kg/s
Density	pound mass per foot^3	$1.601\ 846 \times 10^{1}$	kg/m^3
	pound mass per inch^3	$2.767\ 990 \times 10^{4}$	kg/m^3
	slugs per foot^3	$5.153\ 788 \times 10^{2}$	kg/m^3
Force	pound force	$4.448\ 222$	N
Pressure	atmosphere (760 mm hg)	$1.013\ 25 \times 10^{5}$	Pa
	inch of mercury (60° F)	$3.376\ 85 \times 10^{3}$	Pa
Pressure, stress	pound-force per inch^2	$6.894\ 757 \times 10^{3}$	Pa
Tear strength, adhesion	pound-force per inch	$1.751\ 268 \times 10^{2}$	N/m
Bending moment	pound-force inch	$1.129\ 848 \times 10^{-1}$	N · m
Energy	Btu*	$1.055\ 056 \times 10^{3}$	J
Power	electric horsepower	$7.460\ 000 \times 10^{2}$	W
Thermal Conductivity	Btu* inch per second foot^2 °F	$5.192\ 204 \times 10^{2}$	W/m · K
Specific heat	Btu* per pound-mass °F	$4.186\ 800 \times 10^{3}$	J/kg · K
Dynamic viscosity	pound-force second per foot^2	$4.788\ 026 \times 10^{1}$	Pa · s
Kinematic viscosity	foot^2 per second	$9.290\ 304 \times 10^{-2}$	m^2/s
Illuminance	footcandle	$1.076\ 391 \times 10^{1}$	lx

*International Table.

Table A5. Frequently used metric to SI conversion factors.

Quantity	Metric System	=	Multiplied by	=	SI
Length	micron		$1.000\ 000 \times 10^{-6}$		m
	angstrom		$1.000\ 000 \times 10^{-10}$		m
Area	barn		$1.000\ 000 \times 10^{-28}$		m^2
Volume	liter		$1.000\ 000 \times 10^{-3}$		m^3
Density	grams per cm^3		$1.000\ 000 \times 10^{3}$		kg/m^3
Force	dyne		$1.000\ 000 \times 10^{-5}$		N
Pressure	torr (mm hg, 0°C)		$1.333\ 22 \times 10^{2}$		Pa
	bar		$1.000\ 000 \times 10^{5}$		Pa
Energy	calorie (International Table)		4.186 800		J
	erg		$1.000\ 000 \times 10^{-7}$		J
	electron volt		$1.602\ 19 \times 10^{-19}$		J
Dynamic viscosity	poise		$1.000\ 000 \times 10^{-1}$		Pa · s
Kinematic viscosity	stoke		$1.000\ 000 \times 10^{-4}$		m^2/s
Magnetic flux	maxwell		$1.000\ 000 \times 10^{-8}$		Wb
Magnetic field strength	oersted		$7.957\ 747 \times 10^{1}$		A/m
Illuminance	phot		$1.000\ 000 \times 10^{4}$		lx
Roentgen as defined (Radiation dose absorbed)	rad		$1.000\ 000 \times 10^{-2}$		J/kg

very large or very small quantities are written in the technical literature as a number greater than 1 and less than 10 multiplied by a suitable power of 10, e.g. 3×10^6 V for 3,000,000 V. SI facilitates expression of these quantities in written and oral communication by the use of the prefixes given in Table A3. Thus 3×10^6 V becomes 3 megavolts or simply 3 MV. Although contained in Table A3, the use of centi, deci, deka and hecto is discouraged. As a result, adjacent prefixes are related by a factor of 1000. The basic number is then between 1 and 1000, e.g. 300 V and not 3 hV.

Table A4 gives conversion factors for a number of useful quantities. The first column is the physical quantity and the second column gives the customary units in the U.S. system. Multiplying by the factor in the third column yields the physical quantity expressed in SI units. As an example, consider a resin with an elastic modulus of 5×10^5 psi. Multiplying by the pressure or stress conversion factor of 6.894 757×10^3 gives 3.4473785 GPa. This example illustrates two problems.

The first is the difficulty of thinking in SI; unless one is experienced using SI it would be difficult to even categorize a material with an elastic modulus of 3 GPa, i.e. is it similar to steel or similar to polyethylene? The transition from the U.S. system to SI will be less difficult if a physical quantity is expressed in both systems during the early years of this transition. As an example, technical societies such as the ASME encourage authors to express quantities in the U.S. system of units with the equivalent SI units in parentheses. The next step would be to invert the order and finally omit the U.S. system units altogether.

The second problem is the number of significant figures after the conversion—8 compared to 1—which implies an accuracy greater than that of the given quantity. To be correct, the SI quantity should be rounded to the same number of significant digits as the U.S. quantity, e.g. 3 GPa.

Table A4 doesn't contain the temperature conversion from degree Fahrenheit (°F) in the U.S. system to Kelvin (K) in SI. This conversion is not a simple multiplication but uses the following equation,

$$K = (°F + 459.67)/1.8$$

Where necessary, temperature may be expressed as degrees Celsius or centigrade; the conversion from Kelvin is,

$$°C = K - 273.15$$

Finally, Table A5 contains conversions from commonly used metric units to SI. Although not evident in Table A5, SI considerably simplifies the existing metric system. This is particularly true for electrical units where, in addition to the MKSA system of units, there coexisted CGS (centimeter, gram, second) electrostatic units, and CGS electromagnetic units.

Index

Index